Advances in Delivery Science and Technology

Series Editor
Michael J. Rathbone

More information about this series at: http://www.springer.com/series/8875

Maria José Alonso • Marcos Garcia-Fuentes
Editors

Nano-Oncologicals

New Targeting and Delivery Approaches

Springer

Editors
Maria José Alonso
Center for Research in Molecular Medicine and Chronic Diseases (CIMUS)
Department of Pharmacy and Pharmaceutical Technology and the Health Research Institute of Santiago de Compostela (IDIS)
Universidad de Santiago de Compostela
Santiago de Compostela, Spain

Marcos Garcia-Fuentes
Center for Research in Molecular Medicine and Chronic Diseases (CIMUS)
Department of Pharmacy and Pharmaceutical Technology and the Health Research Institute of Santiago de Compostela (IDIS)
Universidad de Santiago de Compostela
Santiago de Compostela, Spain

ISSN 2192-6204 ISSN 2192-6212 (electronic)
ISBN 978-3-319-34485-0 ISBN 978-3-319-08084-0 (eBook)
DOI 10.1007/978-3-319-08084-0
Springer Cham Heidelberg New York Dordrecht London

Printed on acid-free paper

Springer is part of Springer Science+Business Media (www.springer.com)

Preface

There is an increasing interest in the scientific international community to exploit nanodevices for the delivery of drugs. Indeed, it is now generally accepted that nanomedicine may significantly improve the therapeutic index of many drugs by: (1) protecting the drug from the degradation/metabolization; (2) promoting drug transport through biological barriers (i.e., endothelia and epithelia); (3) increasing the intracellular diffusion of many drugs with poor intracellular penetration; (4) controlling the drug release; and (5) targeting the diseased area, therefore increasing the pharmacological activity and decreasing the drug's side effects and toxicity.

Since the toxicological issue represents the major limitation of most of the anticancer drugs, it is not surprising that many nanomedicines currently on the market are related to the oncology field (i.e., Doxil®, Abraxane®, etc.). To deliver therapeutic agents to tumor cells in vivo, one must overcome the problems of drug resistance at the tumor level because of physiological barriers (noncellular-based mechanisms), drug resistance at the cellular level (cellular mechanisms), and distribution, biotransformation, and clearance of anticancer drugs in the body. The academic research has highlighted that some nanomedicine formulations may overcome various anticancer drug resistance mechanisms (i.e., multidrug resistance, inhibition of nucleoside transporters, etc.) which represent an important medical need. In this view, some nanodrugs have actually reached the phase III clinical trials, like Livatag® for treatment of the resistant hepatocarcinoma. Other approaches, albeit less advanced, include the development of multifunctional nanoparticles combining both therapeutic and imaging functionalities or even nanodevices containing two or more drugs oriented toward different cancer targets (i.e., tumor vasculature and cancer cells). The above considerations illustrate the important contribution of nanomedicines in oncology.

Nano-Oncologicals: New Targeting and Delivery Approaches, edited by Professor Maria José Alonso and Professor Marcos Garcia-Fuentes, represents an important and complete piece of information on the more advanced research and applications in the nano-oncological field. Overcoming the biological barriers for optimizing anticancer drug delivery in tumors is an important challenge that can be addressed by taking advantage of the enhanced permeability and retention (EPR) effect. But inside of the tumor tissue, nano-sized exosome vesicles that are capable of transferring intracellular proteins or nucleic acids contents to the tumoral micro-environment may also represent a promising strategy to deliver antitumor agents to target cancer cells. This is explained in the first part of the book, "Biological Barriers in Cancer." The second part answers the major question of how to efficiently target the tumor tissue by using either polymer nanoparticles, liposomes or micelles, whereas a special chapter deals with the use of nanocarriers for photodynamic therapy. Apart from the use of conventional chemotherapy, researchers have also studied new strategies to enhance the immune response against cancer, including antigen-pulsed dendritic cells, recombinant DNA and viral vaccines, and a number of novel immune adjuvants targeting immune competent cells. In this view, the application of nanotechnology to immunosuppression or immunomodulation also represents a powerful tool to improve cancer immunotherapy and is discussed in detail in the third part of this book.

The following part deals with the use of small interfering RNA (siRNA) or microRNA (miRNA) as a tumor suppressor, when deletions or mutations occur in diverse human malignancies. However, since the delivery of these small therapeutic nucleic acids remains a major challenge because of enzymatic degradation and poor intracellular penetration, the use of polymeric micelles, liposomes, or nanoparticles provides a decisive advantage over the administration of those biomacromolecules in an aqueous solution. The efficient combination of a therapeutic agent with an imaging molecule in a single nanomedicine and the extreme versatility of the nano-theranostic platform could contribute to the development of individualized treatment protocols offering the opportunity for personalized nanomedicine. This is nicely addressed in the fifth part of the book, with a complete overview on the impressive progresses of nano-based approaches for diagnosis, imaging, and therapy of cancers with many illustrative examples arising from recent literature. Two other chapters are focused on more specific nanotheranostic systems made of multifunctional gold nanoparticles or microbubbles. The last part of the book discusses the conditions for the translation of anticancer nanomedicines into the clinic, including the toxicological issues and other important regulatory aspects. Finally, a case study is given concerning a new double-strand RNA–based nanocomplex with potent anticancer activity.

This book assembles the major aspects of nano-oncology. It represents an essential reference to a broad scientific community, including physico-chemists, pharmaceutical technologists, pharmacologists, oncologists, and biomedical scientists in general. I hope that this contribution to the drug delivery field will give rise to positive discussions between all scientists, regardless of their age, position, or experience.

Villebon-sur-Yvette, France Patrick Couvreur

About the Editors

Maria José Alonso is Professor of Biopharmaceutics and Pharmaceutical Technology at the University of Santiago de Compostela (USC) in Spain. She has worked as a visiting scientist at the University of Paris-Sud, France, the University of Angers, France, and the Massachusetts Institute of Technology, US. She has made critical contributions to the design of novel nanostructures for the targeted delivery of drugs and vaccines and to the understanding of the interaction of nanoparticles with biological barriers. She is part of the scientific boards of a number of societies and Director-At-Large of the Controlled Release Society. She is also a member of the editorial board of eight prestigious journals in the field of drug delivery.

Marcos Garcia-Fuentes is an Associate Professor of Pharmacy at Spain's University of Santiago de Compostela (USC), where he also obtained his Ph.D. He was a visiting scholar at Purdue University in Indiana, USA, and a Marie Curie postdoctoral scientist at ETH Zurich in Switzerland. He was recognized with a competitive contract from the regional government for the reintegration of excellent researchers. Dr. Garcia-Fuentes has worked in the design of biomaterials and nanostructures for medical applications. His current focus is on nanomedicines and biomaterials for modulating progenitor cells and cancer stem cells.

Contents

Part I Biological Barriers in Cancer

1 **Passive vs. Active Targeting: An Update of the EPR Role in Drug Delivery to Tumors** 3
Jaydev R. Upponi and Vladimir P. Torchilin

2 **The "EX" and the "SOMA": How They Communicate** 47
Pinar Kanlikilicer, Mohammed Helmy Rashed, Bulent Ozpolat, Cristian Rodriguez-Aguayo, Anil K. Sood, George A. Calin, and Gabriel Lopez-Berestein

Part II Tumor Targeting

3 **Polymeric Nanocarriers for Cancer Therapy** 67
Peng Mi and Nobuhiro Nishiyama

4 **Targeted Liposomes and Micelles as Carriers for Cancer Therapy** 95
Weiyue Lu, Changyou Zhan, and Huimin Hou

5 **Photodynamic Therapy for Cancer: Principles, Clinical Applications, and Nanotechnological Approaches** 123
Claudia Conte, Francesca Ungaro, Antonino Mazzaglia, and Fabiana Quaglia

Part III Targeting the Immune System

6 **Cancer Vaccines: Fundamentals and Strategies** 163
Javier Briones

7 Cancer Immune Modulation and Immunosuppressive Cells: Current and Future Therapeutic Approaches ... 187
Maria Stella Sasso, Vincenzo Bronte, and Ilaria Marigo

8 Nanotechnology Approaches for Cancer Immunotherapy and Immunomodulation ... 215
María de la Fuente, Robert Langer, and Maria José Alonso

Part IV Gene Therapy

9 RNAi-Based Nano-Oncologicals: Delivery and Clinical Applications ... 245
Katharina Müller and Ernst Wagner

10 Nano-Platforms for Tumor-Targeted Delivery of Nucleic Acid Therapies ... 269
Arun K. Iyer, Shanthi Ganesh, and Mansoor M. Amiji

Part V Nanotheranostics

11 Multifunctional Gold Nanocarriers for Cancer Theranostics: From Bench to Bedside and Back Again? ... 295
João Conde, Furong Tian, Pedro V. Baptista, and Jesús M. de la Fuente

12 Microbubbles as Theranostics Agents ... 329
Tuan Pham, Carl Beigie, Yoonjee Park, and Joyce Y. Wong

13 Multimodal Nano-Systems for Cancer Diagnosis, Imaging, and Therapy ... 351
Amit Singh, Adwait J. Oka, Purva Pandya, and Mansoor M. Amiji

Part VI Translational Aspects of Nano-Oncologicals

14 Nanotoxicology: Towards Safety by Design ... 391
Fernando Torres Andón and Bengt Fadeel

15 Regulatory Aspects of Oncologicals: Nanosystems Main Challenges ... 425
Rogério S. Gaspar, Helena F. Florindo, Liana C. Silva, Mafalda A. Videira, M. Luísa Corvo, Bárbara F. Martins, and Beatriz Silva-Lima

16 BO-110, a dsRNA-Based Anticancer Agent ... 453
Mercedes Pozuelo-Rubio, Damià Tormo Carulla, Andrés P. Gagete, and María Soengas

Index ... 471

Contributors

Maria José Alonso Center for Research in Molecular Medicine and Chronic Diseases (CIMUS), Department of Pharmacy and Pharmaceutical Technology, Health Research Institute of Santiago de Compostela (IDIS), Universidad de Santiago de Compostela, Santiago de Compostela, Spain

Mansoor M. Amiji Department of Pharmaceutical Sciences, School of Pharmacy, Bouvé College of Health Sciences, Northeastern University, Boston, MA, USA

Fernando Torres Andón Division of Molecular Toxicology, Institute of Environmental Medicine, Karolinska Institutet, Stockholm, Sweden

Pedro V. Baptista Research Centre for Human Molecular Genetics, Science and Technology Faculty, New University of Lisbon, Caparica, Portugal

Carl Beigie Department of Biomedical Engineering, Boston University, Boston, MA, USA

Javier Briones Hematology Service, Hospital Sant Pau, Barcelona, Spain

Vincenzo Bronte Department of Pathology, Verona University Hospital, Verona, Italy

George A. Calin Department of Experimental Therapeutics and the Center for RNA Interference and Non-coding RNAs, The University of Texas M.D. Anderson Cancer Center, Houston, TX, USA

Damià Tormo Carulla Bioncotech Therapeutics, Paterna, Spain

João Conde Instituto de Nanociencia de Aragon (INA), Universidad de Zaragoza, Zaragoza, Spain

Research Centre for Human Molecular Genetics, Science and Technology Faculty, New University of Lisbon, Caparica, Portugal

Massachusetts Institute of Technology, Harvard–MIT Biomedical Engineering Center, Cambridge, MA, USA

Claudia Conte Department of Pharmacy, University of Napoli Federico II, Napoli, Italy

Jesús M. de la Fuente Instituto de Nanociencia de Aragon (INA), Universidad de Zaragoza, Zaragoza, Spain

María de la Fuente Nano-oncologicals Laboratory, Translational Medical Oncology, Health Research Institute of Santiago de Compostela (IDIS) and Clinical University Hospital of Santiago de Compostela (CHUS), Santiago de Compostela, Spain

Bengt Fadeel Division of Molecular Toxicology, Institute of Environmental Medicine, Karolinska Institutet, Stockholm, Sweden

Helena F. Florindo Faculdade de Farmácia, Universidade de Lisboa, Lisboa, Portugal

Andrés P. Gagete Bioncotech Therapeutics, Paterna, Spain

Shanthi Ganesh Dicerna, Inc., Watertown, MA, USA

Department of Pharmaceutical Sciences, School of Pharmacy, Bouvé College of Health Sciences, Northeastern University, Boston, MA, USA

Rogério S. Gaspar Faculdade de Farmácia, Universidade de Lisboa, Lisboa, Portugal

Huimin Hou National Pharmaceutical Engineering Research Center, Shanghai Institute of Industrial Pharmaceutics, Shanghai, China

Arun K. Iyer Department of Pharmaceutical Sciences, School of Pharmacy, Bouvé College of Health Sciences, Northeastern University, Boston, MA, USA

Pinar Kanlikilicer Department of Experimental Therapeutics, The University of Texas M.D. Anderson Cancer Center, Houston, TX, USA

Robert Langer Department of Chemical, Biological and Mechanical Engineering, David H. Koch Institute for Integrative Cancer Research at MIT, Massachusetts Institute of Technology, Boston, MA, USA

Gabriel Lopez-Berestein Department of Experimental Therapeutics and the Center for RNA Interference and Non-coding RNAs, The University of Texas M.D. Anderson Cancer Center, Houston, TX, USA

Weiyue Lu Department of Pharmaceutics, School of Pharmacy, Fudan University & Key Laboratory of Smart Drug Delivery (Fudan University), Ministry of Education, Shanghai, China

M. Luísa Corvo Faculdade de Farmácia, Universidade de Lisboa, Lisboa, Portugal

Ilaria Marigo Istituto Oncologico Veneto, IOV-IRCCS, Padova, Italy

Bárbara F. Martins Faculdade de Farmácia, Universidade de Lisboa, Lisboa, Portugal

Antonino Mazzaglia CNR-ISMN Institute of Nanostructured Materials, Department of Chemical Sciences, University of Messina, Messina, Italy

Peng Mi Polymer Chemistry Division, Chemical Resources Laboratory, Tokyo Institute of Technology, Yokohama, Japan

Katharina Müller Pharmaceutical Biotechnology, Center for System-Based Drug Research, Center for Nanoscience (CeNS), and Nanosystems Initiative Munich (NIM), Ludwig-Maximilians University, Munich, Germany

Nobuhiro Nishiyama Polymer Chemistry Division, Chemical Resources Laboratory, Tokyo Institute of Technology, Yokohama, Japan

Adwait J. Oka Department of Pharmaceutical Sciences, School of Pharmacy, Bouvé College of Health Sciences, Northeastern University, Boston, MA, USA

Bulent Ozpolat Department of Experimental Therapeutics, The University of Texas M.D. Anderson Cancer Center, Houston, TX, USA

Purva Pandya Department of Pharmaceutical Sciences, School of Pharmacy, Bouvé College of Health Sciences, Northeastern University, Boston, MA, USA

Yoonjee Park, Ph.D. Department of Biomedical Engineering, Boston University, Boston, MA, USA

Tuan Pham Department of Biomedical Engineering, Boston University, Boston, MA, USA

Mercedes Pozuelo-Rubio Bioncotech Therapeutics, Paterna, Spain

Fabiana Quaglia Department of Pharmacy, University of Napoli Federico II, Napoli, Italy

Mohammed Helmy Rashed Department of Experimental Therapeutics, The University of Texas M.D. Anderson Cancer Center, Houston, TX, USA

Faculty of Pharmacy, Department of Pharmacology and Toxicology, Al-Azhar University, Cairo, Egypt

Cristian Rodriguez-Aguayo Department of Cancer Biology and the Center for RNA Interference and Non-coding RNAs, The University of Texas M.D. Anderson Cancer Center, Houston, TX, USA

Maria Stella Sasso Faculdade de Farmácia, Universidade de Lisboa, Lisboa, Italy

Liana C. Silva Faculdade de Farmácia, Universidade de Lisboa, Lisboa, Portugal

Beatriz Silva-Lima Faculdade de Farmácia, Universidade de Lisboa, Lisboa, Portugal

Amit Singh Department of Pharmaceutical Sciences, School of Pharmacy, Bouvé College of Health Sciences, Northeastern University, Boston, MA, USA

María Soengas Melanoma Laboratory, Molecular Oncology Programme, Spanish National Cancer Research Centre (CNIO), Madrid, Spain

Anil K. Sood Departments of Gynecologic Oncology and Cancer Biology and the Center for RNA Interference and Non-coding RNAs, The University of Texas M.D. Anderson Cancer Center, Houston, TX, USA

Furong Tian Focas Research Institute, Dublin Institute of Technology, Dublin, Ireland

Vladimir P. Torchilin Center for Pharmaceutical Biotechnology and Nanomedicine, Northeastern University, Boston, MA, USA

Francesca Ungaro Department of Pharmacy, University of Napoli Federico II, Napoli, Italy

Jaydev R. Upponi Center for Pharmaceutical Biotechnology and Nanomedicine, Northeastern University, Boston, MA, USA

Mafalda A. Videira Faculdade de Farmácia, Universidade de Lisboa, Lisboa, Portugal

Ernst Wagner Pharmaceutical Biotechnology, Center for System-Based Drug Research, Center for Nanoscience (CeNS), and Nanosystems Initiative Munich (NIM), Ludwig-Maximilians University, Munich, Germany

Joyce Y. Wong, Ph.D. Department of Biomedical Engineering, Boston University, Boston, MA, USA

Changyou Zhan Department of Pharmaceutics, School of Pharmacy, Fudan University & Key Laboratory of Smart Drug Delivery (Fudan University), Ministry of Education, Shanghai, China

Laboratory for Biomaterials and Drug Delivery, Department of Anesthesiology, Children's Hospital Boston, Harvard Medical School, Boston, MA, USA

Part I
Biological Barriers in Cancer

Chapter 1
Passive vs. Active Targeting: An Update of the EPR Role in Drug Delivery to Tumors

Jaydev R. Upponi and Vladimir P. Torchilin

1.1 Introduction

The enhanced permeability and retention (EPR) effect is a fundamental vascular phenomenon that has shaped nanotherapeutics and diagnostics of today. From the first liposomes described in the 1960s till date, considerable progress has been made in the field of nano-drug delivery. Due to the vast knowledge and understanding of tumor biology and various challenges and barriers, dramatic progress is being made with the number of nanopreparations under clinical investigation. It has also shaped the current understanding of the passive and active targeting of nanocarriers. Emerging new targets facilitate the development of new ligands for active transport and assist in further discovery of the molecular and subcellular mechanisms of tumor development and identification of new and even better targets.

Targeting nanocarriers to tumor involves both, "passive targeting" and "active targeting"; and the concept of active targeting process is not separate from passive targeting since it only occurs after passive accumulation in tumor tissues. Specific tumor targeting of nanocarriers leads to better pharmacokinetics and pharmacodynamics profiles, enables controlled and sustained release of drugs, improves specificity, increases intracellular delivery and, more importantly, results in lower systemic toxicity (Danhier et al. 2010).

In this chapter, we begin discussing the use of nanocarriers, such as liposomes and micelles, since majority of the active pharmaceutical ingredients (APIs) currently used for cancer treatment are largely cytotoxic, and when administered directly, they are generally distributed nonspecifically within the body resulting in substantial toxicity to normal tissues, hence limiting their clinical application.

J.R. Upponi • V.P. Torchilin (✉)
Center for Pharmaceutical Biotechnology and Nanomedicine, Northeastern University, 140 The Fenway, Room 214, 360 Huntington Avenue, Boston, MA 02115, USA
e-mail: v.torchilin@neu.edu

M.J. Alonso, M. Garcia-Fuentes (eds.), *Nano-Oncologicals: New Targeting and Delivery Approaches*, Advances in Delivery Science and Technology,
DOI 10.1007/978-3-319-08084-0_1

Following this, we discuss the hallmark EPR effect, on which the passive targeting is based, and describe certain factors that affect passive targeting. Later, we discuss active targeting of nanocarriers providing some fundamentals and factors that affect active targeting with some examples. Last, we briefly describe the clinical perspective of actively and passively targeted liposomes and micelles.

1.2 Pharmaceutical Nanocarriers: General Considerations

Currently, a variety of chemotherapeutic agents are available clinically to treat cancer. However, these chemotherapeutic agents pose potential disadvantages such as toxicity or side effects towards normal cells, premature drug degradation resulting in short in vivo half-life, poor bioavailability and thus, repeated administrations (Allen and Cullis 2004). To overcome these issues, it is important to increase the amount of therapeutic active entity at the site of action or in pathological area and to reduce the toxicity by embedding drugs into nontoxic biodegradable carriers (Duncan 2003). To achieve this goal, a variety of drug delivery systems such as synthetic polymers, microcapsules, ghost cells, lipoproteins, liposomes, micelles, and nanoparticles are currently applied or under development (Torchilin 2007) offering a wide range of advantages in delivery of chemotherapeutic agents. Various nanomaterials have also been employed in the development of contrast agents and radiopharmaceuticals intended for applications in the field of imaging (Pauwels and Erba 2007). Of the widely available nanocarriers, polymeric micelles and liposomes are among the most actively investigated.

Micelles are self-assembling colloidal particles that consist of a hydrophobic core and hydrophilic corona (Torchilin 2007). Polymeric micelles belong to a group of amphiphilic colloids, which form spontaneously under certain conditions (concentration and temperature) from amphiphilic molecules. At low concentrations in the aqueous medium, these amphiphilic molecules exist as monomers. As the concentration is increased, these molecules begin to self-assemble into micelles at a concentration known as critical micelle concentration (CMC). The temperature below which amphiphilic molecules exist as unimers and above as aggregates is referred to as critical micelle temperature (CMT) (Torchilin 2001a). The major driving force behind self-association of amphiphilic copolymers is the decrease in the free energy of the system due to removal of hydrophobic fragments from the aqueous surroundings, with the formation of micelle core stabilized with hydrophilic units exposed to water (Martin 1993). In addition, van der Waals interactions amongst the hydrophobic blocks forming the core of the micelles significantly contribute to the stability of polymeric micelles (Jones and Leroux 1999). The characteristic size of polymeric micelles ranges from 5 to 50–100 nm (Torchilin 2002), which makes them ideal candidates for the delivery of therapeutic and contrast agents to the tumor tissue. Because of their small size, micelles are able to extravasate into the interstitium of body compartments with leaky vasculature (tumors and infarcts) by the EPR effect, which will be discussed in the following section.

The other most commonly investigated nanocarrier is liposomes, which are closed artificial phospholipid vesicles with an aqueous phase inside and between the lipid bilayers (Torchilin 2005). Thermodynamics and kinetics of vesicle formation are well documented and reviewed (Guida 2010). The liposome size is determined by its composition and the technique of preparation. For instance, multilamellar vesicles have a size ranging from 500 to 5,000 nm and consist of numerous concentric bilayers. Similarly the large unilamellar vesicles and small unilamellar vesicles range in size from 200 to 800 nm and 50 to 100 nm, respectively. Liposomes can be loaded with a variety of therapeutic agents. Some drugs are encapsulated in the bilayer of the liposome while some are in inner aqueous compartment, and this is depended on the different methods used to prepare liposomes and encapsulation techniques (Torchilin and Weissig 2003). The most commonly used techniques are ultrasonication, gel filtration, and dialysis. Since the application of liposomes is in vivo, stability is imparted by the addition of cholesterol (up to 50 % mol) into the liposomal membrane (Musacchio and Torchilin 2009). Liposomes offer several advantages such as biocompatibility, biodegradability, and low toxicity. Transport of these nanoparticles to the site of action or the tumor can be achieved either via passive targeting or active targeting (see Fig. 1.1). In both cases it is essential to understand the various factors that govern the effective delivery of therapeutic agents, some of which are enlisted in the following sections.

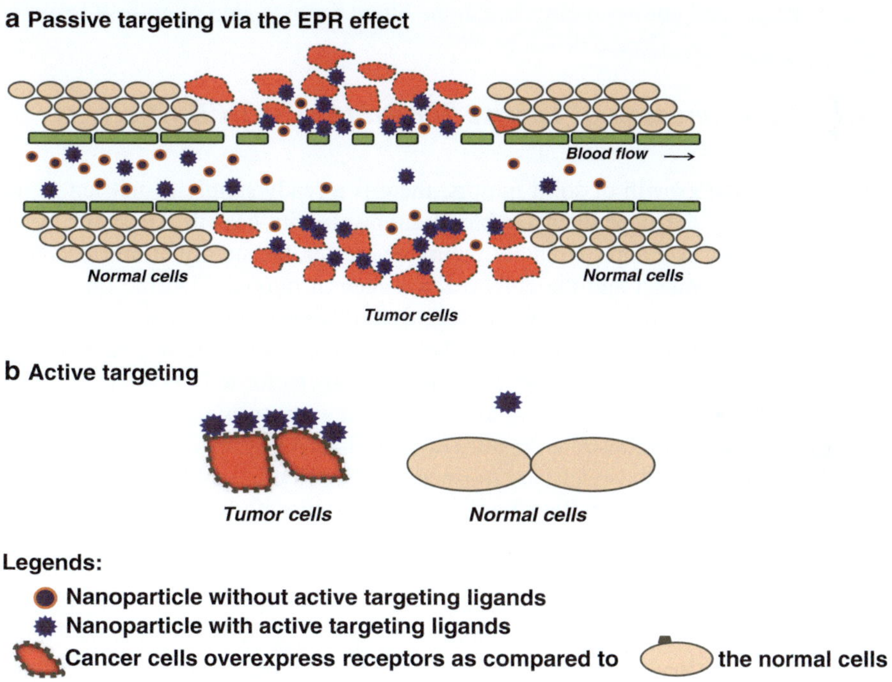

Fig. 1.1 Schematic representation of passive and active targeting of nanoparticles. (**a**) Represents passive targeting: extravasation of nanoparticles, both with and without ligands through the gap junctions of the blood vessels, known as the enhanced permeability and retention effect. (**b**) Represents active targeting. Tumor cells exhibit overexpression of receptors as compared to normal cells. Following extravasation of the actively targeted nanoparticles into the tumor interstitium, the nanoparticles interact with the surface receptors of the tumor cells. Active targeting nanoparticles show greater accumulation of nanoparticles in the tumor tissue due to the combined effect of **a** and **b**

1.3 Passive Targeting of Nanocarriers

It was more than 30 years ago when Matsumura and Maeda first reported the preferential accumulation of macromolecules in tumors and described the EPR effect (Maeda and Matsumura 1989; Matsumura and Maeda 1986). The accumulation of large molecules and even small particles in the tumors is observed due to large pores in the blood vessels resulting in leaky vasculature and also because of compromised lymphatic drainage within the tumor tissue. Smancs or poly(styrene-*co*-maleic acid/half-*n*-butyl ester) (SMA) conjugated with neocarzinostatin (NCS), a 16 kDa polymer conjugate, was the first ever reported antitumor agent to accumulate in the tumor (Matsumura and Maeda 1986). Upon further investigation Matsumura and Maeda reported that proteins less than 30 kDa do not exhibit the EPR effect (Matsumura and Maeda 1986). These early investigations led to advent of a new era of nanocarrier delivery to the tumor via passive targeting. It is important to understand that the EPR effect entails a list of factors that can influence the extravasation and retention of nanocarriers into tumor tissue. Some of the factors that are mentioned in this section are physiological, related to the tumor and its microenvironment itself, while the others are linked to the nanocarrier properties, such as size, surface charge, and ability to circulate in the blood for extended periods of time.

1.3.1 Fundamentals of Passive Targeting: The EPR Effect

During the initial growth of solid tumors, there is a ready supply of nutrients and oxygen and they are dependent on blood supply for its growth. However, after reaching a certain size approximately 2 mm^3 or greater, the supply of nutrients and oxygen is compromised and the cells reach a state of hypoxia. During this stage, cellular hypoxia-inducible factor transcription is increased, thereby activating growth factors such as the vascular-endothelial growth factor (VEGF), the platelet-derived growth factor (PDGF), and the tumor necrosis factor-α (TNF-α) giving rise to new blood vessels, a process known as "angiogenesis" (Folkman 1995). Angiogenesis is well orchestrated in normal conditions, however in tumors irregular shaped, tortuous blood vessels are formed. These blood vessels are disorderly branched, have irregular capillary bed, and are dilated with large interendothelial junctions with irregular basement membrane (Jain and Stylianopoulos 2010) with capillary pore size ranging from 200 to 2,000 nm (Hobbs et al. 1998). This contributes to the enhanced permeation of the EPR effect. The lymphatic function is responsible for draining interstitial fluid from the tissue and returning it to the circulation (Alexander-Bryant et al. 2013). In tumor tissues, the lymphatic vessels are compromised leading to ineffective or poor drainage of fluid within the tumor interstitium. This represents the enhanced retention aspect of the EPR effect.

This phenomenon of extravasation, passive accumulation, or targeting and retention is collectively described as the EPR effect (Matsumura and Maeda 1986; Maeda et al. 1979; Matsumura et al. 1987). When nanoparticles, macromolecules, or drug conjugates with the molecular size greater than 30–40 kDa are injected intravenously

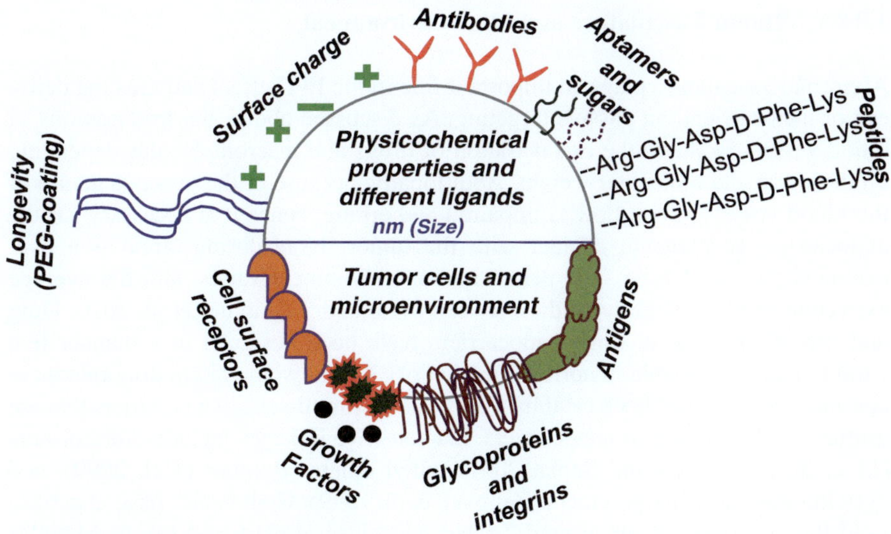

Fig. 1.2 Schematic representation of various factors affecting passive and active targeting of nanoparticles. Figure represents the physicochemical, ligand factors of the nanocarrier and the tumor cellular and microenvironmental factors that affect passive and active targeting

in the abovementioned conditions, they extravasate into the tumor interstitium through large fenestrations, which indicates enhanced permeation. Numerous macromolecules and nanocarriers have shown to accumulate in tumor and infarct tissues via the passive targeting owing to the EPR effect (Hobbs et al. 1998; Torchilin 2011; Torchilin et al. 1992; Abra et al. 2002; Jain 1999).

For a successful delivery of nanocarriers within the tumor vasculature via the EPR effect, it is important to understand the various factors to be considered in the preparation of a suitable nanocarrier. In addition, factors affecting the tumor physiology also need to be taken into consideration. For instance, following passive targeting, liposomes localize in the tumor-residing macrophages within the interstitium surrounding tumor cells (Yuan et al. 1994; Huang et al. 1992). Due to impaired lymphatic drainage, certain types of tumors have high interstitial pressure and large interstitial space; as a result, nanocarriers can have limited distribution within the tumor interstitium (Jain 1989). Hence, it is important to consider the various factors that could affect the delivery of nanocarriers via passive targeting (see Fig. 1.2).

1.3.2 Factors Affecting Passive Targeting via the EPR Effect

Tumor biology plays a critical role in passive targeting. These include the tumor microenvironment and tumor vasculature. Additionally, the physicochemical properties of the nanocarrier that contribute to passive targeting will be discussed in this section.

1.3.2.1 Tumor Vasculature and Microenvironment

Abnormal vasculature plays an important role in the EPR effect and thus the delivery of nanocarriers via passive targeting. As discussed above, the fenestrations of blood vessels facilitate the extravasation of drugs and macromolecules depending on their size and molecular weight. Additionally, because of the tortuous nature of the blood vessels, blood flow is impaired and erratic, hence it is necessary for the nanocarriers to come in contact with the tumor tissue during circulation over extended period of time. The pH of tumor environment varies, and the average extracellular pH is slightly acidic between pH 6 and 7 (Danhier et al. 2010; Ding and Ma 2013). Hence, some nanocarriers have been designed in a manner that would make them stable in normal blood at pH 7.4 but would aid in drug release at acidified pH. This has been established by using stimuli-sensitive polymers that are grafted on the surface of nanocarriers. pH-sensitive linkages include diorthoesters (Li et al. 2005; Guo and Szoka 2001), vinyl ethers (Boomer et al. 2009), and cysteine-cleavable lipopolymers (Zalipsky et al. 1999). Hydrazones have also been used that are stable at pH around 7.5 but hydrolyze at pH 6 and below. Micelles made of poly(ethylene glycol) (PEG)-poly(aspartate hydrazone adriamycin) release drug at lowered pH values facilitating drug delivery and toxicity against cancer cells (Bae et al. 2005). Similarly, pH-sensitive poly(L-histidine) micelles containing doxorubicin showed the release of the drug at pH 6.8 upon the uptake by human ovarian carcinoma (A2780) cells (Gao et al. 2005). *N*-isopropylacrylamide also facilitates liposomal destabilization and drug release at decreased pH values (Deshpande et al. 2013). Such approaches have also been used while designing nanocarriers for active targeting to protect the targeting ligand during circulation.

1.3.2.2 Vascular and Molecular Mediators

In addition to the inherent architecture of the tumor microenvironment, passive targeting of nanocarriers is also affected at the molecular level. The influence of vascular mediators, such as nitric oxide (NO), bradykinins, prostaglandins, matrix metalloproteinase (MMPs), such as collagenase, and other inflammatory cytokines significantly affect extravasation of nanocarriers within the tumor interstitium and have been thoroughly discussed (Maeda et al. 2013; Maeda 2012; Fang et al. 2011). The activation of MMPs in tumor cells causes the extracellular matrix to disintegrate and remodel, facilitating vascular permeability due to the degradation of matrix proteins, and as a result the blood vessels become more leaky (Wu et al. 2001). The MMPs can be activated using NO_2 or a derivative $ONOO^-$ which can decompose to NO (Okamoto et al. 1997). Maeda and group demonstrated increase in the extravasation of Evans Blue following an intradermal injection of $ONOO^-$ (Wu et al. 2001).

NO is generated from L-arginine by nitric oxide synthase (NOS) and is a vital signaling molecule. In tumor tissue, NO is responsible for tumor growth, oxygen supply, and nutrition. Administering nitro derivatives such as nitroglycerine along with nanocarriers potentially increases the EPR effect by activation of the MMPs as described above (Seki et al. 2009). Chemosensitivity of drugs such as docetaxel, cisplatin, and carboplatin was increased in presence of nitroglycerine during the treatment of patients with non-small cell lung cancer (NSCLC) (Yasuda et al. 2006a, b). Bradykinins and kininase inhibitors, such as the angiotensin-converting enzyme (ACE), also potentiate the permeability of the tumor vasculature and enhance the EPR effect (Maeda et al. 1988, 2013; Maeda and Yamamoto 1996; Matsumura et al. 1988, 1991). In addition, other molecular and vascular mediators include TNF-α, tumor growth factor, interferon-α, and interleukins that significantly contribute to the delivery of macromolecules via the EPR effect. For example, intravenous administration of TNF resulted in a three to sixfold increase in EL4 tumor uptake of Evans Blue/Albumin, adenovirus, and long-circulating polymer-coated adenovirus by activating the Rho A/Rho kinase pathway, which subsequently polymerizes the actin fibers within the endothelial cells, thereby causing actin stress (Seki et al. 2011; Koss et al. 2006). Inflammatory cytokines such as interleukin-1 and interferon-α can increase the endothelial-inducible NOS and prolong NO release upon stimulation of the bradykinin receptors (Ignjatovic et al. 2004). VEGF is involved in the upregulation of NO generation system which enhances the EPR effect as discussed above (Maeda et al. 2003).

1.3.2.3 Physicochemical Properties of Nanocarriers

Apart from the inherent molecular and vascular factors that affect the passive accumulation of nanocarriers, the different properties of the nanocarrier itself can influence its accumulation. These include the size of the nanocarrier, its surface charge, and its ability to sustain prolonged circulation in the blood (Bertrand and Leroux 2012; Alexis et al. 2008). These individual parameters can affect the biodistribution, extravasation, and intratumoral diffusion and hence the overall pharmacokinetic behavior of the formulation.

1.3.2.3.1 Size

The carriers have to possess an adequate size in order to extravasate through the leaky pores of the neovasculature and for its diffusion through the tumor interstitium. The cutoff size of the permeabilized vasculature varies from case to case in rather broad limits from 200 to 800 nm (Hobbs et al. 1998; Torchilin 2011; Yuan et al. 1995), and, hence, it is essential to the control size of the nanocarrier for effective EPR-mediated drug delivery. In recent reports, it had been stated that the EPR effect is most effective for nanocarriers particle size from 10 to 500 nm (Torchilin 2011; Maeda et al. 2013; Fang et al. 2011). It has also been observed that smaller molecular weight macromolecules of 3 to 10 kDa and with the size 2 to 3 nm extravasate faster as compared to 70 and 2,000 kDa molecules with diameters 7 and 25 nm, respectively (Dreher et al. 2006). However, in such cases, it was reported that molecules with smaller weight and particle size would diffuse out from the tumor interstitium. The EPR effect was observed in a variety of nanocarrier systems such as liposomes, micelles, polymer drug conjugates, DNA polyplexes, and immunoglobulins up to sizes of approximately 700 to 800 kDa and higher (Maeda et al. 1979; Fang et al. 2011; Seymour et al. 1995).

1.3.2.3.2 Surface Charge

In addition to the size and molecular weight, the surface charge of nanocarriers significantly influences the EPR effect (Bertrand and Leroux 2012; Alexis et al. 2008; Baselga 2006). Surface charge of the nanocarriers affects the penetration and cellular association within large tumors. It has been well documented that the blood vessels are negatively charged on the luminal surface due to the presence of carboxylate sugars and sulfate groups (Maeda et al. 2009). On the other hand, highly positively charged nanocarriers can be opsonized and cleared from the circulation (Xiao et al. 2011; He et al. 2010; Levchenko et al. 2002; Chonn et al. 1991; Nishida et al. 1991; Scherphof and Kamps 2001; Salvador-Morales et al. 2009), whereas those with a negative surface charge can have variable effects depending on the overall charge (Arvizo et al. 2011; Roux et al. 2003; Yamamoto et al. 2001; Gabizon and Papahadjopoulos 1988; Peer and Margalit 2004a). Hence, studies suggest a net positive charge demonstrates enhanced tumor accumulation (Hu-Lieskovan et al. 2005). Positive charge on the nanocarrier surface favors the interaction with the tumor blood vessels thereby inhibiting their entry back into the systemic circulation. Additionally, after endocytosis the nanocarrier could escape from the endosomes and enter a cell's cytoplasm through disruptive interaction of the cationic lipid with endosomal membranes (Ding and Ma 2013; Hafez et al. 2001).

For example, positively charged liposomes containing the lipid 1,2-diacyl-trimethylammonium propane (DOTAP) have shown high tumor accumulation compared to neutral or negative liposomes (Campbell et al. 2002; Schmitt-Sody et al. 2003; Krasnici et al. 2003). Similar effect was observed with positively charged sterically stabilized colloidal macromolecules (Ho et al. 2010; Meng et al. 2011). In a different study, positively charged liposomes showed lower accumulation in spleen and higher association with tumor blood vessels when compared with negatively charged liposomes in vivo (Campbell et al. 2002). In addition, high antitumor effect was observed with neutral PEGylated doxorubicin-loaded liposomes as compared to cationic liposomes (Zhao et al. 2011). Phosphatidylcholine (PC)/cholesterol (Chol) liposomes containing 6 % mol of a charged lipids [stearylamine (SA) or phosphatidic acid (PA) or phosphatidyl serine (PS)] were prepared with varying amounts of poly(ethylene glycol)–phosphatidylethanolamine (PEG–PE) (700 and 500). The clearance of positively charged SA-liposomes was inhibited by both PEG–PE 750 and PEG–PE 5000, whereas the clearance of negatively charged PA-liposomes was inhibited by PEG–PE 5000 alone (Levchenko et al. 2002). Recently, it has been reported that the surface charges represented by zwitterionic materials also influence nonspecific interactions with endothelial cells (Han et al. 2013).

Although the molecular size and surface properties of a nanocarrier might be critical, they alone are not sufficient for the best accumulation. Longevity in the circulation is an important factor affecting the EPR effect.

1.3.2.3.3 Longevity

Longevity is essential to maintain a required level of the nanocarrier in the circulation for an extended period of time. In addition to passive accumulation, increased circulation can also help achieve better active targeting due to the large number of passages made through the target (Torchilin 1996, 2010). The clearance by circulating blood components, such as the macrophages, and subsequently the reticuloendothelial system (RES) contributes to fast removal of the nanocarrier. In order to protect these nanocarriers from the capture and clearance, a steric protection is required (Napper 1983; Needham et al. 1992).

This is achieved by using various polymers, grafted on the surface of the nanocarriers that provide steric protection against opsonins in the biological environment and reduce the clearance by the RES (Torchilin 2001b; Weissig et al. 1998). The most popular polymer is PEG–lipid conjugates [e.g., methoxy polyethylene glycol (mPEG)–distearoylphosphatidylethanolamine (DSPE)], which have been shown to significantly prolong the blood circulation of the nanocarriers (Torchilin 2011). The mechanism includes shielding the surface charges, increased hydrophilicity, and formation of a polymeric layer around the carrier surface, which enhances the repulsion between the blood components and the nanocarriers (Gabizon and Papahadjopoulos 1992).

Molecular weight of the PEG and the amount of the PEG used to sterically stabilize the nanocarrier influence the blood circulation time (Lasic et al. 1991). Thus, higher molecular weight PEG, such as PEG-1900 and PEG-5000, imparted higher circulation to nanocarriers than low molecular weight PEG, such as PEG-750 and PEG-120 (Allen et al. 1991). A variety of long-circulating nanocarriers are prepared by encapsulating different drugs such as cisplatin, adriamycin, and vincristine (Huang et al. 1994; Kim et al. 2001). PEGylated liposomal doxorubicin (Doxil®) reported a sixfold increase in the circulation time compared to free doxorubicin (Bogner and Goebel 1995; Gabizon et al. 1995; Siegal et al. 1995; Safra et al. 2000; Gabizon 2001). When polymeric micelles were functionalized with PEG–PE, their circulation half-life was also significantly increased in mice, rats, and rabbits (Lukyanov et al. 2002). However, PEG–PE micelles have a shorter circulation time as compared to PEG-coated liposomes because of their small size (facilitation extravasation) and dissociation in the blood (Klibanov et al. 1990). Although PEG remains a gold standard for surface modification and protection, it has been reported to induce activation of complement system (Moein Moghimi et al. 2006). Alternatively, other grafting molecules such as polyvinyl alcohol (Takeuchi et al. 2001), L-amino acid-based biodegradable polymer–lipid conjugates (Metselaar et al. 2003), poly[*N*-(2-hydroxypropyl)methacrylamide] (Whiteman et al. 2001), and poly-*N*-vinylpyrrolidones (Torchilin et al. 2001a) have been reported for steric protection and longevity. Additionally, hyaluronic acid (HA) was reported to provide resistance to protein adsorption when coated on poly(D,L-lactic-*co*-glycolic acid) (PLGA) surfaces (Lord et al. 2009; Croll et al. 2006) and prolonged circulation times in vivo, when surface modified over liposomes loaded with mitomycin C (MMC) (Peer and Margalit 2004b).These formulations showed 7–70-fold longer circulation times than liposomes without HA or free MMC in three tumor models: C57BL/6 bearing metastatic clone cells (B16F10.9), BALB/c bearing colon adenocarcinoma (C-26 solid tumors), and metastatic Lewis lung carcinoma (D122) cells. Other nanocarriers such as chitosan/triphosphate nanoparticles have also been reported that use HA as a coating agent (Almalik et al. 2013). In addition to the abovementioned nanocarriers, some proteins and small molecules were also delivered via passive targeting. L-asparaginase modified with PEG showed an increased circulation time of 5.7 days as compared to 1.2 days for the non-modified enzyme (Asselin 1999).

1.3.3 Nanocarriers for Passive Targeting

In addition to the inherent "EPR effect" exploited for passive targeting, we can now appreciate the various physicochemical factors that affect the EPR-based passive targeting. It has repeatedly shown that long-circulating pharmaceutical liposomes and micelles, and other polymeric nanoparticles, can accumulate in various pathological areas with affected vasculature via the EPR effect and have been used for delivery of drugs into tumors via passive accumulation. One of the most widely used chemotherapeutic agents, paclitaxel, has been shown to accumulate in tumors much better than its commercial formulation Taxol®, when loaded into micelles made of PEG-*b*-poly(4-phenyl-1-butanoate)-L-aspartamide conjugates (Hamaguchi et al. 2005). This study reported approximately a 100-fold increase in the plasma area under the curve (AUC) resulting in a 25-fold improved drug accumulation and increase in antitumor activity in mice bearing C-26 tumors. Additionally, many similar reports have been shown to improve the delivery of chemotherapeutic agents using polymeric micelles (Torchilin 2010). However, the problems with drug delivery using polymeric micelles for EPR-mediated passive targeting are usually associated with rapid drug release from the micelles and challenges with respect to intracellular delivery of the drug (Torchilin 2011; Aliabadi and Lavasanifar 2006). This can be overcome by chemically conjugating the drug with hydrophobic blocks of micelle-forming material (Kang et al. 2005; Yuan et al. 2005; Shuai et al. 2004; Lavasanifar et al. 2002). Another widely studied chemotherapeutic agent, doxorubicin, is incorporated into long-circulating PEG-coated liposomes for highly effective EPR-based tumor therapy. For instance, Doxil® and Caelyx® are used for the treatment of solid tumors in patients with breast carcinoma metastases, ovarian cancer, and have shown subsequent survival improvement (Perez et al. 2002).

Additionally, PEGylated interferon, hematopoietic growth factors, lymphokines, and cytokine inhibitors act as the EPR-based agents (Abuchowski et al. 1984;

Bukowski et al. 2002; Eliason 2006). Drug complexes with hydroxypropylmethacrylate (HPMA) polymer are also delivered to tumors via the EPR effect (Kopecek et al. 2000; Peterson et al. 2003). Although there are many macromolecules and nanocarriers delivered via passive targeting (Maeda et al. 2009; Upadhyay et al. 2009; Greish 2007; Modi et al. 2006; Iyer et al. 2006; Luo and Prestwich 2002; Kopecek et al. 2001), it suffers from several disadvantages. Primarily, since the accumulation of nanocarriers is dependent on the cutoff size of the fenestrations within the neovasculature, these cutoff sizes can vary with the variations in tumor type and stage (Hobbs et al. 1998; Bae 2009). Additionally, the diffusion within the tumor interstitium can be hindered by the high interstitial fluid pressure (Jain 1994), which can inhibit the distribution of the nanocarrier within the tumor tissue. It is important to mention, however, that in many pathological situations the integrity of vascular endothelium remains unaffected, and there is no opportunity for EPR (Torchilin 2010). To circumvent these challenges and make the nanocarriers accumulate more effectively in the tumor, systems combining passive targeting and active targeting can be engineered.

1.4 Active Targeting of Nanocarriers

In addition to passive targeting, successful delivery of nanocarriers to the site of action can be achieved using active targeting. A concept once known as magic bullet by the visionary Paul Ehrlich (Strebhardt and Ullrich 2008), active targeting utilizes ligands such as antibodies, peptides, etc., on the surface of the nanoparticles, which facilitates their uptake by a specific diseased or cancerous cell. The homogenous overexpression of certain proteins and molecules on the surface of the cancer cells, or in and around the tumor interstitium also known as receptors, acts as an anchor for interaction with or attachment to the ligands that are functionalized on the surface of the nanocarrier (Peer et al. 2007; Kamaly et al. 2012; Shi et al. 2011; Cheng et al. 2012; Koshkaryev et al. 2013). The two most common strategies of targeted drug delivery are (1) the engineering of stimuli-sensitive nanosystems that will respond to the tumor or disease microenvironment and the changes in the pathological area in a way that would trigger release of the drug and (2) the classic receptor-mediated targeting where the nanoparticle is decorated with ligands that bind specific receptors of interest.

The targeted nanocarrier should be in close proximity with the receptor to achieve optimal benefit of this approach. Hence, it is important to understand the fundamentals and various factors involved in the interaction of the targeted nanocarrier with the receptor of interest without affecting the overall pharmacokinetics of the formulation (Alexis et al. 2008; Byrne et al. 2008). The recognition of the target can occur on various levels, such as on a whole organ, or on a cellular level, or on a level of individual cellular components. The most universal form of target recognition is at the molecular level, based on the identification of organ or tissue compounds or antigens that are specific to the organ of interest (Torchilin 2000).

In addition to the size, surface, and longevity, which are discussed earlier, the choice of the targeting ligand plays a critical role in the development of targeted nanocarriers (Bertrand and Leroux 2012; Monopoli et al. 2012; Gu et al. 2008; Jiang et al. 2008; Valencia et al. 2011). The design of the nanoparticle, the conjugation chemistry used in ligand binding, and the various types of ligands used all contribute to a successful targeted therapy. In this section, the fundamentals of active targeting, factors, and various strategies of targeted nanocarrier delivery systems are discussed.

1.4.1 Fundamentals and Concepts

The most important element in producing a pharmacological effect when using active drug targeting is the ability of the ligand to identify or recognize the receptor of interest. These receptors belong to various categories of molecular entities, which are overexpressed by tumor cells or tumor vasculature compared to the normal cells (Yu et al. 2010; des Rieux et al. 2013). Likewise, different ligands are available, such as antibodies, proteins, peptides, sugars, and nucleic acids, which specifically bind their respective receptors. It is essential to have a sufficient number of ligand molecules on the surface of the nanocarrier for effective interaction, ensuring an increased avidity of the nanocarrier to the receptor site (Wang et al. 2010). In addition, the biodistribution of the targeted nanocarrier will also contribute to the interaction of the ligand with the receptor target (Kamaly et al. 2012; Wu and Chu 2013).

Another fundamental aspect for achieving effective drug delivery is the ability of the nanocarrier with ligand to be in close vicinity of the recognition molecule (receptor) (Florence 2012). Additionally, tumor penetration could be hindered because of the "binding-site barrier" (Danhier et al. 2010), i.e., accumulation of the targeted nanocarriers on the periphery of the tumor, which can be challenging to achieve active targeting into the tumor mass. Hence, careful consideration must be given to the design of the nanocarrier.

Similar to passive targeting, the targeted nanoparticles also accumulate owing to the EPR effect (Lammers et al. 2012; Taurin et al. 2012). In addition, targeting nanocarriers to the receptors located in the extravascular space of the tumor (Bertrand et al. 2014) can overcome some of the abovementioned challenges. However, the nanocarriers have to be carefully optimized and engineered for prolonged circulation (Koshkaryev et al. 2013). One has to also keep in mind that unwanted immune response could be generated with the use of certain protein/antibody ligands (Park et al. 2001; Harding et al. 1997; Benhar et al. 1994). In such cases, replacing large immunoglobulin (IgG) molecules with small peptides or antibody fragments offer a better alternative since they are less immunogenic than a complete IgG molecule (Torchilin 2009; Zeng et al. 2013). Following successful targeting, it is essential for the nanocarrier to release the drug intracellularly (Bartlett et al. 2007; Farokhzad et al. 2006a; Kirpotin et al. 2006). This is achieved through vesicular transport- or receptor-mediated endocytosis or other pathways that have been extensively reviewed and documented (Biswas and Torchilin 2014; Wang et al. 2012; Zhao and Rodriguez 2013; Perche and Torchilin 2013; Sawant and Torchilin 2012a).

1.4.2 *Factors Affecting Active Targeting*

Since active targeting employs nanoparticles used for passive targeting with the exception of surface modification with ligands, the factors affecting passive targeting also apply to actively targeted nanocarriers. More so, the surface properties of the nanocarriers are drastically affected in terms of zeta potential, size, and the presence of ligands (Campbell et al. 2002). Additionally, the choice of nanoparticle used affects the accumulation of the nanocarrier at the site of action, for example, micelles versus liposomes. However the most important governing factor for targeted delivery of nanoparticles is the ligand/receptor density and the preparation of the targeted nanocarrier.

1.4.2.1 Different Chemistry of the Ligand Attachment to Nanocarriers

Methods of functionalizing ligands on the nanocarriers can greatly influence the yield, the density and thus, the overall treatability of the nanocarrier. One of the chemical conjugation strategies often used is a simple one-step method of attaching the distal terminus of the PEG block with any ligand containing amino groups via *p*-nitrophenylcarbonyl (*p*NP) (Torchilin et al. 2001b, c). The distal end of the PEG–lipid can be functionalized using different chemistries depending on the ligand and application of the nanocarrier. This technique was successfully used for conjugating various monoclonal antibodies such as anti-nucleosome 2C5 antibody (Torchilin et al. 2003; Gao et al. 2003). The ligand–PEG conjugate can be mixed with different liposome-forming materials, and the concentration of the required ligand can be varied depending on the yield.

Alternatively, "post-insertion technique" can be used to prepare ligand–nanocarrier conjugates (Ishida et al. 1999). In this technique, the ligand–PEG–PE conjugates are co-incubated with plain or PEGylated nanocarrier ensuring positioning of the ligands on the outer surface of the nanocarrier with high yields (Sawant and Torchilin 2012b). In a study reported, post-insertion of DSPE-PEG-2000 in irinotecan-loaded liposomes showed higher plasma concentration and slower drug release in rats, when compared to pre-insertion technique (Yoshino et al. 2012; Nakamura et al. 2012). In some cases, the combination of the above two chemistries has also been reported. For example, post-insertion technique was used to prepare immuno-Doxil which was modified with (*p*NP)–PEG–PE-modified anticancer 2C5 monoclonal antibody (Torchilin et al. 2001c).

1.4.2.2 Receptor and Ligand Density

Apart from the ligand chemistry, the ligand density on the nanocarrier and the receptor density play an important role in active targeting. Ideally, a high receptor density with high specificity at the target site would be ideal. However, overexpressed receptors density may vary depending on the disease condition, hence it is much more favorable to rely on designing nanocarriers with optimum ligand density to interact with even sparsely expressed receptor binding sites. For example, PLGA nanoparticles were prepared by controlling the ligand cyclic peptide, cLABL, to optimize binding with the target, intercellular cell adhesion molecule-1 (ICAM-1), in human alveolar basal epithelial cells (A549) (Fakhari et al. 2011). Additional studies have reported that increase in the number of ligands or ligand density increased cell uptake of the nanocarriers in vitro (Gao et al. 2013; Stefanick et al. 2013).

However, this is not always the case. It has been observed that increasing the ligand density has negative effects on cellular uptake and binding (Elias et al. 2013). This can be due to the presence of ligands in close proximity with one another resulting in a competition of the ligands for a particular receptor, impeding their interaction, and leading to poor or no pharmacological response (Stefanick et al. 2013). Additionally, oversaturating the nanocarrier with ligands could result in poor design of nanocarrier due to improper ligand orientation rendering it less effective. The distribution of ligand is largely dependent on the number of ligand grafting sites, the conjugation efficiency, and the type of ligand used (Elias et al. 2013; Hakem et al. 2010). For example, high-efficiency yields are reported while using click conjugation technique (Hein et al. 2008). In some case, high ligand density has shown to increase uptake by the macrophages. For instance, PLGA-PEG-RGD nanoparticles composed of 74 % PLGA-PEG-Folate showed an 8.5-fold increase in phagocytosis by the macrophages in vitro (Valencia et al. 2011). The high uptake can also be attributed to the low hydrophilicity of folate, thereby inducing opsonization and presence of folate receptors on the surface of the macrophages (Jiang et al. 2008; Ghandehari et al. 2001). Additionally, the use of other hydrophobic ligands may result in an increased uptake by the macrophages (Valencia et al. 2011).

Since antibodies are one of the most widely used ligands for active targeting, it has been shown that increasing the concentration of antibody moieties increases the interaction with the receptor sites (Hakem et al. 2010). When poly(ethylene glycol)-based nanoparticles were conjugated with transferrin receptor antibody (NP-OKT9), the targeting efficiency was dependent on ligand density (Wang et al. 2010). This phenomenon is observed until a saturation point when no further increase in the number of ligands has any significant effect (Liu et al. 2010a; Zhou et al. 2007), owing to the differences in the binding affinities (Orlova et al. 2006; Park et al. 2000; Dixon et al. 1992). In addition to the ligand density, active targeting can be greatly affected by the diffusion across the tumor interstitium (Thurber and Weissleder 2011; Rudnick and Adams 2009; Fujimori et al. 1989) when using high-affinity ligands (Juweid et al. 1992; Weinstein and van Osdol 1992; van Osdol et al. 1991; Fujimori et al. 1990).

Another factor that significantly affects the active targeting of nanocarriers is the receptor itself. In some instances, it has been observed that although receptor binding sites are overexpressed, they exist as receptor clusters (Elias et al. 2013). HER2/neu receptor is a good example of such receptors, which exists in clusters of various sizes (Kaufmann et al. 2011). Similar effects have also been observed with transferrin receptors (Smart et al. 1996), folate receptors (Liu et al. 2010b), ICAM-1 (Fakhari et al. 2011), and integrins such as the $\alpha_4\beta_3$ (Cluzel et al. 2005). Each ligand has a unique behavior with regard to its binding to specific receptor sites. We will take a closer look at the different receptors and ligands used for active targeting in the following section.

1.4.3 Ligands Used for Active Targeting

A plethora of ligands are available for active targeting of nanocarriers. These depend on the choice of target and the nanoparticle used. Factors, such as stability, ease of preparation, and to some extent cost, are considered in the choice of ligands. Summarized are few of the commonly used ligands in active targeting.

1.4.3.1 Antibodies and Their Fragments

Antibodies are large molecules with an average molecular weight of 150 kDa and with a hydrodynamic radius ranging from 15 to 20 nm. They are glycoproteins and are highly known for their specificity and high binding affinity to the receptor sites. The F(ab′)2 portion of the antibody, also known as the active site, is responsible in eliciting any antibody reaction or to carry out any functions of the antibody. Either the entire antibody molecule or just the F(ab′)2 fragments are used for active targeting. Fully humanized chimeric antibodies with minimum immunogenicity have led to the development of monoclonal antibody-targeted therapies, such as bevacizumab, trastuzumab, cetuximab, and rituximab (Kamaly et al. 2012; Leserman et al. 1980; Heath et al. 1980; Warenius et al. 1981). A variety of nanocarriers such as micelles and liposomes are prepared using antibody ligands for active targeting (Kirpotin et al. 2006; Torchilin et al. 2003; Gao et al. 2003; Sawant and Torchilin 2012b; Skidan et al. 2008; Sawant et al. 2008; Elbayoumi et al. 2007; Roby et al. 2006; Yamada et al. 2008; Mamot et al. 2005).

1.4.3.2 Peptides

Peptides are sequences of amino acids (typically below 50), which exist either in a linear or a cyclic form. Depending on the length of the amino acid sequence they vary in their molecular size and stability. For example, small chain sequence and simple structure contribute to their stability (Bertrand et al. 2014). This greatly influences the conjugation technique during the preparation of the nanocarrier. Due to the advancement in the phage display techniques (Needels et al. 1993; Lam et al. 1991) a variety of peptides have been developed to target new cell-specific domain as targeting moieties (Pasqualini and Ruoslahti 1996; McGuire et al. 2009). The $\alpha_v\beta_3$ integrin receptors are highly expressed on the surface of osteoclasts, angiogenic endothelial cells, and some solid tumors (Millard et al. 2011). They have been extensively studied using the RGD (arginine–glycine–aspartic acid) peptide, which shows high specificity (Kamaly et al. 2012; Shi et al. 2011). Other peptides, such as the cell-penetrating peptides, have also been used in active intracellular targeting of nanocarriers for organelle targeting (Deshpande et al. 2013; Biswas and Torchilin 2014).

1.4.3.3 Sugars

Numerous cancer cells express lectins on the cell surface that can be actively targeted using sugar molecules. Lactose, mannose, and galactose are commonly used targeting ligands (Zubieta et al. 2006; Ohannesian et al. 1995). For instance, hepatocytes in liver cancer overexpress asialoglycoprotein receptors that can be targeted using galactose (Zubieta et al. 2006). Following slight chemical modification, sugar moieties can be attached to the surface of the nanocarrier. Galactosamine was conjugated with *N*-(2-hydroxypropyl) methacrylamide copolymer for treatment of liver cancer (Davis et al. 2008). Additionally, 6-ascorbate-PEG-1,2-distearoyl-*sn*-*glycero*-3-phosphoethanolamine (6-ascorbate-PEG-PE) was used to target glioma cells in vitro (Salmaso et al. 2009). However, when using a sugar moiety as a ligand, a challenge could be faced due to the low binding affinity of carbohydrates, which can be overcome by increasing ligand density and conjugating multiple ligands on the nanocarrier to achieve multivalent interactions (Managit et al. 2003).

1.4.3.4 Aptamers

Aptamers are a class of nucleic acids that include single-stranded oligonucleotides like DNA, RNA or xeno-nucleic acids (Xiao and Farokhzad 2012). They interact with target receptors with high affinity by hydrophobic, electrostatic interactions and hydrogen bonding (Lapointe et al. 2006; Zhu et al. 2011). Compared to antibodies, aptamers are smaller in size, which allows for better tissue penetration (Farokhzad et al. 2006b). Due to the advancement in the technology and the systematic evolution of ligands by exponential enrichment (SELEX), a wide library of aptamers specifically recognizing targets have been developed (Fang and Tan 2010). In addition to their specificity, they are relatively easy to synthesize and modify as compared to antibodies. In the case of targets that are poorly immunogenic (Edwards et al. 2010; Kang et al. 2010), aptamers offer even greater specificity than antibodies. They have been used for the targeting of receptors such as VEGF (Shangguan et al. 2006; Daniels et al. 2003) and prostrate-specific aptamers receptor. For example, biocompatible and biodegradable poly(D,L-lactic-*co*-glycolic acid)-*block*-*poly*(ethylene glycol) (PLGA-*b*-PEG) nanoparticles encapsulating docetaxel were surface functionalized with the A10 2′-fluoropyrimidine RNA aptamers (Farokhzad et al. 2004, 2006a). The use of the aptamer-targeted nanomedicines resulted in 100 % survival of prostate epithelial (LNCaP) cancer xenograft nude mice versus 57 % survival using the nontargeted preparation,

1.4.4 Target Receptors Involved in Active Targeting

From the previous section it can be concluded that many receptors are available to target cancer cells. Still, there are two main approaches to active targeting: (1) to target the overexpressed proteins on the surface of the cells and (2) to target the tumor microenvironment. In the former case, ligands, such as antibodies, are commonly used to target the cell surface antigens (Allen 2002). Both approaches are employed to achieve significant payload at the target site (Byrne et al. 2008). In this section, various receptors overexpressed either on the cell surface, tumor endothelium, or microenvironment are discussed.

1.4.4.1 The Epidermal Growth Factor Receptor

Epidermal growth factor receptor (EGFR) belongs to the family of ErbB-tyrosine kinase receptors. Activation of EGFR stimulates important processes such as growth and progression of tumor including angiogenesis, cell proliferation, invasion, and metastasis. EGFR is overexpressed in a variety of cancer types (Lurje and Lenz 2009) of which the human epidermal receptor-2 is reported to be overexpressed in 14–19 % of patients with breast cancer (Acharya et al. 2009; Scaltriti and Baselga 2006; Baselga et al. 1996; Slamon et al. 1987, 1989).

1.4.4.2 Folate Receptor

Folate receptor binds the vitamin folic acid that is essential for nucleotide base synthesis and cell survival. Additionally, it is also required in carbon metabolism. These receptors are overexpressed particularly on ovarian cancer cells along with other cancer types. Two isoforms of folate receptors are observed, namely folate receptor—α which is overexpressed on more than 40 % of human cancers, and folate receptor—β which is expressed on malignant cells of hematopoietic origin and activated macrophages (Low and Kularatne 2009).

1.4.4.3 Transferrin Receptor

Transferrin is a serum glycoprotein and plays a vital role in regulation of cell growth. Additionally, it is a key component in iron homeostasis where transferrin transports iron across the cell membrane via the receptor-mediated endocytosis. Cancer cells express transferrin receptors to a higher extent than normal cells. Due to their extracellular surface accessibility, they are attractive targets for active cancer therapy (Pourbaghi-Masouleh and Hosseini 2013; Kim et al. 2010; Cho et al. 2008; Sahoo et al. 2004). In addition, they are overexpressed by many tumors in response to their increased metabolic demand (Deshpande et al. 2013; Liechty and Peppas 2012).

1.4.4.4 Glycoproteins

Carbohydrates that are attached to glycoproteins expressed on the cell surface make an attractive receptor for binding with lectins. The carbohydrate–lectin interaction is very specific and was utilized for targeting colon cancer using nanocarriers coupled with lectins (Minko 2004).

1.4.4.5 The VEGF Receptors

VEGF plays an important role in tumor angiogenesis and neovascularization (Danhier et al. 2010). It is a highly specific mitogen for endothelial cells. It binds to the two VEGF receptors, VEGF receptor-1 and VEGF receptor-2 that are expressed on vascular endothelial cells (Matsumura et al. 1988, 1991). Signal transduction involves binding to tyrosine kinase receptors and results in endothelial cell proliferation, migration, and new vessel formation (Hoeben et al. 2004). The driving force for the upregulation of VEGF is the tumor hypoxia and oncogene expression. VEGF and other growth factors interact with receptors result in enhanced angiogenesis. Various approaches targeting VEGF inhibit neovascularization, resulting in tumor cell death due to lack of oxygen and nutrients. To achieve angiogenesis, two approaches are carried out (a) targeting VEGF to inhibit ligand binding to VEGFR and (b) targeting VEGFR to decrease VEGF binding to its receptor (Byrne et al. 2008; Veikkola et al. 2000; Carmeliet 2005). Nanocarriers targeted using with anti-VEGFR-2 showed delay in tumor growth (Chen et al. 2006).

1.4.4.6 The Vascular Cell Adhesion Molecule-1

Vascular cell adhesion molecule-1 (VCAM-1) is a transmembrane glycoprotein that is expressed on the surface of the endothelial tumor cells and is responsible for cell-to-cell adhesion, which is crucial during the process during angiogenesis. VCAM-1 is overexpressed on the surface of a variety of cancers, such as leukemia, melanoma, nephroblastoma, and renal, gastric, lung, and breast cancer, and is targeted using anti-VCAM antibody (Dienst et al. 2005).

1.4.4.7 Integrins

Integrins are glycoproteins that play a key role in binding with extracellular matrix by interacting with proteins, such as osteopontin, thrombospondin, vibronectin, fibronectin, and fibrinogen (Desgrosellier and Cheresh 2010). The $\alpha_4\beta_3$ integrin receptor is a heterodimer containing the α- and β-subunits. The $\alpha_4\beta_3$ integrin receptors are overexpressed on the endothelial cells of many tumors but poorly expressed in endothelial cells of normal organs (Byrne et al. 2008).

1.4.5 Nanocarriers for Active Transport

Various nanocarriers have been employed for active targeting of drugs to the tumor tissue and, as discussed above, a library of ligands are available to target specific receptor sites. Among the various ligands, antibodies are the most diverse and broadly used ligands for active transport. PEG–PE micelles have also been modified with a variety of tumor-specific monoclonal antibodies (Torchilin et al. 2003; Skidan et al. 2008; Sawant et al. 2008; Elbayoumi et al. 2007; Roby et al. 2006). For instance, 2C5 antibody-modified mixed immunomicelles showed effective binding to MCF-7 (breast adenocarcinoma), LLC (Lewis lung carcinoma), EL4 (T lymphoma), and BT-20 cells when compared to antibody-free micelles (Gao et al. 2003). Similar results were observed when mixed immunomicelles containing vitamin E were used in various cell lines (Gao et al. 2003; Sawant et al. 2008; Mu et al. 2005). Unfortunately, there have been some disadvantages to the use of antibody as targeting ligand. Antibody-tagged liposome molecules are reported to have been opsonized by circulating or liver macrophages via the Fc receptors. Additionally, use of whole antibody molecule has been shown to elicit antibody-dependent or complement-mediated cellular cytotoxicity. To overcome these challenges, Fab′ antibody fragments are used. Fab′ fragment of monoclonal antibody HER2 when surface immobilized on doxorubicin-loaded PEG–liposomes demonstrated increased antitumor effect in breast cancer xenograft models compared to nontargeted PEGylated liposomes (Park et al. 2001, 2002; Mamot et al. 2005; Carter et al. 1992). Other nanocarriers have also been prepared specifically to target the tumor interstitium. The accumulation of anti-VCAM-1 and Fab′-coupled liposomes loaded with celecoxib was greatly increased by HUVECs when activated using TNF-α which showed induced expression of VCAM-1 (Kang et al. 2011).

In a different study, EGFR-conjugated micelles prepared from diblock copolymers of methoxy poly(ethylene glycol)-*block*-poly(δ-valerolactone) with a fluorescent dye showed intracellular uptake in MDA-MB-468 breast cancer which overexpressed EGF receptors (Zeng et al. 2006). Other ligands, such as folic acid, have been conjugated to the surface of nanocarriers (Liang et al. 2009; Esmaeili et al. 2008; Kim et al. 2005; Stella et al. 2000). Thus, folic acid-polyethylene glycol-polyethyleneiminepoly(caprolactone) (FA-PEG-PEI-PCL) was used for the delivery of siRNA (Zou et al. 2012). Folic acid–coated anionic liposomes containing doxorubicin demonstrated increased cellular uptake and higher toxicity in human nasopharyngeal carcinoma KB cells that overexpressed folate receptor, compared to their nontargeted liposome (Watanabe et al. 2012). In a recent study, a thermosensitive and folate-functionalized poly(ethylene oxide)-*b*-poly(propylene oxide)-*b*-*poly*(ethylene oxide)-poly(*N*-isopropylacrylamide-*co*-hydroxyethyl methacrylate) (FA-Pluronic-PNH) pluronic micelles loaded with doxorubicin demonstrated increased cellular uptake in human cervical cancer (HeLa) and human lung cancer (A549) cell lines (Yang et al. 2013). Similarly, other folate-conjugated nanocarriers have also been reported (Nair et al. 2013; Zhang et al. 2013; Yu et al. 2013; Chen et al. 2013; Zhu et al. 2013). Another alternative to targeted therapy is the use of

RGD ligands to target the $\alpha_4\beta_3$ integrins. Cationic nanoparticles that were surface modified with RGD ligands selectively delivered genes to angiogenic blood vessels which led to apoptosis and consequently regression of primary and metastatic tumor (Hood et al. 2002). RGD was also attached to polymeric micelles for targeted delivery of doxorubicin (Nasongkla et al. 2004). Cyclic RGD peptide-modified liposomal system showed higher cellular uptake of doxorubicin in human glioblastoma (U87MG) cell line when compared to unmodified liposomes (Chen et al. 2012). RGD peptide-mediated chitosan-based polymeric micelles showed enhance cytotoxicity against human hepatocellular carcinoma (BEL-7402) cell line in the presence of doxorubicin and compared to nontargeted micelles (Cai et al. 2011). Other micellar formulations were prepared using block copolymers such as poly(ethylene glycol)-*block*-polycation carrying ethylenediamine units which were conjugated with peptide (cRGD), and used in vivo for gene delivery in rat carotid artery (Kagaya et al. 2012). Other tumor-targeted delivery systems using RGD ligands are discussed (Knudsen et al. 2012; Du et al. 2011; Garg et al. 2009).

1.5 Passively and Actively Targeted Nanocarriers: Clinical Progress

Significant progress has been made in the area of nanocarrier delivery systems using passive and active targeting. This has led to the emergence of certain nanocarrier candidates that have been approved to be used in the clinic or are under clinical investigation. In this section we discuss the preclinical and clinical status of passively and actively tumor-targeted nanoformulations.

1.5.1 Passively Targeted Liposomes and Micelles

Various nanocarrier platforms for drug delivery via active and passive transport are undergoing clinical studies or approved to be used for various kinds of cancer diseases. Mentioned here (see Table 1.1) are various passively targeted nanocarriers lacking of targeting ligands, wherein their pharmacokinetic and biodistribution parameters are purely based on their physicochemical properties and their prolonged blood circulation and interaction with the site of action (van der Meel et al. 2013). Although a plethora of nanocarriers are under clinical investigation, our analysis here is focused specifically on the use of liposomes and micelles. Among the most popular nanomedicines is the PEGylated liposomal doxorubicin or Doxil® that received the US Food and Drug Administration (FDA) approval in 1995 to treat Kaposi sarcoma. This formulation contained a lipid bilayer membrane in a "liquid-ordered" phase composed of phosphatidylcholine and cholesterol (Barenholz 2012). Similarly, Caelyx® was approved in 1997 to be used in Europe. Some years later, Doxil/Caelyx were also approved for the treatment of ovarian cancer, multiple myeloma, and metastatic

Table 1.1 Passively targeted liposomes and micelles under clinical or preclinical application

Nanocarrier	Drug	Product name	Indication	Status	References
Liposomes					
PEGlyated liposomes	Doxorubicin	Doxil®	Metastatic breast cancer, ovarian cancer, Kaposi sarcoma	Approved/marketed	Kim et al. (2005)
Liposomes	Doxorubicin	Myocet®	Breast cancer (combination with cyclophosphamide)	Approved/marketed in EU[a]	Danhier et al. (2010)
PEGlyated liposomes	Doxorubicin	Doxil® and Caelyx®	Ovarian cancer, multiple myeloma, metastatic breast cancer	Approved/marketed	Danhier et al. (2010) and Stella et al. (2000)
Liposomes	Daunorubicin	DaunoXome®	Kaposi sarcoma	Approved/marketed	Zhang et al. (2013) and Yu et al. (2013)
Liposomes	Vincristine	Onco-TCS®	Non-Hodgkin lymphoma	Approved/marketed	
Liposomes	Cytarabine	DepoCyt®	Lymphomatous meningitis	Approved/marketed	Chen et al. (2013), Zhu et al. (2013), and Hood et al. (2002)
Temperature-sensitive liposomes	Doxorubicin	Thermodox®	Hepatocellular carcinoma and its recurrence	Phase III clinical trials ongoing NCT00617981[b]	Nasongkla et al. (2004) www.clinicaltrials.gov
Nanoliposome	Irinotecan	NL CPT-11	Recurring high-grade glioma	Phase I clinical trial currently recruiting NCT00734682[b]	Chen et al. (2012), Cai et al. (2011), Kagaya et al. (2012), Knudsen et al. (2012), and Du et al. (2011) www.clinicaltrials.gov
Liposomes	Irinotecan HCl and floxuridine	CPX-1	Advanced colorectal carcinoma	Phase II clinical trials complete NCT00361842[b]	
Liposomes	Irinotecan	MM-398	Metastatic pancreatic cancer	Phase III clinical trials ongoing NCT01494506[b]	Nasongkla et al. (2004) www.clinicaltrials.gov
Liposomes	Irinotecan and cyclophosphamide	MM-398	Pediatric solid tumors	Phase I clinical trial currently recruiting NCT02013336[b]	

(continued)

Table 1.1 (continued)

Nanocarrier	Drug	Product name	Indication	Status	References
Micelles					
Polymeric micelles	Paclitaxel	Genexol-PM®	Breast cancer and non-small cell lung cancer	Approved/marketed in Korea	Lao et al. (2013) and O'Brien et al. (2004)
Polymeric micelles	Paclitaxel	Genexol-PM®	Metastatic or recurrent breast cancer	Phase III clinical trial currently recruiting NCT00876486[b]	Nasongkla et al. (2004) and Zhang et al. (2008) www.clinicaltrials.gov
Polymeric micelles	Paclitaxel in combination with doxorubicin	NK911	Solid tumors	Phase I study in Japan[a]	Lammers et al. (2008)
Polymeric micelles	Paclitaxel	NK105	Metastatic breast cancer	Phase III trial currently recruiting NCT01644890[b]	Kripp and Hofheinz (2008) and Gaviani et al. (2013) www.clinicaltrials.gov
Polymeric micelles	Cisplatin	NC-6004 with gemcitabine	Advanced or metastatic pancreatic cancer	Phase I clinical trials complete NCT00910741[b]	Benesch and Urban (2008) www.clinicaltrials.gov
Polymeric micelles	Oxaliplatin	NC-4016	Advanced cancers and lymphoma	Phase I trial currently recruiting NCT01999491[b]	
Polymeric micelles	Paclitaxel	Paclical®	Epithelial ovarian cancer, primary peritoneal cancer, Fallopian tube cancer	Phase III no updated status. NCT00989131[b]	www.clinicaltrials.gov

Clinical trial identifier can be searched on http://www.clinicaltrials.gov
[a]Indicates not marketed in the USA
[b]Indicates ClinicalTrials.gov identifier

breast cancer. Myocet®, a PEGylated liposomal doxorubicin formulation, was approved in Europe and used to treat breast cancer in combination with cyclophosphamide (Danhier et al. 2010). Doxil® formulation was designed to increase blood circulation time. The half-life time of free doxorubicin was increased from 0.2 h to 2.5 h and 55 h for Myocet® and Doxil®, respectively (Hofheinz et al. 2005). Additionally, Doxil® has shown lower cardiotoxicity compared to free doxorubicin (Safra et al. 2000; Anders et al. 2013; Lao et al. 2013; O'Brien et al. 2004; Moreira da Silva and Moreira da Silva 1975). In addition to Myocet®, the FDA approved other non-PEGylated liposomal formulations of daunorubicin (DaunoXome®) and vincristine (Onco-TCS®) for the treatment of Kaposi sarcoma and aggressive non-Hodgkin lymphoma, respectively (Zhang et al. 2008; Lammers et al. 2008). DepoCyt®, a non-PEGylated, sustained-release liposomal formulation of cytarabine, was approved for lymphomatous meningitis in 1999 (Kripp and Hofheinz 2008; Gaviani et al. 2013; Benesch and Urban 2008).

In addition to the abovementioned marketed formulations, various other liposomal preparations are under clinical trials or undergoing extensive preclinical investigation. Thus, Thermodox®, a temperature-sensitive liposomal formulation of doxorubicin, is being tested in phase III trials for the treatment of hepatocellular carcinoma and its recurrence (Svenson 2012). At temperatures of 42 °C, the formulation becomes leaky due to the presence of lysolipids and releases the drug. An Irinotecan-containing nanoliposome formulation, NL CPT-11, is being tested for recurring high-grade gliomas in phase I trials. Additionally, liposomal formulation of Irinotecan HCl in combination with floxuridine (CPX-1) completed phase II clinical trials against advanced colorectal carcinoma (Batist et al. 2009; Bayne et al. 2009; Tardi et al. 2007, 2009; Mayer et al. 2006). MM-398, a liposomal formulation of irinotecan, is undergoing phase III clinical investigation in patients with metastatic pancreatic cancer and phase I study in pediatric solid tumors along with cyclophosphamide (Svenson 2012). Preclinically, many liposomal nanocarriers are being evaluated for the treatment of metastatic pancreatic cancer, and their EPR-mediated passive targeting capacity has been assessed using various imaging techniques (Tagami et al. 2011a, b; Negussie et al. 2011; de Smet et al. 2011; Deckers and Moonen 2010).

Similarly, a large number of micellar nanocarrier platforms have been evaluated clinically. Genexol-PM®, a cremophor-free polymeric micelle of paclitaxel prepared using PEGylated poly(D,L-lactide) copolymer, obtained approval in 2007 to be used in Korea for the treatment of breast cancer and NSCLC (Wacker 2013; Chen et al. 2011). Genexol-PM® increased the maximum tolerated dose (MTD) of paclitaxel up to 300 mg/m^2 when administered every 3 weeks in patients with breast cancer (Montana et al. 2011). In the USA, Genexol-PM® is undergoing phase III clinical trials for safety evaluation compared to Genexol® with Cremophor® EL in patients with metastatic or recurrent breast cancer (Svenson 2012; Lee et al. 2008). NK911, a polymeric micelle formulation containing doxorubicin conjugated to methoxy-PEG-poly(aspartic acid) block copolymer, has undergone phase I study in patients with solid tumors (Matsumura et al. 2004a). Using similar block copolymer, NK105 was prepared using paclitaxel and is being investigated in Phase III trial on patients with metastatic breast cancer (Hamaguchi et al. 2005; Matsumura 2011).

NC-6004, a micelle formulation of cisplatin/gemcitabine loaded into methoxy-PEG-poly(glutamic acid) micelles, is used for advanced or metastatic pancreatic cancer, and a similar formulation, NC-4016 containing oxaliplatin, is now in phase I clinical trials for advanced cancers and lymphoma (Uchino et al. 2005). Additionally, Paclical®, a novel nanoparticle formulation composed of patented excipient XR-17 Paclitaxel (Cremophor® EL), was studied in a phase III study in patients with recurrent epithelial cancer, primary peritoneal cancer, or fallopian tube cancer.

1.5.2 Actively Targeted Liposomes and Micelles

All the nanocarriers described above are intended to passively target to the tumor site via the EPR effect by taking advantage of the leaky vasculature and poor or reduced lymphatic drainage. Many nanocarriers that can actively target the tumor by using various ligands are also under clinical investigation (Table 1.2); however, there are no commercially available targeted nanocarriers. Doxil®/Caelyx® liposomes surface-functionalized with cetuximab, an EGFR antagonist has completed phase I clinical studies in patients with advance solid tumors (Mamot et al. 2012). MM-302, a HER2-targeted (trastuzumab) PEGylated liposomal doxorubicin formulation in combination with cyclophosphamide, is undergoing phase I safety evaluation study in patients with human epidermal growth factor receptor (HER2)+advanced breast cancer (ABC) (Wickham and Futch 2012). Anti-transferrin receptor (TfR) linked to a liposomal complex (SG-T53) was developed to restore the normal human tumor suppressor gene (p53) and is undergoing phase I clinical trials in patients with advanced solid tumors (Senzer et al. 2013). Similarly, MBP-426, a Tf-conjugated liposomal formulation of oxaliplatin, completed phase I clinical trials in patients with gastric, gastroesophageal, or esophageal adenocarcinoma (Senzer et al. 2009). MCC-465, a liposomal doxorubicin functionalized with human monoclonal GAH antibodies that specifically target to GAH overexpressed stomach cancer tissues, has completed phase I studies in patients with metastatic or recurrent stomach cancer (Hosokawa et al. 2003; Matsumura et al. 2004b). In addition to the abovementioned targeted liposomes under clinical trials, a large number of actively targeted micelles and liposomes are undergoing preclinical studies using various chemotherapeutic agents and targeting ligands (Kirpotin et al. 2006; Park et al. 2002; Ying et al. 2010; Danhier et al. 2009; Murphy et al. 2008; Yoo and Park 2004; Gabizon et al. 2003; Cheng et al. 2008; Hu et al. 2008; Hatakeyama et al. 2007; Xiong et al. 2005; Li et al. 2004; Kondo et al. 2004; Pastorino et al. 2003). For example, polymeric micelles prepared from self-assembled diblock copolymer of PLGA and PEG encapsulating doxorubicin were functionalized with ligands towards folate receptor and showed increased accumulation of doxorubicin in tumor tissue in human squamous cell carcinoma (KB cells) xenografted nude mouse model (Yoo and Park 2004). In addition to therapy, nanocarriers are also being investigated for diagnostic purposes. Especially, the application of magnetic nanoparticles is of great potential for the use in magnetic resonance imaging (Cheng et al. 2013; Hilger and Kaiser 2012; Goodwill et al. 2012; Tu and Louie 2012; Yigit et al. 2012).

Table 1.2 Actively targeted liposomes under clinical or preclinical application

Nanocarrier	Drug	Targeting ligand	Product name	Indication	Status	References
Liposomes						
Doxil®/Caelyx® liposomes	Doxorubicin	Cetuximab	–	Advance solid tumors	Phase I clinical studies complete NCT01702129[a]	Svenson (2012) www.clinicaltrials.gov
PEGylated liposome	Doxorubicin	Trastuzumab	MM-302 with cyclophosphamide	Human epidermal growth factor receptor (HER2) + advanced breast cancer (ABC)	Phase I clinical trial currently recruiting NCT01304797[a]	Batist et al. (2009) www.clinicaltrials.gov
Liposomal complex	In combination with docetaxel	Anti-transferrin (scFv)	SG-T53	Advanced solid tumors	Phase I clinical trial currently recruiting NCT00470613[a]	Bayne et al. (2009) www.clinicaltrials.gov
Liposomal formulation	Oxaliplatin	Anti-transferrin	MBP-426	Gastric, gastroesophageal or esophageal-adenocarcinoma	Phase I clinical trial complete NCT00355888[a]	Tardi et al. (2009) www.clinicaltrials.gov
Immunoliposome	Doxorubicin	Monoclonal GAH antibodies	MCC-465	Metastatic or recurrent stomach cancer	Phase I clinical trial complete	Tardi et al. (2007) and Mayer et al. (2006)

Clinical trial identifier can be searched on http://www.clinicaltrials.gov

[a]Indicates ClinicalTrials.gov identifier

From liposomal Doxil® to micellar Genexol-PM, the above examples clearly indicate the significant progress made in the field of nanomedicine, and these advancements are expected to continue. Many of the approved formulations are used in combination with the existing treatment modalities such as radiation therapy or conventional chemotherapy, such as in combined modality anticancer therapy. Several other polymer–drug conjugates, nanoparticles and antibody–drug conjugates micelles, liposomes, etc. have been evaluated clinically. Oncaspar, a PEGylated protein, has already been approved for the treatment of leukemia. Polymer therapeutics, such as and Opaxio, are in or will be in late-stage clinical trials.

While many of the formulations are in clinical trials today, most of them fail to make past the FDA and/or EMA approval either due to the overinterpretation and/or the misunderstanding of their tumor-targeted drug delivery. Additionally, lack of image-guided studies to evaluate the personalized nanomedicine-based therapeutic effects, extensive emphasis on preclinical animal studies, might be a hindrance of nanomedicines to the clinics (Lammers et al. 2012).

1.6 Conclusion and Future Perspective

Many of the chemotherapeutic anticancer agents that are routinely used are hydrophobic, have a large volume of distribution, and are toxic towards healthy tissues. They show poor accumulation in tumor tissue and in tumor cells because of high interstitial fluid pressure, presence of drug efflux pumps, etc. To overcome these barriers and for a successful antitumor effect, nanocarriers are used to ferry the anticancer agents more selectively to pathological sites ("site-specific drug delivery") and/or by guiding them away from potentially endangered healthy tissues ("site-avoidance drug delivery"); nanomedicine formulations aim to improve the balance between the efficacy and the toxicity of systemic (chemo-)therapeutic interventions (Lammers et al. 2012).

Amongst the many nanocarriers described in this chapter, we have mainly focused on liposomes and micelles. A wide variety of polymers such as PEG–PE have been extensively used to encapsulate the chemotherapeutic agents into the micelle core, similarly, PEG–liposomes that are prepared with cholesterol are used as a different drug-carrier platform. Nanocarriers offer many advantages such as protecting the chemotherapeutic anticancer agents from premature degradation; preventing drugs from prematurely interacting with the biological environment; enhancing absorption of the drugs into a selected tissue (e.g., solid tumor), controlling the biodistribution, pharmacokinetic parameters of the drug; and improving intracellular penetration (Peer et al. 2007).

Upon intravenous administration of the nanocarriers, they predominantly accumulate at the site of action via passive targeting or active targeting that is described in this chapter. By exploiting the pathophysiology of the tumor tissue, due to the tortuous vasculature with fenestrations and poor lymphatic drainage, the nanocarriers of certain cutoff size are able to extravasate into the tumor mass via the EPR effect, which is arguably the most important strategy for improving the delivery of low-molecular-weight chemotherapeutic agents to tumors, and it is generally referred to as "passive drug targeting" (Lammers et al. 2012). On the contrary, "active drug targeting" uses targeting ligands, such as antibodies and peptides, which is specific to the receptors that are overexpressed at the target site. Examples of various liposomes and micelles that deliver drugs via passive and active targeting are discussed.

Although the primary mode of tumor accumulation relies on the EPR-mediated passive extravasation, active drug targeting is implemented to improve target cell recognition and target cell uptake, and not to improve overall tumor accumulation. Following extravasation, the nanocarrier has to reach the cancer cells, bind to it, and get internalized in order to elicit a therapeutic response, hence, active targeting is essential component. However, if the tumor vasculature is not leaky, the tumor cells are covered with smooth muscle cells or pericytes, the tumor matrix is dense, and if the interstitial fluid pressure is high, it is likely that active targeting to cancer cells will not at all lead to any benefit over passive targeting (Lammers et al. 2012). Nevertheless, in nanocarriers that possess a targeting moiety and are actively targeted, their cell uptake is more efficient as compared to passively targeted nanocarriers. For example, liposomes targeted with HER2 showed 20 times more efficient uptake than untargeted liposomes (Kirpotin et al. 2006). Additionally, taking advantage of the tumor microenvironment, many stimuli-responsive nanocarriers are used to mediated drug targeting to tumors. The active targeting moiety is either pH sensitive or temperature sensitive. For example, Thermodox® is a temperature-sensitive PEGylated liposome containing doxorubicin. The challenges of such complex stimuli-sensitive nanocarriers are that it can be difficult to formulate to obtain very specific response with regard to their stimuli and thereby release the drug. This can be overcome by using an external stimuli which will enable a more effective and selective delivery of chemotherapeutic agent to the site of action. The external stimuli can be of a mechanical origin such as fiber-optic catheters, for light-exposure in deep-seated tumors (Lammers et al. 2012).

Since cancer represents a heterogeneous population of various diseases, choice of an appropriate nanocarrier or a targeting ligand is not obvious. There is a lot of intra- and inter-tumor- and intra- and interpatient-variability that needs to be accounted for. Formulation and targeting strategies are determined experimentally on a case-by-case basis, which is laborious, time consuming, and not cost effective. Hence, by combining various treatment strategies together with imaging modalities a more personalized form of treatment can be offered. For instance, in addition to using a chemotherapeutic agent, and using a targeting moiety towards the overexpressed receptor, it might be necessary to target the tumor vasculature or the extracellular matrix surrounding the tumor microenvironment; such can be applied to a variety of solid tumors. Additionally, image-guided insights can help prescreen patients and assign them to receive, e.g., liposome- and polymer-based nanomedicine treatments, in order to identify which tumors are amenable to passive and active drug targeting and which are not and to thereby predict which patients are likely to respond to such targeted therapeutic interventions and which are not (Peer et al. 2007).

Culmination of passive targeting, active targeting, and personalized form of medicine could lead to oral anticancer chemotherapy. Considering it has always been and will continue to be the most important administration route. It may change the current treatment regimen of chemotherapy in terms of its sustained plasma drug concentration, convenient and flexible administration, low cost, and improved patient compliance (Peer et al. 2007). While this vision seems distant, moving forward a step closer, personalized and theranostic nanomedicine together with noninvasive imaging techniques can contribute significantly to realizing the potential of nanomedicine.

References

Abra RM, Bankert RB, Chen F, Egilmez NK, Huang K, Saville R, Slater JL, Sugano M, Yokota SJ (2002) The next generation of liposome delivery systems: recent experience with tumor-targeted, sterically-stabilized immunoliposomes and active-loading gradients. J Liposome Res 12:1–3

Abuchowski A, Kazo GM, Verhoest CR Jr, Van Es T, Kafkewitz D, Nucci ML, Viau AT, Davis FF (1984) Cancer therapy with chemically modified enzymes. I. Antitumor properties of polyethylene glycol-asparaginase conjugates. Cancer Biochem Biophys 7:175–186

Acharya S, Dilnawaz F, Sahoo SK (2009) Targeted epidermal growth factor receptor nanoparticle bioconjugates for breast cancer therapy. Biomaterials 30:5737–5750

Alexander-Bryant AA, Vanden Berg-Foels WS, Wen X (2013) Bioengineering strategies for designing targeted cancer therapies (chapter one). In: Kenneth DT, Paul BF (eds) Advances in cancer research. Academic, pp 1–59

Alexis F, Pridgen E, Molnar LK, Farokhzad OC (2008) Factors affecting the clearance and biodistribution of polymeric nanoparticles. Mol Pharm 5:505–515

Aliabadi HM, Lavasanifar A (2006) Polymeric micelles for drug delivery. Expert Opin Drug Deliv 3:139–162

Allen TM (2002) Ligand-targeted therapeutics in anticancer therapy. Nat Rev Cancer 2:750–763

Allen TM, Cullis PR (2004) Drug delivery systems: entering the mainstream. Science 303:1818–1822

Allen TM, Hansen C, Martin F, Redemann C, Yau-Young A (1991) Liposomes containing synthetic lipid derivatives of poly(ethylene glycol) show prolonged circulation half-lives in vivo. Biochim Biophys Acta 1066:29–36

Almalik A, Karimi S, Ouasti S, Donno R, Wandrey C, Day PJ, Tirelli N (2013) Hyaluronic acid (HA) presentation as a tool to modulate and control the receptor-mediated uptake of HA-coated nanoparticles. Biomaterials 34:5369–5380

Anders CK, Adamo B, Karginova O, Deal AM, Rawal S, Darr D, Schorzman A, Santos C, Bash R, Kafri T, Carey L, Miller CR, Perou CM, Sharpless N, Zamboni WC (2013) Pharmacokinetics and efficacy of PEGylated liposomal doxorubicin in an intracranial model of breast cancer. PLoS One 8:e61359

Arvizo RR, Miranda OR, Moyano DF, Walden CA, Giri K, Bhattacharya R, Robertson JD, Rotello VM, Reid JM, Mukherjee P (2011) Modulating pharmacokinetics, tumor uptake and biodistribution by engineered nanoparticles. PLoS One 6:e24374

Asselin BL (1999) The three asparaginases: comparative pharmacology and optimal use in childhood leukemia. Adv Exp Med Biol 457:621–629

Bae YH (2009) Drug targeting and tumor heterogeneity. J Control Release 133:2–3

Bae Y, Nishiyama N, Fukushima S, Koyama H, Yasuhiro M, Kataoka K (2005) Preparation and biological characterization of polymeric micelle drug carriers with intracellular pH-triggered drug release property: tumor permeability, controlled subcellular drug distribution, and enhanced in vivo antitumor efficacy. Bioconjug Chem 16:122–130

Barenholz Y (2012) Doxil(R)–the first FDA-approved nano-drug: lessons learned. J Control Release 160:117–134

Bartlett DW, Su H, Hildebrandt IJ, Weber WA, Davis ME (2007) Impact of tumor-specific targeting on the biodistribution and efficacy of siRNA nanoparticles measured by multimodality in vivo imaging. Proc Natl Acad Sci U S A 104:15549–15554

Baselga J (2006) Targeting tyrosine kinases in cancer: the second wave. Science 312:1175–1178

Baselga J, Tripathy D, Mendelsohn J, Baughman S, Benz CC, Dantis L, Sklarin NT, Seidman AD, Hudis CA, Moore J, Rosen PP, Twaddell T, Henderson IC, Norton L (1996) Phase II study of weekly intravenous recombinant humanized anti-p185HER2 monoclonal antibody in patients with HER2/neu-overexpressing metastatic breast cancer. J Clin Oncol 14:737–744

Batist G, Gelmon KA, Chi KN, Miller WH Jr, Chia SK, Mayer LD, Swenson CE, Janoff AS, Louie AC (2009) Safety, pharmacokinetics, and efficacy of CPX-1 liposome injection in patients with advanced solid tumors. Clin Cancer Res 15:692–700

Bayne WF, Mayer LD, Swenson CE (2009) Pharmacokinetics of CPX-351 (cytarabine/daunorubicin HCl) liposome injection in the mouse. J Pharm Sci 98:2540–2548

Benesch M, Urban C (2008) Liposomal cytarabine for leukemic and lymphomatous meningitis: recent developments. Expert Opin Pharmacother 9:301–309

Benhar I, Padlan EA, Jung SH, Lee B, Pastan I (1994) Rapid humanization of the Fv of monoclonal antibody B3 by using framework exchange of the recombinant immunotoxin B3(Fv)-PE38. Proc Natl Acad Sci U S A 91:12051–12055

Bertrand N, Leroux JC (2012) The journey of a drug-carrier in the body: an anatomo-physiological perspective. J Control Release 161:152–163

Bertrand N, Wu J, Xu X, Kamaly N, Farokhzad OC (2014) Cancer nanotechnology: the impact of passive and active targeting in the era of modern cancer biology. Adv Drug Deliv Rev 66:2–25

Biswas S, Torchilin VP (2014) Nanopreparations for organelle-specific delivery in cancer. Adv Drug Deliv Rev 66:26–41

Bogner JR, Goebel FD (1995) Efficacy of DOX-SL (Stelath liposomal doxorubicin) in the treatment of advanced AIDS-related Kaposi sarcoma. In: Lasic DD, Martin FJ (eds) Stealth liposomes. CRC, Boca Raton, pp 267–278

Boomer JA, Qualls MM, Inerowicz HD, Haynes RH, Patri VS, Kim JM, Thompson DH (2009) Cytoplasmic delivery of liposomal contents mediated by an acid-labile cholesterol-vinyl ether-PEG conjugate. Bioconjug Chem 20:47–59

Bukowski R, Ernstoff MS, Gore ME, Nemunaitis JJ, Amato R, Gupta SK, Tendler CL (2002) Pegylated interferon alfa-2b treatment for patients with solid tumors: a phase I/II study. J Clin Oncol 20:3841–3849

Byrne JD, Betancourt T, Brannon-Peppas L (2008) Active targeting schemes for nanoparticle systems in cancer therapeutics. Adv Drug Deliv Rev 60:1615–1626

Cai LL, Liu P, Li X, Huang X, Ye YQ, Chen FY, Yuan H, Hu FQ, Du YZ (2011) RGD peptide-mediated chitosan-based polymeric micelles targeting delivery for integrin-overexpressing tumor cells. Int J Nanomedicine 6:3499–3508

Campbell RB, Fukumura D, Brown EB, Mazzola LM, Izumi Y, Jain RK, Torchilin VP, Munn LL (2002) Cationic charge determines the distribution of liposomes between the vascular and extravascular compartments of tumors. Cancer Res 62:6831–6836

Carmeliet P (2005) VEGF as a key mediator of angiogenesis in cancer. Oncology 69(Suppl 3):4–10

Carter P, Presta L, Gorman CM, Ridgway JB, Henner D, Wong WL, Rowland AM, Kotts C, Carver ME, Shepard HM (1992) Humanization of an anti-p185HER2 antibody for human cancer therapy. Proc Natl Acad Sci U S A 89:4285–4289

Chen J, Wu H, Han D, Xie C (2006) Using anti-VEGF McAb and magnetic nanoparticles as double-targeting vector for the radioimmunotherapy of liver cancer. Cancer Lett 231:169–175

Chen H, Khemtong C, Yang X, Chang X, Gao J (2011) Nanonization strategies for poorly water-soluble drugs. Drug Discov Today 16:354–360

Chen Z, Deng J, Zhao Y, Tao T (2012) Cyclic RGD peptide-modified liposomal drug delivery system: enhanced cellular uptake in vitro and improved pharmacokinetics in rats. Int J Nanomedicine 7:3803–3811

Chen H, Zhang T, Zhou Z, Guan M, Wang J, Liu L, Zhang Q (2013) Enhanced uptake and cytotoxity of folate-conjugated mitoxantrone-loaded micelles via receptor up-regulation by dexamethasone. Int J Pharm 448:142–149
Cheng C, Wei H, Zhu JL, Chang C, Cheng H, Li C, Cheng SX, Zhang XZ, Zhuo RX (2008) Functionalized thermoresponsive micelles self-assembled from biotin-PEG-*b*-P(NIPAAm-*co*-HMAAm)-*b*-PMMA for tumor cell target. Bioconjug Chem 19:1194–1201
Cheng Z, Al Zaki A, Hui JZ, Muzykantov VR, Tsourkas A (2012) Multifunctional nanoparticles: cost versus benefit of adding targeting and imaging capabilities. Science 338:903–910
Cheng W, Ping Y, Zhang Y, Chuang KH, Liu Y (2013) Magnetic resonance imaging (MRI) contrast agents for tumor diagnosis. J Healthc Eng 4:23–45
Cho K, Wang X, Nie S, Chen ZG, Shin DM (2008) Therapeutic nanoparticles for drug delivery in cancer. Clin Cancer Res 14:1310–1316
Chonn A, Cullis PR, Devine DV (1991) The role of surface charge in the activation of the classical and alternative pathways of complement by liposomes. J Immunol 146:4234–4241
Cluzel C, Saltel F, Lussi J, Paulhe F, Imhof BA, Wehrle-Haller B (2005) The mechanisms and dynamics of (alpha)v(beta)3 integrin clustering in living cells. J Cell Biol 171:383–392
Croll TI, O'Connor AJ, Stevens GW, Cooper-White JJ (2006) A blank slate? Layer-by-layer deposition of hyaluronic acid and chitosan onto various surfaces. Biomacromolecules 7:1610–1622
Danhier F, Vroman B, Lecouturier N, Crokart N, Pourcelle V, Freichels H, Jerome C, Marchand-Brynaert J, Feron O, Preat V (2009) Targeting of tumor endothelium by RGD-grafted PLGA-nanoparticles loaded with paclitaxel. J Control Release 140:166–173
Danhier F, Feron O, Preat V (2010) To exploit the tumor microenvironment: passive and active tumor targeting of nanocarriers for anti-cancer drug delivery. J Control Release 148:135–146
Daniels DA, Chen H, Hicke BJ, Swiderek KM, Gold L (2003) A tenascin-C aptamer identified by tumor cell SELEX: systematic evolution of ligands by exponential enrichment. Proc Natl Acad Sci U S A 100:15416–15421
Davis ME, Chen ZG, Shin DM (2008) Nanoparticle therapeutics: an emerging treatment modality for cancer. Nat Rev Drug Discov 7:771–782
de Smet M, Heijman E, Langereis S, Hijnen NM, Grull H (2011) Magnetic resonance imaging of high intensity focused ultrasound mediated drug delivery from temperature-sensitive liposomes: an in vivo proof-of-concept study. J Control Release 150:102–110
Deckers R, Moonen CT (2010) Ultrasound triggered, image guided, local drug delivery. J Control Release 148:25–33
des Rieux A, Pourcelle V, Cani PD, Marchand-Brynaert J, Preat V (2013) Targeted nanoparticles with novel non-peptidic ligands for oral delivery. Adv Drug Deliv Rev 65:833–844
Desgrosellier JS, Cheresh DA (2010) Integrins in cancer: biological implications and therapeutic opportunities. Nat Rev Cancer 10:9–22
Deshpande PP, Biswas S, Torchilin VP (2013) Current trends in the use of liposomes for tumor targeting. Nanomedicine 8:1509–1528
Dienst A, Grunow A, Unruh M, Rabausch B, Nor JE, Fries JW, Gottstein C (2005) Specific occlusion of murine and human tumor vasculature by VCAM-1-targeted recombinant fusion proteins. J Natl Cancer Inst 97:733–747
Ding HM, Ma YQ (2013) Controlling cellular uptake of nanoparticles with pH-sensitive polymers. Sci Rep 3:2804
Dixon KH, Mulligan T, Chung KN, Elwood PC, Cowan KH (1992) Effects of folate receptor expression following stable transfection into wild type and methotrexate transport-deficient ZR-75-1 human breast cancer cells. J Biol Chem 267:24140–24147
Dreher MR, Liu W, Michelich CR, Dewhirst MW, Yuan F, Chilkoti A (2006) Tumor vascular permeability, accumulation, and penetration of macromolecular drug carriers. J Natl Cancer Inst 98:335–344
Du H, Cui C, Wang L, Liu H, Cui G (2011) Novel tetrapeptide, RGDF, mediated tumor specific liposomal doxorubicin (DOX) preparations. Mol Pharm 8:1224–1232
Duncan R (2003) The dawning era of polymer therapeutics. Nat Rev Drug Discov 2:347–360

Edwards KA, Wang Y, Baeumner AJ (2010) Aptamer sandwich assays: human alpha-thrombin detection using liposome enhancement. Anal Bioanal Chem 398:2645–2654
Elbayoumi TA, Pabba S, Roby A, Torchilin VP (2007) Antinucleosome antibody-modified liposomes and lipid-core micelles for tumor-targeted delivery of therapeutic and diagnostic agents. J Liposome Res 17:1–14
Elias DR, Poloukhtine A, Popik V, Tsourkas A (2013) Effect of ligand density, receptor density, and nanoparticle size on cell targeting. Nanomedicine 9:194–201
Eliason JF (2006) PEGilated proteins in immunotherapy of cancer. Delivery of protein and peptide drugs in cancer. Imperial College Press, London, pp 111–126
Esmaeili F, Ghahremani MH, Ostad SN, Atyabi F, Seyedabadi M, Malekshahi MR, Amini M, Dinarvand R (2008) Folate-receptor-targeted delivery of docetaxel nanoparticles prepared by PLGA-PEG-folate conjugate. J Drug Target 16:415–423
Fakhari A, Baoum A, Siahaan TJ, Le KB, Berkland C (2011) Controlling ligand surface density optimizes nanoparticle binding to ICAM-1. J Pharm Sci 100:1045–1056
Fang X, Tan W (2010) Aptamers generated from cell-SELEX for molecular medicine: a chemical biology approach. Acc Chem Res 43:48–57
Fang J, Nakamura H, Maeda H (2011) The EPR effect: unique features of tumor blood vessels for drug delivery, factors involved, and limitations and augmentation of the effect. Adv Drug Deliv Rev 63:136–151
Farokhzad OC, Jon S, Khademhosseini A, Tran TN, Lavan DA, Langer R (2004) Nanoparticle-aptamer bioconjugates: a new approach for targeting prostate cancer cells. Cancer Res 64:7668–7672
Farokhzad OC, Cheng J, Teply BA, Sherifi I, Jon S, Kantoff PW, Richie JP, Langer R (2006a) Targeted nanoparticle-aptamer bioconjugates for cancer chemotherapy in vivo. Proc Natl Acad Sci U S A 103:6315–6320
Farokhzad OC, Karp JM, Langer R (2006b) Nanoparticle-aptamer bioconjugates for cancer targeting. Expert Opin Drug Deliv 3:311–324
Florence AT (2012) "Targeting" nanoparticles: the constraints of physical laws and physical barriers. J Control Release 164:115–124
Folkman J (1995) Angiogenesis in cancer, vascular, rheumatoid and other disease. Nat Med 1:27–31
Fujimori K, Covell DG, Fletcher JE, Weinstein JN (1989) Modeling analysis of the global and microscopic distribution of immunoglobulin G, F(ab')2, and Fab in tumors. Cancer Res 49:5656–5663
Fujimori K, Covell DG, Fletcher JE, Weinstein JN (1990) A modeling analysis of monoclonal antibody percolation through tumors: a binding-site barrier. J Nucl Med 31:1191–1198
Gabizon AA (2001) Pegylated liposomal doxorubicin: metamorphosis of an old drug into a new form of chemotherapy. Cancer Invest 19:424–436
Gabizon A, Papahadjopoulos D (1988) Liposome formulations with prolonged circulation time in blood and enhanced uptake by tumors. Proc Natl Acad Sci U S A 85:6949–6953
Gabizon A, Papahadjopoulos D (1992) The role of surface charge and hydrophilic groups on liposome clearance in vivo. Biochim Biophys Acta 1103:94–100
Gabizon A, Huang A, Martin FJ, Barenholz Y (1995) Doxorubicin encapsulated in polyethylene-glycol coated liposomes: initial clinical pharmacokinetic studies in solid tumors. In: Lasic DD, Martin FJ (eds) Stealth liposomes. CRC, Boca Raton, pp 245–256
Gabizon A, Horowitz AT, Goren D, Tzemach D, Shmeeda H, Zalipsky S (2003) In vivo fate of folate-targeted polyethylene-glycol liposomes in tumor-bearing mice. Clin Cancer Res 9:6551–6559
Gao Z, Lukyanov AN, Chakilam AR, Torchilin VP (2003) PEG-PE/phosphatidylcholine mixed immunomicelles specifically deliver encapsulated taxol to tumor cells of different origin and promote their efficient killing. J Drug Target 11:87–92
Gao ZG, Lee DH, Kim DI, Bae YH (2005) Doxorubicin loaded pH-sensitive micelle targeting acidic extracellular pH of human ovarian A2780 tumor in mice. J Drug Target 13:391–397

Gao H, Yang Z, Zhang S, Cao S, Shen S, Pang Z, Jiang X (2013) Ligand modified nanoparticles increases cell uptake, alters endocytosis and elevates glioma distribution and internalization. Sci Rep 3:2534

Garg A, Tisdale AW, Haidari E, Kokkoli E (2009) Targeting colon cancer cells using PEGylated liposomes modified with a fibronectin-mimetic peptide. Int J Pharm 366:201–210

Gaviani P, Corsini E, Salmaggi A, Lamperti E, Botturi A, Erbetta A, Milanesi I, Legnani F, Pollo B, Silvani A (2013) Liposomal cytarabine in neoplastic meningitis from primary brain tumors: a single institutional experience. Neurol Sci 34:2151–2157

Ghandehari H, Sharan R, Rubas W, Killing WM (2001) Molecular modeling of arginine-glycine-aspartic acid (RGD) analogs: relevance to transepithelial transport. J Pharm Pharm Sci 4:32–41

Goodwill PW, Saritas EU, Croft LR, Kim TN, Krishnan KM, Schaffer DV, Conolly SM (2012) X-space MPI: magnetic nanoparticles for safe medical imaging. Adv Mater 24:3870–3877

Greish K (2007) Enhanced permeability and retention of macromolecular drugs in solid tumors: a royal gate for targeted anticancer nanomedicines. J Drug Target 15:457–464

Gu F, Zhang L, Teply BA, Mann N, Wang A, Radovic-Moreno AF, Langer R, Farokhzad OC (2008) Precise engineering of targeted nanoparticles by using self-assembled biointegrated block copolymers. Proc Natl Acad Sci U S A 105:2586–2591

Guida V (2010) Thermodynamics and kinetics of vesicles formation processes. Adv Colloid Interface Sci 161:77–88

Guo X, Szoka FC Jr (2001) Steric stabilization of fusogenic liposomes by a low-pH sensitive PEG–diortho ester–lipid conjugate. Bioconjug Chem 12:291–300

Hafez IM, Maurer N, Cullis PR (2001) On the mechanism whereby cationic lipids promote intracellular delivery of polynucleic acids. Gene Ther 8:1188–1196

Hakem IF, Leech AM, Johnson JD, Donahue SJ, Walker JP, Bockstaller MR (2010) Understanding ligand distributions in modified particle and particlelike systems. J Am Chem Soc 132:16593–16598

Hamaguchi T, Matsumura Y, Suzuki M, Shimizu K, Goda R, Nakamura I, Nakatomi I, Yokoyama M, Kataoka K, Kakizoe T (2005) NK105, a paclitaxel-incorporating micellar nanoparticle formulation, can extend in vivo antitumour activity and reduce the neurotoxicity of paclitaxel. Br J Cancer 92:1240–1246

Han HS, Martin JD, Lee J, Harris DK, Fukumura D, Jain RK, Bawendi M (2013) Spatial charge configuration regulates nanoparticle transport and binding behavior in vivo. Angew Chem Int Ed Engl 52:1414–1419

Harding JA, Engbers CM, Newman MS, Goldstein NI, Zalipsky S (1997) Immunogenicity and pharmacokinetic attributes of poly(ethylene glycol)-grafted immunoliposomes. Biochim Biophys Acta 1327:181–192

Hatakeyama H, Akita H, Ishida E, Hashimoto K, Kobayashi H, Aoki T, Yasuda J, Obata K, Kikuchi H, Ishida T, Kiwada H, Harashima H (2007) Tumor targeting of doxorubicin by anti-MT1-MMP antibody-modified PEG liposomes. Int J Pharm 342:194–200

He C, Hu Y, Yin L, Tang C, Yin C (2010) Effects of particle size and surface charge on cellular uptake and biodistribution of polymeric nanoparticles. Biomaterials 31:3657–3666

Heath TD, Fraley RT, Papahdjopoulos D (1980) Antibody targeting of liposomes: cell specificity obtained by conjugation of F(ab')2 to vesicle surface. Science 210:539–541

Hein CD, Liu XM, Wang D (2008) Click chemistry, a powerful tool for pharmaceutical sciences. Pharm Res 25:2216–2230

Hilger I, Kaiser WA (2012) Iron oxide-based nanostructures for MRI and magnetic hyperthermia. Nanomedicine 7:1443–1459

Ho EA, Ramsay E, Ginj M, Anantha M, Bregman I, Sy J, Woo J, Osooly-Talesh M, Yapp DT, Bally MB (2010) Characterization of cationic liposome formulations designed to exhibit extended plasma residence times and tumor vasculature targeting properties. J Pharm Sci 99:2839–2853

Hobbs SK, Monsky WL, Yuan F, Roberts WG, Griffith L, Torchilin VP, Jain RK (1998) Regulation of transport pathways in tumor vessels: role of tumor type and microenvironment. Proc Natl Acad Sci U S A 95:4607–4612

Hoeben A, Landuyt B, Highley MS, Wildiers H, Van Oosterom AT, De Bruijn EA (2004) Vascular endothelial growth factor and angiogenesis. Pharmacol Rev 56:549–580
Hofheinz RD, Gnad-Vogt SU, Beyer U, Hochhaus A (2005) Liposomal encapsulated anti-cancer drugs. Anticancer Drugs 16:691–707
Hood JD, Bednarski M, Frausto R, Guccione S, Reisfeld RA, Xiang R, Cheresh DA (2002) Tumor regression by targeted gene delivery to the neovasculature. Science 296:2404–2407
Hosokawa S, Tagawa T, Niki H, Hirakawa Y, Nohga K, Nagaike K (2003) Efficacy of immunoliposomes on cancer models in a cell-surface-antigen-density-dependent manner. Br J Cancer 89:1545–1551
Hu Z, Luo F, Pan Y, Hou C, Ren L, Chen J, Wang J, Zhang Y (2008) Arg-Gly-Asp (RGD) peptide conjugated poly(lactic acid)-poly(ethylene oxide) micelle for targeted drug delivery. J Biomed Mater Res A 85:797–807
Huang SK, Mayhew E, Gilani S, Lasic DD, Martin FJ, Papahadjopoulos D (1992) Pharmacokinetics and therapeutics of sterically stabilized liposomes in mice bearing C-26 colon carcinoma. Cancer Res 52:6774–6781
Huang SK, Stauffer PR, Hong K, Guo JW, Phillips TL, Huang A, Papahadjopoulos D (1994) Liposomes and hyperthermia in mice: increased tumor uptake and therapeutic efficacy of doxorubicin in sterically stabilized liposomes. Cancer Res 54:2186–2191
Hu-Lieskovan S, Heidel JD, Bartlett DW, Davis ME, Triche TJ (2005) Sequence-specific knockdown of EWS-FLI1 by targeted, nonviral delivery of small interfering RNA inhibits tumor growth in a murine model of metastatic Ewing's sarcoma. Cancer Res 65:8984–8992
Ignjatovic T, Stanisavljevic S, Brovkovych V, Skidgel RA, Erdos EG (2004) Kinin B1 receptors stimulate nitric oxide production in endothelial cells: signaling pathways activated by angiotensin I-converting enzyme inhibitors and peptide ligands. Mol Pharmacol 66:1310–1316
Ishida T, Iden DL, Allen TM (1999) A combinatorial approach to producing sterically stabilized (Stealth) immunoliposomal drugs. FEBS Lett 460:129–133
Iyer AK, Khaled G, Fang J, Maeda H (2006) Exploiting the enhanced permeability and retention effect for tumor targeting. Drug Discov Today 11:812–818
Jain RK (1989) Delivery of novel therapeutic agents in tumors: physiological barriers and strategies. J Natl Cancer Inst 81:570–576
Jain RK (1994) Barriers to drug delivery in solid tumors. Sci Am 271:58–65
Jain RK (1999) Transport of molecules, particles, and cells in solid tumors. Annu Rev Biomed Eng 1:241–263
Jain RK, Stylianopoulos T (2010) Delivering nanomedicine to solid tumors. Nat Rev Clin Oncol 7:653–664
Jiang W, Kim BY, Rutka JT, Chan WC (2008) Nanoparticle-mediated cellular response is size-dependent. Nat Nanotechnol 3:145–150
Jones M, Leroux J (1999) Polymeric micelles – a new generation of colloidal drug carriers. Eur J Pharm Biopharm 48:101–111
Juweid M, Neumann R, Paik C, Perez-Bacete MJ, Sato J, van Osdol W, Weinstein JN (1992) Micropharmacology of monoclonal antibodies in solid tumors: direct experimental evidence for a binding site barrier. Cancer Res 52:5144–5153
Kagaya H, Oba M, Miura Y, Koyama H, Ishii T, Shimada T, Takato T, Kataoka K, Miyata T (2012) Impact of polyplex micelles installed with cyclic RGD peptide as ligand on gene delivery to vascular lesions. Gene Ther 19:61–69
Kamaly N, Xiao Z, Valencia PM, Radovic-Moreno AF, Farokhzad OC (2012) Targeted polymeric therapeutic nanoparticles: design, development and clinical translation. Chem Soc Rev 41:2971–3010
Kang N, Perron ME, Prud'homme RE, Zhang Y, Gaucher G, Leroux JC (2005) Stereocomplex block copolymer micelles: core-shell nanostructures with enhanced stability. Nano Lett 5:315–319
Kang H, O'Donoghue MB, Liu H, Tan W (2010) A liposome-based nanostructure for aptamer directed delivery. Chem Commun 46:249–251

Kang DI, Lee S, Lee JT, Sung BJ, Yoon JY, Kim JK, Chung J, Lim SJ (2011) Preparation and in vitro evaluation of anti-VCAM-1-Fab'-conjugated liposomes for the targeted delivery of the poorly water-soluble drug celecoxib. J Microencapsul 28:220–227

Kaufmann R, Muller P, Hildenbrand G, Hausmann M, Cremer C (2011) Analysis of Her2/neu membrane protein clusters in different types of breast cancer cells using localization microscopy. J Microsc 242:46–54

Kim ES, Lu C, Khuri FR, Tonda M, Glisson BS, Liu D, Jung M, Hong WK, Herbst RS (2001) A phase II study of STEALTH cisplatin (SPI-77) in patients with advanced non-small cell lung cancer. Lung Cancer 34:427–432

Kim SH, Jeong JH, Chun KW, Park TG (2005) Target-specific cellular uptake of PLGA nanoparticles coated with poly(L-lysine)-poly(ethylene glycol)-folate conjugate. Langmuir 21:8852–8857

Kim K, Kim JH, Park H, Kim YS, Park K, Nam H, Lee S, Park JH, Park RW, Kim IS, Choi K, Kim SY, Park K, Kwon IC (2010) Tumor-homing multifunctional nanoparticles for cancer theragnosis: simultaneous diagnosis, drug delivery, and therapeutic monitoring. J Control Release 146:219–227

Kirpotin DB, Drummond DC, Shao Y, Shalaby MR, Hong K, Nielsen UB, Marks JD, Benz CC, Park JW (2006) Antibody targeting of long-circulating lipidic nanoparticles does not increase tumor localization but does increase internalization in animal models. Cancer Res 66:6732–6740

Klibanov AL, Maruyama K, Torchilin VP, Huang L (1990) Amphipathic polyethyleneglycols effectively prolong the circulation time of liposomes. FEBS Lett 268:235–237

Knudsen NO, Schiffelers RM, Jorgensen L, Hansen J, Frokjaer S, Foged C (2012) Design of cyclic RKKH peptide-conjugated PEG liposomes targeting the integrin alpha(2)beta(1) receptor. Int J Pharm 428:171–177

Kondo M, Asai T, Katanasaka Y, Sadzuka Y, Tsukada H, Ogino K, Taki T, Baba K, Oku N (2004) Anti-neovascular therapy by liposomal drug targeted to membrane type-1 matrix metalloproteinase. Int J Cancer 108:301–306

Kopecek J, Kopeckova P, Minko T, Lu Z (2000) HPMA copolymer-anticancer drug conjugates: design, activity, and mechanism of action. Eur J Pharm Biopharm 50:61–81

Kopecek J, Kopeckova P, Minko T, Lu ZR, Peterson CM (2001) Water soluble polymers in tumor targeted delivery. J Control Release 74:147–158

Koshkaryev A, Sawant R, Deshpande M, Torchilin V (2013) Immunoconjugates and long circulating systems: origins, current state of the art and future directions. Adv Drug Deliv Rev 65:24–35

Koss M, Pfeiffer GR 2nd, Wang Y, Thomas ST, Yerukhimovich M, Gaarde WA, Doerschuk CM, Wang Q (2006) Ezrin/radixin/moesin proteins are phosphorylated by TNF-alpha and modulate permeability increases in human pulmonary microvascular endothelial cells. J Immunol 176:1218–1227

Krasnici S, Werner A, Eichhorn ME, Schmitt-Sody M, Pahernik SA, Sauer B, Schulze B, Teifel M, Michaelis U, Naujoks K, Dellian M (2003) Effect of the surface charge of liposomes on their uptake by angiogenic tumor vessels. Int J Cancer 105:561–567

Kripp M, Hofheinz RD (2008) Treatment of lymphomatous and leukemic meningitis with liposomal encapsulated cytarabine. Int J Nanomedicine 3:397–401

Lam KS, Salmon SE, Hersh EM, Hruby VJ, Kazmierski WM, Knapp RJ (1991) A new type of synthetic peptide library for identifying ligand-binding activity. Nature 354:82–84

Lammers T, Hennink WE, Storm G (2008) Tumour-targeted nanomedicines: principles and practice. Br J Cancer 99:392–397

Lammers T, Kiessling F, Hennink WE, Storm G (2012) Drug targeting to tumors: principles, pitfalls and (pre-) clinical progress. J Control Release 161:175–187

Lao J, Madani J, Puertolas T, Alvarez M, Hernandez A, Pazo-Cid R, Artal A, Anton Torres A (2013) Liposomal Doxorubicin in the treatment of breast cancer patients: a review. J Drug Deliv 2013:456409

Lapointe J, Truong TQ, Falstrault L, Brissette L (2006) Differential abilities of mouse liver parenchymal and nonparenchymal cells in HDL and LDL (native and oxidized) association and cholesterol efflux. Biochem Cell Biol 84:250–256

Lasic DD, Martin FJ, Gabizon A, Huang SK, Papahadjopoulos D (1991) Sterically stabilized liposomes: a hypothesis on the molecular origin of the extended circulation times. Biochim Biophys Acta 1070:187–192

Lavasanifar A, Samuel J, Kwon GS (2002) The effect of fatty acid substitution on the in vitro release of amphotericin B from micelles composed of poly(ethylene oxide)-*block*-poly(*N*-hexyl stearate-L-aspartamide). J Control Release 79:165–172

Lee KS, Chung HC, Im SA, Park YH, Kim CS, Kim SB, Rha SY, Lee MY, Ro J (2008) Multicenter phase II trial of Genexol-PM, a Cremophor-free, polymeric micelle formulation of paclitaxel, in patients with metastatic breast cancer. Breast Cancer Res Treat 108:241–250

Leserman LD, Barbet J, Kourilsky F, Weinstein JN (1980) Targeting to cells of fluorescent liposomes covalently coupled with monoclonal antibody or protein A. Nature 288:602–604

Levchenko TS, Rammohan R, Lukyanov AN, Whiteman KR, Torchilin VP (2002) Liposome clearance in mice: the effect of a separate and combined presence of surface charge and polymer coating. Int J Pharm 240:95–102

Li L, Wartchow CA, Danthi SN, Shen Z, Dechene N, Pease J, Choi HS, Doede T, Chu P, Ning S, Lee DY, Bednarski MD, Knox SJ (2004) A novel antiangiogenesis therapy using an integrin antagonist or anti-Flk-1 antibody coated 90Y-labeled nanoparticles. Int J Radiat Oncol Biol Phys 58:1215–1227

Li W, Huang Z, MacKay JA, Grube S, Szoka FC Jr (2005) Low-pH-sensitive poly(ethylene glycol) (PEG)-stabilized plasmid nanolipoparticles: effects of PEG chain length, lipid composition and assembly conditions on gene delivery. J Gene Med 7:67–79

Liang B, He ML, Chan CY, Chen YC, Li XP, Li Y, Zheng D, Lin MC, Kung HF, Shuai XT, Peng Y (2009) The use of folate-PEG-grafted-hybranched-PEI nonviral vector for the inhibition of glioma growth in the rat. Biomaterials 30:4014–4020

Liechty WB, Peppas NA (2012) Expert opinion: responsive polymer nanoparticles in cancer therapy. Eur J Pharm Biopharm 80:241–246

Liu J, Weller GE, Zern B, Ayyaswamy PS, Eckmann DM, Muzykantov VR, Radhakrishnan R (2010a) Computational model for nanocarrier binding to endothelium validated using in vivo, in vitro, and atomic force microscopy experiments. Proc Natl Acad Sci U S A 107: 16530–16535

Liu AP, Aguet F, Danuser G, Schmid SL (2010b) Local clustering of transferrin receptors promotes clathrin-coated pit initiation. J Cell Biol 191:1381–1393

Lord MS, Pasqui D, Barbucci R, Milthorpe BK (2009) Protein adsorption on derivatives of hyaluronic acid and subsequent cellular response. J Biomed Mater Res A 91:635–646

Low PS, Kularatne SA (2009) Folate-targeted therapeutic and imaging agents for cancer. Curr Opin Chem Biol 13:256–262

Lukyanov AN, Gao Z, Mazzola L, Torchilin VP (2002) Polyethylene glycol-diacyllipid micelles demonstrate increased accumulation in subcutaneous tumors in mice. Pharm Res 19:1424–1429

Luo Y, Prestwich GD (2002) Cancer-targeted polymeric drugs. Curr Cancer Drug Targets 2:209–226

Lurje G, Lenz HJ (2009) EGFR signaling and drug discovery. Oncology 77:400–410

Maeda H (2012) Macromolecular therapeutics in cancer treatment: the EPR effect and beyond. J Control Release 164:138–144

Maeda H, Matsumura Y (1989) Tumoritropic and lymphotropic principles of macromolecular drugs. Crit Rev Ther Drug Carrier Syst 6:193–210

Maeda H, Yamamoto T (1996) Pathogenic mechanisms induced by microbial proteases in microbial infections. Biol Chem Hoppe Seyler 377:217–226

Maeda H, Takeshita J, Kanamaru R (1979) A lipophilic derivative of neocarzinostatin. A polymer conjugation of an antitumor protein antibiotic. Int J Pept Protein Res 14:81–87

Maeda H, Matsumura Y, Kato H (1988) Purification and identification of [hydroxyprolyl3]bradykinin in ascitic fluid from a patient with gastric cancer. J Biol Chem 263:16051–16054

Maeda H, Fang J, Inutsuka T, Kitamoto Y (2003) Vascular permeability enhancement in solid tumor: various factors, mechanisms involved and its implications. Int Immunopharmacol 3: 319–328

Maeda H, Bharate GY, Daruwalla J (2009) Polymeric drugs for efficient tumor-targeted drug delivery based on EPR-effect. Eur J Pharm Biopharm 71:409–419

Maeda H, Nakamura H, Fang J (2013) The EPR effect for macromolecular drug delivery to solid tumors: improvement of tumor uptake, lowering of systemic toxicity, and distinct tumor imaging in vivo. Adv Drug Deliv Rev 65:71–79

Mamot C, Drummond DC, Noble CO, Kallab V, Guo Z, Hong K, Kirpotin DB, Park JW (2005) Epidermal growth factor receptor-targeted immunoliposomes significantly enhance the efficacy of multiple anticancer drugs in vivo. Cancer Res 65:11631–11638

Mamot C, Ritschard R, Wicki A, Stehle G, Dieterle T, Bubendorf L, Hilker C, Deuster S, Herrmann R, Rochlitz C (2012) Tolerability, safety, pharmacokinetics, and efficacy of doxorubicin-loaded anti-EGFR immunoliposomes in advanced solid tumours: a phase 1 dose-escalation study. Lancet Oncol 13:1234–1241

Managit C, Kawakami S, Nishikawa M, Yamashita F, Hashida M (2003) Targeted and sustained drug delivery using PEGylated galactosylated liposomes. Int J Pharm 266:77–84

Martin A (1993) Physical pharmacy. Lippincott Williams and Wilkins, Philadelphia

Matsumura Y (2011) Preclinical and clinical studies of NK012, an SN-38-incorporating polymeric micelles, which is designed based on EPR effect. Adv Drug Deliv Rev 63:184–192

Matsumura Y, Maeda H (1986) A new concept for macromolecular therapeutics in cancer chemotherapy: mechanism of tumoritropic accumulation of proteins and the antitumor agent smancs. Cancer Res 46:6387–6392

Matsumura Y, Oda T, Maeda H (1987) General mechanism of intratumor accumulation of macromolecules: advantage of macromolecular therapeutics. Gan To Kagaku Ryoho 14:821–829

Matsumura Y, Kimura M, Yamamoto T, Maeda H (1988) Involvement of the kinin-generating cascade in enhanced vascular permeability in tumor tissue. Jpn J Cancer Res 79:1327–1334

Matsumura Y, Maruo K, Kimura M, Yamamoto T, Konno T, Maeda H (1991) Kinin-generating cascade in advanced cancer patients and in vitro study. Jpn J Cancer Res 82:732–741

Matsumura Y, Hamaguchi T, Ura T, Muro K, Yamada Y, Shimada Y, Shirao K, Okusaka T, Ueno H, Ikeda M, Watanabe N (2004a) Phase I clinical trial and pharmacokinetic evaluation of NK911, a micelle-encapsulated doxorubicin. Br J Cancer 91:1775–1781

Matsumura Y, Gotoh M, Muro K, Yamada Y, Shirao K, Shimada Y, Okuwa M, Matsumoto S, Miyata Y, Ohkura H, Chin K, Baba S, Yamao T, Kannami A, Takamatsu Y, Ito K, Takahashi K (2004b) Phase I and pharmacokinetic study of MCC-465, a doxorubicin (DXR) encapsulated in PEG immunoliposome, in patients with metastatic stomach cancer. Ann Oncol 15:517–525

Mayer LD, Harasym TO, Tardi PG, Harasym NL, Shew CR, Johnstone SA, Ramsay EC, Bally MB, Janoff AS (2006) Ratiometric dosing of anticancer drug combinations: controlling drug ratios after systemic administration regulates therapeutic activity in tumor-bearing mice. Mol Cancer Ther 5:1854–1863

McGuire MJ, Li S, Brown KC (2009) Biopanning of phage displayed peptide libraries for the isolation of cell-specific ligands. Methods Mol Biol 504:291–321

Meng H, Xue M, Xia T, Ji Z, Tarn DY, Zink JI, Nel AE (2011) Use of size and a copolymer design feature to improve the biodistribution and the enhanced permeability and retention effect of doxorubicin-loaded mesoporous silica nanoparticles in a murine xenograft tumor model. ACS Nano 5:4131–4144

Metselaar JM, Bruin P, de Boer LW, de Vringer T, Snel C, Oussoren C, Wauben MH, Crommelin DJ, Storm G, Hennink WE (2003) A novel family of L-amino acid-based biodegradable polymer-lipid conjugates for the development of long-circulating liposomes with effective drug-targeting capacity. Bioconjug Chem 14:1156–1164

Millard M, Odde S, Neamati N (2011) Integrin targeted therapeutics. Theranostics 1:154–188

Minko T (2004) Drug targeting to the colon with lectins and neoglycoconjugates. Adv Drug Deliv Rev 56:491–509

Modi S, Prakash Jain J, Domb AJ, Kumar N (2006) Exploiting EPR in polymer drug conjugate delivery for tumor targeting. Curr Pharm Des 12:4785–4796

Moein Moghimi S, Hamad I, Bunger R, Andresen TL, Jorgensen K, Hunter AC, Baranji L, Rosivall L, Szebeni J (2006) Activation of the human complement system by cholesterol-rich and

PEGylated liposomes-modulation of cholesterol-rich liposome-mediated complement activation by elevated serum LDL and HDL levels. J Liposome Res 16:167–174
Monopoli MP, Aberg C, Salvati A, Dawson KA (2012) Biomolecular coronas provide the biological identity of nanosized materials. Nat Nanotechnol 7:779–786
Montana M, Ducros C, Verhaeghe P, Terme T, Vanelle P, Rathelot P (2011) Albumin-bound paclitaxel: the benefit of this new formulation in the treatment of various cancers. J Chemother 23:59–66
Moreira da Silva L, Moreira da Silva L (1975) Prognosis of neurotic patients under ambulatory care. Arq Neuropsiquiatr 33:376–381
Mu L, Elbayoumi TA, Torchilin VP (2005) Mixed micelles made of poly(ethylene glycol)-phosphatidylethanolamine conjugate and d-alpha-tocopheryl polyethylene glycol 1000 succinate as pharmaceutical nanocarriers for camptothecin. Int J Pharm 306:142–149
Murphy EA, Majeti BK, Barnes LA, Makale M, Weis SM, Lutu-Fuga K, Wrasidlo W, Cheresh DA (2008) Nanoparticle-mediated drug delivery to tumor vasculature suppresses metastasis. Proc Natl Acad Sci U S A 105:9343–9348
Musacchio T, Torchilin VP (2009) Lipid-based delivery systems: liposomes and lipid-core micelles—properties and applications. Biodrug delivery systems: fundamentals, applications, and clinical development. Informa Healthcare, pp 260–292
Nair KL, Jagadeeshan S, Nair SA, Kumar GS (2013) Folic acid conjugated delta-valerolactone-poly(ethylene glycol) based triblock copolymer as a promising carrier for targeted doxorubicin delivery. PLoS One 8:e70697
Nakamura K, Yamashita K, Itoh Y, Yoshino K, Nozawa S, Kasukawa H (2012) Comparative studies of polyethylene glycol-modified liposomes prepared using different PEG-modification methods. Biochim Biophys Acta 1818:2801–2807
Napper DH (1983) Polymeric stabilization of colloidal dispersions. Academic, London/New York
Nasongkla N, Shuai X, Ai H, Weinberg BD, Pink J, Boothman DA, Gao J (2004) cRGD-functionalized polymer micelles for targeted doxorubicin delivery. Angew Chem Int Ed Engl 43:6323–6327
Needels MC, Jones DG, Tate EH, Heinkel GL, Kochersperger LM, Dower WJ, Barrett RW, Gallop MA (1993) Generation and screening of an oligonucleotide-encoded synthetic peptide library. Proc Natl Acad Sci U S A 90:10700–10704
Needham D, McIntosh TJ, Lasic DD (1992) Repulsive interactions and mechanical stability of polymer-grafted lipid membranes. Biochim Biophys Acta 1108:40–48
Negussie AH, Yarmolenko PS, Partanen A, Ranjan A, Jacobs G, Woods D, Bryant H, Thomasson D, Dewhirst MW, Wood BJ, Dreher MR (2011) Formulation and characterisation of magnetic resonance imageable thermally sensitive liposomes for use with magnetic resonance-guided high intensity focused ultrasound. Int J Hyperthermia 27:140–155
Nishida K, Mihara K, Takino T, Nakane S, Takakura Y, Hashida M, Sezaki H (1991) Hepatic disposition characteristics of electrically charged macromolecules in rat in vivo and in the perfused liver. Pharm Res 8:437–444
O'Brien ME, Wigler N, Inbar M, Rosso R, Grischke E, Santoro A, Catane R, Kieback DG, Tomczak P, Ackland SP, Orlandi F, Mellars L, Alland L, Tendler C, Group CBCS (2004) Reduced cardiotoxicity and comparable efficacy in a phase III trial of pegylated liposomal doxorubicin HCl (CAELYX/Doxil) versus conventional doxorubicin for first-line treatment of metastatic breast cancer. Ann Oncol 15:440–449
Ohannesian DW, Lotan D, Thomas P, Jessup JM, Fukuda M, Gabius HJ, Lotan R (1995) Carcinoembryonic antigen and other glycoconjugates act as ligands for galectin-3 in human colon carcinoma cells. Cancer Res 55:2191–2199
Okamoto T, Akaike T, Nagano T, Miyajima S, Suga M, Ando M, Ichimori K, Maeda H (1997) Activation of human neutrophil procollagenase by nitrogen dioxide and peroxynitrite: a novel mechanism for procollagenase activation involving nitric oxide. Arch Biochem Biophys 342:261–274
Orlova A, Magnusson M, Eriksson TL, Nilsson M, Larsson B, Hoiden-Guthenberg I, Widstrom C, Carlsson J, Tolmachev V, Stahl S, Nilsson FY (2006) Tumor imaging using a picomolar affinity HER2 binding affibody molecule. Cancer Res 66:4339–4348

Park BW, Zhang HT, Wu C, Berezov A, Zhang X, Dua R, Wang Q, Kao G, O'Rourke DM, Greene MI, Murali R (2000) Rationally designed anti-HER2/neu peptide mimetic disables P185HER2/neu tyrosine kinases in vitro and in vivo. Nat Biotechnol 18:194–198

Park JW, Kirpotin DB, Hong K, Shalaby R, Shao Y, Nielsen UB, Marks JD, Papahadjopoulos D, Benz CC (2001) Tumor targeting using anti-her2 immunoliposomes. J Control Release 74:95–113

Park JW, Hong K, Kirpotin DB, Colbern G, Shalaby R, Baselga J, Shao Y, Nielsen UB, Marks JD, Moore D, Papahadjopoulos D, Benz CC (2002) Anti-HER2 immunoliposomes: enhanced efficacy attributable to targeted delivery. Clin Cancer Res 8:1172–1181

Pasqualini R, Ruoslahti E (1996) Organ targeting in vivo using phage display peptide libraries. Nature 380:364–366

Pastorino F, Brignole C, Marimpietri D, Cilli M, Gambini C, Ribatti D, Longhi R, Allen TM, Corti A, Ponzoni M (2003) Vascular damage and anti-angiogenic effects of tumor vessel-targeted liposomal chemotherapy. Cancer Res 63:7400–7409

Pauwels EK, Erba P (2007) Towards the use of nanoparticles in cancer therapy and imaging. Drug News Perspect 20:213–220

Peer D, Margalit R (2004a) Tumor-targeted hyaluronan nanoliposomes increase the antitumor activity of liposomal doxorubicin in syngeneic and human xenograft mouse tumor models. Neoplasia 6:343–353

Peer D, Margalit R (2004b) Loading mitomycin C inside long circulating hyaluronan targeted nano-liposomes increases its antitumor activity in three mice tumor models. Int J Cancer 108:780–789

Peer D, Karp JM, Hong S, Farokhzad OC, Margalit R, Langer R (2007) Nanocarriers as an emerging platform for cancer therapy. Nat Nanotechnol 2:751–760

Perche F, Torchilin VP (2013) Recent trends in multifunctional liposomal nanocarriers for enhanced tumor targeting. J Drug Deliv 2013:705265

Perez AT, Domenech GH, Frankel C, Vogel CL (2002) Pegylated liposomal doxorubicin (Doxil) for metastatic breast cancer: the Cancer Research Network, Inc., experience. Cancer Invest 20(Suppl 2):22–29

Peterson CM, Shiah JG, Sun Y, Kopeckova P, Minko T, Straight RC, Kopecek J (2003) HPMA copolymer delivery of chemotherapy and photodynamic therapy in ovarian cancer. Adv Exp Med Biol 519:101–123

Pourbaghi-Masouleh M, Hosseini V (2013) Amorphous calcium phosphate nanoparticles could function as a novel cancer therapeutic agent by employing a suitable targeted drug delivery platform. Nanoscale Res Lett 8:449

Roby A, Erdogan S, Torchilin VP (2006) Solubilization of poorly soluble PDT agent, *meso*-tetraphenylporphin, in plain or immunotargeted PEG-PE micelles results in dramatically improved cancer cell killing in vitro. Eur J Pharm Biopharm 62:235–240

Roux E, Lafleur M, Lataste E, Moreau P, Leroux JC (2003) On the characterization of pH-sensitive liposome/polymer complexes. Biomacromolecules 4:240–248

Rudnick SI, Adams GP (2009) Affinity and avidity in antibody-based tumor targeting. Cancer Biother Radiopharm 24:155–161

Safra T, Muggia F, Jeffers S, Tsao-Wei DD, Groshen S, Lyass O, Henderson R, Berry G, Gabizon A (2000) Pegylated liposomal doxorubicin (doxil): reduced clinical cardiotoxicity in patients reaching or exceeding cumulative doses of 500 mg/m^2. Ann Oncol 11:1029–1033

Sahoo SK, Ma W, Labhasetwar V (2004) Efficacy of transferrin-conjugated paclitaxel-loaded nanoparticles in a murine model of prostate cancer. Int J Cancer 112:335–340

Salmaso S, Pappalardo JS, Sawant RR, Musacchio T, Rockwell K, Caliceti P, Torchilin VP (2009) Targeting glioma cells in vitro with ascorbate-conjugated pharmaceutical nanocarriers. Bioconjug Chem 20:2348–2355

Salvador-Morales C, Zhang L, Langer R, Farokhzad OC (2009) Immunocompatibility properties of lipid-polymer hybrid nanoparticles with heterogeneous surface functional groups. Biomaterials 30:2231–2240

Sawant RR, Torchilin VP (2012a) Multifunctional nanocarriers and intracellular drug delivery. Curr Opin Solid State Mater Sci 16:269–275

Sawant RR, Torchilin VP (2012b) Challenges in development of targeted liposomal therapeutics. AAPS J 14:303–315

Sawant RR, Sawant RM, Torchilin VP (2008) Mixed PEG-PE/vitamin E tumor-targeted immunomicelles as carriers for poorly soluble anti-cancer drugs: improved drug solubilization and enhanced in vitro cytotoxicity. Eur J Pharm Biopharm 70:51–57

Scaltriti M, Baselga J (2006) The epidermal growth factor receptor pathway: a model for targeted therapy. Clin Cancer Res 12:5268–5272

Scherphof GL, Kamps JA (2001) The role of hepatocytes in the clearance of liposomes from the blood circulation. Prog Lipid Res 40:149–166

Schmitt-Sody M, Strieth S, Krasnici S, Sauer B, Schulze B, Teifel M, Michaelis U, Naujoks K, Dellian M (2003) Neovascular targeting therapy: paclitaxel encapsulated in cationic liposomes improves antitumoral efficacy. Clin Cancer Res 9:2335–2341

Seki T, Fang J, Maeda H (2009) Enhanced delivery of macromolecular antitumor drugs to tumors by nitroglycerin application. Cancer Sci 100:2426–2430

Seki T, Carroll F, Illingworth S, Green N, Cawood R, Bachtarzi H, Subr V, Fisher KD, Seymour LW (2011) Tumour necrosis factor-alpha increases extravasation of virus particles into tumour tissue by activating the Rho A/Rho kinase pathway. J Control Release 156:381–389

Senzer NN, Matsuno K, Yamagata N, Fujisawa T, Wasserman E, Sutherland W, Sharma S, Phan A (2009) MBP-426, a novel liposome-encapsulated oxaliplatin, in combination with 5-FU/leucovorin (LV): phase I results of a phase I/II study in gastro-esophageal adenocarcinoma, with pharmacokinetics. In: AACR-NCI-EORTC international conference: molecular targets and cancer therapeutics, Molecular Cancer Therapeutics, Boston, 2009

Senzer N, Nemunaitis J, Nemunaitis D, Bedell C, Edelman G, Barve M, Nunan R, Pirollo KF, Rait A, Chang EH (2013) Phase I study of a systemically delivered p53 nanoparticle in advanced solid tumors. Mol Ther 21:1096–1103

Seymour LW, Miyamoto Y, Maeda H, Brereton M, Strohalm J, Ulbrich K, Duncan R (1995) Influence of molecular weight on passive tumour accumulation of a soluble macromolecular drug carrier. Eur J Cancer 31A:766–770

Shangguan D, Li Y, Tang Z, Cao ZC, Chen HW, Mallikaratchy P, Sefah K, Yang CJ, Tan W (2006) Aptamers evolved from live cells as effective molecular probes for cancer study. Proc Natl Acad Sci U S A 103:11838–11843

Shi J, Xiao Z, Kamaly N, Farokhzad OC (2011) Self-assembled targeted nanoparticles: evolution of technologies and bench to bedside translation. Acc Chem Res 44:1123–1134

Shuai X, Merdan T, Schaper AK, Xi F, Kissel T (2004) Core-cross-linked polymeric micelles as paclitaxel carriers. Bioconjug Chem 15:441–448

Siegal T, Horowitz A, Gabizon A (1995) Doxorubicin encapsulated in sterically stabilized liposomes for the treatment of a brain tumor model: biodistribution and therapeutic efficacy. J Neurosurg 83:1029–1037

Skidan I, Dholakia P, Torchilin V (2008) Photodynamic therapy of experimental B-16 melanoma in mice with tumor-targeted 5,10,15,20-tetraphenylporphin-loaded PEG-PE micelles. J Drug Target 16:486–493

Slamon DJ, Clark GM, Wong SG, Levin WJ, Ullrich A, McGuire WL (1987) Human breast cancer: correlation of relapse and survival with amplification of the HER-2/neu oncogene. Science 235:177–182

Slamon DJ, Godolphin W, Jones LA, Holt JA, Wong SG, Keith DE, Levin WJ, Stuart SG, Udove J, Ullrich A et al (1989) Studies of the HER-2/neu proto-oncogene in human breast and ovarian cancer. Science 244:707–712

Smart EJ, Mineo C, Anderson RG (1996) Clustered folate receptors deliver 5-methyltetrahydrofolate to cytoplasm of MA104 cells. J Cell Biol 134:1169–1177

Stefanick JF, Ashley JD, Kiziltepe T, Bilgicer B (2013) A systematic analysis of peptide linker length and liposomal polyethylene glycol coating on cellular uptake of peptide-targeted liposomes. ACS Nano 7:2935–2947

Stella B, Arpicco S, Peracchia MT, Desmaele D, Hoebeke J, Renoir M, D'Angelo J, Cattel L, Couvreur P (2000) Design of folic acid-conjugated nanoparticles for drug targeting. J Pharm Sci 89:1452–1464
Strebhardt K, Ullrich A (2008) Paul Ehrlich's magic bullet concept: 100 years of progress. Nat Rev Cancer 8:473–480
Svenson S (2012) Clinical translation of nanomedicines. Curr Opin Solid State Mater Sci 16:287–294
Tagami T, Ernsting MJ, Li SD (2011a) Optimization of a novel and improved thermosensitive liposome formulated with DPPC and a Brij surfactant using a robust in vitro system. J Control Release 154:290–297
Tagami T, Ernsting MJ, Li SD (2011b) Efficient tumor regression by a single and low dose treatment with a novel and enhanced formulation of thermosensitive liposomal doxorubicin. J Control Release 152:303–309
Takeuchi H, Kojima H, Yamamoto H, Kawashima Y (2001) Evaluation of circulation profiles of liposomes coated with hydrophilic polymers having different molecular weights in rats. J Control Release 75:83–91
Tardi PG, Gallagher RC, Johnstone S, Harasym N, Webb M, Bally MB, Mayer LD (2007) Coencapsulation of irinotecan and floxuridine into low cholesterol-containing liposomes that coordinate drug release in vivo. Biochim Biophys Acta 1768:678–687
Tardi P, Johnstone S, Harasym N, Xie S, Harasym T, Zisman N, Harvie P, Bermudes D, Mayer L (2009) In vivo maintenance of synergistic cytarabine:daunorubicin ratios greatly enhances therapeutic efficacy. Leuk Res 33:129–139
Taurin S, Nehoff H, Greish K (2012) Anticancer nanomedicine and tumor vascular permeability; where is the missing link? J Control Release 164:265–275
Thurber GM, Weissleder R (2011) Quantitating antibody uptake in vivo: conditional dependence on antigen expression levels. Mol Imaging Biol 13:623–632
Torchilin VP (1996) How do polymers prolong circulation times of liposomes. J Liposome Res 9:99–116
Torchilin VP (2000) Drug targeting. Eur J Pharm Sci 11(Suppl 2):S81–S91
Torchilin VP (2001a) Structure and design of polymeric surfactant-based drug delivery systems. J Control Release 73:137–172
Torchilin VP (2001b) Diacyl lipid-polyethene glycols as pharmaceutical carriers. In: 28th international symposium on controlled release of bioactive materials, Controlled Release Society, Inc., San Diego, 2001, pp 57–58
Torchilin VP (2002) PEG-based micelles as carriers of contrast agents for different imaging modalities. Adv Drug Deliv Rev 54:235–252
Torchilin VP (2005) Recent advances with liposomes as pharmaceutical carriers. Nat Rev Drug Discov 4:145–160
Torchilin VP (2007) Micellar nanocarriers: pharmaceutical perspectives. Pharm Res 24:1–16
Torchilin V (2009) Multifunctional and stimuli-sensitive pharmaceutical nanocarriers. Eur J Pharm Biopharm 71:431–444
Torchilin VP (2010) Passive and active drug targeting: drug delivery to tumors as an example. Handb Exp Pharmacol (197):3–53.
Torchilin V (2011) Tumor delivery of macromolecular drugs based on the EPR effect. Adv Drug Deliv Rev 63:131–135
Torchilin VP, Weissig V (2003) Liposomes: a practical approach. Practical approach series, vol 264. Oxford University Press, Oxford/New York, pp xxiii, 396 p
Torchilin VP, Klibanov AL, Huang L, O'Donnell S, Nossiff ND, Khaw BA (1992) Targeted accumulation of polyethylene glycol-coated immunoliposomes in infarcted rabbit myocardium. FASEB J 6:2716–2719
Torchilin VP, Levchenko TS, Whiteman KR, Yaroslavov AA, Tsatsakis AM, Rizos AK, Michailova EV, Shtilman MI (2001a) Amphiphilic poly-*N*-vinylpyrrolidones: synthesis, properties and liposome surface modification. Biomaterials 22:3035–3044
Torchilin VP, Rammohan R, Weissig V, Levchenko TS (2001b) TAT peptide on the surface of liposomes affords their efficient intracellular delivery even at low temperature and in the presence of metabolic inhibitors. Proc Natl Acad Sci U S A 98:8786–8791

Torchilin VP, Levchenko TS, Lukyanov AN, Khaw BA, Klibanov AL, Rammohan R, Samokhin GP, Whiteman KR (2001c) *p*-Nitrophenylcarbonyl-PEG-PE-liposomes: fast and simple attachment of specific ligands, including monoclonal antibodies, to distal ends of PEG chains via *p*-nitrophenylcarbonyl groups. Biochim Biophys Acta 1511:397–411

Torchilin VP, Lukyanov AN, Gao Z, Papahadjopoulos-Sternberg B (2003) Immunomicelles: targeted pharmaceutical carriers for poorly soluble drugs. Proc Natl Acad Sci U S A 100:6039–6044

Tu C, Louie AY (2012) Nanoformulations for molecular MRI. Wiley Interdiscip Rev Nanomed Nanobiotechnol 4:448–457

Uchino H, Matsumura Y, Negishi T, Koizumi F, Hayashi T, Honda T, Nishiyama N, Kataoka K, Naito S, Kakizoe T (2005) Cisplatin-incorporating polymeric micelles (NC-6004) can reduce nephrotoxicity and neurotoxicity of cisplatin in rats. Br J Cancer 93:678–687

Upadhyay KK, Agrawal HG, Upadhyay C, Schatz C, Le Meins JF, Misra A, Lecommandoux S (2009) Role of block copolymer nanoconstructs in cancer therapy. Crit Rev Ther Drug Carrier Syst 26:157–205

Valencia PM, Hanewich-Hollatz MH, Gao W, Karim F, Langer R, Karnik R, Farokhzad OC (2011) Effects of ligands with different water solubilities on self-assembly and properties of targeted nanoparticles. Biomaterials 32:6226–6233

van der Meel R, Vehmeijer LJ, Kok RJ, Storm G, van Gaal EV (2013) Ligand-targeted particulate nanomedicines undergoing clinical evaluation: current status. Adv Drug Deliv Rev 65:1284–1298

van Osdol W, Fujimori K, Weinstein JN (1991) An analysis of monoclonal antibody distribution in microscopic tumor nodules: consequences of a "binding site barrier". Cancer Res 51:4776–4784

Veikkola T, Karkkainen M, Claesson-Welsh L, Alitalo K (2000) Regulation of angiogenesis via vascular endothelial growth factor receptors. Cancer Res 60:203–212

Wacker M (2013) Nanocarriers for intravenous injection – the long hard road to the market. Int J Pharm 457:50–62

Wang J, Tian S, Petros RA, Napier ME, Desimone JM (2010) The complex role of multivalency in nanoparticles targeting the transferrin receptor for cancer therapies. J Am Chem Soc 132: 11306–11313

Wang T, Upponi JR, Torchilin VP (2012) Design of multifunctional non-viral gene vectors to overcome physiological barriers: dilemmas and strategies. Int J Pharm 427:3–20

Warenius HM, Galfre G, Bleehen NM, Milstein C (1981) Attempted targeting of a monoclonal antibody in a human tumour xenograft system. Eur J Cancer Clin Oncol 17:1009–1015

Watanabe K, Kaneko M, Maitani Y (2012) Functional coating of liposomes using a folate- polymer conjugate to target folate receptors. Int J Nanomedicine 7:3679–3688

Weinstein JN, van Osdol W (1992) The macroscopic and microscopic pharmacology of monoclonal antibodies. Int J Immunopharmacol 14:457–463

Weissig V, Lasch J, Erdos G, Meyer HW, Rowe TC, Hughes J (1998) DQAsomes: a novel potential drug and gene delivery system made from Dequalinium. Pharm Res 15:334–337

Whiteman KR, Subr V, Ulbrich K, Torchilin VP (2001) Poly(HPMA)-coated liposomes demonstrate prolonged circulation in mice. J Liposome Res 11:153–164

Wickham T, Futch K (2012) A phase I study of MM-302, a HER2-targeted liposomal doxorubicin, in patients with advanced, HER2- positive breast cancer. In: Thirty-fifth annual CTRC-AACR San Antonio Breast Cancer Symposium, Cancer Research, San Antonio, 2012

Wu J, Chu C-C (2013) Water insoluble cationic poly(ester amide)s: synthesis, characterization and applications. J Mater Chem B 1:353–360

Wu J, Akaike T, Hayashida K, Okamoto T, Okuyama A, Maeda H (2001) Enhanced vascular permeability in solid tumor involving peroxynitrite and matrix metalloproteinases. Jpn J Cancer Res 92:439–451

Xiao Z, Farokhzad OC (2012) Aptamer-functionalized nanoparticles for medical applications: challenges and opportunities. ACS Nano 6:3670–3676

Xiao K, Li Y, Luo J, Lee JS, Xiao W, Gonik AM, Agarwal RG, Lam KS (2011) The effect of surface charge on in vivo biodistribution of PEG-oligocholic acid based micellar nanoparticles. Biomaterials 32:3435–3446

Xiong XB, Huang Y, Lu WL, Zhang X, Zhang H, Nagai T, Zhang Q (2005) Intracellular delivery of doxorubicin with RGD-modified sterically stabilized liposomes for an improved antitumor efficacy: in vitro and in vivo. J Pharm Sci 94:1782–1793

Yamada A, Taniguchi Y, Kawano K, Honda T, Hattori Y, Maitani Y (2008) Design of folate-linked liposomal doxorubicin to its antitumor effect in mice. Clin Cancer Res 14:8161–8168

Yamamoto Y, Nagasaki Y, Kato Y, Sugiyama Y, Kataoka K (2001) Long-circulating poly(ethylene glycol)-poly(D, L-lactide) block copolymer micelles with modulated surface charge. J Control Release 77:27–38

Yang C, Zhao H, Yuan H, Yu R, Lan M (2013) Preparation and characterization of thermosensitive and folate functionalized Pluronic micelles. J Nanosci Nanotechnol 13:6553–6559

Yasuda H, Nakayama K, Watanabe M, Suzuki S, Fuji H, Okinaga S, Kanda A, Zayasu K, Sasaki T, Asada M, Suzuki T, Yoshida M, Yamanda S, Inoue D, Kaneta T, Kondo T, Takai Y, Sasaki H, Yanagihara K, Yamaya M (2006a) Nitroglycerin treatment may enhance chemosensitivity to docetaxel and carboplatin in patients with lung adenocarcinoma. Clin Cancer Res 12: 6748–6757

Yasuda H, Yamaya M, Nakayama K, Sasaki T, Ebihara S, Kanda A, Asada M, Inoue D, Suzuki T, Okazaki T, Takahashi H, Yoshida M, Kaneta T, Ishizawa K, Yamanda S, Tomita N, Yamasaki M, Kikuchi A, Kubo H, Sasaki H (2006b) Randomized phase II trial comparing nitroglycerin plus vinorelbine and cisplatin with vinorelbine and cisplatin alone in previously untreated stage IIIB/IV non-small-cell lung cancer. J Clin Oncol 24:688–694

Yigit MV, Moore A, Medarova Z (2012) Magnetic nanoparticles for cancer diagnosis and therapy. Pharm Res 29:1180–1188

Ying X, Wen H, Lu WL, Du J, Guo J, Tian W, Men Y, Zhang Y, Li RJ, Yang TY, Shang DW, Lou JN, Zhang LR, Zhang Q (2010) Dual-targeting daunorubicin liposomes improve the therapeutic efficacy of brain glioma in animals. J Control Release 141:183–192

Yoo HS, Park TG (2004) Folate receptor targeted biodegradable polymeric doxorubicin micelles. J Control Release 96:273–283

Yoshino K, Nakamura K, Terajima Y, Kurita A, Matsuzaki T, Yamashita K, Isozaki M, Kasukawa H (2012) Comparative studies of irinotecan-loaded polyethylene glycol-modified liposomes prepared using different PEG-modification methods. Biochim Biophys Acta 1818:2901–2907

Yu B, Tai HC, Xue W, Lee LJ, Lee RJ (2010) Receptor-targeted nanocarriers for therapeutic delivery to cancer. Mol Membr Biol 27:286–298

Yu J, Xie X, Wu J, Liu Y, Liu P, Xu X, Yu H, Lu L, Che X (2013) Folic acid conjugated glycol chitosan micelles for targeted delivery of doxorubicin: preparation and preliminary evaluation in vitro. J Biomater Sci Polym Ed 24:606–620

Yuan F, Leunig M, Huang SK, Berk DA, Papahadjopoulos D, Jain RK (1994) Microvascular permeability and interstitial penetration of sterically stabilized (stealth) liposomes in a human tumor xenograft. Cancer Res 54:3352–3356

Yuan F, Dellian M, Fukumura D, Leunig M, Berk DA, Torchilin VP, Jain RK (1995) Vascular permeability in a human tumor xenograft: molecular size dependence and cutoff size. Cancer Res 55:3752–3756

Yuan X, Harada A, Yamasaki Y, Kataoka K (2005) Stabilization of lysozyme-incorporated polyion complex micelles by the omega-end derivatization of poly(ethylene glycol)-poly(alpha, beta-aspartic acid) block copolymers with hydrophobic groups. Langmuir 21:2668–2674

Zalipsky S, Qazen M, Walker JA 2nd, Mullah N, Quinn YP, Huang SK (1999) New detachable poly(ethylene glycol) conjugates: cysteine-cleavable lipopolymers regenerating natural phospholipid, diacyl phosphatidylethanolamine. Bioconjug Chem 10:703–707

Zeng F, Lee H, Allen C (2006) Epidermal growth factor-conjugated poly(ethylene glycol)-*block*-poly(delta-valerolactone) copolymer micelles for targeted delivery of chemotherapeutics. Bioconjug Chem 17:399–409

Zeng J, Zhang J, Tanaka T, Rabbitts T (2013) Single domain antibody fragments as drug surrogates targeting protein–protein interactions inside cells. Antibodies 2:306–320

Zhang L, Gu FX, Chan JM, Wang AZ, Langer RS, Farokhzad OC (2008) Nanoparticles in medicine: therapeutic applications and developments. Clin Pharmacol Ther 83:761–769

Zhang L, Gong F, Zhang F, Ma J, Zhang P, Shen J (2013) Targeted therapy for human hepatic carcinoma cells using folate-functionalized polymeric micelles loaded with superparamagnetic iron oxide and sorafenib in vitro. Int J Nanomedicine 8:1517–1524

Zhao G, Rodriguez BL (2013) Molecular targeting of liposomal nanoparticles to tumor microenvironment. Int J Nanomedicine 8:61–71

Zhao W, Zhuang S, Qi XR (2011) Comparative study of the in vitro and in vivo characteristics of cationic and neutral liposomes. Int J Nanomedicine 6:3087–3098

Zhou Y, Drummond DC, Zou H, Hayes ME, Adams GP, Kirpotin DB, Marks JD (2007) Impact of single-chain Fv antibody fragment affinity on nanoparticle targeting of epidermal growth factor receptor-expressing tumor cells. J Mol Biol 371:934–947

Zhu CL, Lu CH, Song XY, Yang HH, Wang XR (2011) Bioresponsive controlled release using mesoporous silica nanoparticles capped with aptamer-based molecular gate. J Am Chem Soc 133:1278–1281

Zhu H, Cao J, Cui S, Qian Z, Gu Y (2013) Enhanced tumor targeting and antitumor efficacy via hydroxycamptothecin-encapsulated folate-modified *N*-succinyl-*N*′-octyl chitosan micelles. J Pharm Sci 102:1318–1332

Zou S, Cao N, Cheng D, Zheng R, Wang J, Zhu K, Shuai X (2012) Enhanced apoptosis of ovarian cancer cells via nanocarrier-mediated codelivery of siRNA and doxorubicin. Int J Nanomedicine 7:3823–3835

Zubieta MR, Furman D, Barrio M, Bravo AI, Domenichini E, Mordoh J (2006) Galectin-3 expression correlates with apoptosis of tumor-associated lymphocytes in human melanoma biopsies. Am J Pathol 168:1666–1675

Chapter 2
The "EX" and the "SOMA": How They Communicate

Pinar Kanlikilicer, Mohammed Helmy Rashed, Bulent Ozpolat, Cristian Rodriguez-Aguayo, Anil K. Sood, George A. Calin, and Gabriel Lopez-Berestein

2.1 Introduction

Exosomes were first thought to function as removers of intracellular debris. Over the past few years, accumulating evidence has shown that these small vesicles (30–100 nm in diameter) are capable of transferring internal cellular material such as RNAs, microRNAs (miRNAs), and proteins to the outside of the cell and

P. Kanlikilicer • B. Ozpolat
Department of Experimental Therapeutics, The University of Texas M.D. Anderson Cancer Center, 1515 Holcombe Blvd, Houston, TX 77030, USA

M.H. Rashed
Department of Experimental Therapeutics, The University of Texas M.D. Anderson Cancer Center, 1515 Holcombe Blvd, Houston, TX 77030, USA

Faculty of Pharmacy, Department of Pharmacology and Toxicology, Al-Azhar University, Cairo, Egypt

C. Rodriguez-Aguayo
Department of Cancer Biology and the Center for RNA Interference and Non-coding RNAs, The University of Texas M.D. Anderson Cancer Center, Houston, TX, USA

A.K. Sood
Departments of Gynecologic Oncology and Cancer Biology and the Center for RNA Interference and Non-coding RNAs, University of Texas M.D. Anderson Cancer Center, Houston, TX, USA

G.A. Calin • G. Lopez-Berestein (✉)
Department of Experimental Therapeutics and the Center for RNA Interference and Non-coding RNAs, The University of Texas M.D. Anderson Cancer Center, 1515 Holcombe Blvd, Houston, TX 77030, USA
e-mail: glopez@mdanderson.com

M.J. Alonso, M. Garcia-Fuentes (eds.), *Nano-Oncologicals: New Targeting and Delivery Approaches*, Advances in Delivery Science and Technology,
DOI 10.1007/978-3-319-08084-0_2

Table 2.1 Size distribution of different extracellular vesicles[a]

Extracellular vesicle	Size	Morphology	Cell surface markers	Centrifugal force to collect
Apoptotic bodies	800–5 μm	Round shaped	TSP, C3b	2,000*g*
Microvesicles	200–800 nm	Diverse in shape	ARF6, VCAMP3	12,200*g*
Exosome	40–100 nm	Cup-shaped	CD63, CD9	120,000*g*

[a]Modified from Crescitelli et al. (2013) and Akers et al. (2013)

may function as an intercellular communication system (Valadi et al. 2007). The cup-shaped exosome structures are formed within the cell and participate in endocytic pathways before being taken up into a multivesicular body. The multivesicular body fuses with the plasma membrane, releasing its cargo of exosomes into the extracellular space. Microvesicles and exosomes differ in many aspects, such as morphology, biogenesis, and release (Table 2.1). Crescitelli et al. (2013) showed that apoptotic bodies and multivesicles to be more diverse in their morphology than exosomes. Exosomes are uniform in structure, size, and endocytic origin, whereas microvesicles range from 200 nm to greater than 1 μm in diameter and originate from many different cell types.

The protein content of exosomes is located mainly in the cytosol, the membrane of the endocytic compartments, or the plasma membrane. These proteins comprise cytosolic proteins such as annexin and Rab, signal transduction enzymes such as tyrosine kinases, heat shock proteins, and large amounts of tetraspin molecules such as CD9, CD81, CD82, CD83, and CD63. Exosomes also include antigen-presenting proteins such as MHC class I and MHC class II, integrins, immunoglobulin family members, and cell surface peptidases (Thery et al. 2002). Recently, RNA content of exosomes was analyzed by next generation sequencing in two different breast cancer cells in vitro (Jenjaroenpun et al. 2013). Exosomes were found to be enriched in fragmented ribosomal RNA (rRNA) and other non-coding RNAs. In this study, they also found out that mRNA content of exosomes reflects the mRNA content of the host breast cancer cells. An miRNA array study revealed 121 miRNAs present in exosomes (Valadi et al. 2007).

Exosomes are secreted by many cell types, including immune cells, neurons, tumor cells, and stem cells (Fruehbeis et al. 2013; Raposo et al. 1996; Hegmans et al. 2004; Arslan et al. 2013). Individual cell types may produce different amount of exosomes. For instance, SKOV-3 ovarian cancer cells released 2.7-fold more exosomes compared to OVCAR-3 ovarian cancer cells (Kobayashi et al. 2014). In an in vivo setting, Takahashi et al. visualize the localization of exosomes in mice using bioluminescent imaging. They showed that exosomes were removed from the blood circulation in a few minutes after administration. They detected little luciferase activity in serum after 4 h. Liver and lung were the first two organs showing exosome-derived signals (Takahashi et al. 2013).

Since exosomes can be released by cancer cells into the microenvironment and the body fluids, they most likely participate in extracellular communications; however, the underlying mechanisms of this participation are not well understood.

2.2 Exosome Communication

Various communication mechanisms have been proposed in the literature for eukaryotic cells. One of these mechanisms is based on fusion, whereby exosomes fuse with the plasma membrane (Parolini et al. 2009). Exosomal lipid content and structure most likely enhances its capacity for fusing with other membranes.

When the multivesicular body fuses directly with the plasma membrane, its intraluminal vesicles are released into the extracellular environment as exosomes. The formation of intraluminal vesicles is initiated in early endosomes. Endosomes are classified as early, late, and recycling stages. It involves the transport of endocytic materials. Each group of endosomes has different protein and nucleic acid contents. Endosomes are capable of docking and fusing with the membrane with the help of some membrane trafficking and coordinating proteins such as clathrin, adaptor protein complexes, actin cytoskeletons, SNAREs (soluble *N-ethylmaleimide-sensitive* factor attachment protein receptors), and Rabs. The clathrin-mediated endocytosis pathway, together with the action of the early enzyme adaptor proteins, governs the best-characterized cargo-carrying mechanism. Actin provides the force necessary for vesicle uptake into the cells in this model of communication. AP1, AP2, AP3, and Golgi-localized gamma-adaptin ear–containing ARF-binding proteins (GGA) are the adaptor proteins that have major roles in the formation of clathrin-coated vesicles (Traub 2003; Robinson 2004). The localization of adaptor proteins varies. AP1 is localized on the trans-Golgi network and endosomes, whereas AP2 is found at the plasma membrane. AP3 homolog protein is also mostly localized on endosomes (Robinson 2004). GGA is structurally very different than the other major adaptor proteins.

The distribution of the adaptor proteins was suggested to play a role in cargo selectivity (Robinson 2004). Another adaptor protein, clathrin assembly lymphoid myeloid leukemia (CALM), was shown to bind directly to the arginine-contributing soluble *N*-ethylmaleimide-sensitive factor (NSF) attachment protein (SNAP) receptor proteins (R-SNAREs) such as VAMP8, VAMP3, and VAMP2 in HeLa-M cells and HEK293 cells (Miller et al. 2011; Koo et al. 2011). Mutation of CALM:SNARE interface residues resulted in inhibition of R-SNARE binding and endocytosis (Miller et al. 2011). Exosome uptake via clathrin-mediated endocytosis has been demonstrated previously in many cell lines, such as ovarian tumor cells and dendritic cells (Morelli et al. 2004; Escrevente et al. 2011).

Fusion of exosomes with the plasma membrane takes place at the final step of exosome secretion. Membrane fusion is crucial for exocytic and endocytic trafficking of signaling molecules. SNARE protein complexes have been suggested to be the key molecules in membrane fusion (Palfreyman and Jorgensen 2008). The SNARE proteins can be grouped into several small protein families (Fasshauer et al. 1998), synaptobrevin (R-SNARE), syntaxin (vesicle SNARE or v-SNARE), and SNAP-25 are the individual SNARE motifs that come together to form a four-helix bundle during membrane fusion. Basically, the function of SNARE proteins is to bring the vesicle membrane and the cell membrane into close proximity so that the fusion can take place (Fig. 2.1).

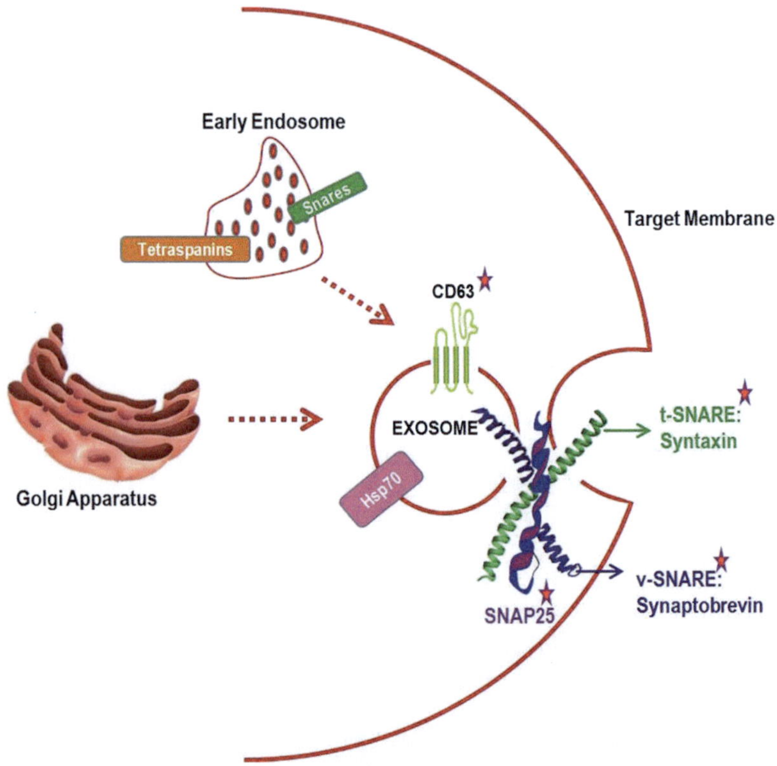

Fig. 2.1 Schematic representation of exosomal fusion in the cell. SNARE protein complexes (t-SNARE, v-SNARE, and SNAP-25) are involved in the fusion of the exosomal cargo

SNAREs have been thought to encode for specificity of membrane fusion, since it is thought that each intracellular fusion is mediated by a specific set of SNARE proteins (Table 2.2). There are more than 35 SNARE proteins. Differences in localization of each SNARE sub-family in the cell compartments have been suggested to drive the fusion selectivity. Scales et al. analyzed whether vesicle trafficking is determined by the specific pairing of the SNARE proteins. They demonstrated that inhibition or rescue of exocytosis of norepinephrine occurs only with a specific SNARE complex (Scales et al. 2000). McNew et al. tested in yeast the compartmental specificity of the plasma membrane, vacuolar, and endoplasmic reticulum (ER)–Golgi target SNARE (t-SNARE) complexes fused with possible combinations of non-syntaxin v-SNAREs, which were reconstituted into liposomes. They identified the pairs of v-SNAREs and t-SNAREs that are able to fuse significantly in a specific cellular compartment (Table 2.2; McNew et al. 2000). Moreover, they also identified four non-syntaxin v-SNAREs (Snc1, Snc2, Bet1, and Nyv1) that are able to promote fusion by binding a t-SNARE located in a different compartment.

These findings suggest that the specificity of exosome target cell communication may be a molecular level, rather than a system level, phenomenon.

Table 2.2 Membrane compartmental organization of SNARE proteins within the cell[a]

Mammal			
Localization	t-SNARE[b]	R-SNARE	SNAP
Plasma membrane	Syn1,2	Vamp 2,8	SNAP-23,25
Early endosomes	Syn7,13	Vamp 3,8	Syn8
Late endosomes	Syn7	Vamp 7,8	Syn8,VTI1b
Endoplasmic reticulum–Golgi	Syn5	Sec22b	GS27,BET1
Trans-Golgi network	Syn 13,16	Vamp4	VTI1a,Syn6

Yeast		
Localization	t-SNARE	Candidate v-SNAREs
Golgi	t-Sed5/Bos1,Sec22	Sft1 Gos1
Vacuole	t-Vam3/Vam7,Vit1	v-Nyv-1
Plasma membrane	t-Sso1/Sec9c	
Endoplasmic reticulum–Golgi		v-Bet Bos1 Ykt6 Sec22
Golgi/PM		v-Snc1 v-Snc2
Golgi/endosome		Tlg1
Vac/Golgi/endosome		Vti1

[a]Modified from Scheller (2013), Scales et al. (2000), and McNew et al. (2000)
[b]*t-SNARE* target SNARE, *v-SNARE* vesicle SNARE, *PM* plasma membrane, *Vac* vacuole

Graff et al. examined the existence of SNARE proteins in endocrine tumors because these proteins are widely present in endocrine cells (Capella et al. 1995; Graff et al. 2001). SNAP-25 and syntaxin1 in high-grade malignant small cell lung carcinomas (SCLCs), large cell carcinomas, adenocarcinomas, and squamous cell carcinomas of the lung were analyzed in a study evaluating SNAREs as differentiation markers for tumor identification. They found that SCLCs consistently expressed two of the major SNAREs; at the transcriptional level, however, the presence of mRNAs encoding for SNAP-25 and syntaxin1 was shown in the SCLC cell lines but not in the adenocarcinoma cell line (Graff et al. 2001).

Rab GTPases and their regulators have important roles in membrane trafficking. Rab5 and Rab7 were shown to be required for transport of cargo from the late endosomes to lysosomes for degradation (Stenmark 2009). Different populations of Rab proteins were abundant in different type of endosomes. The Rab5, Rab4 and Rab5, and Rab4 and Rab11 protein populations were abundant in early endosomes, whereas late endosomes were enriched with Rab7 and Rab9 proteins (Stenmark 2009). Likewise, exosome release is regulated by different Rab proteins in different cell populations. Exosome secretion was shown to be regulated by Rab35 in the central nervous system (Hsu et al. 2010), whereas Rab27a and Rab27b were shown to control exosome release in HeLa cells (Ostrowski et al. 2010). The presence of

CD63-positive, endosomal compartments in Rab27A-depleted cells, and perinuclear localization of CD63 in Rab27B-depleted cells suggest that Rab27A plays a role in MVE fusion with the plasma membrane, whereas Rab27B participate in the trafficking of MVEs and efflux proteins.

Cell–cell communication by exosomes leads to the acquisition of some other proteins within the cells. Exosomes were found to be rich in glycoproteins6 (Escrevente et al. 2011) and to contain high concentrations of ceramide (Trajkovic et al. 2008). Vallhov et al. (2011) found that glycoprotein glycol protein 350 mediates the specific binding of exosomes to B cells. Sphingolipid ceramide transfers exosome-associated domains into the endosomal lumens. Heparan sulfate proteoglycans was shown to be the key mediator for internalization of exosomes and their functional activity (Christianson et al. 2013).

2.2.1 Exosomes and Plasticity

Exosome-mediated intercellular communication may be a comprehensive phenomenon that includes plasticity. The way exosomes communicate with the target cells may depend on the experience-dependent properties present in cells. In neurons, exosome release was shown to be regulated in an activity-dependent manner by glutamatergic synaptic activity in brain (Lachenal et al. 2011). The synaptic plasticity–disrupting activity of Aβ in the brains of Alzheimer disease patients was also shown to be neutralized by exosomes, indicating their role in synaptic plasticity (An et al. 2013).

Savina et al. (2003) demonstrated that monensin regulates exosome release in K562 cells in a calcium-dependent manner. Monensin, which acts on acidic intracellular organelles such as endosomes and lysosomes, exchanges Na^+ ions with H^+ ions and thus alters the intracellular Ca^{2+} flux. Many intracellular trafficking reactions, such as membrane fusion (Mayer 1999), exocytosis (Erxleben et al. 1997), and ER-to-Golgi transport (Beckers and Balch 1989), depend on Ca^{2+} ions. Membrane fusion is initiated by the action of SNARE proteins, which are regulated in a Ca^{2+}-dependent manner (Di Giovanni et al. 2010). Therefore, since exosomes fuse with the plasma membrane to communicate with other cells, either the Ca^2+ flux of the cell or a complex of SNARE proteins might be the rate limiting step in the communication process of exosomes.

Parolini et al. (2009) suggested that exosome-to-cell fusion is a more frequent event in tumor tissues than in normal tissues. The transformation of a normal cell into a cancer cell is also associated with change of Ca^{2+} flux within the cell (Capiod et al. 2007). Hence, Ca^{2+}-dependent plasticity might determine the exosome-to-cell fusion characteristics of a cancer cell. Moreover, Gross et al. (2012) demonstrated that Wnt proteins, which are involved in many vital cell functions such as cell proliferation, survival, and migration, are located on the surface of exosomes. They also demonstrated that Wnt transport to endosomal compartments on the surface of exosomes requires the R-SNARE protein Ykt6. Hence, the regulation of Wnt signaling, one of the most important signaling pathways for cancer (Anastas and Moon 2013), might also be affected by Ca^{2+}-dependent plasticity of exosomes.

2.3 Exosomes and Cancer

The role of exosomes in cancer is not fully understood. While, there is evidence indicating that exosomes stimulate antitumor response by presenting tumor antigens to dendritic cells, they also modulate immunosuppressive functions, thus contributing to cancer pathogenesis. Exosomes secreted from cancer cells have been suggested to play important roles in angiogenesis, invasion, and metastasis, thus promoting tumor development (Christianson et al. 2013; Skog et al. 2008). Lung cancer microvesicles and glioblastoma-derived microvesicles in brain endothelial cells were shown to play roles in angiogenesis (Janowska-Wieczorek et al. 2005; Skog et al. 2008). Exosomes were shown to promote tumor invasiveness via the stimulation of matrix metalloproteinase 1 (MMP1) in patients with gastrointestinal stromal tumor (Atay et al. 2014). The promotion of metastases by exosomes has been previously demonstrated in melanoma (Peinado et al. 2012). They pointed out the premetastatic niche formation by showing the endothelial permeability caused by B16-F10 metastatic melanoma cells in the lung. Exosomes increased the metastatic behavior of primary melanoma tumors through the up-regulation of receptor tyrosine kinase MET. In breast cancer, fibroblast-secreted exosomes were demonstrated to promote breast cancer cell motility via Wnt induced signaling (Luga et al. 2012). Luga et al. also showed that coinjection of breast cancer cells with CD81 positive fibroblasts stimulates metastasis in orthotopic mouse models of breast cancer. Exosomes have been shown to target and modulate non-transformed cells in premetastatic organs in metastatic rat adenocarcinoma in vivo (Rana et al. 2013). Another study demonstrated that melanoma exosomes travel to sentinel lymph nodes (Hood et al. 2011). Thus, many studies indicate that cancer cells may benefit from exosomal homing mechanisms to communicate signals to distant cells and tissues.

Exosome secretion has also been linked to chemoresistance (Corcoran et al. 2012). Doxorubicin and other small molecules such as cisplatin were shown to accumulate in vesicles, resulting in exosome shedding (Shedden et al. 2003; Chen et al. 2006). Once cisplatin enters the exosomal pathway, cisplatin-resistant cells exclude it (Safaei et al. 2005). Corcoran et al. (2012) found a significant increase in resistance to docetaxel in docetaxel-sensitive parent cells that were exposed to exosomes isolated from docetaxel-resistant cancer cells. Survivin, which was shown to be released from cancer cells via exosomes, was suggested to suppress the efficacy of protein irradiation (Khan et al. 2011).

Tumor-released exosomes elicit antitumor immune responses by promoting antigen-presenting cells. The differentiation of circulating monocytes into functional dendritic cells was shown to be deteriorated by tumor exosomes; moreover, exosomes mediate myeloid suppressive activity on T-cell functions, such as proliferation (Valenti et al. 2006). Some tumor-released exosomes cause apoptosis in activated T cells by expressing tumor necrosis factor-related apoptosis-inducing ligand FasL (Kim et al. 2005).

Exosomes may interfere with many vital signaling pathways through the phosphorylation of proteins. One of the biological processes affected in this way is the mesenchymal stem cells (MSCs) transition to carcinoma-associated fibroblasts

(CAFs). MSCs to CAFs transition is known to promote tumor growth (Quante et al. 2011). Gu et al. (2012) showed the presence of TGF-β in gastric cancer derived exosomes and suggested that exosomes promote MSCs to CAFs transition by the stimulation of the phosphorylation of Smad-2/3 through the TGF-β signaling pathway. In another study investigating the mechanism of exosome uptake revealed that non-clathrin-dependent, lipid raft-mediated endocytosis mediates the exosome uptake (Svensson et al. 2013). Exosome treatment results in 2–4.5-fold induction of many lipid raft associated proteins such as p-FAK, the heat-shock protein p-HSP27, and p-ERK1/2 and its downstream target p-MSK1/2 (Svensson et al. 2013).

Many factors such as inflammation, hypoxia, and other microenvironmental conditions contribute to tumor growth and proliferation through exosome interaction. Microenvironment is one of the key factors in tumor malignancy (Luciani et al. 2004; Webb et al. 1999). Parolini et al. (2009) demonstrated that acidic conditions promote membrane fusion. Fusion between buffered exosomes and buffered cells was shown to be less than that between acidic exosomes and an acidic microenvironment. Metastatic melanoma cells were shown to be significantly more acidic than samples derived from primary melanoma cells (Lugini et al. 2006). Hypoxia, which promotes tumor progression, was demonstrated to enhance the release of exosomes in breast cancer cell lines (King et al. 2012). Ovarian tumor-derived exosomes were shown to induce pro-inflammatory cytokines in macrophages (Atay et al. 2013). The study suggested that fibronectins expressed by tumor exosomes account for the pro-inflammatory microenvironment observed in cancer, whereas thermal and oxidative stress was found to enhance the release of immunosuppressive exosomes in leukemia/lymphoma cells (Hedlund et al. 2011).

Some proteins secreted from exosomes are responsible for antitumor responses. Heat shock protein (Hsp) is one of the major proteins figuring in exosome-related antitumor response. In human hepatocellular carcinoma cells, resistance to anticancer drugs was shown to enhance the release of exosomes by Hsp (Lv et al. 2012). p53, which is mutated or lost in most tumors, promotes exosome production. p53 activation after DNA damage led to an increase in the amount of proteins that are secreted into the medium via exosomes (Yu et al. 2006). PTEN gene, which is usually localized in the cytoplasm and nucleus, was found to be secreted in exosomes (Putz et al. 2012). In recipient fibroblasts, PTEN hinders pAkt signaling, which leads to inhibition of cell proliferation. Thus, tumor suppressor genes can be used to target tumor bearing cancer cells, and induce immune responses by exosome-mediated delivery.

2.4 Exosome-Mediated Drug Delivery

Having the capacity to carry genetic material, exosomes are promising tools for drug delivery. Drugs delivered by an exosomal delivery vehicle are more stable and more highly concentrated in the blood than drugs delivered by other such carriers (Sun et al. 2010). Exosomes have been studied as drug carriers for many pathological conditions such as cancer and inflammatory diseases (Sun et al. 2010; Ohno et al. 2013).

Exosomes provide target specificity, directing the drug only to abnormal cells, and thus decreasing the drug's toxicity. Most of the cancer chemotherapeutic agents act on tumor cells by inducing DNA damage. While normal cells can escape from these agents by inducing cell cycle arrest as a response to DNA damage, treatments that are too aggressive or prolonged can damage normal cells. Therefore, innovative treatments for cancer must have substantially less toxicity than current therapeutic agents.

Exosomes generated from monocyte-derived dendritic cells, which contain MHC-I and MHC-II antigens, were shown to be potent activators of T cells through antigen presentation (Hsu et al. 2003). Exosomal delivery was shown to enhance the anti-inflammatory activity of curcumin in a mouse model of lipopolysaccharide-induced septic shock (Sun et al. 2010). In this study, in order to incorporate curcumin into exosomes, curcumin–EL-4-derived exosome mixture was prepared at 22 °C and then, sucrose gradient centrifugation was applied. The fraction corresponding to a yellowish band was designated as the exosomal curcumin. This study demonstrated that curcumin delivered by exosomes is more stable and achieves higher concentrations in the blood than curcumin administered by other means. By targeting the inflammatory cells only, undesired off-target effects are eliminated.

Exosomes have been used successfully for in vivo and in vitro delivery of small-interfering RNA (siRNA) (El-Andaloussi et al. 2012). siRNA delivery was achieved in the brains of mice by using neuron-specific RVG peptide3-targeted exosomes through loading of exogenous siRNA into exosomal membrane protein Lamb2 (Alvarez-Erviti et al. 2011). They performed exosome loading using 12 μg exosomes and either 400 nmol or 12 μg of siRNA for cell culture and in vivo injections. The exosome–siRNA mixtures were electroporated in a cuvette. For the in vivo experiments, 150 μg exosomes-encapsulated siRNA resuspended in 80 μl of 5 % glucose was injected i.v to the C57BL/6 male 8- to 10-week-old mice.

In vitro exosome-mediated siRNA delivery was demonstrated by knockdown of the RAD51 transcript in both HeLa and HT1080 cells. The exosomes caused selective silencing of the RAD51 gene in the target cancer cells, resulting in cell death (Shtam et al. 2013). In this study, siRNAs were loaded into exosomes by two approaches: chemical treatment and electroporation. For the chemical treatment, 2 μmol/ml siRNA was first mixed with Lipofectamine and incubated for 10 min at room temperature. Then, the resulting mixture was mixed and incubated with 300 μl of exosome suspension for an additional 30 min at room temperature. The resulting exosome–siRNA mixture was purified by filtration. For electroporation, 300 μl of exosomes either derived from HeLa, HT1080, or ascitic fluids was diluted in cytomix transfection buffer and then mixed with siRNAs against RAD51 or RAD52. The mixtures were transferred into ice cold 0.4-cm cuvettes and electroporated. Exosome-mediated delivery of siRNA was also demonstrated in human mononuclear blood cells (Wahlgren et al. 2012). In that study, siRNA was introduced into human exosomes by electroporation, and these exosomes caused selective silencing of the MAPK1 gene in monocytes and lymphocytes. In another study, nanoparticles were used to deliver siRNA in conjunction with exosomes (van den Boorn et al. 2011). Exosomes have also been used to efficiently deliver antitumor miRNA to cancer cells. Ohno et al. (2013) intravenously injected exosomes targeting EGFR to epidermal growth factor receptor (EGFR)-expressing breast cancer cells. Let-7a loaded GE11-positive exosomes efficiently delivered their cargo to EGFR-expressing tumors. However, there are a number of challenges that need to be overcome. So far, the incorporation efficiencies of functional proteins or nucleic acids into exosomes are low. Incorporation of multiple proteins is a complicated process. In addition, exosomal components that are required for the assembly of functional drug delivery systems have not been identified.

A novel study used an exosome mimetic to deliver chemotherapeutics for the treatment of malignant tumors (Jang et al. 2013). These authors showed that exosome-mimetic nanovesicles that targeted endothelial cells were produced from either macrophages or monocytes. They also found that the exosome-mimetic nanovesicles loaded with doxorubicin reduced tumor growth to the same extent as 20-fold higher doses of free drug and were similarly effective in treating tumors in vivo as exosomes loaded with doxorubicin.

2.5 Exosomes as Biomarkers

Extensive miRNA profiling of circulating tumor exosomes has been undertaken. Samples taken from body fluids, either urine or plasma, revealed many exosomal proteins or miRNAs involved in pathological conditions such as breast cancer, prostate cancer, ovarian cancer, and liver disease due to high alcohol consumption (Wittmann and Jaeck 2010; Corcoran et al. 2011; Li et al. 2009; Bala et al. 2012).

Quantitative reverse-transcriptase polymerase chain reaction (RT-PCR) studies that compared miRNA expression in tumor and normal tissues revealed many circulating miRNAs in the serum that can be used as diagnostic markers for various cancer types. miRNA profiles for several types of cancers also have been analyzed. For instance, altered levels of miR-155 were found in the plasma of patients with B-cell lymphoma, breast, ovarian, or pancreatic cancer (Lawrie et al. 2008; Zhu et al. 2009; Wang et al. 2009; Resnick et al. 2009). miR-21 is also not only highly abundant in patients with B-cell lymphoma (Chen 2005) but also is elevated in the plasma of patients with pancreatic, gastric, or ovarian cancer (Resnick et al. 2009; Wang et al. 2009; Tsujiura et al. 2010). However, the abundance of miRNAs in exosomes may not always correlate with the amount of miRNAs of the cells from which the exosomes originate. In ovarian cancer cells, the let-7 miRNAs were found to be more abundant in OVCAR-3 cell than SKOV-3 cells, whereas let-7 family miRNAs were found to be more abundant in exosomes from SKOV-3 than OVCAR-3 cells. miR-200 was only identified in OVCAR-3 cells and corresponding exosomes from this cell line (Kobayashi et al. 2014).

Some exosome-based diagnostic assays are commercially available for clinical use; these include ExoTEST (HansaBioMed, Estonia) and EXO50-60-70 (Exosome Diagnostics Inc., New York, NY, USA) for use in plasma, serum, and cerebrospinal fluid (CSF) specimens. ExoTEST is a sandwich ELISA developed to capture and quantify exosomes in human plasma and other body fluids, as well as in cell culture supernatants. This method couples an exosomal protein (such as tetraspin molecule CDC63) with a tumor-specific marker (Logozzi et al. 2009). EXO50 enables the extraction of RNA from microvesicles in plasma/serum (Exo50), CSF (Exo60), and urine (Exo70).

2.6 Isolation of Exosomes

Exosomes are isolated from body fluids such as plasma, serum, or ascites, mainly by one of the three approaches: (1) ultracentrifugation, (2) size exclusion chromatography, and (3) antibody-coupled magnetic beads (DynaBeads). These methods isolate exosomes for downstream processing such as identification of specific RNA populations by quantitative RT-PCR profiling or identification and analysis of specific marker proteins by microarray profiling. A fourth method, ExoQuick precipitation, is performed with a commercial kit. ExoQuick precipitation was found to yield greater quantity and purity of exosomal RNA and proteins than the other three methods (Taylor and Gercel-Taylor 2008).

The ultracentrifugation method, as the name implies, is based on centrifugation of exosomes at high speeds such as 100,000*g* for 1 h. After the first centrifugation, the pellet is resuspended with appropriate buffer and then subjected to centrifugation again at 100,000*g* for an additional hour. The resulting pellet contains exosomes. For size exclusion chromatography, aliquots of body fluid samples are applied to a 2 % agarose-based gel column. The column is eluted with phosphate-buffered saline solution at a certain flow rate and monitored at 280 nm, and fractions are collected. The void volume fractions are pooled and subjected to centrifugation at 100,000*g* for 1 h (Taylor and Gercel-Taylor 2008). The DynaBeads method is based on coupling of magnetic microbeads with anti-epithelial cell adhesion molecule (EpCAM). Exosomes from epithelial tumors express EpCAM on their surface. Therefore, tumor-derived exosomes can be specifically isolated by modified magnetic activated cell sorting (MACS) using EpCAM. For this method, serum samples are incubated with anti-EpCAM coupled to magnetic microbeads. An LD microcolumn is placed in the magnetic field of a MACS separator and the column is rinsed. The magnetic immune complex is applied onto the column and then washed. The specifically selected exosomes can be recovered by removing the column from the separator. The magnetically labeled exosomes are obtained by applying the plunger. The isolated exosomes/microbeads are subjected to centrifugation to separate the microbeads from the exosomes. The exosomes (supernatant) are then subjected to centrifugation at a high speed.

2.7 Conclusion

The underlying mechanisms of exosome interaction with the body environment are not well understood. Plasticity might be involved, in conjunction with the known communication mechanisms to explain intracellular cell trafficking. Whether the specificity of exosomes is determined at the system level or molecular level is still being interrogated. Specificity may be determined at the biogenesis of the vesicle, when the vesicle coat is formed by the interaction of clathrin and its corresponding effector protein; these proteins are present in various subunits and are localized to different compartments of the cell. Rab localization in the cell may also bring specificity to exosome release. Moreover, SNARE proteins, which were shown to be functional only if assembled with the correct subunits (Scales et al. 2000), may provide a model for understanding this specificity. Calcium ions, which regulate membrane fusion, might be the key regulator of exosome–cell communication.

The potential use of exosomal miRNA (exRNA) may open a new era of cancer therapeutics. As we improve our understanding of cell–exosome–cell communication and the role of exRNA in cellular and body functions, this may become a reality. Nevertheless, the success for both diagnosis and treatment of the cancer needs more understanding of the interactions of nucleic acid or protein contents in tumor microenvironment and the mechanisms explaining how they migrate to distant cells. Thus, this young area of research needs to mature as more exosomal diagnostic and therapeutic approaches go into applications. How exosomes communicate within the microenvironmental structures and distant sites is a challenge at present. Elucidation of such mechanisms will lead to a better understanding of cellular communication and the potential of using exosomes as delivery systems.

Acknowledgments *Financial Support.* This study was supported by the Department of Defense grant W81XWH-09-1-0212; the NIH grants U54CA143837, U54CA151668, RC2GM092599, and P50 CA083639; the CPRIT grant RP121071 from the State of Texas; the RGK Foundation; the Gilder Foundation; and the Ernest Cockrell Jr. Distinguished Endowed Chair. The costs of publication of this article were defrayed in part by the payment of page charges. This article must therefore be hereby marked advertisement in accordance with 18 U.S.C. Section 1734 solely to indicate this fact.

References

Akers JC, Gonda D, Kim R, Carter BS, Chen CC (2013) Biogenesis of extracellular vesicles (EV): exosomes, microvesicles, retrovirus-like vesicles, and apoptotic bodies. J Neurooncol 113:1–11

Alvarez-Erviti L, Seow Y, Yin H, Betts C, Lakhal S, Wood MJA (2011) Delivery of siRNA to the mouse brain by systemic injection of targeted exosomes. Nat Biotechnol 29:341–345

An K, Klyubin I, Kim Y, Jung JH, Mably AJ, O'Dowd ST, Lynch T, Kanmert D, Lemere CA, Finan GM, Park JW, Kim T-W, Walsh DM, Rowan MJ, Kim J-H (2013) Exosomes neutralize synaptic-plasticity-disrupting activity of Abeta assemblies in vivo. Mol Brain 6:47

Anastas JN, Moon RT (2013) WNT signalling pathways as therapeutic targets in cancer. Nat Rev Cancer 13:11–26

Arslan F, Lai RC, Smeets MB, Akeroyd L, Choo A, Aguor ENE, Timmers L, van Rijen HV, Doevendans PA, Pasterkamp G, Lim SK, de Kleijn DP (2013) Mesenchymal stem cell-derived exosomes increase ATP levels, decrease oxidative stress and activate PI3K/Akt pathway to enhance myocardial viability and prevent adverse remodeling after myocardial ischemia/reperfusion injury. Stem Cell Res 10:301–312

Atay S et al (2013) Ovarian cancer-derived exosomal fibronectin induces pro-inflammatory IL-1 beta. Exosomes Microvesicles 1:1–10

Atay S, Banskota S, Crow J, Sethi G, Rink L, Godwin AK (2014) Oncogenic KIT-containing exosomes increase gastrointestinal stromal tumor cell invasion. Proc Natl Acad Sci U S A 111:711–716

Bala S, Petrasek J, Mundkur S, Catalano D, Levin I, Ward J, Alao H, Kodys K, Szabo G (2012) Circulating microRNAs in exosomes indicate hepatocyte injury and inflammation in alcoholic, drug-induced, and inflammatory liver diseases. Hepatology 56:1946–1957

Beckers CJM, Balch WE (1989) Calcium and GTP – essential components in vesicular trafficking between the endoplasmic-reticulum and Golgi-apparatus. J Cell Biol 108:1245–1256

Capella C, Heitz PU, Hofler H, Solcia E, Kloppel G (1995) Revised classification of neuroendocrine tumors of the lung, pancreas and gut. Virchows Arch 425:547–560

Capiod T, Shuba Y, Skryma R, Prevarskaya N (2007) Calcium signalling and cancer cell growth. Subcell Biochem 45:405–427

Chen CZ (2005) MicroRNAs as oncogenes and tumor suppressors. N Engl J Med 353:1768–1771

Chen KG, Valencia JC, Lai B, Zhang G, Paterson JK, Rouzaud F, Berens W, Wincovitch SM, Garfield SH, Leapman RD, Hearing VJ, Gottesman MM (2006) Melanosomal sequestration of cytotoxic drugs contributes to the intractability of malignant melanomas. Proc Natl Acad Sci U S A 103:9903–9907

Christianson HC, Svensson KJ, van Kuppevelt TH, Li J-P, Belting M (2013) Cancer cell exosomes depend on cell-surface heparan sulfate proteoglycans for their internalization and functional activity. Proc Natl Acad Sci U S A 110:17380–17385

Corcoran C, Friel AM, Duffy MJ, Crown J, O'Driscoll L (2011) Intracellular and extracellular microRNAs in breast cancer. Clin Chem 57:18–32

Corcoran C, Rani S, O'Brien K, O'Neill A, Prencipe M, Sheikh R, Webb G, Mcdermott R, Watson W, Crown J, O'Driscoll L (2012) Docetaxel-resistance in prostate cancer: evaluating associated phenotypic changes and potential for resistance transfer via exosomes. PLoS One 7

Crescitelli R, Lasser C, Szabo TG, Kittel A, Eldh M, Dianzani I, Buzas EI, Lotvall J (2013) Distinct RNA profiles in subpopulations of extracellular vesicles: apoptotic bodies, microvesicles and exosomes. J Extracell Vesicles 2

di Giovanni J, Iborra C, Maulet Y, Leveque C, El Far O, Seagar M (2010) Calcium-dependent regulation of SNARE-mediated membrane fusion by calmodulin. J Biol Chem 285:23665–23675

El-Andaloussi S, Lee Y, Lakhal-Littleton S, Li J, Seow Y, Gardiner C, Alvarez-Erviti L, Sargent IL, Wood MJA (2012) Exosome-mediated delivery of siRNA in vitro and in vivo. Nature Protoc 7:2112–2126

Erxleben C, Klauke N, Flotenmeyer M, Blanchard MP, Braun C, Plattner H (1997) Microdomain Ca^{2+} activation during exocytosis in Paramecium cells. Superposition of local subplasmalemmal calcium store activation by local Ca^{2+} influx. J Cell Biol 136:597–607

Escrevente C, Keller S, Altevogt P, Costa J (2011) Interaction and uptake of exosomes by ovarian cancer cells. BMC Cancer 11:108

Fasshauer D, Sutton RB, Brunger AT, Jahn R (1998) Conserved structural features of the synaptic fusion complex: SNARE proteins reclassified as Q- and R-SNAREs. Proc Natl Acad Sci U S A 95:15781–15786

Fruehbeis C, Froehlich D, Kuo WP, Amphornrat J, Thilemann S, Saab AS, Kirchhoff F, Moebius W, Goebbels S, Nave K-A, Schneider A, Simons M, Klugmann M, Trotter J, Kraemer-Albers E-M (2013) Neurotransmitter-triggered transfer of exosomes mediates oligodendrocyte-neuron communication. PLoS Biol 11:e1001604

Graff L, Castrop F, Bauer M, Hofler H, Gratzl M (2001) Expression of vesicular monoamine transporters, synaptosomal-associated protein 25 and syntaxin1: a signature of human small cell lung carcinoma. Cancer Res 61:2138–2144

Gross JC, Chaudhary V, Bartscherer K, Boutros M (2012) Active Wnt proteins are secreted on exosomes. Nat Cell Biol 14:1036–1045

Gu J, Qian H, Shen L, Zhang X, Zhu W, Huang L, Yan Y, Mao F, Zhao C, Shi Y, Xu W (2012) Gastric cancer exosomes trigger differentiation of umbilical cord derived mesenchymal stem cells to carcinoma-associated fibroblasts through TGF-beta/Smad pathway. PLoS One 7:e52465

Hedlund M, Nagaeva O, Kargl D, Baranov V, Mincheva-Nilsson L (2011) Thermal- and oxidative stress causes enhanced release of NKG2D ligand-bearing immunosuppressive exosomes in leukemia/lymphoma T and B cells. PLoS One 6:e16899

Hegmans J, Bard MPL, Hemmes A, Luider TM, Kleijmeer MJ, Prins JB, Zitvogel L, Burgers SA, Hoogsteden HC, Lambrecht BN (2004) Proteomic analysis of exosomes secreted by human mesothelioma cells. Am J Pathol 164:1807–1815

Hood JL, San RS, Wickline SA (2011) Exosomes released by melanoma cells prepare sentinel lymph nodes for tumor metastasis. Cancer Res 71:3792–3801

Hsu DH, Paz P, Villaflor G, Rivas A, Mehta-Damani A, Angevin E, Zitvogel L, Le Pecq JB (2003) Exosomes as a tumor vaccine: enhancing potency through direct loading of antigenic peptides. J Immunother 26:440–450

Hsu C, Morohashi Y, Yoshimura S-I, Manrique-Hoyos N, Jung S, Lauterbach MA, Bakhti M, Gronborg M, Moebius W, Rhee J, Barr FA, Simons M (2010) Regulation of exosome secretion by Rab35 and its GTPase-activating proteins TBC1D10A-C. J Cell Biol 189:223–232

Jang SC, Kim OY, Yoon CM, Choi DS, Roh TY, Park J, Nilsson J, Lötvall J, Kim YK, Gho YS (2013) Bioinspired exosome-mimetic nanovesicles for targeted delivery of chemotherapeutics to malignant tumors. ACS Nano 7:7698–7710

Janowska-Wieczorek A, Wysoczynski M, Kijowski J, Marquez-Curtis L, Machalinski B, Ratajczak J, Ratajczak MZ (2005) Microvesicles derived from activated platelets induce metastasis and angiogenesis in lung cancer. Int J Cancer 113:752–760

Jenjaroenpun P, Kremenska Y, Nair VM, Kremenskoy M, Joseph B, Kurochkin IV (2013) Characterization of RNA in exosomes secreted by human breast cancer cell lines using next-generation sequencing. PeerJ 1:e201

Khan S, Jutzy JMS, Aspe JR, Mcgregor DW, Neidigh JW, Wall NR (2011) Survivin is released from cancer cells via exosomes. Apoptosis 16:1–12

Kim JW, Wieckowski E, Taylor DD, Reichert TE, Watkins S, Whiteside TL (2005) Fas ligand-positive membranous vesicles isolated from sera of patients with oral cancer induce apoptosis of activated T lymphocytes. Clin Cancer Res 11:1010–1020

King HW, Michael MZ, Gleadle JM (2012) Hypoxic enhancement of exosome release by breast cancer cells. BMC Cancer 12:421

Kobayashi M, Salomon C, Tapia J, Illanes SE, Mitchell MD, Rice GE (2014) Ovarian cancer cell invasiveness is associated with discordant exosomal sequestration of Let-7 miRNA and miR-200. J Transl Med 12:4

Koo SJ, Markovic S, Puchkov D, Mahrenholz CC, Beceren-Braun F, Maritzen T, Dernedde J, Volkmer R, Oschkinat H, Haucke V (2011) SNARE motif-mediated sorting of synaptobrevin by the endocytic adaptors clathrin assembly lymphoid myeloid leukemia (CALM) and AP180 at synapses. Proc Natl Acad U S A 108:13540–13545

Lachenal G, Pernet-Gallay K, Chivet M, Hemming FJ, Belly A, Bodon G, Blot B, Haase G, Goldberg Y, Sadoul R (2011) Release of exosomes from differentiated neurons and its regulation by synaptic glutamatergic activity. Mol Cell Neurosci 46:409–418

Lawrie CH, Gal S, Dunlop HM, Pushkaran B, Liggins AP, Pulford K, Banham AH, Pezzella F, Boultwood J, Wainscoat JS, Hatton CSR, Harris AL (2008) Detection of elevated levels of tumour-associated microRNAs in serum of patients with diffuse large B-cell lymphoma. Br J Haematol 141:672–675

Li J, Sherman-Baust CA, Tsai-Turton M, Bristow RE, Roden RB, Morin PJ (2009) Claudin-containing exosomes in the peripheral circulation of women with ovarian cancer. BMC Cancer 9:244

Logozzi M, De Milito A, Lugini L, Borghi M, Calabro L, Spada M, Perdicchio M, Marino ML, Federici C, Iessi E, Brambilla D, Venturi G, Lozupone F, Santinami M, Huber V, Maio M, Rivoltini L, Fais S (2009) High levels of exosomes expressing CD63 and caveolin-1 in plasma of melanoma patients. PLoS One 4:e5219

Luciani F, Spada M, De Milito A, Molinari A, Rivoltini L, Montinaro A, Marra M, Lugini L, Logozzi M, Lozupone F, Federici C, Lessi E, Parmiani G, Arancia G, Belardelli F, Fais S (2004) Effect of proton pump inhibitor pretreatment on resistance of solid tumors to cytotoxic drugs. J Natl Cancer Inst 96:1702–1713

Luga V, Zhang L, Viloria-Petit AM, Ogunjimi AA, Inanlou MR, Chiu E, Buchanan M, Hosein AN, Basik M, Wrana JL (2012) Exosomes mediate stromal mobilization of autocrine Wnt-PCP signaling in breast cancer cell migration. Cell 151:1542–1556

Lugini L, Matarrese P, Tinari A, Loznpone F, Federici C, Iessi E, Gentile M, Luciani F, Parmiani G, Rivoltini L, Malorni W, Fais S (2006) Cannibalism of live lymphocytes by human metastatic but not primary melanoma cells. Cancer Res 66:3629–3638

Lv L-H, Wan Y-L, Lin Y, Zhang W, Yang M, Li G-L, Lin H-M, Shang C-Z, Chen Y-J, Min J (2012) Anticancer drugs cause release of exosomes with heat shock proteins from human hepatocellular carcinoma cells that elicit effective natural killer cell antitumor responses in vitro. J Biol Chem 287:15874–15885

Mayer A (1999) Intracellular membrane fusion: SNAREs only? Curr Opin Cell Biol 11:447–452

Mcnew JA, Parlati F, Fukuda R, Johnston RJ, Paz K, Paumet F, Sollner TH, Rothman JE (2000) Compartmental specificity of cellular membrane fusion encoded in SNARE proteins. Nature 407:153–159

Miller SE, Sahlender DA, Graham SC, Hoening S, Robinson MS, Peden AA, Owen DJ (2011) The molecular basis for the endocytosis of small R-SNAREs by the clathrin adaptor CALM. Cell 147:1118–1131

Morelli AE, Larregina AT, Shufesky WJ, Sullivan MLG, Stolz DB, Papworth GD, Zahorchak AF, Logar AJ, Wang ZL, Watkins SC, Falo LD, Thomson AW (2004) Endocytosis, intracellular sorting, and processing of exosomes by dendritic cells. Blood 104:3257–3266

Ohno S-I, Takanashi M, Sudo K, Ueda S, Ishikawa A, Matsuyama N, Fujita K, Mizutani T, Ohgi T, Ochiya T, Gotoh N, Kuroda M (2013) Systemically injected exosomes targeted to EGFR deliver antitumor microRNA to breast cancer cells. Mol Ther 21:185–191

Ostrowski M, Carmo NB, Krumeich S, Fanget I, Raposo G, Savina A, Moita CF, Schauer K, Hume AN, Freitas RP, Goud B, Benaroch P, Hacohen N, Fukuda M, Desnos C, Seabra MC, Darchen F, Amigorena S, Moita LF, Thery C (2010) Rab27a and Rab27b control different steps of the exosome secretion pathway. Nat Cell Biol 12:19–30

Palfreyman MT, Jorgensen EM (2008) Roles of SNARE proteins in synaptic vesicle fusion. In: Wang Z-W (ed) Molecular mechanisms of neurotransmitter release. Humana, Totowa, pp 35–59

Parolini I, Federici C, Raggi C, Lugini L, Palleschi S, De Milito A, Coscia C, Iessi E, Logozzi M, Molinari A, Colone M, Tatti M, Sargiacomo M, Fais S (2009) Microenvironmental pH is a key factor for exosome traffic in tumor cells. J Biol Chem 284:34211–34222

Peinado H, Aleckovic M, Lavotshkin S, Matei I, Costa-Silva B, Moreno-Bueno G, Hergueta-Redondo M, Williams C, Garcia-Santos G, Ghajar CM, Nitadori-Hoshino A, Hoffman C, Badal K, Garcia BA, Callahan MK, Yuan J, Martins VR, Skog J, Kaplan RN, Brady MS, Wolchok JD, Chapman PB, Kang Y, Bromberg J, Lyden D (2012). Melanoma exosomes educate bone marrow progenitor cells toward a pro-metastatic phenotype through MET. Nat Med 18:883–891

Putz U, Howitt J, Doan A, Goh C-P, Low L-H, Silke J, Tan S-S (2012) The tumor suppressor PTEN is exported in exosomes and has phosphatase activity in recipient cells. Sci Signal 5:ra70

Quante M, Tu SP, Tomita H, Gonda T, Wang SSW, Takashi S, Baik GH, Shibata W, Diprete B, Betz KS, Friedman R, Varro A, Tycko B, Wang TC (2011) Bone marrow-derived myofibroblasts contribute to the mesenchymal stem cell niche and promote tumor growth. Cancer Cell 19:257–272

Rana S, Malinowska K, Zoller M (2013) Exosomal tumor microRNA modulates premetastatic organ cells. Neoplasia 15:281–295

Raposo G, Nijman HW, Stoorvogel W, Leijendekker R, Harding CV, Melief CJM, Geuze HJ (1996) B lymphocytes secrete antigen-presenting vesicles. J Exp Med 183:1161–1172

Resnick KE, Alder H, Hagan JP, Richardson DL, Croce CM, Cohn DE (2009) The detection of differentially expressed microRNAs from the serum of ovarian cancer patients using a novel real-time PCR platform. Gynecol Oncol 112:55–59

Robinson MS (2004) Adaptable adaptors for coated vesicles. Trends Cell Biol 14:167–174

Safaei R, Larson BJ, Cheng TC, Gibson MA, Otani S, Naerdemann W, Howell SB (2005) Abnormal lysosomal trafficking and export of cisplatin in drug-resistant ovarian carcinoma cells. Mol Cancer Ther 4:1595–1604

Savina A, Furlan M, Vidal M, Colombo MI (2003) Exosome release is regulated by a calcium-dependent mechanism in K562 cells. J Biol Chem 278:20083–20090
Scales SJ, Chen YA, Yoo BY, Patel SM, Doung YC, Scheller RH (2000) SNAREs contribute to the specificity of membrane fusion. Neuron 26:457–464
Scheller RH (2013) In search of the molecular mechanism of intracellular membrane fusion and neurotransmitter release. Nat Med 19:1232–1235
Shedden K, Xie XT, Chandaroy P, Chang YT, Rosania GR (2003) Expulsion of small molecules in vesicles shed by cancer cells: association with gene expression and chemosensitivity profiles. Cancer Res 63:4331–4337
Shtam T, Kovalev R et al (2013) Exosomes are natural carriers of exogenous siRNA to human cells in vitro. Cell Commun Signal 11:88
Skog J, Wuerdinger T, Van Rijn S, Meijer DH, Gainche L, Sena-Esteves M, Curry WT Jr, Carter BS, Krichevsky AM, Breakefield XO (2008) Glioblastoma microvesicles transport RNA and proteins that promote tumour growth and provide diagnostic biomarkers. Nat Cell Biol 10:1470–1476
Stenmark H (2009) Rab GTPases as coordinators of vesicle traffic. Nat Rev Mol Cell Biol 10:513–525
Sun D, Zhuang X, Xiang X, Liu Y, Zhang S, Liu C, Barnes S, Grizzle W, Miller D, Zhang H-G (2010) A novel nanoparticle drug delivery system: the anti-inflammatory activity of curcumin is enhanced when encapsulated in exosomes. Mol Ther 18:1606–1614
Svensson KJ, Christianson HC, Wittrup A, Bourseau-Guilmain E, Lindqvist E, Svensson LM, Morgelin M, Belting M (2013) Exosome uptake depends on ERK1/2-heat shock protein 27 signaling and lipid raft-mediated endocytosis negatively regulated by caveolin-1. J Biol Chem 288:17713–17724
Takahashi Y, Nishikawa M, Shinotsuka H, Matsui Y, Ohara S, Imai T, Takakura Y (2013) Visualization and in vivo tracking of the exosomes of murine melanoma B16-BL6 cells in mice after intravenous injection. J Biotechnol 165:77–84
Taylor DD, Gercel-Taylor C (2008) MicroRNA signatures of tumor-derived exosomes as diagnostic biomarkers of ovarian cancer. Gynecol Oncol 110:13–21
Thery C, Zitvogel L, Amigorena S (2002) Exosomes: composition, biogenesis and function. Nat Rev Immunol 2:569–579
Trajkovic K, Hsu C, Chiantia S, Rajendran L, Wenzel D, Wieland F, Schwille P, Bruegger B, Simons M (2008) Ceramide triggers budding of exosome vesicles into multivesicular endosomes. Science 319:1244–1247
Traub LM (2003) Sorting it out: AP-2 and alternate clathrin adaptors in endocytic cargo selection. J Cell Biol 163:203–208
Tsujiura M, Ichikawa D, Komatsu S, Shiozaki A, Takeshita H, Kosuga T, Konishi H, Morimura R, Deguchi K, Fujiwara H, Okamoto K, Otsuji E (2010) Circulating microRNAs in plasma of patients with gastric cancers. Br J Cancer 102:1174–1179
Valadi H, Ekstrom K, Bossios A, Sjostrand M, Lee JJ, Lotvall JO (2007) Exosome-mediated transfer of mRNAs and microRNAs is a novel mechanism of genetic exchange between cells. Nat Cell Biol 9:654–659
Valenti R, Huber V, Filipazzi P, Pilla L, Sovena G, Villa A, Corbelli A, Fais S, Parmiani G, Rivoltini L (2006) Human tumor-released microvesicles promote the differentiation of myeloid cells with transforming growth factor-beta-mediated suppressive activity on T lymphocytes. Cancer Res 66:9290–9298
Vallhov H, Gutzeit C, Johansson SM, Nagy N, Paul M, Li Q, Friend S, George TC, Klein E, Scheynius A, Gabrielsson S (2011) Exosomes containing glycoprotein 350 released by EBV-transformed B cells selectively target B cells through CD21 and block EBV infection in vitro. J Immunol 186:73–82
Van Den Boorn JG, Schlee M, Coch C, Hartmann G (2011) SiRNA delivery with exosome nanoparticles. Nat Biotechnol 29:325–326
Wahlgren J, Karlson TDL, Brisslert M, Sani FV, Telemo E, Sunnerhagen P, Valadi H (2012) Plasma exosomes can deliver exogenous short interfering RNA to monocytes and lymphocytes. Nucleic Acids Res 40:e130

Wang J, Chen J, Chang P, Leblanc A, Li D, Abbruzzesse JL, Frazier ML, Killary AM, Sen S (2009) MicroRNAs in plasma of pancreatic ductal adenocarcinoma patients as novel blood-based biomarkers of disease. Cancer Prev Res (Phila) 2:807–813

Webb SD, Sherratt JA, Fish RG (1999) Mathematical modelling of tumour acidity: regulation of intracellular pH. J Theor Biol 196:237–250

Wittmann J, Jaeck H-M (2010) Serum microRNAs as powerful cancer biomarkers. Biochim Biophys Acta 1806:200–207

Yu X, Harris SL, Levine AJ (2006) The regulation of exosome secretion: a novel function of the p53 protein. Cancer Res 66:4795–4801

Zhu W, Qin W, Atasoy U, Sauter ER (2009) Circulating microRNAs in breast cancer and healthy subjects. BMC Res Notes 2:89

Part II
Tumor Targeting

Chapter 3
Polymeric Nanocarriers for Cancer Therapy

Peng Mi and Nobuhiro Nishiyama

3.1 Introduction

Cancer has become one of the top threats to people's health with a rapid disease rate and a huge number of lethal worldwide; thus, the control of cancer has become an urgent mission but still quite challenging. Many efforts have been made for cancer treatments, including surgical treatment, radiotherapy, and chemotherapy. Chemotherapy is a common method for cancer treatment accompanying with other cancer treatment methods, but is limited by unfavorable pharmacokinetics of small molecular therapeutic compounds. Conventional anticancer drugs exhibit fast clearance in circulation and systemically disperse in the body, resulting limitation in their efficacy and considerable toxicity to normal organs. Therefore, it is desirable to develop "dream drugs" targeting only the disease sites. Loading therapeutic compounds to nanocarriers has emerged as one approach to overcome the limitations of conventional therapeutic compounds, expecting reduced side effects, enhanced therapeutic efficacy, and improved dosing compliance (i.e., decrease the frequency and dose of administration) (Langer 1998).

In the recent 20 years, with the advance in nanotechnology, nanoscaled drug vehicles were designed and developed to deliver therapeutic compounds to tumor sites in a cancer-selective manner, demonstrating their promises in clinical application (Ferrari 2005; Davis et al. 2008; Peppas 2013; Peer et al. 2007; Kataoka et al. 2001; Cho et al. 2008; Matsumura and Kataoka 2009). Generally, polymeric nanoparticles with a size around 100 nm, which are formed from natural or synthetic polymers, have been considered as a platform to transport drugs for cancer

P. Mi (✉) • N. Nishiyama
Polymer Chemistry Division, Chemical Resources Laboratory, Tokyo Institute of Technology, R1-11, 4259 Nagatsuta, Midori-ku, Yokohama 226-8503, Japan
e-mail: mi@bmw.t.u-tokyo.ac.jp

M.J. Alonso, M. Garcia-Fuentes (eds.), *Nano-Oncologicals: New Targeting and Delivery Approaches*, Advances in Delivery Science and Technology,
DOI 10.1007/978-3-319-08084-0_3

Table 3.1 Polymer-based nanoparticles for cancer therapy

Identity	Bioactive compound	Type of carrier	Type of target	Status	References
BIND-014	Docetaxel	Polymeric nanoparticle, target prostate-specific membrane antigen	Active target	Phase I	Kim et al. (2004)
NK012	SN-38	Polymeric micelle, PEG-poly(glutamic acid) conjugate	Passive target	Phase II	Hamaguchi et al. (2005)
NC-6004	Cisplatin	Polymeric micelle, PEG-poly(glutamic acid), 30 nm	Passive target	Phase III	Nishiyama et al. (2003a)
NK105	Paclitaxel	PEG-polyaspartate modified with 4-phenyl-butanol	Passive target	Phase III	Nakanishi et al. (2001)
NC-4016	DACHPt	Polymeric micelle, PEG-poly(glutamic acid), 30 nm	Passive target	Phase I	Cabral et al. (2011b)
NC-6300/K-912	Epirubicin	Polymeric micelle, PEG-polyaspartate modified with hydrazide group	Passive target	Phase I	Takahashi et al. (2013)
Livatag	Doxorubicin	Doxorubicin polyalkylcyanoacrylate nanoparticles	Passive target	Phase III	Svenson (2012)
SP1049C	Doxorubicin	Polymeric micelle, P-glycoprotein-targeting pluronics	Passive target	Phase II	Valle et al. (2011)
CALAA-01	siRNA	PEGylated with adamantane, cyclodextrin-based polymer	Active target	Phase I/II	Davis et al. (2010)
XYOTAX	Paclitaxel	Polymeric nanoparticle, paclitaxel poliglumex	Passive target	Phase III	Cho et al. (2008)
Abraxane	Taxol	Polymeric nanoparticle, albumin-Taxol	Passive target	Approved	Davis et al. (2008)
Genexol-PM	Paclitaxel	Polymeric micelle, PEG-poly(D,L-lactide)	Passive target	Approved	Deng et al. (2012)

targeting (Duncan 2006). Polymeric nanoparticles with a payload of imaging probes, low-molecular-weight drugs or emerging biomedicines, for instance, proteins (antibodies) and small interfering RNA (siRNA) could significantly optimize the results of cancer diagnosis and therapy (Cabral et al. 2011a; Ferrari 2005). The therapeutic compounds are incorporated into the core of polymeric nanoparticles, protecting the drugs from metabolism by enzymes and avoiding interaction with healthy tissue and organs (Brewer et al. 2011). Polymeric nanoparticles can improve drug circulation half-life to maintain a therapeutic drug level as a result of decreased renal secretion and hindered uptake by the reticuloendothelial system (RES). The surface coating a poly(ethylene glycol) (PEG) shell to protect from interaction with plasma proteins is important for avoiding their recognition by the RES (Kaul and Amiji 2002; Jokerst et al. 2011; Li and Huang 2009; Otsuka et al. 2012). Polymeric nanoparticles could passively accumulate and retain in the malignant tissues characterized by the leaky vasculature and impaired lymphatic drainage, which is known as the enhanced permeability and retention (EPR) effect (Matsumura and Maeda 1986; Chauhan and Jain 2013). In the tumor microenvironment, polymeric nanoparticles could actively target tumor cells by decorating targeting ligands on their surface, which could specifically bind with various receptors overexpressed on each type of tumor cells (Davis et al. 2008; Byrne et al. 2008; Kamaly et al. 2012). Until polymeric nanoparticles access to the cancer sites, the encapsulated therapeutic compounds can be released from vehicles in a sustained manner within a therapeutic window or in response to internal or external stimulus unique to the target sites. The constituent polymers of nanoparticle are generally degraded and gradually disappear after the delivery, which might make polymeric nanoparticles as an attractive carrier for drug delivery (Jeong et al. 1999; Soppimath et al. 2001). Thus, polymeric nanoparticles have been attracting great interests from both academic and industry, so that several polymeric nanoparticle-based anticancer drugs are under the clinical study or have already been approved for clinical use (Table 3.1).

In this chapter, we will focus on the design and development of polymeric nanoparticles incorporating therapeutic compounds for cancer therapy. The way of design and preparation of polymeric nanoparticles for drug delivery and how those nanoparticles overcome multiple biological barriers are highlighted. Finally, the strategies of applying polymeric nanoparticle-based drug delivery system for cancer therapy are summarized.

3.2 Design and Preparation of Polymeric Nanoparticles for Cancer Therapy

3.2.1 Design and Structure of Polymeric Nanoparticles

Polymeric nanoparticles, mostly with a size arrange from 10 to 100 nm, as a platform for drug delivery system should satisfy several requirements. Thus, from the initial design of nanoparticles, many factors need to be considered, such as safety,

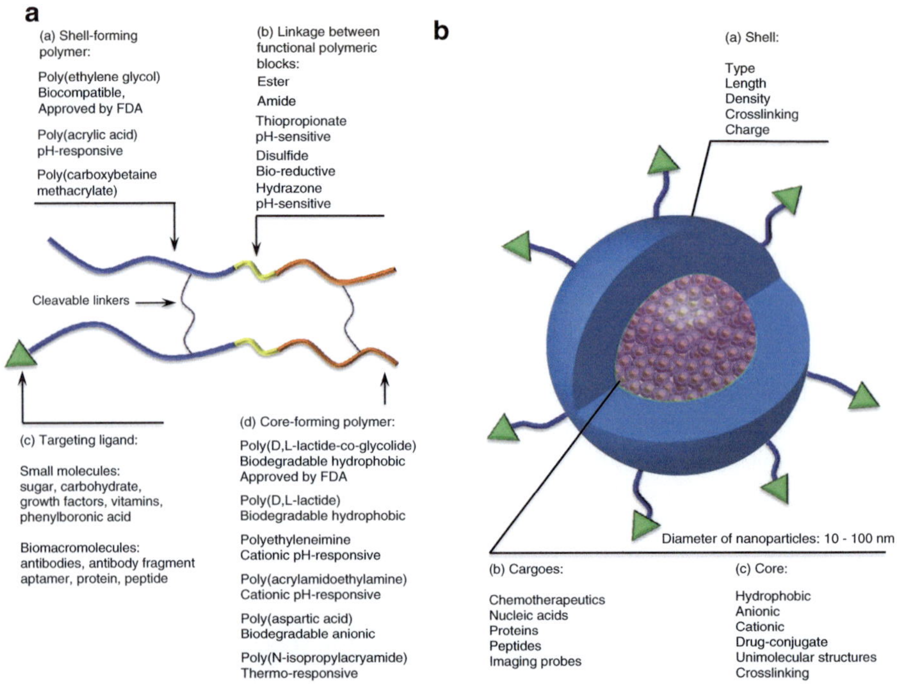

Fig. 3.1 Various types of building polymeric blocks (**a**) and the composition of multifunctional nanoparticles (**b**) for drug delivery

biocompatibility, stability, and even the loading efficacy (Fig. 3.1) (Elsabahy and Wooley 2012). A myriad of polymeric blocks with different biochemical properties are designed and synthesized for building nanoparticles, and some of those polymers have been already approved by the US Food and Drug Administration (FDA) for clinical applications, such as PEG (Otsuka et al. 2012) and poly(D,L-lactide-*co-glycolide*) acid (PLGA) (Danhier et al. 2012) due to their biodegradability and biocompatibility, and some other polymers are under preclinical testing or in clinical trial. Therapeutic compounds are generally incorporated in the core of polymeric nanoparticles and shielded with protecting polymers. Hydrophobic, cationic, or environment-responsive polymeric blocks are usually used to form the core of polymeric nanoparticles loaded with delivering cargoes. The incorporated therapeutic compounds could be released from the core of nanoparticles at specific sites, being triggered by external or internal stimulus such as pH, temperature, redox potential, enzymatic or oxidative reactions, and so on. The shield polymer layer of nanoparticles is generally constructed with neutral, hydrophilic, and flexible polymer blocks such as PEG, which can avoid the interaction with enzymes and proteins in bloodstream and the recognition by RES in several organs, showing the prolonged blood circulation (stealth property). In order to achieve specific and enhanced therapeutic efficacy, targeting ligands are decorated on the outer shell of polymeric nanoparticles, such as antibody and its fragments, peptide, and other molecules which specifically interact with tumor cells while sparing normal cells. Moreover, introduction of cleavable linkages for drug conjugation can improve the loading efficacy and the stability of polymeric nanoparticles until arriving at the targets. Indeed, a wide range of therapeutic components including imaging contrast agents, anticancer drugs, nucleic acids, or antibodies could be simultaneously incorporated into single platform nanoparticles, achieving multifunctions such as diagnosis and therapy (Zhang et al. 2007; Saad et al. 2008; Kaida et al. 2010). But co-delivering of multicomponents in the same nanoparticles may make it difficult to control the dose of each component. Also, the synergistic effects on potential side effects should be considered in the co-delivery strategies.

3.2.2 Methods for Preparing Polymeric Nanoparticles

Polymers, which fulfill the requirements for biomedical applications and assemble to nanoparticles, are obtained from natural resources, including chitosan, alginate, albumin, and gelatin, and synthetic polymers such as PLA, PLGA and poly(epsilon-caprolactone) (PCL) homopolymers and PEG-PLA, and PEG-PLGA and PEG-PCL block copolymers. The therapeutic compounds are dissolved, entrapped, encapsulated, or conjugated to the matrix of polymers described above, and depending on the method of preparation, nanoparticles, nanospheres, or nanocapsules can be obtained. A number of methods are applied to prepare size-controlled and well morphology-defined polymeric nanoparticles, including self-assembly, emulsification, nanoprecipitation, and mechanical method (Soppimath et al. 2001; Vauthier and Bouchemal 2009; Zhang et al. 2011; Rao and Geckeler 2011). If defined with the start materials for preparation, there are two routes to obtain polymeric nanoparticles: one route starts with polymers and drugs or drug-conjugated polymers to produce polymeric nanoparticles and the other route starts with monomers to polymerize by the interfacial or radical reaction during preparation procedures.

Self-assembly is a procedure of building various complex architectures in nature. Synthetic polymers from a wide range of monomers can also self-assemble and construct polymeric nanoparticles with variable size and shape (Simon 2013; Grzelczak et al. 2010). Amphiphilic block copolymers, which are composed of two distinct polymer chains with a large solubility difference, are widely used to form nanostructures in an aqueous medium (Kataoka et al. 2001). The self-assembly of polymeric nanoparticles is generally proceeded through the combination of intermolecular forces, including hydrophobic interaction, electrostatic interaction, metal complexation, and hydrogen bonding of constituent polymers. The hydrophilic polymeric blocks with a flexible nature are forming the shell of polymeric nanoparticles, and the highly dense shell could achieve effective steric stabilization propensities. In this way, anticancer drugs (Kataoka et al. 2001; Uchino et al. 2005; Plummer et al. 2011; Nishiyama and Kataoka 2006; Nishiyama et al. 2003a), nucleic acid (Miyata et al. 2012; Kataoka and Harashima 2001; Nishiyama et al. 2005), antibodies (Lee et al. 2007, 2008, 2009, 2010; Lee and Kataoka 2009), photosensitizers (Nishiyama et al. 2003b, 2005, 2006, 2007, 2009; Ideta et al. 2005; Jang et al. 2006; Zhang et al. 2003), and imaging agents (Cabral et al. 2011a; Mi et al. 2013; Kaida et al. 2010) could be incorporated into polymeric nanoparticles through the self-assembly in aqueous media. Besides, polymers can form nanoparticles by the layer-by-layer methods, in which polycationic and polyanionic polymers are alternately self-assembled through the electrostatic interactions (Ai et al. 2003; Poon et al. 2011; Hammond 2012). Self-assembly could be used to prepare bulk amount of polymeric nanoparticles for drug delivery, and this procedure is easily applied for industry because of its spontaneous manufacturing procedure, easy size, and morphology control. Furthermore, self-assembly also exists in many preparation methods including nanoprecipitation, emulsification/solvent evaporation, salt-out method, etc., and has been found to be appropriate for several applications in nanotechnology, including detergents, paints, electronics, cosmetics, lubricants, tissue engineering, and drug delivery, according to the function of nanostructures.

Emulsification is a traditional technique for pharmaceutical engineering of formulation, and it also could be utilized for preparing polymeric nanoparticles (nanoemulsions) in the way of single emulsion (oil in water: O/W or water in oil: W/O) and double emulsion (W/O/W) (Fig. 3.2). Polymeric nanoparticles made by emulsification mainly depend on the conditions of polymer blocks, drugs, and the degree of miscibility of the oil phase with the water phase. In the single emulsion (O/W) method, the loading drugs can be dispersed in organic phase containing polymer or surfactant and mixed with aqueous phase containing stabilizer to make O/W emulsions. Polymeric nanoparticles could be obtained after solvent evaporation, and it could be disposed in aqueous solvent, for instance, PLA-PEG and PLGA-PEG nanoparticles are obtained by this method (Avgoustakis 2004). Different from single emulsion method, double emulsion (W/O/W) is generally used to encapsulate

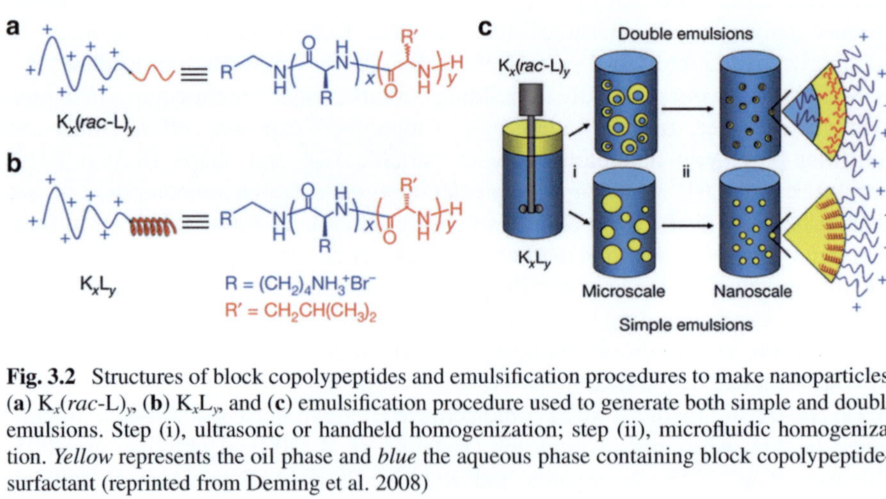

Fig. 3.2 Structures of block copolypeptides and emulsification procedures to make nanoparticles: (**a**) $K_x(rac\text{-}L)_y$, (**b**) K_xL_y, and (**c**) emulsification procedure used to generate both simple and double emulsions. Step (i), ultrasonic or handheld homogenization; step (ii), microfluidic homogenization. *Yellow* represents the oil phase and *blue* the aqueous phase containing block copolypeptide–surfactant (reprinted from Deming et al. 2008)

hydrophilic drugs. The first step is to disperse drugs in aqueous phase and then mix with organic phase containing hydrophobic polymers and stabilizers in each phase to make W/O emulsion. Then, the W/O emulsion is dispersed in water phase containing hydrophilic polymers and stabilizers to make W/O/W emulsion, and polymeric nanoparticles finally obtained by solvent evaporation, such as PLA nanoparticles incorporating BSA were obtained by this method (Wang and Uludag 2008). Recent research demonstrated that amphiphilic copolymers or single-component block copolypeptides can be utilized as stabilizer or surfactant to form nanoparticles without conventional ingredients of stabilizer and surfactant (Deming et al. 2008). The emulsification method requires individual two phases (organic phase and aqueous phase) containing polymers and drugs for nanoparticle preparation, involving additives of surfactants or stabilizers, which are important for controlling the size and morphology and enhancing the stability of obtained products. However, this method makes the purification as an important procedure to remove the organic solvent, extra stabilizers and surfactants because some of those show potential toxicity.

Nanoprecipitation is also called solvent displacement method, which is initialed by Fessi et al. (1989) for the preparation of indomethacin-encapsulated PLA nanocapsules. In this method, hydrophobic drugs and polymers are generally dissolved in organic solvent and then dispersed into bulk mount of aqueous solution under stirring. The rapid disperse of organic phase to aqueous phase results in the decrease of interfacial tension between the two phases, which form small organic drops containing polymers and drugs by precipitation and self-assembly in aqueous solvent. Nanoparticles or nanospheres could be obtained after removal of the organic solvent (Mishra et al. 2010), such as cisplatin- and docetaxel-co-loaded PLGA and PLA polymeric nanoparticles are prepared by nanoprecipitation for co-delivering anticancer drugs to prostate cancer cells with synergistic cytotoxicity (Kolishetti et al. 2010).

Particle replication in non-wetting templates (PRINT) is a mechanical approach capable of producing monodisperse nanoparticles incorporating drugs with well-defined size, shape, and modulus (Fig. 3.3) (Perry et al. 2011). In this method, firstly cast solution containing polymers and drugs on a polyethylene terephthalate (PET) substrate using a Mayer rod to help disperse and remove the solvent by heat. Then, the obtained drug and polymer layer are covered with a perfluoropolyether elastomeric mold with different size and shape of cavities on it, and nanoparticles are fabricated by passing through a heated nip and split. Finally, nanoparticles are harvested by contacting with a high-energy film and passing through the heated nip without splitting, cool down. This method also could be used to prepare layer-by-layer functionalized nanoparticles (Morton et al. 2013). PRINT method for preparing polymeric nanoparticles with high throughput continues producing high-yield and well-morphology control, but this method may have higher requirements of equipment and templates used during manufacturing procedures.

Microfluidics is another mechanical method used for preparing nanoparticles, which could prepare nanoparticles by single emulsion or double emulsion from aqueous phase and organic phase in microfluidic devices (Capretto et al. 2013; Rhee et al. 2011; Karnik et al. 2008). The size and morphology could be defined by handling the flow parameters of microfluidic devices, and double emulsion could be

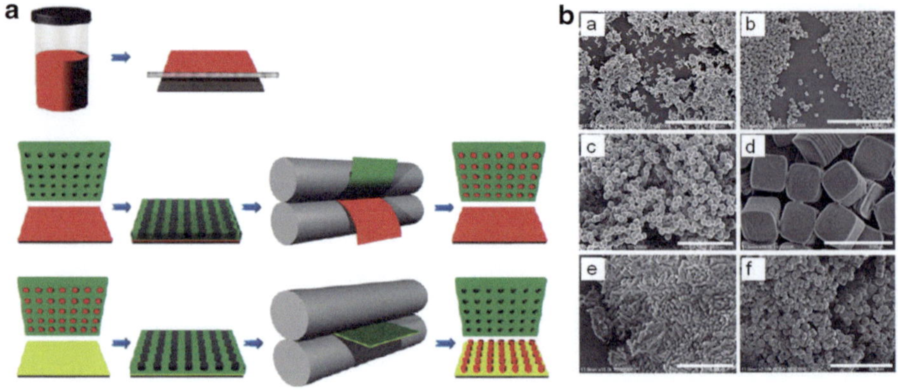

Fig. 3.3 Schematic illustration of the PRINT process to produce polymeric nanoparticles (**a**) and obtain polymeric nanoparticles (**b**), such as 80×320 nm cylinders (*a*), 200×200 nm cylinders (*b*), 1 μm sphere (*c*), 2 μm cubes with ridges (*d*), 200×600 nm cylinders (*e*), and 3 μm particles with center fenestrations (*f*) (reprinted from Perry et al. 2011)

obtained by the linking of multistage microfluidic devices (Utada et al. 2007). Higher drug encapsulation could be achieved by this method without increasing the size of nanoparticles, and this method offers an easy way to control the manufacturing procedures, which may finally control the size or shape of obtained polymeric nanoparticles. Recent progresses in microfluidics are expected to improve the preparation techniques of polymeric nanoparticles and accelerate their transition to clinical test. Although many of these microfluidic systems are still under design and development, they have the potential to become widely applied, as they are economical and reproducible, have modifications available, and can be integrated with other technologies (Valencia et al. 2012).

3.3 Polymeric Nanoparticles Targeting Cancer

3.3.1 Polymeric Nanoparticles for Passive Cancer Targeting

After systemic administration, polymeric nanoparticles incorporating therapeutic compounds can gradually enter the tumor sites from the leaky tumor vessels and retain there due to the aforementioned EPR effect, of which the procedure is called the passive targeting. Conventional free therapeutic compounds such as cisplatin, doxorubicin, and nucleic acids rapidly disperse to the whole body and cannot accumulate in the tumor tissue effectively. Passive targeting is a general way for the tumor targeting using nanoparticles, and the first passively targetable nanocarriers reached clinical trial in the 1980s and were commercialized in the middle of 1990s (Peer et al. 2007). Later, a myriad of nanocarriers based on this strategy have been applied for cancer-targeted therapy and diagnosis. Many efforts have been made to improve the targeting abilities of nanoparticles to malignant tumors.

In the passive targeting, several factors such as size, morphology, and surface properties should be important for the design and development of polymeric nanoparticles. Many researches revealed how these factors could influence on the tumor target ability of the nanoparticles, and "golden rules" from those researches pave the way of developing "ideal" polymeric nanoparticles and translating from bench to bedside.

The size of polymeric nanoparticles is a dominant parameter that can affect the cellular uptake rate, thereby directly affecting the blood circulation and tumor accumulation of the nanoparticles. There are two major endocytic procedures by which particles enter cancer cells, i.e., phagocytosis and pinocytosis (Conner and Schmid 2003). Particles with a large size (>1 μm) are generally internalized by phagocytosis, which is a specific process to typical phagocytic cells, such as macrophages, neutrophils, or dendritic cells. Polymeric nanoparticles enter cells mainly through pinocytosis, which can occur either via the adsorptive pinocytosis induced by non-specific adsorption of nanoparticles to the cell membrane or via the receptor-mediated endocytosis of nanoparticles based on the interaction of the ligand-installed nanoparticles with the receptors on the cell surface. Recent research revealed that nanoparticles could control the cellular functions, and potentially kill cancer cells, by virtue of their size alone even without drugs (Ferrari 2008; Jiang et al. 2008). They found that nanoparticles with a size range of 40–50 nm exhibit the most effective interactions with breast cancer cells, while nanoparticles with a size of 200 and 10 nm cannot be effectively taken up by breast cancer cells. Moreover, nanoparticles decorated with Herceptin, a drug binding to ErbB2 receptors on breast cancer cells, showed higher cellular uptake than those without surface modification.

Regarding the size of nanoparticles for the passive targeting, the upper size limitation for extravasation into solid tumors has been suggested to be approximately 400 nm, and it is generally observed that nanoparticles with a size smaller than 200 nm can accumulate effectively within tumor tissues (Torchilin 2007). Recently, Kataoka et al. demonstrated that nanoparticles with the size smaller than 100 nm are suitable for the passive targeting of hypopermeable cancers such as pancreatic cancers. They studied the accumulation and therapeutic effects of the anticancer drug (1,2-diaminocyclohexane) platinum (II) (DACHPt) incorporating polymeric nanoparticles (DACHPt/m) with varying sizes such as 30, 50, 70, and 100 nm against two types of tumor models with different permeability, hyperpermeable murine colon adenocarcinoma 26 (C26) model and hypopermeable pancreatic adenocarcinoma BxPC3 model (Fig. 3.4) (Cabral et al. 2011b). The DACHPt/m with different sizes were obtained by self-assembly of DACHPt with the carboxylic groups of poly(glutamic acid) (P(Glu)) in PEG–*b*-P(Glu) and P(Glu), and the drug could be released in chloride ion-rich environments by ligand exchange (Fig. 3.4a). Consequently, all DACHPt-loaded micelles showed penetrability and significant antitumor activity against C26 tumors in vivo, but only the 30 nm-sized micelles showed deep penetration and significant antitumor activity against BxPC3 pancreatic tumors. Other researches also revealed that nanoparticles with a smaller size could penetrate into deeper positions of tumor tissues (Fig. 3.4b, c) (Popovic et al. 2010).

In the past decades, the sophisticated polymer design has provided the possibility to constitute polymeric nanoparticles with various morphologies via self-assembly,

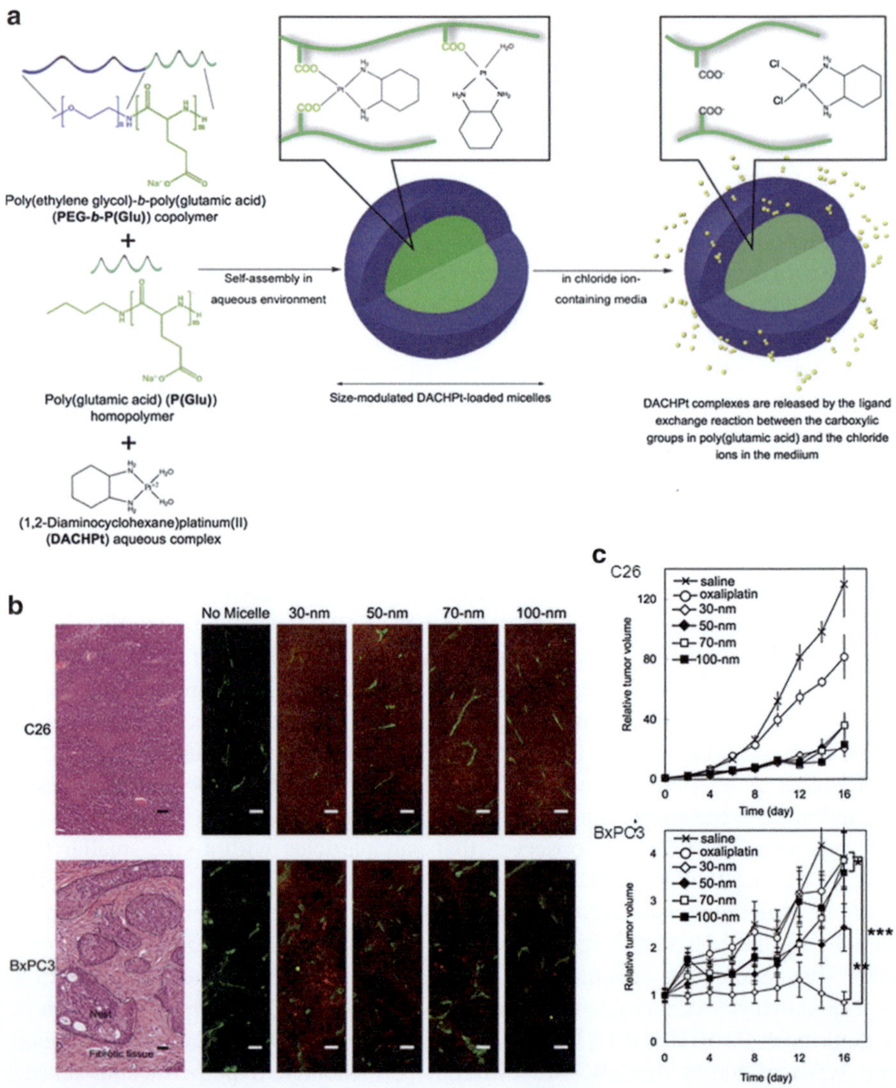

Fig. 3.4 (**a**) Schematic showing DACHPt/m formed through the interaction between DACHPt and the carboxylic groups of poly(glutamic acid) (*green*) in PEG–*b*-P(Glu) and P(Glu). In aqueous media containing chloride ions, DACHPt (*yellow circles*) is released from the micelles through ligand exchange. (**b**) Microdistribution of fluorescently labeled DACHPt/m (*red color*) of varying sizes in tumors; the blood vessels were marked with PECAM-1 to indicate the extensive distribution of blood vessels in C26 tumors and Alexa 488 secondary antibody (*green*). (**c**) Antitumor activities of DACHPt/m with different size against C26 and BxPC3 tumors (reprinted from Cabral et al. 2011b)

even though the majority of nanoparticles in drug delivery display spherical morphology. Recent study reported that morphology of polymeric nanoparticles, also noted as shape and dimensions, is another factor that could impact the behaviors as drug vehicles (Venkataraman et al. 2011). Besides, spherical nanoparticles and other morphologies such as cylindrical or vesicular architectures are obtained by designing the polymer structure and their preparation procedure. It is generally found that spherical nanoparticles enter cells to a greater extent than do elongated cylindrical ones. And, the nonspherical shape of nanoparticles has been identified as one of the key factors that affect biological behaviors of anticancer drugs such as cellular uptake and biodistribution. For example, Geng et al. (2007) constructed filamentous micelles (filomicelles) from the biodegradable PEG-*b*-PCL and the nonbiodegradable PEG-*b*-polyethylethylene and compared their biological properties with spherical micelles of similar chemistry (Fig. 3.6). Inert filomicelles, with a length of 18 μm and a cylinder diameter of 20–60 nm, persisted in the circulation for a prominently long time, up to 1 week after intravenous injection, which is about ten times longer than their spherical counterparts, due to the reduced rate of phagocytosis and clearance by the phagocytic system. The clearance of filomicelles also shows a strong dependence on the micellar length. For L_o up to 8 μm, longer filomicelles exhibit longer half-life of circulation after intravenous injection. In a fluidic flow environment, smaller micelles and vesicles are captured more by cells than long filomicelles, which flow pass the cells easily only with occasionally uptake of a fragment. Interestingly, filomicelles have a higher accumulation in tumor sites via the EPR effect due to its long circulation property and it has better performance for deep penetration through tumor tissues. Finally, filomicelles incorporating paclitaxel demonstrated much better antitumor efficacy than control, and longer size of filomicelles exhibited better antitumor activity than shorter ones. Other researchers such as Wooley et al. (Zhang et al. 2008) and DeSimone et al. (Gratton et al. 2008) discovered similar phenomena that the morphology of polymeric nanoparticles has high influence on the cell uptake, i.e., smaller spherical nanoparticles were internalized by the cells more rapidly than longer cylindrical nanoparticles, and nanoparticles with the shape but with different ratio between height and diameter showed different cellular uptake. However, "stick" polymeric nanoparticles with extended circulation half-life require enough flexibility or deformability and maintain the length of one dimension less than 200 nm to avoid blocking in the spleen or other tissues.

Polymeric nanoparticles with long circulation property could achieve high accumulation in tumor sites by the EPR effect, and, therefore, the surface characterization of nanoparticles is important for this besides the cellular bioavailability of cancer cells (Verma and Stellacci 2010). The charge property of nanoparticles can greatly affect the availability of nanoparticles at many aspects, including physical stability, interactions with extracellular matrix, stability during circulation, cellular uptake, and endocytosis (Verma and Stellacci 2010). Charged nanoparticles are easy to aggregate or interact with components of the extracellular matrix, resulting in their cell uptake ratio. After systemic administration, the charged nanoparticles are easy to undergo opsonization and aggregation with proteins in blood vessels,

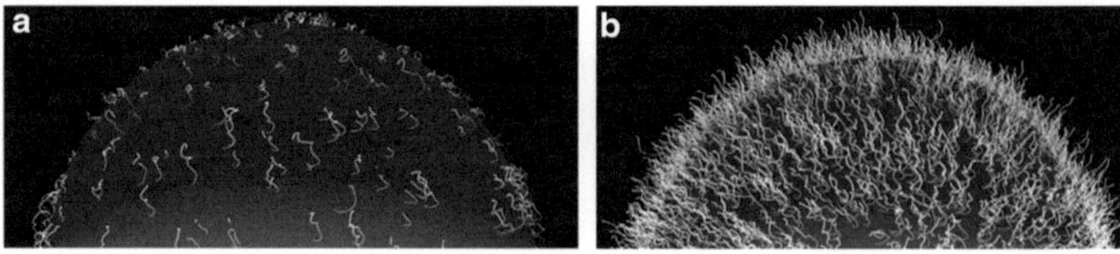

Fig. 3.5 Schematic diagram of PEG configurations on the upper hemisphere of a polymeric nanoparticle. (**a**) Low surface coverage of PEG chains leads to the "mushroom" configuration where most of the chains are located closer to the particle surface. (**b**) High surface coverage and lack of mobility of the PEG chains lead to the "brush" configuration where most of the chains are extended away from the surface (reprinted from Owens and Peppas 2006)

leading to fast clearance from circulation. Therefore, nanoparticles with a hydrophilic, neutral shell are generally acceptable for designing drug delivery system to achieve longer half-life by avoiding recognition by the RES. PEG is a neutrally charged polymer with biocompatible properties, thus providing suitable material for drug delivery (Otsuka et al. 2012; Jokerst et al. 2011). Drugs or drug-loaded polymeric matrix core is generally in the shield of PEG layer in polymeric nanoparticles, but the density of PEG on the particle surface is important (Fig. 3.5) (Owens and Peppas 2006). Low PEG density on the surface of nanoparticles sometimes failed to cover the entire core, and PEG chains are located closer to the particle surface, making the shell thinner. High-PEG-density surface using longer PEG chains (generally the molecular weight of PEG is larger than 2,000) can make the shell thicker, allowing long circulation of polymeric nanoparticles, because most of PEG chains "stand" on the surface of nanoparticles.

3.3.2 Polymeric Nanoparticles for Active Cancer Targeting

Although nanoparticles can access to tumor sites in a passive way of extravasation from the leaky tumor vasculatures, the cellular uptake of nanoparticles depends on tumor cells. In this regard, considerable efforts have been devoted to develop polymeric nanoparticles that could selectively and even specifically target cancer cells, of which procedure is called active targeting. A common approach of designing active targeting nanoparticles is to functionalize targeting ligands on the surface of nanoparticles. Nanoparticles presenting ligands on the surfaces have been designed to enhance their selective binding to specific receptors overexpressed on the target cells. This approach is beneficial in terms of enhancing the possibility of cancer cell uptake of nanoparticles.

There are two ways of decorating ligands on the surface of nanoparticles. One way is to prepare polymeric nanoparticles directly from targeting ligand-conjugated polymers (Gu et al. 2008; Deshayes et al. 2013; Vachutinsky et al. 2011; Oba et al. 2007; Bae et al. 2005, 2007). The final amount of ligands on the surface of polymeric

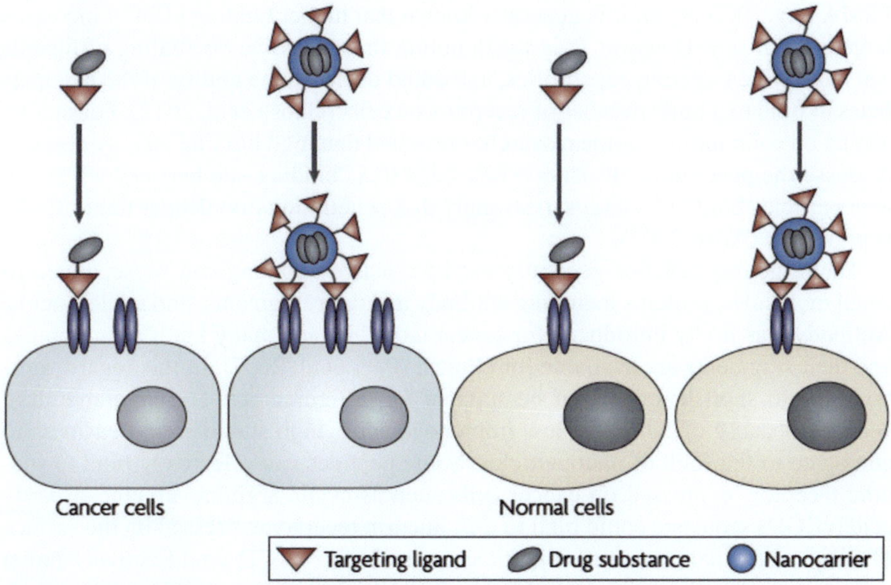

Fig. 3.6 Nanoparticles with numerous targeting ligands can provide multivalent binding to the surface of cancer cells with high receptor density (reprinted from Davis et al. 2008)

nanoparticles can be managed by mixing different ratios of polymers with ligands and those without ligands during fabrication procedures. In this way, it needs to ensure that almost all of the ligands have finally appeared on the outer surface of nanoparticles, and it is not suitable for ligands with hydrophobicity or ligands that could easily interact with drugs or the core matrixes. Another choice is to decorate ligands on the surface of preformed nanoparticles and finally achieve different ratio of ligands by controlling the reactions (Kolishetti et al. 2010; Miura et al. 2013). This method could avoid side interactions of ligands during preparation procedures, and it is a promising way to decorate all of the ligands on the outer surface of drug delivery vehicles.

The amount of targeting ligands on the surface of nanoparticles is a key parameter for their cellular uptake, as the surface density of the receptor is low on normal cell and much higher on cancer cells; therefore, nanoparticles with multi-ligands on the surface have a higher affinity to cancer cells than normal cells by ligand–receptor interaction (Fig. 3.6) (Davis et al. 2008). The final ligands ratio on the shell could be controlled by proper method as described above. For instance, aptamers (Apts)-conjugated PLGA-*b*-PEG polymers were firstly synthesized, and then mixed with different ratio of PLGA-*b*-PEG polymers to prepare docetaxel-incorporated nanoparticles, which were used for targeting prostate cancer (PCa) cells by binding to prostate-specific membrane antigen (PSMA) (Gu et al. 2008). The PSMA-targeting nanoparticles exhibited increased cellular uptake with higher amount of Apts density on the surface of nanoparticles when exposing to PCa cells that express the PSMA (LNCaP), while there was quite low cellular uptake of control cells lacking

PSMA (e.g., PC3 cells). It is generally known that higher binding affinity increases targeting efficacy. However, it is worth noting that excessive decorating of ligands can reduce the targeting capabilities, and could decrease the ability of the nanoparticles to bind to a finite number of receptors on cells (Ghosh et al. 2012). For in vivo testing on solid tumors, some researches revealed that high binding affinity also can decrease the penetration of nanoparticles due to a "binding-site barrier," where the nanoparticles bind to its target so strongly that penetration into deeper tumor tissue is prevented (Allen 2002).

The targeting moieties generally used for active targeting can be separated to small molecules, proteins including antibody and their fragments, and nucleic acids. Antibody was firstly introduced for cancer targeting, and many kinds of antibodies and their fragments are available for clinical (Peer et al. 2007). In this regard, antibodies with short length might be suitable for designing actively targetable drug vehicles because of relatively low immunogenicity, high stability, and easiness to conjugate to the shell of nanoparticles. Some peptides can selectively bind to specific receptors expressed on cancer cells such as cyclic arginine–glycine–aspartic acid (cRGD) sequence could bind to $\alpha_v\beta_3$ integrin receptor expressed on the surface of some cancer cells, and nanoparticles coated with cRGD could actively target solid tumors for chemotherapy (Miura et al. 2013) and gene therapy (Oba et al. 2007; Vachutinsky et al. 2011). Other peptides like LyP-1- and iRGD-modified nanoparticles also have high affinity to tumor cells and penetration in tumors, demonstrating promising active targeting ability for cancer therapy (Park et al. 2010; Agemy et al. 2011).

Apts, single-strand of DNA or RNA oligonucleotides, holding the properties of small size, low immunity, and reproducible for synthesis, are used as targeting ligands for active targeting as shown in Fig. 3.9 and discussed above. Despite the difficulty to produce antibodies at large quantities, Apts could be obtained in a high-throughput way. However, as oligonucleotides, Apts might lack the thermal stability and degrade with exonuclease or endonuclease, which may affect their targeting efficacy.

In addition to biomacromolecules such as antibodies and Apts, some small molecules are also useful as targeting ligands for active targeting, such as folic acid (folate), carbohydrate, growth factors, vitamins, and phenylboronic acid (PBA). Carbohydrates, which are displayed more frequently than protein oncogene markers in all tumor cells due to aberrant glycosylation, provide an exceptional target for active targeting. Sialic acid (SA) is one type of carbohydrates, which are overexpressed on cancer cells and closely related to cancer progression and metastasis. Recent study has demonstrated the active-targeting ability of PBA for selective recognition of SA overexpressed on the surface of cancer cells (Fig. 3.7) (Matsumoto et al. 2010; Deshayes et al. 2013). PBA-installed polymeric micelles incorporating the anticancer drugs DACHPt (PBA-DACHPt/m) exhibited higher cellular uptake by B16F10 murine melanoma cells. Further in vivo study revealed that PBA-DACHPt/m had higher retention in tumor tissues at 48 h post-intravenous injection and significant antitumor activity against both of orthotopic and lung metastasis models of melanoma. Other small molecules, such as folate, also demonstrated

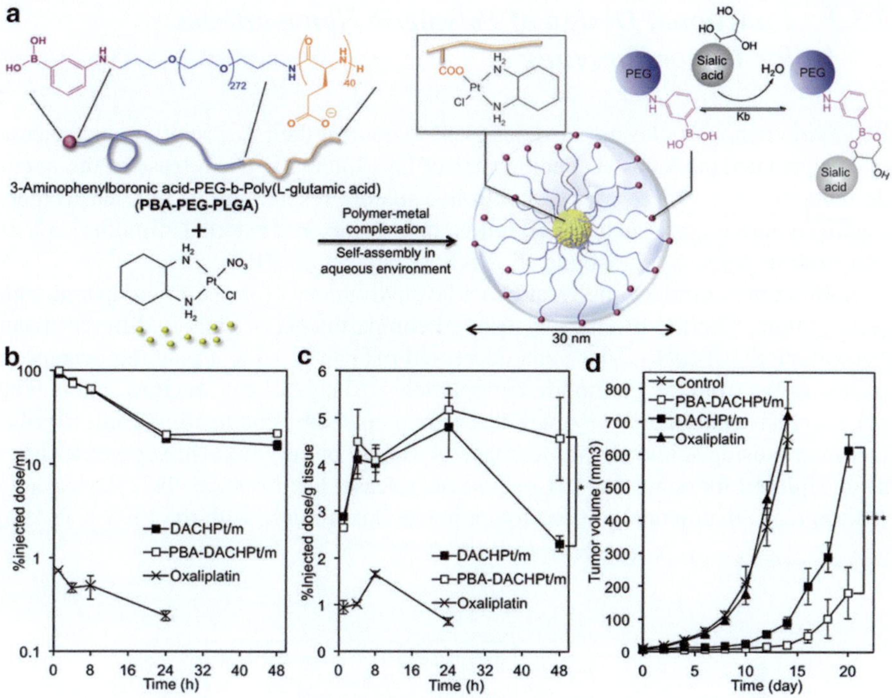

Fig. 3.7 (**a**) Preparation of PBA-installed DACHPt-loaded micelles (PBA-DACHPt/m), (**b**) plasma clearance, (**c**) tumor accumulation and (**d**) antitumor activity against orthotopic B16F10 tumors with oxaliplatin, DACHPt/m, and PBA-DACHPt/m (reprinted from Deshayes et al. 2013)

enhanced tumor targeting for effective cancer therapy (Werner et al. 2011; Bae et al. 2005, 2007).

The benefits of using small molecules as targeting moieties are due to the easily accessible synthetic route, facile coupling to drug delivery systems, and a wide range of candidates available. However, the targeting selectivity or specificity may not be an issue compared to antibodies. For example, SA is also present on red blood cells and the luminal surfaces of vascular endothelium, which may decrease the targeting efficacy of polymeric nanoparticles with ligands for SA targeting.

Considering the insufficient targeting ability by utilizing one type of targeting ligands, nanoparticles with two or more kinds of ligands for recognizing different receptors on cancer cells were introduced as a new targeting strategy, which are also recognized as multi-targeting systems (TMS). Targeting ligands of TMS could interact with multi-types of receptors overexpressed on the surface of one type of cancer cells or multi-types of cancer cells. This strategy would enhance the cellular recognition and internalization. For instance, the combination of transferrin and antibody on polymeric nanoparticles can improve their in vitro targeting efficiency and cytotoxicity, achieving higher tumor accumulation and enhanced therapeutic efficacy compared to single ligand-targeted ones (Sawant et al. 2013; Saul et al. 2006).

3.3.3 Functional Design of Polymeric Nanoparticles for Tumor Targeting

Polymeric nanoparticles are also designed to control their dissociation and release of encapsulated payloads at specific sites or time intervals. The release of therapeutic compounds could be remote controlled mainly responding to internal triggers (endogenous triggers) such as pH, redox, or enzyme, and external stimulus such as temperature, light, or ultrasound (Fig. 3.8) (Cheng et al. 2013).

Solid tumors produce the weakly acidic environments (pH ~6.8) compared with solid tumors. The pH of endo-/lysosomal compartments of cells is different from physiological pH (pH ~7.4). Such decreased pH can provide a potential trigger to release drugs from the polymeric nanoparticles (Maxfield and McGraw 2004). The pH-responsive polymeric nanoparticles were prepared by conjugating drugs to polymer matrix using acidic cleavable bonds. A variety of polymeric nanoparticles have been explored for achieving pH-responsive release. For instance, the pH-cleavable linkage (e.g., hydrazone) is used to conjugate doxorubicin with the PEG-b-P(Asp)

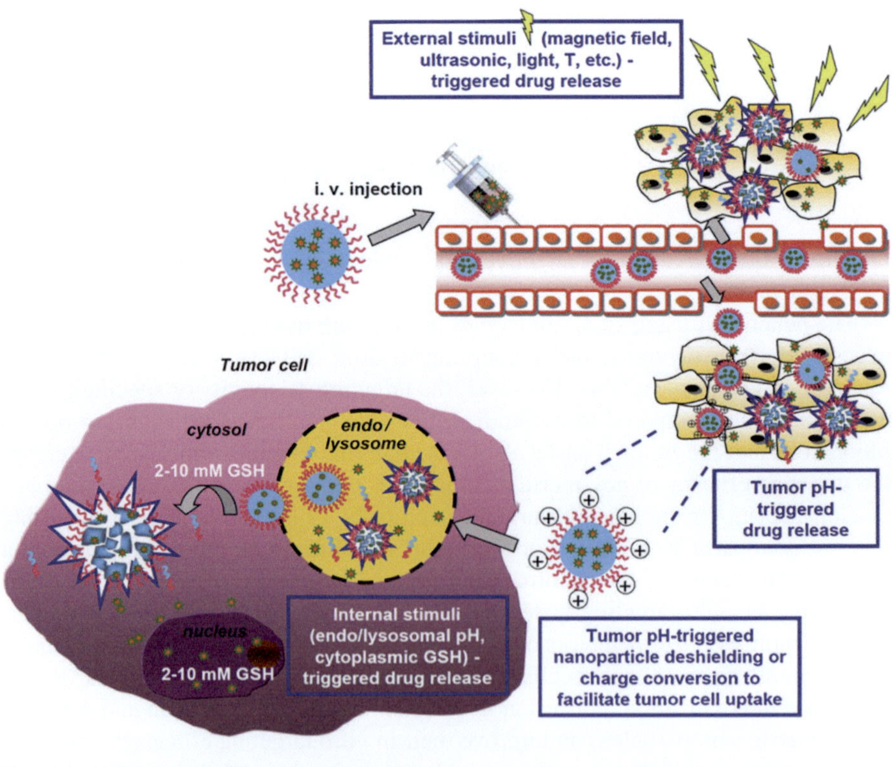

Fig. 3.8 Scheme of polymeric nanoparticles as emerging controlled drug release systems (reprinted from Cheng et al. 2013)

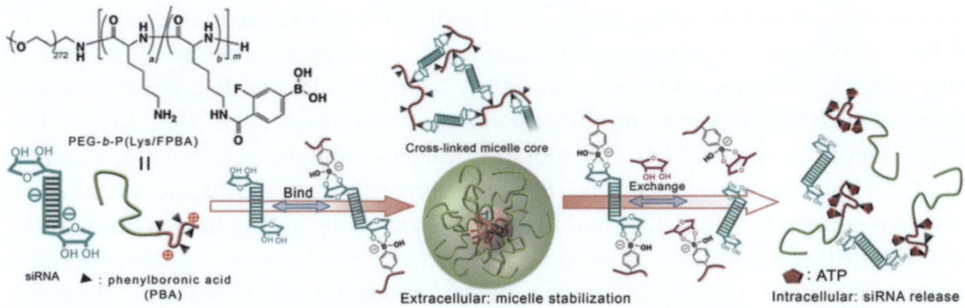

Fig. 3.9 Schematic illustration of the phenylboronic acid-modified block copolymers for ATP-responsive siRNA delivery (reprinted from Naito et al. 2012)

copolymer, allowing the release of doxorubicin in the endo-/lysosomes (Bae et al. 2005). Moreover, the pH-responsive nanoparticles coated with the folate ligand could achieve dual functions for drug delivery, i.e., folate ligands can enhance the cellular uptake, and pH-responsive function could release doxorubicin selectively in the cancer cells (Bae et al. 2007).

Besides pH-responsive strategy, polymeric nanoparticles responding to other triggers were developed as well. For instance, redox-responsive polymeric nanoparticles were prepared by disulfide cross-linking of micellar core incorporating packaged plasmid DNA (pDNA), and this system can dissociate to release pDNA in response to intracellular reductive condition (Miyata et al. 2004). Disulfide cross-linked micelles achieved ~100-fold higher gene transfection efficacy compared to non-cross-linked ones which were not stable at physiological ionic strength. Moreover, recently, the approach of PBA-assisted polyplex micelles is reported, where siRNAs serve as cross-linker via reversible complex formation between PBA and 1,2- or 1,3-*cis*-diols on a ribose ring. The polyplex micelles can be tailored to exhibit dramatic structural disruption, accompanying with the release of siRNAs in response to a change of ATP concentration (Fig. 3.9) (Naito et al. 2012). Nanoparticles also can release drugs with the stimuli of light, mainly due to the light irradiation that changed the properties of light-responsive agents incorporated inside nanoparticles (Tong et al. 2012; Yan et al. 2011; Zhao 2012).

3.4 Strategies for Cancer Therapy Using Polymeric Nanoparticles

3.4.1 Primary Cancer

Polymeric nanoparticles were initially designed and developed to treat primary tumors through the EPR effect. Polymeric nanoparticles are generally tested on subcutaneous tumor models to verify its targeting ability and antitumor activities. Subcutaneous tumor models could be prepared by inoculating different kinds of cancer cells subcutaneously on mice. It is a simple way to characterize the toxicity, blood circulation, tumor accumulation, and therapeutic effects of polymeric nanoparticles (Kwon and Kataoka 1995). Nanoparticles, with passive targeting (Jang et al. 2006) or active targeting functions (Vachutinsky et al. 2011; Miura et al. 2013; Deshayes et al. 2013), incorporating various therapeutic compounds such as siRNA (Kataoka and Harashima 2001) or anticancer drugs (Kataoka et al. 2001; Kaida et al. 2010), could be tested in a simple way using subcutaneous tumor models. For instance, cisplatin-incorporated polymeric micelle was tested on C26 subcutaneous tumor models on CDF1 mice, demonstrating significant antitumor efficacy (Nishiyama et al. 2003a). Another research based on HT29 or drug-resistant HT29 subcutaneous tumor models revealed that DACHPt/m nanoparticles can deliver drugs to the nearby nucleus of cancer cells, exhibiting higher antitumor activity even against drug-resistant tumors (Murakami et al. 2011). Some other researches tested on subcutaneous tumor models also revealed that actively targetable drug delivery systems showed higher antitumor activities than passive drug delivery systems (Vachutinsky et al. 2011; Deshayes et al. 2013; Miura et al. 2013).

Polymeric nanoparticles are also tested on orthotopic tumor models, wherever tumor grows in the organs usually happened in clinical, such as liver cancer, lung cancer, stomach cancer, etc. Therefore, most of those tumor models are inside the body, which are not suitable for investigation. Recently, imaging function was introduced to nanoparticles for visible drug delivery and monitoring of the therapeutic procedures by imaging methods, which are termed "theranostics" (Cabral et al. 2011a). For instance, MRI contrast agent diethylenetriaminepentaacetic acid gadolinium (III) (Gd-DTPA) and anticancer drug co-loaded polymeric nanoparticles have been applied to treat orthotopic pancreatic tumor (BxPC3) (Kaida et al. 2010). The nanoparticles demonstrated significant antitumor effects compared with free drugs, and the information of location and size of pancreatic tumor could be diagnosed by MRI.

Pancreatic tumors can spontaneously grow by utilizing transgenic mice, and this tumor models may be useful for preclinical evaluation of nanoparticles. According to recent study, DACHPt/m nanoparticles have enhanced the antitumor activity than oxaliplatin, and more importantly nanoparticles prolonged the survival of mice with spontaneous murine pancreatic tumors and prevented peritoneal metastasis in the body (Cabral et al. 2013).

3.4.2 Metastatic Cancer

Metastatic cancer is a clinical description for the spread of cancer cells from the primary tumor sites to distant organs, establishing secondary tumor sites. Detachment of cancer cells from primary tumor sites and their circulation in the blood allow the cells to be arrested in organs such as lungs, liver, lymph nodes, kidneys, brain, colon, skin, and bones, where they can extravasate and proliferate (Chambers et al. 2002; Chiang and Massague 2008; Fidler 2003). Despite advances in the understanding of the pathway of cancer cells to initiate metastasis and progress in early diagnosis, surgical treatments, new drugs for chemotherapy, and irradiation technologies, metastasis is responsible for as much as 90 % of cancer-associated mortality (Chaffer and Weinberg 2011). The treatment of metastasis is one of the greatest challenges in the decades, because it is limited by many factors, such as difficulty to find out all the metastatic tumor sites and remove all cancer cells by surgery and drug resistance to chemotherapy. Even worse, for many patients, by the time cancer is found, metastasis has already occurred, for example, over 80 % of patients diagnosed with lung cancer associated with metastatic cancer. Therefore, improved therapy of metastatic cancer is strongly demanded. Currently, chemotherapy, hormonal therapy, and radiation serve palliative purposes in the metastatic treatments, and some of them offer a modest but statistically significant extension of survival (Steeg 2006). Recent achievements in nanotechnology especially in polymeric nanoparticles, for instance, early detection of metastasis, targeting small disseminated malignant cells, and effective particle trafficking to cancer cells, highlight new approaches to treat metastatic cancers (Schroeder et al. 2012).

In these decades, more and more researches in the field of nanomedicine are focusing on developing nanoparticles for diagnosis and chemotherapy of metastatic cancers. For metastatic cancer treatment and therapy, diagnosis is essentially important to find metastatic lesions by some imaging methods and provide critical information about locations and stages of metastasis. Those information is important for some treatments, such as directed radiotherapy, require precise tumor location, and the treatment plan depends on the stage of metastasis. Nanoparticles also were developed for cancer diagnosis to provide more precise diagnostic results of metastasis for further clinical treatments. Ultrasmall superparamagnetic iron oxide nanoparticles (USPIOs), such as Combidex, could help in the successful detection of lymph node metastases in patients with prostate cancer by magnetic resonance imaging (MRI) in clinical (Harisinghani et al. 2003). Iron oxide nanoparticles decorated with some tumor cell-specific ligands, for instance, cRGD peptide, could target $\alpha_v\beta_3$ integrin expressed on the surface of some kinds of cancer cells for noninvasive diagnosis of primary tumor and liver metastasis (Peiris et al. 2012). Other inorganic nanoparticles, for instance, quantum dots (QDs) and silica nanoparticles, also demonstrated high reliability for metastatic cancer detection (Voura et al. 2004; Tang et al. 2012), such as aptamer-functionalized, 25 nm silica nanoparticles loading fluorescence dye and radionuclide (e.g., ^{64}Cu) for dual near-infrared (NIR) and positron emission tomography (PET) imaging of lymph node metastasis. Identification of normal and cancer cells on its nature of different cell surfaces is a general strategy, and different cell types, such as normal, cancerous, and metastatic cells, could be distinguished in a rapid and effective way (Bajaj et al. 2009).

The properties of polymeric nanoparticles, such as size and surface charge, have major influence on its target ability to specific organs with metastasis, which is illustrated in Table 3.2. The size of nanoparticle affects its ability to target specific organs, while surface properties of nanoparticles, especially with specific ligand conjugation, enhance the targeting efficacy to tumor cells. Several studies revealed how those factors affect nanoparticles' biodistribution and targeting efficacy. For

Table 3.2 Properties of polymeric nanoparticles for targeting specific organs

Target organs	Size of nanoparticle	Surface property	Mechanical properties	References
Brain	5–100 nm, smaller size leads to higher uptake efficiency	Lipophilic surface or lipid surface, ligand decorate, neutral charge	Leukocytes uptake nanoparticles and transport to brain metastasis sites	Miura et al. (2013)
Lung	100 nm or above makes nanoparticles to be arrested in lung capillaries	Positive surface charge, surface with ligand decorate	Inhaled particles with low density (<0.4 g/cm^3) and large size (>2.5 mm) are also retained in the lung	Xu et al. (2012)
Liver	Around 100 nm or smaller nanoparticles target hepatocytes, above 100 nm will uptake by Kupffer cells	Ligand decorating on the surface of nanoparticles	Nanoparticles are easily entrapped by macrophages in the liver	Peiris et al. (2012)
Lymph node	3–40 nm, iv administration or local administration	Non-cationic, sugar-based nanoparticles	Nanoparticles can be taken up by macrophage or leukocytes and trafficked to lymph nodes	Harisinghani et al. (2003)

example, 20 nm particles could access to lymph node metastasis, while 200 nm particles failed to target lymph node metastasis (Tang et al. 2012). Polymeric nanoparticles with a size of 30 nm, incorporating DACHPt, also demonstrated significant antitumor activity against primary gastric cancer and its lymph node metastasis (Rafi et al. 2012). Lymph nodes play an important role in the immune response, thus metastasis is easily initiated in lymph nodes by the spread of cancer cells through the lymph network, causing further metastatic spread and poor patient survival (Moore et al. 1998). Therefore, the targeting lymph node should be primarily important to prevent metastasis and extend survival period.

Chemotherapy of brain metastatic cancer has been limited by a lack of effective method to deliver therapeutic compounds to the lesions inside the brain due to the barriers of BBB, blood–cerebrospinal fluid barrier, and blood–tumor barrier, further limiting effective chemotherapy (Lesniak and Brem 2004). Lipid nanoparticles could passively target the brain with their lipophilic features by the procedures of endocytosis (Kaur et al. 2008; Andrieux and Couvreur 2009). In most of the cases, polymeric nanoparticle surface decorated with peptide ligand could deliver chemotherapeutic agents to the brain after administration, for instance, cyclic RGD-linked polymeric micelles can deliver platinum anticancer drugs specifically to glioblastoma through the BBB by intravenous injection (Miura et al. 2013). Moreover, glycopeptide-installed polymeric nanoparticles could reach the target also after intraperitoneal, intranasal, and even oral administrations (Tosi et al. 2013). Polysorbate-80-coated poly(butylcyanoacrylate) (PBCA) nanoparticles could pass the BBB by receptor-mediated endocytosis of the brain endothelial cells and deliver doxorubicin to brain tumors (Koo et al. 2006). The targeting peptide-coated polymeric nanoparticles loading imaging compounds such as MRI probe or dyes also could pass through BBB for diagnosis of brain metastasis (Veiseh et al. 2009; Qiao et al. 2012). Therefore, the receptor-mediated transport of drugs across the BBB will greatly enhance the therapy efficacy of brain metastasis in the near future.

Polymeric nanoparticles are also applied to target other kinds of metastasis, such as lung metastasis and liver metastasis. Liver is a common organ with metastasis, as vascular is widely dispersed in liver tissue with low shear rates and accessible capillaries, which provide friendly "soil" for migrating cancer cells to "seed" (Fidler 2003). In spite of most of systemically administrated polymeric nanoparticles exhibit high uptake by the liver, generally it is captured in macrophages (Kupffer cells) that reside within and near vascular. Regarding this mechanism, some polymeric nanoparticles deliver doxorubicin to generate gradient drug concentrations for a massive and prolonged diffusion of drugs towards the neoplastic tissue (Chiannilkulchai et al. 1990). Besides anticancer drugs, polymeric nanoparticles comprised of a core of high-molecular-weight linear polyethylenimine (LPEI) complexed with DNA and surrounded by a shell of PEG demonstrated effective gene therapy of malignant disease with liver and lung metastasis (Yang et al. 2013). The active targeting through the endothelial fenestrations to the liver metastasis has been facilitated via integration of ligand to the surface of nanoparticles for diagnosis and chemotherapy (Peiris et al. 2012; Zhao et al. 2013). Some polymeric nanoparticles have been further developed from beach to bedside, and recently FDA approved for paclitaxel albumin-stabilized nanoparticle formulation, Abraxane®, to treat metastatic pancreatic cancer and its liver metastasis. Like the liver, lung metastasis can be accessed by passive and target delivery of therapeutic agents with polymeric nanoparticles. PLGA nanoparticles have been utilized to co-deliver paclitaxel and Stat3 siRNA to overcome cellular resistance in lung cancer cells (Su et al. 2012). Delivering hedgehog pathway inhibitor HPI-1 with polymeric nanoparticles can inhibit systemic metastasis including lung metastasis in an orthotopic model of human hepatocellular carcinoma (Xu et al. 2012). Polymeric nanoparticles existing affinity ligands outside such as PBA (Deshayes et al. 2013) and peptide (Sarfati et al. 2011; Wang et al. 2013) demonstrate effective therapy effect against lung metastasis or prohibit lung metastasis.

3.5 Challenges and Future Perspectives

In these decades, polymeric nanoparticles with different types and targeting abilities have been designed and developed for delivering therapeutic compounds to malignant tumor or metastatic tumors, and some of these formulations have been tested in preclinical, clinical, or approved for clinical applications. Great efforts on polymeric nanoparticle-based drug delivery systems have improved pharmacokinetics, restrained side effects, increased targeting ability, enhanced cellular uptake, and finally significantly promoted the anticancer activities surpassing free therapeutic compounds. The exploration of polymeric nanocarriers has acquired the achievements including synthesized several materials for the application of drug delivery, designed many approaches for nanoparticle preparation, developed nanoparticles with different physicochemical properties and targeting abilities, revealed several impact factors of nanocarriers for drug delivery, found transporting mechanisms of nanocarriers in tumors tissues or cancer cells, and translated polymeric nanoparticles from bench to clinical for cancer therapy.

Respecting progresses on polymeric nanoparticle-based nanomedicines has shown hopeful light on cancer therapy; there are still some practical limitations that may influence their final applications, which need to be considered during the development of novel drug delivery systems. The limitations are varying from: (1) lack of proper safe, biodegradable, and biocompatible polymeric materials for drug delivery, and only several kinds of polymers are available in clinical until now; (2) advanced techniques for particle design and preparation are required for production in a large scale, including facile preparation and purification; (3) demanding to screen more cancer cells or tumor microenvironments specific ligands for active drug delivery; (4) optimization of the biophysicochemical properties of polymeric nanoparticles to meet problems of low loading efficacy, unreliable stability, short half-life, weak targeting efficacy to tumors, difficult uptake by cancer cells specifically, poor penetrability in tumor tissues, and controlled release; (5) polymeric nanoparticles should be designed to avoid drug resistance, which was found to be a common phenomenon of most anticancer drugs; (6) efficient gene delivery systems are required to deliver nucleic acids for gene therapy; and (7) translating more developed nanoparticle from bench to clinical for cancer therapy, as until now there are only tens of polymeric nanoparticles in clinical trial, regarding hundreds of patents issued and thousands of scientific papers published yearly.

We have the confidence that more powerful and wonderful polymeric nanoparticle-based drug delivery systems will be designed and developed in the near future with the rapid scientific and technological advancements in the fields ranging from chemistry, materials, physics, biology to oncology. We are optimistically viewing that more and more nanocarriers will be translated from bench to clinical trial and will be finally approved for clinical applications to treat patients with cancer, to control cancer from progressing and metastasis, to extend their survival periods, or even to recover their health. Furthermore, nanoparticles could be further broadened its applications to treat other medical problems, such as cardiovascular disease and vaccines.

References

Agemy L, Friedmann-Morvinski D, Kotamraju VR et al (2011) Targeted nanoparticle enhanced proapoptotic peptide as potential therapy for glioblastoma. Proc Natl Acad Sci U S A 108(42):17450–17455

Ai H, Jones SA, Lvov YM (2003) Biomedical applications of electrostatic layer-by-layer nanoassembly of polymers, enzymes, and nanoparticles. Cell Biochem Biophys 39(1):23–43

Allen TM (2002) Ligand-targeted therapeutics in anticancer therapy. Nat Rev Cancer 2(10):750–763

Andrieux K, Couvreur P (2009) Polyalkylcyanoacrylate nanoparticles for delivery of drugs across the blood-brain barrier. Wiley Interdiscip Rev Nanomed Nanobiotechnol 1(5):463–474

Avgoustakis K (2004) Pegylated poly(lactide) and poly(lactide-*co*-glycolide) nanoparticles: preparation, properties and possible applications in drug delivery. Curr Drug Deliv 1(4):321–333

Bae Y, Jang WD, Nishiyama N et al (2005) Multifunctional polymeric micelles with folate-mediated cancer cell targeting and pH-triggered drug releasing properties for active intracellular drug delivery. Mol Biosyst 1(3):242–250

Bae Y, Nishiyama N, Kataoka K (2007) In vivo antitumor activity of the folate-conjugated pH-sensitive polymeric micelle selectively releasing adriamycin in the intracellular acidic compartments. Bioconjug Chem 18(4):1131–1139

Bajaj A, Miranda OR, Kim IB et al (2009) Detection and differentiation of normal, cancerous, and metastatic cells using nanoparticle-polymer sensor arrays. Proc Natl Acad Sci U S A 106(27):10912–10916

Brewer E, Coleman J, Lowman A (2011) Emerging technologies of polymeric nanoparticles in cancer drug delivery. J Nanomater 2011:1–10

Byrne JD, Betancourt T, Brannon-Peppas L (2008) Active targeting schemes for nanoparticle systems in cancer therapeutics. Adv Drug Deliv Rev 60(15):1615–1626

Cabral H, Nishiyama N, Kataoka K (2011a) Supramolecular nanodevices: from design validation to theranostic nanomedicine. Acc Chem Res 44(10):999–1008

Cabral H, Matsumoto Y, Mizuno K et al (2011b) Accumulation of sub-100nm polymeric micelles in poorly permeable tumours depends on size. Nat Nanotechnol 6:815–823

Cabral H, Murakami M, Hojo H et al (2013) Targeted therapy of spontaneous murine pancreatic tumors by polymeric micelles prolongs survival and prevents peritoneal metastasis. Proc Natl Acad Sci U S A 110(28):11397–11402

Capretto L, Carugo D, Mazzitelli S et al (2013) Microfluidic and lab-on-a-chip preparation routes for organic nanoparticles and vesicular systems for nanomedicine applications. Adv Drug Deliv Rev 65(11–12):1496–1532

Chaffer CL, Weinberg RA (2011) A perspective on cancer cell metastasis. Science 331(6024):1559–1564

Chambers AF, Groom AC, MacDonald IC (2002) Dissemination and growth of cancer cells in metastatic sites. Nat Rev Cancer 2(8):563–572

Chauhan VP, Jain RK (2013) Strategies for advancing cancer nanomedicine. Nat Mater 12(11):958–962

Cheng R, Meng FH, Deng C et al (2013) Dual and multi-stimuli responsive polymeric nanoparticles for programmed site-specific drug delivery. Biomaterials 34(14):3647–3657

Chiang AC, Massague J (2008) Molecular basis of metastasis. N Engl J Med 359(26):2814–2823

Chiannilkulchai N, Ammoury N, Caillou B et al (1990) Hepatic tissue distribution of doxorubicin-loaded nanoparticles after i.v. administration in reticulosarcoma M 5076 metastasis-bearing mice. Cancer Chemother Pharmacol 26(2):122–126

Cho K, Wang X, Nie S et al (2008) Therapeutic nanoparticles for drug delivery in cancer. Clin Cancer Res 14(5):1310–1316

Conner SD, Schmid SL (2003) Regulated portals of entry into the cell. Nature 422(6927):37–44

Danhier F, Ansorena E, Silva JM et al (2012) PLGA-based nanoparticles: an overview of biomedical applications. J Control Release 161(2):505–522

Davis ME, Chen ZG, Shin DM (2008) Nanoparticle therapeutics: an emerging treatment modality for cancer. Nat Rev Drug Discov 7(9):771–782

Davis ME, Zuckerman JE, Choi CH et al (2010) Evidence of RNAi in humans from systemically administered siRNA via targeted nanoparticles. Nature 464(7291):1067–1070

Deming TJ, Hanson JA, Chang CB et al (2008) Nanoscale double emulsions stabilized by single-component block copolypeptides. Nature 455(7209):85–88

Deng C, Jiang Y, Cheng R et al (2012) Biodegradable polymeric micelles for targeted and controlled anticancer drug delivery: promises, progress and prospects. Nano Today 7:467–480

Deshayes S, Cabral H, Ishii T et al (2013) Phenylboronic acid-installed polymeric micelles for targeting sialylated epitopes in solid tumors. J Am Chem Soc 135(41):15501–15507

Duncan R (2006) Polymer conjugates as anticancer nanomedicines. Nat Rev Cancer 6(9):688–701

Elsabahy M, Wooley KL (2012) Design of polymeric nanoparticles for biomedical delivery applications. Chem Soc Rev 41(7):2545–2561

Ferrari M (2005) Cancer nanotechnology: opportunities and challenges. Nat Rev Cancer 5(3):161–171

Ferrari M (2008) Nanogeometry: beyond drug delivery. Nat Nanotechnol 3(3):131–132

Fessi H, Puisieux F, Devissaguet JP et al (1989) Nanocapsule formation by interfacial polymer deposition following solvent displacement. Int J Pharm 55(1):R1–R4

Fidler IJ (2003) The pathogenesis of cancer metastasis: the 'seed and soil' hypothesis revisited. Nat Rev Cancer 3(6):453–458

Geng Y, Dalhaimer P, Cai S et al (2007) Shape effects of filaments versus spherical particles in flow and drug delivery. Nat Nanotechnol 2(4):249–255

Ghosh D, Lee Y, Thomas S et al (2012) M13-templated magnetic nanoparticles for targeted in vivo imaging of prostate cancer. Nat Nanotechnol 7(10):677–682

Gratton SE, Ropp PA, Pohlhaus PD et al (2008) The effect of particle design on cellular internalization pathways. Proc Natl Acad Sci U S A 105(33):11613–11618

Grzelczak M, Vermant J, Furst EM et al (2010) Directed self-assembly of nanoparticles. ACS Nano 4(7):3591–3605

Gu F, Zhang L, Teply BA et al (2008) Precise engineering of targeted nanoparticles by using self-assembled biointegrated block copolymers. Proc Natl Acad Sci U S A 105(7):2586–2591

Hamaguchi T, Matsumura Y, Suzuki M et al (2005) NK105, a paclitaxel-incorporating micellar nanoparticle formulation, can extend in vivo antitumour activity and reduce the neurotoxicity of paclitaxel. Br J Cancer 92(7):1240–1246

Hammond PT (2012) Building biomedical materials layer-by-layer. Mater Today 15(5):196–206

Harisinghani MG, Barentsz J, Hahn PF et al (2003) Noninvasive detection of clinically occult lymph-node metastases in prostate cancer. N Engl J Med 348(25):2491–2499

Ideta R, Tasaka F, Jang WD et al (2005) Nanotechnology-based photodynamic therapy for neovascular disease using a supramolecular nanocarrier loaded with a dendritic photosensitizer. Nano Lett 5(12):2426–2431

Jang WD, Nakagishi Y, Nishiyama N et al (2006) Polyion complex micelles for photodynamic therapy: incorporation of dendritic photosensitizer excitable at long wavelength relevant to improved tissue-penetrating property. J Control Release 113(1):73–79

Jeong B, Choi YK, Bae YH et al (1999) New biodegradable polymers for injectable drug delivery systems. J Control Release 62(1–2):109–114

Jiang W, Kim BYS, Rutka JT et al (2008) Nanoparticle-mediated cellular response is size-dependent. Nat Nanotechnol 3(3):145–150

Jokerst JV, Lobovkina T, Zare RN et al (2011) Nanoparticle PEGylation for imaging and therapy. Nanomedicine 6(4):715–728

Kaida S, Cabral H, Kumagai M et al (2010) Visible drug delivery by supramolecular nanocarriers directing to single-platformed diagnosis and therapy of pancreatic tumor model. Cancer Res 70(18):7031–7041

Kamaly N, Xiao Z, Valencia PM et al (2012) Targeted polymeric therapeutic nanoparticles: design, development and clinical translation. Chem Soc Rev 41(7):2971–3010

Karnik R, Gu F, Basto P et al (2008) Microfluidic platform for controlled synthesis of polymeric nanoparticles. Nano Lett 8(9):2906–2912

Kataoka K, Harashima H (2001) Gene delivery systems: viral vs. non-viral vectors. Adv Drug Deliv Rev 52(3):151

Kataoka K, Harada A, Nagasaki Y (2001) Block copolymer micelles for drug delivery: design, characterization and biological significance. Adv Drug Deliv Rev 47(1):113–131

Kaul G, Amiji M (2002) Long-circulating poly(ethylene glycol)-modified gelatin nanoparticles for intracellular delivery. Pharm Res 19(7):1061–1067
Kaur IP, Bhandari R, Bhandari S et al (2008) Potential of solid lipid nanoparticles in brain targeting. J Control Release 127(2):97–109
Kim TY, Kim DW, Chung JY et al (2004) Phase I and pharmacokinetic study of Genexol-PM, a cremophor-free, polymeric micelle-formulated paclitaxel, in patients with advanced malignancies. Clin Cancer Res 10(11):3708–3716
Kolishetti N, Dhar S, Valencia PM et al (2010) Engineering of self-assembled nanoparticle platform for precisely controlled combination drug therapy. Proc Natl Acad Sci U S A 107(42):17939–17944
Koo YEL, Reddy GR, Bhojani M et al (2006) Brain cancer diagnosis and therapy with nanoplatforms. Adv Drug Deliv Rev 58(14):1556–1577
Kwon GS, Kataoka K (1995) Block-copolymer micelles as long-circulating drug vehicles. Adv Drug Deliv Rev 16(2–3):295–309
Langer R (1998) Drug delivery and targeting. Nature 392:5–10
Lee Y, Kataoka K (2009) Biosignal-sensitive polyion complex micelles for the delivery of biopharmaceuticals. Soft Matter 5(20):3810–3817
Lee Y, Fukushima S, Bae Y et al (2007) A protein nanocarrier from charge-conversion polymer in response to endosomal pH. J Am Chem Soc 129(17):5362–5363
Lee Y, Miyata K, Oba M et al (2008) Charge-conversion ternary polyplex with endosome disruption moiety: a technique for efficient and safe gene delivery. Angew Chem Int Ed 47(28):5163–5166
Lee Y, Ishii T, Cabral H et al (2009) Charge-conversional polyionic complex micelles-efficient nanocarriers for protein delivery into cytoplasm. Angew Chem Int Ed 48(29):5309–5312
Lee Y, Ishii T, Kim HJ et al (2010) Efficient delivery of bioactive antibodies into the cytoplasm of living cells by charge-conversional polyion complex micelles. Angew Chem Int Ed 49(14):2552–2555
Lesniak MS, Brem H (2004) Targeted therapy for brain tumours. Nat Rev Drug Discov 3(6):499–508
Li SD, Huang L (2009) Nanoparticles evading the reticuloendothelial system: role of the supported bilayer. Biochim Biophys Acta 1788(10):2259–2266
Matsumoto A, Cabral H, Sato N et al (2010) Assessment of tumor metastasis by the direct determination of cell-membrane sialic acid expression. Angew Chem Int Ed 49(32):5494–5497
Matsumura Y, Kataoka K (2009) Preclinical and clinical studies of anticancer agent-incorporating polymer micelles. Cancer Sci 100(4):572–579
Matsumura Y, Maeda H (1986) A new concept for macromolecular therapeutics in cancer chemotherapy: mechanism of tumoritropic accumulation of proteins and the antitumor agent smancs. Cancer Res 46(12):6387–6392
Maxfield FR, McGraw TE (2004) Endocytic recycling. Nat Rev Mol Cell Biol 5(2):121–132
Mi P, Cabral H, Kokuryo D et al (2013) Gd-DTPA-loaded polymer-metal complex micelles with high relaxivity for MR cancer imaging. Biomaterials 34(2):492–500
Mishra B, Patel BB, Tiwari S (2010) Colloidal nanocarriers: a review on formulation technology, types and applications toward targeted drug delivery. Nanomedicine 6(1):9–24
Miura Y, Takenaka T, Toh K et al (2013) Cyclic RGD-linked polymeric micelles for targeted delivery of platinum anticancer drugs to glioblastoma through the blood-brain tumor barrier. ACS Nano 7(10):8583–8592
Miyata K, Kakizawa Y, Nishiyama N et al (2004) Block catiomer polyplexes with regulated densities of charge and disulfide cross-linking directed to enhance gene expression. J Am Chem Soc 126(8):2355–2361
Miyata K, Nishiyama N, Kataoka K (2012) Rational design of smart supramolecular assemblies for gene delivery: chemical challenges in the creation of artificial viruses. Chem Soc Rev 41(7):2562–2574
Moore A, Sergeyev N, Bredow S et al (1998) A model system to quantitate tumor burden in locoregional lymph nodes during cancer spread. Invasion Metastasis 18(4):192–197

Morton SW, Herlihy KP, Shopsowitz KE et al (2013) Scalable manufacture of built-to-order nanomedicine: spray-assisted layer-by-layer functionalization of PRINT nanoparticles. Adv Mater 25(34):4707–4713

Murakami M, Cabral H, Matsumoto Y et al (2011) Improving drug potency and efficacy by nanocarrier-mediated subcellular targeting. Sci Transl Med 3(64):64ra62.

Naito M, Ishii T, Matsumoto A et al (2012) A phenylboronate-functionalized polyion complex micelle for ATP-triggered release of siRNA. Angew Chem Int Ed 51(43):10751–10755

Nakanishi T, Fukushima S, Okamoto K et al (2001) Development of the polymer micelle carrier system for doxorubicin. J Control Release 74(1–3):295–302

Nishiyama N, Kataoka K (2006) Current state, achievements, and future prospects of polymeric micelles as nanocarriers for drug and gene delivery. Pharmacol Ther 112(3):630–648

Nishiyama N, Okazaki S, Cabral H et al (2003a) Novel cisplatin-incorporated polymeric micelles can eradicate solid tumors in mice. Cancer Res 63(24):8977–8983

Nishiyama N, Stapert HR, Zhang GD et al (2003b) Light-harvesting ionic dendrimer porphyrins as new photosensitizers for photodynamic therapy. Bioconjug Chem 14(1):58–66

Nishiyama N, Iriyama A, Jang WD et al (2005) Light-induced gene transfer from packaged DNA enveloped in a dendrimeric photosensitizer. Nat Mater 4(12):934–941

Nishiyama N, Arnida, Jang WD et al (2006) Photochemical enhancement of transgene expression by polymeric micelles incorporating plasmid DNA and dendrimer-based photosensitizer. J Drug Target 14(6):413–424

Nishiyama N, Jang WD, Kataoka K (2007) Supramolecular nanocarriers integrated with dendrimers encapsulating photosensitizers for effective photodynamic therapy and photochemical gene delivery. New J Chem 31(7):1074–1082

Nishiyama N, Morimoto Y, Jang WD et al (2009) Design and development of dendrimer photosensitizer-incorporated polymeric micelles for enhanced photodynamic therapy. Adv Drug Deliv Rev 61(4):327–338

Oba M, Fukushima S, Kanayama N et al (2007) Cyclic RGD peptide-conjugated polyplex micelles as a targetable gene delivery system directed to cells possessing alpha(v)beta(3) and alpha(v) beta(5) integrins. Bioconjug Chem 18(5):1415–1423

Otsuka H, Nagasaki Y, Kataoka K (2012) PEGylated nanoparticles for biological and pharmaceutical applications. Adv Drug Deliv Rev 64:246–255

Owens DE 3rd, Peppas NA (2006) Opsonization, biodistribution, and pharmacokinetics of polymeric nanoparticles. Int J Pharm 307(1):93–102

Park JH, von Maltzahn G, Xu MJ et al (2010) Cooperative nanomaterial system to sensitize, target, and treat tumors. Proc Natl Acad Sci U S A 107(3):981–986

Peer D, Karp JM, Hong S et al (2007) Nanocarriers as an emerging platform for cancer therapy. Nat Nanotechnol 2(12):751–760

Peiris PM, Toy R, Doolittle E et al (2012) Imaging metastasis using an integrin-targeting chain-shaped nanoparticle. ACS Nano 6(10):8783–8795

Peppas NA (2013) Historical perspective on advanced drug delivery: how engineering design and mathematical modeling helped the field mature. Adv Drug Deliv Rev 65(1):5–9

Perry JL, Herlihy KP, Napier ME et al (2011) PRINT: a novel platform toward shape and size specific nanoparticle theranostics. Acc Chem Res 44(10):990–998

Plummer R, Wilson RH, Calvert H et al (2011) A phase I clinical study of cisplatin-incorporated polymeric micelles (NC-6004) in patients with solid tumours. Br J Cancer 104(4):593–598

Poon Z, Chang D, Zhao XY et al (2011) Layer-by-layer nanoparticles with a pH-Sheddable layer for in vivo targeting of tumor hypoxia. ACS Nano 5(6):4284–4292

Popovic Z, Liu W, Chauhan VP et al (2010) A nanoparticle size series for in vivo fluorescence imaging. Angew Chem Int Ed 49(46):8649–8652

Qiao RR, Jia QJ, Huwel S et al (2012) Receptor-mediated delivery of magnetic nanoparticles across the blood-brain barrier. ACS Nano 6(4):3304–3310

Rafi M, Cabral H, Kano MR et al (2012) Polymeric micelles incorporating (1,2-diaminocyclohexane) platinum (II) suppress the growth of orthotopic scirrhous gastric tumors and their lymph node metastasis. J Control Release 159(2):189–196

Rao JP, Geckeler KE (2011) Polymer nanoparticles: preparation techniques and size-control parameters. Prog Polym Sci 36(7):887–913
Rhee M, Valencia PM, Rodriguez MI et al (2011) Synthesis of size-tunable polymeric nanoparticles enabled by 3D hydrodynamic flow focusing in single-layer microchannels. Adv Mater 23(12):H79–H83
Saad M, Garbuzenko OB, Minko T (2008) Co-delivery of siRNA and an anticancer drug for treatment of multidrug-resistant cancer. Nanomedicine 3(6):761–776
Sarfati G, Dvir T, Elkabets M et al (2011) Targeting of polymeric nanoparticles to lung metastases by surface-attachment of YIGSR peptide from laminin. Biomaterials 32(1):152–161
Saul JM, Annapragada AV, Bellamkonda RV (2006) A dual-ligand approach for enhancing targeting selectivity of therapeutic nanocarriers. J Control Release 114(3):277–287
Sawant RR, Jhaveri AM, Koshkaryev A et al (2013) The effect of dual ligand-targeted micelles on the delivery and efficacy of poorly soluble drug for cancer therapy. J Drug Target 21(7):630–638
Schroeder A, Heller DA, Winslow MM et al (2012) Treating metastatic cancer with nanotechnology. Nat Rev Cancer 12(1):39–50
Simon U (2013) Nanoparticle self-assembly bonding them all. Nat Mater 12(8):694–696
Soppimath KS, Aminabhavi TM, Kulkarni AR et al (2001) Biodegradable polymeric nanoparticles as drug delivery devices. J Control Release 70(1–2):1–20
Steeg PS (2006) Tumor metastasis: mechanistic insights and clinical challenges. Nat Med 12(8):895–904
Su WP, Cheng FY, Shieh DB et al (2012) PLGA nanoparticles codeliver paclitaxel and Stat3 siRNA to overcome cellular resistance in lung cancer cells. Int J Nanomedicine 7:4269–4283
Svenson S (2012) Clinical translation of nanomedicines. Curr Opin Solid State Mater Sci 16(6):287–294
Takahashi A, Yamamoto Y, Yasunaga M et al (2013) NC-6300, an epirubicin-incorporating micelle, extends the antitumor effect and reduces the cardiotoxicity of epirubicin. Cancer Sci 104(7):920–925
Tang L, Yang XJ, Dobrucki LW et al (2012) Aptamer-functionalized, ultra-small, monodisperse silica nanoconjugates for targeted dual-modal imaging of lymph nodes with metastatic tumors. Angew Chem Int Ed 51(51):12721–12726
Tong R, Hemmati HD, Langer R et al (2012) Photoswitchable nanoparticles for triggered tissue penetration and drug delivery. J Am Chem Soc 134(21):8848–8855
Torchilin VP (2007) Targeted pharmaceutical nanocarriers for cancer therapy and imaging. AAPS J 9(2):E128–E147
Tosi G, Ruozi B, Belletti D et al (2013) Brain-targeted polymeric nanoparticles: in vivo evidence of different routes of administration in rodents. Nanomedicine 8(9):1373–1383
Uchino H, Matsumura Y, Negishi T et al (2005) Cisplatin-incorporating polymeric micelles (NC-6004) can reduce nephrotoxicity and neurotoxicity of cisplatin in rats. Br J Cancer 93(6):678–687
Utada AS, Chu LY, Fernandez-Nieves A et al (2007) Dripping, jetting, drops, and wetting: the magic of microfluidics. MRS Bull 32(9):702–708
Vachutinsky Y, Oba M, Miyata K et al (2011) Antiangiogenic gene therapy of experimental pancreatic tumor by sFlt-1 plasmid DNA carried by RGD-modified crosslinked polyplex micelles. J Control Release 149(1):51–57
Valencia PM, Farokhzad OC, Karnik R et al (2012) Microfluidic technologies for accelerating the clinical translation of nanoparticles. Nat Nanotechnol 7(10):623–629
Valle JW, Armstrong A, Newman C et al (2011) A phase II study of SP1049C, doxorubicin in P-glycoprotein-targeting pluronics, in patients with advanced adenocarcinoma of the esophagus and gastroesophageal junction. Invest New Drugs 29(5):1029–1037
Vauthier C, Bouchemal K (2009) Methods for the preparation and manufacture of polymeric nanoparticles. Pharm Res 26(5):1025–1058
Veiseh O, Sun C, Fang C et al (2009) Specific targeting of brain tumors with an optical/magnetic resonance imaging nanoprobe across the blood-brain barrier. Cancer Res 69(15):6200–6207

Venkataraman S, Hedrick JL, Ong ZY et al (2011) The effects of polymeric nanostructure shape on drug delivery. Adv Drug Deliv Rev 63(14–15):1228–1246

Verma A, Stellacci F (2010) Effect of surface properties on nanoparticle-cell interactions. Small 6(1):12–21

Voura EB, Jaiswal JK, Mattoussi H et al (2004) Tracking metastatic tumor cell extravasation with quantum dot nanocrystals and fluorescence emission-scanning microscopy. Nat Med 10(9):993–998

Wang G, Uludag H (2008) Recent developments in nanoparticle-based drug delivery and targeting systems with emphasis on protein-based nanoparticles. Expert Opin Drug Deliv 5(5):499–515

Wang ZH, Yu Y, Dai WB et al (2013) A specific peptide ligand-modified lipid nanoparticle carrier for the inhibition of tumor metastasis growth. Biomaterials 34(3):756–764

Werner ME, Karve S, Sukumar R et al (2011) Folate-targeted nanoparticle delivery of chemo- and radiotherapeutics for the treatment of ovarian cancer peritoneal metastasis. Biomaterials 32(33):8548–8554

Xu Y, Chenna V, Hu C et al (2012) Polymeric nanoparticle-encapsulated hedgehog pathway inhibitor HPI-1 (NanoHHI) inhibits systemic metastases in an orthotopic model of human hepatocellular carcinoma. Clin Cancer Res 18(5):1291–1302

Yan B, Boyer JC, Branda NR et al (2011) Near-infrared light-triggered dissociation of block copolymer micelles using upconverting nanoparticles. J Am Chem Soc 133(49):19714–19717

Yang J, Hendricks W, Liu GS et al (2013) A nanoparticle formulation that selectively transfects metastatic tumors in mice. Proc Natl Acad Sci U S A 110(36):14717–14722

Zhang GD, Harada A, Nishiyama N et al (2003) Polyion complex micelles entrapping cationic dendrimer porphyrin: effective photosensitizer for photodynamic therapy of cancer. J Control Release 93(2):141–150

Zhang L, Radovic-Moreno AF, Alexis F et al (2007) Co-delivery of hydrophobic and hydrophilic drugs from nanoparticle-aptamer bioconjugates. ChemMedChem 2(9):1268–1271

Zhang K, Fang H, Chen Z et al (2008) Shape effects of nanoparticles conjugated with cell-penetrating peptides (HIV Tat PTD) on CHO cell uptake. Bioconjug Chem 19(9):1880–1887

Zhang J, Wu LB, Chan HK et al (2011) Formation, characterization, and fate of inhaled drug nanoparticles. Adv Drug Deliv Rev 63(6):441–455

Zhao Y (2012) Light-responsive block copolymer micelles. Macromolecules 45(9):3647–3657

Zhao C, Feng Q, Dou Z et al (2013) Local targeted therapy of liver metastasis from colon cancer by galactosylated liposome encapsulated with doxorubicin. PLoS One 8(9):e73860

Chapter 4
Targeted Liposomes and Micelles as Carriers for Cancer Therapy

Weiyue Lu, Changyou Zhan, and Huimin Hou

List of Abbreviations

9-NC	9-Nitro-camptothecin
CMC	Critical micellar concentration
CPT	Camptothecin
DHPC	Diheptanolyphosphatidylcholine
DPPC	1,2-Dipalmitoyl-*sn*-glycero-2-phosphocholine
DPPGOG	1,2-Dipalmitoyl-*sn*-glycero-3-phosphoglyceroglycerol
ECM	Extracellular matrix
EGFR	Epidermal growth factor receptors
ELP	Elastin-like polypeptide
EPR	Enhanced permeability and retention
FRV	Freeze-dried rehydration vesicle
LCST	Low critical solution temperature

W. Lu (✉)
Department of Pharmaceutics, School of Pharmacy, Fudan University & Key Laboratory of Smart Drug Delivery (Fudan University), Ministry of Education, 826 Zhangheng Road, Shanghai 201203, China
e-mail: wylu@shmu.edu.cn

C. Zhan
Department of Pharmaceutics, School of Pharmacy, Fudan University & Key Laboratory of Smart Drug Delivery (Fudan University), Ministry of Education, 826 Zhangheng Road, Shanghai 201203, China

Laboratory for Biomaterials and Drug Delivery, Department of Anesthesiology, Children's Hospital Boston, Harvard Medical School, Boston, MA 02115, USA

H. Hou
National Pharmaceutical Engineering Research Center, Shanghai Institute of Industrial Pharmaceutics, Shanghai 201203, China

M.J. Alonso, M. Garcia-Fuentes (eds.), *Nano-Oncologicals: New Targeting and Delivery Approaches*, Advances in Delivery Science and Technology,
DOI 10.1007/978-3-319-08084-0_4

MDR	Multidrug resistance
MLV	Multilamellar large vesicle
MMPs	Matrix metalloproteinases
MPS	Mononuclear phagocyte system
PEG-*b*-PDLA	Poly(ethylene glycol)-*b*-poly(D-lactic acid)
PEG-*b*-PLLA	Poly(ethylene glycol)-*b*-poly(L-lactic acid)
PEG-DSPE	Poly(ethyl glycol)-1,2-distearoyl-*sn*-glycero-3-phosphoethanolamine
PEG-PCL	Poly(ethylene glycol)-*b*-poly(ε-caprolactone)
PFP	Perfluoropentane
pHLIP	Low pH inserting peptide
PIC	Polyion complex micelle
PNIPPAm	Poly(N-isopropylacrylamide)
RES	Reticuloendothelial system
REV	Reverse-phase evaporation vesicle
sPLA2	Secretory phospholipase A2
SUV	Small unilamellar vesicle
TEM	Transmission electron microscopy
TfR	Transferrin receptor
uPA	Urokinase plasminogen activator
uPAR	Urokinase plasminogen activator receptor
VEGFR	Vascular endothelial growth factor receptor

4.1 Introduction

In recent years, the medical application of nanotechnologies, usually termed nanomedicine, has given a crucial impulse to the development of drug discovery and delivery. In particular, nanomedicine plays a pivotal role in the treatments of cancer owing to the presence of numerous oncologic characteristics, e.g., enhanced permeability and retention (EPR) effect in solid tumors (Lindner et al. 2004), and overexpression of receptors or antigens on tumor cells. Nanomedicine also expands the application of antineoplastics by elevating solubility of hydrophobic payloads, stabilizing the degradable agents (such as proteins, peptides, and genetic materials), and enhancing intracellular drug delivery efficiency. Among the numerous classes of materials employed for cancer therapy, self-assembly systems such as liposomes and polymeric micelles have attracted extensive attention because of the flexibility of drug loading and ease of functionalization, among many others. In this chapter, we emphasize on the targeted liposomes and polymeric micelles for cancer therapy.

Liposomes, which were first described by Alec Bangham in 1961 (Bangham and Horne 1964), are artificially prepared vesicles composed of lipid bilayer. Liposomes can be classified according to their lamellarity, where unilamellar vesicles comprise one lipid bilayer and contain large aqueous internal phase, and multilamellar vesicles consist of several concentric lipid bilayer with micron size. Liposomes form spontaneously when certain lipids are hydrated in aqueous media. Biocompatibility of lipids broadens the biomedical applications of liposomes. As drug carrier, liposomes are ideal delivery systems for encapsulating different kinds of payload with high capacity. They can entrap hydrophilic (or amphiphilic) pharmaceutical agents in their internal aqueous compartment and hydrophobic (or amphiphilic) drugs into the lipid bilayer. Liposomes are not only drug carriers to assist delivery but also protecting barriers that prevent the pharmaceutical agents from inactivating effects. Liposomes can be taken up by cells through specific pathways, such as membrane fusion and endocytosis, providing unique opportunity to accomplish intracellular drug delivery (Torchilin 2005). During the past several decades, considerable progress has been witnessed and some liposomal formulations have been in clinical use for cancer therapy (O'Brien et al. 2004; Boulikas 2009; Zhang et al. 2008; Gelmon et al. 1999; Guaglianone et al. 1994), and numerous liposome-based formulations are still on the way from bench to bedside. In contrast to the formation of lipid bilayer in liposomes, polymeric micelles form monolayer in selective solvents from amphiphilic polymers that consist of hydrophilic and hydrophobic monomer units. They have typical core-shell architecture with size ranging from 5 to 100 nm (Nicolas et al. 2013), restricting the uptake by mononuclear phagocyte system (MPS) (Kataoka et al. 2000). Polymeric micelles have been reported as drug carriers since the 1980s (Hubert et al. 1984) where the hydrophobic core serves as a reservoir of various drugs (such as chemotherapeutic agents) and the hydrophilic shell stabilizes the structure in aqueous media, making them appropriate candidates for systemic drug delivery. The core may consist of hydrophilic polymers with chemical conjugation of hydrophobic moieties (Kataoka et al. 2000; Yokoyama et al. 1998) or form through the association of two oppositely charged polyions (polyion complex micelles, PIC) (Oberoi et al. 2011; Oishi et al. 2005). Compared with the traditional surfactant micelles, polymeric micelles are generally more stable, registering a remarkably lower critical micellar concentration (CMC) and a slower dissociation rate (Kataoka et al. 2001). Above the CMC, polymers exist as micelles in equilibrium with a small population of single chains while below the CMC, they exist as unimer in solution. It has been demonstrated, however, micelles remain stable for extended periods of time below CMC due to the unique properties of the core-forming blocks (Mikhail and Allen 2009). CMC defines the thermodynamic stability of micelles, a very important parameter in the drug delivery application of polymeric micelles because of the extreme dilution by circulating blood after intravenous injection. Fortunately, the kinetic stability of polymeric micelles is relatively high. Some micelles with "solid" core and the glass transition temperature above physiological temperature may survive for many hours and even days upon dilution below the CMC (Kwon and Okano 1999). In addition, numerous strategies have been developed to improve in vivo stability, including shell crosslinking (O'Reilly et al. 2006; Kim et al. 2010a) and the design of unimolecular micelles (Zhu et al. 2011a). Monodisperse stereocomplex block copolymer micelles, which were obtained through the self-assembly of equimolar mixtures of poly(ethylene glycol)-*b*-poly(L-lactic acid) (PEG-*b*-PLLA) and poly(ethylene glycol)-*b*-poly(D-lactic acid) (PEG-*b*-PDLA), exhibited kinetic stability and redispersion properties superior to micelles prepared with isotactic or racemic polymers alone (Kang et al. 2005).

Liposomes emerge as versatile vehicles for loading a broad spectrum of drugs with high stability, while micelles are normally designed for the delivery of hydrophobic drugs and charged agents (PIC micelle). Liposomes have also been capable of providing multimodal drug release kinetics and delivering drug combinations for boosting anticancer effects. However, micelles are valuable candidates in comparison with liposome when designing small-size (smaller than 50 nm) nanocarriers for specific applications (such as for enhancing tumor penetration).

Cancer remains one of the most life-threatening diseases; however, substantial improvements have been achieved owing to better understanding of cancer biology and emerging diagnostic techniques (Carmeliet and Jain 2000; Peer et al. 2007). Current cancer treatments include surgical resection, radiotherapy, and chemotherapy. Since chemotherapeutic agents are always toxic to healthy tissues and cause severe side effects to patients, it is extremely desirable to develop tumor targeting drug delivery systems that facilitate drug accumulation in tumor tissue to boost antitumor effects and decrease drug distribution in healthy tissues to minimize side effects. The concept of "magic bullet" proposed by Paul Ehrlich in 1906 envisioned selectively delivery of active agents to the organs in question (Strebhardt and Ullrich 2008). Nanomedicines designed as reservoirs of chemotherapeutic agents have been extensively exploited to passively and/or actively target tumor and/or tumor-related tissues. In contrast to normal vessels, tumor vessels are structurally and functionally abnormal. They are highly disorganized and have numerous endothelial fenestrae and discontinuous or absent basement membrane, making them leaky and highly permeable. Whereas free drugs may non-specifically diffuse, nanocarriers can take advantage of vessel leakiness and extravasate into the tumor tissue, where poor lymphatic drainage is beneficial for retaining the accumulated nanocarriers and facilitating drug release in the vicinity of tumor cells. One way to enhance tumor targeting efficiency is to decorate on the surface of nanocarriers tumor homing ligand, which specifically binds the overexpressed tumor markers (antigens or receptors). The ligands bind to tumor cell surface after extravasation, tremendously expediting accumulation of chemotherapeutics in tumor. Multidrug resistance (MDR) occurs because of overexpression of protein transporters, e.g., drug efflux pump (P-glycoprotein), on tumor cells. Active targeting may work as effective strategy to circumvent MDR by unique receptor-mediated endocytosis mechanism. Liposome- or micelle-based drug delivery systems that are easy to be functionalized have been extensively investigated to maximize the benefit from active targeting strategy.

4.2 Liposome and Micelle: Preparation and Functionalization

4.2.1 Preparation Methods

Various methods have been employed for the preparation of different classes of liposome. Among those, thin-film hydration is an exceptionally simple method to prepare multilamellar large vesicles (MLVs), one of the most widely used liposomes (Bangham et al. 1974). In brief, the mixture of lipids is dissolved in the organic solvents and rotary evaporated to form a thin film on the wall of a round-bottom flask under reduced pressure. After the removal of residual organic solvents, aqueous buffer is added and the thin film is hydrated at a temperature above the gel-liquid crystal transition temperature (Tc) of lipids or above the Tc of the highest melting component in the mixture by shaking, rotating, or vortex (Olson et al. 1979). For the preparation of small unilamellar vesicles (SUVs), classical procedures include sonicating a dispersion of phospholipids with either a bath type sonicator or a probe sonicator immersed into liposome dispersion (Johnson et al. 1971; Huang 1969). Reverse-phase evaporation vesicles (REVs) are capable of entrapping most of the water-soluble substances (Szoka and Papahadjopoulos 1978). High aqueous space-to-lipid ratio of REVs also enables high encapsulation efficiency of hydrophilic therapeutic agents. To prepare REVs, phospholipids are dissolved in organic solvents, such as diethylether, chloroform, or mixtures. The aqueous material is added directly to this phospholipid-solvent mixture. A homogeneous emulsion forms after sonication for a short period. The organic solvents are removed under reduced pressure, resulting in formation of a viscous gel-like intermediate phase. At a critical point, the gel collapses and spontaneously forms liposomal dispersion when residual solvent is removed by continued rotary evaporation under reduced pressure. Detergent removal is another essentially mild method for the preparation of phospholipid vesicles encapsulating sensitive protein, which results in formation of unilamellar vesicles. The detergent is removed from mixed micelle solution formed by detergent and phospholipids using different methods such as dilution (Schurtenberger et al. 1984), dialysis (Milsmann et al. 1978), gel filtration (Brunner et al. 1976), and absorption. Brunner et al. (1976) reported that the treatment of EPC with sodium cholate in a 1:2 molar ratio followed by gel filtration to remove detergent resulted in the formation of a homogeneous population of unilamellar vesicles with a mean diameter of 30 nm. Freeze-dried rehydration vesicles (FRVs) are formed from preformed vesicles to achieve extreme high entrapment efficiencies, even for macromolecules. To prepare liposomes with high pressure homogenization, there are three different ways: (1) to homogenize well-prepared liposomes, (2) to transfer the phospholipids and buffer to SUV, and (3) to homogenize during injecting organic lipid solution into the aqueous phase (Mayhew et al. 1984; Brandl et al. 1990).

Depending on the physicochemical properties, polymeric micelles are mainly prepared by two classes of procedure. The first one is direct dissolution where copolymers along with drugs are dissolved in aqueous solution. This method is mainly employed for copolymers with less hydrophobicity owing to the premise of water solubility. The Pluronic triblock copolymer, which consists of hydrophilic ethylene oxide and hydrophobic propylene oxide blocks, can self-assemble in aqueous media at the concentration above CMC and be used for drug delivery. Pluronic P85 has been selected to encapsulate carboplatin, an active chemotherapeutic agent compromised by side effects and drug resistance (Exner et al. 2005). The carboplatin-Pluronic P85 micelles formed by dissolving 1 % polymer along with various concentrations of drug could robustly enhance antitumor effect of carboplatin against the DHB/K12/TRb rat colorectal carcinoma cell line. Direct dissolution is also used to prepare PIC micelle. Muir et al. have developed long-circulating, cyclic RGD peptide-conjugated polymeric micelles incorporating (1,2-diaminocyclohexane) platinum (II) (DACHPt), the parent complex of the potent anticancer drug oxaliplatin, through the metal complex formation-driven self-assembly of poly(ethyl glycol)-*b*-poly(L-glutamic acid) and DACHPt in an aqueous solution. In another research, soluble poly(L-lysine) quantitatively grafted with 3-fluoro-4-carboxyphenylboronic acid condensed siRNA in aqueous medium for ATP-triggered release (Naito et al. 2012). The second class of micellization procedure applies to amphiphilic copolymers that are not readily water soluble. In this case, an organic solvent or mixed organic solvent is usually needed to dissolve copolymer and drug, and micelle forms by the removal of organic solvent. For water-miscible organic solvents (such as methanol, ethanol, acetone, acetonitrile, DMF, and DMSO), the copolymer/drug mixture can be dialyzed against water, whereby the slow removal of organic solvent facilitates micellization and drug loading. Alternatively, solvent-casting method enables the formation of thin film of copolymer/drug mixture, and rehydration with aqueous medium initiates micellization. PEG-PLA micelles have been widely used as drug delivery systems. When the PLA block is short, PEG-PLA copolymer can be dissolved in acetonitrile to form thin film by vacuum evaporation, and rehydration with aqueous medium yields micelles with the size ranging from 20 to 50 nm (Zhan et al. 2010a, b, 2011, 2012a). However, rehydration of the thin film results in aggregation when the PLA block is too long. Physical entrapment of hydrophobic drugs can be further achieved through oil-in-water (O/W) emulsion process. Hu et al. (2009) reported the synthesis of PEG_{3K}-PLA_{50K} nanocarrier by using a water-in-oil-in-water (W/O/W) emulsion process. PEG-PLA and drug dissolved in dichloromethane were emulsified in water by sonication, and the formed O/W primary emulsion was further emulsified in sodium cholate solution. The final PEG-PLA micelle was obtained by diluting with sodium cholate solution and evaporating dichloromethane.

4.2.2 Functionalization

Functionalization on the periphery of liposomes and micelles is needed to synthesize active tumor targeting drug delivery systems. Active ligands that specifically recognize the overexpressed receptors or antigens in tumor and/or tumor-related tissues can inspire the active targeting drug delivery toward tumors, thus boosting the antitumor effects with minimal side effects. On the premise of maintaining the bioactivity of ligand, various strategies, including conjugating chemistry (covalent or non-covalent) and modification mode (prior to or after the preparation of nanocarriers), have been employed for the functionalization.

As for liposomes, additional amphiphilic materials, such as poly(ethyl glycol)-1,2-distearoyl-*sn*-glycero-3-phosphoethanolamine (PEG-DSPE) (Yan et al. 2011, 2012a), are needed for the conjugation of active ligands. The hydrophobic moiety, DSPE, is used to insert into the lipid bilayer, and the other distal of PEG is available for modification. Lee and Low (1995) prepared folate modified liposome for tumor targeting drug delivery. *N*-succinyl DSPE was conjugated with folate-PEG-NH_2 in the presence of dicyclohexylcarbodiimide and pyridine. The purified folate-PEG-DSPE was then mixed with lipid to form liposome. Cholesterol, usually an indispensible composition of liposome, has also been used for liposome functionalization. Tang et al. (2013) designed a tumor-targeting liposome by co-modification with transferrin and TAT peptide. The specific ligand transferrin and cationic cell penetrating peptide TAT were conjugated with cholesterol through a PEG spacer, and the obtained amphiphilic conjugates were mixed with lipids to form liposomes. The targeting moieties are presented on the particle surface owing to the hydrophilicity of PEG spacer. Compared with liposome, micelle is much easy to be functionalized because the active ligands can be directly linked to the copolymer.

For the self-assembly systems, active ligands can be either modified prior to the formation of nanocarriers, or directly on the surface of preformed nanocarriers in aqueous solution. The choice of synthetic route is usually governed by at least three factors: (1) Size of active ligand. If the active ligand is small molecule or small peptide, conjugation to copolymers may not affect the hydrophilic/hydrophobic balance. When the active ligand is polypeptide or antibody, the conjugation of such bulky ligand prior to the formation of nanocarriers would destroy the balance. In this case, ligands are usually modified on the surface of formed liposomes and micelles. (2) Stability of the ligands. The use of organic solvent in formulating step may denature polypeptides and antibodies that possess three-dimensional confirmations, and therefore attenuate the bioactivity or even deactivate them. (3) Conjugating chemistry. The prepared nanocarriers are usually suspended in neutral aqueous solution. If organic solvent or acidic/basic condition is needed to conduct the conjugation, it will destroy the self-assembly system and lead to the leakiness of payloads, restricting the application of directly modification on the particle surface.

For the uncertainty of reaction yield, it is not easy to control the ligand density presented on the particle surface when ligand modification performed after the formulation of nanocarriers. The advantage of conjugating active ligand prior to the formation lies in the feasibility to control the reaction yield and to tune the ligand density on the particle surface. However, drawbacks include the possibilities of destroying the self-assembly property and deactivating the ligands owing to the use of organic solvent and/or high temperature during formulating steps. Conjugation after nanocarrier formation may bring difficulties on purification and chemical characterization.

The preparation of ligand modified liposome and micelle with envisioned properties requires the chemical conjugation of small molecules, peptides, polypeptides, or antibodies, among many others. A variety of conjugating strategies have been employed for ligand modification, including the formation of amide bond (Wang et al. 2011) between activated carboxyl group and amino group, disulfide bond (Yan et al. 2012b) via the reaction between pyridyldithiol and thiol, Michael addition between thiol or amino group and maleimide group (Zhan et al. 2010a, 2012b), and "click" chemistry (Parrish et al. 2005). In some case, non-covalent linkage has also been exploited for ligand modification, such as biotin-streptavidin interaction (Loughrey et al. 1987). The detailed description of those strategies has been illustrated in previous review (Nicolas et al. 2013), and we will not discuss in this chapter.

4.3 Active Tumor-Targeting Drug Delivery for Cancer Therapy

4.3.1 Design of Active Tumor-Targeting Liposomes and Micelles

Rationale behind the design of tumor targeting nanocarriers is to increase antitumor efficiency of payloads, while reducing systemic side effects. Numerous investigations on cancer biology have illustrated the features of tumor cells and tumor vasculature, leading to the development of active tumor-targeting drug delivery systems. A variety of receptors and antigens play pivotal roles in the development and progression of human cancers, and the overexpression provides targets for the design of tumor homing drug delivery.

Ligands are usually attached at the surface of nanocarriers to bind corresponding receptors or antigens overexpressed on tumor cells and/or tumor vasculature. For cancer cells, active targeting of internalization-prone cell-surface receptors is preferable to improve the cellular uptake of nanocarriers, especially for the payloads that take effect intracellularly. Active targeting is particularly attractive for the intracellular delivery of macromolecular drugs, such as genetic materials and proteins. Compared to elevated tumor accumulation, direct internalization into target cells is responsible for enhanced antitumor efficiency of targeting nanocarriers. To date, numerous internalization-prone receptors, including folate receptors, transferrin receptor (TfR), and EGFR, have been exploited to expedite tumor targeting drug delivery. Folate receptors are cysteine-rich cell-surface glycoproteins that bind folic acid or folate conjugated nanocarriers with high avidity. Although at very low levels in most tissues, folate receptors are expressed at high levels in numerous cancers to meet the folate demand of rapidly dividing cells under low folate conditions (Chen et al. 2013). The folate receptors are well-known tumor markers that can facilitate intracellular delivery via receptor-mediated endocytosis. To deliver doxorubicin loaded micelle into KB cells, folic acid was separately conjugated with PEG-PLGA to produce folate-PEG-PLGA. Folate decorated micelles were prepared by mixing doxorubicin conjugated PLGA-PEG (DOX-PLGA-PEG), folate-PEG-PLGA, and free doxorubicin. The mixtures were dissolved in acetone in the presence of triethylamine, followed by directly dispersing into aqueous phase with gentle stirring for 24 h. The micelles encapsulating a high loading amount of doxorubicin demonstrated superior cellular uptake over free drug and micelles without folate modification (Yoo and Park 2004), indicating high efficiency of receptor-mediated endocytosis induced by folate receptors. Han et al. (2009) prepared folate-conjugated PEG-DSPE micelle by mixing PEG-DSPE and folate-PEG-DSPE for tumor targeting delivery of 9-nitro-camptothecin (9-NC). The optimal molar ratio between folate-PEG-DSPE and PEG-DSPE was 1:100, by which the formed folate modified micelles could avoid the uptake of macrophage and boost the antitumor effects of 9-NC by receptor-mediated endocytosis. When conjugated onto the surface of liposome, folate also inspired the targeting delivery of encapsulated oligonucleotides to folate receptors-positive tumor cells in vitro and in vivo (Leamon et al. 2003).

TfR is a 180-kDa dimeric transmembrane glycoprotein overexpressed on cancer and leukemia cells. Transferrin, an 80-kDa glycoprotein, is the natural occurring ligand of TfR and is internalized by receptor-mediated endocytosis. Besides, antibodies against TfR have also attracted widely attention to inspire tumor targeting drug delivery. The possibility of using transferrin for developing an intracellular drug delivery system has been explored by conjugating transferrin on the surface of liposome with an average diameter of 100–130 nm (Ishida et al. 2001). Transmission electron microscopy (TEM) has been employed to investigate the endocytotic uptake pathway of transferrin-PEG-liposome into colon-26 cells, demonstrating that the liposomes could bind specific cell surface receptors and be internalized through receptor-mediated endocytosis. In vivo biodistribution studies demonstrated that transferrin-PEG-liposome retained the capability of specific receptor binding and triggering receptor-mediated endocytosis after extravasation, leading to significant tumor accumulation. Transferrin modified liposomes have also been exploited as drug delivery systems to circumvent MDR. The overexpression of P-glycoprotein has been associated with the development of MDR in cancer cells. To tackle this problem, TfR-mediated endocytosis presents a unique pathway for circumventing the P-gp efflux and facilitating the intracellular delivery of liposomes into MDR cells. When conjugated on the surface of liposomes consisting of egg-PC and cholesterol, transferrin markedly enhanced intracellular delivery and increased cytotoxicity of encapsulated doxorubicin on MDR cells by 3.5-fold higher than free doxorubicin (Kobayashi et al. 2007).

Due to the dynamic flow environment of the bloodstream, binding with high avidity usually appears to be preferable (Danhier et al. 2010). However, it is evident in solid tumors that high binding affinity attenuates the penetration of nanocarriers due to the presence of "binding-site barrier," where the targeting nanocarriers bind to their targets so strong after extravasation and further penetration is restricted (Adams et al. 2001). Relatively low dissociation affinity ligands can rapidly diffuse into tumor by using sequential binding to and dissociation from unoccupied target receptor or antigen. Under the same condition, ligands with high binding avidity will durably occupy the target receptor or antigen pool that can be reached in a single diffusion step. Ligands that subsequently enter the tumor from blood vessel will be unable to reach free receptor or antigen, and the high interstitial pressure will force them back out of the tumor.

Besides tumor cell targeting, tumor endothelium targeting provides an alternative pathway for active tumor-targeting drug delivery. As suggested by Judah Folkman in 1971, tumor growth might be negatively regulated by preventing tumors from recruiting new blood vessels (Folkman 1972). Anti-angiogenesis provides a crucial paradigm in cancer therapy by attacking the oxygen and nutrients supply. In this strategy, nanocarriers can directly target and kill the tumor endothelial cells. Compared to the strategy of tumor cell targeting, the advantages of tumor endothelium targeting lie in that binding is directly carried out after intravenous injection and no extravasation is needed. Additionally, most of endothelial cell markers, such as integrins and vascular endothelial growth factor receptor (VEGFR), are expressed whatever the tumor type, involving a ubiquitous approach and an eventual broad application spectrum. Tumor vasculature is a particularly suitable target for cancer therapy because it consists of nonmalignant endothelial cells that are genetically stable and unlikely to mutate into drug-resistant variants (Arap et al. 1998). Cyclic RGD peptide (cRGD), a specific ligand of integrin $\alpha_v\beta_3$ and $\alpha_v\beta_5$, has been successfully conjugated on the surface of poly(ethylene glycol)-*b*-poly(ε-caprolactone) (PEG-PCL) to enable tumor endothelial cells targeting (Nasongkla et al. 2004). Flow cytometry and confocal laser scanning microscopy have been employed to study the uptake of micelles into SLK tumor endothelial cells after 2 h of incubation. Higher density of cRGD on the micellar surface led to higher level of cellular internalization. A maximum of 30-fold enhancement has been achieved with 76 % cRGD conjugated PEG-PCL in the micelle. In another report, a peptide with the sequence ATWLPPR has demonstrated high binding affinity for the VEGF receptor, which is overexpressed on the surface of tumor endothelial cells. After incubation with HUVEC cells for 1 h at 4 °C, peptide modified PEG-liposome displayed a tenfold higher binding as compared to unmodified PEG-liposome (Janssen et al. 2003).

4.3.2 *Ligands for Active Tumor-Targeting Drug Delivery*

Active tumor targeting drug delivery promises to expand the therapeutic windows of drugs by increasing delivery to target sites as well as the target-nontarget tissue ratio. To date, different ligands, including small molecules, peptides, and antibodies and aptamers, have been identified to inspire tumor targeting drug delivery. These ligands target tumor cells and/or tumor endothelial cells to elevate accumulation in tumor sites or to expedite intracellular delivery. To choose an appropriate ligand, a couple of considerations should be taken into account. First, the targeted antigen or receptor should have a high density on the surface of the target cells compared with the non-target cells. Second, it is crucial that the corresponding receptor can trigger receptor-mediated endocytosis, especially for the delivery of cell membrane impermeable cargos. Third, immunogenicity is a potential limitation, in particular with the use of antibody, for the application of ligand-inspired tumor targeting drug delivery.

4.3.2.1 Small-Molecule-Type Ligand

As active ligands for tumor targeting drug delivery, small molecules are preferable for their favorable pharmacokinetic properties and reduced probability of immunogenicity allowing for repeated administration. In addition, convenient availability and low cost offer the potential of wide application of small-molecule-type ligands. Small-molecule ligands are applicable for fabricating liposome- and micelle-based tumor targeting drug delivery systems with relatively simple conjugating chemistry and tiny influence on self-assembly. Folic acid, a naturally occurring small-molecule-type ligand, can specifically target folate receptors. When conjugated on the surface of micelles (Yoo and Park 2004; Park et al. 2005; Zhao and Yung 2008) or liposomes

(Lee and Low 1995; Leamon et al. 2003; Lee and Huang 1996) via γ-carboxyl, folic acid retains its receptor binding properties and triggers the receptor-mediated endocytosis process (Sudimack and Lee 2000), further expediting the targeting delivery of non-permeable drugs or circumventing the multi-drug resistance (Goren et al. 2000; Kim et al. 2008).

Anisamide is another small-molecule ligand that is widely used for tumor targeting. Sigma receptors are well-known membrane-bound proteins, which show high affinity for neuroleptics (Walker et al. 1990). These receptors are expressed on normal tissues and the physiological role is largely unknown so far (Banerjee et al. 2004). Various human tumors, including melanoma, non-small cell lung carcinoma, breast cancer of neural origin, and prostate cancer, overexpress sigma receptors, suggestive of the prospect of using receptor-binding ligand for tumor targeting drug delivery (Vilner et al. 1995). By modifying anisamide on the surface, Leaf Huang's group has synthesized various liposome-based nanocarriers for tumor-targeting delivery of siRNA (Li et al. 2008, 2012) and doxorubicin (Banerjee et al. 2004).

Numerous successful applications of small-molecule-type ligand have been presented for inspiring tumor targeting drug delivery; however, broadening usage has been restricted by the identification of small-molecule-type ligand with high specificity and affinity. In addition, covalently conjugating with small-molecule ligand is risky to attenuate the binding avidity with corresponding receptor owing to small binding interface between ligand and receptor.

4.3.2.2 Peptide-Type Ligand

Peptide-type ligands have attracted much attention for the development of tumor targeting drug delivery owing to the availability of facile design techniques. Structure-guided peptide design is applicable to design specific ligand for the corresponding receptor by mimicking the protein–protein interaction. Urokinase plasminogen activator receptor (uPAR) is overexpressed on a variety of cancer cells, such as those of the pancreas (Cantero et al. 1997), the prostate (Gavrilov et al. 2001) and the breast (Costantini et al. 1996). Urokinase plasminogen activator (uPA) (Siehler et al. 1999) together with uPAR mediates various biological activities at cell surface, including plasminogen activation, extracellular matrix (ECM) remodeling, growth factor activation, and initiation of intracellular signaling. Crystallographic studies of the uPAR–uPA conjugate have revealed that the binding region of uPA for uPAR is localized at the tip of a β-hairpin loop within the growth factor domain, representing amino residues 12–32 (Huai et al. 2006). Within this tip are two looped structures, and the second loop, U11, appears to be the primary uPAR binding motif with an equilibrium dissociation constant, K_d, of 1.3–1.4 μM (Wang et al. 2009). Considering the binding affinity and the receptor-mediated endocytosis properties of the uPAR–uPA complex, U11 possesses the potential for targeting nanocarriers to various cancers.

Alternatively, selection of phage display libraries was widely used to isolate peptides that home specifically tumor cells or tumor blood vessels. Endothelial cells within solid tumor express several receptors that are absent or barely detectable in established vessels. Ruoslahti's group has applied in vivo selection of phage peptide libraries to identify tumor targeting ligands by injecting phage peptide libraries into the circulation of nude mice bearing human breast carcinoma xenograft. Recovery of phage from tumors led to the identification of RGD peptide, NGR peptide, and GSL peptide (Arap et al. 1998). The RGD peptide, which binds selectively to $\alpha_v\beta_3$ and $\alpha_v\beta_5$ integrins, has been widely used for the fabrication of tumor targeting drug delivery systems (Zhan et al. 2010a; Xiong et al. 2005; Dubey et al. 2004). NGR peptide has been identified as a cell adhesion motif (Koivunen et al. 1994). As the NGR sequence appears in the cell-binding region of the fibronectin molecule, this sequence could contribute to the specific recognition of fibronectin $\alpha_v\beta_1$(Koivunen et al. 1994). Recently, another intriguing peptide, termed tumor penetrating peptide (iRGD), has been identified by in vivo phage display with multistep binding and penetration mechanism (Sugahara et al. 2009). The iRGD peptide accumulates at the surface of α_v integrin-expressing endothelial and other cells in tumor and RGD motif mediates integrin binding. After that, the peptide is cleaved by cell surface-associated protease(s) to expose the cryptic CendR element at the C terminus to mediate neuropilin-1 binding, resulting in penetration of cells and tissues. Either mixing or conjugating with the nanocarriers, iRGD can inspire tumor targeting drug delivery with enhanced penetration efficiency (Sugahara et al. 2010; Zhu et al. 2011b; Feron 2010).

Peptide-type ligands hold extensive potential for tumor targeting drug delivery; however, the enzymatic environment in blood plasma tremendously threatens the stability of these short peptides, thus attenuating their targeting efficiency. Peptidomimetics, such as D-peptides that are composed of D-amino acids and display resistance to proteolytic degradation, provide a unique method to identify tumor targeting ligands with high stability and efficiency. Li et al. (2013) described the design of retro-inverso peptide of SP5 peptide, which can recognize tumor neovasculature but not normal blood vessels, leading to a stable peptide-type ligand for tumor endothelium targeting.

4.3.2.3 Antibody-Type Ligand

Monoclonal antibody inspired tumor targeting was described by Milstein in 1981 (Warenius et al. 1981), and the feasibility of antibody-based tumor targeting has been clinically demonstrated over the past decades (Peer et al. 2007). The first angiogenesis inhibitor for treating colorectal cancer, bevacizumab (Avastin), an anti-VEGF mAb that inhibits the factor responsible for the growth of new blood vessels, was approved in 2004. To date, hundreds of delivery systems based on antibodies or their fragments are in preclinical and clinical trials. Antibodies can be used in their native state or as fragments for tumor targeting. However, the whole antibody is preferable owing to the presence of two binding sites within a single molecule to give rise to higher binding avidity and the ability to maintain stability during long-term storage. Antibody fragments, such as antigen-binding fragments, dimmers of antigen-binding fragments, single-chain fragment variables, and other engineered fragments, are considered safer for systemic administration owing to reduced non-specific binding. Phage display libraries that involve a high throughput approach can be used to rapidly identify antibodies or their fragments with high binding affinity to antigens. To construct tumor targeting drug delivery systems, antibody-modified liposomes (immunoliposome) and micelles (immunomicelle) have demonstrated extensive potential. A sterically stabilized immunoliposome encapsulating paclitaxel has been developed to promote the efficiency of intracellular delivery through receptor-mediated endocytosis (Yang et al. 2007). Thiolated Herceptin, which recognizes human epidermal growth factor receptor-2, was successfully conjugated onto the reactive maleimide of PEG-DSPE as well as being incorporated in the liposome bilayers. Confocal microscopy studies displayed substantially higher cellular uptake than the PEGylated liposome in two breast cancer cell lines overexpressing human epidermal growth factor receptor 2.

4.3.2.4 Aptamer-Type Ligand

Aptamers are oligonucleic acids such as DNA or RNA that bear unique 3-dimensional conformations capable of binding target non-nucleic acid target molecules, such as peptides, proteins, drugs, or even whole cells, with high avidity and specificity (Cho et al. 2008). They are isolated and chemically synthesized from combinatorial oligonucleotide libraries by a process known as in vitro systematic evolution of ligands by exponential enrichment (SELEX) (Daniels et al. 2003). Based on their specific binding pockets for the target molecules, aptamers possess high potential for tumor targeting drug delivery. AS1411, a 26-mer DNA aptamer, was identified to have high binding affinity to nucleolin (Soundararajan et al. 2008). Cao et al. (Cao et al. 2009) prepared cholesterol modified AS1411 with a 12-T-base spacer and inserted this cholesterol tagged AS1411 into lipid bilayer. The extra 12-T bases were presumable to extend the binding domain away from liposome surface and ensured its bioactivity to nucleolin. Nucleolin overexpression on the plasma membrane has been observed in various human diseases such as breast cancers, providing a potential target for cancer cell targeting drug delivery. When encapsulating cisplatin, this aptamer functionalized liposome could effectively facilitate cancer cell specific drug delivery and significantly enhance antitumor effects of the payload. A DNA aptamer that is specific to Ramos cells (a B cell lymphoma cell line), termed TDO5, has been linked onto the surface of PEG-lipid micelles. To prove the potential application of the aptamer-functionalized micelle in biological living systems, the authors mimicked a tumor site in bloodstream by immobilizing tumor cells onto the surface of a flow channel device (Wu et al. 2010). When flushed through the channel, micelles demonstrated selective recognition under flow circulation in human whole-blood sample.

4.4 Microenvironment-Responsive Tumor Targeting Drug Delivery

Tumor consists of tumor cells and non-malignant host stromal cells, including endothelial cells, peri-vascular cells, fibroblasts, myofibroblasts, macrophages, lymphocytes, dendritic cells, and mast cells embedded within a protein-rich ECM and interstitial fluid. The overexpression of various genes is affected by the interactions among those cells, surrounding matrix, and their local microenvironment. The products encoded by these genes, in turn, give rise to abnormal organization, structure, and function of tumor vasculature. These abnormalities are also responsible for the high vascular permeability and special microenvironment characterized by hypoxia and acidosis (Fukumura and Jain 2007). Based on the numerous differences compared with normal tissue including vascular abnormalities, oxygenation, perfusion, pH, and metabolic states, researchers have elaborated a variety of therapeutic strategies for tumor targeting drug delivery.

4.4.1 pH-Sensitive Tumor Targeting Drug Delivery

The extracellular pH (pH_e) of normal tissues and blood pH are kept constant at 7.4, which is slightly higher than their intracellular pH (pH_i at 7.2). However, the pH gradient in most tumors determined by invasive and noninvasive methods is reversed. The measured pH_e of human and animal solid tumors is usually below pH 7.2. Tumor vasculature is often inadequate to supply the nutritional needs of the expanding population of tumor cells, leading to the deficiency of oxygen and many other nutrients. The production of lactic acid under anaerobic conditions and the hydrolysis of ATP in an energy-deficient environment contribute to the acidic microenvironment which has been found in many types of tumor (Tannock and Rotin 1989). The resulting pH gradient between intracellular and extracellular tumor cells but also between tumor mass and surrounding tissues presents a potential source for tumor targeting.

pH-sensitive copolymers have demonstrated to be promising for developing tumor targeting drug delivery systems. Poly(β-amino ester) contains tertiary amine and its pK_b value is about 6.5 (Lynn et al. 2001; Potineni et al. 2003). It is soluble below pH 6.5 but insoluble above pH 6.5 in aqueous media. By attaching with a hydrophilic moiety, PEG, the copolymer PEG-poly(β-amino ester) forms self-assembled polymeric micelles with a sharp micellization/demicellization transition at the tumoral acidic pH value (Kim et al. 2006). Using a simple solvent casting method, PEG-poly(β-amino ester) could easily form micelles and efficiently encapsulate fluorescent dyes and anticancer drugs for cancer imaging and therapy (Min et al. 2010). The encapsulated dyes and drugs displayed rapid release in weakly acidic aqueous solution owing to the sharp transition of poly(β-amino ester).

In MDA-MB231 human breast tumor-bearing mice, PEG-poly(β-amino ester) delivered significantly higher dose of dye into the solid tumor when compared with PEG-PLA micelle. The pH-targeting characteristic of this micelle also contributed to the enhancement of antitumor effect of camptothecin (CPT). Compared to free CPT and CPT encapsulated PEG-PLA micelles, PEG-poly(β-amino ester) micelles encapsulating camptothecin significantly increased therapeutic efficacy with minimum side effects in breast tumor-bearing mice.

Poly(L-histidine) is another promising biomaterial for the construction of pH-sensitive polymeric micelles. The imidazole ring (pK_b ~6.5) of poly(L-histidine) has lone pairs of electrons on the unstructured nitrogen that confer pH-dependent amphoteric properties (Lee et al. 2007a). Polymeric micelles with particle diameter of 80 nm have been constructed as a tumor pH-specific anticancer drug carrier. The solution of PLA-PEG-poly(L-histidine) triblock copolymer in DMSO was dialyzed against pH 8.0 aqueous solution, assuming that poly(L-histidine) block was embedded into the hydrophobic core to form a flower-like morphology. Due to the ionization of poly(L-histidine) at a slightly acidic pH, the deformation of the micellar core gave rise to the pH-sensitivity of the micelles. However, the co-presence of pH-insensitive lipophilic PLA block in the core prevented disintegration of the micelles and caused swelling/aggregation in response to pH change. The fluorescence probe study and micellar size change confirmed that the micelles were not dissociated but rather swollen/aggregated. The polarity of pyrene retained in the micelles increased as pH decreased from 7.4 to 6.6, indicating a change to more hydrophilic environment in the micelles. The size increased up to 580 nm at pH 6.6 from 80 nm at pH 7.4, and the transmittance of micellar solution increased with decreasing pH. Interestingly, the subsequent decline of pyrene polarity below pH 6.6 suggested re-self-assembly of the block copolymers, most likely forming a PLA block core while poly(L-histidine) block relocating back to the surface. Consequently, pH-dependent physical changes of PLA-*b*-PEG-*b*-poly(L-histidine) micelles provide a mechanism for triggering drug release from the micelles by small change in pH (pH 7.2–6.5).

The design of pH-sensitive peptide broadens the application of pH-triggered tumor targeting drug delivery. Low pH inserting peptide (pHLIP) is a 36-residue polypeptide containing the sequence of the C-helix of the integral membrane protein bacteriorhodopsin. It exhibits significant solubility in aqueous buffers free of both detergents and denaturants and contains two aspartic acid residues in the membrane-spanning region. At neutral pH, the peptide associates with lipid bilayers in a non-helical and presumably peripheral confirmation. With acidic aqueous buffer (pH is lower than 7.0), pHLIP reversibly inserts into the bilayer as a transbilayer α-helix (Hunt et al. 1997). At physiological pH, the soluble form of pHLIP is favored, whereas at acidic pH the transmembrane α-helix predominates. The pHLIP affinity for membrane at pH 5.0 is 20-fold higher than that at pH 8.0 (Andreev et al. 2007). As a pH sensitive peptide, pHLIP has been successfully applied for intracellular drug delivery of cell impermeable molecules through a disulfide bond at its C-termini (Thevenin et al. 2009). Besides, it can also be formulated with imaging probes (Vavere et al. 2009; Shan 2004; Macholl et al. 2012; Daumar et al. 2012) and nanocarriers (Zhao et al. 2013; Yao et al. 2013a, b) for tumor diagnosis and therapy. When presented on the surface of PEGylated liposomes, pHLIP tremendously enhanced membrane fusion and lipid exchange in a pH-dependent fashion, leading to increase of cellular uptake and payload release and inhibition of cell proliferation by liposomes containing ceramide (Yao et al. 2013a).

Intriguingly, some shielding drug delivery systems have also been presented by employing the pH-triggered cleavage. Sawant et al. (2006) designed a smart liposomal drug delivery system, in which long chain PEG grafted lipid with a hydrazone linker has been fabricated into a TAT modified liposome. Considering that the hydrazone is stable at pH 7.4, long PEG chain is beneficial to prolonged circulation. When reached tumor site, hydrazone was cleaved at relatively low pH and the shielded TAT was activated and exposed to tumor cells to facilitate intracellular delivery. Positively charged nanocarriers are believed to be beneficial to intracellular drug delivery. To improve its selectivity to tumor cells, pH-sensitive polymer-doxorubicin micelle has been developed and the pH-triggering tumor-targeting delivery has been studied. Diblock copolymer monomethoxyl poly(ethylene glycol)-*b*-poly-(allyl ethylene phosphate) prepared by open-ring polymerization was grafted with cysteamine to reverse the negative charge in the backbone into positively charged amino group, which is partially converted to sulfhydryl groups by a reaction with 2-iminothiolane. After the conjugation of doxorubicin through an acid-labile hydrazone bond and the modification of free amino groups with 2,3-dimethylmaleic anhydride (DMMA) to obtain the pH-dependent charge conversion, this polymer-drug conjugate was self-assembled into micelle-like nanocarrier (Du et al. 2011), which was capable of reversing its surface charge from negative to positive at tumor extracellular pH to facilitate cell internalization. Subsequently, the further increased acidity in endosome and lysosome promoted doxorubicin release from the endocytosed drug carriers.

The strategy of pH-sensitive tumor targeting provides a unique method for tumor-homing drug delivery. However, the heterogeneity of acidity in tumor microenvironment restricts the targeting efficacy. The responsive pH for tumor targeting in most cases is at or below pH 6.5, which is usually lower than that of tumor extracellular pH (6.8–7.2). To employ the pH-triggering tumor targeting, novel polymeric materials that can precisely recognize the narrow pH change are desirable.

4.4.2 Proteases-Sensitive Tumor Targeting Drug Delivery

The progression of malignant cancer depends on the successful acquisition of numerous important capabilities, including expanding growth of tumor cells, evasion of programmed cell death, development of sustained angiogenic response, as well as infiltration and invasion into the connective tissues (Stetler-Stevenson and Yu 2001). A number of studies have demonstrated that successful invasive tumor or endothelial cells must orchestrate a close cooperation among cellular adhesion molecules, cytoskeletal elements, ECM degrading proteases, and regulatory molecules. For example, matrix metalloproteinases (MMPs) are a family of zinc-containing endopeptidases and capable of cleaving most ECM proteins. Considerable body of evidence illustrates increased MMPs expression in cancer tissues relative to corresponding and adjacent normal tissues. The upregulation of MMPs in tumor microenvironment provides a unique opportunity for tumor-targeting drug delivery.

The substrate peptides of MMPs have attracted tremendous attention to trigger the MMPs responsiveness. In a recent study, such peptide substrate was conjugated as spacer between PEG and lipid, and the MMPs-cleavable conjugate was fabricated into galactosylated liposomes for hepatocellular carcinoma targeting drug delivery (Terada et al. 2006). Since PEG chain shielded the galactose moiety on the liposomal surface, the uptake by normal hepatocytes could be inhibited by steric hindrance. Furthermore, PEG chain also reduced the liposome uptake by reticulo-endothelial system (RES) and extended the blood circulation of liposome. In tumor site, PEG chain was removed from the liposomal surface by MMPs cleavage and the galactose was exposed to the cell surface, inspiring the receptor mediated endocytosis and accomplishing tumor targeting drug delivery. Another similar research has constructed MMPs responsive micelle for the tumor targeting delivery of doxorubicin (Lee et al. 2007b). PEGylated peptide-doxorubicin conjugate, which was subject to MMPs cleavage, was chemically synthesized. In vivo antitumor studies indicated that such micelles significantly inhibited tumor growth but with relatively low toxicity to normal tissues.

Secretory phospholipase A_2 ($sPLA_2$), which is overexpressed in inflammatory and tumor tissues, has been extensively utilized to realize tumor targeting drug delivery. PLA_2 is an enzyme that catalyzes the hydrolysis of the ester *sn*-2-acyl chain of phospholipids, and the products of the enzymatic cleavage are free fatty acids and 1-acyl-lysophospholipids (Sofou 2007). Polymer covered liposomes consisting of diacyl phospholipids can be converted into lysolipids by site-specific hydrolysis in the tumor tissue as a consequence of the elevated levels of $sPLA_2$. The disruption of lipid bilayer triggered rapid site-specific release of encapsulated chemotherapeutics (Jorgensen et al. 1999; Andresen et al. 2005). Alternatively, liposome fabricated with lipid prodrugs of the anticancer ether lipids, which are stable in bloodstream and display less cytotoxicity than free drug, could accumulate in tumor tissues owing to EPR effects. Overexpressed $sPLA_2$ catalyzed the prodrug and induced drug release in the target sites.

4.5 Triggerable Tumor Targeting Drug Delivery

With better understanding of various physical phenomena and the development of novel technology, numerous external stimuli triggering strategies, including light, heat, ultrasound, and magnet, have been employed in the design of tumor targeting drug delivery systems.

4.5.1 Light-Triggered Tumor Targeting Drug Delivery

The energetic confinement available from modern laser system has expedited the utilization in a number of biomedical applications, including tumor diagnosis and therapy. An underdeveloped opportunity, however, is the use of laser excitation methods for activating physical, chemical, or biochemical processes such as photoisomerization, photopolymerization, and photocleavage.

4.5.1.1 Light-Triggered Liposome for Tumor Targeting Drug Delivery

PEGylated liposomes are long-circulating nanocarriers due to reduced rates of opsonization and uptake by the RES, leading to their preferential accumulation in tumors through EPR effect. A persistent problem in the application of PEGylated liposomes, however, is the slow release after their deposition at the target site (Shum et al. 2001). The rationale behind the design of light-triggered liposome lies in the sharp transition of the liposome stability, leading to rapid release at target site. The use of light to stimulate the release of payloads from liposome is attractive by possibly controlling the spatial and temporal delivery of the radiation. Bis-SorbPC is a photosensitive lipid, in which polymerization can be triggered by UV. When PEGylated liposome is composed in part of bis-SorbPC, it can effectively encapsulate water soluble compounds and release them upon exposure to UV light in the presence of oxygen. The cooperation of bis-SorbPC did not change the long-circulating profile of PEGylated liposome, while the observed increase in liposome membrane permeability was about 200-fold at high photoconversion of the monomeric bis-SorbPC. The photoinitiated destabilization increased liposome permeability, making it possible to release entrapped agents at target sites in a manner of tens of minutes to hours, rather than over the course of weeks (Bondurant et al. 2001).

Tissue penetration and phototoxicity, however, limited the extensive application of UV for in vivo triggering. Alternatively, longer-wavelength light is preferable for the further utilization of photo-triggered tumor targeting drug delivery. Semi-synthetic plasmenylcholine liposome encapsulating calcein and a membrane-bound sensitizer has been prepared by extrusion. Considering the known effect of plasmalogen photooxidation on membrane permeability, irradiation of air-saturated liposome solution enhanced membrane permeability toward calcein and Mn^{2+} and promoted membrane fusion process compared with non-irradiated or anaerobic controls. 100 % of encapsulated calcein was released in less than 20 min by shining 300 mW laser with 800 nm wavelength, two orders of magnitude of observed release rate higher than that of liposome consisting of egg lecithin (Thompson et al. 1996).

4.5.1.2 Light-Triggered Micelle for Tumor Targeting Drug Delivery

Micellization of copolymers depends on the hydrophilic/hydrophobic balance; however, the photoinitiated switch from hydrophobicity to hydrophilicity provides a rational method to disintegrate the micellar system and release cargo (Fomina et al. 2012).

By attaching a hydrophobic light-sensitive 2-diazonaphthoquinone to a PEG chain, an amphiphilic molecule was constructed to form micelle with low CMC and capability of hydrophobic drug encapsulation (Goodwin et al. 2005). Upon the irradiation by UV or near infrared (Vijayalakshmi et al. 2003) light (via absorption of one or two photons), 2-diazonaphthoquinone underwent a Wolff rearrangement to form a hydrophilic carboxylic acid, leading to dissolving of micelles incorporated 2-diazonaphthoquinone and quick release of encapsulated hydrophobic drugs. Similar results have been collected from PEG-poly(2-nitrobenzyl methacrylate) micelle. 2-Nitrobenzyl moieties were detached from the amphiphilic chain and photolysis transformed the hydrophobic block into hydrophilic poly(methacrylic acid), triggering the quick release of payloads (Jiang et al. 2006). Alternatively, photocleavage can also be used to control the bioactivity of targeting ligand. Dvir et al. (2010) reported a novel and simple proof-of-concept of nanocarriers for selective targeting upon irradiation. YIGSR peptide is an amino acid sequence in laminin that is crucial for adhesion to integrin β1 on the cell membrane. The biological activity of YIGSR can be greatly attenuated by mutation or deletion of tyrosine. To design a photo-triggered peptide, the side chain of tyrosine in YIGSR peptide was caged with 4,5-dimethoxy-2-nitrobenzyl (DMNB), which could be rapidly removed by shining UV light. When modified on the surface of nanocarriers, caged YIGSR peptide displayed the capability of binding cells selectively upon illumination.

Photocleavage provides applicable method to control micellization, leading to quick drug release in the target sites. Photoisomerization, however, broadens the application of photo-triggered drug delivery by reversibly adjusting the target structures. An intriguing design by Tong et al. (2012) demonstrated the synthesis of photoswitchable micelles for triggered tissue penetration and tumor targeting drug delivery. The micelles composed of spiropyran and lipids were formulated in aqueous solution. Nitrobenzopyran and indoline moiety with orthogonal orientation in spiropyran absorb in the UV spectrum independently, inducing ring-opening in the pyran to form merocyanine. While in the dark, the zwitterionic merocyanine form was less stable than the hydrophobic spiropyran form and underwent spontaneous ring-closing back to spiropyran. Upon photo-triggering with UV light, the volume of micelles comprising spiropyran, which underwent reversible photoisomerization, changed from 150 to 40 nm. PEGylated lipid enabled repetitive dosing from a single administration and the shrinkage of micelle size enhanced penetration in tumor tissues.

4.5.2 Thermo-Triggered Tumor Targeting Drug Delivery

4.5.2.1 Thermo-Responsive Liposome for Tumor Targeting Drug Delivery

Hyperthermia increases the efficiency of various chemotherapeutic agents and is administrated as an adjunct to chemotherapy for the treatment of cancer patients. Thermo-responsive liposome, which can be destabilized by heating, is promising to facilitate rapid drug release in target sites, such as tumors (Weinstein et al. 1979; Tacker and Anderson 1982). In combination with 1.2-dipalmitoyl-*sn*-glycero-3-phosphocholine and 1.2-distearoyl-*sn*-glycero-3-phosphocholine, the novel lipid 1,2-dipalmitoyl-*sn*-glycero-3-phosphoglyceroglycerol (DPPGOG), which is closely related to the naturally occurring 1,2-dipalmitoyl-*sn*-glycero-3-phosphoglycerol, provides long-circulating temperature-sensitive liposomes with favorable properties under mildly hyperthermic conditions (41–42 °C). DPPGOG facilitated temperature-triggered drug release from these liposomes (diameter, 175 nm) and led to a substantially prolonged plasma half-life for the encapsulated drug with $t_{1/2}$=9.6 h in hamsters and $t_{1/2}$=5.0 h in rats. Quantitative fluorescence microscopy of amelanotic melanoma grown in the transparent dorsal skin fold chamber of hamsters demonstrated favorable drug accumulation in heated tissue after i.v. application of these liposomes (42 °C for 1 h). The mean area under the curve for tissue drug concentration increased by more than sixfold with the application of the novel liposome compared with non-liposomal drug delivery (Lindner et al. 2004). A novel approach has been described by Chen and Regen (2005). Liposomes containing 1,2-dipalmitoyl-sn-glycero-2-phosphocholine (DPPC) have been fabricated with a pore-forming conjugates to amplify the release of encapsulated drugs when applying hyperthermia. Alternatively, Brij surfactant could also be incorporated into the DPPC liposome by the thin film hydration or the post-insertion method with an optimal range of 1–8 mol%. The optimal Brij surfactant, Brij78, has been verified to improve thermosensitive drug delivery compared with the lyso-lipid temperature sensitive liposomes consisting of DPPC, MSPC, and PEG-DSPE in a cell culture system (Tagami et al. 2011).

4.5.2.2 Thermo-Responsive Micelle for Tumor Targeting Drug Delivery

Developing thermo-responsive polymeric micelles as tumor targeting nanocarriers has recently attracted extensive attention. Such polymeric micelles react with a sharp change of properties in response to a small exterior change of temperature (Rapoport 2007). The thermo-responsive fragment, which has a low critical solution temperature (LCST), can be incorporated to either micelle core or shell. Block copolymers of PEG as a hydrophilic block and poly(*N*-isopropylacrylamide) (PNIPAAm) or poly(NIPAAm-*co*-*N*-(2-hydroxypropyl)methacrylamide-dilactate) as a thermosensitive block were able to self-assemble into micelles, whose size in aqueous solutions can be regulated by heating rate of the polymer solution. Particularly, stable micelles with a size around 50–70 nm were formed when the cloud point of the block polymer was extremely rapid passed, making them suitable carriers for tumor targeting drug delivery (Neradovic et al. 2004).

Besides the thermo-responsive block of PNIPAAm and pluronic polymers, elastin-like polypeptide (ELP) displays suitable LCST for biomedical applications. ELPs are biopolymers composed of a Val-Pro-Gly-Xaa-Gly pentapeptide repeat (where the “guest residue” Xaa is any amino acid except Pro) derived from a structural motif found in mammalian elastin (Meyer et al. 2001). ELPs undergo an inverse temperature phase transition in response to an increase in temperature. They are soluble in aqueous solutions at temperature below their transition temperature but aggregate at the temperature above their transition temperature (Dreher et al. 2008). ELPs that contain hydrophobic amino acids in the fourth amino acid position, such as tyrosine, display a conformational transition from random coil to repetitive type II β turns at temperature well below 37 °C. The blocks contain a charged glutamic acid; however, persist as random coil in physiological temperature range. Two amphiphilic diblock polypeptides with the sequence of VPGEG[(VPGVG)(VPGEG)(VPGVG)(VPGEG)(VPGVG)]$_x$-[C_4G_3]-[(IPGVG)$_2$VPGYG(IPGVG)$_2$]$_y$VPGYG were synthesized and self-assembled into micellar structures with multiple cysteine residues incorporated at the core-shell surface with the CMC ranging from 2 to 4 μM (Kim et al. 2010b). At low temperature (5 °C), the intensity of scattered light was very low, indicating the presence of only free chains in solution. Above 25 °C, well-defined micellar structure was clearly observed as evidenced by a significant increase in scattered light intensity and a dramatic decrease in the polydispersity index. Additionally, the formation of disulfide bond has demonstrated pivotal role in stabilizing the micelles, suggestive of the responsiveness to reductive microenvironment.

4.5.3 Ultrasound-Triggered Tumor Targeting Drug Delivery

Liposomes containing small amount of air have potential to carry pharmaceutical agents and their acoustic activity could enable them to respond to ultrasound stimulation by releasing cargos. Liposome composed of phosphatidylcholine, phosphatidylethanolamine, phosphatidylglycerol, and cholesterol was made acoustically active by hydrating a lipid film, sonication, freezing in the presence of mannitol, lyophilization, and rehydration. The procedure for generating acoustically active liposome was compatible with an encapsulation efficiency of 15 % or more. The presence of mannitol during freeze-drying was essential not only for generation of acoustic activity but also for efficient encapsulation. Ultrasound-triggered release was achieved by applying 1 MHz ultrasound at 2 W/cm^2 for 10 s. The inclusion of 4 % diheptanolyphosphatidylcholine (DHPC) elevated the sensitivity of liposome to ultrasound stimulation and led to very efficient stimulated release of contents (1/3 released in 10 s, 2/3 released in six such applications). Release of contents was highly correlated with the loss of air induced either by ultrasound or rapid pressure reduction. These encapsulation and triggered release techniques are highly efficient, and hence may be applicable to tumor targeting drug delivery (Huang and MacDonald 2004). The strategy of ultrasound triggering can also be applied in micelle-based drug delivery system. At room temperature, the formulations consisted of the mixtures of polymeric micelles and nanoemulsion droplets formed by perfluoropentane (PFP), which were stabilized by the same biodegradable block copolymer that formed the micelles. At physiological temperatures, the nanodroplets converted into stable nano/microbubbles that could survive for many hours. The encapsulated doxorubicin was confined to the microbubble walls formed by the bubble-stabilizing block copolymer. When the microbubbles are sonicated in the presence of cells, the drug is effectively transferred from the microbubble surface to the interior of the cells owing to the microbubble collapse in the process of inertial cavitation. The ultrasound triggering release also induced a dramatic enhancement of intracellular drug uptake by the tumor cells and effective tumor regression in vivo (Rapoport et al. 2007). In addition, continuous wave ultrasound exerts heating effects, which is useful in combination with the thermo-responsive liposomes and micelles.

4.5.4 Magnet-Triggered Tumor Targeting Drug Delivery

To control the growth of primary tumors effectively with systemic chemotherapy, magnet triggered tumor targeting drug delivery provides a potential modality to elevate drug accumulation in tumors. Magnetic liposomes with diameter of 146 nm have been prepared by the reverse-phase evaporation method (Kubo et al. 2001). Magnet-triggered tumor targeting efficiency was studied with Syrian male hamsters inoculated with osteosarcoma, Os515, in the right hind limb at 7 days after inoculation. One day prior to the animal study, either a permanent magnet (with magnetic force) or non-magnetic alloy (without magnetic force) was implanted in the center of the tumors. Treatment with magnetic adriamycin liposomes under magnetic force showed significantly greater antitumor activity than intravenous administration of adriamycin solution or that of magnetic adriamycin liposomes without magnetic force. Adriamycin administered as magnetic liposomes eliminated weight loss of hamsters, one of the side effects produced by adriamycin. Histological examination notably demonstrated the selective accumulation of magnetite particles in tumor blood vessels, suggesting that this systemic chemotherapy could effectively control the primary tumor owing to the high targeting efficiency guided by magnetic triggering.

The magnetic gradient decreases with the distance to the target, however, the main limitation of magnet triggered tumor targeting drug delivery relates to the strength of the external field that can be applied in vivo. In addition, it must be noted that the magnetic nanocarriers accumulate not only at the desired site but also throughout the cross-section from the external source to the depth making the effective field limit (Arruebo et al. 2007). Finally, state-of-the-art magnetic drug delivery seems mainly applicable to well-defined tumors, thereby it remains a challenge in the treatment of metastatic neoplasms and small tumors in the early stage of their growth.

4.6 Conclusion

More effective delivery of active targeting nanocarriers has resulted in the development of novel methods to treat cancer. The picture of the ideal tumor targeting drug delivery system can be pointed back to the original vision of Paul Ehrlich's "magic bullet" more than 100 years ago, where an anticancer agent introduced into the body is capable of being concentrated in diseased tissues while healthy organs untouched. Many different strategies have been employed for designing tumor targeting liposomes and micelles over the years. However, cancer is an extremely heterogeneous disease and the treatment may involve a multifaceted approach. The development of cancer biology and emerging of new technologies will catalyze the identification of multi-functional nanocarriers with greater customization to achieve more efficacious anticancer response.

References

Adams GP et al (2001) High affinity restricts the localization and tumor penetration of single-chain fv antibody molecules. Cancer Res 61(12):4750–4755

Andreev OA et al (2007) Mechanism and uses of a membrane peptide that targets tumors and other acidic tissues in vivo. Proc Natl Acad Sci U S A 104(19):7893–7898

Andresen TL, Jensen SS, Jorgensen K (2005) Advanced strategies in liposomal cancer therapy: problems and prospects of active and tumor specific drug release. Prog Lipid Res 44(1):68–97

Arap W, Pasqualini R, Ruoslahti E (1998) Cancer treatment by targeted drug delivery to tumor vasculature in a mouse model. Science 279(5349):377–380

Arruebo M, Fernandez-Pacheco R, Ibarra MR, Santamaria J (2007) Magnetic nanoparticles for drug delivery. Nano Today 2(3):22–32

Banerjee R, Tyagi P, Li S, Huang L (2004) Anisamide-targeted stealth liposomes: a potent carrier for targeting doxorubicin to human prostate cancer cells. Int J Cancer 112(4):693–700

Bangham AD, Horne RW (1964) Negative staining of phospholipids and their structural modification by surface-active agents as observed in the electron microscope. J Mol Biol 8:660–668

Bangham AD, Hill MW, Miller NGA (1974) Preparation and use of liposomes as models of biological membranes. Plenum

Bondurant B, Mueller A, O'Brien DF (2001) Photoinitiated destabilization of sterically stabilized liposomes. Biochim Biophys Acta 1511(1):113–122

Boulikas T (2009) Clinical overview on Lipoplatin: a successful liposomal formulation of cisplatin. Expert Opin Investig Drugs 18(8):1197–1218

Brandl M, Bachmann D, Drechsler M, Bauer KH (1990) Liposome preparation by a new high pressure homogenizer Gaulin Micron Lab 40. Drug Dev Ind Pharm 16(14):2167–2197

Brunner J, Skrabal P, Hauser H (1976) Single bilayer vesicles prepared without sonication. Physico-chemical properties. Biochim Biophys Acta 455(2):322–331

Cantero D et al (1997) Enhanced expression of urokinase plasminogen activator and its receptor in pancreatic carcinoma. Br J Cancer 75(3):388–395

Cao ZH et al (2009) Reversible cell-specific drug delivery with aptamer-functionalized liposomes. Angew Chem Int Ed Engl 48(35):6494–6498

Carmeliet P, Jain RK (2000) Angiogenesis in cancer and other diseases. Nature 407(6801):249–257

Chen WH, Regen SL (2005) Thermally gated liposomes. J Am Chem Soc 127(18):6538–6539

Chen C et al (2013) Structural basis for molecular recognition of folic acid by folate receptors. Nature 500(7463):486–489

Cho KJ, Wang X, Nie SM, Chen Z, Shin DM (2008) Therapeutic nanoparticles for drug delivery in cancer. Clin Cancer Res 14(5):1310–1316

Costantini V et al (1996) Combined overexpression of urokinase, urokinase receptor, and plasminogen activator inhibitor-1 is associated with breast cancer progression – an immunohistochemical comparison of normal, benign, and malignant breast tissues. Cancer 77(6):1079–1088

Danhier F, Feron O, Preat V (2010) To exploit the tumor microenvironment: passive and active tumor targeting of nanocarriers for anti-cancer drug delivery. J Control Release 148(2):135–146

Daniels DA, Chen H, Hicke BJ, Swiderek KM, Gold L (2003) A tenascin-C aptamer identified by tumor cell SELEX: systematic evolution of ligands by exponential enrichment. Proc Natl Acad Sci U S A 100(26):15416–15421

Daumar P et al (2012) Efficient (18)F-labeling of large 37-amino-acid pHLIP peptide analogues and their biological evaluation. Bioconjug Chem 23(8):1557–1566

Dreher MR et al (2008) Temperature triggered self-assembly of polypeptides into multivalent spherical micelles. J Am Chem Soc 130(2):687–694

Du JZ, Du XJ, Mao CQ, Wang J (2011) Tailor-made dual pH-sensitive polymer-doxorubicin nanoparticles for efficient anticancer drug delivery. J Am Chem Soc 133(44):17560–17563

Dubey PK, Mishra V, Jain S, Mahor S, Vyas SP (2004) Liposomes modified with cyclic RGD peptide for tumor targeting. J Drug Target 12(5):257–264

Dvir T, Banghart MR, Timko BP, Langer R, Kohane DS (2010) Photo-targeted nanoparticles. Nano Lett 10(1):250–254

Exner AA, Krupka TM, Scherrer K, Teets JM (2005) Enhancement of carboplatin toxicity by Pluronic block copolymers. J Control Release 106(1–2):188–197

Feron O (2010) Tumor-penetrating peptides: a shift from magic bullets to magic guns. Sci Transl Med 2(34)

Folkman J (1972) Anti-angiogenesis: new concept for therapy of solid tumors. Ann Surg 175(3):409–416

Fomina N, Sankaranarayanan J, Almutairi A (2012) Photochemical mechanisms of light-triggered release from nanocarriers. Adv Drug Deliv Rev 64(11):1005–1020

Fukumura D, Jain RK (2007) Tumor microenvironment abnormalities: causes, consequences, and strategies to normalize. J Cell Biochem 101(4):937–949

Gavrilov D, Kenzior O, Evans M, Calaluce R, Folk WR (2001) Expression of urokinase plasminogen activator and receptor in conjunction with the ets family and AP-1 complex transcription factors in high grade prostate cancers. Eur J Cancer 37(8):1033–1040

Gelmon KA et al (1999) Phase I study of liposomal vincristine. J Clin Oncol 17(2):697–705

Goodwin AP, Mynar JL, Ma YZ, Fleming GR, Frechet JMJ (2005) Synthetic micelle sensitive to IR light via a two-photon process. J Am Chem Soc 127(28):9952–9953

Goren D et al (2000) Nuclear delivery of doxorubicin via folate-targeted liposomes with bypass of multidrug-resistance efflux pump. Clin Cancer Res 6(5):1949–1957

Guaglianone P et al (1994) Phase I and pharmacologic study of liposomal daunorubicin (DaunoXome). Invest New Drugs 12(2):103–110

Han X et al (2009) 9-NC-loaded folate-conjugated polymer micelles as tumor targeted drug delivery system: preparation and evaluation in vitro. Int J Pharm 372(1–2):125–131

Hu KL et al (2009) Lactoferrin-conjugated PEG-PLA nanoparticles with improved brain delivery: in vitro and in vivo evaluations. J Control Release 134(1):55–61

Huai Q et al (2006) Structure of human urokinase plasminogen activator in complex with its receptor. Science 311(5761):656–659

Huang C (1969) Studies on phosphatidylcholine vesicles. Formation and physical characteristics. Biochemistry 8(1):344–352

Huang SL, MacDonald RC (2004) Acoustically active liposomes for drug encapsulation and ultrasound-triggered release. Biochim Biophys Acta 1665(1–2):134–141

Hubert B, Helmut R, Brigitta S (1984) Watersoluble polymers in medicine. Die Angewandte Makromolekulare Chemie 123(1):457–485

Hunt JF, Rath P, Rothschild KJ, Engelman DM (1997) Spontaneous, pH-dependent membrane insertion of a transbilayer alpha-helix. Biochemistry 36(49):15177–15192

Ishida O et al (2001) Liposomes bearing polyethyleneglycol-coupled transferrin with intracellular targeting property to the solid tumors in vivo. Pharm Res 18(7):1042–1048

Janssen AP et al (2003) Peptide-targeted PEG-liposomes in anti-angiogenic therapy. Int J Pharm 254(1):55–58

Jiang JQ, Tong X, Morris D, Zhao Y (2006) Toward photocontrolled release using light-dissociable block copolymer micelles. Macromolecules 39(13):4633–4640

Johnson SM, Bangham AD, Hill MW, Korn ED (1971) Single bilayer liposomes. Biochim Biophys Acta 233(3):820–826

Jorgensen K, Vermehren C, Mouritsen OG (1999) Enhancement of phospholipase A2 catalyzed degradation of polymer grafted PEG-liposomes: effects of lipopolymer-concentration and chain-length. Pharm Res 16(9):1491–1493

Kang N et al (2005) Stereocomplex block copolymer micelles: core-shell nanostructures with enhanced stability. Nano Lett 5(2):315–319

Kataoka K et al (2000) Doxorubicin-loaded poly(ethylene glycol)-poly(beta-benzyl-L-aspartate) copolymer micelles: their pharmaceutical characteristics and biological significance. J Control Release 64(1–3):143–153

Kataoka K, Harada A, Nagasaki Y (2001) Block copolymer micelles for drug delivery: design, characterization and biological significance. Adv Drug Deliv Rev 47(1):113–131

Kim MS et al (2006) pH-responsive PEG-poly(beta-amino ester) block copolymer micelles with a sharp transition. Macromol Rapid Commun 27(6):447–451

Kim D, Lee ES, Oh KT, Gao ZG, Bae YH (2008) Doxorubicin-loaded polymeric micelle overcomes multidrug resistance of cancer by double-targeting folate receptor and early endosomal pH. Small 4(11):2043–2050

Kim S, Shi Y, Kim JY, Park K, Cheng JX (2010a) Overcoming the barriers in micellar drug delivery: loading efficiency, in vivo stability, and micelle-cell interaction. Expert Opin Drug Deliv 7(1):49–62

Kim W, Thevenot J, Ibarboure E, Lecommandoux S, Chaikof EL (2010b) Self-assembly of thermally responsive amphiphilic diblock copolypeptides into spherical micellar nanoparticles. Angew Chem Int Ed Engl 49(25):4257–4260

Kobayashi T et al (2007) Effect of transferrin receptor-targeted liposomal doxorubicin in P-glycoprotein-mediated drug resistant tumor cells. Int J Pharm 329(1–2):94–102

Koivunen E, Wang B, Ruoslahti E (1994) Isolation of a highly specific ligand for the alpha 5 beta 1 integrin from a phage display library. J Cell Biol 124(3):373–380

Kubo T et al (2001) Targeted systemic chemotherapy using magnetic liposomes with incorporated adriamycin for osteosarcoma in hamsters. Int J Oncol 18(1):121–125

Kwon GS, Okano T (1999) Soluble self-assembled block copolymers for drug delivery. Pharm Res 16(5):597–600

Leamon CP, Cooper SR, Hardee GE (2003) Folate-liposome-mediated antisense oligodeoxynucleotide targeting to cancer cells: evaluation in vitro and in vivo. Bioconjug Chem 14(4):738–747

Lee RJ, Huang L (1996) Folate-targeted, anionic liposome-entrapped polylysine-condensed DNA for tumor cell-specific gene transfer. J Biol Chem 271(14):8481–8487

Lee RJ, Low PS (1995) Folate-mediated tumor cell targeting of liposome-entrapped doxorubicin in vitro. Biochim Biophys Acta 1233(2):134–144

Lee ES, Oh KT, Kim D, Youn YS, Bae YH (2007a) Tumor pH-responsive flower-like micelles of poly(L-lactic acid)-b-poly (ethylene glycol)-b-poly(L-histidine). J Control Release 123(1):19–26

Lee GY, Park K, Kim SY, Byun Y (2007b) MMPs-specific PEGylated peptide-DOX conjugate micelles that can contain free doxorubicin. Eur J Pharm Biopharm 67(3):646–654

Li SD, Chono S, Huang L (2008) Efficient oncogene silencing and metastasis inhibition via systemic delivery of siRNA. Mol Ther 16(5):942–946

Li J, Yang Y, Huang L (2012) Calcium phosphate nanoparticles with an asymmetric lipid bilayer coating for siRNA delivery to the tumor. J Control Release 158(1):108–114

Li Y et al (2013) Potent retro-inverso D-peptide for simultaneous targeting of angiogenic blood vasculature and tumor cells. Bioconjug Chem 24(1):133–143

Lindner LH et al (2004) Novel temperature-sensitive liposomes with prolonged circulation time. Clin Cancer Res 10(6):2168–2178

Loughrey H, Bally MB, Cullis PR (1987) A non-covalent method of attaching antibodies to liposomes. Biochim Biophys Acta 901(1):157–160

Lynn DM, Amiji MM, Langer R (2001) pH-responsive polymer microspheres: rapid release of encapsulated material within the range of intracellular pH financial support was provided by the NSF (Cooperative Agreement No. ECC9843342 to the MIT Biotechnology Process Engineering Center), the NIH (GM26698), and the Department of the Army (Cooperative Agreement DAMD 17-99-2-9-001 to the Center for Innovative Minimally Invasive Therapy). D.M.L. wishes to thank the NIH for a Postdoctoral Fellowship (NRSA Fellowship No. 1 F32 GM20227-01). Scanning electron microscopy and confocal microscopy images were acquired by William Fowle at the Northeastern University Center for Electron Microscopy. Dr. David Putnam, David Ting, and Tommy Thomas are thanked for helpful discussions. Angew Chem Int Ed Engl 40(9):1707–1710

Macholl S et al (2012) In vivo pH imaging with (99m)Tc-pHLIP. Mol Imaging Biol 14(6):725–734
Mayhew E, Lazo R, Vail WJ, King J, Green AM (1984) Characterization of liposomes prepared using a microemulsifier. Biochim Biophys Acta 775(2):169–174
Meyer DE, Kong GA, Dewhirst MW, Zalutsky MR, Chilkoti A (2001) Targeting a genetically engineered elastin-like polypeptide to solid tumors by local hyperthermia. Cancer Res 61(4):1548–1554
Mikhail AS, Allen C (2009) Block copolymer micelles for delivery of cancer therapy: transport at the whole body, tissue and cellular levels. J Control Release 138(3):214–223
Milsmann MH, Schwendener RA, Weder HG (1978) The preparation of large single bilayer liposomes by a fast and controlled dialysis. Biochim Biophys Acta 512(1):147–155
Min KH et al (2010) Tumoral acidic pH-responsive MPEG-poly(beta-amino ester) polymeric micelles for cancer targeting therapy. J Control Release 144(2):259–266
Naito M et al (2012) A phenylboronate-functionalized polyion complex micelle for ATP-triggered release of siRNA. Angew Chem Int Ed Engl 51(43):10751–10755
Nasongkla N et al (2004) cRGD-functionalized polymer micelles for targeted doxorubicin delivery. Angew Chem Int Ed Engl 43(46):6323–6327
Neradovic D, Soga O, Van Nostrum CF, Hennink WE (2004) The effect of the processing and formulation parameters on the size of nanoparticles based on block copolymers of poly(ethylene glycol) and poly(*N*-isopropylacrylamide) with and without hydrolytically sensitive groups. Biomaterials 25(12):2409–2418
Nicolas J, Mura S, Brambilla D, Mackiewicz N, Couvreur P (2013) Design, functionalization strategies and biomedical applications of targeted biodegradable/biocompatible polymer-based nanocarriers for drug delivery. Chem Soc Rev 42(3):1147–1235
O'Brien ME et al (2004) Reduced cardiotoxicity and comparable efficacy in a phase III trial of pegylated liposomal doxorubicin HCl (CAELYX/Doxil) versus conventional doxorubicin for first-line treatment of metastatic breast cancer. Ann Oncol 15(3):440–449
O'Reilly RK, Hawker CJ, Wooley KL (2006) Cross-linked block copolymer micelles: functional nanostructures of great potential and versatility. Chem Soc Rev 35(11):1068–1083
Oberoi HS, Laquer FC, Marky LA, Kabanov AV, Bronich TK (2011) Core cross-linked block ionomer micelles as pH-responsive carriers for *cis*-diamminedichloroplatinum(II). J Control Release 153(1):64–72
Oishi M, Nagasaki Y, Itaka K, Nishiyama N, Kataoka K (2005) Lactosylated poly(ethylene glycol)-siRNA conjugate through acid-labile beta-thiopropionate linkage to construct pH-sensitive polyion complex micelles achieving enhanced gene silencing in hepatoma cells. J Am Chem Soc 127(6):1624–1625
Olson F, Hunt CA, Szoka FC, Vail WJ, Papahadjopoulos D (1979) Preparation of liposomes of defined size distribution by extrusion through polycarbonate membranes. Biochim Biophys Acta 557(1):9–23
Park EK, Kim SY, Lee SB, Lee YM (2005) Folate-conjugated methoxy poly(ethylene glycol)/poly(epsilon-caprolactone) amphiphilic block copolymeric micelles for tumor-targeted drug delivery. J Control Release 109(1–3):158–168
Parrish B, Breitenkamp RB, Emrick T (2005) PEG- and peptide-grafted aliphatic polyesters by click chemistry. J Am Chem Soc 127(20):7404–7410
Peer D et al (2007) Nanocarriers as an emerging platform for cancer therapy. Nat Nanotechnol 2(12):751–760
Potineni A, Lynn DM, Langer R, Amiji MM (2003) Poly(ethylene oxide)-modified poly(beta-amino ester) nanoparticles as a pH-sensitive biodegradable system for paclitaxel delivery. J Control Release 86(2–3):223–234
Rapoport N (2007) Physical stimuli-responsive polymeric micelles for anti-cancer drug delivery. Prog Polym Sci 32(8–9):962–990
Rapoport N, Gao Z, Kennedy A (2007) Multifunctional nanoparticles for combining ultrasonic tumor imaging and targeted chemotherapy. J Natl Cancer Inst 99(14):1095–1106
Sawant RM et al (2006) "SMART" drug delivery systems: double-targeted pH-responsive pharmaceutical nanocarriers. Bioconjug Chem 17(4):943–949

Schurtenberger P, Mazer N, Waldvogel S, Kanzig W (1984) Preparation of monodisperse vesicles with variable size by dilution of mixed micellar solutions of bile salt and phosphatidylcholine. Biochim Biophys Acta 775(1):111–114

Shan L (2004) Cy5.5-labeled pH low insertion peptide (pHLIP). Molecular Imaging and Contrast Agent Database (MICAD), Bethesda (MD)

Shum P, Kim JM, Thompson DH (2001) Phototriggering of liposomal drug delivery systems. Adv Drug Deliv Rev 53(3):273–284

Siehler S, Zupanc GK, Seuwen K, Hoyer D (1999) Characterisation of the fish sst3 receptor, a member of the SRIF1 receptor family: atypical pharmacological features. Neuropharmacology 38(3):449–462

Sofou S (2007) Surface-active liposomes for targeted cancer therapy. Nanomedicine (Lond) 2(5):711–724

Soundararajan S, Chen WW, Spicer EK, Courtenay-Luck N, Fernandes DJ (2008) The nucleolin targeting aptamer AS1411 destabilizes bcl-2 messenger RNA in human breast cancer cells. Cancer Res 68(7):2358–2365

Stetler-Stevenson WG, Yu AE (2001) Proteases in invasion: matrix metalloproteinases. Semin Cancer Biol 11(2):143–152

Strebhardt K, Ullrich A (2008) Paul Ehrlich's magic bullet concept: 100 years of progress. Nat Rev Cancer 8(6):473–480

Sudimack J, Lee RJ (2000) Targeted drug delivery via the folate receptor. Adv Drug Deliv Rev 41(2):147–162

Sugahara KN et al (2009) Tissue-penetrating delivery of compounds and nanoparticles into tumors. Cancer Cell 16(6):510–520

Sugahara KN et al (2010) Coadministration of a tumor-penetrating peptide enhances the efficacy of cancer drugs. Science 328(5981):1031–1035

Szoka F Jr, Papahadjopoulos D (1978) Procedure for preparation of liposomes with large internal aqueous space and high capture by reverse-phase evaporation. Proc Natl Acad Sci U S A 75(9):4194–4198

Tacker JR, Anderson RU (1982) Delivery of antitumor drug to bladder cancer by use of phase transition liposomes and hyperthermia. J Urol 127(6):1211–1214

Tagami T, Ernsting MJ, Li SD (2011) Optimization of a novel and improved thermosensitive liposome formulated with DPPC and a Brij surfactant using a robust in vitro system. J Control Release 154(3):290–297

Tang J et al (2013) Synergistic targeted delivery of payload into tumor cells by dual-ligand liposomes co-modified with cholesterol anchored transferrin and TAT. Int J Pharm 454(1):31–40

Tannock IF, Rotin D (1989) Acid pH in tumors and its potential for therapeutic exploitation. Cancer Res 49(16):4373–4384

Terada T, Iwai M, Kawakami S, Yamashita F, Hashida M (2006) Novel PEG-matrix metalloproteinase-2 cleavable peptide-lipid containing galactosylated liposomes for hepatocellular carcinoma-selective targeting. J Control Release 111(3):333–342

Thevenin D, An M, Engelman DM (2009) pHLIP-mediated translocation of membrane-impermeable molecules into cells. Chem Biol 16(7):754–762

Thompson DH, Gerasimov OV, Wheeler JJ, Rui YJ, Anderson VC (1996) Triggerable plasmalogen liposomes: improvement of system efficiency. Biochim Biophys Acta 1279(1):25–34

Tong R, Hemmati HD, Langer R, Kohane DS (2012) Photoswitchable nanoparticles for triggered tissue penetration and drug delivery. J Am Chem Soc 134(21):8848–8855

Torchilin VP (2005) Recent advances with liposomes as pharmaceutical carriers. Nat Rev Drug Discov 4(2):145–160

Vavere AL et al (2009) A novel technology for the imaging of acidic prostate tumors by positron emission tomography. Cancer Res 69(10):4510–4516

Vijayalakshmi P, Srivastava KK, Poornima B, Nirmalan P (2003) Visual outcome of cataract surgery in children with congenital rubella syndrome. J AAPOS 7(2):91–95

Vilner BJ, John CS, Bowen WD (1995) Sigma-1 and sigma-2 receptors are expressed in a wide variety of human and rodent tumor cell lines. Cancer Res 55(2):408–413

Walker JM et al (1990) Sigma receptors: biology and function. Pharmacol Rev 42(4):355–402
Wang M, Lowik DWPM, Miller AD, Thanou M (2009) Targeting the urokinase plasminogen activator receptor with synthetic self-assembly nanoparticles. Bioconjug Chem 20(1):32–40
Wang YG et al (2011) Materializing sequential killing of tumor vasculature and tumor cells via targeted polymeric micelle system. J Control Release 149(3):299–306
Warenius HM, Galfre G, Bleehen NM, Milstein C (1981) Attempted targeting of a monoclonal antibody in a human tumour xenograft system. Eur J Cancer Clin Oncol 17(9):1009–1015
Weinstein JN, Magin RL, Yatvin MB, Zaharko DS (1979) Liposomes and local hyperthermia: selective delivery of methotrexate to heated tumors. Science 204(4389):188–191
Wu YR, Sefah K, Liu HP, Wang RW, Tan WH (2010) DNA aptamer-micelle as an efficient detection/delivery vehicle toward cancer cells. Proc Natl Acad Sci U S A 107(1):5–10
Xiong XB et al (2005) Enhanced intracellular delivery and improved antitumor efficacy of doxorubicin by sterically stabilized liposomes modified with a synthetic RGD mimetic. J Control Release 107(2):262–275
Yan ZQ et al (2011) LyP-1-conjugated doxorubicin-loaded liposomes suppress lymphatic metastasis by inhibiting lymph node metastases and destroying tumor lymphatics. Nanotechnology 22(41)
Yan ZQ et al (2012a) LyP-1-conjugated PEGylated liposomes: a carrier system for targeted therapy of lymphatic metastatic tumor. J Control Release 157(1):118–125
Yan HH et al (2012b) Two-order targeted brain tumor imaging by using an optical/paramagnetic nanoprobe across the blood brain barrier. ACS Nano 6(1):410–420
Yang T et al (2007) Preparation and evaluation of paclitaxel-loaded PEGylated immunoliposome. J Control Release 120(3):169–177
Yao L, Daniels J, Wijesinghe D, Andreev OA, Reshetnyak YK (2013a) pHLIP (R)-mediated delivery of PEGylated liposomes to cancer cells. J Control Release 167(3):228–237
Yao L et al (2013b) pHLIP peptide targets nanogold particles to tumors. Proc Natl Acad Sci U S A 110(2):465–470
Yokoyama M et al (1998) Incorporation of water-insoluble anticancer drug into polymeric micelles and control of their particle size. J Control Release 55(2–3):219–229
Yoo HS, Park TG (2004) Folate receptor targeted biodegradable polymeric doxorubicin micelles. J Control Release 96(2):273–283
Zhan C et al (2010a) Cyclic RGD conjugated poly(ethylene glycol)-*co*-poly(lactic acid) micelle enhances paclitaxel anti-glioblastoma effect. J Control Release 143(1):136–142
Zhan C, Yan Z, Xie C, Lu W (2010b) Loop 2 of Ophiophagus hannah toxin b binds with neuronal nicotinic acetylcholine receptors and enhances intracranial drug delivery. Mol Pharm 7(6): 1940–1947
Zhan C et al (2011) Micelle-based brain-targeted drug delivery enabled by a nicotine acetylcholine receptor ligand. Angew Chem Int Ed Engl 50(24):5482–5485
Zhan C et al (2012a) Co-delivery of TRAIL gene enhances the anti-glioblastoma effect of paclitaxel in vitro and in vivo. J Control Release 160(3):630–636
Zhan CY et al (2012b) Cyclic RGD-polyethylene glycol-polyethylenimine for intracranial glioblastoma-targeted gene delivery. Chem Asian J 7(1):91–96
Zhang L et al (2008) Nanoparticles in medicine: therapeutic applications and developments. Clin Pharmacol Ther 83(5):761–769
Zhao HZ, Yung LYL (2008) Selectivity of folate conjugated polymer micelles against different tumor cells. Int J Pharm 349(1–2):256–268
Zhao Z et al (2013) A controlled-release nanocarrier with extracellular pH value driven tumor targeting and translocation for drug delivery. Angew Chem Int Ed Engl 52(29):7487–7491
Zhu S et al (2011a) RGD-modified PEG-PAMAM-DOX conjugate: in vitro and in vivo targeting to both tumor neovascular endothelial cells and tumor cells. Adv Mater 23(12):H84–H89
Zhu ZS et al (2011b) The effect of hydrophilic chain length and iRGD on drug delivery from poly(epsilon-caprolactone)-poly(*N*-vinylpyrrolidone) nanoparticles. Biomaterials 32(35): 9525–9535

Chapter 5
Photodynamic Therapy for Cancer: Principles, Clinical Applications, and Nanotechnological Approaches

Claudia Conte, Francesca Ungaro, Antonino Mazzaglia, and Fabiana Quaglia

5.1 Principles of Photodynamic Therapy

Photodynamic therapy (PDT) is based on photochemical processes between light and an exogenous photosensitizer (PS) localized at disease level. These components, tolerated singly by the cells, generate cytotoxic oxygen-based molecular species in proper dosage and concentration. Mechanistically it consists in three phases: excitation of PS, generation of toxic oxygen species, and cell death (Fig. 5.1). In the first phase, light of an appropriate wavelength, usually visible (VIS) or near-infrared (NIR), excites the PS molecules from the ground state to the excited singlet state. The radiation wavelength is usually chosen to coincide with the maximum absorption wavelength of the drug molecule. In analogy with many other fluorescent molecules, PS can, at this stage, decay to its ground state with concomitant emission of light in the form of fluorescence. On the other hand, the excited singlet state PS may also undergo a process known as intersystem crossing whereby the spin of the excited electron inverts to form a relatively more stable and long-lived excited triplet-state that has electrons which spin in a parallel conformation in the two highest occupied molecular orbitals. High quantum efficiency for this transition is a key characteristic of a good PS. The PS in triplet state can either decay to the ground state or transfer electron/energy to the surroundings through two types of reactions. A type 1 process

C. Conte • F. Ungaro • F. Quaglia (✉)
Department of Pharmacy, University of Napoli Federico II,
Via Domenico Montesano 49, 80131 Napoli, Italy
e-mail: quaglia@unina.it

A. Mazzaglia
CNR-ISMN Institute of Nanostructured Materials, c/o Dept of Chemical Sciences of the University of Messina, V.le F. Stagno D'Alcontres 31, 98166 Messina, Italy

M.J. Alonso, M. Garcia-Fuentes (eds.), *Nano-Oncologicals: New Targeting and Delivery Approaches*, Advances in Delivery Science and Technology,
DOI 10.1007/978-3-319-08084-0_5

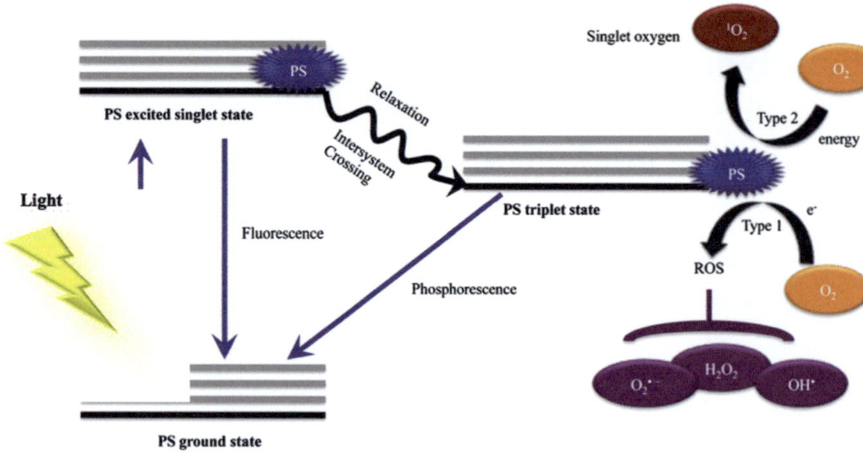

Fig. 5.1 Schematic illustration of a photodynamic reaction. The PS initially absorbs a photon that excites it to the short-lived singlet state. This can decay by fluorescence emission, by non-radiative relaxation with emission of heat, or by intersystem crossing to the long-lived triplet state. This triplet PS can interact with molecular oxygen in two pathways, type 1 and type 2, leading to the formation of oxygen radicals and singlet oxygen ($^1O_2$), respectively (adapted from Gupta et al. 2013 with permission from Elsevier)

can occur whereby PS reacts directly with an organic molecule in a cellular microenvironment, acquiring or losing a hydrogen atom or electron to form a radical. Subsequent autoxidation of the reduced PS produces a superoxide anion radical ($O_2^{\cdot-}$). Dismutation or one electron reduction of $O_2^{\cdot-}$ in the presence of two protons gives hydrogen peroxide (H_2O_2), which in turn can undergo one-electron reduction to a powerful and virtually indiscriminate oxidant hydroxyl radical (HO^-). In a type 2 process the transfer of energy to molecular oxygen leads to the formation of singlet oxygen ($^1O_2$) which initiates oxidation of susceptible substrates.

Both type 1 and type 2 reactions can occur simultaneously and competitively, and the ratio between these processes depends on the type of PS used, as well as the concentrations of substrate and oxygen. Type 2 reaction, however, appears to play a central role in cytotoxicity, because of the highly efficient interaction of the $^1O_2$ species with various biomolecules. Singlet oxygen, in fact, is believed to be the main cytotoxic agent in PDT (Agostinis et al. 2011). Singlet oxygen species have a lifetime of less than 3.5 μs in aqueous environment and can diffuse only 0.01 to 0.02 μm during this period. Nevertheless, it should be taken into account that singlet oxygen senses the inherent heterogeneity of cell environment and its lifetime can be consequently affected (Kuimova et al. 2008). Therefore, the initial extent of the damage is limited to the site of concentration of the PS (Robertson et al. 2009). This is usually the mitochondria, plasma membrane, Golgi apparatus, lysosomes, endosomes, and endoplasmic reticulum (ER). The nucleus and nuclear membrane are usually spared and DNA damage is rare. Net ionic charge (from −4 to +4), hydrophobicity, and the degree of asymmetry of PS are reported to play a role in cell uptake and intracellular localization.

5.2 Light Sources and Light Delivery

No single light source is ideal for all PDT indications, even with the same PS. The choice of light source should therefore be based on PS absorption, disease, cost, and size. Furthermore, the clinical efficacy of PDT is dependent on dosimetry: total light dose, light exposure time, light delivery mode (single vs. fractionated or even metronomic), and fluence rate (intensity of light delivery).

The effective excitation light magnitude is determined by the combination of optical absorption and scattering properties of the tissue. Absorption is largely due to endogenous tissue chromophores such as hemoglobin, myoglobin, and cytochromes (Fig. 5.2).

The optical scattering of a tissue decreases with wavelength. For the spectral range of 450–1,750 nm, tissue scattering is, in general, more prevalent than absorption, although for the range of 450–600 nm, melanin and hemoglobin provide significant absorption, while water plays a similar role for wavelength >1,350 nm. Therefore, the optimal optical window for PDT, as well as for optical imaging, is in the NIR spectral region (600–1,300 nm), where the scattering and absorption by tissue are minimized and, therefore, the longest penetration depth can be achieved. Within this optical window, the longer is the wavelength, the deeper is the penetration depth. However, light up to only approximately 800 nm can extensively generate singlet oxygen, because longer wavelengths have insufficient energy to initiate a photodynamic reaction (Sandell and Zhu 2011). Recently the PDT effectiveness of new porphyrin dimers activated at longer wavelengths was demonstrated (Balaz et al. 2009; Dahlstedt et al. 2009).

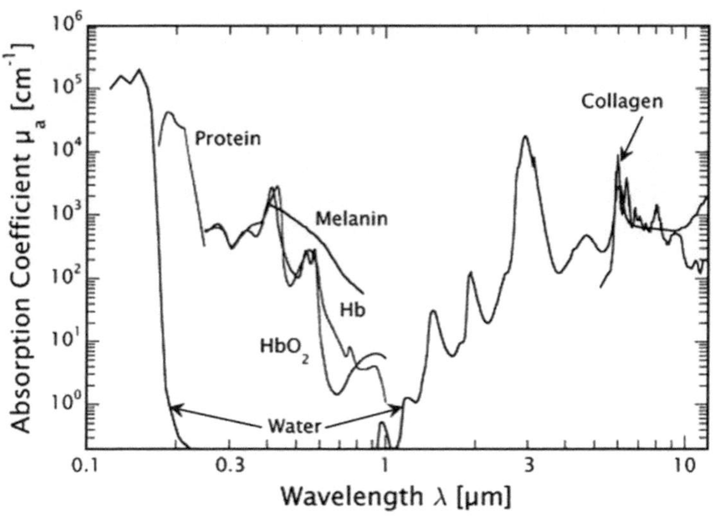

Fig. 5.2 Optical absorption coefficients of principal tissue chromophores in the human body (reproduced from Vogel and Venugopalan 2003 with permission from the American Chemical Society)

Table 5.1 Optical penetration depth (mm) as a function of light wavelength of some tissues (reproduced from Lee and Kopelman 2011 with permission from Springer)

	Wavelength of light						
Tissue	630 nm	632.8 nm	665 nm	675 nm	780 nm	835 nm	1,064 nm
Blood		0.19		0.28	0.42	0.51	
Mammary tissue		2.59		2.87	3.12	3.54	
Mammary carcinoma		2.87		3.14	3.62	4.23	
Mammary carcinoma in C3H/HEJ mice	2.0		2.3				3.7
Brain (postmortem)		0.92		1.38	2.17	2.52	
Brain	1.6						
Brain tumor	3.1						
Colon		2.48		2.73	2.91		
Lung		0.81		1.09	1.86	2.47	
Lung carcinoma		1.68		2.01	2.82	3.89	

The ideal light source for PDT must exhibit suitable spectral characteristics that coincide with the maximum absorption wavelength range of the PS applied in order to generate enough radical oxygen species (ROS) to produce a cytotoxic effect. Nevertheless, as previously explained, PDT treatment is generally carried out with red light at higher penetration also for those PS which exibit maximum absorption in the blue region of spectrum.

Currently approved PS absorb in the visible spectral regions below 700 nm, where light penetration into the skin is only a few millimeters, clinically limiting PDT to treating topical lesions. Optical penetration depth of selected healthy and tumor tissues is reported in Table 5.1.

Laser systems are widely used for treating dermatological conditions (Barolet 2008). The gold vapor laser (628 nm), the argon ion-pumped dye laser (630 or 635 nm), and the copper vapor-pumped dye laser constitute the most popular systems. They allow the selection of a wavelength that has a maximal effective tissue penetration of approximately 10 mm and have been used in combination with all types of PS. The laser beams can be launched into an optical fiber applicator, enabling light to be delivered directly into internal tumors. However, these techniques are relatively expensive, require specialized supporting staff, and are space-consuming.

It is likely that such systems will eventually be replaced by laser diode arrays which are very convenient since they can be easily handled, require only a single phase supply, and are also relatively inexpensive. Commercially available incoherent light sources, such as incandescent or arc lamps, have been used in topical PDT by several groups mainly for the treatment of large lesions (Warren et al. 2009). Because coherence of light is not necessary for PDT, such sources offer the advantage of being less expensive and easier to handle. The most popular of them is the filtered slide projector, which excludes light below 600 nm with glass filters, minimizing the emission of shorter wavelengths. Due to strong absorption by hemoglobin, it could lead to the production of heat-induced erythema. Despite this, unfiltered white light has also been employed, and recently, professional incoherent lamp like PDT 1200 lamp has been developed for PDT.

5.3 PDT-Mediated Tumor Destruction

PDT treatment is connected to the modification of different signal transduction cellular pathways, in relation to calcium expression levels, lipid metabolism, tyrosine kinase expression, cell adhesion molecules, and cytokines. In particular, PDT treatment induces an acute stress reaction leading to changes in cellular metabolism which may result in apoptosis, necrosis, or cell survival (Robertson et al. 2009).

Three distinct mechanisms have been recognized which contribute to reduction or disappearance of tumors after PDT treatment (Fig. 5.3).

In the first case, ROS that are generated by PDT can kill tumor cells directly by apoptosis and/or necrosis if the PS has been taken up by tumor cells (Oleinick et al. 2002). Alternatively, PDT is able to damage the tumor-associated vasculature, thus leading to tumor death via lack of oxygen and nutrients (Krammer 2001). Finally, the acute inflammation as well as the release of cytokines and stress response proteins induced in the tumor by PDT can lead to an invasion of leukocytes that can both contribute to tumor destruction as well as stimulate the immune system to recognize and destroy tumor cells (Castano et al. 2006; Pizova et al. 2012; van

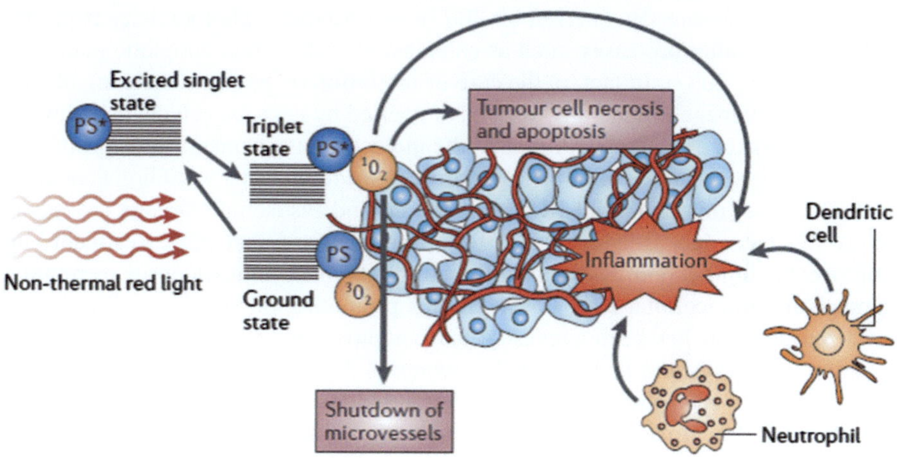

Fig. 5.3 The mechanism of tumor attack in PDT. PS absorbs light and an electron moves to the first short-lived excited singlet state. This is followed by intersystem crossing, in which the excited electron changes its spin and produces a longer-lived triplet state. The PS triplet transfers energy to ground state triplet oxygen, which produces reactive singlet oxygen. Singlet oxygen can directly kill tumor cells by induction of necrosis and/or apoptosis, can cause destruction of tumor vasculature, and produces an acute inflammatory response that attracts leukocytes such as dendritic cells and neutrophils (reproduced from Castano et al. 2006 with permission from the American Chemical Society)

Duijnhoven et al. 2003). The relative importance of each mechanism for the overall tumor response is yet to be defined and so requires further research. It is clear, however, that the combination of all these components in PDT is required for optimum long-term tumor regression, especially of tumors that may have metastasized.

5.3.1 Mechanisms of Cell Death in PDT

Although PDT can induce many cellular and molecular signaling pathway events in cells, the final effect is the induction of cell death through the activation of three main cell death pathways: apoptosis, necrosis, and autophagy (Reiners Jr et al. 2010).

The mode and the extent of cell death is related to different elements, including the concentration, the physiochemical properties and the subcellular location of the PS, the concentration of oxygen, the wavelength and intensity of the light, as well as the cell type. For instance, it is recognized that lower doses of PDT lead to more apoptotic cells, while higher doses lead to proportionately more necrotic cells (Castano et al. 2005).

The major death modality in cells responding to PDT is generally apoptosis, which is mediated by the activation of the pro-apoptotic members of the family Bcl (Kessel and Castelli 2001; Xue et al. 2001). With mitochondria-associated PS, photodamage to membrane-bound Bcl-2 can induce the subsequent release of caspase activators such as cytochrome c, or other proapoptotic molecules. In particular, cell death after PDT treatment has been associated to high levels of caspase 3 and 9 within the cell lysate (Buytaert et al. 2007). Nevertheless, phototoxicity is propagated also by other proteases, such as calpains, as well as non-apoptotic pathways (Buytaert et al. 2007). In fact, in the case of inhibition or genetic deficiency of caspases, photodamage is generally delayed or exerted by necrotic cell death pathways (Vanlangenakker et al. 2008). Although the molecular mechanisms underlying programmed necrosis are not completely clarified, specific events including activation of RIP kinase, mitochondrial ROS production, and lysosomal damage are recurrently involved. In particular, it has been demonstrated that PDT treatment can improve the mitochondrial permeability through inner mitochondria membrane damage and intracellular Ca^{2+} overload, thus promoting necrotic as well as apoptotic cell death. It has been suggested that apoptosis and necrosis share common initiation pathways and that the final outcome is dependent on the presence of an active caspase (Nakagawa et al. 2005). This implies that apoptosis inhibition reorients cells to necrosis, so that cells sufficiently damaged by PDT are destroyed, regardless of the mechanism involved (Buytaert et al. 2007).

The last cell death pathway induced by PDT is associated to the stimulation of autophagy by various stress signals including oxidative stress and ROS production as primary damaging agents. However, autophagy can have not only a pro-death role after cancer therapies but also a cytoprotective effect (Kessel 2006). There is a correlation between autophagy and the other cell death pathways. For instance, accumulation of ROS-damaged cytoplasmic components may then potentiate phototoxicity in apoptosis-competent cells (Reiners Jr et al. 2010). However, a better understanding of the interplay between autophagy, apoptosis, and necrosis in tumor response will be a requisite to optimize application of PDT in cancer treatment (Buytaert et al. 2007).

After PDT, cancer cells usually develop a cytoprotective mechanism in order to avoid cytotoxic effects such as the production of antioxidant molecules (e.g., some amino acids, glutathione, vitamin E) and of enzymes that can detoxify ROS. For example, superoxide dismutase (SOD) overexpression or treatment with SOD mimetics has been shown to counteract the cytotoxic effect of PDT (Golab et al. 2003). A further cytoprotective mechanism is based on the expression of proteins which are involved in particular signaling pathways that can regulate PDT-induced apoptosis or participate in the repair of lesions caused by oxidative stress. Amid them, different stress-related transcription factors induced by PDT are widely known, including the nuclear factor κB (NF-κB), an activator protein-1 (AP-1), a hypoxia-inducible factor-1α (HIF), or the nuclear factor (erythroid-derived 2)-like 2 (Nrf_2) (Oleinick et al. 2002). Furthermore, it has been demonstrated that PDT upregulates heme oxygenase-1 expression depending on Nrf_2 nuclear accumulation and induces the expression of various heat shock proteins (HSPs) exerting a protective role (Weiss et al. 2012). For example, transfection of tumor cells with the HSP gene increased the survival of tumor cells after PDT. In particular, the increase of HSP60 and HSP70 levels in tumor cells are inversely correlated with sensitivity to the PDT thanks to their ability to bind the damaged proteins (Hanlon et al. 2001; Nonaka et al. 2004). Many HSPs play a critical role also in the regulation of pro-survival pathways, thus increasing, for example, the ubiquitination of carbonylated proteins and preventing the formation of toxic protein aggregates (Szokalska et al. 2009).

5.3.2 Antivascular Effects of PDT

Depending on the type of PS and the PS-light interval administered, PDT can provoke several vascular effects, including microvasculature collapse as well as high tissue hypoxia, complete blood flow stasis, hemorrhage, and, in some cases, the formation of platelet aggregates in larger vessels.

Although microvascular damage and hypoxia after PDT contribute to greater tumor response, reduction in oxygen during treatment can limit tumor control by inducing the production of proangiogenic markers such as the vascular endothelial growth factor (VEGF), the cyclooxygenase-2 (COX-2), the matrix metalloproteinase (MMP), and other cytokines, creating enhanced environment for tumor recurrence (Bhuvaneswari et al. 2009). All these phenomena described at vessel level are directly correlated to the photodamage of the endothelium, which is reported as the primary target for PDT in vivo (Fig. 5.4). It is widely known, in fact, that endothelial cells are characterized by a high sensitivity to PDT treatment (Gomer et al. 1988).

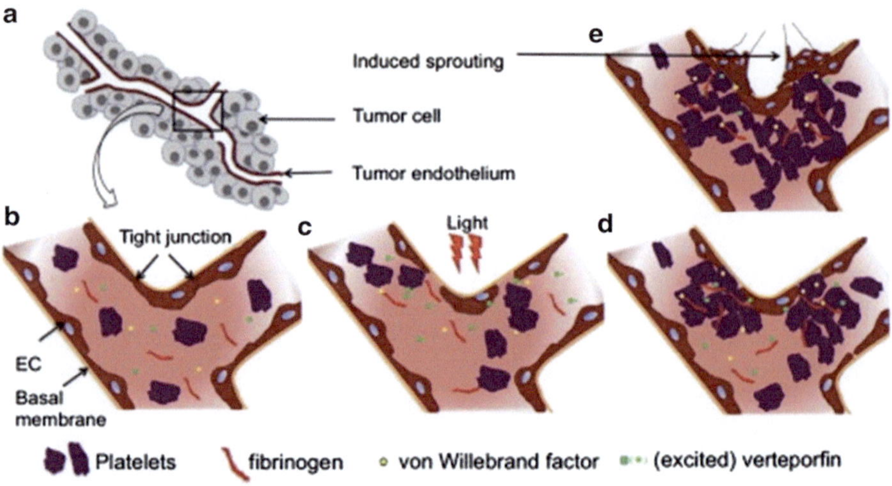

Fig. 5.4 Tumor endothelial responses after PDT leading to blood flow stasis. (**a**) Tumor blood vessel before PDT. (**b–e**) Magnification of the junction of the tumor vessel. (**b**) Before PDT endothelial cells are tightly attached to the basement membrane of the vessel wall, lining the blood vessel. Endothelial cells are connected through tight junctions. (**c**) After injection of a PS and light exposure, cellular stress inside the endothelial cells results in disruption of tight junctions, partial retraction, and detachment from the vessel wall. (**d**) Blood gets in contact with the vessel wall collagen and the clotting cascade is initiated, ultimately leading, through the interaction with fibrinogen, to the formation of a stabilized thrombus, leading to obstruction of the vessel. (**e**) Due to the angiogenic switch, endothelial cell proliferation, migration, and sprout formation are observed (reproduced from Weiss et al. 2012 with permission from the American Chemical Society)

The difference in sensitivity between the endothelial cells and the other proliferating tumor cells is accompanied by a greater PS accumulation in the endothelial cells. Endothelial cell responses to sublethal doses of PDT may also contribute to vascular changes observed in tissue (Fingar 1996; Fingar et al. 1992; Pizova et al. 2012).

5.3.3 *PDT and the Immune Response*

Numerous preclinical and clinical studies have demonstrated that PDT can influence the adaptive immune response in different ways: some regimens result in potentiation of adaptive immunity, whereas others lead to immunosuppression. However, the precise mechanism leading to potentiation versus suppression is unclear; nevertheless, it seems as though the effect of PDT on the immune system is dependent on the PS type, the treatment regimen, and the area treated. Furthermore, recent findings suggest that clinical antitumor PDT can increase antitumor immunity (Gollnick et al. 2006).

First of all, PDT frequently provokes a strong acute inflammatory reaction observed as localized edema at the targeted site. This reaction starts with the generation of important alarm/danger signals, also called damage-associated molecular patterns (DAMPs) or cell death-associated molecular patterns (CDAMPs), at the treated site that can be detected by the innate immunity (Korbelik 2006). As consequence, PDT induces oxidative stress which triggers a vast array of signal pathways via Toll-like receptors (TLRs). This includes expression of HSPs, NF-κB, and AP-1 that can then induce expression of immunoregulatory and proinflammatory proteins such as interleukins (IL-1α, -1β, -2, -6, -8, -11, -12, -15), tumor necrosis factor (TNF), chemokines, and interferons (IFN-α/β). Further, photo-oxidative degradation of membrane lipids and generation of arachidonic acid metabolites, which are themselves potent inflammatory mediators, can cause a rapid and strong inflammatory reaction (Castano et al. 2006). These processes together with the release of histamine and serotonin from damaged vasculature cause infiltration of the tumor site by diverse populations of immune cells (neutrophils, mast cells, and macrophages) that become activated and engaged in tumor cell destruction (Krosl et al. 1995). However, a key event appears to be PDT induced complement activation.

Photosensitization induces the innate immunity response with subsequent development of adaptive immunity. In particular, PDT activates both humoral and cell-mediated antitumor immunity. PDT-induced acute local and systemic inflammation is postulated to culminate in the maturation and activation of dendritic cells (DCs) (Sousa 2004). Activated DCs then migrate to tumor draining lymph nodes, where they stimulate T-cell activation. Generation of CD8+ effector and memory T cells is frequently, but not always, dependent upon the presence and activation of CD4+ T cells. PDT-induced antitumor immunity may or may not depend on CD4+ T cells and may be augmented by natural killer cells. Nevertheless, in some cases, certain PDT regimens have been shown to systemically suppress immune reactivity (Gollnick et al. 2006).

5.4 Treatment of Cancer Through PDT

A PDT treatment is a two-stage process where a PS is administered in the body locally or by intravenous injection. After a certain period, PS accumulates in cancer cells and is activated by application of light at level of diseased area where biological effects occur. PDT efficacy depends on multiple factors related to PS photochemical features (singlet oxygen production yield, light penetration depth), PS pharmacokinetics (blood protein interaction, clearance) and consequent tissue/cell distribution, properties of the tumor (vasculature, oxygenation level) as well as light parameters (light dose, drug-to-light interval). Hence, therapeutic success depends on the optimization of all these factors as described in the following.

5.4.1 Photosensitizers in Cancer Treatment

PS are generally classified as porphyrinoids and non-porphyrinoids. Amid porphyrinoids-based PS, first, second, and third generation PS are reported.

The first generation agent hematoporphyrin (Hp) was isolated from hemoglobin of dried blood using concentrated sulfuric acid in 1841 by Scherer. Purification and solubilization of Hp are the chemical processes leading to the preparation of HpD (HP derivative), a very complex mixture of several components, whereby approximately 50 % is identifiable as oligomeric ematoporphyrins and protoporphyrins, which have a low in vivo photosensitizing activity. Further chemical purification yielded to the preparation of porfimer sodium (Photofrin®), a lyophilized concentrated form of monomeric and oligomeric hematoporphyrin derivatives. Photofrin® is characterized by an absorption band at 630 nm (corresponding to a penetration of about 5–10 mm) and a low molar extinction coefficient which in turn demands high concentrations of Photofrin® and light to obtain adequate tumor eradication, and a long half-life of 452 h, causing long-lasting photosensitivity. The time delay between drug delivery and the time it takes to maximize the tumor to normal cell uptake within the target tissue determines the correct time for light application. Photofrin®-mediated PDT involves iv administration of PS followed by irradiation (100–200 J/cm^2 of red light) 24–48 h later. During this period Photofrin® is cleared from a number of tissues and remains concentrated at target site (Allison et al. 2010).

Second generation porphyrins have been developed with the aim to alleviate certain problems associated with first-generation molecules such as prolonged skin photosensitization and suboptimal tissue penetration. Amid alternative strategies to induce photosensitization, the use of the endogenous PS protoporphyrin IX (PpIX), precursor in the biosynthesis of heme, has been proposed. By providing exogenous 5-aminolevulinic acid (5-ALA) to cells, the production rate of PpIX, as well as of reactive oxygen, is increased, especially in cancer cells. 5-ALA is now approved for the treatment of actinic keratosis (AK) (Levulan®) and is in clinical trials for other types of cancer (Nokes et al. 2013). Due to its poor ability to cross the skin, lipophilic derivatives have been proposed such as 5-methyl-aminolevulinate (Metvix®) and hexyl ester of 5-ALA (Hexvix®) (Fukuda et al. 2006). However, various chemical modifications of the tetrapyrrolic ring of the porphyrins characterize the different groups of the second-generation PS (Kreimer-Birnbaum 1989). They have high absorption coefficients, satisfactory singlet oxygen quantum yields as well as absorption peaks in the IR (660–700 nm) or NIR (700–850 nm) regions. The serum half-life of these compounds is short and tissue accumulation is improved and occurs quickly (within 1–6 h after injection). Thus, the treatment can be carried out on the same day as the administration of the drug. In addition, the risk of burns by accidental sun exposure is low because clearance from normal tissues is rapid. Finally, toxicity to skin and internal organs in the absence of light (the so-called "dark" toxicity) is absent or minimal. PS currently under investigation belong to two chemical groups: the chlorins and the phthalocyanines (Pc) (Garland et al. 2009).

Chlorins are reduced porphyrins such as the mono-L-aspartyl chlorine e6 (NPe6) and chlorin e6, hydrophilic compounds with some amphiphilicity and similar photobiological properties. They have been found effective in treating basal cell carcinoma (BCC) and squamous cell carcinoma (SCC). Tin etiopurpurin dichloride ($SnET_2$) (Purlytin) and *meta*-tetrahydroxylphenyl chlorin (*m*-THPC, Temoporfin) (Foscan®) are hydrophobic and their use for dermatological indications has been investigated extensively. Finally, bacteriochlorins might be particularly useful for the PDT of pigmented tumors.

The joining of four benzenes or naphthalene rings to the β-pyrrolic positions of porphyrins and the substitution of the methylene-bridge carbons with nitrogen produce Pc and naphthalocyanines, respectively. The presence of Al (III), Zn (II), Si (IV), Ru (II), and other diamagnetic metal ions with axial ligands gives hexacoordination and guarantees a satisfactory yield of singlet oxygen generation, thus decreasing the tendency to form PS self-aggregates and inducing high photodynamic efficiency and reduced phototoxic side effects (Garland et al. 2009).

In addition, other PS with different modifications of the porphyrin structure are available. Synthetic porphycenes and isomers of porphines are efficient generators of singlet oxygen and inducers of tumor regression. For instance, 9-acetoxy-2,7,12,17-tetrakis-(beta-methoxyethyl)-porphycene (ATMPn) is a chemically pure substance with fast pharmacokinetics and superior photodynamic properties in vitro and in vivo as compared to Photofrin® (Abels et al. 1997; Leunig et al. 1993). Texaphyrins are synthetic, water-soluble expanded-ring porphyrin analogues, in which a phenyl ring replaces one pyrrole ring. They can be easily complexed with large metal cations, such as Ln (III), Lu (III) or Gd (III), to give metal complexes photoactives in vivo with high selectivity and an efficient generation of singlet oxygen. The lack of significant persistent skin phototoxicity is another outstanding characteristic of these compounds. Photosensitivity lasts up to 6 weeks, but strict avoidance of sunlight is needed only for the first 2 weeks (Sessler and Miller 2000). Tetra-sulfonatophenylporphine (TPPS) is a hydrophilic substituted porphyrin that is 25–30 times more effective than HpD and Photofrin® (Calzavara-Pinton et al. 2007; O'Connor et al. 2009).

Non-porphyrin derivates include hypericin, a naturally occurring compound from *Hypericum* plants that shows maximum absorption in the UV (330 nm) and visible (550 and 588 nm) light range, high singlet oxygen production, minimal dark toxicity, and high clearance from the body after administration. Hypocrellins, methylene blue, azure C, methylene violet, thionine, methylene green, Nile blue A, and rhodamine 123 are potential PS for PDT.

When considering the ideal properties that a PS should possess, some features can be easily manipulated, whereas some others cannot be easily controlled, thus prompting to novel formulation approaches based on nanotechnologies.

Each of the currently commercially available PS has specific characteristics, but none of them is an ideal agent. For example, most of the PS are hydrophobic and can aggregate very easily in aqueous media which can affect their photophysical, chemical, and biological properties (Allison and Sibata 2010).

Second generation PS show lower toxicity, but most of them exhibit poor solubility in aqueous media, preventing intravenous delivery into the bloodstream and affecting their efficacy and tumor selectivity. Thus, it becomes a key fundamental to develop more performing compounds with improved deliverability.

Currently, research efforts have been focused on the development of third generation PS, characterized by a higher specificity to target cells, thus resulting in minimized accumulation in healthy tissues. The first approach is based on the conjugation of PS with a targeting component, such as an antibody directed against the tumor antigens, in order to promote the localization and the accumulation of the drug at the diseased site (Staneloudi et al. 2007; Nowis et al. 2005).

The PS bioconjugate is then able to (specifically) photodynamically inactivate tumor cells expressing the tumor-associated antigen, minimizing healthy cell localization and concomitant damage. An alternative approach in the development of third generation PS would be their delivery through nanoscale carrier (nanoPDT).

5.4.2 Clinical PDT for Cancer

PDT has been utilized for pre-neoplastic and neoplastic diseases in a wide variety of organ systems, including skin, genitourinary, esophagus, prostate, bile duct, pancreas, head and neck, and brain. Several medicines have been approved or are currently in clinical trials (Table 5.2).

Skin cancers and precancers have been among the first to be studied in PDT due to their accessibility to PS and external light. In the definitive setting, PDT is currently approved for the treatment of AK, BCC, and SCC.

Successful results for PDT of nonhyperkeratotic AK have been achieved with systemically administered porfimer sodium as well as topically applied ALA and methyl-ALA (MAL). Twenty-eight randomized clinical trials (RCTs) that reported the use of PDT in the treatment of AK have been identified (Lim 2002), and aggregated data indicate better rates of complete response and better cosmetic results with PDT than with the other treatments (Biel 2010).

Thirteen RCTs on superficial and nodular BCC have been reported, comparing ALA-PDT with surgical excision, cryotherapy, or placebo. In particular, for superficial BCC, the outcome after PDT appears similar to surgery or cryotherapy, whereas for nodular (deep) BCC, PDT is less effective than surgery for lesion clearance. Finally, PDT can substantially reduce the size of large SCC tumors reducing morbidity and increasing overall curative response (Kostron 2010).

Regarding ophthalmic disease, the only clinically approved PS for PDT treatment is Verteporfin (Visudyne®). In particular, it has been approved for the treatment of age-related macular degeneration (AMD) worldwide since 2000. AMD therapy involves iv administration of verteporfin followed by activation through an ophthalmoscope equipped with a 690 nm diode laser while the PS is still in the general circulation. Several RCTs are currently under way to evaluate the efficacy of AMD PDT for use with other PS, such as $SnET_2$, motexafin lutetium, and NPe6 (Eljamel 2010).

Table 5.2 PS for PDT of cancer (adapted from Anand et al. 2012)

Trade name	Photosensitizer	Structure	Excitation λ (nm)	Cancer indication
Photosensitizers in clinical trials[a]				
Hexvix; Benzvix	ALA esters (hexyl-ALA/benzyl-ALA)	Porphyrin precursor	635	Bladder, gastrointestinal, and skin
Lutex	Lutetium texaphyrin/motexafin lutetium	Texaphyrin	732	Brain, breast, cervical, and prostate
Pc4	Silicon phthalocyanine 4	Phthalocyanine	675	Cutaneous T cell lymphoma
Photochlor	2-(1-hexyloxyethyl)-2-devinylpyropheophorbide-alpha (HPPH)	Chlorin	665	Basal cell carcinoma, esophagus, head and neck, and lung
Purlytin	Tin ethyl etiopurpurin (SnEt2)	Chlorin	664	Breast, prostate, Kaposi's sarcoma and skin cancer
Tookad	Padoporfin	Bacteriochlorin	762	Prostate cancer
Verteporfin	Benzoporphyrin derivative-monoacid ring A (BPD-MA)	Porphyrin	690	Basal cell carcinoma (also approved as Visudyne™ for age-related macular degeneration, a non-cancer condition)
WST-11	Padeliporfin	Bacteriochlorin	753	Prostate cancer
Clinically approved photosensitizers[b]				
Foscan	*Meta-tetrahydroxyphenylchlorin* (*m*-THPC, Temoporfin)	Chlorin	652	Cervical cancer (Japan), esophageal cancer and dysplasia (Canada, EU, USA, Japan), gastric cancer (Japan)
Fotoditazin	Chlorin e6 + chlorin p6	Chlorin	660	Skin cancer (Russia, S. Korea)
Fotolon	Chlorin e6 + polyvinylpyrrolidone	Chlorin	660	Skin, vulva, cervix, oral cancer (Russia)
Laserphyrin	Taporfin sodium (Talaporfin, mono-(L)-aspartylchlorin-e6, MACE, NPe6, LS11)	Chlorin	664	Lung cancer (Canada, EU, Japan, USA), head and neck cancer (EU), lung cancer (Japan)
Levulan	5-Aminolevulinic acid (ALA)	Porphyrin precursor	635	Actinic keratosis (Canada, USA)
Metvix, Metvixia	Methylester of 5-ALA	Porphyrin precursor	635	Actinic keratosis (Canada, USA)
Photofrin	Porfimer sodium; also called hematoporphyrin derivative (HPD)	Porphyrin	630	Basal cell skin cancer and squamous cell carcinoma in situ (EU), bladder cancer (Canada)
Photosense	Aluminum sulfonated phthalocyanines	Phthalocyanine	675	Skin, vulva, oral, esophagus and stomach; breast metastases (Russia)

[a]Compounds in the upper table are still under investigation. Also, note that many clinically approved PS in the lower table are being investigated beyond their originally approved indications

[b]Compounds, diseases, and countries in which clinical uses for each disease indication are fully approved

In the field of head and neck cancer, over 1,500 patients have been treated with PDT (Simone et al. 2012) with Photofrin®, 5-ALA, Foscan®, and Photochlor by systemic delivery; in particular, Foscan® was approved in Europe in 2001 for the palliative treatment of patients with advanced head and neck cancer who have exhausted other treatment options. Furthermore, various formulations of porfimer sodium, 5-ALA, and Temoporfin are currently undergoing intensive clinical investigation as an adjunctive treatment for brain tumors, such as glioblastoma multiforme, anaplastic astrocytoma, malignant ependymomas or meningiomas, melanoma and lung cancer brain metastasis, and recurrent pituitary adenomas (Minnich et al. 2010). Currently, PS are being evaluated both as intraoperative diagnostic tools by means of photodetection (PD) and fluorescence guided resection (FGR) as well as during PDT as an adjunctive therapeutic modality. The most recently published trials that employed PD, FGR, and PDT provided additional encouraging results, but the initial delay in tumor progression did not translate to extended overall survival (Friedberg 2011).

PDT is increasingly being used to treat cancers of the airways and other tumors in the thoracic cavity, especially non-small cell lung carcinoma (NSCLC) (Wiedmann and Caca 2004). Different RCTs based on talaporfin or porfimer sodium-mediated PDT showed good results and complete response rate in patients with early stage lung cancer or for whom surgery is not feasible (Ortner 2011).

In gastroenterology, endoscopically accessible premalignant or malignant lesions located within the esophagus, the stomach, the bile duct, or the colorectum with a high surgical risk have become suitable targets of endoscopic PDT (Bown et al. 2002). Photofrin®-PDT has been approved for obstructing esophageal cancer, early-stage esophageal cancer, and Barrett's esophagus in several countries, as an alternative to esophagectomy because these are superficial and large mucosal areas that are easily accessible for light. Recent pilot studies have demonstrated that endoscopic Photofrin®-PDT is also effective in the palliative treatment of hilar cholangiocarcinoma (Anand et al. 2012), for early duodenal and ampullary cancers as well as for advanced adenomas.

Due to advances in light applicators, the interstitial PDT is now becoming a practical option for solid lesions, including those in parenchymal organs such as the liver and pancreas (Parhi et al. 2012). Talaporfin-mediated PDT may have efficacy in treating hepatocellular carcinoma, whereas Foscan® looked promising in the treatment of pancreatic cancer (Kessel and Erickson 1992). In the case of prostate cancer, Foscan®, Tookad Tookad, and Lutex Tookad, looked minimally invasive alternatives to surgery or radiotherapy, reducing the risk of the post-surgical side effects of incontinence and impotence. Bladder cancer tends to be a superficial condition, and for this reason it is proposed that a superficial treatment mediated with ALA or its ester derivatives, by intravesical instillation, may be a preferable means for local therapy (Zuluaga and Lange 2008).

The last PDT application refers to the treatment of gynecological cancers. For cervical intraepithelial neoplasia, PDT based on chlorine e6 (Fotolon®) or hexyl-ALA offers a nonscarring alternative to cone biopsy. For vulvar intraepithelial neoplasia, use of Foscan® or ALA may ameliorate the need for radical mutilating surgery. Similarly, penile intraepithelial neoplasia and anal intraepithelial neoplasia have been treated with ALA-based PDT, sometimes with complete clearance. Extramammary Paget's disease responds to PDT with porfimer sodium or ALA (Labib et al. 1991).

5.4.3 PDT Combined to Other Treatment Modalities in Cancer

At present, combination of different chemotherapeutic drugs in a chemotherapy regimen is an attractive strategy for effective cancer treatment. In a clinical setting the treated patients were found to fail the experiences of single agent chemotherapy, because it is limited to act on specific cancer survival pathways and showed low response rates and relapse of tumor. To improve the therapeutic potential of cancer chemotherapy, it is essential to establish alternative approaches that could provide a solution to the problems involved in single drug chemotherapy. To this end, much attention has been given to combination approaches for a better long-term prognosis and to decrease side effects associated with high doses of single drug treatment. Unlike single agent therapy, combination therapy can modulate different signaling pathways, maximizing the therapeutic effect by overcoming toxicity and, moreover, can overcome the mechanisms of multi-drug resistance (MDR).

The use of combination therapy for cancer treatment has been well established in recent years and its advantages are illustrated below. One of the prime benefits of combination therapies is the potential for providing synergistic effects. In combination therapy the overall therapeutic benefit of the drugs in combination was found to be greater than the sum of the effects of the drugs individually. These advantages have driven drug discovery efforts toward the search for combination therapies. The best drug combination with maximal antitumor efficacy can be calculated by multiple drug effect/combination index isobologram analysis, an effective way to demonstrate that drugs are working synergistically. The prime mechanism of synergistic effect following combinational drug treatment could act on the same or different signaling pathways to achieve more-favorable outcomes at a lower dose with equal or increased efficacy (Parhi et al. 2012; Lehar et al. 2009).

Combinations of various therapeutic modalities with non-overlapping toxicities are among the commonly used strategies to improve the therapeutic index of treatments in modern oncology. Two general approaches may increase the antitumor effectiveness of PDT: (1) sensitization of tumor cells to PDT and (2) interference with cytoprotective molecular responses triggered by PDT in surviving tumor or stromal cells. Any interactions between PDT and PDT-sensitizing agents will be confined to the illuminated area. Therefore, the potentiated toxicity of the combinations is not systemic. This should be of special importance in elderly or debilitated patients who poorly tolerate more intensive therapeutic regimens. Moreover, considering its unique singlet oxygen-dependent cytotoxic effects, PDT can be safely combined with other antitumor treatments without the risk of inducing cross-resistance (Kessel and Erickson 1992). However, there have been few studies on combinations of PDT with standard antitumor regimens published to date

(Zuluaga and Lange 2008). PDT can be used in combination with surgery as a neoadjuvant, adjuvant, or repetitive adjuvant treatment, preferably image-guided fluorescence to confine illumination to the most suspicious lesions. PDT has also been successfully combined with radiotherapy and chemotherapy (Golab et al. 2003; Szokalska et al. 2009).

Another approach to promote PDT efficacy involves increased PS delivery or impaired loss from tumor cells. The first approach involves conjugation of PS to various tumor-targeting molecules. This may be important in the treatment of tumors where large surface areas are illuminated and hence increased tumor selectivity is desired. The use of compounds that impair PS efflux has also been demonstrated to effectively sensitize tumor cells to PDT, although such approaches seem to be limited to those PS that are the substrates of outward transport systems such as ABCG2 (Liu et al. 2007).

The development of novel target-specific antitumor drugs has enabled examination of a number of concept-based combinations that by various molecular mechanisms sensitize tumor cells to the cytotoxic effects of PDT. Proteins are major targets for oxidative reactions because they constitute nearly 70 % of the dry weight of cells. Oxidized proteins can be refolded by molecular chaperones such as HSPs. Inefficient restoration of their structure leads to accumulation of misfolded proteins and their aggregation, which precipitates cell death. Accumulation of damaged or misfolded proteins within endoplasmic reticulum (ER) triggers a process called ER stress, which can be ameliorated by unfolded protein response or can lead to cell death. Therapeutic approaches that interfere with refolding or removal of oxidized proteins can be used to sensitize tumor cells to PDT. For example, modulation of HSP function with geldanamycin, a HSP90 inhibitor, sensitizes tumor cells to PDT (Ferrario et al. 2007). Bortezomib, a proteasome inhibitor successfully used in the treatment of hematological disorders, potentiates the cytotoxic effects of PDT by aggravation of ER stress (Szokalska et al. 2009). Moreover, several apoptosis-modulating factors such as rapamycin, Bcl-2 antagonists, ursodeoxycholic acid, or ceramide analogues have been shown to increase PDT-mediated cancer cell death (Zuluaga and Lange 2008) (Table 5.3).

5.4.4 Drawbacks in PDT

The efficacy of a PDT treatment depends on multiple factors related to PS photochemical and physico-chemical properties (singlet oxygen production efficiency, tissue penetration of excitation light), PS biodistribution in the body, localization in a specific compartment and dose at target tissue, and the light parameters (light dose, fluence rate, interval between administration and light exposure). Obviously, cancer tissue characteristics (vascularization, oxygenation level) play an important role in determining the therapeutic outcome of PDT.

The low extinction coefficients of PS require the administration of relatively large amounts of drug to obtain a satisfactory therapeutic response. Furthermore, several hydrophobic PS tend to aggregate in physiological conditions and as a

Table 5.3 Combinations of PDT and various therapeutic modalities in cancer treatment (adapted from Agostinis et al. 2011)

Drug or treatment modality	Outcome/results
Chemotherapeutics and novel anticancer drugs	
Anthracyclines	Doxorubicin improves PDT-mediated tumor growth control in mice
Platinum compounds	Cisplatin potentiates antitumor activity of PDT in mice
Antimetabolites	Methotrexate enhances in vitro cytotoxicity of PDT with ALA by upregulation of protoporphyrin IX production
Microtubule inhibitors	Vincristine administered prior to or immediately after PDT improves its antitumor activity in mice
DNA methyltransferase inhibitors	5-Azadeoxycytidine prolongs survival of PDT-treated animals and improves tumor growth control
Proteasome inhibitors	Bortezomib enhances PDT-mediated ER stress in cancer cells in vitro and significantly delays post-PDT tumor regrowth in mice
Radiotherapy: Two-way enhancement of antitumor effects: PDT sensitizes cancer cells to radiotherapy and radiotherapy increases anticancer efficacy of PDT, prolonged tumor growth control induced by combined treatment	
Drugs modulating arachidonic acid cascade	
COX-2 inhibitors	COX-2 inhibitors (such as NS-398, nimesulide, or celecoxib) potentiate antitumor effects of PDT, possibly through indirect antiangiogenic effects
LOX inhibitors	MK-886 sensitizes tumor cells to PDT-mediated killing
Agents increasing PS accumulation in tumor cells	
Vitamin D	Increases ALA-induced protoporphyrin IX accumulation and thus potentiates PDT cytotoxicity in vitro
Imatinib	Increases intracellular accumulation of second-generation PS and thus potentiates PDT cytotoxicity in vitro and in vivo
Lipid-lowering drugs	Lovastatin improves in vitro LDL binding and porfimer sodium uptake by cancer cells
Salicylate	Enhancement of PDT efficacy via increased PS uptake by tumor cells
Approaches increasing oxygen delivery to tumor cells	
EPO	EPO improves chemotherapy-induced anemia and restores antitumor efficacy of PDT in mice118; however, EPO might also inhibit direct PDT-mediated cytotoxicity toward certain cancer cells
Hyperbaric oxygen	Increased antitumor effects of PDT in mice and in advanced pleural tumors in humans
Hyperthermia	In various treatment regimens, hyperthermia potentiates antitumor efficacy of PDT in vitro and in animal models the short time interval between these 2 treatment modalities might increase normal tissue injury via vascular effects
Targeting cytoprotective mechanisms and increasing of radical formation in cancer cells	
Disruption of heme degradation pathway	Targeting of HO˙ with selective inhibitors and siRNA as well as an siRNA-mediated knockdown of ferrochelatase or chelation of iron ions potentiate antitumor effects of PDT
Inhibition of SOD 2	2-Methoxyestradiol, a natural SOD inhibitor, enhances PDT cytotoxicity in vitro and improves antitumor effects of PDT in mice
NO synthase inhibition	Improved tumor response to PDT in mice
HSP90 modulation	Interference with HSP90 client proteins binding using a geldanamycin derivative improves responsiveness to PDT both in vitro and in vivo
Targeting of tumor vasculature	
Antiangiogenic treatment	Anti-VEGF or anti-VEGFR monoclonal antibodies, matrix metalloproteinase inhibitor (prinomastat), TNP-470, and other antiangiogenic agents as well as adenovirus-driven IL-12 expression potentiate antitumor effects of PDT in mice

consequence to produce singlet oxygen with very low yields. This state is one of the determining factors, which can hinder the efficacy of the drug in vivo by decreasing its bioavailability and limiting its capacity to absorb light.

The absorption maximum of PS falls at relatively short wavelengths leading to a poor tissue penetration of light. This has prompted toward development of alternative strategies to improve quantum yields of singlet oxygen such as two-photon induced excitation. This strategy combines the energy of two photons (in the range 780–950 nm) where tissues have maximum transparency to light but where the energy of one photon is not high enough to produce singlet oxygen.

Selectivity remains another key issue in PDT. A PDT treatment can be considered to be selective in that the toxicity to tumor tissue is induced by the local activation of the PS, while normal tissue not exposed to light are spared. Second generation PS show improved selectivity and clearance rate from the body so as increased therapeutic efficiency and mostly alleviated toxicity due to post-PDT photosensitization are experienced. However, this problem is far to be solved and prompts toward finding of third-generation PS.

PS can be administered by either topic or intravenous administration. For topical administration there is no need to delay light application except for drugs that need metabolic pathways to become active (such as 5-ALA). For systemic administration, PS location and extent of PS accumulation in a target tissue depend on post-injection time. At time shorter than PS half-life, the drug predominantly stays in the vascular compartment of the tumor while at longer time, PS can accumulate in extravascular sites due to interstitial diffusion. Therefore drug-light interval may play a crucial role for the therapeutic outcome. Concerning administration, most PS are hydrophobic. This lipophilic nature may be an important factor affecting the preferential accumulation in cellular hydrophobic loci since these molecules must be able to get into the cells by crossing lipid membranes. Due to their minute solubility in water, iv administration of PS is greatly hampered, thus requiring suitable vehicles for both topical and intravenous delivery.

5.5 Nanocarriers in PDT

Nanotechnology offers a great opportunity in PDT based on the concept that a PS formulated in a nanocarrier can alter its pharmacokinetics enhancing the treatment ability to target and kill cancer cells of diseased tissues/organs while affecting as few healthy cells as possible (Chatterjee et al. 2008; Chen 2008). Nanocarrier can also facilitate PS entry inside cells and potentiate photodynamic effects. Finally, nanocarriers offer the opportunity of multimodality since they can be designed to transport one or more drugs/species with different mechanisms of cancer cell killing and can incorporate an imaging agent useful to track the system in the body. Depending on the properties of the carrier (size, surface, shape), large variations in pharmacokinetics, cell uptake and subcellular distribution of PS can occur resulting in major therapeutical implications.

To drive PS distribution in the body after intravenous injection, sterically stabilized nanocarriers with a biomimetic coating (stealth nanocarriers) can be largely beneficial. These systems, generally constituted of a polyethylene glycol (PEG) shell, withstand phagocytosis in mononuclear phagocyte system (MPS) and recognition by the immune system, thus showing an increased blood circulation (Gref et al. 2000; d'Angelo et al. 2014; Huynh et al. 2010). Long-circulating nanocarriers potentially accumulate in pathological sites with compromised leaky vasculature such as that found in tumors (passive targeting). Defective vascular architecture coupled with poor lymphatic drainage is common to many kinds of solid tumors and has been exploited to promote drug accumulation in tumor site (the so-called enhanced permeability and retention, EPR, effect) (Fig. 5.5) (Maeda 2012; Maeda et al. 2013).

A further development of this concept implies exploitation of differences between cancer and normal cells. Surface decoration of nanocarriers with specific ligands able to interact with cell surface structures overexpressed on either endothelial cells of tumor capillary or cancer cells (particular antigens, peptide receptors, folate, transferrin, and integrin surface receptors) is a fascinating option for improving the specificity of anticancer treatments (active targeting) (Fig. 5.5). Nevertheless, "stimuli-sensitive" nanocarriers can trigger release of drug cargo in response to environmental conditions in a tumor (lower pH, presence of specific enzymes) (Wang and Thanou 2010).

Another advantage of nanocarriers with sustained release properties relies on the prolongation of drug half-life in the circulation and at target site, which could in theory reduce the number of administrations with consequent improvement of the

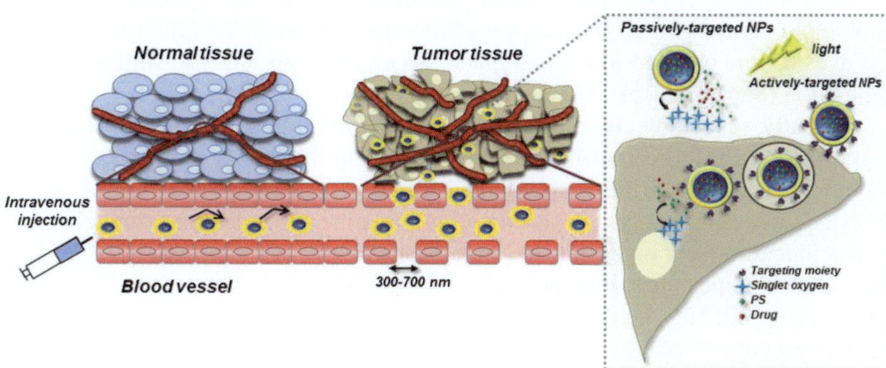

Fig. 5.5 Accumulation of nanocarriers in solid tumors after intravenous injection. Long-circulating nanocarriers potentially accumulate in a tumor site through enhanced permeability and retention (EPR) effect (passive targeting). Nanocarriers can be decorated with ligands able to interact with cell surface receptors overexpressed on either endothelial cells of tumor capillary or cancer cells (active targeting). Following extravasation, light can activate the PS before and/or after its release from the nanocarrier located in the extracellular tumor matrix. Finally, light can activate the PS directly inside the cellular compartments, after nanocarrier uptake has occurred. In some cases an anticancer drug can be combined with PS to increase cytotoxicity

patient compliance. Improved delivery to cancer cells results in an optimized pharmacological response and milder toxicity profile. It has been suggested also that drug accumulation and slow release inside tumor cells can be useful to circumvent MDR.

The ideal drug delivery system for PDT should enable the selective accumulation of the PS within the diseased tissue and the delivery of therapeutic concentrations of PS to the target site with little or no uptake by non-target cells. The carrier should be able also to incorporate the PS without loss or alteration of its photoactivity. Another reason for using nanocarriers is to provide an environment where the PS can be administered in a monomeric form and maintain its photochemical properties also in an in vivo setting. Poor solubility of PS can benefit of the inclusion in a nanocarrier which can also drive PS with a tendency to aggregation in aqueous environment toward their monomeric state. Also biological stability of some PS can be improved by incorporation in certain nanocarriers.

As mentioned before, commonly used PS are activated with visible light with low penetration depth, limiting PDT to the treatment of superficial lesions. To overcome this limitation, Förster resonance energy transfer (FRET) has been suggested as a possible approach to improve generation of singlet oxygen under NIR irradiation (Sapsford et al. 2006a). FRET is a non-radiative energy transfer from a photo-excited donor to an acceptor molecule after absorption of a higher energy photon. Thus in FRET a donor with high absorption coefficient and fluorescence quantum efficiency is excited with NIR light and activates an acceptor (PS) at a wavelength suitable to generate singlet oxygen (Chen et al. 2008). NPs designed for FRET have to take into account that a nanoscopic critical distance is required between donor and acceptor. On the other hand, two-photon absorption-induced excitation is based on the absorption of two NIR photons providing the same energy level to a single photon of visible light. To this purpose, either new PS can be designed or two-photon dyes can be coupled to PS and make use of FRET to increase singlet oxygen generation (Sapsford et al. 2006a, b).

Currently, several nanosized carriers, made of different materials have been proposed in nanoPDT (Fig. 5.6). Each type of system highlights pros and cons and should be selected on the basis of delivery requirements. In the following, we describe some nanocarriers proposed for PDT treatment that demonstrated potential in preclinical models together with emerging concepts for their advanced design. These include liposomes, polymer nanoparticles (NPs) and micelles, and inorganic NPs (silica, gold NPs, etc.). Two-photon excitation or optical upconversion can be used instead of one-photon excitation to increase tissue penetration at longer wavelengths. In some cases NPs itself can act as a PS as in the case of quantum dots (QDs).

5.5.1 Liposomes

Liposomes have been extensively studied as drug delivery systems for both local and systemic applications. Liposomes are biodegradable lipid vesicles composed by one (monolamellar) or more (multilamellar) bilayers surrounding an aqueous inner core.

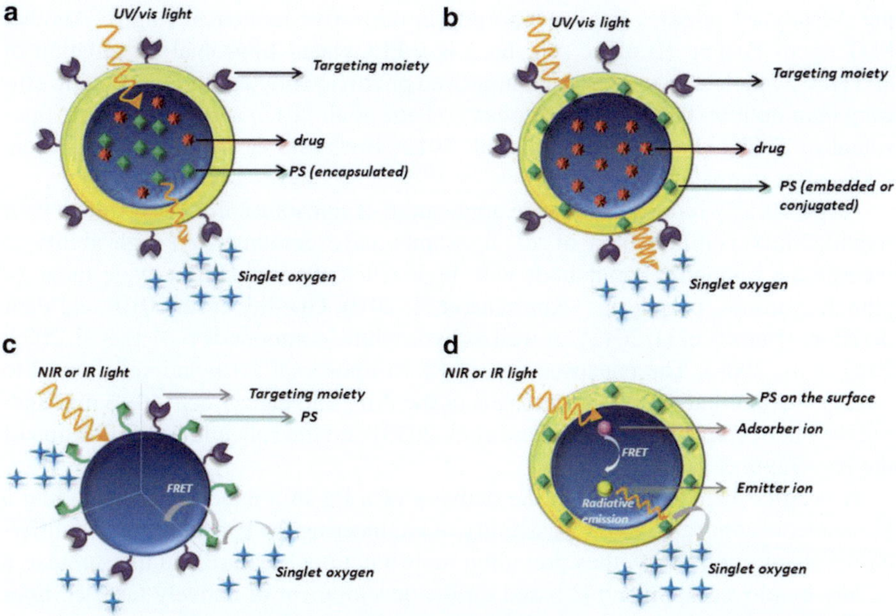

Fig. 5.6 Sketch representation of some types of NPs for PDT. PS can be entrapped (**a**) inside the core of NPs, alone, or in association with a conventional anticancer drugs or (**b**) exposed on their external shell through non-covalent interactions or chemical conjugation. PS is then activated by specific UV/vis irradiation, thus producing cytotoxic singlet oxygen. A further approach is represented by semiconductor nanocrystals including quantum dots (**c**) with intrinsic photophysical properties and able to absorb IR and NIR light. After activation, QDs can transfer energy directly to surrounding oxygen or to PS eventually linked on the surface, with the final production of singlet oxygen. Upconvertion NPs (**d**) can generate higher energy light from lower energy radiation through the use of transition metal, lanthanide, or actinide ions contained in the core matrix. After activation of the adsorber ion, a non-radiative energy is produced and activates an emitter ion. A consequent radiative emission can finally activate the PS

Phospholipids and cholesterol are commonly used for assembling liposomes able to incorporate both hydrophilic and lipophilic compounds. In the field of PDT for cancer, liposomes have been intensively employed as biocompatible carrier systems of PS since the late 1990s (Jiang et al. 1997, 1998), especially due to their propensity to entrap lipophilic PS in the lipid layers, thus preventing their fast aggregation in aqueous environments (Chen et al. 2005; Derycke and de Witte 2004; Skupin-Mrugalska et al. 2013). Furthermore, the high versatility of this carrier is underlined by the suitability to encapsulate also hydrophilic PS in the internal aqueous core. Lipid properties, lipid composition, liposome preparation method, size, targeting ligands, and triggered release strategies have been taken into account in order to optimize PDT anticancer efficiency, depending also on the final administration route.

Visudyne® is a successful example of non-PEGylated liposomal formulation approved by the FDA for PDT AMD, ocular histoplasmosis, and pathologic myopia. Visudyne® releases the benzoporphyrin derivative monoacid ring A. Another PDT agent FosPeg (Biolitec AG, Inc.) is a PEGylated liposomal formulation of *m*-THPC, which has demonstrated improved pharmacokinetics and therapeutic efficacy than both *m*-THPC alone (Foscan®) (Petri et al. 2012) and conventional liposomal *m*-THPC (Foslip®) (Bovis et al. 2012; Buchholz et al. 2005; Pegaz et al. 2006) after intravenous injection.

As previously reported, the wide application of liposomes in PDT is due to their remarkable versatility. First of all, liposomes have demonstrated high ability to incorporate lipophilic compounds with high efficiency including a wide range of phthalocyanines, porphyrins (Kuntsche et al. 2010; Guelluy et al. 2010) and their derivates (Sutoris et al. 2013), as well as hydrophilic compounds (Casas et al. 2002; Fang et al. 2008). The encapsulation of PS in liposomal formulations allowed to modify the internalization mechanisms of the drug into cancer cells, thus modulating the endocytic pathways (Soriano et al. 2013) and the subcellular localization of the PS (Gupta et al. 2011).

A peculiar benefit related to the delivery of a PS in a nanocarrier, including a liposome, is connected to the possibility of engineering the carrier system at different levels of complexity, thus providing multifunctional features. In this context, a widely employed approach is based on the development of actively targeted liposomes through surface decoration with functional ligands targeted to cancer cells. In so doing, a bifunctional selectivity against the tumor site is achieved, thus combining the selectivity of the PDT based on the irradiation of a specific pathological area associated with specificity of the targeted surface. As an example, folate-conjugated liposomes were used to deliver AlPc tetrasulfonate ($AlPcS_4$) (Qualls and Thompson 2001), *m*-THPC (Moret et al. 2013), and Zn tetraphenylporphyrin (ZnTPP) (Garcia-Diaz et al. 2011). Transferrin conjugated PEG-liposomes were also tested to improve cellular endocytosis of the carrier as well as cytotoxicity (Derycke and de Witte 2002; Gijsens et al. 2002). Another strategy relies in the delivery of PS through liposomes targeted to specific tumor cells with antibodies (Mir et al. 2013).

Finally, as previously described, in order to enhance the anticancer effect of PDT, research has recently focused on the combination of multifunctional carrier systems for the delivery of PS and other treatment modality. In the field of liposomes, a recent study showed that the entrapment of ZnPc and cucurbituril complexed in the same liposomal formulation allowed an improvement of the cytotoxic effect of the system against melanoma cells, thus demonstrating that PDT and magnetohyperthermia are more effective in combination (Bolfarini et al. 2012).

5.5.2 Nanoparticles

A variety of materials of synthetic and natural origin have been used to develop PS-loaded NPs. Most of them are polymeric and PS are generally physically entrapped in the matrix. Other NPs are inorganic and PS can be either physically introduced (silica) or covalently bound to NP surface.

The design of NPs is very versatile (Fig. 5.6). In their simplest design, NPs are inert and simply carry a PS. In a more complex design, light-harvesting NPs (e.g., two-photon absorption NPs) (Dayal and Burda 2008) and upconverting NPs (Chatterjee and Yong 2008) use different strategies to expand the applicable excitation wavelengths for PDT. Some NPs have dual functions and act also as PS themselves. For example, quantum dot (QD)-based NPs with tunable absorption and size hold potential as new PS (Bakalova et al. 2004). Indeed, they can transfer energy to PS or directly to oxygen and generate singlet oxygen (Samia et al. 2003, 2006).

There are also some designs where NPs are made of PS (Hu et al. 2009) or PS-containing building blocks. Multifunctional NPs have been built to provide besides PDT, delivery of chemotherapeutics, image contrast enhancement (fluorescence imaging and MRI), or other therapies (hyperthermia or radiotherapy).

Depending on base material, some NPs are biodegradable and others are not. PS can be loaded in NPs through encapsulation, covalent linkage, or post-loading. Encapsulation relies on physical entrapment of PS in NP matrix on the basis of hydrophobic or electrostatic interactions and hydrogen bond formation. Aggregation of PS inside the matrix has to be controlled in order to avoid loss of photodynamic efficiency. In the post-loading method, PS are added to preformed NPs by equilibrium in solution. The latter method is simple to perform although NPs suffer of premature PS leaching which can be a drawback for in vivo application. Covalent binding of PS to NPs is difficult to attain and requires either attachment of PS to monomers that are then polymerized or self-assembled in NPs or to post-modification of preformed NPs. Advantages of this strategy consist in preventing PS leaching from NPs and avoid PS aggregation in biological environment.

It is worth of note that release of PS from NPs is not considered determinant since molecular oxygen can penetrate NPs and produced singlet oxygen can diffuse out of NPs to induce photodynamic reactions. In such cases, PDT efficiency depends on NP type (size and oxygen permeability of the matrix) as demonstrated by Tang et al. (2005).

Photochemical properties are assessed through in-depth examination of light-absorbing properties and ability to produce singlet oxygen through chemical probes (water soluble anthracene-9,10-dipropionic acid disodium salt) or direct evaluation by NIR phosphorescence.

5.5.2.1 Polymeric NPs

Biodegradable polymeric NPs have received tremendous attention for delivering antineoplastic conventional agents for their capacity of high drug loading, the possibility of controlling the drug release, and the existence of a large variety of materials and manufacturing processes allowing to tune their overall properties. Investigation into biodegradable polymeric NPs for use in PDT began as early as 1990 with efforts to enhance carrier capacity and control drug release.

In 1991, Labib et al. reported the encapsulation of ZnPc tetrasulfonate ($ZnPcS_4$) or AlPc into poly(isobutyl cyanoacrylate) or poly(ethylbutylcyanoacrylate) nanocapsules or nanospheres (Labib et al. 1991). Almost all subsequent research have utilized poly(lactic acid) (PLA) and poly(lactic-*co*-glycolic acid) (PLGA) NPs due to their better safety profile (degradation products are water and carbon dioxide) and controlled degradation rate.

In vitro and in vivo photodynamic activities of verteporfin-loaded PLGA NPs intended for iv administration were investigated (Konan-Kouakou et al. 2005). Two types of verteporfin-loaded PLGA NPs (167 and 370 nm in diameter) were prepared (Konan et al. 2003). A higher photocytotoxic effect was evident in the case of smaller sized NPs due to a higher extent of intracellular uptake. Furthermore, in vivo studies on rhabdomyosarcoma-bearing DBA/2 mice demonstrated that verteporfin-loaded small NPs effectively controlled tumor growth for 20 days with early light irradiation times following drug administration. The efficacy of *m*-THPP loaded PLGA and PLA NPs was investigated on the chick embryo model (Vargas et al. 2004). It was proven that PDT-induced vascular occlusion of *m*THPP was enhanced when encapsulated into NPs due to a longer residence time inside the vasculature. In another study, Ricci-Junior and Marchetti (2006) reported the preparation, characterization, and phototoxicity of PLGA NPs delivering ZnPc which showed a strong phototoxicity on P388-D1 cells. A similar PLGA nanosystem entrapping ZnPc was developed and tested on tumor-bearing mice (Fadel et al. 2010). Other PS that have been entrapped inside PLGA NPs include indocyanine green (ICG) (Saxena et al. 2006) and hypericin (Zeisser-Labouebe et al. 2006) which have the potential to be used for both diagnostic and therapeutic purposes (Chatterjee et al. 2008). Recently, hypocrellin was entrapped in PLGA NPs and used for PDT treatment of SCC growing subcutaneously in syngeneic mice (Korbelik et al. 2012).

Use of PEGylated NPs prepared from PLA, PLGA, and polycaprolactone (PCL) covalently linked with a PEG segment has been recently attempted with the aim to increase circulation time of the nanocarrier after iv injection and promote tumor accumulation through EPR effect. Temoporfin-loaded core-shell PEG-PLGA NPs were prepared and in vivo activity assessed on athymic nude-Foxn1 mice (Rojnik et al. 2012). Despite numerous studies on PLGA NPs, an open issue relies in the actual mechanism underlying potentiated effect of PS when delivered through NPs. Indeed, if PS are released from NPs outside the cells, sustained release can alter the amount of PS available to cells. On the other hand, strong association of PS to NPs can result in improved PS uptake and peculiar localization inside cells. In this last case, photoactivity of PS-loaded NPs and ability to generate singlet oxygen under light exposure can be strongly impaired by the presence of a polymer shield, resulting in very low quantum yield of the whole system. It is worth of note that PEGylation results also in a poor cell uptake of NPs as compared to non-PEGylated NPs, thus resulting in predominant extracellular effects.

Also polysaccharides (chitosan, alginate) and proteins (albumin) were tested for PS incorporation and delivery. *m*THPP was incorporated in biodegradable modified chitosan NPs and in vitro photocytotoxicity and cellular uptake investigated (Reza et al. 2011). Chemical conjugation of the PS chlorin e6 and PpIX to modified chitosans allowed to control release rate of chlorin e6 and to achieve consistent antitumor effects (Lee et al. 2011a, b).

Polyacrylamide (PAA) NPs are hydrogel-like nanostructure prepared by polymerization of a nanoemulsion template. PAA NPs can be prepared from biodegradable and nonbiodegradable cross-linkers and decorated on the surface with targeting ligands. PAA NPs encorporating different PS have been tested as potential anticancer treatment (Koo et al. 2006; Reddy et al. 2006; Tang et al. 2005, 2008; Wu et al. 2009). PAA NPs delivering Photofrin® and surface-decorated with F3 peptide were demonstrated to be active both in vitro and in vivo after iv injection (Reddy et al. 2006). Improved survival was found only in the animal group treated with targeted NPs where 40 % of animal were tumor-free after 6 months from PDT treatment.

The idea to combine two drugs with different mechanisms of action and pharmacokinetics in a nanocarrier with well-tailored properties can allow a control over anticancer drug/PS biological fate and promote co-localization in the same area of the body. This approach is rather recent and demonstrated to be a promising strategy to overcome tumor drug resistance in a mice tumor model treated with doxorubicin in combination with the PS methylene blue (Khdair et al. 2010). In particular, surfactant-polymer hybrid NPs formulated using an anionic surfactant, Aerosol-OT, and a naturally occurring polysaccharide polymer, sodium alginate, were used for synchronized delivery of the two drugs. NP-mediated combination treatment resulted in enhanced tumor (mammary adenocarcinoma) accumulation of both doxorubicin and methylene blue, significant inhibition of tumor cell proliferation, and increased induction of apoptosis. More recently, core shell nanoassemblies based on PEG–PCL diblock copolymers were prepared for the delivery of docetaxel and ZnPc inspired by combination therapy of cancer. These systems showed superior antitumor activity as compared to the free drugs in an mice model of orthotopic amelanotic melanoma (Conte et al. 2013).

To implement multimodality of a cancer therapy, very promising results have been obtained also by combining PDT to conventional chemotherapy and other light-activated therapies. Indeed, cytotoxic effects induced by nitric oxide (NO) (Fukumura et al. 2006) combined with PDT represent a very promising strategy in view of a multimodal cancer treatment due to a potential to attack biological substrates of different nature, to avoid multiple drug resistance and to improve selectivity of therapy. Finally, since NO release is independent on O_2 availability, it can potentially very well complement PDT at the onset of hypoxic conditions. Interesting results in this field have been obtained by employing light-activated nanoassemblies delivering nitric oxide, PS, and conventional anticancer drugs (Fraix et al. 2013a, b; Kandoth et al. 2012; Sortino 2010; Swaminathan et al. 2014).

5.5.2.2 Polymeric Micelles

Polymeric micelles are supramolecular spherical nanostructures that have been attracting intensive interest due to their ability to accommodate lipophilic drugs in the core and to provide a shell at biological interface. They are formed from amphiphilic block copolymers (ABCs) which are materials comprising hydrophilic and hydrophobic regions with opposite affinities for an aqueous solvent. ABCs with proper building block chemistry, molecular weight, hydrophilic/hydrophobic balance can form spontaneously supramolecular aggregates in aqueous media driven by the decrease of free energy in the system because of the removal of hydrophobic fragments from the aqueous environment. Their lipophilic core is highly suited for the delivery of hydrophobic PS and their size is small enough to benefit of EPR effect.

A variety of biocompatible polymeric micelles have been studied to deliver hydrophobic PS. Sibata et al. (2004) developed PEG-phosphatidyl ethanolamine (PE)-based micelles entrapping ZnPc which demonstrated enhanced fluorescence quantum yields, longer lifetime of triplet excited state, and good stability of the entrapped drug. Following this study, Roby et al. (2006) described the application of similar micelles as well as tumor-targeted immunomicelles (with anticancer monoclonal 2C5 antibody) delivering the poorly water soluble meso-tetraphenylporphyrin.

Other diblock copolymer-based micelles such as polyethylene glycol PEG–PCL have great potential to deliver hydrophobic PS (Knop et al. 2009). Li et al. (2007) studied PpIX-loaded PEG–PCL micelles for PDT. Compared to free PpIX, PpIX-loaded micelles showed a higher cellular uptake as well as a better PDT effect upon irradiation in RIF-1 cells.

Kataoka group studied polyion complex micelles of PEG-poly(L-lysine) block copolymers with an anionic dendrimer Pc (DPc) (Jang et al. 2006). The PDT effect of DPc was two orders of magnitude higher than the free DPc at the same irradiation time. In vivo PDT efficacy of dPc-loaded micelles (Nishiyama et al. 2009) was higher than free DPc and Photofrin®. Furthermore, the skin phototoxicity of the DPc-loaded micelles was significantly reduced after white light irradiation.

Stimuli-sensitive delivery systems are based on the concept that drug cargo can be released in response to an internal factor (pH variation in a specific compartment) or an external factor (temperature increase). Since tumors usually have a pH lower than healthy tissues, pH-responsive micelles made of poly(2-ethyl-2-oxazoline)-*b*-poly(D-L-lactide) entrapping *m*-THPC were developed to deliver PS at pH around 5 and suppress release at pH 7.4. Nevertheless, micelles displayed in vivo PDT activity similar to that of free *m*-THPC while strongly attenuating skin phototoxicity (Shieh et al. 2010).

Thermosensitive biodegradable micelles based on methoxy poly(ethyleneglycol)-*block*-poly(*N*-(2-hydroxypropyl)methacrylamide-dilactate) with a hydrophobic solketal-substituted Pc (Si(sol)2Pc) have been investigated (Rijcken et al. 2007). PS concentration controlled the aggregation state and resulting photodynamic efficiency of micelles. Micelles loaded with low concentration of Si(sol)2Pc had a similar PDT effect compared to the free PS with 10 % serum.

Two-photon absorbing block copolymer micelles have been developed recently and demonstrated to transfer energy and enhance the singlet oxygen generation of a hydrophobic porphyrin (Chen et al. 2007).

5.5.2.3 Cyclodextrin-Based Nanoassemblies

Self-assembled NPs are formed by proper selection of monomer building blocks that spontaneously form a well-oriented three-dimensional structure incorporating PS. One of the topics that some of us developed is the design of self-assembled systems formed by PS and cyclodextrins (CDs) as third generation PS for PDT (Mazzaglia 2011). CDs comprise a family of biocompatible cyclic oligosaccharide nanocages, which offer both in monomeric and in aggregate form functional scaffolds to bind PS guests by supramolecular interactions, preserving their photodynamic properties (Mazzaglia et al. 2012). PS embedded in cationic amphiphilic CD (aCD) nanoassemblies are effective in inducing photodynamic damage in cancer cells. A peculiarity of this strategy relies on the preservation of a considerably high quantum yield of singlet oxygen when PS is entrapped in the aCD nanocarrier with respect to its free form in aqueous medium (Sortino et al. 2006). A further interesting aspect of these amphiphilic nanomaterials is that the PS units are entangled in the aCD network but not included in the interior or proximity of the CD cavity. This offers the opportunity to exploit the empty cavity of the NPs for the accommodation of additional guests.

On these basis, a multifunctional photoactive molecular constructs possessing simultaneous photogeneration of singlet oxygen and NO and imaging capacities in living cells has been recently developed exploiting the different affinity of an anionic PS and a tailor-made NO photodonor incorporating an adamantan portion (an excellent CD encapsulator) for the diverse compartments of aCD nanoassemblies. The combination of the dual photodynamic action and the imaging capacities in one single nanostructure, together with its biocompatibility, makes this supramolecular construct an appealing candidate for applications in cancer research (Kandoth et al. 2012). In line with the latter findings, our group designed an emitting nanoassembly composed of a novel aCD functionalized with a covalently appended fluorophore and an anionic PS. This nanoassembly internalizes effectively in tumor cells, allowing simultaneously the detection of the carrier and the PS (Mazzaglia et al. 2011).

5.5.2.4 Silica NPs

Silica NPs are certainly very promising systems in drug delivery. Exquisite control over size, shape, and porosity, very low (less than 50 nm) and uniform size can be attained through appropriate control of preparation parameters. Both non-covalent encapsulation and covalent conjugation have been employed as strategies to immobilize PS inside or on the surface of silica NPs. The porous structure with large surface area and pore volume can promote drug loading and protect cargo from degradation in the body. Furthermore, the surface can be easily modified to attain proper functionalization.

The first paper reporting the use of ceramic-based NPs as a novel drug-carrier system for PDT used silica-based spherical particles encapsulating the anticancer drug 2-devinyl-2-(1-hexyloxyethyl) pyropheophorbide (HPPH) alpha (Roy et al. 2003). Following this pioneering paper, several other PS have been entrapped inside silica and demonstrated to give higher quantum yields of singlet oxygen generation as compared to free PS.

Energy transferring organically modified silica (ORMOSIL) NPs for two-photon PDT were developed where an aggregated two-photon fluorescent dye (9,10-bis[4′-(4″-aminostyryl)styryl] anthracene) efficiently converted the NIR light energy and transferred it to HPPH. To avoid premature release in the bloodstream, covalent bonding of PS molecules onto ORMOSIL NPs creates more stable formulations (Ohulchanskyy et al. 2007; Kim et al. 2007).

5.5.2.5 Gold NPs

Gold NPs (AuNPs) are particularly appealing in PDT due to their biocompatibility, stability, size control, and ease of surface functionalization. Both covalent and non-covalent approaches have been applied to transport PS (Cheng et al. 2008; Hone et al. 2002).

For covalent linking, most of the PS have to be modified with thiols for direct attachment on Au NPs. In this configuration, their premature release can be hampered. Furthermore, if AuNPs and PS are very close, the excited states of the PS can be quenched due to energy transfer and, therefore, photodynamic effects can be lowered (Fan et al. 2003; Huang and Murray 2002).

PDT outcomes are strongly affected by the bond type between PS and NP surface. Cheng et al. (2010) prepared AuNPs conjugates loaded with a Pc and tested their PDT efficacy in HeLa cancer cells. Pharmacokinetics and phototherapeutic properties of a free Pc (C11Pc) and Au NP-bound C11Pc in a murine tumor model (amelanotic melanoma) were investigated (Camerin et al. 2010). It was shown that while labile amino adsorption to the AuNP surface allows efficient PS release into cancer cells and efficient PDT, a covalent thiol bond leads to PS delivery into cell suppressing PDT effect.

Recently, Khaing Oo et al. (2012) studied the size-dependent enhancement of ROS production by decorating Au NP of varying size with PpIX. In vitro study on the ROS formation produced by PpIX-conjugated AuNP in human breast cancer cells (MDA-MB-231) revealed size-dependent enhancement of intracellular ROS formation, this effect being greatly dependent on cellular uptake of AuNPs.

Interaction of PS and AuNPs on the basis of electrostatic interactions is highly desirable but very complicate to achieve. Recent findings from Burda group have demonstrated that PEGylated Au nanoconjugates can embed the PS Pc_4 presumably due to interaction between PS and Au surface (electrostatic, hydrophobic) as well as a hydrotropic effect of PEG chains (Cheng et al. 2010). High efficient in vitro PDT activity without internalization of AuNPs was found. Fast (within minutes) tumor accumulation of Pc_4 was found in vivo resulting in tumor shrinkage without evident side effects (Chen et al. 2014). Gold core can be used also for plasmonic imaging and hyperthermal treatment. This latter strategy, named photothermal therapy of tumors (PTT), uses the ability of AuNPs to absorb light in the visible and NIR regions. This is due to coherent oscillation of free electrons in the electromagnetic field of light with particular frequency. This process is known as surface plasmon resonance oscillation decays through scattering or releasing heat, which makes imaging and PTT feasible. Thus, PTT joined with PDT can generate from the same nanoplatform heat and ROS under visible or NIR light.

AuNPs, nanorods, nanocages, and nanoshells, free or coupled with other colloids and decorated with receptor targeting groups, are useful platforms to link chemotherapeutics or PS, thus improving the efficacy and selectivity of intracellular-delivery with overall healing effect (Dreaden et al. 2012). Most of these nanosystems are not photoactivated with a single radiation due to the difference of light absorption wavelengths between AuNPs and PS. Khlebtsov et al. described novel composite NPs consisting of a gold–silver nanocage core and a mesoporous silica shell functionalized with the PS Yb2,4-dimethoxyhematoporphyrin (Yb-HP) which can be accumulated in mice bearing Ehrlich carcinoma tumors. The nanocomposite generated singlet oxygen upon 630 nm excitation and produced heat under laser irradiation at the plasmon resonance wavelength (750–800 nm). The simultaneous diagnostic and PDT effects are assessed by IR-Luminescence, due to absorption in the tissue-transparent window of Yb^{3+} (900–1,060 nm), furnishing to this systems attractive theranostic properties (Khlebtsov et al. 2011).

Based on this concept, a supramolecular hybrid based on spontaneous assembly in aqueous medium of AuNPs, aCDs, and an anionic porphyrin showed an enhanced therapeutic action as compared to PS alone incorporated in aCD by irradiating with two light source AuNPs and PS, respectively (Trapani et al. 2013). By matching the absorption wavelengths of inorganic nanoplatforms with the excitation wavelengths of PS, simultaneous PDT and PTT upon single-laser irradiation can be performed. As intriguing examples with high potential to act as bimodal phototherapeutics, a complex of gold-nanorod (AuNR) with aluminum $AlPcS_4$ generated highly effective PDT/PTT dual therapy with a further benefit for NIR fluorescence imaging of tumor sites (Jang et al. 2011). AuNR and ICG-encapsulated chitosan hybrid nanospheres were successfully prepared and used for PTT and PDT combined therapy by triggering with a single NIR irradiation (Chen et al. 2013).

5.5.2.6 Quantum Dots

QDs are semiconductor nanocrystals with intrinsic photophysical properties and able to absorb light from UV to IR region depending on size and composition. High emission quantum yields again in UV–NIR regions and high photostability in a specific size range (1–6 nm) are found. QDs can be used to deliver ionizing radiation in a specific location in the body, thus circumventing the low penetration depth of light needed to activate PS. Under NIR irradiation, QDs can transfer energy to surrounding oxygen, with consequent cytotoxic effects.

Several recent papers have explored their potential as PS in their own. In the first paper mentioning this possibility, the authors discovered that semiconductor QDs alone can generate singlet oxygen without a mediating PS molecule in toluene probably because of the intercalation of dissolved oxygen molecules within the nonpolar layer at the QD surface (Samia et al. 2003). The authors also showed the first example of FRET from CdSe QDs to the attached PS.

To circumvent the inefficiency of QDs alone to generate singlet oxygen with good yields (as compared to second generation PS), several attempts have been made to covalently conjugate PS to CdSe/ZnS via organic bridges (Shi et al. 2006) or by electrostatic interactions. Recently, AlPc conjugated with amine-dihydrolipoic acid coated QDs by an electrostatic binding was able to destroy cancer cells via FRET-mediated PDT (Li et al. 2012).

Issues related to the use of QDs in PDT and imaging mainly rely on elimination from the body (QDs < 5.5 nm are effectively cleared through renal excretion) and release of inherently toxic cadmium ions.

5.5.2.7 Other Nanoparticulate Systems

Self-lighting PDT is a new approach to cancer treatment through a combination of radiation therapy and PDT. Upon exposure to ionizing radiation such as X-rays, scintillation luminescence will emit from NPs and activate the PS; as a consequence, singlet oxygen is produced to enhance the killing of cancer cells by ionizing radiation. Supplementing conventional radiation therapy (that can damage healthy tissues as well) with PDT will enable the use of lower doses of radiation. When NPs have persistent luminescence, e.g., BaFBr:Eu^+, Mn^+ NPs, short X-ray exposures are followed by prolonged PS excitation. The period of afterglow is increased in vivo because of higher ambient temperatures (Chen and Zhang 2006). However, direct application in biological systems has not been reported yet.

Luminescent materials with triplet excitation states emit light of higher energy than exciting radiation (anti-Stokes emission) by different mechanisms, including simultaneous two-photon absorption and upconversion (Auzel 2004). Simultaneous two-photon absorption requires a single entity where the transition from ground to an excited electronic state is brought about by simultaneous absorption of two low energy photons whose combined energy is sufficient to induce the transition. Quantum mechanically, this takes place through the attainment of a virtual intermediate state on absorption of the first photon. Upconversion relies on sequential discrete absorption and luminescence steps where at least two metastable entities (usually ions) are involved, the first serving as an excitation reservoir, and the second as the emitting state. It is generally a very efficient process, does not require coherent radiation, and involves real (as opposed to virtual) intermediate states. Anti-Stokes emissions for upconversion processes are found to exceed excitation energies by 10–100 times kT. Since both the mechanisms allow excitation with low energy light and since this is known to penetrate tissues deeply, NPs excited by each process have been used to activate PS. The common advantage is in extending the ability of PDT to reach tumor sites at several centimeters below the skin/mucosal surface. The ability to observe the emission from the NPs to some depth within tissues can have enormous impact on the diagnosis and monitoring of some tumors. NIR light, for example, can penetrate to a considerable extent in soft fibro-fatty tissues like breast, and upconversion NPs can potentially be used for therapy as well as monitoring of tumors over time.

Upconversion NPs are modified nanometer-sized composites which generate higher energy light from lower energy radiation, usually near-infrared or infrared, through the use of transition metal, lanthanide, or actinide ions embedded into a solid state host. The role of the NPs here can be compared to a nanotransducer. NPs are by themselves unable to generate singlet oxygen species from dissolved oxygen and require the attachment of an appropriate PS with an excitation band matching the emission of the NPs. A variety of core materials and dopants have been demonstrated for upconversion particles with actual/possible biological applications. The first reported use of upconversion NPs in PDT used $NaYF_4$:Yb^{3+}, Er^{3+}+ NPs coated with a porous, thin layer of silica doped with merocyanine-540 PS and functionalized with a tumor targeting antibody (Zhang et al. 2007). Then, upconversion ZnPc/PEI/$NaYF_4$ NPs conjugated with folic acid producing red/green emission with NIR excitation both in vitro and in vivo were developed (Chatterjee et al. 2008).

5.6 Conclusions

PDT has demonstrated a great potential in managing different cancer types, has been used for localized superficial or endoluminal malignant and premalignant conditions, and has now expanded to the treatment of solid tumors. Potentiating PDT effects and coupling PDT to other treatment modalities are considered promising strategies to fight tumors. Nanotechnologies can support this evolution allowing engineering of multimodal nanoplatforms useful for therapeutic, diagnostic, and theranostic applications in cancer. Nanocarriers design should take into account the basic design rules (size, surface) to overcome biological barriers (MPS recognition, extravasation at tumor level, cell accumulation, subcellular localization) together with optimization of photochemical properties. To this respect, quantum yield of singlet oxygen and oxygen radicals as well as strategies able to expand applicability of PDT by harvesting light in the NIR (FRET-based, two-photon absorption, upconversion, self-lighting nanocarriers) can allow overcoming typical limitations of PDT. Rational combination of building elements in a single nanoplatform can also couple PDT to other treatment modalities (conventional chemotherapy, PPT, radiotherapy) or imaging (MRI, fluorescence) and propel the application of PDT to the forefront of diagnosis and therapy of cancer. Toxicological issues remain a concern mainly for inorganic materials that needs to be addressed in the next future.

References

Abels C, Szeimies RM, Steinbach P, Richert C, Goetz AE (1997) Targeting of the tumor microcirculation by photodynamic therapy with a synthetic porphycene. J Photochem Photobiol B 40(3):305–312

Agostinis P, Berg K, Cengel KA et al (2011) Photodynamic therapy of cancer: an update. CA Cancer J Clin 61(4):250–281

Allison RR, Sibata CH (2010) Oncologic photodynamic therapy photosensitizers: a clinical review. Photodiagnosis Photodyn Ther 7(2):61–75

Allison RR, Bagnato VS, Sibata CH (2010) Future of oncologic photodynamic therapy. Future Oncol 6(6):929–940

Anand S, Ortel BJ, Pereira SP, Hasan T, Maytin EV (2012) Biomodulatory approaches to photodynamic therapy for solid tumors. Cancer Lett 326(1):8–16

Auzel F (2004) Upconversion and anti-Stokes processes with f and d ions in solids. Chem Rev 104(1):139–173

Bakalova R, Ohba H, Zhelev Z, Ishikawa M, Baba Y (2004) Quantum dots as photosensitizers? Nat Biotechnol 22(11):1360–1361

Balaz M, Collins HA, Dahlstedt E, Anderson HL (2009) Synthesis of hydrophilic conjugated porphyrin dimers for one-photon and two-photon photodynamic therapy at NIR wavelengths. Org Biomol Chem 7(5):874–888

Barolet D (2008) Light-emitting diodes (LEDs) in dermatology. Semin Cutan Med Surg 27(4): 227–238

Bhuvaneswari R, Gan YY, Soo KC, Olivo M (2009) The effect of photodynamic therapy on tumor angiogenesis. Cell Mol Life Sci 66(14):2275–2283

Biel MA (2010) Photodynamic therapy of head and neck cancers. Methods Mol Biol 635:281–293

Bolfarini GC, Siqueira-Moura MP, Demets GJ, Morais PC, Tedesco AC (2012) In vitro evaluation of combined hyperthermia and photodynamic effects using magnetoliposomes loaded with cucurbituril zinc phthalocyanine complex on melanoma. J Photochem Photobiol B 115:1–4

Bovis MJ, Woodhams JH, Loizidou M et al (2012) Improved in vivo delivery of *m*-THPC via pegylated liposomes for use in photodynamic therapy. J Control Release 157(2):196–205

Bown SG, Rogowska AZ, Whitelaw DE et al (2002) Photodynamic therapy for cancer of the pancreas. Gut 50(4):549–557

Buchholz J, Kaser-Hotz B, Khan T et al (2005) Optimizing photodynamic therapy: in vivo pharmacokinetics of liposomal meta-(tetrahydroxyphenyl)chlorin in feline squamous cell carcinoma. Clin Cancer Res 11(20):7538–7544

Buytaert E, Dewaele M, Agostinis P (2007) Molecular effectors of multiple cell death pathways initiated by photodynamic therapy. Biochim Biophys Acta 1776(1):86–107

Calzavara-Pinton PG, Venturini M, Sala R (2007) Photodynamic therapy: update 2006. Part 1: Photochemistry and photobiology. J Eur Acad Dermatol Venereol 21(3):293–302

Camerin M, Magaraggia M, Soncin M et al (2010) The in vivo efficacy of phthalocyanine-nanoparticle conjugates for the photodynamic therapy of amelanotic melanoma. Eur J Cancer 46(10): 1910–1918

Casas A, Perotti C, Saccoliti M et al (2002) ALA and ALA hexyl ester in free and liposomal formulations for the photosensitisation of tumour organ cultures. Br J Cancer 86(5):837–842

Castano AP, Demidova TN, Hamblin MR (2005) Mechanisms in photodynamic therapy: part two-cellular signaling, cell metabolism and modes of cell death. Photodiagnosis Photodyn Ther 2(1):1–23

Castano AP, Mroz P, Hamblin MR (2006) Photodynamic therapy and anti-tumour immunity. Nat Rev Cancer 6(7):535–545

Chatterjee DK, Yong Z (2008) Upconverting nanoparticles as nanotransducers for photodynamic therapy in cancer cells. Nanomedicine (Lond) 3(1):73–82

Chatterjee DK, Fong LS, Zhang Y (2008) Nanoparticles in photodynamic therapy: an emerging paradigm. Adv Drug Deliv Rev 60(15):1627–1637

Chen W (2008) Nanoparticle fluorescence based technology for biological applications. J Nanosci Nanotechnol 8(3):1019–1051

Chen W, Zhang J (2006) Using nanoparticles to enable simultaneous radiation and photodynamic therapies for cancer treatment. J Nanosci Nanotechnol 6(4):1159–1166

Chen B, Pogue BW, Hasan T (2005) Liposomal delivery of photosensitising agents. Expert Opin Drug Deliv 2(3):477–487

Chen CY, Tian Y, Cheng YJ et al (2007) Two-photon absorbing block copolymer as a nanocarrier for porphyrin: energy transfer and singlet oxygen generation in micellar aqueous solution. J Am Chem Soc 129(23):7220–7221

Chen H, Kim S, He W et al (2008) Fast release of lipophilic agents from circulating PEG-PDLLA micelles revealed by in vivo Förster resonance energy transfer imaging. Langmuir 24(10): 5213–5217

Chen R, Wang X, Yao X et al (2013) Near-IR-triggered photothermal/photodynamic dual-modality therapy system via chitosan hybrid nanospheres. Biomaterials 34(33):8314–8322

Chen Y, Samia AC, Li J et al (2014) Tuning the delivery and PDT efficacy of a cancer drug via the bond to the gold nanoparticle vector. Langmuir

Cheng Y, Samia C, Meyers JD et al (2008) Highly efficient drug delivery with gold nanoparticle vectors for in vivo photodynamic therapy of cancer. J Am Chem Soc 130(32):10643–10647
Cheng Y, Samia AC, Li J et al (2010) Delivery and efficacy of a cancer drug as a function of the bond to the gold nanoparticle surface. Langmuir 26(4):2248–2255
Conte C, Ungaro F, Maglio G et al (2013) Biodegradable core-shell nanoassemblies for the delivery of docetaxel and Zn(II)-phthalocyanine inspired by combination therapy for cancer. J Control Release 167(1):40–52
d'Angelo I, Conte C, Miro A, Quaglia F, Ungaro F (2014) Core-shell nanocarriers for cancer therapy. Part I: biologically oriented design rules. Expert Opin Drug Deliv 11:283–297
Dahlstedt E, Collins HA, Balaz M et al (2009) One- and two-photon activated phototoxicity of conjugated porphyrin dimers with high two-photon absorption cross sections. Org Biomol Chem 7(5):897–904
Dayal S, Burda C (2008) One- and two-photon induced QD-based energy transfer and the influence of multiple QD excitations. Photochem Photobiol Sci 7(5):605–613
Derycke AS, de Witte PA (2002) Transferrin-mediated targeting of hypericin embedded in sterically stabilized PEG-liposomes. Int J Oncol 20(1):181–187
Derycke AS, de Witte PA (2004) Liposomes for photodynamic therapy. Adv Drug Deliv Rev 56(1):17–30
Dreaden EC, Alkilany AM, Huang X, Murphy CJ, El-Sayed MA (2012) The golden age: gold nanoparticles for biomedicine. Chem Soc Rev 41(7):2740–2779
Eljamel S (2010) Photodynamic applications in brain tumors: a comprehensive review of the literature. Photodiagnosis Photodyn Ther 7(2):76–85
Fadel M, Kassab K, Fadeel DA (2010) Zinc phthalocyanine-loaded PLGA biodegradable nanoparticles for photodynamic therapy in tumor-bearing mice. Lasers Med Sci 25(2):283-72
Fan C, Wang S, Hong JW et al (2003) Beyond superquenching: hyper-efficient energy transfer from conjugated polymers to gold nanoparticles. Proc Natl Acad Sci U S A 100(11):6297–6301
Fang YP, Tsai YH, Wu PC, Huang YB (2008) Comparison of 5-aminolevulinic acid-encapsulated liposome versus ethosome for skin delivery for photodynamic therapy. Int J Pharm 356(1–2): 144–152
Ferrario A, Rucker N, Wong S, Luna M, Gomer CJ (2007) Survivin, a member of the inhibitor of apoptosis family, is induced by photodynamic therapy and is a target for improving treatment response. Cancer Res 67(10):4989–4995
Fingar VH (1996) Vascular effects of photodynamic therapy. J Clin Laser Med Surg 14(5):323–328
Fingar VH, Wieman TJ, Wiehle SA, Cerrito PB (1992) The role of microvascular damage in photodynamic therapy: the effect of treatment on vessel constriction, permeability, and leukocyte adhesion. Cancer Res 52(18):4914–4921
Fraix A, Goncalves AR, Cardile V et al (2013a) A multifunctional bichromophoric nanoaggregate for fluorescence imaging and simultaneous photogeneration of RNOS and ROS. Chem Asian J 8(11):2634–2641
Fraix A, Kandoth N, Manet I et al (2013b) An engineered nanoplatform for bimodal anticancer phototherapy with dual-color fluorescence detection of sensitizers. Chem Commun (Camb) 49(40):4459–4461
Friedberg JS (2011) Photodynamic therapy for malignant pleural mesothelioma: the future of treatment? Expert Rev Respir Med 5(1):49–63
Fukuda H, Casas A, Batlle A (2006) Use of ALA and ALA derivatives for optimizing ALA-based photodynamic therapy: a review of our experience. J Environ Pathol Toxicol Oncol 25(1–2):127–143
Fukumura D, Kashiwagi S, Jain RK (2006) The role of nitric oxide in tumour progression. Nat Rev Cancer 6(7):521–534
Garcia-Diaz M, Nonell S, Villanueva A et al (2011) Do folate-receptor targeted liposomal photosensitizers enhance photodynamic therapy selectivity? Biochim Biophys Acta 1808(4): 1063–1071

Garland MJ, Cassidy CM, Woolfson D, Donnelly RF (2009) Designing photosensitizers for photodynamic therapy: strategies, challenges and promising developments. Future Med Chem 1(4):667–691

Gijsens A, Derycke A, Missiaen L et al (2002) Targeting of the photocytotoxic compound AlPcS4 to Hela cells by transferrin conjugated PEG-liposomes. Int J Cancer 101(1):78–85

Golab J, Nowis D, Skrzycki M et al (2003) Antitumor effects of photodynamic therapy are potentiated by 2-methoxyestradiol. A superoxide dismutase inhibitor. J Biol Chem 278(1):407–414

Gollnick SO, Owczarczak B, Maier P (2006) Photodynamic therapy and anti-tumor immunity. Lasers Surg Med 38(5):509–515

Gomer CJ, Rucker N, Murphree AL (1988) Differential cell photosensitivity following porphyrin photodynamic therapy. Cancer Res 48(16):4539–4542

Gref R, Luck M, Quellec P et al (2000) 'Stealth' corona-core nanoparticles surface modified by polyethylene glycol (PEG): influences of the corona (PEG chain length and surface density) and of the core composition on phagocytic uptake and plasma protein adsorption. Colloids Surf B Biointerfaces 18(3–4):301–313

Guelluy PH, Fontaine-Aupart MP, Grammenos A et al (2010) Optimizing photodynamic therapy by liposomal formulation of the photosensitizer pyropheophorbide-a methyl ester: in vitro and ex vivo comparative biophysical investigations in a colon carcinoma cell line. Photochem Photobiol Sci 9(9):1252–1260

Gupta S, Dwarakanath BS, Chaudhury NK et al (2011) In vitro and in vivo targeted delivery of photosensitizers to the tumor cells for enhanced photodynamic effects. J Cancer Res Ther 7(3):314–324

Gupta A, Avci P, Sadasivam M et al (2013) Shining light on nanotechnology to help repair and regeneration. Biotechnol Adv 31:607–631

Hanlon JG, Adams K, Rainbow AJ, Gupta RS, Singh G (2001) Induction of Hsp60 by photofrin-mediated photodynamic therapy. J Photochem Photobiol B 64(1):55–61

Hone DC, Walker PI, Evans-Gowing R et al (2002) Generation of cytotoxic singlet oxygen via phthalocyanine stabilized gold nanoparticles: a potential delivery vehicle for photodynamic therapy. Langmuir 18(8):2985–2987

Hu Z, Pan Y, Wang J et al (2009) Meso-tetra (carboxyphenyl) porphyrin (TCPP) nanoparticles were internalized by SW480 cells by a clathrin-mediated endocytosis pathway to induce high photocytotoxicity. Biomed Pharmacother 63(2):155–164

Huang T, Murray RW (2002) Quenching of Ru(bpy)3 fluorescence by binding to Au nanoparticles. Langmuir 18:7077–7081

Huynh NT, Roger E, Lautram N, Benoit JP, Passirani C (2010) The rise and rise of stealth nanocarriers for cancer therapy: passive versus active targeting. Nanomedicine (Lond) 5(9):1415–1433

Jang WD, Nakagishi Y, Nishiyama N et al (2006) Polyion complex micelles for photodynamic therapy: incorporation of dendritic photosensitizer excitable at long wavelength relevant to improved tissue-penetrating property. J Control Release 113(1):73–79

Jang B, Park JY, Tung CH, Kim IH, Choi Y (2011) Gold nanorod-photosensitizer complex for near-infrared fluorescence imaging and photodynamic/photothermal therapy in vivo. ACS Nano 5(2):1086–1094

Jiang F, Lilge L, Logie B, Li Y, Chopp M (1997) Photodynamic therapy of 9L gliosarcoma with liposome-delivered photofrin. Photochem Photobiol 65(4):701–706

Jiang F, Lilge L, Grenier J et al (1998) Photodynamic therapy of U87 human glioma in nude rat using liposome-delivered photofrin. Lasers Surg Med 22(2):74–80

Kandoth N, Vittorino E, Sciortino MT et al (2012) A cyclodextrin-based nanoassembly with bimodal photodynamic action. Chemistry 18(6):1684–1690

Kessel D (2006) Death pathways associated with photodynamic therapy. Med Laser Appl 21(4):219–224

Kessel D, Castelli M (2001) Evidence that bcl-2 is the target of three photosensitizers that induce a rapid apoptotic response. Photochem Photobiol 74(2):318–322

Kessel D, Erickson C (1992) Porphyrin photosensitization of multi-drug resistant cell types. Photochem Photobiol 55(3):397–399

Khaing Oo MK, Yang Y, Hu Y et al (2012) Gold nanoparticle-enhanced and size-dependent generation of reactive oxygen species from protoporphyrin IX. ACS Nano 6(3):1939–1947

Khdair A, Chen D, Patil Y et al (2010) Nanoparticle-mediated combination chemotherapy and photodynamic therapy overcomes tumor drug resistance. J Control Release 141(2):137–144

Khlebtsov B, Panfilova E, Khanadeev V et al (2011) Nanocomposites containing silica-coated gold-silver nanocages and Yb-2,4-dimethoxyhematoporphyrin: multifunctional capability of IR-luminescence detection, photosensitization, and photothermolysis. ACS Nano 5(9): 7077–7089

Kim S, Ohulchanskyy TY, Pudavar HE, et al (2007) Organically modified silica nanoparticles co-encapsulating photosensitizing drug and aggregation-enhanced two-photon absorbing fluorescent dye aggregates for two photon photodynamic therapy. J Am Chem Soc 129(9):2669–2675

Knop K, Mingotaud AF, El-Akra N, Violleau F, Souchard JP (2009) Monomeric pheophorbide(a)-containing poly(ethyleneglycol-*b*-epsilon-caprolactone) micelles for photodynamic therapy. Photochem Photobiol Sci 8(3):396–404

Konan YN, Cerny R, Favet J et al (2003) Preparation and characterization of sterile sub-200 nm meso-tetra(4-hydroxylphenyl)porphyrin-loaded nanoparticles for photodynamic therapy. Eur J Pharm Biopharm 55(1):115–124

Konan-Kouakou YN, Boch R, Gurny R, Allemann E (2005) In vitro and in vivo activities of verteporfin-loaded nanoparticles. J Control Release 103(1):83–91

Koo YE, Reddy GR, Bhojani M et al (2006) Brain cancer diagnosis and therapy with nanoplatforms. Adv Drug Deliv Rev 58(14):1556–1577

Korbelik M (2006) PDT-associated host response and its role in the therapy outcome. Lasers Surg Med 38(5):500–508

Korbelik M, Madiyalakan R, Woo T, Haddadi A (2012) Antitumor efficacy of photodynamic therapy using novel nanoformulations of hypocrellin photosensitizer SL052. Photochem Photobiol 88(1):188–193

Kostron H (2010) Photodynamic diagnosis and therapy and the brain. Methods Mol Biol 635:261–280

Krammer B (2001) Vascular effects of photodynamic therapy. Anticancer Res 21(6B):4271–4277

Kreimer-Birnbaum M (1989) Modified porphyrins, chlorins, phthalocyanines, and purpurins: second-generation photosensitizers for photodynamic therapy. Semin Hematol 26(2):157–173

Krosl G, Korbelik M, Dougherty GJ (1995) Induction of immune cell infiltration into murine SCCVII tumor by photofrin-based photodynamic therapy. Br J Cancer 71(3):549–555

Kuimova MK, Yahioglu G, Ogilby PR (2008) Singlet oxygen in a cell: spatially dependent lifetimes and quenching rate constants. J Am Chem Soc 131(1):332–340

Kuntsche J, Freisleben I, Steiniger F, Fahr A (2010) Temoporfin-loaded liposomes: physicochemical characterization. Eur J Pharm Sci 40(4):305–315

Labib A, Lenaerts V, Chouinard F et al (1991) Biodegradable nanospheres containing phthalocyanines and naphthalocyanines for targeted photodynamic tumor therapy. Pharm Res 8(8): 1027–1031

Lee YE, Kopelman R (2011) Polymeric nanoparticles for photodynamic therapy. Methods Mol Biol 726:151–178

Lee SJ, Koo H, Jeong H et al (2011a) Comparative study of photosensitizer loaded and conjugated glycol chitosan nanoparticles for cancer therapy. J Control Release 152(1):21–29

Lee SJ, Koo H, Lee DE et al (2011b) Tumor-homing photosensitizer-conjugated glycol chitosan nanoparticles for synchronous photodynamic imaging and therapy based on cellular on/off system. Biomaterials 32(16):4021–4029

Lehar J, Krueger AS, Avery W et al (2009) Synergistic drug combinations tend to improve therapeutically relevant selectivity. Nat Biotechnol 27(7):659–666

Leunig M, Richert C, Gamarra F et al (1993) Tumour localisation kinetics of photofrin and three synthetic porphyrinoids in an amelanotic melanoma of the hamster. Br J Cancer 68(2):225–234

Li B, Moriyama EH, Li F et al (2007) Diblock copolymer micelles deliver hydrophobic protoporphyrin IX for photodynamic therapy. Photochem Photobiol 83(6):1505–1512

Li L, Zhao JF, Won N et al (2012) Quantum dot-aluminum phthalocyanine conjugates perform photodynamic reactions to kill cancer cells via fluorescence resonance energy transfer (FRET). Nanoscale Res Lett 7(1):386

Lim JI (2002) Photodynamic therapy for choroidal neovascular disease: photosensitizers and clinical trials. Ophthalmol Clin North Am 15(4):473–478, vii

Liu W, Baer MR, Bowman MJ et al (2007) The tyrosine kinase inhibitor imatinib mesylate enhances the efficacy of photodynamic therapy by inhibiting ABCG2. Clin Cancer Res 13(8):2463–2470

Maeda H (2012) Macromolecular therapeutics in cancer treatment: the EPR effect and beyond. J Control Release 164(2):138–144

Maeda H, Nakamura H, Fang J (2013) The EPR effect for macromolecular drug delivery to solid tumors: improvement of tumor uptake, lowering of systemic toxicity, and distinct tumor imaging in vivo. Adv Drug Deliv Rev 65(1):71–79

Mazzaglia A (2011) In: Bilensoy E (ed) Cyclodextrins in pharmaceutics, cosmetics, and biomedicine: current and future industrial applications. Wiley, New York, pp 343–361

Mazzaglia A, Valerio A, Micali N et al (2011) Effective cell uptake of nanoassemblies of a fluorescent amphiphilic cyclodextrin and an anionic porphyrin. Chem Commun 47(32): 9140–9142

Mazzaglia A, Sciortino MT, Kandoth N, Sortino S (2012) Cyclodextrin-based nanoconstructs for photoactivated therapies. J Drug Deliv Sci Technol 22(3):235–242

Minnich DJ, Bryant AS, Dooley A, Cerfolio RJ (2010) Photodynamic laser therapy for lesions in the airway. Ann Thorac Surg 89(6):1744–1748

Mir Y, Elrington SA, Hasan T (2013) A new nanoconstruct for epidermal growth factor receptor-targeted photo-immunotherapy of ovarian cancer. Nanomedicine 9(7):1114–1122

Moret F, Scheglmann D, Reddi E (2013) Folate-targeted PEGylated liposomes improve the selectivity of PDT with *meta*-tetra(hydroxyphenyl)chlorin (*m*-THPC). Photochem Photobiol Sci 12(5):823–834

Nakagawa T, Shimizu S, Watanabe T et al (2005) Cyclophilin D-dependent mitochondrial permeability transition regulates some necrotic but not apoptotic cell death. Nature 434(7033):652–658

Nishiyama N, Nakagishi Y, Morimoto Y et al (2009) Enhanced photodynamic cancer treatment by supramolecular nanocarriers charged with dendrimer phthalocyanine. J Control Release 133(3):245–251

Nokes B, Apel M, Jones C, Brown G, Lang JE (2013) Aminolevulinic acid (ALA): photodynamic detection and potential therapeutic applications. J Surg Res 181(2):262–271

Nonaka M, Ikeda H, Inokuchi T (2004) Inhibitory effect of heat shock protein 70 on apoptosis induced by photodynamic therapy in vitro. Photochem Photobiol 79(1):94–98

Nowis D, Makowski M, Stoklosa T et al (2005) Direct tumor damage mechanisms of photodynamic therapy. Acta Biochim Pol 52(2):339–352

O'Connor AE, Gallagher WM, Byrne AT (2009) Porphyrin and nonporphyrin photosensitizers in oncology: preclinical and clinical advances in photodynamic therapy. Photochem Photobiol 85(5):1053–1074

Ohulchanskyy TY, Roy I, Goswami LN et al (2007) Organically modified silica nanoparticles with covalently incorporated photosensitizer for photodynamic therapy of cancer. Nano Lett 7(9):2835–2842

Oleinick NL, Morris RL, Belichenko I (2002) The role of apoptosis in response to photodynamic therapy: what, where, why, and how. Photochem Photobiol Sci 1(1):1–21

Ortner MA (2011) Photodynamic therapy for cholangiocarcinoma. Lasers Surg Med 43(7):776–780

Parhi P, Mohanty C, Sahoo SK (2012) Nanotechnology-based combinational drug delivery: an emerging approach for cancer therapy. Drug Discov Today 17(17–18):1044–1052

Pegaz B, Debefve E, Ballini JP et al (2006) Photothrombic activity of *m*-THPC-loaded liposomal formulations: pre-clinical assessment on chick chorioallantoic membrane model. Eur J Pharm Sci 28(1–2):134–140

Petri A, Yova D, Alexandratou E, Kyriazi M, Rallis M (2012) Comparative characterization of the cellular uptake and photodynamic efficiency of Foscan(R) and Fospeg in a human prostate cancer cell line. Photodiagnosis Photodyn Ther 9(4):344–354

Pizova K, Tomankova K, Daskova A et al (2012) Photodynamic therapy for enhancing antitumour immunity. Biomed Pap Med Fac Univ Palacky Olomouc Czech Repub 156(2):93–102

Qualls MM, Thompson DH (2001) Chloroaluminum phthalocyanine tetrasulfonate delivered via acid-labile diplasmenylcholine-folate liposomes: intracellular localization and synergistic phototoxicity. Int J Cancer 93(3):384–392

Reddy GR, Bhojani MS, McConville P et al (2006) Vascular targeted nanoparticles for imaging and treatment of brain tumors. Clin Cancer Res 12(22):6677–6686

Reiners JJ Jr, Agostinis P, Berg K, Oleinick NL, Kessel D (2010) Assessing autophagy in the context of photodynamic therapy. Autophagy 6(1):7–18

Reza SM, Tabatabaie RM, Maharramov A, Ali RM (2011) Synthesis and in vitro studies of biodegradable modified chitosan nanoparticles for photodynamic treatment of cancer. Int J Biol Macromol 49(5):1059–1065

Ricci-Junior E, Marchetti JM (2006) Zinc(II) phthalocyanine loaded PLGA nanoparticles for photodynamic therapy use. Int J Pharm 310(1–2):187–195

Rijcken CJ, Hofman JW, van Zeeland F, Hennink WE, van Nostrum CF (2007) Photosensitiser-loaded biodegradable polymeric micelles: preparation, characterisation and in vitro PDT efficacy. J Control Release 124(3):144–153

Robertson CA, Evans DH, Abrahamse H (2009) Photodynamic therapy (PDT): a short review on cellular mechanisms and cancer research applications for PDT. J Photochem Photobiol B 96(1):1–8

Roby A, Erdogan S, Torchilin VP (2006) Solubilization of poorly soluble PDT agent, meso-tetraphenylporphin, in plain or immunotargeted PEG-PE micelles results in dramatically improved cancer cell killing in vitro. Eur J Pharm Biopharm 62(3):235–240

Rojnik M, Kocbek P, Moret F et al (2012) In vitro and in vivo characterization of temoporfin-loaded PEGylated PLGA nanoparticles for use in photodynamic therapy. Nanomedicine (Lond) 7(5):663–677

Roy I, Ohulchanskyy TY, Pudavar HE et al (2003) Ceramic-based nanoparticles entrapping water-insoluble photosensitizing anticancer drugs: a novel drug-carrier system for photodynamic therapy. J Am Chem Soc 125(26):7860–7865

Samia AC, Chen X, Burda C (2003) Semiconductor quantum dots for photodynamic therapy. J Am Chem Soc 125(51):15736–15737

Samia AC, Dayal S, Burda C (2006) Quantum dot-based energy transfer: perspectives and potential for applications in photodynamic therapy. Photochem Photobiol 82(3):617–625

Sandell JL, Zhu TC (2011) A review of in-vivo optical properties of human tissues and its impact on PDT. J Biophotonics 4(11–12):773–787

Sapsford KE, Berti L, Medintz IL (2006a) Materials for fluorescence resonance energy transfer analysis: beyond traditional donor-acceptor combinations. Angew Chem Int Ed Engl 45(28):4562–4588

Sapsford KE, Pons T, Medintz IL, Mattoussi H (2006b) Biosensing with luminescent semiconductor quantum dots. Sensors (Basel) 6(8):925–953

Saxena V, Sadoqi M, Shao J (2006) Polymeric nanoparticulate delivery system for Indocyanine green: biodistribution in healthy mice. Int J Pharm 308(1–2):200–204

Sessler JL, Miller RA (2000) Texaphyrins: new drugs with diverse clinical applications in radiation and photodynamic therapy. Biochem Pharmacol 59(7):733–739

Shi L, Hernandez B, Selke M (2006) Singlet oxygen generation from water-soluble quantum dot-organic dye nanocomposites. J Am Chem Soc 128(19):6278–6279

Shieh MJ, Peng CL, Chiang WL et al (2010) Reduced skin photosensitivity with meta-tetra(hydroxyphenyl) chlorin-loaded micelles based on a poly(2-ethyl-2-oxazoline)-b-poly(D,L-lactide) diblock copolymer in vivo. Mol Pharm 7(4):1244–1253

Sibata MN, Tedesco AC, Marchetti JM (2004) Photophysicals and photochemicals studies of zinc(II) phthalocyanine in long time circulation micelles for photodynamic therapy use. Eur J Pharm Sci 23(2):131–138

Simone CB, Friedberg JS, Glatstein E et al (2012) Photodynamic therapy for the treatment of non-small cell lung cancer. J Thorac Dis 4(1):63–75

Skupin-Mrugalska P, Piskorz J, Goslinski T et al (2013) Current status of liposomal porphyrinoid photosensitizers. Drug Discov Today 18(15–16):776–784

Soriano J, Villanueva A, Stockert JC, Canete M (2013) Vehiculization determines the endocytic internalization mechanism of Zn(II)-phthalocyanine. Histochem Cell Biol 139(1):149–160

Sortino S (2010) Light-controlled nitric oxide delivering molecular assemblies. Chem Soc Rev 39(8):2903–2913
Sortino S, Mazzaglia A, Monsù Scolaro L et al (2006) Nanoparticles of cationic amphiphilic cyclodextrins entangling anionic porphyrins as carrier-sensitizer system in photodynamic cancer therapy. Biomaterials 27(23):4256–4265
Sousa CRE (2004) Activation of dendritic cells: translating innate into adaptive immunity. Curr Opin Immunol 16(1):21–25
Staneloudi C, Smith KA, Hudson R et al (2007) Development and characterization of novel photosensitizer : scFv conjugates for use in photodynamic therapy of cancer. Immunology 120(4):512–517
Sutoris K, Rakusan J, Karaskova M et al (2013) Novel topical photodynamic therapy of prostate carcinoma using hydroxy-aluminum phthalocyanine entrapped in liposomes. Anticancer Res 33(4):1563–1568
Swaminathan S, Garcia-Amoros J, Fraix A et al (2014) Photoresponsive polymer nanocarriers with multifunctional cargo. Chem Soc Rev 43:4167–4178
Szokalska A, Makowski M, Nowis D et al (2009) Proteasome inhibition potentiates antitumor effects of photodynamic therapy in mice through induction of endoplasmic reticulum stress and unfolded protein response. Cancer Res 69(10):4235–4243
Tang W, Xu H, Kopelman R, Philbert MA (2005) Photodynamic characterization and in vitro application of methylene blue-containing nanoparticle platforms. Photochem Photobiol 81(2):242–249
Tang W, Xu H, Park EJ, Philbert MA, Kopelman R (2008) Encapsulation of methylene blue in polyacrylamide nanoparticle platforms protects its photodynamic effectiveness. Biochem Biophys Res Commun 369(2):579–583
Trapani M, Romeo A, Parisi T et al (2013) Supramolecular hybrid assemblies based on gold nanoparticles, amphiphilic cyclodextrin and porphyrins with combined phototherapeutic action. RSC Adv 3:5607–5614
van Duijnhoven FH, Aalbers RI, Rovers JP, Terpstra OT, Kuppen PJ (2003) The immunological consequences of photodynamic treatment of cancer, a literature review. Immunobiology 207(2):105–113
Vanlangenakker N, Vanden Berghe T, Krysko DV, Festjens N, Vandenabeele P (2008) Molecular mechanisms and pathophysiology of necrotic cell death. Curr Mol Med 8(3):207–220
Vargas A, Pegaz B, Debefve E et al (2004) Improved photodynamic activity of porphyrin loaded into nanoparticles: an in vivo evaluation using chick embryos. Int J Pharm 286(1–2):131–145
Vogel A, Venugopalan V (2003) Mechanisms of pulsed laser ablation of biological tissues. Chem Rev 103(2):577–644
Wang M, Thanou M (2010) Targeting nanoparticles to cancer. Pharmacol Res 62(2):90–99
Warren CB, Karai LJ, Vidimos A, Maytin EV (2009) Pain associated with aminolevulinic acid-photodynamic therapy of skin disease. J Am Acad Dermatol 61(6):1033–1043
Weiss A, den Bergh Hv, Griffioen AW, Nowak-Sliwinska P (2012) Angiogenesis inhibition for the improvement of photodynamic therapy: the revival of a promising idea. Biochim Biophys Acta 1826(1):53–70
Wiedmann MW, Caca K (2004) General principles of photodynamic therapy (PDT) and gastrointestinal applications. Curr Pharm Biotechnol 5(4):397–408
Wu J, Xu H, Tang W et al (2009) Eradication of bacteria in suspension and biofilms using methylene blue-loaded dynamic nanoplatforms. Antimicrob Agents Chemother 53(7):3042–3048
Xue LY, Chiu SM, Oleinick NL (2001) Photochemical destruction of the Bcl-2 oncoprotein during photodynamic therapy with the phthalocyanine photosensitizer Pc 4. Oncogene 20(26): 3420–3427
Zeisser-Labouebe M, Lange N, Gurny R, Delie F (2006) Hypericin-loaded nanoparticles for the photodynamic treatment of ovarian cancer. Int J Pharm 326(1–2):174–181
Zhang P, Steelant W, Kumar M, Scholfield M (2007) Versatile photosensitizers for photodynamic therapy at infrared excitation. J Am Chem Soc 129(15):4526–4527
Zuluaga MF, Lange N (2008) Combination of photodynamic therapy with anti-cancer agents. Curr Med Chem 15(17):1655–1673

Part III
Targeting the Immune System

Chapter 6
Cancer Vaccines: Fundamentals and Strategies

Javier Briones

6.1 Introduction

Cancer vaccines have been the subject of intensive research in the last decade. A better understanding of the mechanisms involved in antigen presentation, T cell activation, and tumor-induced immunosuppression has given researchers powerful tools to manipulate the immune system to elicit a specific antitumor immune response. Manipulating the immune system has taken two forms: passive immunotherapy consists of adoptive transfer of monoclonal antibodies or effector T cells specific to tumor cells; in contrast, active immunotherapy, by means of cancer vaccines, promotes a specific immune response against the tumor by stimulating the tumor-bearing host. Although a humoral response mediated by antibodies has clearly a role in the antitumor immune response, most cancer vaccine strategies have focused on induction of a cellular immune response, mainly mediated by T cells ($CD4^+$ and $CD8^+$) and NK cells.

Generation of antitumor immunity involves a number of sequential steps, including antigen presentation, activation of effector cells, and overcoming tumor immunosuppression.

T cells recognize antigens in the form of small peptides loaded onto major histocompatibility complex (MHC) molecules (HLA molecules in humans), which are classified in two types: MHC class I molecules present short peptides (8–11 amino acid-long) and are recognized by $CD8^+$ T cells (Klein and Sato 2000a, b). MHC class II molecules present longer peptides that are recognized by $CD4^+$ T cells.

J. Briones (✉)
Hematology Service, Hospital Sant Pau, Mas Casanovas, 90, 08010 Barcelona, Spain
e-mail: jbriones@santpau.cat

M.J. Alonso, M. Garcia-Fuentes (eds.), *Nano-Oncologicals: New Targeting and Delivery Approaches*, Advances in Delivery Science and Technology,
DOI 10.1007/978-3-319-08084-0_6

Activation of antigen-specific T cells requires the presentation of antigens by antigen-presenting cells (APCs). Dendritic cells (DCs) are the most potent and efficient APCs (Palucka and Banchereau 2012). DCs capture tumor antigens, either delivered exogenously or from apoptotic tumor cells, and migrate to draining lymph nodes where presentation of antigens (on MHC class I and II) to T cells occurs. To elicit an effective T cell response, additional costimulatory signals need to be provided to T cells. Depending on the type of stimuli a $CD8^+$ mediated T cell response is generated, although in the majority of cases the concourse of $CD4^+$ T cells is needed. DCs can also stimulate natural killer (NK) cells and antibody responses, all contributing to tumor immunity (Heath and Carbone 2009).

Finally, tumor-specific T cells must enter the tumor microenvironment to eliminate cancer cells. To this end, effector T cells must overcome the immunosuppressive environment generated by the tumor: secretion of molecules that inhibits T cell function [i.e., transforming growth factor beta (TGFβ), vascular endothelial growth factor (VEGF), indoleamine 2,3-dioxygenase], expression of molecules that downregulates T cell function (i.e., programmed cell death ligand 1; PD-L1), or enable expansion of immunosuppressive regulatory T cells (Tregs).

Major hurdles in developing cancer vaccines include: identification of tumor-specific antigens; development of strategies to generate a robust immune response enough to eradicate the tumor; and overcoming immunosuppressive mechanisms leading to immune response evasion by tumors.

Tumor antigens can be generally categorized in four groups: (1) antigens unique to an individual patient's tumor (i.e., B cell idiotype); (2) antigens derived from genes mutated in tumor cells (i.e., *p53, ras*); (3) tissue-differentiation antigens (i.e., MAGE, BAGE, NY-ESO), and (4) ubiquitous antigens expressed by both, normal and malignant cells (i.e., tyrosinase, MART-1, PSA). While the list of tumor antigens available is extensive (Neller et al. 2008), the number of "true" tumor specific antigens, that is, those expressed only in tumor cells, is very low. Thus, cancer vaccine strategies will be mostly determined by issues such as characterization of a tumor-specific antigen, antigen expression by tumor cells in different patients, and antigen immunogenicity. Finally, the generation of a successful vaccine will depend on a multistep approach, consisting of: (1) choice of a tumor antigen and the form that it takes (i.e., peptide, protein, DNA, or RNA); (2) choice of the adjuvant for delivery of that antigen; (3) choice of activation stimulus to recruit and activate effector T cells and innate immune responses (i.e., NK cells); and (4) overcoming tumor immunosuppression and tumor escape (Table 6.1).

6.2 Peptide/Protein Vaccines

The observation that short peptides loaded onto MHC molecules are recognized by T cells prompted the use of peptides as cancer vaccines in patients with different tumors. The cloning of the first human tumor-associated antigen and the identification of its

Table 6.1 Cancer vaccines modalities

Vaccine	Antigen	Adjuvant
Peptides or protein	Idiotype, melanoma (MART-1, gp100, tyrosinase)	KLH, Montanide
Plasmid DNA	Tumor antigen genes (i.e., idiotype, prostate, melanoma)	Cytokine genes (IL-2, IL-12, IL-15, GM-CSF) Tetanus toxin fragment C
Recombinant virus (adenovirus, poxvirus)	Tumor antigen genes (i.e., idiotype, prostate, melanoma)	Cytokine genes (IL-2, IL-12, GM-CSF) Genes encoding costimulatory molecules (CD40L, B7/ICAM1/LFA3)
Dendritic cells	Pulsed with tumor peptides/ proteins Pulsed with tumor lysates Fused to tumor cells	Transduced with viral vectors encoding: cytokines, costimulatory molecules
Whole tumor cells (autologous, allogeneic)	Multiple known/unknown antigens	GM-CSF, IL-2 genes Costimulatory genes

Abbreviations: *MART-1* melanoma antigen recognized by T cells, *KLH* keyhole limpet hemocyanin, *GM-CSF* granulocyte-macrophage colony stimulating factor

peptide sequence binding to MHC was a definitive step to translate this approach to the clinic (Traversari et al. 1992). Tumor-derived peptides can be synthetically manufactured and administered in combination with an immunological adjuvant to promote antitumor response. This allows activation of DCs that will capture and further present these peptides to T cells which will target tumor cells expressing those peptides. However, many epitopes from the tumor peptides have a low to intermediate binding affinity for the MHC molecule, thus generating low-affinity T cells that have limited antitumor activity. Modifications of the epitope sequence to increase affinity of the peptide for MHC molecules have been done resulting in an enhanced immune response (Kaumaya 2013). However, these epitope enhancement strategies have not resulted in a clinical benefit in the vast majority of vaccinated patients with melanoma (Rosenberg et al. 2004). One potentially promising approach consists of the use of longer peptides (20 amino acids) which are thought to be more efficient at generating effector T cells. This approach proved to be of clinical benefit in patients with human papillomavirus associated neoplasia, but failed to induce a clinically beneficial immune response in patients with other types of solid tumors otherwise vaccinated with peptides derived from oncoproteins (i.e., *p53*). The main advantage of peptide vaccination is the ease of manufacture (Table 6.2); however, peptide-based vaccines require knowledge of the patient's HLA haplotype and matching the antigen sequence. Since the most common HLA in Caucasian population is HLA-A2, which only accounts for roughly 30 % of population, a peptide-based approach will only be suitable for one third of the patients, whose tumors express a tumor antigen for which the immunodominant HLA-binding epitope has been defined, thus drastically limiting the numbers of patients suitable for this approach.

Table 6.2 Comparison of different vaccine strategies

Vaccine	Advantages	Disadvantages
Peptide	Easy to produce	Need to match patient's HLA Low immunogenicity
Protein	Easy to produce	Weak T cell priming
Dendritic cells	High immunogenicity Presentation of multiple antigen formulations (peptide, protein, DNA, RNA)	Production is costly and time consuming Individualized cell processing Different DCs types with different actions
Plasmid DNA	Easy to produce Allow combination of tumor antigen plus costimulation	Low immunogenicity
Recombinant virus	Highly immunogenic Different tumor genes plus stimulation genes	Difficult to manufacture Safety issues Preexisting immunity against several virus strains
Tumor cells (autologous, allogeneic, cell lines)	Express an array of different tumor antigens Cell lines are easy to prepare	Difficult to obtain for some tumor types May include antigens that induce autoimmunity

However, the availability of tumor DNA sequencing may revolutionize the field of immunotherapy with peptides. Genes from a patient's tumor can be sequenced to identify mutant peptides potentially recognized by T cells. Relevant epitopes that could bind to a given HLA could be predicted and used to immunize cancer patients. It is conceivable that in the near future the collective of mutated peptides from a patient's tumor (the so-called mutanome) may be analyzed and these peptides can be generated in a vaccine form with the assumption that may harnessing a stronger T cell response, which in addition will be patient's tumor specific (Heemskerk et al. 2013; Robbins et al. 2013).

The entire protein has also been explored as a vaccine, especially in B-cell tumors (Hsu et al. 1997). The full protein contains a broader profile of immunodominant epitopes that may be presented by DCs eventually leading to an enhanced immune response. However, only half of the patients with B cell tumors vaccinated with the tumor protein (idiotype) coupled to an adjuvant developed an immune response, which had a limited clinical benefit (Ai et al. 2009). Administration of a peptide/protein combined with cytokines (i.e., IL-2, GM-CSF), or Toll-like receptors (TLR) agonists will potentially contribute to increase tumor-specific $CD8^+$ T cell responses.

6.3 DNA Vaccines

DNA vaccines have been explored as an alternative to protein vaccination in cancer. DNA vaccines are vehicles for in vivo transfection and antigen production thus representing a simple method for immune system activation. The concept consists in simply taking a tumor-antigen DNA sequence, cloning it into a bacterial plasmid

DNA, and delivering directly to the patient, so the gene is translated in vivo and presented to the immune system. Since they represent a delivery system that does not require in vitro protein expression, a DNA vaccine can be rapidly produced with PCR technology in a short period of time, which represents a substantial advantage over the vaccines based upon in vitro protein production (Stevenson et al. 2010).

Studies in the last years have partially revealed the mechanisms by which DNA vaccines induce immune responses. DNA vaccines can be delivered via many routes (i.e., skin, muscle, mucosal). The route of inoculation has effects on the type of immune response elicited and on the dose of vaccine needed to induce such response. DNA vaccines are able to stimulate both, humoral (antibody responses) and cellular responses, mediated by $CD4^+$ and $CD8^+$ T cells. When injected into the muscle or skin, the encoded tumor-antigen is expressed by myocytes or keratinocytes. However, these cells do not express costimulatory molecules needed to directly stimulate T cells. A number of studies have shown that the DNA delivered can be found in the draining lymph nodes where it is uptaken by APCs (mostly DCs) which present appropriate peptides to specific $CD4^+$ and $CD8^+$ T cells, a process known as "cross-presentation." Additionally, DCs can phagocyte the protein produced in the muscle or skin cells which will contribute to the stimulation of a specific immune response. Alternatively, depending on the delivery method, DCs can be directly transfected in vivo with the plasmid DNA. In this scenario, intracellular processing of the antigen within the DCs will lead to the generation of mostly not only $CD8^+$ but also $CD4^+$ specific T cells (Rice et al. 2008; Vittes et al. 2011).

Studies in a number of cancer models have shown that DNA vaccines encoding a tumor antigen are poorly immunogenic (Syrengelas and Levy 1999). Additional molecules are required to activate effective immune responses against weak tumor antigens. An important fact that contributes to the stimulation of the immune system, when using a bacterial plasmid DNA, is the existence of unmethylated cytidine-phosphate-guanosine (CpG) oligonucleotide-containing motif sequences (Brightbill and Modlin 2000). In contrast to bacterial DNA, human DNA has very low numbers of CpG motifs that are heavily methylated. Thus, bacterial CpG motifs act as a "danger signal" to activate both the innate and the adaptive immune system. These sequences play an important role in the activation (i.e., enhanced surface expression of costimulatory molecules CD40, CD80/CD86) and maturation of DCs and contribute to the stimulation of T, B, and NK cells through the interaction with TLR, specifically TLR9. Activation of TLR9 by these sequences leads to the release of pro-inflammatory cytokines (i.e., IFN-γ, IL-12, and TNF-α) which contributes to the activation of NK-mediated, and Th1 T cell specific responses. Due to the critical role that TLR stimulation has on the development of a coordinated innate and adaptive immune response against a tumor antigen, CpGs as well as other TLR agonists are being extensively studied as vaccine adjuvants in different cancer models (Bode et al. 2011).

In addition to the CpG motifs, a growing list of immune enhancers is being assessed in the context of fusion genes—tumor antigen gene plus immune activating gene (Table 6.3). The aim of the fusion gene is to increase the level of the immune response against the tumor antigen and also to direct the immune response outcome.

Table 6.3 Novel immune adjuvants for cancer vaccines

Group	Types
Toll-like receptor agonists	CpG containing sequences Flagellin
T cell modulation	
Immune checkpoints inhibition	CTLA4 blockade PD-1 blockade LAG-3 blockade
Stimulatory	Agonist OX40 Agonist 4-1BB Costimulatory molecules (B7/ICAM1/LFA3)
NK/NKT cell activation	Agonist NKG2D α-Galceramide
T regulatory cell depletion	Anti CD25 Anti GITR Anti folate receptor 4

Abbreviations: *CTLA4* cytotoxic T-lymphocyte-associated protein 4, *LAG-3* lymphocyte-activation gene 3, *PD-1* programmed cell death 1, *NKG2D* natural killer activating receptor, *GITR* glucocorticoid-induced tumor necrosis factor

The list of immune enhancers is extensive, and includes cytokines (i.e., GM-CSF, IL-2, IL-12), chemokines (i.e., MIP1α), and T cell-costimulatory molecules (CD80, CD86, CD40L). Fusion genes can be developed to redirect the immune response, for example, towards a Th1 response or the generation of antigen-specific antibodies. To this purpose, genes encoding microbial products, such as the *Clostridium tetani* toxin fragment C, have been incorporated into the DNA vaccine (King et al. 1998). Alternatively, minigenes encoding peptides recognized by MHC class I and II molecules can lead to enhanced $CD8^+$ and $CD4^+$ responses, respectively, against tumor-derived peptides (Rice et al. 2002). These approaches have been tested in animal models with different tumors and have shown that they are able to enhance antibody and T cell specific responses consistent with a Th1 response.

In the last years, the concept of DNA vaccines has been translated into the clinic in patients with solid and hematological cancers. In one clinical trial, patients with B cell lymphoma were vaccinated with a plasmid encoding the tumor-specific idiotype (Timmerman et al. 2002a). DNA vaccines were also tested in patients with colorectal carcinoma, melanoma, and prostate cancer, using carcinoembrionic antigen (CEA), tyrosinase, and prostate-specific antigen (PSA), respectively. Although the vaccine schedules (dose and route) and adjuvants used (i.e., GM-CSF, IL-2) were different between those studies, overall level of antitumor response was modest with no significant clinical responses (Ribas et al. 2003). Further optimization of the DNA vaccine including vaccine design, selection of appropriate tumor antigens and immunostimulatory sequences, and DNA delivery is needed to achieve the major goal of a significant clinical impact in patients with cancer.

6.4 Recombinant Virus Vaccines

The potential use of recombinant viruses as cancer vaccines has been extensively explored in the last years (Harrop and Carroll 2006). Viruses are the most diverse and efficient gene-transfer agents. Their natural cell tropism and biological features can significantly enhance the immunogenicity of antigens carried within them. Recombinant viruses offer several advantages over naked DNA: they are highly immunogenic; they efficiently transduce genes into a variety of cellular types, either in vitro or in vivo, yielding high levels of protein; and high titers can be produced under good manufacturing practice (GMP) conditions in a relatively short period of time.

Adenoviruses are of special interest for cancer vaccines because they meet all of the abovementioned criteria. The concept consists of the use of a replication-defective adenovirus with a particular tumor antigen-encoded gene. The most frequently used adenoviral vector in cancer vaccines is based on the adenovirus type 5. The viral vector was rendered replication-defective by removing the E1 gene, which is required both for adenovirus replication and for viral gene expression. Genes encoding a tumor antigen may be inserted in place of E1 gene with expression driven by the CMV early promoter (Ng and Graham 2002).

In infectious disease animal models, vaccination with adenovirus encoding a model antigen has proved to be efficacious in preventing disease (Shiver and Emini 2004), and a similar approach has been taken for cancer. Vaccination with adenovirus encoding a tumor antigen has shown the ability to stimulate antigen-specific humoral and cellular immunity and to break preexisting immunologic tolerance (Chen et al. 1996). Importantly, immunization with a single dose of the adenovirus encoding a tumor-antigen generates a greater immune response against that antigen, compared with the immune response generated with the corresponding peptide or protein. After intramuscular or intradermal administration, the tumor antigen gene is expressed directly in DCs or cross-presented by other cells to DCs, to promote an antibody and $CD8^{+}$ T cells mediated immune response. In addition, the adenovirus itself generates a strong inflammatory response that functions as a potent adjuvant (mediated in part through TLR receptors on DCs) to enhance the immune response against the tumor antigen-encoded gene.

Studies done in murine B cell lymphoma models have shown that vaccination with a single dose of an adenovirus encoding the idiotype genes was able to stimulate a protective, specific immune response against lymphoma (Timmerman et al. 2001). Importantly, tumor immunity seems to be superior to that afforded by the classical idiotype protein vaccination and naked DNA. The tumor immunity promoted by adenovirus vaccine has also been demonstrated in experimental models of solid tumors, such as melanoma and colon cancer.

In the clinical setting, the most evaluated viral vectors for cancer vaccines are from the poxviridae family (Mastrangelo et al. 2000). They include vaccinia and the avipoxviruses (fowlpox and canarypox; ALVAC). Poxviruses have features similar to adenovirus although, in our experience, rates of transfection of lymphoid cells are higher than with adenovirus. Several cancer vaccine studies have been conducted with poxviruses (Schlom 2012). In one trial, patients with prostate cancer received a poxvirus vector encoding PSA and the costimulatory molecules B7-1, ICAM-1, and LFA-3, along with GM-CSF. Patients receiving the vaccine had a superior overall survival than the control group (patients treated with an empty vector) (Kantoff et al. 2010a).

Despite the potential of viral vector vaccines to induce tumor immunity, several disadvantages need to be overcome before these vaccines will be used in the clinic. First, the inherent immunogenicity of the vector that results in the generation of neutralizing antibodies which may prevent re-administration of the vector to boost the tumor immune response. The high prevalence of preexisting immunity in humans to several viral vectors will likely reduce their immunogenicity in the clinical setting. In fact, studies in melanoma patients vaccinated with adenovirus encoding the melanoma associated antigens gp100 or MART-1 have shown high titers of neutralizing antibodies in the majority of the patients (Rosenberg et al. 1998). This contributed, at least partially, to the failure of generating an antitumor response in the vaccinated patients. A similar problem might be observed when using recombinant poxviruses. Second, a theoretical concern is based upon the immunodominance concept (Chen and McCluskey 2006); in this regard, strong viral antigens may redirect the immune response thereby inhibiting the responses to the weak tumor antigen, thus, reducing the antitumor efficacy of the viral vaccine.

6.5 Dendritic Cell Vaccines

Presentation of an antigen to T cells is one of the crucial steps to stimulate an efficient immune response against that antigen. DCs are the most important APCs and play a crucial role in both the induction and regulation of the immune response. DCs are bone marrow-derived cells expressing surface class I and II MHC molecules, along with a number of adhesion and costimulatory molecules needed to stimulate naïve T cells. In addition, DCs are able to produce a variety of chemokines and immunostimulatory cytokines that contribute to the final activation and expansion of effector cells, including T, B, and NK cells. In addition to cellular immune response, DCs also have an important role in regulating humoral immunity. DCs directly interact with B cells and regulate $CD4^+$ T cell differentiation and expansion, all of which contributes to antibody response (Palucka and Banchereau 2012).

In the early phase of the immune response process, DCs are in an immature state and express an array of surface receptors specialized in antigen uptake. In this situation, DCs can uptake antigens from bacteria, viruses, dead or dying tumor cells, proteins, and immunocomplexes, through a mechanism of phagocytosis and endocytosis. The antigens are processed into peptides, coupled to MHC molecules, and expressed in the membrane for recognition of antigen-specific $CD4^+$ and $CD8^+$ T cells (Trombetta and Mellman 2005). When compared with other APCs, such as macrophages or B cells, DCs are extremely efficient at antigen presentation and can elicit very low numbers of T cells to respond. After capturing the antigen, DCs undergo another process called maturation. This process transforms DCs into cells specialized for T cell stimulation. Maturation is characterized by reduced antigen uptake, enhanced ability to migrate to lymphoid tissues, secretion of cytokines, and activation of antigen-specific T cells. A number of molecules released by damaged tissues and tumors may stimulate the maturation process in which several receptor types, such as TLRs, are involved. Mature DCs express high levels of class I and II MHC molecules, adhesion and costimulatory molecules CD80, CD86, ICAM-1, and CD40, all devoted to efficiently activate T cells.

A number of experimental studies have demonstrated that DCs can capture tumor antigens that are released from tumor cells and present these antigens to T cells in the tumor-draining lymph nodes. This results in the generation of tumor-specific CD8+ T cells that contribute to tumor eradication. In line with this, DCs have been considered an important part of different cancer vaccine strategies (Ueno et al. 2011). To this end, the first step is to provide DCs with tumor-specific antigens. This can be achieved by culturing in vitro DCs that have been obtained from the patient with the tumor-specific antigen plus an adjuvant; alternatively, DCs can be induced in vivo to uptake the tumor-specific antigen.

The list of possibilities includes the use of DCs cocultured in vitro with tumor-derived peptides or proteins (also called "pulsed" DCs) or tumor lysates or DCs fused to tumor cells. Alternatively, DCs can be transfected in vitro with DNA, or RNA encoding a tumor-antigen, or transduced with recombinant viruses encoding the antigen (Table 6.1).

A number of DCs cancer vaccine trials have been reported (Schuler 2010). The first was reported in patients with B cell lymphoma (Hsu et al. 1996). Based on preclinical studies, a small pilot trial was conducted more than 15 years ago at Stanford University. In this study, cellular immune responses and clinical responses were noted after the administration of ex vivo idiotype-pulsed DCs to four patients with relapsed, advanced lymphoma. Encouraged by these results, the study was further extended by including 35 patients (Timmerman et al. 2002b). Sixty-five percent of the patients developed either a humoral or cellular anti-idiotype response. More important, among ten relapsed patients, one had a partial response (i.e., at least 50 % reduction of tumor burden) and three had a complete response lasting more than 4 years. Concerning the 25 patients vaccinated incomplete remission after chemotherapy, 70 % of the patients remained progression-free after vaccination, at a median of 4 years after completion of the treatment.

DCs vaccines have also been studied in patients with solid tumors, including melanoma, prostate and colon cancer. In the majority of these studies, patients were vaccinated with ex vivo generated DCs pulsed with different tumor-associated peptides. However, although cellular immune responses were seen, no significant clinical responses were detected in the majority of patients, with perhaps the exception of prostate cancer patients. In one trial, metastatic prostate cancer patients received DCs that were cultured in vitro with a fusion protein of prostatic acid phosphatase and GM-CSF (Kantoff et al. 2010b). Patients receiving the vaccine had an improvement in overall survival compared with the control group (patients receiving unpulsed DCs).

These data were deemed significant by the US Food and Drug Administration in a patient population that has very few, if any, effective therapeutic options and lead to the first cellular immunotherapy product approved for cancer patients.

To enhance the antitumor vaccine efficacy, DCs have been genetically modified. DCs can be ex vivo transduced with recombinant viruses encoding tumor antigen genes (Humrich and Jenne 2003). This strategy can result in prolonged presentation of multiple epitopes by various class I and II MHC molecules, with the ability to stimulate high-avidity T cells capable of recognizing different tumor antigens. In line with this, experimental studies have shown that DCs transduced with recombinant viruses encoding a tumor antigen stimulate a potent $CD8^{+}$ T cell response able to eradicate established tumors. The antitumor response efficacy can be further enhanced by the transfer into DCs of genes encoding cytokines (i.e., IL-12 or IL-15) or costimulatory molecules (i.e., CD80, CD86, ICAM-1, or LFA-3).

The previously discussed approaches are based on the characterization of tumor antigens. However, as previously discussed, specific tumor antigens are currently not known for the majority of tumors. To overcome this problem, another strategy has been developed, consisting of the use of DCs stimulated with whole tumor cells. Thus, vaccination with DCs cocultured in vitro with a tumor lysate elicits an immune response capable of eradicating established tumors in animal models. This approach may be more efficient than peptide/protein vaccination, at least in some experimental tumor models (Gatza and Okada 2002).

Another strategy takes advantage of the ability of DCs to be fused to tumor cells. This approach ensures proper presentation of an array of tumor antigens coupled to adequate stimulation of antigen-specific T cells. The potential advantage of this method relies on the fact that several relevant tumor antigens, instead of just one, may be presented to the effector T cells along with appropriate costimulation, thus potentially increasing the frequency of different antigen-specific T cells. This strategy has proved to be effective in a number of solid tumors and hematological cancers (Gong et al. 2000; Raje et al. 2004). The system can be further improved by using adjuvants such as CpGs or genetic modification with immunostimulatory genes. Thus, fused DCs and tumor cells transduced with a recombinant adenovirus encoding CD40L elicit an immune response able to eradicate established lymphoma (Alvarez et al. 2010). A clinical trial with fusions of DCs with tumor cells has been conducted in patients with multiple myeloma and other solid tumors such as renal, brain, and breast cancer (Rosenblatt et al. 2013).

Another approach to deliver antigens to DCs relies on the use of antibodies. In vivo delivery of antigens can be done by using an antibody targeting a DC receptor fused to an antigen protein. Administration of an antibody directed to a DC surface receptor (i.e., DC-SIGN) coupled to a tumor antigen elicits potent antigen-specific $CD4^{+}$ and $CD8^{+}$ T cells (Kretz-Rommel et al. 2007). However, targeting to some DC receptors polarizes T cells into suppressor cells, which in turn might negatively affect tumor specific effector T cells. The efficacy of this approach also may be different depending on the DC subtype targeted. Thus, a deeper knowledge of the immune outcome after targeting DC surface receptors is needed before this approach can be translated into the clinic as a cancer vaccine.

Clinical trials of DC vaccines rely on the generation of large numbers of clinical grade human DCs. Currently, two different approaches may be used: (1) ex vivo differentiation of DCs from CD34⁺ hematopoietic progenitor cells or from peripheral blood monocytes; and (2) purification of immature DC precursors obtained from peripheral blood. In addition, maturation of DCs can be accomplished with a variety of stimuli, including, TNF-α, or CD40 ligation yielding different results. The need for ex vivo culture to generate DCs in addition to the lack of standardized methods for obtaining DCs, assessing maturation status, and loading antigens constitutes a significant limitation for the broad application of a DC-based vaccine in cancer patients.

6.6 Tumor Cell Vaccines

A completely different approach to generate a cancer vaccine consists of the use of tumor cells to induce a host immune response. An advantage of the use of whole tumor cells as a vaccine is that characterization of the tumor antigen is no longer required since they represent a source of a broad spectrum of tumor antigens by themselves. Thus, with this strategy, it can be generated a polyvalent immune response against several tumor antigens, in contrast to a response against a single antigen when a peptide or a single gene is used for vaccination.

For an appropriate antigen-specific T cell activation, T cells are required to interact with APCs, recognition of the peptide-MHC complexes and costimulation of those T cells. This later step is critical for generation of an effective T cell immune response against a predefined antigen (Liebowitz et al. 1998). In fact, recognition of antigens by T cells in absence of proper costimulation results in T cell anergy, rather than activation. Among the costimulatory molecules involved in this process, the B7 family proteins—mainly B7.1 (CD80) and B7.2 (CD86)—are of great importance for T cell stimulation. The expression of these molecules is tightly controlled and is restricted to APCs, including DCs and B cells. Both, CD80 and CD86 interact with the CD28 molecule on activated T cells to further enhance its activation. In addition, other molecules expressed by APCs, such as ICAM-1 and LFA-3, interact with different receptors on T cells to promote cellular adhesion and activation. Experimental data have shown that tumor cells are very inefficient at presenting antigens to APCs for the stimulation of a T cell response. However, genetic manipulation of the tumor cells may increase their immunogenicity.

Tumor cells can be genetically modified, by DNA transfection or by use of recombinant viruses, to secrete cytokines and chemokines that contribute to the activation of DCs and T cells. Several studies in tumor animal models have shown that vaccination with irradiated tumor cells engineered to express cytokines (i.e., IL-2, IL-12, IL-15) stimulates T cells that are able to recognize and kill unmanipulated tumor cells, thus providing systemic immunity (Ochoa et al. 2013). Alternatively, tumor cells can be transduced with genes directly acting on DCs. GM-CSF has been one of the cytokines most extensively studied. In murine models, vaccination with irradiated tumor cells transduced with a recombinant virus encoding GM-CSF provides a protective systemic immunity mediated by both CD4+ and CD8+ T cells (Levitsky et al. 1996). This approach proved to be very robust in generating an antitumor immune response and was moved to the clinic (Soiffer et al. 2003). Patients with melanoma and prostate cancer were vaccinated with autologous tumor cells transduced with a retrovirus encoding human GM-CSF. Although an extensive inflammatory infiltration with T cells and necrosis have been detected in the tumors of most of those patients, a very few clinical responses were recorded, and clinical activity of this approach has not been proven. However, a meta-analysis of more than 100 immunotherapy trials in several solid tumor types revealed that patients immunized with whole-tumor cells vaccines had significantly higher rates of clinical responses than patients immunized with molecularly defined tumor antigens (Neller et al. 2008).

In the clinical scenario, the use of autologous tumor cells for vaccination is hampered by the long, cumbersome, and expensive manufacturing process. Moreover, since they are a personalized vaccine, obtaining and preparing the tumor cells for each patient may be extremely difficult, limiting its broad application for clinical use. To avoid these problems, allogeneic tumor cell lines containing shared tumor antigens have been used. Cancer cell lines can be ex vivo transduced with genes encoding cytokines (i.e., GM-CSF, IL-2) and used to immunize the patients. These vaccines are easier to produce than autologous tumor cells and have been tested in melanoma and prostate cancer patients. However, despite early promising results, clinical efficacy was not shown in randomized trials (Copier and Dalgleish 2010).

Another novel approach to stimulate antitumor immunity focuses on the ability of some tumors to present antigens to T cells. In contrast to solid tumors, B cell lymphomas are able to present antigens to T cells but the low expression of costimulatory molecules renders these cells inefficient for T cell stimulation (Schultze et al. 1995). A number of methods have been developed to turn tumor cells into powerful APCs capable of stimulating T cells (Schultze et al. 1997). A CD40 agonistic monoclonal antibody can activate DCs leading to cross-presentation of tumor antigens and generation of a CD8+ T cell response (Sotomayor et al. 1999). Alternatively, direct transfer to the tumor of costimulatory molecules (i.e., B7, ICAM-1, and LFA-3) by using recombinant viruses may enhance their immunogenicity, and these cells can be used as a vaccine (Grosenbach et al. 2003). This concept has been tested in lymphoma models where it has been shown to induce protective systemic antitumor immunity (Briones et al. 2003). Alternatively, a recombinant virus encoding a tumor antigen gene combined to genes encoding those costimulatory molecules has been shown to induce antigen-specific T cells with higher avidity and more effective at killing tumor cells (Yang et al. 2005).

Another approach extensively studied involves the CD40-CD40 ligand (CD40L) system which has a critical role in T cell dependent immune responses. CD40 is a member of the tumor necrosis factor receptor family that is expressed by APCs (including DCs and B cells) and interacts with its ligand, CD40L, expressed on activated T cells (van Kooten and Banchereau 2000). Stimulation of CD40 induces upregulation of MHC class II and costimulatory molecules (i.e., CD80/CD86,

ICAM-1, 4-1BBL) on lymphoma cells leading to efficient activation of tumor-specific T cells. In B cell lymphoma models, vaccination with tumor cells transduced with a recombinant virus encoding the CD40L gene stimulates a systemic immune response against unmanipulated lymphoma cells (Briones et al. 2002). This concept has been translated into the clinic in patients with hematological cell malignancies (Wierda et al. 2010; Rousseau et al. 2006). In one approach, autologous tumor B cells were ex vivo transduced with a recombinant adenovirus encoding CD40L and reinfused to the patients with no other concomitant treatments. Another approach has used autologous leukemic cells mixed with fibroblasts transduced with a recombinant virus encoding CD40L and IL-2. In these studies, tumor-specific T cells have been detected along with clinical responses, which suggest that CD40-based vaccines may be useful in hematological malignancies.

6.7 NKT Cell Based Vaccines

Studies in the last 10 years have focused on the role of Natural Killer T (NKT) cells in the immune system. NKT cells constitute a particular type of T cells that share some features with NK cells (Bendelac et al. 2007). Although NKT cells are a heterogeneous population, type I NKT cells (the most studied subtype) express a limited array of αβT cell receptors (TCR) (i.e., the αchain Vα24-Jα18 paired with particular TCR-Vβ chains (i.e., the Vβ11 in humans)). In contrast to T cells, NKT cells do not recognize peptides on the MHC molecules. Type I NKT cells are characterized by their ability to recognize lipid antigens through the CD1d receptor, expressed on APCs. Although they represent minor population in the immune system, NKT cells have a critical role in immunosurveillance and in tumor immunology, since they coordinate both, the innate (DCs and NK cells) and the adaptive (B and T cells) immune responses.

α-Galactosylceramide (α-GalCer) is the prototypic lipid recognized by type I NKT cells (Kinjo et al. 2005). After administration of α-GalCer, APCs (either DCs or B cells) may present this antigen to NKT cells. This results in activation of NKT cells by upregulation of CD40L. CD40 cross-linking induces DCs to upregulate CD40, MHC molecules, and B7 costimulatory molecules, and to secrete IL-12. This, in turn, enhances NKT cell activation and cytokine production. Activated NKT cells produce Th1 (IFN-γ and IL-12) and Th2 (IL-4) cytokines which contribute to the generation of a T cell and B cell mediated immune response. Thus, killing of tumor cells after NKT cell activation is produced by two different mechanisms: (1) direct killing of $CD1d^+$ tumor cells after recognition of lipid tumor antigens; (2) cross-priming of DCs mediated by IL-12 and enhancement of NK cells and antigen-specific $CD4^+$ and $CD8^+$ T cell response (Cerundolo et al. 2009).

DCs have an important role in the antitumor response mediated by NKT cells. A number of experimental studies have shown that administration of DCs cocultured with CD1d analogs (i.e., α-GalCer) results in activation and expansion of NKT cells resulting in the induction of a strong T cell immune response capable of eradicating established tumors. Other studies have further refined the efficacy of this approach by combining α-GalCer loaded DCs with tumor cells and showing a greater induction of antitumor immunity (Chang et al. 2005; Liu et al. 2005; Alvarez E and Briones J; unpublished data).

An alternative approach to activate NKT cells may involve the use of tumor cells. Because most human leukemias and B-cell lymphomas express the CD1d receptor, they can induce the activation of NKT cells. Experimental studies in lymphoma models have shown that vaccination with B lymphoma cells loaded with α-GalCer generates specific T cell-mediated antilymphoma immunity (Chung et al. 2007). This approach has recently been taken to the clinic in patients with multiple myeloma (Richter et al. 2013).

6.8 Strategies to Improve Cancer Vaccines Efficacy

Despite all the advances in tumor immunology, the clinical impact of cancer vaccines is far from ideal. In the last years, a better understanding of the tumor's mechanisms to evade the immune system has led to the development of new strategies to enhance the efficacy of the cancer vaccines.

Despite the expression of foreign (mutated) antigens, which are potentially highly immunogenic, tumors grow progressively supporting the concept that tumor cells escape the immune recognition. Several mechanisms for tumor escape have been proposed, including those related to the inherent instability of the tumor cells and others that may be shared by many normal cells (Gajewski et al. 2013). The latter include: (1) the lack of expression of costimulatory molecules; (2) the secretion of proteins with immune inhibitory capacity by tumor cells; and (3) the induction of a T cell population with suppressor activity.

Low or absent expression of costimulatory molecules may be involved, at least in part, in the inhibition of the immune system since, as previously mentioned, recognition of tumor peptides by T cells in the absence of costimulation generates T cell anergy (Schultze et al. 1995). In line with this, direct stimulation of T cells through activating receptors may enhance T cell function and improved tumor killing. There are a number of activating receptors that can be targets for agonistic antibodies, some of which are already in the clinic: OX40, 4-1BB, CD27, and CD40.

OX40 is a costimulatory molecule that promotes T cell proliferation, survival, and enhanced function and migration. In addition, OX40 also enhances memory T cell development (Chen et al. 1999). Interestingly, activation of OX40 inhibits proliferation of regulatory T (Treg) cells (Piconese et al. 2008). Studies in experimental tumor models have shown that administration of OX40 agonists could eradicate established tumors, and that this response was $CD4^+$ and $CD8^+$ T cell dependent (Jensen et al. 2010). Alternatively, vaccines with tumor cells transduced with OX40L stimulate strong antitumor-specific T cell responses. Currently, clinical studies with agonistic OX40 antibodies are being conducted in patients with solid tumors (Curti et al. 2013).

4-1BB is another costimulatory molecule extensively studied (Wang et al. 2009). A member of the tumor necrosis factor receptor family, 4-1BB is expressed on activated T cells and NK cells. Its ligand is expressed on APCs and macrophages. Targeting 4-1BB on T cells induces activation of $CD8^+$ T cells and NK cells, stimulating tumor antigen responses. 4-1BB stimulation can reverse tolerance of $CD8^+$ T cells and promote antitumor memory T cell survival (Wilcox et al. 2004). Several clinical trials are currently investigating the role of 4-1BB agonistic antibodies in patients with melanoma and B cell lymphoma (Weinberg et al. 2011).

Downregulation of MHC molecules and/or tumor antigens has also been associated with reduced ability to be targeted by T cells, providing a scenario for tumor escape from the immune system. Tumors may produce different molecules with immunosuppressive properties. Thus, tumor secretion of VEGF inhibits the antigen processing and maturation capacity of DCs (Osada et al. 2008). In addition, TGFβ and IL-10 have a strong inhibitory effect on antigen-specific T cells as well as DCs.

Immunosuppressive regulatory cells have been the object of much attention in the last years. These cells play a physiological role in the induction of immune tolerance to self-antigens preventing autoimmunity. However, there is increasing evidence that, in cancer patients, these cells are largely responsible for preventing effective antitumor immune responses.

The best characterized immunoregulatory cells are the $CD4^+$ Treg cells (Roncarolo and Battaglia 2007). This cell population constitutes a functionally distinct lineage of T cells involved in the maintenance of peripheral tolerance in vivo. Tregs represent 5–10 % of peripheral $CD4^+$ T cells and are characterized by the expression of CD25, the transcription factor forkhead box P3 (Foxp3), and glucocorticoid-induced tumor necrosis factor (GITR). Treg cells can induce immunosuppression by different mechanisms. Treg cells may suppress the proliferation and cytokine production of antigen-specific T cells through direct contact with the responding T cell. Alternatively, Treg cells produce high amounts of immunosuppressive cytokines such as IL-10 and TGFβ with inhibitory effect on DCs and cytotoxic $CD8^+$ T cells. High levels of TGFβ in the tumor microenvironment may also enhance the conversion of Foxp3-negative $CD4^+$ effector T cells into Treg cells. In addition, Treg cells also restrain the antitumor activities of NKT cells. Several studies in cancer patients have shown significant increases in regulatory T cell populations both in the tumor and in peripheral blood, which negatively correlate with clinical outcome (Preston et al. 2013; Blatner et al. 2012).

The impact of Treg cells on the development of an effective antitumor immune response led to the evaluation of Treg cell depletion as a strategy to enhance the immune response against cancer. Studies in animal tumor models have shown that depletion of Treg cells by administration of monoclonal antibodies enhanced the effect of a tumor vaccine (Marabelle et al. 2013). This approach has been taken to the clinic in patients receiving a cancer vaccine with similar results. Administration of monoclonal antibodies targeting CD25 or chemotherapy agents, such as cyclophosphamide, prior to antitumor vaccination contributes to Treg cell depletion and augmentation of tumor antigen-specific T cell responses (Morse et al. 2008; Litzinger et al. 2007).

Myeloid-derived suppressor cells (MDSCs) are another heterogeneous population of immature myeloid cells (i.e., granulocyte and monocyte lineage) with many immunosuppressive functions (Talmadge and Gabrilovich 2013). MDSCs have been shown to accumulate in the peripheral blood and tumors of patients with different cancer types. MDSCs mediate their immunosuppressive activity through different mechanisms: production of arginase and nitric oxide which induce $CD4^+$ T cell apoptosis; induction of Treg cells, and promotion of immature DCs through IL-10. Attempts to reduce or eliminate these immunosuppressive cells are the subject of intensive research (Albeituni et al. 2013). However, to date, the systemic depletion of any of regulatory cell subsets is hampered by the fact that none of the regulatory cell subsets express currently known unique markers that can be targeted without affecting nonregulatory effector cells.

A very promising novel strategy for reducing antitumor immunoregulation is based upon the specific blocking of the molecules involved in T cell immunosuppression, a concept that evolved in the last years as the "immune checkpoint blockade" (Bluestone and Small 2012). The list of molecules that can be potentially targetable is rapidly growing and can be expressed either on the cell surface of tumor cells or effector T and NK cells (Fig. 6.1). So far, two molecules have received enormous attention and have been focus of intensive research to increase the quality of the immune response after vaccination.

The first of these negative immune regulator molecules studied is cytotoxic T-lymphocyte-associated protein 4 (CTLA-4) (Chambers et al. 2001). CTLA-4 is upregulated in conventional effector T cells during the antigen stimulation and contributes to inhibit T cell function and limit T cell proliferation by binding to the costimulatory molecules CD80/CD86. CTLA-4 is also expressed in Treg cells where it induces the production of immunosuppressive molecules by DCs (i.e., indoleamine 2,3-dioxygenase) promoting T cell anergy. Ipilimumab is a monoclonal antibody that

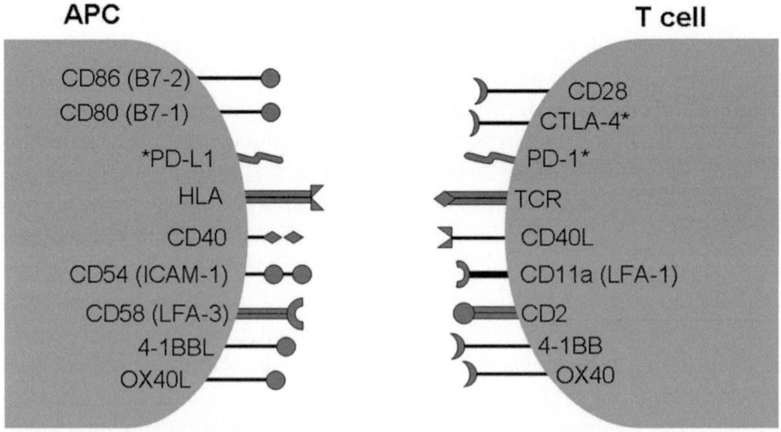

Fig. 6.1 Costimulatory and inhibitory (*) interactions between antigen presenting cell (*APC*) and T cell. *TCR* T cell receptor, *HLA* human leukocyte antigen

antagonizes CTLA-4, blocking its interaction with CD80/CD86. CTLA-4 blockade also promotes apoptosis of Treg cells contributing to an enhancement of the antitumor T cell responses. In the clinical scenario, ipilimumab as single therapy was able to induce long-term remissions in a group of heavily pre-treated melanoma patients (Hodi et al. 2010). Although the clinical benefit occurred in a small percentage of patients, these results, in an otherwise very advanced cancer patients with no therapeutic options, were considered of great impact. In experimental studies, CTLA-4 blockade was shown to augment the efficacy of a cellular vaccine consisting of tumor cells transduced with GM-CSF, with an enhancement of the $CD8^+$ T cell responses (Li et al. 2009).

A second inhibitory molecule is the programmed cell death-1 (PD-1), which is expressed by activated $CD4^+$ and $CD8^+$ T cells, B and NK cells (Keir et al. 2008). Its ligand (PD-L1) is expressed on a variety of tumors and APCs. Ligation of PD-1 on T cells is associated with decreased proliferation and cytokine secretion and, ultimately, T cell exhaustion and death. Blocking PD-1 may restore T cell function and contributes to the enhancement of the T cell responses (Chen and Mellman 2013).

In the clinical setting, administration of a monoclonal antibody blocking PD-1 as single therapy resulted in complete responses in patients with solid tumors and hematological malignancies (Brahmer et al. 2012).

Similar to what it has been shown for CTLA-4, PD-1 blockade has been shown to improve the immune response after vaccination in cancer models, by increasing the number and function of antigen-specific $CD8^+$ T cells (Duraiswamy et al. 2013; Kamphorst and Ahmed 2013). Thus, combination of PD-1 blockade with a tumor vaccine is being tested in several clinical trials involving both, cellular and peptide vaccines. Data from these trials will provide critical information for the design of combinatorial immunotherapy for the treatment of cancer patients.

6.9 Nanotechnology Applications to Cancer Vaccines

A critical issue on the development of cancer vaccines is the efficient presentation of tumor antigens to T cells. While this can be achieved with the use of ex vivo generated DCs or recombinant DNA-viral vaccines encoding tumor antigens, all these strategies are labor extensive and costly, and pose a formidable challenge for administration to cancer patients under clinically accepted GMPs conditions. In this regard, nanotechnology, through their broad range of immunological applications, has arisen as a promising strategy to develop cancer vaccines (Smith et al. 2013).

Nanovaccines are nanoscale complexes that can be designed to deliver antigens and/or adjuvants to DCs or T cells to generate an immune response. Nanovaccines can be designed to tailor immune responses by specific targeting of immune cells. Thus, nanoparticles (NPs) can be decorated with antibodies directed to DCs or with adjuvants that interact with specific receptors on DCs (i.e., TLR ligands, or lectin receptor ligands) enhancing the immune response (Paulis et al. 2013). Multiple immunomodulators and antigens can be loaded onto a single NP, and the dose and rate of antigen delivery can be adjusted (Ali et al. 2014). In addition, NPs can be designed to deliver chemokines or cytokines to recruit APCs first and subsequently present tumor antigens (Ali et al. 2009). This strategy proved to be effective in murine models of cancer. Interestingly, the size and shape of the NPs can be modulated so they can differentially target distinct DCs populations; small NPs (up to 200 nm) target selected populations of DCs that better migrate to lymph nodes to present antigens to T cells (Manolova et al. 2008). Furthermore, NPs can penetrate tumors so they can be designed to target different subsets of cells (i.e., cancer stem cells) or the tumor microenvironment.

Since T cells are the most important effector cells in the immune response, adoptive therapy with ex vivo generated antigen specific T cells has become a promising strategy to treat cancer patients. In this regard, NPs can be designed to contain cytokines promoting T cell function and survival (i.e., IL-15 or IL-21), so they can be ex vivo tethered to T cells before administration to patients (Stephan et al. 2010).

Nanovaccines are versatile complexes that have a broad array of immunological properties which make them promising candidates as cancer vaccines. Importantly, the ease of manufacturing and high quality production under GMP conditions, in addition to long-term storage properties, are of great value at the time of translation of these strategies into clinical trials for cancer patients.

6.10 Conclusions

The considerable progresses made in the knowledge of the biology of DCs as well as effector T cell immunology clearly open the avenues for the development of greatly improved clinical protocols in cancer immunotherapy. The design of better adjuvants and the identification of new tumor targets will be critical for the advancement in the field. While passive immunotherapies involving monoclonal antibodies or antigen-specific T cells are relatively straightforward, they have the limitation of the requirement for a known epitope surface expression. This can be overcome by active immunotherapy with cellular vaccines (i.e., DCs, recombinant viruses, and whole tumor cells) which may induce a T cell response against intracellular antigens. Importantly, recent advances in the knowledge of T cell inhibitory molecules have led to the development of blocking antibodies that will likely enhance the antitumor effect of these cancer vaccines.

Finally, the recent development of biomaterials and nanoparticles has shown an enormous potential for cancer immunotherapy. Particles vaccines may have advantages over cellular or viral-based vaccines: They can incorporate multiple immune modulators, including antigens and adjuvants; they can target specific cell populations and subcellular compartments, which may increase antigen-delivery to effector cells, thus reducing toxicity while improving the immune response against tumor cells. The use of nanoparticles with immunological properties may represent a promising strategy for the development of cancer vaccines.

References

Ai WZ, Tibshirani R, Taidi B, Czerwinski D, Levy R (2009) Anti-idiotype antibody response after vaccination correlates with better overall survival in follicular lymphoma. Blood 113(23): 5743–5746

Albeituni SH, Ding C, Yan J (2013) Hampering immune suppressors: therapeutic targeting of myeloid-derived suppressor cells in cancer. Cancer J 19(6):490–501

Ali OA, Huebsch N, Cao L, Dranoff G, Mooney DJ (2009) Infection-mimicking materials to program dendritic cells in situ. Nat Mater 8(2):151–158

Ali OA, Verbeke C, Johnson C, Sands W, Lewin SA, White D, Doherty E, Dranoff G, Mooney DJ (2014) Identification of immune factors regulating anti-tumor immunity using polymeric vaccines with multiple adjuvants. Cancer Res 74(6):1670–1681

Alvarez E, Moga E, Barquinero J, Sierra J, Briones J (2010) Dendritic and tumor cell fusions transduced with adenovirus encoding CD40L eradicate B-cell lymphoma and induce a Th17-type response. Gene Ther 17(4):469–477

Bendelac A, Savage PB, Teyton L (2007) The biology of NKT cells. Annu Rev Immunol 25: 297–336

Blatner NR, Mulcahy MF, Dennis KL, Scholtens D, Bentrem DJ, Phillips JD, Ham S, Sandall BP, Khan MW, Mahvi DM, Halverson AL, Stryker SJ, Boller AM, Singal A, Sneed RK, Sarraj B, Ansari MJ, Oft M, Iwakura Y, Zhou L, Bonertz A, Beckhove P, Gounari F, Khazaie K (2012) Expression of RORγt marks a pathogenic regulatory T cell subset in human colon cancer. Sci Transl Med 4(164):164ra159

Bluestone JA, Small EJ (2012) The future of cancer treatment: will it include immunotherapy? Cancer Cell 22(1):7–8

Bode C, Zhao G, Steinhagen F, Kinjo T, Klinman DM (2011) CpG DNA as a vaccine adjuvant. Expert Rev Vaccines 10(4):499–511

Brahmer JR, Tykodi SS, Chow LQ, Hwu WJ, Topalian SL, Hwu P, Drake CG, Camacho LH, Kauh J, Odunsi K, Pitot HC, Hamid O, Bhatia S, Martins R, Eaton K, Chen S, Salay TM, Alaparthy S, Grosso JF, Korman AJ, Parker SM, Agrawal S, Goldberg SM, Pardoll DM, Gupta A, Wigginton JM (2012) Safety and activity of anti-PD-L1 antibody in patients with advanced cancer. N Engl J Med 366(26):2455–2465

Brightbill HD, Modlin RL (2000) Toll-like receptors: molecular mechanisms of the mammalian immune response. Immunology 101(1):1–10

Briones J, Timmerman J, Levy R (2002) In vivo antitumor effect of CD40L-transduced tumor cells as a vaccine for B-cell lymphoma. Cancer Res 62(11):3195–3199

Briones J, Timmerman JM, Panicalli DL, Levy R (2003) Antitumor immunity after vaccination with B lymphoma cells overexpressing a triad of costimulatory molecules. J Natl Cancer Inst 95(7):548–555

Cerundolo V, Silk JD, Masri SH, Salio M (2009) Harnessing invariant NKT cells in vaccination strategies. Nat Rev Immunol 9:28–38

Chambers CA, Kuhns MS, Egen JG, Allison JP (2001) CTLA-4-mediated inhibition in regulation of T cell responses: mechanisms and manipulation in tumor immunotherapy. Annu Rev Immunol 19:565–594

Chang DH, Osman K, Connolly J, Kukreja A, Krasovsky J, Pack M, Hutchinson A, Geller M, Liu N, Annable R, Shay J, Kirchhoff K, Nishi N, Ando Y, Hayashi K, Hassoun H, Steinman RM, Dhodapkar MV (2005) Sustained expansion of NKT cells and antigen-specific T cells after injection of alpha-galactosyl-ceramide loaded mature dendritic cells in cancer patients. J Exp Med 201:1503–1517

Chen W, McCluskey J (2006) Immunodominance and immunodomination: critical factors in developing effective CD8+ T-cell-based cancer vaccines. Adv Cancer Res 95:203–247

Chen DS, Mellman I (2013) Oncology meets immunology: the cancer-immunity cycle. Immunity 39(1):1–10

Chen PW, Wang M, Bronte V, Zhai Y, Rosenberg SA, Restifo NP (1996) Therapeutic antitumor response after immunization with a recombinant adenovirus encoding a model tumor-associated antigen. J Immunol 156(1):224–231

Chen AI, McAdam AJ, Buhlmann JE, Scott S, Lupher ML Jr, Greenfield EA, Baum PR, Fanslow WC, Calderhead DM, Freeman GJ, Sharpe AH (1999) Ox40-ligand has a critical costimulatory role in dendritic cell: T cell interactions. Immunity 11(6):689–698

Chung Y, Qin H, Kang CY, Kim S, Kwak LW, Dong C (2007) An NKT-mediated autologous vaccine generates CD4 T-cell dependent potent antilymphoma immunity. Blood 110:2013–2019

Copier J, Dalgleish A (2010) Whole-cell vaccines: a failure or a success waiting to happen? Curr Opin Mol Ther 12(1):14–20

Curti BD, Kovacsovics-Bankowski M, Morris N, Walker E, Chisholm L, Floyd K, Walker J, Gonzalez I, Meeuwsen T, Fox BA, Moudgil T, Miller W, Haley D, Coffey T, Fisher B, Delanty-Miller L, Rymarchyk N, Kelly T, Crocenzi T, Bernstein E, Sanborn R, Urba WJ, Weinberg AD (2013) OX40 is a potent immune stimulating target in late stage cancer patients. Cancer Res 73(24):7189–7198 [Epub 2013 Oct 31]

Duraiswamy J, Kaluza KM, Freeman GJ, Coukos G (2013) Dual blockade of PD-1 and CTLA-4 combined with tumor vaccine effectively restores T-cell rejection function in tumors. Cancer Res 73(12):3591–3603

Gajewski TF, Schreiber H, Fu YX (2013) Innate and adaptive immune cells in the tumor microenvironment. Nat Immunol 14(10):1014–1022

Gatza E, Okada CY (2002) Tumor cell lysate-pulsed dendritic cells are more effective than TCR Id protein vaccines for active immunotherapy of T cell lymphoma. J Immunol 169(9):5227–5235

Gong J, Avigan D, Chen D, Wu Z, Koido S, Kashiwaba M, Kufe D (2000) Activation of antitumor cytotoxic T lymphocytes by fusions of human dendritic cells and breast carcinoma cells. Proc Natl Acad Sci U S A 97(6):2715–2718

Grosenbach DW, Schlom J, Gritz L, Gómez Yafal A, Hodge JW (2003) A recombinant vector expressing transgenes for four T-cell costimulatory molecules (OX40L, B7-1, ICAM-1, LFA-3) induces sustained CD4+ and CD8+ T-cell activation, protection from apoptosis, and enhanced cytokine production. Cell Immunol 222(1):45–57

Harrop R, Carroll MW (2006) Viral vectors for cancer immunotherapy. Front Biosci 11:804–817

Heath WR, Carbone FR (2009) Dendritic cell subsets in primary and secondary T cell responses at body surfaces. Nat Immunol 10(12):1237–1244

Heemskerk B, Kvistborg P, Schumacher TN (2013) The cancer antigenome. EMBO J 32(2):194–203

Hodi FS, O'Day SJ, McDermott DF, Weber RW, Sosman JA, Haanen JB, Gonzalez R, Robert C, Schadendorf D, Hassel JC, Akerley W, van den Eertwegh AJ, Lutzky J, Lorigan P, Vaubel JM, Linette GP, Hogg D, Ottensmeier CH, Lebbé C, Peschel C, Quirt I, Clark JI, Wolchok JD, Weber JS, Tian J, Yellin MJ, Nichol GM, Hoos A, Urba WJ (2010) Improved survival with ipilimumab in patients with metastatic melanoma. N Engl J Med 363(8):711–723

Hsu FJ, Benike C, Fagnoni F, Liles TM, Czerwinski D, Taidi B, Engleman EG, Levy R (1996) Vaccination of patients with B-cell lymphoma using autologous antigen-pulsed dendritic cells. Nat Med 2(1):52–58

Hsu FJ, Caspar CB, Czerwinski D, Kwak LW, Liles TM, Syrengelas A, Taidi-Laskowski B, Levy R (1997) Tumor-specific idiotype vaccines in the treatment of patients with B-cell lymphoma: long-term results of a clinical trial. Blood 89(9):3129–3135

Humrich J, Jenne L (2003) Viral vectors for dendritic cell-based immunotherapy. Curr Top Microbiol Immunol 276:241–259

Jensen SM, Maston LD, Gough MJ, Ruby CE, Redmond WL, Crittenden M, Li Y, Puri S, Poehlein CH, Morris N, Kovacsovics-Bankowski M, Moudgil T, Twitty C, Walker EB, Hu HM, Urba WJ, Weinberg AD, Curti B, Fox BA (2010) Signaling through OX40 enhances antitumor immunity. Semin Oncol 37(5):524–532

Kamphorst AO, Ahmed R (2013) Manipulating the PD-1 pathway to improve immunity. Curr Opin Immunol 25(3):381–388

Kantoff PW, Schuetz TJ, Blumenstein BA, Glode LM, Bilhartz DL, Wyand M, Manson K, Panicali DL, Laus R, Schlom J, Dahut WL, Arlen PM, Gulley JL, Godfrey WR (2010a) Overall survival analysis of a phase II randomized controlled trial of a Poxviral-based PSA-targeted immunotherapy in metastatic castration-resistant prostate cancer. J Clin Oncol 28(7):1099–1105

Kantoff PW, Higano CS, Shore ND, Berger ER, Small EJ, Penson DF, Redfern CH, Ferrari AC, Dreicer R, Sims RB, Xu Y, Frohlich MW, Schellhammer PF; IMPACT Study Investigators (2010) Sipuleucel-T immunotherapy for castration-resistant prostate cancer. N Engl J Med 363(5):411–422

Kaumaya PT (2013) Bridging oncology and immunology: expanding horizons with innovative peptide vaccines and peptidomimetics. Immunotherapy 5(11):1159–1163

Keir ME, Butte MJ, Freeman GJ, Sharpe AH (2008) PD-1 and its ligands in tolerance and immunity. Annu Rev Immunol 26:677–704

King CA, Spellerberg MB, Zhu D, Rice J, Sahota SS, Thompsett AR, Hamblin TJ, Radl J, Stevenson FK (1998) DNA vaccines with single-chain Fv fused to fragment C of tetanus toxin induce protective immunity against lymphoma and myeloma. Nat Med 4(11):1281–1286

Kinjo Y, Wu D, Kim G, Xing GW, Poles MA, Ho DD, Tsuji M, Kawahara K, Wong CH, Kronenberg M (2005) Recognition of bacterial glycosphingolipids by natural killer T cells. Nature 434:520–525

Klein J, Sato A (2000) The HLA system. First of two parts. N Engl J Med 343(10):702–709

Klein J, Sato A (2000) The HLA system. Second of two parts. N Engl J Med 343(11):782–763

Kretz-Rommel A, Qin F, Dakappagari N, Torensma R, Faas S, Wu D, Bowdish KS (2007) In vivo targeting of antigens to human dendritic cells through DC-SIGN elicits stimulatory immune responses and inhibits tumor growth in grafted mouse models. J Immunother 30(7):715–726

Levitsky HI, Montgomery J, Ahmadzadeh M, Staveley-O'Carroll K, Guarnieri F, Longo DL, Kwak LW (1996) Immunization with granulocyte-macrophage colony-stimulating factor-transduced, but not B7-1-transduced, lymphoma cells primes idiotype-specific T cells and generates potent systemic antitumor immunity. J Immunol 156(10):3858–3865

Li B, VanRoey M, Wang C, Chen TH, Korman A, Jooss K (2009) Anti-programmed death-1 synergizes with granulocyte macrophage colony-stimulating factor—secreting tumor cell immunotherapy providing therapeutic benefit to mice with established tumors. Clin Cancer Res 15(5):1623–1634

Liebowitz DN, Lee KP, June CH (1998) Costimulatory approaches to adoptive immunotherapy. Curr Opin Oncol 10(6):533–541

Litzinger MT, Fernando R, Curiel TJ, Grosenbach DW, Schlom J, Palena C (2007) IL-2 immunotoxin denileukin diftitox reduces regulatory T cells and enhances vaccine-mediated T-cell immunity. Blood 110(9):3192–3201

Liu K, Idoyaga J, Charalambous A, Fujii S, Bonito A, Mordoh J, Wainstok R, Bai XF, Liu Y, Steinman RM (2005) Innate NKT lymphocytes confer superior adaptive immunity via tumor-capturing dendritic cells. J Exp Med 202:1507–1516

Manolova V, Flace A, Bauer M, Schwarz K, Saudan P, Bachmann MF (2008) Nanoparticles target distinct dendritic cell populations according to their size. Eur J Immunol 38(5):1404–1413

Marabelle A, Kohrt H, Sagiv-Barfi I, Ajami B, Axtell RC, Zhou G, Rajapaksa R, Green MR, Torchia J, Brody J, Luong R, Rosenblum MD, Steinman L, Levitsky HI, Tse V, Levy R (2013) Depleting tumor-specific Tregs at a single site eradicates disseminated tumors. J Clin Invest 123(6):2447–2463

Mastrangelo MJ, Eisenlohr LC, Gomella L, Lattime EC (2000) Poxvirus vectors: orphaned and underappreciated. J Clin Invest 105(8):1031–1034

Morse MA, Hobeika AC, Osada T, Serra D, Niedzwiecki D, Lyerly HK, Clay TM (2008) Depletion of human regulatory T cells specifically enhances antigen-specific immune responses to cancer vaccines. Blood 112(3):610–618

Neller MA, López JA, Schmidt CW (2008) Antigens for cancer immunotherapy. Semin Immunol 20(5):286–295

Ng P, Graham FL (2002) Construction of first-generation adenoviral vectors. Methods Mol Med 69:389–414

Ochoa MC, Mazzolini G, Hervas-Stubbs S, de Sanmamed MF, Berraondo P, Melero I (2013) Interleukin-15 in gene therapy of cancer. Curr Gene Ther 13(1):15–30

Osada T, Chong G, Tansik R, Hong T, Spector N, Kumar R, Hurwitz HI, Dev I, Nixon AB, Lyerly HK, Clay T, Morse MA (2008) The effect of anti-VEGF therapy on immature myeloid cell and dendritic cells in cancer patients. Cancer Immunol Immunother 57(8):1115–1124

Palucka K, Banchereau J (2012) Cancer immunotherapy via dendritic cells. Nat Rev Cancer 12(4):265–277

Paulis LE, Mandal S, Kreutz M, Figdor CG (2013) Dendritic cell-based nanovaccines for cancer immunotherapy. Curr Opin Immunol 25(3):389–395

Piconese S, Valzasina B, Colombo MP (2008) OX40 triggering blocks suppression by regulatory T cells and facilitates tumor rejection. J Exp Med 205(4):825–839

Preston CC, Maurer MJ, Oberg AL, Visscher DW, Kalli KR, Hartmann LC, Goode EL, Knutson KL (2013) The Ratios of CD8(+) T Cells to CD4(+)CD25(+) FOXP3(+) and FOXP3(−) T Cells Correlate with Poor Clinical Outcome in Human Serous Ovarian Cancer. PLoS One 8(11):e80063

Raje N, Hideshima T, Davies FE, Chauhan D, Treon SP, Young G, Tai YT, Avigan D, Gong J, Schlossman RL, Richardson P, Kufe DW, Anderson KC (2004) Tumour cell/dendritic cell fusions as a vaccination strategy for multiple myeloma. Br J Haematol 125(3):343–352

Ribas A, Butterfield LH, Glaspy JA, Economou JS (2003) Current developments in cancer vaccines and cellular immunotherapy. J Clin Oncol 21(12):2415–2432

Rice J, Buchan S, Stevenson FK (2002) Critical components of a DNA fusion vaccine able to induce protective cytotoxic T cells against a single epitope of a tumor antigen. J Immunol 169(7):3908–3913

Rice J, Ottensmeier CH, Stevenson FK (2008) DNA vaccines: precision tools for activating effective immunity against cancer. Nat Rev Cancer 8(2):108–120

Richter J, Neparidze N, Zhang L, Nair S, Monesmith T, Sundaram R, Miesowicz F, Dhodapkar KM, Dhodapkar MV (2013) Clinical regressions and broad immune activation following combination therapy targeting human NKT cells in myeloma. Blood 121(3):423–430

Robbins PF, Lu YC, El-Gamil M, Li YF, Gross C, Gartner J, Lin JC, Teer JK, Cliften P, Tycksen E, Samuels Y, Rosenberg SA (2013) Mining exomic sequencing data to identify mutated antigens recognized by adoptively transferred tumor-reactive T cells. Nat Med 19(6):747–752

Roncarolo MG, Battaglia M (2007) Regulatory T-cell immunotherapy for tolerance to self antigens and alloantigens in humans. Nat Rev Immunol 7(8):585–598

Rosenberg SA, Zhai Y, Yang JC, Schwartzentruber DJ, Hwu P, Marincola FM, Topalian SL, Restifo NP, Seipp CA, Einhorn JH, Roberts B, White DE (1998) Immunizing patients with metastatic melanoma using recombinant adenoviruses encoding MART-1 or gp100 melanoma antigens. J Natl Cancer Inst 90(24):1894–1900

Rosenberg SA, Yang JC, Restifo NP (2004) Cancer immunotherapy: moving beyond current vaccines. Nat Med 10(9):909–915

Rosenblatt J, Avivi I, Vasir B, Uhl L, Munshi NC, Katz T, Dey BR, Somaiya P, Mills H, Campigotto F, Weller E, Joyce R, Levine JD, Tzachanis D, Richardson P, Laubach J, Raje N, Boussiotis V, Yuan YE, Bisharat L, Held V, Rowe J, Anderson K, Kufe D, Avigan D (2013) Vaccination with dendritic cell/tumor fusions following autologous stem cell transplant induces immunologic and clinical responses in multiple myeloma patients. Clin Cancer Res 19(13):3640–3648

Rousseau RF, Biagi E, Dutour A, Yvon ES, Brown MP, Lin T, Mei Z, Grilley B, Popek E, Heslop HE, Gee AP, Krance RA, Popat U, Carrum G, Margolin JF, Brenner MK (2006) Immunotherapy of high-risk acute leukemia with a recipient (autologous) vaccine expressing transgenic human CD40L and IL-2 after chemotherapy and allogeneic stem cell transplantation. Blood 107(4): 1332–1341

Schlom J (2012) Therapeutic cancer vaccines: current status and moving forward. J Natl Cancer Inst 104(8):599–613

Schuler G (2010) Dendritic cells in cancer immunotherapy. Eur J Immunol 40(8):2123–2130

Schultze JL, Cardoso AA, Freeman GJ, Seamon MJ, Daley J, Pinkus GS, Gribben JG, Nadler LM (1995) Follicular lymphomas can be induced to present alloantigen efficiently: a conceptual model to improve their tumor immunogenicity. Proc Natl Acad Sci U S A 92(18):8200–8204

Schultze JL, Michalak S, Seamon MJ, Dranoff G, Jung K, Daley J, Delgado JC, Gribben JG, Nadler LM (1997) CD40-activated human B cells: an alternative source of highly efficient antigen presenting cells to generate autologous antigen-specific T cells for adoptive immunotherapy. J Clin Invest 100(11):2757–2765

Shiver JW, Emini EA (2004) Recent advances in the development of HIV-1 vaccines using replication-incompetent adenovirus vectors. Annu Rev Med 55:355–372

Smith DM, Simon JK, Baker JR Jr (2013) Applications of nanotechnology for immunology. Nat Rev Immunol 8:592–605

Soiffer R, Hodi FS, Haluska F, Jung K, Gillessen S, Singer S, Tanabe K, Duda R, Mentzer S, Jaklitsch M, Bueno R, Clift S, Hardy S, Neuberg D, Mulligan R, Webb I, Mihm M, Dranoff G (2003) Vaccination with irradiated, autologous melanoma cells engineered to secrete granulocyte-macrophage colony-stimulating factor by adenoviral-mediated gene transfer augments antitumor immunity in patients with metastatic melanoma. J Clin Oncol 21(17): 3343–3350

Sotomayor EM, Borrello I, Tubb E, Rattis FM, Bien H, Lu Z, Fein S, Schoenberger S, Levitsky HI (1999) Conversion of tumor-specific CD4+ T-cell tolerance to T-cell priming through in vivo ligation of CD40. Nat Med 5(7):780–787

Stephan MT, Moon JJ, Um SH, Bershteyn A, Irvine DJ (2010) Therapeutic cell engineering with surface-conjugated synthetic nanoparticles. Nat Med 16(9):1035–1041

Stevenson FK, Ottensmeier CH, Rice J (2010) DNA vaccines against cancer come of age. Curr Opin Immunol 22(2):264–270

Syrengelas AD, Levy R (1999) DNA vaccination against the idiotype of a murine B cell lymphoma: mechanism of tumor protection. J Immunol 162(8):4790–4797

Talmadge JE, Gabrilovich DI (2013) History of myeloid-derived suppressor cells. Nat Rev Cancer 13(10):739–752

Timmerman JM, Caspar CB, Lambert SL, Syrengelas AD, Levy R (2001) Idiotype-encoding recombinant adenoviruses provide protective immunity against murine B-cell lymphomas. Blood 97(5):1370–1377

Timmerman JM, Singh G, Hermanson G, Hobart P, Czerwinski DK, Taidi B, Rajapaksa R, Caspar CB, Van Beckhoven A, Levy R (2002a) Immunogenicity of a plasmid DNA vaccine encoding chimeric idiotype in patients with B-cell lymphoma. Cancer Res 62(20):5845–5852

Timmerman JM, Czerwinski DK, Davis TA, Hsu FJ, Benike C, Hao ZM, Taidi B, Rajapaksa R, Caspar CB, Okada CY, van Beckhoven A, Liles TM, Engleman EG, Levy R (2002b) Idiotype-pulsed dendritic cell vaccination for B-cell lymphoma: clinical and immune responses in 35 patients. Blood 99(5):1517–1526

Traversari C, van der Bruggen P, Luescher IF, Lurquin C, Chomez P, Van Pel A, De Plaen E, Amar-Costesec A, Boon T (1992) A nonapeptide encoded by human gene MAGE-1 is recognized on HLA-A1 by cytolytic T lymphocytes directed against tumor antigen MZ2-E. J Exp Med 176(5):1453–1457

Trombetta ES, Mellman I (2005) Cell biology of antigen processing in vitro and in vivo. Annu Rev Immunol 23:975–1028

Ueno H, Klechevsky E, Schmitt N, Ni L, Flamar AL, Zurawski S, Zurawski G, Palucka K, Banchereau J, Oh S (2011) Targeting human dendritic cell subsets for improved vaccines. Semin Immunol 23(1):21–27

van Kooten C, Banchereau J (2000) CD40-CD40 ligand. J Leukoc Biol 67(1):2–17

Vittes GE, Harden EL, Ottensmeier CH, Rice J, Stevenson FK (2011) DNA fusion gene vaccines induce cytotoxic T-cell attack on naturally processed peptides of human prostate-specific membrane antigen. Eur J Immunol 41(8):2447–2456

Wang C, Lin GH, McPherson AJ, Watts TH (2009) Immune regulation by 4-1BB and 4-1BBL: complexities and challenges. Immunol Rev 229(1):192–215

Weinberg AD, Morris NP, Kovacsovics-Bankowski M, Urba WJ, Curti BD (2011) Science gone translational: the OX40 agonist story. Immunol Rev 244(1):218–231

Wierda WG, Castro JE, Aguillon R, Sampath D, Jalayer A, McMannis J, Prussak CE, Keating M, Kipps TJ (2010) A phase I study of immune gene therapy for patients with CLL using a membrane-stable, humanized CD154. Leukemia 24(11):1893–1900

Wilcox RA, Tamada K, Flies DB, Zhu G, Chapoval AI, Blazar BR, Kast WM, Chen L (2004) Ligation of CD137 receptor prevents and reverses established anergy of CD8+ cytolytic T lymphocytes in vivo. Blood 103(1):177–184

Yang S, Hodge JW, Grosenbach DW, Schlom J (2005) Vaccines with enhanced costimulation maintain high avidity memory CTL. J Immunol 175(6):3715–3723

Chapter 7
Cancer Immune Modulation and Immunosuppressive Cells: Current and Future Therapeutic Approaches

Maria Stella Sasso, Vincenzo Bronte, and Ilaria Marigo

7.1 Introduction

7.1.1 *Role of the Immune System in Cancer Pathology*

According to the theory of cancer immunosurveillance/immunoediting, the immune system is able to detect and destroy newborn neoplastic cells preventing the growth of incipient tumors (Schreiber et al. 2011). Therefore, tumors have to overcome the immune defense to reach a clinically detectable stage. The tumor ability of evading immune destruction has recently been proposed as an emerging hallmark of cancer (Hanahan and Weinberg 2011), stressing the increasing consideration that this concept has received in the last decade.

Clinical epidemiologic data support the existence of an anti-tumor immune response at least in certain human cancers. Among the strongest evidence is the existence of a positive correlation between a dense T lymphocyte tumor infiltrate and improved patient prognosis, which has been reported in different human tumors (Bindea et al. 2010). Remarkably, a study dated 2006 on colorectal cancer showed that immune cells type, density, and location in tumors dramatically influenced patient survival and these parameters represented better prognosis predictors than classical staging methods based on histopathology (Galon et al. 2006).

M.S. Sasso
Department of Surgery, Oncology and Gastroenterology, Oncology and Immunology Section, University of Padova, 35128 Padova, Italy

V. Bronte
Department of Pathology, Verona University Hospital, 37134 Verona, Italy

I. Marigo (✉)
Istituto Oncologico Veneto, IOV-IRCCS, Via Gattamelata, 64, 35128 Padova, Italy
e-mail: ilaria.marigo@ioveneto.it

M.J. Alonso, M. Garcia-Fuentes (eds.), *Nano-Oncologicals: New Targeting and Delivery Approaches*, Advances in Delivery Science and Technology,
DOI 10.1007/978-3-319-08084-0_7

Other data supporting the relevance of anti-tumor immune responses in humans come from the observation of an increased incidence of cancers, including forms of non-viral origin, in transplanted patients undergoing immunosuppressive treatments (Engels et al. 2011; Jiang et al. 2010; Moloney et al. 2006; Schrem et al. 2013; Vajdic et al. 2006) and human immunodeficiency virus (HIV)-infected individuals (Chaturvedi et al. 2007; Kirk et al. 2007; Shiels and Engels 2012). Moreover, induced immunosuppression has been reported to cause the growth of previously undetectable tumors derived from donor tissues in transplantation cases (Desai et al. 2012; MacKie et al. 2003).

Along with the anti-tumor role described so far, the immune system has been shown to play an opposite and direct pro-tumor function. Tumors are usually characterized by an inflammatory microenvironment, which enables cancer growth and progression by supplying growth factors, pro-angiogenic mediators, extracellular matrix modifying enzymes, and signaling molecules, all favoring cancer cell invasion and metastatization (Hanahan and Weinberg 2011). Innate immune cells including granulocytes, monocytes, macrophages, and dendritic cells (DCs) are key players of tumor-associated inflammation. Moreover, innate immune cells, including macrophages, DCs, and more immature cells, have been shown to exert a crucial immunosuppressive activity on T cell-mediated anti-tumor response, as discussed below.

7.1.2 Immunosuppressive and Tumor-Promoting Immune Populations

Myeloid cells represent the innate arm of the immune system, which is the first line of defense against pathogens and is required to activate and sustain the subsequent antigen-specific adaptive immune response. Moreover, myeloid cells contribute to tissue homeostasis by removing dying cells and debris and participating in tissue remodeling.

Tumors can alter the normal myeloid cell differentiation and function by releasing multiple soluble factors, able to act both systemically and locally. Different cytokines, like granulocyte-macrophage colony stimulating factor (GM-CSF), granulocyte colony stimulating factor (G-CSF), macrophage colony stimulating factor (M-CSF), stem cell factor (SCF), and vascular endothelial growth factor (VEGF), have been shown to promote cancer-induced myelopoiesis. Pro-inflammatory mediators released by both tumor cells and tumor-infiltrating immune cells, as interleukin-1β (IL-1β), IL-6, IL-4, IL-10, IL-13, S100 calcium binding protein A8 and A9 (S100A8 and S100A9), interferon gamma (IFN-γ), induce then the activation of immunosuppressive and tumor-promoting programs in myeloid cells (Gabrilovich et al. 2012).

One of the main consequences of these alterations is the accumulation of an heterogeneous population of immature, highly immunosuppressive myeloid cells termed myeloid-derived suppressor cells (MDSCs). Moreover, terminally differentiated populations, such as DCs and macrophages, which are normally devoted to antigen presentation, pathogen killing, and pro-inflammatory cytokines secretion, become either poorly immunostimulating or overtly immunosuppressive. Beside cells of myeloid origin, a particular subset of T lymphocytes, named T regulatory cells (Tregs), has been shown to contribute to tumor-induced immunosuppression. Tregs and tumor-promoting myeloid populations are described singularly below, with a particular attention to MDSCs and macrophages, whose targeting will be the focus of this chapter.

7.1.2.1 Tumor-Associated Macrophages

Macrophages are classically thought to differentiate from circulating monocytes, following their recruitment into peripheral tissues. In contrast with this paradigm, recent studies in mouse models have shown that certain tissue resident macrophage populations descend from primitive yolk-sack-derived macrophages or from fetal liver monocytes and are completely maintained by self-renewing during the adult life (Guilliams et al. 2013; Hoeffel et al. 2012; Schulz et al. 2012). Circulating monocytes poorly contribute to the maintenance of tissue-resident macrophages under steady state conditions, while they dramatically expand during inflammation and give rise to inflammation-associated macrophage and DCs (Shi and Pamer 2011; Wynn et al. 2013). Notably, in mouse tumor models, Ly6C^{high} inflammatory monocytes have been shown to be the direct precursors of tumor-associated macrophages (TAMs) (Movahedi et al. 2010).

Macrophages are phenotypically defined by the presence/absence of specific cell surface markers. They are broadly identified as CD11b^{+}, F4/80^{+}, CD115^{+}, Ly6G^{-} cells in mice and as CD68^{+}, CD163^{+}, CD16^{+}, CD312^{+}, and CD115^{+} cells in humans. However, different subsets of macrophages exist, which can be identified by a finer phenotypic characterization, together with functional and gene expression assays (Qian and Pollard 2010).

One of the first macrophage classifications distinguishes between M1 (classically activated) and M2 (alternatively activated) macrophages. The M1 and M2 states are induced by different environmental factors, including cytokines and toll-like receptor (TLR) ligands, and are characterized by specific functional properties. M1 macrophages have stronger antigen-presenting and killing abilities, higher production of reactive oxygen and nitrogen species, express higher levels of pro-inflammatory cytokines, and promote a “Th1-oriented” T cell response (Sica and Mantovani 2012). Th1 lymphocytes are, in turn, required for the activation of cytotoxic T cells that are crucial for tumor cell recognition and elimination (Vesely et al. 2011). M2 macrophages are characterized by efficient phagocytic activity, high expression of scavenging surface molecules, low production of pro-inflammatory cytokines, and high levels of the anti-inflammatory cytokine IL-10, and are typically involved in either the resolution of inflammation or maintenance of smoldering chronic inflammation (Sica and Mantovani 2012).

Although in most cases the M1/M2 distinction is not clear-cut and some cells may bear features of both types, there is strong evidence that TAMs, at least in advanced tumors, are mainly M2-oriented and have poor anti-tumor functions (Sica and Mantovani 2012). In addition, different pro-tumor properties have been ascribed to TAMs, involving both immune and non-immune mechanisms (Gabrilovich et al. 2012). According to this, TAMs are a negative prognostic factor in human cancers (DeNardo et al. 2011; Steidl et al. 2010).

A first immune mechanism by which TAMs promote cancer progression is the loss of an anti-tumor M1-oriented response. Since macrophage polarization in sustained by positive feedback loops, the existence of a predominant M2 orientation further prevents the differentiation of M1 macrophages. For example, M2 macrophages support the development of Th2 cells, which secrete cytokines (as IL-4) that further drive M2 macrophage differentiation (DeNardo et al. 2009).

Moreover, TAMs are directly immunosuppressive through multiple mechanisms. First, tumor-associated monocytes/macrophages may express the membrane molecule programmed death ligand 1 (PD-L1), which inhibits activated T cells expressing the PD1 receptor (Kuang et al. 2009). In human ovarian cancer, TAMs attract immunosuppressive Tregs through the secretion of CC-chemokine ligand 22 (CCL22) (Curiel et al. 2004). In addition, mouse TAMs express the enzyme arginase 1 (ARG1), which locally depletes L-arginine thus impairing T cell activation and proliferation (Rodriguez et al. 2004), as further discussed below.

Non-immune, pro-tumor functions include promotion of tumor cell invasion, stimulation of angiogenesis, support of seeding and growth of metastatic cells (Qian and Pollard 2010). TAMs express high level of different proteases, which support tumor cell mobility and invasion through the degradation and remodeling of the extracellular matrix (Kessenbrock et al. 2010). Moreover, macrophages directly stimulate tumor cell migration, through the secretion of paracrine factors like the epithelial growth factor (Wyckoff et al. 2004) and favor the ingress of invasive cells into blood vessels (Wyckoff et al. 2007). Macrophages have been shown to contribute to tumor-induced neoangiogenesis, which is required for continuous tumor growth and progression to malignancy. A major contribution is due to the production of VEGF, both by direct secretion (Lin et al. 2006, 2007) and by proteolytic mobilization of extracellular depots (Kessenbrock et al. 2010).

The mechanisms underlining the process of metastatization are still largely unknown; however, recent studies have suggested that myeloid cells are critically required for the formation of metastasis (Qian and Pollard 2010). Tumor-derived soluble factors, released systemically, can induce the accumulation of myeloid cells at distant sites, thus increasing the efficiency of metastatic cell seeding at those sites, defined pre-metastatic niches (Hiratsuka et al. 2006; Kaplan et al. 2005). The presence of macrophages at the pre-metastatic niche may increase the efficiency of circulating tumor cell extravasation and favor the subsequent growth of metastatic lesions (Qian et al. 2009).

7.1.2.2 Myeloid-Derived Suppressor Cells

Cancer is associated with abnormal myeloid cell differentiation that leads to the accumulation of immature myeloid cells with immunosuppressive functions, the MDSCs. MDSC expansion has been reported in several mouse tumor models

(Gabrilovich and Nagaraj 2009) and in different human cancers (Montero et al. 2012). In mice, immunosuppressive myeloid cells accumulate both in peripheral lymphoid organs, mainly in the spleen, and at the tumor site (Gabrilovich and Nagaraj 2009; Gallina et al. 2006), although evidence suggests that they may reach their full immunosuppressive potential only within the tumor microenvironment (Gabrilovich and Nagaraj 2009). Moreover, at the tumor site MDSCs may differentiate into TAMs, thus contributing to increase this tumor promoting population (Kusmartsev and Gabrilovich 2005).

The spleen is a unique site of cancer-associated extramedullary hematopoiesis in mice (Bronte and Pittet 2013). In mouse models, cancer induces the proliferation and expansion of splenic myeloid progenitors, which sustain the accumulation of immunosuppressive myeloid cells (Bronte et al. 2000; Cortez-Retamozo et al. 2012; Ugel et al. 2012). There is some evidence that similar splenic hematopoiesis may occur in humans, since myeloid precursor expansion has been reported in the spleen of patients with invasive cancers (Bronte and Pittet 2013; Cortez-Retamozo et al. 2012). Notably, in mice, the spleen has been reported to be essential for the induction of systemic T cell tolerance to tumor antigens, preventing the onset of a proficient anti-tumor immune response (Ugel et al. 2012).

In mice, MDSCs include monocytic and polymorphonuclear (or granulocytic) populations, whose phenotype resembles the phenotype of inflammatory monocytes and neutrophils, respectively. Murine monocytic (M)-MDSCs are $CD11b^{+}Gr1^{int}LY6C^{high}Ly6G^{-}CD49d^{+}$, while polymorphonuclear (PMN)-MDSCs are $CD11b^{+}Gr1^{high}Ly6C^{low}Ly6G^{+}CD49d^{-}$ (Gabrilovich et al. 2012). In tumor bearing mice, PMN-MDSCs are the most abundant population, but M-MDSCs have stronger immunosuppressive abilities on a per cell basis (Dolcetti et al. 2010; Youn et al. 2008). Splenic $Ly6C^{high}$ M-MDSCs have been shown to be a determinant for the induction of immune tolerance to tumor antigens (Ugel et al. 2012), stressing the relevance of this cell subset in cancer-associated immunosuppression.

Human MDSCs have been shown to be highly heterogeneous and express different surface markers in different cancer types, depending both on the stage of myeloid development and on the differentiation context provided by factors secreted by cancer cells. As a general classification, human MDSCs belong to three broad subsets, namely M-MDSCs, PMN-MDSCs, and immature MDSCs (Filipazzi et al. 2012; Montero et al. 2012).

M-MDSCs are identified by the expression of the CD14 monocyte marker, plus common myeloid markers as CD11b and CD33. In addition, M-MDSC are often defined as interleukin 4 receptor α (IL4Rα) positive and/or human leukocyte antigen (HLA)-DR low-negative cells. PMN-MDSCs, as M-MSDC, express the myeloid antigens CD11b and CD33, but are distinguished by the expression of granulocyte markers such as CD15 and/or CD66b. Immature MDSCs have been identified within the lineage negative (Lin^{-}) peripheral blood cell fraction (Lin^{-} cells defined as cells lacking the expression of lineage differentiation antigens, including CD3, CD14, CD19, and CD56 or CD57) and are $CD11b^{+}CD33^{+}HLA\text{-}DR^{-}$ cells (human MDSC phenotypes have been extensively reviewed in Filipazzi et al. 2012; Montero et al. 2012).

MDSCs exploit different mechanisms to inhibit T cell recruitment, activation, and proliferation. A well-described mechanism is the depletion of L-arginine from the local microenvironment through the expression of the enzyme ARG1. Amino acid starvation is sensed by T cells resulting in the suppression of cell growth and proliferation. Notably, L-arginine depletion inhibits the expression of the ζ-chain of the T cell receptor (TCR) in T cells, thus preventing their activation (Bronte and Zanovello 2005). MDSC cell metabolism also results in the local depletion of cysteine, which is required for T cell activation and function (Srivastava et al. 2010).

Inducible nitric oxide synthase (iNOS/NOS2) is a second L-arginine converting enzyme, critically involved in MDSC suppressive activity. Nitric oxide (NO) produced by iNOS is able to block the signaling pathway downstream of the IL-2 receptor, thus impairing T cell proliferation, and high NO levels can constitute a direct pro-apoptotic signal for exposed cells (Bronte and Zanovello 2005). However, the strongest iNOS immunosuppressive function is exerted in the presence of a concomitant ARG1 activity in the same microenvironment, as it happens in the tumor microenvironment. L-arginine depletion shifts iNOS activity from NO production to superoxide anion (O_2^-) production; the superoxide anion in turn reacts with NO to produce peroxynitrite ($ONOO^-$) (Bronte and Zanovello 2005). Peroxynitrite is a potent nitrating agent that can nitrate TCR and CD8 molecules on the surface of T cells, thus impairing antigen-specific recognition (Nagaraj et al. 2007). Moreover, peroxynitrite nitrates CC-chemokine ligand 2 (CCL2) in the tumor microenvironment, preventing T cell recruitment mediated by this chemokine (Molon et al. 2011). In addition to reactive nitrogen species (RNS), MDSCs also have an increased production of reactive oxygen species (ROS), mainly H_2O_2, which contributes to impair T cell functionality (Bronte and Zanovello 2005).

MDSC-dependent immune subversion also involves MDSC-mediated expansion of other immunosuppressive populations, such as Tregs and M2 macrophages, through multiple and not yet fully elucidated mechanisms (Pan et al. 2010; Serafini et al. 2008; Sinha et al. 2007).

7.1.2.3 Dendritic Cells

DCs are the most potent antigen-presenting cells (APCs) with the unique ability to activate naïve T lymphocytes (Hivroz et al. 2012). Most DCs are of myeloid origin and derive from myeloid-committed bone marrow precursors. Classical DCs (cDCs) in lymphoid organs and some DC populations in non-lymphoid tissues differentiate from pre-cDCs, which are released from the bone marrow in the blood circulation. DCs in non-lymphoid tissues may also derive from circulating monocytes, especially during inflammation (Geissmann et al. 2010; Liu and Nussenzweig 2010). Plasmacytoid DCs (pDCs) are a particular subset of DCs, specialized in responding to viral infections by producing large amounts of type I interferon. pDCs differentiate in the bone marrow from myeloid precursors and migrate to peripheral lymphoid organs (Geissmann et al. 2010; Idoyaga and Steinman 2011). To induce T cell activation, DCs need first to "mature" in presence of environmental signals derived from pathogens or damaged cells and tissues, which are referred as pathogen-associated molecular patterns (PAMPs) and damage-associated molecular patterns (DAMPs), respectively. DC maturation is associated to the upregulation of costimulatory molecules, cytokine secretion, and higher antigen presenting and activating abilities (Reis e Sousa 2006).

In cancer, the previously described abnormal myeloid cell differentiation driven by high level growth factors and cytokines (including VEGF, GM-CSF, M-CSF, IL-6, IL-10) is also associated with an impaired terminal DC differentiation and with the prevalence of immature DCs, characterized by low costimulatory molecule expression and poor activating abilities (Gabrilovich 2004). Moreover, tumor microenvironmental factors such as hypoxia, low pH, increased level of lactate and adenosine also affect DC biology and contribute to the dysfunctional state of tumor-associated DCs (Gabrilovich et al. 2012).

Besides having reduced stimulating abilities, tumor-conditioned DCs may also exert an active suppression on T cells immunity. One of the major mechanisms of DC-mediated immunosuppression is the expression of the tryptophan-metabolizing enzyme indoleamine 2,3-dioxygenase (IDO) (Lee et al. 2003; Munn and Mellor 2013; Munn et al. 2004). Tryptophan depletion in the microenvironment induces signaling events in T cells leading to anergy and apoptosis; moreover, tryptophan catabolytes produced by IDO have direct immunosuppressive effects on T cells (Platten et al. 2012). In DCs, it has been shown that IDO expression and function are modulated by a number of factors, including IFN-γ, transforming growth factor β (TGFβ) and adenosine levels (Hanks et al. 2013; Jasperson et al. 2009; Novitskiy et al. 2008; Pallotta et al. 2011). IDO may also be upregulated following the interaction of DCs with Tregs (Fallarino et al. 2003; Grohmann et al. 2002). Tumor-associated DCs were also reported to express ARG1 (Norian et al. 2009), the immunosuppressive enzyme upregulated in MDSCs and macrophages (see previous sections).

Even though DCs might exert active immunosuppression in cancer, the main immunomodulating approach to target DCs is based on inducing their maturation and activation, rather than developing strategies to eliminate the immunosuppressive subsets. Functional DCs are indeed strictly required for the induction of a proficient anti-tumor adaptive response, and their activation is one of the central aims of cancer vaccine research. Nanotechnology-based approaches aimed at activating DCs will be described in the context of vaccination in this book.

7.1.2.4 T Regulatory Cells

Tregs are a particular population of T lymphocytes able to constrain the immune response. Tregs found in peripheral sites, including the tumor microenvironment, comprise subsets with different origins. Naturally occurring Tregs (nTregs) are generated in the thymus during the normal development of the immune system and are characterized by a high-level expression of the IL-2 receptor α chain (CD25) and the transcription factor forkhead box protein 3 (FOXP3). FOXP3 expression, in addition of being a phenotypic marker, plays a critical role in nTreg differentiation and maintenance (Josefowicz and Rudensky 2009). nTregs may be recruited from lymphoid organs to the tumor sites under the influence of chemokines like CC-chemokine ligand 22 (CCL22) (Curiel et al. 2004) and proliferate locally. Moreover, $CD25^{+}FOXP3^{+}$ Tregs can be induced in peripheral sites (inducible Tregs, iTregs) through the conversion of conventional $CD4^{+}$ T cells in presence of cytokine stimuli such as IL-2 and TGF-β (Curotto de Lafaille and Lafaille 2009), which may be abundant in the tumor microenvironment. Also subsets of $FOXP3^{-}$ $CD4^{+}$ or $CD8^{+}$ T regulatory cells exist, some of which may be induced in the tumor microenvironment in the presence of immunosuppressive cytokines such as IL-10 and contact with tumor-conditioned DCs (Zou 2006).

Tregs are necessary for the maintenance of the immune tolerance against self-antigens and commensal microorganisms, and loss-of-function mutations in the FOXP3 gene cause severe autoimmune disorders with early onset in both mice and humans (Wildin et al. 2001). However, in cancer Tregs usually play a negative role since they have been shown to contribute to the failure of naturally occurring anti-tumor immunity. Treg depletion by anti-CD25 antibodies was shown to be associated with reduced tumor growth and improved anti-tumor immune response in different mouse tumor models (Zou 2006). Moreover, an increased frequency of Tregs in peripheral blood has been reported for different human cancers (Zou 2006). In these studies, Tregs were shown to be immunosuppressive ex vivo, thus supporting a functional role in human cancer pathology.

Different mechanisms of Treg mediated immunosuppression have been identified so far (von Boehmer 2005). These include the secretion of immunosuppressive cytokines, such as IL-10 and TGF-β, as well as cell contact-dependent mechanisms. The inhibitory receptor cytotoxic T-lymphocyte-associated protein 4 (CTLA-4) expressed on Tregs binds CD80/CD86 molecules on the surface of T effector cells provoking a direct suppression of T cell activity. In addition, CTLA-4 binds the same molecules on APCs and induces the expression of the IDO immunosuppressive enzyme. Treg cells might also be able to directly kill other T cells through granzyme and perforin secretion. Furthermore Tregs are thought to compete with effector T cells for environmental IL-2, since Tregs express a higher level of the high-affinity IL-2 receptor, as said above. Since IL-2 is important for T cell proliferation, its consumption by expanded Treg populations might contribute to impair T cell function (von Boehmer 2005).

Although different strategies to manipulate Tregs in cancer have been developed, ranging from low dose chemotherapy to targeted molecular inhibitors (von Boehmer and Daniel 2013), therapeutic approaches aimed at depleting Tregs will be not discussed in this chapter, since to date there are still few nanotechnology applications in this field. Barriers to the exploitation of nanovehicles to Treg targeting include their limited uptake ability compared to professional phagocytes, like macrophages, and the absence of cell-specific surface markers. The CD25 antigen, which is considered as a major Treg surface marker, is actually not expressed by certain induced Treg subsets, as noted above. Moreover, this antigen is also expressed on the surface of activated effector T cells, although at lower amounts (Sakaguchi 2004). Therefore, CD25 targeting may cause off-target effects on the "good" effector lymphocytes (Jacobs et al. 2010).

7.1.3 Cancer Immunotherapy and Immunosuppressive Cells

The rationale behind cancer immunotherapy is to exploit the selective cytotoxicity of adaptive immunity (primarily dependent on $CD8^+$ T lymphocytes) to design anti-cancer therapies with high specificity and low off-target toxicity. To date, different approaches have been developed to potentiate the host immune response against tumor cells, evaluated at clinical and preclinical levels. Approaches that have been studied in clinical trials include antibody-based therapies, cancer vaccines, and the adoptive transfer of tumor-specific T cells (adoptive cell therapy, ACT).

Ipilimumab, a blocking antibody targeting the inhibitory molecule CTLA-4 on T cell surface, was shown to produce survival benefits in patients with late-stage, metastatic melanoma with no other treatment options, and the drug was approved by the FDA in 2011 (Hodi et al. 2010; Sliwkowski and Mellman 2013). By blocking an inhibitory receptor, this antibody promotes prolonged activation of T cells despite the presence of constraining signals. Moreover, the efficacy of anti-CTLA-4 antibodies may be in part also due to interference with Treg-mediated immunosuppression (Zou 2006) (see the Sect. 1.2.4).

Currently, antibodies targeting programmed death-1 (PD-1) receptor, another T cell negative regulator, or its ligand PD-L1 are in phase of clinical evaluation. A phase 1 clinical trial evaluating the antitumor activity and safety of a PD-1 blocking antibody reported an objective response rate of 1/4–1/5 in patients with non-small-cell lung cancer, melanoma, or renal cell cancer (Topalian et al. 2012). Moreover, the immunohistochemical analysis of tumor specimens from a subgroup of patients revealed the existence of a relationship between therapy efficacy and PDL-1 expression in tumors, since none of the patients with PD-L1 negative tumors responded to the anti-PD1 treatment (Topalian et al. 2012). This result suggests the possibility to select patients eligible for PD-1 targeted immunotherapy on the base of the presence/absence of PD-L1 upregulation in the tumor mass. In parallel, also an antibody targeting PD-L1 was clinically evaluated and shown to induce durable tumor regression or prolonged stabilization of disease in patients with advanced solid tumors (Brahmer et al. 2012), stressing the importance of inhibiting the PD-1/PD-L1 immunosuppressive pathway in cancer immunotherapy.

Therapeutic cancer vaccines are designed to immunize patients against tumor-associated antigens in order to boost host immune response against the disease. Differently, ACT directly supplies the patients with tumor-specific lymphocytes of exogenous origin, usually obtained ex vivo from patient-derived blood samples through passages of in vitro selection and expansion (Restifo et al. 2012).

While cancer vaccines have obtained limited results in late-stage clinical trials so far (Klebanoff et al. 2011), ACT was shown to mediate cancer regression in advanced melanoma patients with high response rates (around 50 %) and cases of durable regression (Restifo et al. 2012). Although these results on melanoma represent a milestone in the history of cancer immunotherapy, they leave important challenges unsolved. First, the reason why only some patients respond to the therapy remains unclear. Moreover, comparable results in other forms of cancer are still lacking, at least in part because of the lower immunogenicity of most human tumors compared to melanoma and the consequent difficulty in isolating naturally occurring, tumor-specific lymphocytes.

To overcome this barrier, different approaches to generate genetically engineered, tumor-specific lymphocytes are currently under preclinical or clinical evaluation (Restifo et al. 2012).

The presence of an immunosuppressive tumor microenvironment has been progressively recognized as a critical factor underlying the failure of cancer immunotherapy and in particular of ACT approaches. The depletion/alteration of immunosuppressive cells has been proposed as one of the mechanisms explaining the efficacy of lympho-myeloablating regimens in increasing the clinical response to ACT (Gattinoni et al. 2006). The advantage in using chemotherapy and/or total body irradiation to partially deplete the host immune system prior to ACT is supported by studies in both mouse models (Wrzesinski et al. 2010) and human patients (Dudley et al. 2002, 2008). The reduction of specific immunosuppressive populations, including MDSCs (Ghansah et al. 2013; Ugel et al. 2012) and Tregs (Ghiringhelli et al. 2004), following treatment with selected chemotherapy agents, has been shown to increase the efficacy of ACT and cancer vaccines in rodent models.

Of note, in the clinical trials on melanoma patients reported above, therapeutic efficacy was achieved through the combination of ACT with preconditioning lympho-myeloablating regimens and IL-2 treatment, the latter being aimed at increasing the proliferation and survival of transferred T cells. However, these adjuvant treatments may cause severe toxicity, thus limiting the applicability of the therapy to patients with good performance status (Restifo et al. 2012).

The development of more effective and widely applicable immunotherapies is therefore dependent on the availability of new adjuvant treatments with good efficacy and low toxicity. The search for highly selective strategies to target and neutralize immunosuppressive cells may provide a great advantage in this respect.

7.2 Targeting TAMs

7.2.1 Liposome Interactions with Macrophages

Liposomes are the most common vehicles employed to target monocytes and macrophages. Liposomes are nanometric vesicles composed of one or more phospholipid bilayers surrounding an aqueous core. These vesicles may have positive, neutral, or negative surface charge, depending on the composition.

A common in vivo fate of liposomes is opsonization and phagocytosis by mononuclear phagocytes (monocytes and macrophages). Opsonization consists in the binding to the surface of foreign bodies (as pathogens and synthetic particles) of molecules (termed opsonins) that can make them recognizable to phagocytic cells. Main opsonins are immunoglobulins (IgG and IgM) and complement fragments, which are abundant in the blood stream. The constant portion of absorbed immunoglobulins is recognized by Fc receptors, which are highly expressed by macrophages. Immunoglobulins usually recognize specific antigenic molecules, but they can also be absorbed on the surface of circulating particles through non-specific interactions, such as hydrophobic interactions (Hillaireau and Couvreur 2009).

Complement fragments acting as opsonins are produced by the activation of the complement system, which can be triggered both by antigen-bound antibodies (classical pathway) and by the direct binding of complement proteins to the pathogen surface (alternative and lectin pathways). Highly charged liposomes have been shown to activate the complement more strongly than neutral ones, by both the classical and the alternative pathways (Chonn et al. 1991; Devine et al. 1994).

Circulating liposomes are typically easily opsonized and phagocytized by monocytes and macrophages, unless they have a very small size (lower than 50–100 nm) and possess a hydrophilic coating able to repel opsonins, such as a polyethyleneglycol (PEG) shell (Hillaireau and Couvreur 2009). The internalization rate of weakly charged PEGylated nanoparticles by macrophages has been suggested to positively correlate with particle size, with significantly reduced particle uptake below 100 nm (Yu et al. 2012). This non-specific, size-dependent internalization involves, at least in vitro, the macrophage scavenger receptor A (CD204) (Yu et al. 2012).

Highly charged liposomes, besides having a higher rate of opsonization, also exert stronger direct interactions with the macrophage membrane. Positively charged liposomes electrostatically interact with the negatively charged cell surface, thus facilitating cell uptake. However, delivery nanosystems based on cationic lipids (including cationic liposomes, and cationic-lipid/nucleic acid lipoplexes) have been reported to have different drawbacks, which limit their in vivo application: first, they may cause cell toxicity (Lv et al. 2006); moreover, their behavior may be modified in presence of serum, due to electrostatic interactions with anionic serum protein. Highly cationic liposomes were shown to severely aggregate in presence of serum, and in vivo liposome–serum interactions were found to cause reduced circulation time of the nanovehicles and rapid accumulation in the lungs, while other sites, including the spleen and the tumor, were poorly reached (Zhao et al. 2011).

Liposomes containing negatively charged phospholipids, like phosphatidylserine and phosphatidylglycerol, are preferentially uptaken by macrophages, at least in part due to the engagement of scavenger receptors on their membrane (Kelly et al. 2011).

Liposomes loaded with bisphosphonates (primarily clodronate) have been used for "suicide" killing of macrophages since the end of the 1980s (Van Rooijen 1989). Macrophages internalize bisphosphonate-loaded liposomes and destroy the lipidic shell during the endosomal digestion, thus allowing the release of the drug. More recently, liposomes loaded with clodronate have been shown to successfully deplete monocytes/macrophages in different mouse tumor models, resulting in the reduction of primary tumor growth, impaired angiogenesis and decreased metastasis (Gazzaniga et al. 2007; Halin et al. 2009; Hiraoka et al. 2008; Kimura et al. 2007; Miselis et al. 2008; Zeisberger et al. 2006). In addition, bisphosphonates and aminobisphosphonates can regulate both Matrix Metalloprotease-9 (MMP-9) function and its expression in monocytic-macrophagic cells (Giraudo et al. 2004; Valleala et al. 2003). MMP-9 downregulation by the amino-bisphosphonate zoledronate was shown to reduce tumor infiltration by myeloid cells (including F4/80^{+} macrophages) and negatively impact MDSC expansion in a mouse breast cancer model (Melani et al. 2007). This immunomodulatory effect appears to rely on a MMP-9/VEGF regulatory axis: MMP-9 activity in the tumor microenvironment (mainly dependent on MMP-9 expression by bone marrow derived cells) increases the availability of

VEGF, which in turn stimulates bone marrow hematopoiesis. Amino-bisphosphonates interfere with this circuit, thus controlling tumor associated aberrant hematopoiesis (Melani et al. 2007).

7.2.2 TAM-Targeted Liposomes and Nanoparticles

Although liposomes are naturally uptaken by macrophages their targeting efficiency can be increased by adding ligands that bind to macrophage-specific surface molecules. Moreover, the synthesis of nanovehicles selectively targeting TAMs may allow to face the major challenge of interfering with TAM function without damaging other macrophage populations, which are instead required for normal homeostatic and immune functions.

A first approach is based on liposome or nanoparticle mannosylation, which consists in incorporating mannoside groups that target the mannose receptor (MR) in particle formulations. The MR is a C-type lectin receptor, primarily expressed on the surface of macrophages and DCs. This receptor recognizes both endogenous ligands and pathogen-associated ligands, as the glucide mannose expressed on the surface of many microorganisms (Gazi and Martinez-Pomares 2009). Liposome mannosylation has been repeatedly shown to improve macrophage targeting, both in vitro and in vivo (Kelly et al. 2011). The functionalization of polymeric nanoparticles for nucleic acid delivery with surface mannose has been reported to increase siRNA uptake by a human monocytic/macrophagic cell line and by mouse primary macrophages (Yu et al. 2013). In this work, MR-targeting nanoparticles optimized for nucleic acid delivery were obtained by combining three functionally distinct polymeric blocks: a core pH-sensitive block (capable of disrupting endosomes at low pH), a cationic block for condensation of nucleic acids, and an external azide-displaying block for conjugation of targeting molecules (Yu et al. 2013).

MR ligands are particularly suitable for TAM targeting, since this receptor is highly expressed in M2 macrophages compared to M1 macrophages (Mantovani et al. 2002). Mannosylated nanoparticles have been employed to specifically target TAMs in vivo, with reduced internalization by phagocytes in non-tumoral sites, by taking advantage of mannosylation combined with an acid-sensitive PEG shell (Zhu et al. 2013). Nanoparticle PEGylation was carried out by employing PEG-containing amphiphilic molecules that spontaneously hydrolyzed at acid pH. In the acidic extracellular tumor microenvironment (~pH 6.8) the PEG shell was shed, exposing the MR ligands (Zhu et al. 2013).

A different approach is based on liposome or nanoparticle coupling with antibodies targeting TAM specific antigens. The hemoglobin scavenger receptor (CD163), as the mannose receptor, is overexpressed by M2 macrophages (Mantovani et al. 2002). PEG-coated liposomes coupled with anti-CD163 antibodies have been shown to efficiently target human and mouse macrophages in vitro (Etzerodt et al. 2012) and might be employed for TAM targeting in vivo.

Cell-specific targeting can be achieved also by the use of peptides. Different amino acid sequences have been shown to improve liposome uptake by monocytes and neutrophils (Kelly et al. 2011). Recently, an M2 macrophage-targeting peptide, named M2pep, was reported (Cieslewicz et al. 2013). In vitro, M2pep preferentially binds to M2 macrophages rather than M1 macrophages and DCs. Moreover, in a mouse in vivo model, this peptide was shown to target tumor-infiltrating macrophages, but not liver and spleen macrophages. Although the ability of this peptide to mediate the uptake of nanovehicles remains to be assessed, it represents a promising candidate for the design of TAM-specific delivery systems.

7.2.3 Prevent Macrophage Generation by Monocyte Targeting

As reported in Sect. 1.2.1, TAMs are primarily derived from inflammatory monocytes ($Ly6C^{high}$ monocytes in mice) that are recruited to the tumor mass (Movahedi et al. 2010). It is, therefore, possible to impair TAM generation by targeting monocytes, either in blood circulation or in lymphoid organs.

In mice, $Ly6C^{high}$ monocyte recruitment to inflammation sites depends on the chemokine/chemokine receptor axis monocyte chemoattractant protein-1 (MCP-1)/CC-chemokine receptor type 2 (CCR2) (Shi and Pamer 2011). This signaling axis also is likely involved in the recruitment of human $CD14^{+}CD16^{-}$ monocytes, which share phenotypic features and homing potential with mouse $Ly6C^{high}$ inflammatory monocytes (Geissmann et al. 2003). To block CCR2 signaling, a small interfering RNA (siRNA) targeting CCR2 mRNA (siCCR2) loaded into optimized lipid nanoparticles was recently developed (Leuschner et al. 2011). This nanosystem was shown to rapidly reach mouse spleen and bone marrow after intravenous injection and efficiently knock down CCR2 mRNA in splenic $Ly6C^{high}$ monocytes. Moreover, repeated treatments with siCCR2 resulted in a reduced accumulation of TAMs in two different mouse tumor models, likely due to impaired monocyte recruitment at the tumor site (Leuschner et al. 2011).

Monocytes can be also selectively targeted by specific chemotherapy drugs administered at appropriate doses. Trabectedin, a recently approved chemotherapy agent, was shown to induce selective apoptosis in human and mouse monocytes, but not in other leukocytes such as granulocytes and lymphocytes (Germano et al. 2013). Prolonged treatment with trabectedin, at clinically relevant doses, caused the depletion of circulating and splenic $Ly6C^{high}$ monocytes in four different mouse tumor models and eventually resulted in the reduction of TAMs (Germano et al. 2013). The impact of trabectedin on monocytes and TAMs was further confirmed in soft-tissue sarcoma patients, supporting the clinical relevance of this finding (Germano et al. 2013).

Inflammatory monocytes in mice are commonly identified as $Ly6C^{high}CCR2^{+}$ mononuclear myeloid cells (Shi and Pamer 2011). This phenotype also identifies M-MDSCs (see Sect. 1.2.2), which are functionally defined by their immunosuppressive activity (Gallina et al. 2006; Ugel et al. 2012). It is then likely that these two myeloid populations, at least in part, overlap each other, and MDSC-targeting strategies (discussed in the next paragraph) also affect inflammatory monocytes and vice versa. For instance, low-dose chemotherapy with selected anti-proliferative drugs, which has been reported to target M-MDSCs in the spleen (Ugel et al. 2012), might also affect TAM precursors and ultimately reduce the number of TAMs. Nevertheless, further studies are required to clarify this possibility.

7.3 Targeting MDSCs

MDSC targeting may (1) inhibit the mechanisms of immunosuppression, (2) reduce MDSC number by either killing preexistent cells or preventing their generation, and (3) force the conversion of immature immunosuppressive cells into mature APCs.

7.3.1 Inhibition of Immunosuppressive Mechanisms

The impairment of MDSC-mediated immunosuppression has been achieved in preclinical mouse models by employing drugs that exert a double inhibitory effect on iNOS and ARG1, such as phosphodiesterase-5 (PDE5) inhibitors (Serafini et al. 2006), nitroaspirin (De Santo et al. 2005), and AT38 (Molon et al. 2011). PDE5 inhibitors, such as sildenafil, reduce iNOS and ARG1 expression through an incompletely elucidated mechanism, possibly involving mRNA destabilization dependent on increased cGMP levels (Serafini et al. 2006). Nitroaspirin is a modified aspirin molecule, covalently linked to a NO-releasing group. The NO release exerts a negative feedback inhibition on both iNOS activity and expression, while the salicylic portion of the drug is responsible of ARG1 inhibition, probably due to interference with the signaling pathways that induce the enzyme upregulation (De Santo et al. 2005). AT38 has been developed by modification of the nitroaspirin molecule to increase the efficiency of iNOS and ARG1 inhibition and obtain more potent in vivo effects (Molon et al. 2011). The use of nanocarriers to selectively deliver these drugs to the tumor site, while reducing their accumulation in off-target organs, could help to decrease the side effects associated to chronic administration and facilitate the translation to the clinic.

7.3.2 Depletion of MDSCs

MDSC generation can be prevented by acting on the signaling pathways that drive their expansion and activation. The extracellular signaling molecules involved in this process are multiple and redundant, including different cytokines and growth factors (Gabrilovich et al. 2012). However, most of these signals converge in the activation of transcription factors, some of which belong to the signal transducer and activator of transcription (STAT) family (Gabrilovich et al. 2012). For example, STAT3 activation has a primary role in promoting MDSC proliferation and survival, hence its inhibition by peptides and chemical compounds may be exploited to interfere with cancer-induced MDSC accumulation (Ugel et al. 2009). Moreover, STAT3 signaling has been reported to control MDSC function, by upregulating ARG1 expression (Vasquez-Dunddel et al. 2013).

In addition, STAT3 is constitutively activated in different human cancers and is therefore a suitable molecular target for cancer therapy (Yu et al. 2007; Yue and Turkson 2009). STAT3 inhibitors may then exert an anti-cancer activity in two ways: by directly impairing tumor cell proliferation and survival and by reducing the expansion of tumor-promoting MDSCs.

To date, several STAT3 inhibitors have been developed and reported to exert anti-cancer activity in preclinical tumor models, but most of them still lack pharmacological and toxicological properties sufficiently good for clinic applications (Yue and Turkson 2009). Among the most promising approaches there are nucleic-acid-based STAT3 inhibitors, including antisense RNA, small interfering RNA (siRNA), and decoy oligodeoxynucleotide (ODN) (Sen and Grandis 2012). These systems provide high selectivity and specificity, but their in vivo efficacy is limited by low stability, due to nuclease degradation, and poor cellular uptake (Sen and Grandis 2012). The in vivo performance of STAT3-targeting oligonucleotides would hence take advantage of delivery nanosystems able to protect them from degradation and increase the uptake rate by both tumor and myeloid cells.

As mentioned above, different strategies may be used to affect STAT3 pathway. Curcumin, a natural polyphenol found in the rhizomes of *Curcuma longa* (turmeric), and its synthetic derivatives have been reported to inhibit STAT3 activation (Bharti et al. 2003; Bill et al. 2012; Wang et al. 2012; Yang et al. 2012). Curcumin can be delivered after encapsulation in exosomes. Exosomes are membrane-bound nanometric vesicles (30–100 nm) naturally released from normal, dead, and neoplastic cells, which are present in bodily fluids and can be used as delivery vehicles (Zhang and Grizzle 2011). Exosomes derived from the murine lymphoma cell line EL-4 were mixed with curcumin to prepare exosomal curcumin (Sun et al. 2010). This curcumin formulation was administered during lipopolysaccharide (LPS)-induced septic shock and caused a pathology reduction by inducing apoptosis of $CD11b^+$ $Gr\text{-}1^+$ myeloid cells in mouse lungs (Sun et al. 2010). Moreover, hydrazinocurcumin-loaded nanoparticles have been reported to inhibit STAT3 activation and reeducate in vitro conditioned RAW264.7 murine macrophages, skewing them from an M2 to an M1 phenotype (Zhang et al. 2013).

MDSC number has been shown to be reduced by low-dose chemotherapy with selected cytotoxic drugs, such as sunitinib, sorafenib, bortezomib, gemcitabine, fludarabine, 5-fluorouracyl (Ugel et al. 2012; Vincent et al. 2010), and docetaxel (Kodumudi et al. 2010). Among them, gemcitabine and 5-fluorouracyl have been reported to exert the strongest cytotoxic activity against MDSCs, at doses that have a minimum or absent effect on tumor growth (Ugel et al. 2012; Vincent et al. 2010). Splenic M-MDSCs are particularly sensitive to low-dose chemotherapy, at least in part due to a higher proliferation rate compared to PMN-MDSCs (Ugel et al. 2012). The splenic M-MDSC subset is particularly relevant to tumor-induced immune tolerance, since its depletion by splenectomy or low-dose chemotherapy results in the recovery of cytotoxic T cell function and in an improved response to immunotherapy in mouse tumor models (Ugel et al. 2012).

On the basis of these results, targeting M-MDSCs in the spleen may represent an interesting approach for cancer immunomodulation. Pluronic-stabilized poly(propylene sulfide) nanoparticles (PPS-NPs), administered intradermally to healthy mice, were shown to preferentially distribute to the blood, spleen, and kidney, rather than the liver, likely due to the particle's small size (30 nm) and PEG coating (Kourtis et al. 2013). These nanoparticles strongly associated with monocytes and macrophages in the blood, spleen, and lymph nodes (LNs) of healthy mice and efficiently targeted M-MDSCs in the spleen, tumor, and LNs of tumor-bearing mice (Kourtis et al. 2013). Of note, the analysis of PPS-NP distribution with respect to spleen cytoarchitecture showed a preferential location of the nanoparticles in the splenic red pulp and in the marginal zone of spleen follicles (Kourtis et al. 2013). The spleen marginal zone is particularly interesting when designing MDSC-targeted approaches, since this compartment has been proposed to be a critical site for the interaction between M-MDSC and memory $CD8^+$ T cells, leading to the establishment of the immunological tolerance towards tumor antigens (Ugel et al. 2012).

PPS-NPS and other nanosystems with similar biodistribution proprieties at organ and cell-level could be exploited for the delivery of chemotherapy agents and other immunomodulating moieties to M-MDSCs, within the tumor and peripheral lymphoid organs. Nanosystems with enhanced tropism for the spleen might improve the delivery of MDSC-directed drugs and achieve a significant effect with very low drug doses and minimal side effects. This spleen-targeted approach could be particularly relevant in cases in which reaching tumor sites by nanomedicine appears challenging due to physiologic barriers, for instance, poor tumor vascularization of certain tumor areas that reduces the enhanced permeability and retention (EPR) effect (Fang et al. 2011).

Besides chemotherapy, MDSC elimination by apoptosis can be induced by using IL-4Rα-binding aptamers to block the IL-4 receptor α (IL-4Rα/CD124) on their surface (Roth et al. 2012). RNA aptamers with specific affinity for murine IL-4Rα were generated and shown to bind primarily to M-MDSCs in the spleen, and to M-MDSCs and TAMs in mouse tumors. Such binding resulted in the depletion of targeted cell populations (Roth et al. 2012). The IL-4Rα is highly expressed on MDSCs and TAMs, where it mediates the activation of STAT6 in response to IL-4 and IL-13, resulting in the upregulation of ARG-1 and increased immunosuppressive functions (Gabrilovich et al. 2012). In addition to these effects on cell function, the IL-4/STAT6 axis mediates pro-survival signals in MDSCs, so that its inhibition by IL-4Rα blocking aptamers causes increased MDSC apoptosis (Roth et al. 2012).

IL-4R targeting strategies have also been developed with the aim to directly kill IL-4R expressing tumor cells. A recombinant IL-4 cytotoxin, composed of a circularly permuted IL-4 and a mutated form of *Pseudomonas* exotoxin, was employed as antitumor agent to treat IL-4R expressing ovarian, pancreatic, prostatic, and hematologic tumors in mouse xenograft models (Husain et al. 2003; Kawakami et al. 2005a; Kioi et al. 2005; Shimamura et al. 2007). The same toxin showed antitumor activity in high-grade glioma patients following intratumoral administration (Rand et al. 2000). In these studies the effects of IL-4 *Pseudomonas* exotoxin administration on IL-4R expressing myeloid cells were not evaluated. Moreover, since all preclinical evaluations were performed in immunodeficient mice (required to allow the growth of human tumor xenografts), in those settings the immunological component was not supposed to play a significant role in the therapeutic activity of IL-4 cytotoxin. However, it might be worthwhile to evaluate the effects of IL4-R targeting toxins on IL-4R expressing myeloid cells, as MDSCs and TAMS, in a tumor context. Of note, in murine models of glioma, MDSCs were recently proven to play a critical role in supporting tumor growth and to functionally depend on IL-4R signaling (Kohanbash et al. 2013). In the same work, human glioma samples were shown to be infiltrated by IL-4Rα expressing MDSCs (Kohanbash et al. 2013). These results raise the possibility that the reported effects of IL-4 cytotoxin on glioma patients, mentioned above (Rand et al. 2000), might be in part due to a toxin activity on immunosuppressive myeloid cells, besides the direct tumor cytotoxicity. However, further studies would be needed to clarify this point.

As said so far, IL-4Rα expression is increased in MDSCs and TAMs; nevertheless, this receptor subunit is also expressed by other immune cells, including T and B lymphocytes (Gessner and Rollinghoff 2000). Hence, IL-4R directed approaches, even though more selective than other commonly employed MDSC-targeting methods, still lack full selectivity and consequently are not potentially devoid of side effects. However, combining IL-4Rα targeting with drugs selective for MDSCs or TAMs might amplify specificity.

In addition to the strategies described above, MDSC presence and function can be modulated indirectly by approaches that primarily target different immune cell subsets. As an example, the intra-tumor administration of the CC-chemokine ligand 21 (CCL21), encapsulated into ribonucleopreoteic recombinant nanoparticles, was shown to reduce tumor growth and shape the immune infiltrate in a mouse orthotopic lung cancer model (Kar et al. 2011). CCL21 acts as chemo-attractant for CC-chemokine receptor type 7 (CCR7)-expressing cells, including mature DCs, naïve and memory T cells (Sallusto and Lanzavecchia 2000). Besides enhancing tumor infiltration by these immune populations, CCL21 administration was shown to reduce the percentage of MDSCs and Tregs at the tumor site (Kar et al. 2011). The complex and still incompletely understood network of interaction that characterize the tumor microenvironment may account for these indirect immunomodulating effects, which should be evaluated carefully when examining new immunotherapy approaches.

7.3.3 Conversion of Immunosuppressive Cells into Mature APCs

All-trans-retinoic acid (ATRA), a derivative of vitamin A, has been reported to promote the in vitro differentiation of mouse splenic MDSCs into more mature cells devoid of immunosuppressive activity (Gabrilovich et al. 2001). In mouse tumor models, ATRA was shown to promote the differentiation of immature myeloid cells into mature DCs, macrophages, and granulocytes (Kusmartsev et al. 2003). The reduction in MDSC numbers was associated to the rescue of T cell-mediated antitumor responses and improved efficacy of cancer vaccines (Kusmartsev et al. 2003).

ATRA effects on MDSCs appear to depend on the upregulation of glutathione synthase, which results in a dramatic increase in glutathione (GSH) levels and a consequent decrease in cellular ROS, as direct consequence of GSH antioxidant activity (Nefedova et al. 2007). As mentioned before, MDSCs typically have a high ROS production that contributes to their function (Bronte and Zanovello 2005), thus ROS neutralization reduces the immunosuppressive potential of these cells. Moreover, ROS scavenging appears to push MDSCs towards differentiation in more mature APCs (Kusmartsev and Gabrilovich 2003; Nefedova et al. 2007).

The possible relevance of ATRA as immunomodulating agent in clinical oncology has been stressed in a pilot, phase I study on metastatic renal cell carcinoma (Mirza et al. 2006). In this study, patients having high plasma levels of ATRA after oral administration reported a decrease in circulating immature myeloid cells compared to pre-treatment values (Mirza et al. 2006). This effect was associated to an improved ability of circulating mononuclear cells to stimulate T cell activation in vitro (Mirza et al. 2006).

ATRA is primarily employed in the treatment of acute myeloid leukemia (AML). Nevertheless, its efficacy is limited by a progressive reduction of plasma drug levels following chronic oral administration, mainly because of the induction of enhanced drug catabolism (Muindi et al. 1992; Regazzi et al. 1997; Warrell 1993). Moreover, ATRA is a highly hydrophobic molecule, and its bioavailability after oral administration depends on its release from the formulation and likely on the pH and fatty acid composition of intraluminal bile (Regazzi et al. 1997). All these factors lead to a wide inter-patient variability in ATRA pharmacokinetics and bioavailability after oral administration (Mirza et al. 2006). Of note, ATRA effects on tumor-induced MDSCs were observed only in patients with plasma levels of the drug higher than 150 ng/ml, suggesting the relevance of an appropriate ATRA bioavailability for MDSC targeting (Mirza et al. 2006). Liposomal and polymeric micelle formulations of ATRA, administered intravenously, were shown to maintain higher ATRA plasma levels compared to oral or intravenous administration of the free drug (Kawakami et al. 2005b; Ozpolat et al. 2003). Besides applications in the treatment of AML, the use of nanosystems for ATRA delivery could increase the efficacy of this drug as MDSC-targeted agent and promote its use in the immunotherapy of solid tumors.

Immunological adjuvants are commonly employed either to potentiate or to modulate the immune response to vaccines, typically by promoting the maturation and activation of APCs. In virtue of their modulating activity on the innate immunity, these compounds might also affect MDSC phenotype and function in a tumor context. By examining the effects of a nanoparticulated adjuvant, the very small size proteoliposomes (VSSPs), on mouse MDSCs, a proof of principle of this concept was recently provided (Fernandez et al. 2011). VSSPs were obtained by incorporating gangliosides into the outer membrane protein complex of *Neisseria meningitides* (Estevez et al. 1999). This formulation was tested both in a phase I clinical trial as a vaccine to induce an immune response against poorly immunogenic, tumor-related gangliosides (Carr et al. 2003) and as an adjuvant for immunotherapy protocols (Ramirez et al. 2006; Torrens et al. 2005). VSSPs were subsequently shown to dramatically reduce the immunosuppressive properties of splenic MDSCs in different mouse tumor models (Fernandez et al. 2011). In particular, VSSPs injection into healthy mice induced splenomegaly due to the accumulation of $CD11b^+$ myeloid cells as a consequence of its immunostimulating activity. Nevertheless, these splenic myeloid cells did not exert any relevant immunosuppression on T cells (Fernandez et al. 2011). In addition, the injection of this adjuvant in tumor-bearing mice was able to modulate the phenotype of splenic MDSCs and strongly reduce their ability to impair T cytotoxic cell activation (Fernandez et al. 2011).

A different adjuvant comprising CpG oligonucleotides was similarly shown to induce the differentiation and reduce the immunosuppressive activity of M-MDSCs in tumor-bearing mice (Shirota et al. 2012). Nanoparticles loaded with either CpG or paclitaxel have been employed to specifically target tumor-draining lymph nodes (Thomas et al. 2014). In B16-F10 melanoma-bearing mice, these adjuvant nanoparticles induced DC maturation and increased the $CD8^+/CD4^+$ T cell ratio, as well as the frequency of activated $CD8^+$ T cells (Thomas et al. 2014).

The delivery of CpG oligonucleotides and other immunological adjuvant through nanosystems could be thus considered in future studies as a strategy to interfere with MDSC differentiation and activity, both in peripheral lymphoid organs and at the tumor site.

7.4 Conclusions

To date, there is still a limited application of nanotechnology to the targeting of immunosuppressive and tumor-promoting immune cells. Although a few promising TAM-targeted nanosystems have been developed, full experimental evidence of their in vivo immunological effects, either alone or in combination with other immunotherapy treatments, is still lacking. On the other side, while different pharmacological approaches either to deplete or modulate functionally MDSCs and TAMs have been studied at preclinical level, the combination of these approaches with nanotechnologies for drug delivery has not been fully exploited. Such combination might result in an improvement in terms of increased efficacy and reduced off-target toxicity.

Most of the work in cancer nanomedicine has been oriented to the targeting of the tumor site. Nanosystems able to selectively accumulate within the tumor mass may be indeed highly useful to reach tumor-infiltrating MDSCs and TAMs. However, together with tumor-infiltrating immune cells, myeloid populations outside the tumor, and notably in the spleen, appear to be critical in sustaining tumor-induced immunosuppression. The spleen supports myeloid cell expansions and continuously provides the tumor with new immunosuppressive and tumor-promoting cells (Bronte and Pittet 2013). On the basis of these findings, the spleen appears as a new relevant target for cancer nano-immunotherapy, easier to reach than the tumor mass.

A better integration between researchers working in the fields of nanomedicine and cancer immunology would probably allow to fulfill experimental gaps on both sides and lead to the development of more powerful and selective immune cell targeting strategies. Such integrated work is expected to ultimately lead to the translation of developed approaches to the clinic that to date is still lacking for most MDSC and TAM targeting therapies.

Acknowledgments This work was supported by grants from the Italian Ministry of Education, University and Research (FIRB project RBAP11T3WB_003 and MIUR project 2009NREAT2_003), Italian Association for Cancer Research (AIRC, Grants IG10400, 12182, AGIMM 100005, and 6599), FP7-NMP-2011(Trans-Int), and Euronanomed 2013 (053 NICHE).

References

Bharti AC, Donato N, Aggarwal BB (2003) Curcumin (diferuloylmethane) inhibits constitutive and IL-6-inducible STAT3 phosphorylation in human multiple myeloma cells. J Immunol 171:3863–3871

Bill MA, Nicholas C, Mace TA, Etter JP, Li C, Schwartz EB, Fuchs JR, Young GS, Lin L, Lin J et al (2012) Structurally modified curcumin analogs inhibit STAT3 phosphorylation and promote apoptosis of human renal cell carcinoma and melanoma cell lines. PLoS One 7:e40724

Bindea G, Mlecnik B, Fridman WH, Pages F, Galon J (2010) Natural immunity to cancer in humans. Curr Opin Immunol 22:215–222

Brahmer JR, Tykodi SS, Chow LQ, Hwu WJ, Topalian SL, Hwu P, Drake CG, Camacho LH, Kauh J, Odunsi K et al (2012) Safety and activity of anti-PD-L1 antibody in patients with advanced cancer. N Engl J Med 366:2455–2465

Bronte V, Pittet MJ (2013) The spleen in local and systemic regulation of immunity. Immunity 39:806–818

Bronte V, Zanovello P (2005) Regulation of immune responses by L-arginine metabolism. Nat Rev Immunol 5:641–654

Bronte V, Apolloni E, Cabrelle A, Ronca R, Serafini P, Zamboni P, Restifo NP, Zanovello P (2000) Identification of a CD11b(+)/Gr-1(+)/CD31(+) myeloid progenitor capable of activating or suppressing CD8(+) T cells. Blood 96:3838–3846

Carr A, Rodriguez E, Arango Mdel C, Camacho R, Osorio M, Gabri M, Carrillo G, Valdes Z, Bebelagua Y, Perez R, Fernandez LE (2003) Immunotherapy of advanced breast cancer with a heterophilic ganglioside (NeuGcGM3) cancer vaccine. J Clin Oncol 21:1015–1021

Chaturvedi AK, Pfeiffer RM, Chang L, Goedert JJ, Biggar RJ, Engels EA (2007) Elevated risk of lung cancer among people with AIDS. AIDS 21:207–213

Chonn A, Cullis PR, Devine DV (1991) The role of surface charge in the activation of the classical and alternative pathways of complement by liposomes. J Immunol 146:4234–4241

Cieslewicz M, Tang J, Yu JL, Cao H, Zavaljevski M, Motoyama K, Lieber A, Raines EW, Pun SH (2013) Targeted delivery of proapoptotic peptides to tumor-associated macrophages improves survival. Proc Natl Acad Sci U S A 110:15919–15924

Cortez-Retamozo V, Etzrodt M, Newton A, Rauch PJ, Chudnovskiy A, Berger C, Ryan RJ, Iwamoto Y, Marinelli B, Gorbatov R et al (2012) Origins of tumor-associated macrophages and neutrophils. Proc Natl Acad Sci U S A 109:2491–2496

Curiel TJ, Coukos G, Zou L, Alvarez X, Cheng P, Mottram P, Evdemon-Hogan M, Conejo-Garcia JR, Zhang L, Burow M et al (2004) Specific recruitment of regulatory T cells in ovarian carcinoma fosters immune privilege and predicts reduced survival. Nat Med 10:942–949

Curotto de Lafaille MA, Lafaille JJ (2009) Natural and adaptive foxp3+ regulatory T cells: more of the same or a division of labor? Immunity 30:626–635

De Santo C, Serafini P, Marigo I, Dolcetti L, Bolla M, Del Soldato P, Melani C, Guiducci C, Colombo MP, Iezzi M et al (2005) Nitroaspirin corrects immune dysfunction in tumor-bearing hosts and promotes tumor eradication by cancer vaccination. Proc Natl Acad Sci U S A 102:4185–4190

DeNardo DG, Barreto JB, Andreu P, Vasquez L, Tawfik D, Kolhatkar N, Coussens LM (2009) CD4(+) T cells regulate pulmonary metastasis of mammary carcinomas by enhancing protumor properties of macrophages. Cancer Cell 16:91–102

DeNardo DG, Brennan DJ, Rexhepaj E, Ruffell B, Shiao SL, Madden SF, Gallagher WM, Wadhwani N, Keil SD, Junaid SA et al (2011) Leukocyte complexity predicts breast cancer survival and functionally regulates response to chemotherapy. Cancer Discov 1:54–67

Desai R, Collett D, Watson CJ, Johnson P, Evans T, Neuberger J (2012) Cancer transmission from organ donors-unavoidable but low risk. Transplantation 94:1200–1207

Devine DV, Wong K, Serrano K, Chonn A, Cullis PR (1994) Liposome-complement interactions in rat serum: implications for liposome survival studies. Biochim Biophys Acta 1191:43–51

Dolcetti L, Peranzoni E, Ugel S, Marigo I, Fernandez Gomez A, Mesa C, Geilich M, Winkels G, Traggiai E, Casati A et al (2010) Hierarchy of immunosuppressive strength among myeloid-derived suppressor cell subsets is determined by GM-CSF. Eur J Immunol 40:22–35

Dudley ME, Wunderlich JR, Robbins PF, Yang JC, Hwu P, Schwartzentruber DJ, Topalian SL, Sherry R, Restifo NP, Hubicki AM et al (2002) Cancer regression and autoimmunity in patients after clonal repopulation with antitumor lymphocytes. Science 298:850–854

Dudley ME, Yang JC, Sherry R, Hughes MS, Royal R, Kammula U, Robbins PF, Huang J, Citrin DE, Leitman SF et al (2008) Adoptive cell therapy for patients with metastatic melanoma: evaluation of intensive myeloablative chemoradiation preparative regimens. J Clin Oncol 26:5233–5239

Engels EA, Pfeiffer RM, Fraumeni JF Jr, Kasiske BL, Israni AK, Snyder JJ, Wolfe RA, Goodrich NP, Bayakly AR, Clarke CA et al (2011) Spectrum of cancer risk among US solid organ transplant recipients. JAMA 306:1891–1901

Estevez F, Carr A, Solorzano L, Valiente O, Mesa C, Barroso O, Sierra GV, Fernandez LE (1999) Enhancement of the immune response to poorly immunogenic gangliosides after incorporation into very small size proteoliposomes (VSSP). Vaccine 18:190–197

Etzerodt A, Maniecki MB, Graversen JH, Moller HJ, Torchilin VP, Moestrup SK (2012) Efficient intracellular drug-targeting of macrophages using stealth liposomes directed to the hemoglobin scavenger receptor CD163. J Control Release 160:72–80

Fallarino F, Grohmann U, Hwang KW, Orabona C, Vacca C, Bianchi R, Belladonna ML, Fioretti MC, Alegre ML, Puccetti P (2003) Modulation of tryptophan catabolism by regulatory T cells. Nat Immunol 4:1206–1212

Fang J, Nakamura H, Maeda H (2011) The EPR effect: unique features of tumor blood vessels for drug delivery, factors involved, and limitations and augmentation of the effect. Adv Drug Deliv Rev 63:136–151

Fernandez A, Mesa C, Marigo I, Dolcetti L, Clavell M, Oliver L, Fernandez LE, Bronte V (2011) Inhibition of tumor-induced myeloid-derived suppressor cell function by a nanoparticulated adjuvant. J Immunol 186:264–274

Filipazzi P, Huber V, Rivoltini L (2012) Phenotype, function and clinical implications of myeloid-derived suppressor cells in cancer patients. Cancer Immunol Immunother 61:255–263

Gabrilovich D (2004) Mechanisms and functional significance of tumour-induced dendritic-cell defects. Nat Rev Immunol 4:941–952

Gabrilovich DI, Nagaraj S (2009) Myeloid-derived suppressor cells as regulators of the immune system. Nat Rev Immunol 9:162–174

Gabrilovich DI, Velders MP, Sotomayor EM, Kast WM (2001) Mechanism of immune dysfunction in cancer mediated by immature Gr-1+ myeloid cells. J Immunol 166:5398–5406

Gabrilovich DI, Ostrand-Rosenberg S, Bronte V (2012) Coordinated regulation of myeloid cells by tumours. Nat Rev Immunol 12:253–268

Gallina G, Dolcetti L, Serafini P, De Santo C, Marigo I, Colombo MP, Basso G, Brombacher F, Borrello I, Zanovello P et al (2006) Tumors induce a subset of inflammatory monocytes with immunosuppressive activity on CD8+ T cells. J Clin Invest 116:2777–2790

Galon J, Costes A, Sanchez-Cabo F, Kirilovsky A, Mlecnik B, Lagorce-Pages C, Tosolini M, Camus M, Berger A, Wind P et al (2006) Type, density, and location of immune cells within human colorectal tumors predict clinical outcome. Science 313:1960–1964

Gattinoni L, Powell DJ Jr, Rosenberg SA, Restifo NP (2006) Adoptive immunotherapy for cancer: building on success. Nat Rev Immunol 6:383–393

Gazi U, Martinez-Pomares L (2009) Influence of the mannose receptor in host immune responses. Immunobiology 214:554–561

Gazzaniga S, Bravo AI, Guglielmotti A, van Rooijen N, Maschi F, Vecchi A, Mantovani A, Mordoh J, Wainstok R (2007) Targeting tumor-associated macrophages and inhibition of MCP-1 reduce angiogenesis and tumor growth in a human melanoma xenograft. J Invest Dermatol 127:2031–2041

Geissmann F, Jung S, Littman DR (2003) Blood monocytes consist of two principal subsets with distinct migratory properties. Immunity 19:71–82

Geissmann F, Manz MG, Jung S, Sieweke MH, Merad M, Ley K (2010) Development of monocytes, macrophages, and dendritic cells. Science 327:656–661

Germano G, Frapolli R, Belgiovine C, Anselmo A, Pesce S, Liguori M, Erba E, Uboldi S, Zucchetti M, Pasqualini F et al (2013) Role of macrophage targeting in the antitumor activity of trabectedin. Cancer Cell 23:249–262

Gessner A, Rollinghoff M (2000) Biologic functions and signaling of the interleukin-4 receptor complexes. Immunobiology 201:285–307

Ghansah T, Vohra N, Kinney K, Weber A, Kodumudi K, Springett G, Sarnaik AA, Pilon-Thomas S (2013) Dendritic cell immunotherapy combined with gemcitabine chemotherapy enhances survival in a murine model of pancreatic carcinoma. Cancer Immunol Immunother 62:1083–1091

Ghiringhelli F, Larmonier N, Schmitt E, Parcellier A, Cathelin D, Garrido C, Chauffert B, Solary E, Bonnotte B, Martin F (2004) CD4+CD25+ regulatory T cells suppress tumor immunity but are sensitive to cyclophosphamide which allows immunotherapy of established tumors to be curative. Eur J Immunol 34:336–344

Giraudo E, Inoue M, Hanahan D (2004) An amino-bisphosphonate targets MMP-9-expressing macrophages and angiogenesis to impair cervical carcinogenesis. J Clin Invest 114:623–633

Grohmann U, Orabona C, Fallarino F, Vacca C, Calcinaro F, Falorni A, Candeloro P, Belladonna ML, Bianchi R, Fioretti MC, Puccetti P (2002) CTLA-4-Ig regulates tryptophan catabolism in vivo. Nat Immunol 3:1097–1101

Guilliams M, De Kleer I, Henri S, Post S, Vanhoutte L, De Prijck S, Deswarte K, Malissen B, Hammad H, Lambrecht BN (2013) Alveolar macrophages develop from fetal monocytes that differentiate into long-lived cells in the first week of life via GM-CSF. J Exp Med 210:1977–1992

Halin S, Rudolfsson SH, Van Rooijen N, Bergh A (2009) Extratumoral macrophages promote tumor and vascular growth in an orthotopic rat prostate tumor model. Neoplasia 11:177–186

Hanahan D, Weinberg RA (2011) Hallmarks of cancer: the next generation. Cell 144:646–674

Hanks BA, Holtzhausen A, Evans KS, Jamieson R, Gimpel P, Campbell OM, Hector-Greene M, Sun L, Tewari A, George A et al (2013) Type III TGF-beta receptor downregulation generates an immunotolerant tumor microenvironment. J Clin Invest 123:3925–3940

Hillaireau H, Couvreur P (2009) Nanocarriers' entry into the cell: relevance to drug delivery. Cell Mol Life Sci 66:2873–2896

Hiraoka K, Zenmyo M, Watari K, Iguchi H, Fotovati A, Kimura YN, Hosoi F, Shoda T, Nagata K, Osada H et al (2008) Inhibition of bone and muscle metastases of lung cancer cells by a decrease in the number of monocytes/macrophages. Cancer Sci 99:1595–1602

Hiratsuka S, Watanabe A, Aburatani H, Maru Y (2006) Tumour-mediated upregulation of chemoattractants and recruitment of myeloid cells predetermines lung metastasis. Nat Cell Biol 8:1369–1375

Hivroz C, Chemin K, Tourret M, Bohineust A (2012) Crosstalk between T lymphocytes and dendritic cells. Crit Rev Immunol 32:139–155

Hodi FS, O'Day SJ, McDermott DF, Weber RW, Sosman JA, Haanen JB, Gonzalez R, Robert C, Schadendorf D, Hassel JC et al (2010) Improved survival with ipilimumab in patients with metastatic melanoma. N Engl J Med 363:711–723

Hoeffel G, Wang Y, Greter M, See P, Teo P, Malleret B, Leboeuf M, Low D, Oller G, Almeida F et al (2012) Adult Langerhans cells derive predominantly from embryonic fetal liver monocytes with a minor contribution of yolk sac-derived macrophages. J Exp Med 209:1167–1181

Husain SR, Kawakami K, Kawakami M, Puri RK (2003) Interleukin-4 receptor-targeted cytotoxin therapy of androgen-dependent and -independent prostate carcinoma in xenograft models. Mol Cancer Ther 2:245–254

Idoyaga J, Steinman RM (2011) SnapShot: dendritic cells. Cell 146(660–660):e662

Jacobs JF, Punt CJ, Lesterhuis WJ, Sutmuller RP, Brouwer HM, Scharenborg NM, Klasen IS, Hilbrands LB, Figdor CG, de Vries IJ, Adema GJ (2010) Dendritic cell vaccination in combination with anti-CD25 monoclonal antibody treatment: a phase I/II study in metastatic melanoma patients. Clin Cancer Res 16:5067–5078

Jasperson LK, Bucher C, Panoskaltsis-Mortari A, Mellor AL, Munn DH, Blazar BR (2009) Inducing the tryptophan catabolic pathway, indoleamine 2,3-dioxygenase (IDO), for suppression of graft-versus-host disease (GVHD) lethality. Blood 114:5062–5070

Jiang Y, Villeneuve PJ, Wielgosz A, Schaubel DE, Fenton SS, Mao Y (2010) The incidence of cancer in a population-based cohort of Canadian heart transplant recipients. Am J Transplant 10:637–645

Josefowicz SZ, Rudensky A (2009) Control of regulatory T cell lineage commitment and maintenance. Immunity 30:616–625

Kaplan RN, Riba RD, Zacharoulis S, Bramley AH, Vincent L, Costa C, MacDonald DD, Jin DK, Shido K, Kerns SA et al (2005) VEGFR1-positive haematopoietic bone marrow progenitors initiate the pre-metastatic niche. Nature 438:820–827

Kar UK, Srivastava MK, Andersson A, Baratelli F, Huang M, Kickhoefer VA, Dubinett SM, Rome LH, Sharma S (2011) Novel CCL21-vault nanocapsule intratumoral delivery inhibits lung cancer growth. PLoS One 6:e18758

Kawakami M, Kawakami K, Kioi M, Leland P, Puri RK (2005a) Hodgkin lymphoma therapy with interleukin-4 receptor-directed cytotoxin in an infiltrating animal model. Blood 105:3707–3713

Kawakami S, Opanasopit P, Yokoyama M, Chansri N, Yamamoto T, Okano T, Yamashita F, Hashida M (2005b) Biodistribution characteristics of all-trans retinoic acid incorporated in liposomes and polymeric micelles following intravenous administration. J Pharm Sci 94:2606–2615

Kelly C, Jefferies C, Cryan SA (2011) Targeted liposomal drug delivery to monocytes and macrophages. J Drug Deliv 2011:727241

Kessenbrock K, Plaks V, Werb Z (2010) Matrix metalloproteinases: regulators of the tumor microenvironment. Cell 141:52–67

Kimura YN, Watari K, Fotovati A, Hosoi F, Yasumoto K, Izumi H, Kohno K, Umezawa K, Iguchi H, Shirouzu K et al (2007) Inflammatory stimuli from macrophages and cancer cells synergistically promote tumor growth and angiogenesis. Cancer Sci 98:2009–2018

Kioi M, Takahashi S, Kawakami M, Kawakami K, Kreitman RJ, Puri RK (2005) Expression and targeting of interleukin-4 receptor for primary and advanced ovarian cancer therapy. Cancer Res 65:8388–8396

Kirk GD, Merlo C, O'Driscoll P, Mehta SH, Galai N, Vlahov D, Samet J, Engels EA (2007) HIV infection is associated with an increased risk for lung cancer, independent of smoking. Clin Infect Dis 45:103–110

Klebanoff CA, Acquavella N, Yu Z, Restifo NP (2011) Therapeutic cancer vaccines: are we there yet? Immunol Rev 239:27–44

Kodumudi KN, Woan K, Gilvary DL, Sahakian E, Wei S, Djeu JY (2010) A novel chemoimmunomodulating property of docetaxel: suppression of myeloid-derived suppressor cells in tumor bearers. Clin Cancer Res 16:4583–4594

Kohanbash G, McKaveney K, Sakaki M, Ueda R, Mintz AH, Amankulor N, Fujita M, Ohlfest JR, Okada H (2013) GM-CSF promotes the immunosuppressive activity of glioma-infiltrating myeloid cells through interleukin-4 receptor-alpha. Cancer Res 73:6413–6423

Kourtis IC, Hirosue S, de Titta A, Kontos S, Stegmann T, Hubbell JA, Swartz MA (2013) Peripherally administered nanoparticles target monocytic myeloid cells, secondary lymphoid organs and tumors in mice. PLoS One 8:e61646

Kuang DM, Zhao Q, Peng C, Xu J, Zhang JP, Wu C, Zheng L (2009) Activated monocytes in peritumoral stroma of hepatocellular carcinoma foster immune privilege and disease progression through PD-L1. J Exp Med 206:1327–1337

Kusmartsev S, Gabrilovich DI (2003) Inhibition of myeloid cell differentiation in cancer: the role of reactive oxygen species. J Leukoc Biol 74:186–196

Kusmartsev S, Gabrilovich DI (2005) STAT1 signaling regulates tumor-associated macrophage-mediated T cell deletion. J Immunol 174:4880–4891

Kusmartsev S, Cheng F, Yu B, Nefedova Y, Sotomayor E, Lush R, Gabrilovich D (2003) All-trans-retinoic acid eliminates immature myeloid cells from tumor-bearing mice and improves the effect of vaccination. Cancer Res 63:4441–4449

Lee JR, Dalton RR, Messina JL, Sharma MD, Smith DM, Burgess RE, Mazzella F, Antonia SJ, Mellor AL, Munn DH (2003) Pattern of recruitment of immunoregulatory antigen-presenting cells in malignant melanoma. Lab Invest 83:1457–1466

Leuschner F, Dutta P, Gorbatov R, Novobrantseva TI, Donahoe JS, Courties G, Lee KM, Kim JI, Markmann JF, Marinelli B et al (2011) Therapeutic siRNA silencing in inflammatory monocytes in mice. Nat Biotechnol 29:1005–1010

Lin EY, Li JF, Gnatovskiy L, Deng Y, Zhu L, Grzesik DA, Qian H, Xue XN, Pollard JW (2006) Macrophages regulate the angiogenic switch in a mouse model of breast cancer. Cancer Res 66:11238–11246

Lin EY, Li JF, Bricard G, Wang W, Deng Y, Sellers R, Porcelli SA, Pollard JW (2007) Vascular endothelial growth factor restores delayed tumor progression in tumors depleted of macrophages. Mol Oncol 1:288–302

Liu K, Nussenzweig MC (2010) Origin and development of dendritic cells. Immunol Rev 234:45–54

Lv H, Zhang S, Wang B, Cui S, Yan J (2006) Toxicity of cationic lipids and cationic polymers in gene delivery. J Control Release 114:100–109

MacKie RM, Reid R, Junor B (2003) Fatal melanoma transferred in a donated kidney 16 years after melanoma surgery. N Engl J Med 348:567–568

Mantovani A, Sozzani S, Locati M, Allavena P, Sica A (2002) Macrophage polarization: tumor-associated macrophages as a paradigm for polarized M2 mononuclear phagocytes. Trends Immunol 23:549–555

Melani C, Sangaletti S, Barazzetta FM, Werb Z, Colombo MP (2007) Amino-bisphosphonate-mediated MMP-9 inhibition breaks the tumor-bone marrow axis responsible for myeloid-derived suppressor cell expansion and macrophage infiltration in tumor stroma. Cancer Res 67:11438–11446

Mirza N, Fishman M, Fricke I, Dunn M, Neuger AM, Frost TJ, Lush RM, Antonia S, Gabrilovich DI (2006) All-trans-retinoic acid improves differentiation of myeloid cells and immune response in cancer patients. Cancer Res 66:9299–9307

Miselis NR, Wu ZJ, Van Rooijen N, Kane AB (2008) Targeting tumor-associated macrophages in an orthotopic murine model of diffuse malignant mesothelioma. Mol Cancer Ther 7: 788–799

Molon B, Ugel S, Del Pozzo F, Soldani C, Zilio S, Avella D, De Palma A, Mauri P, Monegal A, Rescigno M et al (2011) Chemokine nitration prevents intratumoral infiltration of antigen-specific T cells. J Exp Med 208:1949–1962

Moloney FJ, Comber H, O'Lorcain P, O'Kelly P, Conlon PJ, Murphy GM (2006) A population-based study of skin cancer incidence and prevalence in renal transplant recipients. Br J Dermatol 154:498–504

Montero AJ, Diaz-Montero CM, Kyriakopoulos CE, Bronte V, Mandruzzato S (2012) Myeloid-derived suppressor cells in cancer patients: a clinical perspective. J Immunother 35:107–115

Movahedi K, Laoui D, Gysemans C, Baeten M, Stange G, Van den Bossche J, Mack M, Pipeleers D, In't Veld P, De Baetselier P, Van Ginderachter JA (2010) Different tumor microenvironments contain functionally distinct subsets of macrophages derived from Ly6C(high) monocytes. Cancer Res 70:5728–5739

Muindi JR, Frankel SR, Huselton C, DeGrazia F, Garland WA, Young CW, Warrell RP Jr (1992) Clinical pharmacology of oral all-trans retinoic acid in patients with acute promyelocytic leukemia. Cancer Res 52:2138–2142

Munn DH, Mellor AL (2013) Indoleamine 2,3 dioxygenase and metabolic control of immune responses. Trends Immunol 34:137–143

Munn DH, Sharma MD, Hou D, Baban B, Lee JR, Antonia SJ, Messina JL, Chandler P, Koni PA, Mellor AL (2004) Expression of indoleamine 2,3-dioxygenase by plasmacytoid dendritic cells in tumor-draining lymph nodes. J Clin Invest 114:280–290

Nagaraj S, Gupta K, Pisarev V, Kinarsky L, Sherman S, Kang L, Herber DL, Schneck J, Gabrilovich DI (2007) Altered recognition of antigen is a mechanism of CD8+ T cell tolerance in cancer. Nat Med 13:828–835

Nefedova Y, Fishman M, Sherman S, Wang X, Beg AA, Gabrilovich DI (2007) Mechanism of all-trans retinoic acid effect on tumor-associated myeloid-derived suppressor cells. Cancer Res 67:11021–11028

Norian LA, Rodriguez PC, O'Mara LA, Zabaleta J, Ochoa AC, Cella M, Allen PM (2009) Tumor-infiltrating regulatory dendritic cells inhibit CD8+ T cell function via L-arginine metabolism. Cancer Res 69:3086–3094

Novitskiy SV, Ryzhov S, Zaynagetdinov R, Goldstein AE, Huang Y, Tikhomirov OY, Blackburn MR, Biaggioni I, Carbone DP, Feoktistov I, Dikov MM (2008) Adenosine receptors in regulation of dendritic cell differentiation and function. Blood 112:1822–1831

Ozpolat B, Lopez-Berestein G, Adamson P, Fu CJ, Williams AH (2003) Pharmacokinetics of intravenously administered liposomal all-trans-retinoic acid (ATRA) and orally administered ATRA in healthy volunteers. J Pharm Pharm Sci 6:292–301

Pallotta MT, Orabona C, Volpi C, Vacca C, Belladonna ML, Bianchi R, Servillo G, Brunacci C, Calvitti M, Bicciato S et al (2011) Indoleamine 2,3-dioxygenase is a signaling protein in long-term tolerance by dendritic cells. Nat Immunol 12:870–878

Pan PY, Ma G, Weber KJ, Ozao-Choy J, Wang G, Yin B, Divino CM, Chen SH (2010) Immune stimulatory receptor CD40 is required for T-cell suppression and T regulatory cell activation mediated by myeloid-derived suppressor cells in cancer. Cancer Res 70:99–108

Platten M, Wick W, Van den Eynde BJ (2012) Tryptophan catabolism in cancer: beyond IDO and tryptophan depletion. Cancer Res 72:5435–5440

Qian BZ, Pollard JW (2010) Macrophage diversity enhances tumor progression and metastasis. Cell 141:39–51

Qian B, Deng Y, Im JH, Muschel RJ, Zou Y, Li J, Lang RA, Pollard JW (2009) A distinct macrophage population mediates metastatic breast cancer cell extravasation, establishment and growth. PLoS One 4:e6562

Ramirez BS, Pestana ES, Hidalgo GG, Garcia TH, Rodriguez RP, Ullrich A, Fernandez LE (2006) Active antimetastatic immunotherapy in Lewis lung carcinoma with self EGFR extracellular domain protein in VSSP adjuvant. Int J Cancer 119:2190–2199

Rand RW, Kreitman RJ, Patronas N, Varricchio F, Pastan I, Puri RK (2000) Intratumoral administration of recombinant circularly permuted interleukin-4-Pseudomonas exotoxin in patients with high-grade glioma. Clin Cancer Res 6:2157–2165

Regazzi MB, Iacona I, Gervasutti C, Lazzarino M, Toma S (1997) Clinical pharmacokinetics of tretinoin. Clin Pharmacokinet 32:382–402

Reis e Sousa C (2006) Dendritic cells in a mature age. Nat Rev Immunol 6:476–483

Restifo NP, Dudley ME, Rosenberg SA (2012) Adoptive immunotherapy for cancer: harnessing the T cell response. Nat Rev Immunol 12:269–281

Rodriguez PC, Quiceno DG, Zabaleta J, Ortiz B, Zea AH, Piazuelo MB, Delgado A, Correa P, Brayer J, Sotomayor EM et al (2004) Arginase I production in the tumor microenvironment by mature myeloid cells inhibits T-cell receptor expression and antigen-specific T-cell responses. Cancer Res 64:5839–5849

Roth F, De La Fuente AC, Vella JL, Zoso A, Inverardi L, Serafini P (2012) Aptamer-mediated blockade of IL4Ralpha triggers apoptosis of MDSCs and limits tumor progression. Cancer Res 72:1373–1383

Sakaguchi S (2004) Naturally arising CD4+ regulatory t cells for immunologic self-tolerance and negative control of immune responses. Annu Rev Immunol 22:531–562

Sallusto F, Lanzavecchia A (2000) Understanding dendritic cell and T-lymphocyte traffic through the analysis of chemokine receptor expression. Immunol Rev 177:134–140

Schreiber RD, Old LJ, Smyth MJ (2011) Cancer immunoediting: integrating immunity's roles in cancer suppression and promotion. Science 331:1565–1570

Schrem H, Kurok M, Kaltenborn A, Vogel A, Walter U, Zachau L, Manns MP, Klempnauer J, Kleine M (2013) Incidence and long-term risk of de novo malignancies after liver transplantation with implications for prevention and detection. Liver Transpl 19:1252–1261

Schulz C, Gomez Perdiguero E, Chorro L, Szabo-Rogers H, Cagnard N, Kierdorf K, Prinz M, Wu B, Jacobsen SE, Pollard JW et al (2012) A lineage of myeloid cells independent of Myb and hematopoietic stem cells. Science 336:86–90

Sen M, Grandis JR (2012) Nucleic acid-based approaches to STAT inhibition. JAKSTAT 1:285–291

Serafini P, Meckel K, Kelso M, Noonan K, Califano J, Koch W, Dolcetti L, Bronte V, Borrello I (2006) Phosphodiesterase-5 inhibition augments endogenous antitumor immunity by reducing myeloid-derived suppressor cell function. J Exp Med 203:2691–2702

Serafini P, Mgebroff S, Noonan K, Borrello I (2008) Myeloid-derived suppressor cells promote cross-tolerance in B-cell lymphoma by expanding regulatory T cells. Cancer Res 68: 5439–5449

Shi C, Pamer EG (2011) Monocyte recruitment during infection and inflammation. Nat Rev Immunol 11:762–774

Shiels MS, Engels EA (2012) Increased risk of histologically defined cancer subtypes in human immunodeficiency virus-infected individuals: clues for possible immunosuppression-related or infectious etiology. Cancer 118:4869–4876

Shimamura T, Royal RE, Kioi M, Nakajima A, Husain SR, Puri RK (2007) Interleukin-4 cytotoxin therapy synergizes with gemcitabine in a mouse model of pancreatic ductal adenocarcinoma. Cancer Res 67:9903–9912

Shirota Y, Shirota H, Klinman DM (2012) Intratumoral injection of CpG oligonucleotides induces the differentiation and reduces the immunosuppressive activity of myeloid-derived suppressor cells. J Immunol 188:1592–1599

Sica A, Mantovani A (2012) Macrophage plasticity and polarization: in vivo veritas. J Clin Invest 122:787–795

Sinha P, Clements VK, Bunt SK, Albelda SM, Ostrand-Rosenberg S (2007) Cross-talk between myeloid-derived suppressor cells and macrophages subverts tumor immunity toward a type 2 response. J Immunol 179:977–983

Sliwkowski MX, Mellman I (2013) Antibody therapeutics in cancer. Science 341:1192–1198

Srivastava MK, Sinha P, Clements VK, Rodriguez P, Ostrand-Rosenberg S (2010) Myeloid-derived suppressor cells inhibit T-cell activation by depleting cystine and cysteine. Cancer Res 70:68–77

Steidl C, Lee T, Shah SP, Farinha P, Han G, Nayar T, Delaney A, Jones SJ, Iqbal J, Weisenburger DD et al (2010) Tumor-associated macrophages and survival in classic Hodgkin's lymphoma. N Engl J Med 362:875–885

Sun D, Zhuang X, Xiang X, Liu Y, Zhang S, Liu C, Barnes S, Grizzle W, Miller D, Zhang HG (2010) A novel nanoparticle drug delivery system: the anti-inflammatory activity of curcumin is enhanced when encapsulated in exosomes. Mol Ther 18:1606–1614

Thomas SN, Vokali E, Lund AW, Hubbell JA, Swartz MA (2014) Targeting the tumor-draining lymph node with adjuvanted nanoparticles reshapes the anti-tumor immune response. Biomaterials 35:814–824

Topalian SL, Hodi FS, Brahmer JR, Gettinger SN, Smith DC, McDermott DF, Powderly JD, Carvajal RD, Sosman JA, Atkins MB et al (2012) Safety, activity, and immune correlates of anti-PD-1 antibody in cancer. N Engl J Med 366:2443–2454

Torrens I, Mendoza O, Batte A, Reyes O, Fernandez LE, Mesa C, Guillen G (2005) Immunotherapy with CTL peptide and VSSP eradicated established human papillomavirus (HPV) type 16 E7-expressing tumors. Vaccine 23:5768–5774

Ugel S, Delpozzo F, Desantis G, Papalini F, Simonato F, Sonda N, Zilio S, Bronte V (2009) Therapeutic targeting of myeloid-derived suppressor cells. Curr Opin Pharmacol 9:470–481

Ugel S, Peranzoni E, Desantis G, Chioda M, Walter S, Weinschenk T, Ochando JC, Cabrelle A, Mandruzzato S, Bronte V (2012) Immune tolerance to tumor antigens occurs in a specialized environment of the spleen. Cell Rep 2:628–639

Vajdic CM, McDonald SP, McCredie MR, van Leeuwen MT, Stewart JH, Law M, Chapman JR, Webster AC, Kaldor JM, Grulich AE (2006) Cancer incidence before and after kidney transplantation. JAMA 296:2823–2831

Valleala H, Hanemaaijer R, Mandelin J, Salminen A, Teronen O, Monkkonen J, Konttinen YT (2003) Regulation of MMP-9 (gelatinase B) in activated human monocyte/macrophages by two different types of bisphosphonates. Life Sci 73:2413–2420

Van Rooijen N (1989) The liposome-mediated macrophage 'suicide' technique. J Immunol Methods 124:1–6

Vasquez-Dunddel D, Pan F, Zeng Q, Gorbounov M, Albesiano E, Fu J, Blosser RL, Tam AJ, Bruno T, Zhang H et al (2013) STAT3 regulates arginase-I in myeloid-derived suppressor cells from cancer patients. J Clin Invest 123:1580–1589

Vesely MD, Kershaw MH, Schreiber RD, Smyth MJ (2011) Natural innate and adaptive immunity to cancer. Annu Rev Immunol 29:235–271

Vincent J, Mignot G, Chalmin F, Ladoire S, Bruchard M, Chevriaux A, Martin F, Apetoh L, Rebe C, Ghiringhelli F (2010) 5-Fluorouracil selectively kills tumor-associated myeloid-derived suppressor cells resulting in enhanced T cell-dependent antitumor immunity. Cancer Res 70:3052–3061

von Boehmer H (2005) Mechanisms of suppression by suppressor T cells. Nat Immunol 6: 338–344

von Boehmer H, Daniel C (2013) Therapeutic opportunities for manipulating T(Reg) cells in autoimmunity and cancer. Nat Rev Immunol 12:51–63

Wang X, Zhang Y, Zhang X, Tian W, Feng W, Chen T (2012) The curcumin analogue hydrazinocurcumin exhibits potent suppressive activity on carcinogenicity of breast cancer cells via STAT3 inhibition. Int J Oncol 40:1189–1195

Warrell RP Jr (1993) Retinoid resistance in acute promyelocytic leukemia: new mechanisms, strategies, and implications. Blood 82:1949–1953

Wildin RS, Ramsdell F, Peake J, Faravelli F, Casanova JL, Buist N, Levy-Lahad E, Mazzella M, Goulet O, Perroni L et al (2001) X-linked neonatal diabetes mellitus, enteropathy and endocrinopathy syndrome is the human equivalent of mouse scurfy. Nat Genet 27:18–20

Wrzesinski C, Paulos CM, Kaiser A, Muranski P, Palmer DC, Gattinoni L, Yu Z, Rosenberg SA, Restifo NP (2010) Increased intensity lymphodepletion enhances tumor treatment efficacy of adoptively transferred tumor-specific T cells. J Immunother 33:1–7

Wyckoff J, Wang W, Lin EY, Wang Y, Pixley F, Stanley ER, Graf T, Pollard JW, Segall J, Condeelis J (2004) A paracrine loop between tumor cells and macrophages is required for tumor cell migration in mammary tumors. Cancer Res 64:7022–7029

Wyckoff JB, Wang Y, Lin EY, Li JF, Goswami S, Stanley ER, Segall JE, Pollard JW, Condeelis J (2007) Direct visualization of macrophage-assisted tumor cell intravasation in mammary tumors. Cancer Res 67:2649–2656

Wynn TA, Chawla A, Pollard JW (2013) Macrophage biology in development, homeostasis and disease. Nature 496:445–455

Yang CL, Liu YY, Ma YG, Xue YX, Liu DG, Ren Y, Liu XB, Li Y, Li Z (2012) Curcumin blocks small cell lung cancer cells migration, invasion, angiogenesis, cell cycle and neoplasia through Janus kinase-STAT3 signalling pathway. PLoS One 7:e37960

Youn JI, Nagaraj S, Collazo M, Gabrilovich DI (2008) Subsets of myeloid-derived suppressor cells in tumor-bearing mice. J Immunol 181:5791–5802

Yu H, Kortylewski M, Pardoll D (2007) Crosstalk between cancer and immune cells: role of STAT3 in the tumour microenvironment. Nat Rev Immunol 7:41–51

Yu SS, Lau CM, Thomas SN, Jerome WG, Maron DJ, Dickerson JH, Hubbell JA, Giorgio TD (2012) Size- and charge-dependent non-specific uptake of PEGylated nanoparticles by macrophages. Int J Nanomedicine 7:799–813

Yu SS, Lau CM, Barham WJ, Onishko HM, Nelson CE, Li H, Smith CA, Yull FE, Duvall CL, Giorgio TD (2013) Macrophage-specific RNA interference targeting via "click", mannosylated polymeric micelles. Mol Pharm 10:975–987

Yue P, Turkson J (2009) Targeting STAT3 in cancer: how successful are we? Expert Opin Investig Drugs 18:45–56

Zeisberger SM, Odermatt B, Marty C, Zehnder-Fjallman AH, Ballmer-Hofer K, Schwendener RA (2006) Clodronate-liposome-mediated depletion of tumour-associated macrophages: a new and highly effective antiangiogenic therapy approach. Br J Cancer 95:272–281

Zhang HG, Grizzle WE (2011) Exosomes and cancer: a newly described pathway of immune suppression. Clin Cancer Res 17:959–964

Zhang X, Tian W, Cai X, Wang X, Dang W, Tang H, Cao H, Wang L, Chen T (2013) Hydrazinocurcumin encapsuled nanoparticles "re-educate" tumor-associated macrophages and exhibit anti-tumor effects on breast cancer following STAT3 suppression. PLoS One 8:e65896

Zhao W, Zhuang S, Qi XR (2011) Comparative study of the in vitro and in vivo characteristics of cationic and neutral liposomes. Int J Nanomedicine 6:3087–3098

Zhu S, Niu M, O'Mary H, Cui Z (2013) Targeting of tumor-associated macrophages made possible by PEG-sheddable, mannose-modified nanoparticles. Mol Pharm 10:3525–3530

Zou W (2006) Regulatory T cells, tumour immunity and immunotherapy. Nat Rev Immunol 6:295–307

Chapter 8
Nanotechnology Approaches for Cancer Immunotherapy and Immunomodulation

María de la Fuente, Robert Langer, and Maria José Alonso

8.1 The Immune System and Cancer

Cancer is, without a doubt, one of the leading causes of mortality in the developed world. The increasing impact of this disease, estimated by the World Health Organization (WHO) as 12 million deaths by 2030, urges the search for improved and powerful therapies. The poor patient survival is directly attributable to widespread metastasis, drug resistance, and the lack of effective treatment strategies (Jemal et al. 2011; Stewart and Wild 2014). Despite the attempts in drug discovery and drug delivery to optimize small drugs-based chemotherapeutics, as well as biomacromolecule-based therapies, the level of success achieved to date is limited and there is a clear need for novel ideas for fighting against cancer.

Cancer-related inflammation and the ability of cancer cells to evade immunological destruction are considered an emerging hallmark of cancer (Colotta et al. 2009; Hanahan and Weinberg 2011), in addition to the six hallmarks previously proposed by Hanahan and Weinberg (2000) as distinctive and complementary

M. de la Fuente
Nano-oncologicals Laboratory, Translational Medical Oncology, Health Research Institute of Santiago de Compostela (IDIS) and Clinical University Hospital of Santiago de Compostela (CHUS), Travesía da Choupana s/n, Santiago de Compostela 15706, Spain

R. Langer
Department of Chemical, Biological and Mechanical Engineering, David H. Koch Institute for Integrative Cancer Research at MIT, Massachusetts Institute of Technology, Boston, USA

M.J. Alonso (✉)
Center for Research in Molecular Medicine and Chronic Diseases (CIMUS), Department of Pharmacy and Pharmaceutical Technology, Health Research Institute of Santiago de Compostela (IDIS), Universidad de Santiago de Compostela, Av. Barcelona s/n, Campus Sur, Santiago de Compostela 15782, Spain
e-mail: mariaj.alonso@usc.es

M.J. Alonso, M. Garcia-Fuentes (eds.), *Nano-Oncologicals: New Targeting and Delivery Approaches*, Advances in Delivery Science and Technology,
DOI 10.1007/978-3-319-08084-0_8

capabilities that enable tumor growth and metastatic dissemination. Consequently, immunotherapy in a broad sense is now regarded as a promising strategy among the various novel anticancer therapies (Hanahan and Weinberg 2011; Seton-Rogers 2012; Vanneman and Dranoff 2012).

Active immunotherapy, or therapeutic vaccination, involves priming of the immune system in order to generate a T-cell response against the tumor-associated antigen (Winter et al. 2011; Boghossian and Von-Delwig 2012; De Pas et al. 2012; Goldinger et al. 2012). The efficacy of this approach can benefit from nonspecific immune activation mediated by immunostimulatory molecules such as Toll-like receptors (TLRs) agonists (Flemming 2011; Hamdy et al. 2011; Silva et al. 2013a).

Associated with this is the concept of immunogenic cell death (ICD), which takes advantage of conventional chemotherapies, since it is well known that certain chemotherapeutics can stimulate tumor-specific immune responses by mediating the immunogenic death of tumoral cells. This is characterized by the release of a set of molecules called damage-associated molecular patterns (DAMPs) (such as calreticulin, HMGB1, and ATP), leading to an immunogenic profile which contributes to efficient antigen presentation and cytotoxic T-cell activation, therefore achieving a vaccine effect in vivo (Kepp et al. 2009; Garg et al. 2010; Galluzzi et al. 2012a).

Passive immunotherapy, based on the use of therapeutic monoclonal antibodies (mAb), has become established as one of the most successful anti-cancer therapies (Weiner et al. 2010; Scott et al. 2012). Most commonly, mAb are given in combination with other anticancer drugs depending on the specific disease and status (Skak et al. 2008; Shuptrine et al. 2012; Vanneman and Dranoff 2012; National Comprehensive Clinical Cancer Network 2013).

Despite the recognized progress achieved with the so-called immunotherapies, it is broadly believed that further improvements are needed. These may come upon consideration of the complex interplay between cancer cells and their immunological microenvironment, which is known to play a critical role in cancer progression and influence the response to cancer therapies (Shiao and Ganesan 2011; Junttila and de Sauvage 2013). Indeed, the progress made in the knowledge of the tumor microenvironment has led to the conclusion that a prominent component of solid tumors is represented by non-tumoral cells, including stromal cells (fibroblasts and endothelial cells) as well as several subtypes of leukocytes [i.e., tumor-associated macrophages (TAMs), dendritic cells (DCs), and myeloid-derived suppressor cells (MDSCs)], which are key initiators of the persistent inflammation in the tumor microenvironment and play a pro-tumorigenic role (Mantovani et al. 2008; Shiao and Ganesan 2011; Vasievich and Huang 2011; Emens et al. 2012).

The improved knowledge of the tumor microenvironment has provided an explanation for the so far limited success of cancer immunotherapy; this has been associated with the nonselective immunosuppression generated, both in the tumor microenvironment and, systemically, in the subsequent fosterage of tumor development and dissemination (Mantovani et al. 2008; Gabrilovich and Nagaraj 2009; Yaguchi et al. 2011; Benencia et al. 2012; Allavena and Mantovani 2012; Junttila and de Sauvage 2013). Several therapeutic strategies that specifically target the immunosuppressive tumor environment have already been described, including the re-education of pro-tumoral immune cells. The efficacy of these novel therapies will largely depend on the availability of a vector that enables the intracellular delivery of gene therapies, or the delivery of specific chemotherapeutics.

8.2 Target Cells in Immunotherapy and Immunomodulation of Cancer (Good Cells and Bad Cells)

Immunocompetent cells are key players in the initiation of adaptive immune responses, and are therefore the target cells for therapeutic vaccination. Immunomodulatory therapies addressed to revert the immunosuppressive tumor microenvironment also target immunocompetent cells, which in this case behave as pro-tumoral cells since they have an altered phenotype and support tumor growth and dissemination. Passive immunotherapy targets cancer cells. A schematic view of the proposed therapeutics, target cells, and expected effects for each modality of therapy is presented in Fig. 8.1.

8.2.1 Target Cells in Therapeutic Vaccination: Immunocompetent Cells (Good Cells)

Active cell immunotherapy involves activating the patient's immune system with the intention of promoting an antigen-specific anti-tumor effect using the body's own immune cells. This therapeutic intervention seeks to create a durable immune response that can protect against tumor recurrence. Therapeutic vaccination involves priming of the immune system in order to generate a T-cell response against the tumor-associated antigen. In this scenario, peripheral blood mononuclear cells, which include lymphocytes [T cells, B cells, and natural killer (NK) cells] and monocytes (precursors to dendritic cells and macrophages), play a critical role. Antigen-presenting cells (APCs), i.e., macrophages and immature dendritic cells, initiate the mechanism of humoral and cellular immunity. They internalize and digest antigens, exposing the resulting peptides at their surface. While this process takes place, dendritic cells undergo a maturation process and migrate to the nearest draining lymph node. The exposure of the antigen onto mature dendritic cells and recognition by T lymphocytes, CD4+ T-helper lymphocytes and CD8+ cytotoxic T lymphocytes, mediate humoral and cellular responses; recognition by helper T cells promotes B cell differentiation into plasma cells and the production of antibodies (humoral responses), while recognition by cytotoxic lymphocytes or killer T cells is responsible for cellular responses (kill the tumor cells). For therapeutic cancer vaccination, an induction of CD8+ T-cell responses against tumor antigens is required (Foged et al. 2012; Silva et al. 2013b). Additionally, antibodies may bind to tumoral APCs, causing them to express more antigens for destruction by macrophages or attack by natural killer cells.

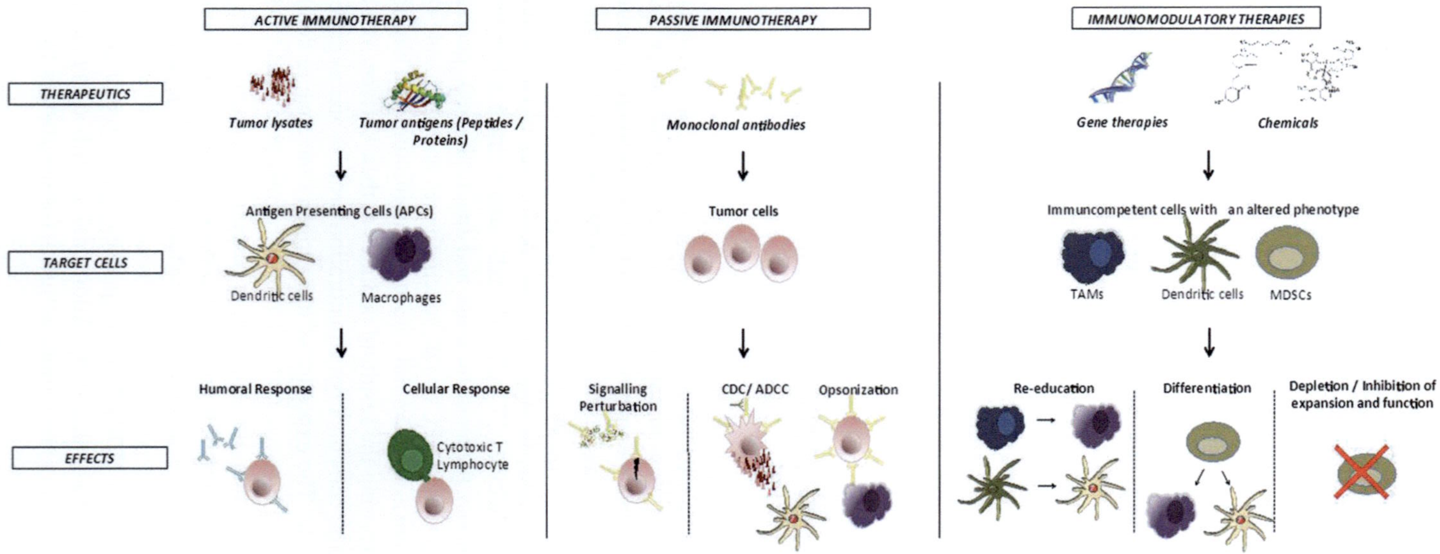

Fig. 8.1 Proposed therapies, target cells, and expected therapeutic effects in immunotherapy and immunomodulation of cancer. *CDC* complement-dependent cytotoxicity, *ADCC* antibody-dependent cell-mediated cytotoxicity, *TAMs* tumor associated macrophages, *MDSCs* myeloid-derived suppressor cells

8.2.2 Target Cells in Passive Immunotherapy: Cancer Cells (Bad Cells)

Passive immunotherapy is based on the delivery of monoclonal antibodies that interact with tumor antigens expressed at the surface of cancer cells, such as growth factor receptors, or with ligands to these receptors, which are secreted by cancer cells (Weiner et al. 2010; Shuptrine et al. 2012; Dienstmann et al. 2012). The main purpose of passive immunotherapy is therefore to block ligand binding and/or signaling through these receptors (signaling perturbation). Complement-dependent cytotoxicity (CDC) and antibody-dependent cell-mediated cytotoxicity (ADCC) are other important contributions to the anti-tumor activity of many therapeutic antibodies. In addition, the hypothesis of the induction of adaptive immunity responses has also been suggested by several groups, as recently reviewed by Shuptrine et al. (2012). Mechanistically, CDC and ADCC lead to tumor cell fragments and the release of tumor cell antigens, which could therefore contribute to efficient antigen presentation and cytotoxic T-cell activation (as described for ICD). Additionally, they can act as opsonins, triggering the phagocytosis of tumor cells by APCs.

8.2.3 Target Cells for Immunomodulatory Therapies: Altered Immunocompetent Cells (Bad Cells)

An altered phenotype of immune cells reverts their role in immunotherapy, since, instead of fighting tumor cells, they promote the conditions known to favor tumor growth and progression, thereby interfering with the efficacy of the therapy. These altered immunosuppressive cells have been identified in the primary tumor, as they form part of the tumor microenvironment (for illustration see Fig. 8.2), and are also

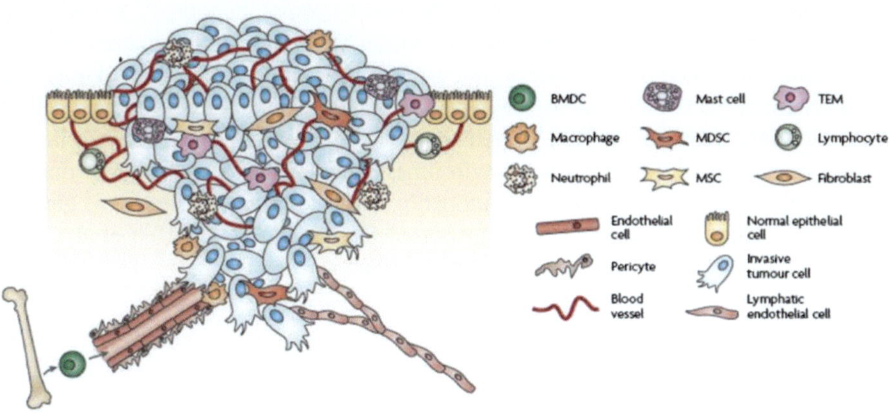

Fig. 8.2 Illustration of the tumor microenvironment. Tumor cells are surrounded by numerous cells including immune cells. *BMDCs* bone marrow-derived cells, *MDSCs* myeloid-derived suppressor cells, *TEMs* TIE2-expressing monocytes, *MSCs* mesenchymal stem cells. Adapted from Joyce and Pollard (2009) with permission

present in metastatic sites, as well as in secondary organs where metastases are subsequently generated, forming what is known as a pre-metastatic niche. Their complex interplay with the tumor is now recognized to play a critical role in the limited success of cancer therapy (Kaplan et al. 2006; Yaguchi et al. 2011; Hanahan and Weinberg 2011; Peinado et al. 2011; Vasievich and Huang 2011; Laoui et al. 2011; Montero and Diaz-Montero 2012; Junttila and de Sauvage 2013). Hence, targeted therapies to manipulate these cells are gaining increasing importance to improve the efficacy of anticancer therapies and address the process of metastasis.

8.2.3.1 Tumor-Associated Macrophages

TAMs represent the major inflammatory component of the stroma, and constitute a driving force in the initiation, proliferation, metastasis, angiogenesis, matrix remodeling, and suppression of adaptive immunity of various tumors (Pollard 2004; Leimgruber et al. 2009; Coffelt et al. 2009; Vasievich and Huang 2011). The tumor milieu strongly influences the recruitment of macrophages and shapes several of their features. Tumor-derived chemotactic factors (CC-chemokines, e.g., CCL2), macrophage colony stimulating factor (M-CSF), and vascular endothelial growth factor (VEGF) actively recruit circulating blood monocytes at the tumor site. CCL2 is probably the most frequently found CC chemokine in tumors, and its levels of expression correlate with the increased infiltration of macrophages. Other CC chemokines produced by tumors, which are related to CCL2, such as CCL7 and CCL8, have also been shown to recruit monocytes. Interestingly, CCL2 production has also been detected in TAMs, indicating the existence of an amplification loop for their recruitment (Sica et al. 2006; Solinas et al. 2009, 2010).

Macrophages are heterogeneous cells, functioning in distinct differentiation forms that can change their functional state in response to microenvironmental cues (Laoui et al. 2011). Indeed, macrophage heterogeneity has been simplified in the macrophage polarization concept where the two extreme phenotypes, the M1 and M2 macrophages, have distinct features. While M1 or pro-inflammatory macrophages are stimulated by bacterial products and T helper type 1 (Th1) cytokines, M2 or alternatively activated macrophages differentiate in microenvironments rich in Th2 cytokines (e.g., IL-4, IL-13) (Allavena and Mantovani 2012). Interestingly, TAMs present in the tumor microenvironment display hallmarks of M2 macrophages and establish a symbiotic relationship with the tumor by means of genetic information and cytokines that are transported to cancer cells (Coffelt et al. 2009; Solinas et al. 2009, 2010; Schmieder et al. 2012).

8.2.3.2 Dendritic Cells

Dendritic cells are professional APCs that play a sentinel role both in the peripheral organs and in the blood. Upon triggering by microbial products, by inflammatory cytokines, or by tissue damage, they move to secondary lymphoid organs where they meet T cells and initiate the inflammatory response. Tumor-infiltrating dendritic cells have also been identified and contribute to the complex mononuclear phagocyte infiltrate of tumors (Vicari et al. 2004; Vasievich and Huang 2011; Laoui et al. 2011; Benencia et al. 2012). Due to their potential capacity to activate tumor-specific T-cell responses, DCs play an important role in cancer immunosurveillance. Interestingly, circulating and tumor-infiltrating DCs from cancer patients appear to be phenotypically and functionally defective. The tumor environment seems to lack angiogenesis-inhibitory myeloid DCs, whereas angiogenesis stimulatory DCs, such as plasmacytoid DCs are present. Several tumor-derived factors have been shown to be responsible for systemic and local DCs defects. Also, DCs are able to actively down-regulate an immune response or to induce immune tolerance by influencing the activity of other cell types (Fricke and Gabrilovich 2006).

8.2.3.3 Myeloid-Derived Suppressor Cells

The name highlights the two strongest features of MDSCs, making clear both their origin and function. MDSCs are a heterogeneous population of myeloid-cell progenitors and precursors of myeloid cells, identified in peripheral lymphoid organs (mainly in the spleen), in peripheral blood, and in the tumor microenvironment. MDSCs have a common biological activity: suppressing the immune response in cancer (Gabrilovich and Nagaraj 2009; Solito et al. 2011; Montero and Diaz-Montero 2012; Gabrilovich et al. 2012).

Tumor cells mediate bone marrow stimulation and the production of MDSCs, and likewise emit signals to recruit these cells, which will be part of the tumor microenvironment. The activation of MDSCs results in the up-regulated expression of immunosuppressive factors such as arginase (ARG1), inducible nitric oxide synthase (iNOS), nitric oxide (NO), and reactive oxygen species (ROS). The interaction of these cells with the primary tumor causes changes in their microenvironment, leading to alterations in signaling cascades, thus enhancing tumor progression (Kaplan et al. 2007; Gabrilovich and Nagaraj 2009; Schmid and Varner 2010; Montero and Diaz-Montero 2012). Moreover, MDSCs play an essential role in tumor invasion and metastasis, and have been associated with the formation of pre-metastatic niches, through the induction of matrix metalloproteinases (MMPs) and chemoattractants that contribute to the generation of an ideal pre-metastatic environment for subsequent colonization by tumor cells (Wels et al. 2008; Peinado et al. 2008, 2011; Joyce and Pollard 2009).

8.3 The Value of Nanotechnology in Cancer Immunotherapy

In the area of cancer therapy, and also in the area of vaccination, nanocarriers have shown a specific promising behavior with regard to their ability to help active compounds to overcome biological barriers and reach their tissue and cell targets. Below, we summarize some key biological features that are critical for their use in cancer immunotherapy.

8.3.1 Tumor Targeting

Nowadays there is significant evidence of the potential for nanotechnology in the development of novel oncological treatments (Wang et al. 2012; Schroeder et al. 2012). Most of the research activity in the field has been oriented to the design of nanocarriers for the passive targeting of small hydrophobic anticancer drugs to the tumor environment. This passive targeting has been mainly attributed to the long circulation and enhanced penetration of the nanocarriers in highly inflamed and vascularized tumor tissue and has been the basis for the clinical development of a number of nanomedicines (Eifler and Thaxton 2011). Currently, there are several nanomedicines on the market which use a variety of nanocarriers, including liposomes (Myocet®, Doxil®, Doxisome®, Daunoxome®, Depocyt®), albumin nanoparticles (Abraxane®), polymeric micelles (Genexol-PM®), and polymer- (Oncaspar®) and monoclonal antibody-drug conjugates (Ontak®, Mylotarg®, Zevalin®, Bexxar®). There is also evidence of hundreds of clinical studies based on the application of nanotechnology to the development of novel anticancer therapies (www.clinicaltrials.gov). Overall, this passive targeting approach has led to a significant reduction in the toxicity of classical cytotoxic drugs and has opened new avenues in cancer treatment.

Nanoparticles can also be designed to overcome systemic barriers and reach the tumoral tissue in a specific manner. Indeed, research into cancer nanomedicine has shown the possibility of improving the efficacy of cytotoxic drugs using an active targeting strategy; by identifying those biomarkers that are aberrantly expressed at the surface of cancer cells, it is possible to prepare surface-functionalized nanoparticles with the ability to recognize cancer cells and, thus, with the potential ability to release their cargo into them. As an example, PLGA-PEG nanoparticles targeting PSMA (prostate specific membrane antigen), for the treatment of prostate cancer, are currently undergoing Phase II clinical trials (Hrkach et al. 2012).

8.3.2 Intratumoral Diffusion

The tumor microenvironment compromises the efficacy of cancer therapies. In particular, the high interstitial pressure and specific extracellular matrix (ECM) represent severe microenvironmental obstacles for the drugs to reach their targets.

During tumor development, fibroblastic stroma progression, also known as stromagenesis, results in significant changes in the surrounding mesenchyme, i.e., remodeling of the ECM, which are believed to promote tumorigenesis. The rapidly proliferating stromal fibroblasts (as well as other cells recruited to the tumor-associated site) undergo morphological changes and begin to express myofibroblastic markers. This myofibroblastic differentiation leads to an increased deposition of ECM proteins such as collagen-I and fibronectin, resulting in a significantly denser tumor microenvironment, which is less penetrable than the normal or non-tumorigenic extracellular environment. As a consequence, diffusion of chemotherapeutics following drug extravasation into the tumor microenvironment is highly restricted. Several experimental approaches have been described for enhancing the efficacy of treatments by improving their access to cancer cells, including the use of nanocarriers, which avoid unspecific interactions of the drugs with extracellular proteins/glycosaminoglycans and allow the co-encapsulation of tumor-associated stromal-depleting drugs to overcome resistance (Cukierman and Khan 2010; Holback and Yeo 2011). Nevertheless, the size threshold and surface properties that are appropriate for tumor penetration still need to be defined.

Nanoparticles could be designed to facilitate transport by adapting to each physiological barrier. As an example, gelatin-coated quantum dots were prepared with a mean size of 100 nm for preferential extravasation from leaky regions of the tumor vasculature. Gelatin would be degraded in the presence of metalloproteases (MMP-2) in the tumor surroundings, thus releasing small nanoparticles (quantum dots with a mean size of 10 nm) for improved penetration into the tumor parenchyma (Wong et al. 2011). The use of specific peptides, such as RGD and Lyp-1, called tumor-homing penetration peptides, has been proposed for nanoparticles' decoration (Ruoslahti 2012). Lipid micelles (Sugahara et al. 2009), PAMAM dendrimers (Waite and Roth 2011), and iron oxide nanoparticles (Roth et al. 2012) decorated with these peptides have shown improved accumulation and penetration into the tumor parenchyma after intravenous administration to mice, thus leading to an improved therapeutic efficacy. Positive results have also been reported for nanocomplexes of siRNA with tandem tumor-penetrating and membrane-translocating peptides (Ren et al. 2012).

8.3.3 Interaction with Immunocompetent Cells

Nanostructured materials have shown interesting potential as adjuvants in preventive and therapeutic vaccination. Particles in the viral size range are known to facilitate antigen uptake by APCs in peripheral tissues, and there is also evidence of their potential to target the antigen directly to the lymph node-associated T-cells. In addition to this targeting capacity, the association of antigens to specific nanocarriers offers attractive features to them, such as the ability to overcome mucosal barriers, protection against degradation, and controlled delivery inside the immunocompetent cells (Gonzalez-Aramundiz et al. 2012). These recognized properties are in alignment with the presence of antigen nanocarriers in the clinical development pathway as well as on the market, as illustrated in a number of review articles (Plebanski et al. 2013; Correia-Pinto et al. 2013). On the other hand, it is currently known that the inherent adjuvanticity of nanocarriers, associated with their interaction with the APCs, can be further enhanced or modulated by combination with immune response modulators, such as TLRs ligands or cytokines (Schlosser et al. 2008; Hamdy et al. 2008; Salem and Weiner 2009; Flemming 2011; Nguyen-Hoai et al. 2012; Rosenthal et al. 2014). As an example, chitosan nanocapsules that associate the recombinant hepatitis B surface antigen and co-encapsulate the TLR7 agonist imiquimod provided a protective effect upon intranasal administration to mice (Vicente et al. 2013b).

Nanostructures can be designed to improve their interaction with the immune system, in such a way that it is possible to obtain the desired immunomodulation. The great diversity of nanostructures that can be synthesized, in terms of composition and structural organization, makes it difficult to generalize about the behavior that a specific prototype might have with regard to its interaction with the immune system. In terms of composition, biodegradable biomaterials with a good safety record have widely been explored for the development of nanovaccines. Poly(lactic-co-glycolic acid) (PLGA) and chitosan nanoparticles have received a great deal of attention for systemic and mucosal vaccination and can elicit antibody and cytotoxic T-cell responses (Csaba et al. 2009; Foged et al. 2012; Ma et al. 2011; Silva et al. 2013b; Vicente et al. 2013a; Wen et al. 2011; Zhang et al. 2011). With respect to physicochemical properties, several studies have been carried out to understand the importance of particle size and superficial charge. Even though there is still some controversy, it is generally assumed that small and large nanoparticles reach the lymph nodes by different mechanisms and interact differently with APCs, with the modulation of these interactions being particularly important for the induction of combined cellular and humoral responses (Shen et al. 2006; Xiang et al. 2006; Li et al. 2011). On the other hand, a cationic surface has been indicated to positively affect particle interactions with immunocompetent cells (Vila et al. 2004; Waeckerle-Men and Groettrup 2005; Prego et al. 2010; Li et al. 2013; Vicente et al. 2013a; Correia-Pinto et al. 2013; Silva et al. 2013a), although this might also represent a hurdle for particle diffusion across specific bio-environments, i.e., intratumoral or subcutaneous tissue.

Overall, the lessons learned from the use of nanostructured biomaterials for vaccination provide us with key elements for the control of the way in which immunoactive molecules can be presented to the immune system and, thus, for their use in oncological immunotherapy. Although there is not yet a recipe for the most appropriate nanocarrier for a specific immunotherapy, well-defined comparative studies are expected to drive progress forward towards the development of novel nano-immunotherapies.

8.4 Specific Applications of Nanotechnology in Immunomodulation and Immunotherapy of Cancer

Nanotechnology approaches have already been explored for the different modalities of immunotherapy and immunomodulation of cancer. In this section, we attempt to summarize the specific achievements in the field, which may help to define the pathway for future developments of novel cancer therapies.

8.4.1 Active Immunotherapy

As indicated above, therapeutic vaccination involves priming of the immune system in order to generate an immune response against a tumor-associated antigen, and seeks to create a durable response that can protect against tumor progression and recurrence (Winter et al. 2011; Boghossian and Von-Delwig 2012; De Pas et al. 2012; Goldinger et al. 2012). Therapeutic vaccination can be proposed to treat either metastatic patients in an advanced setting or disease-free patients in an adjuvant setting. Even if the ideal setting might be the adjuvant one, it is currently accepted that therapeutic vaccination may also be beneficial in advanced disease in combination with other therapeutic options. This as yet hypothetical benefit is justified by the increasing number of proposed vaccines for the treatment of many types of cancers, which are expected to be widely studied in the coming years (Boghossian and Von-Delwig 2012; Vanneman and Dranoff 2012; De Pas et al. 2012; Kvistborg et al. 2013; Silva et al. 2013a). It is important to highlight that Sipuleucel-T (Provenge®), an autologous DC-based vaccine targeting the antigen prostatic acid phosphatase (PAP), is the first vaccine approved for cancer therapy in advanced prostate cancer (autologous DCs are removed from patients, exposed to PAP, and then re-infused back into the body). The approval of this novel vaccine represents a key milestone in the new era of active cancer immunotherapy.

As expected and justified by the comments in Sect. 3, the early development of therapeutic vaccines has been associated with the use of antigen nanocarriers, in particular liposomes. Indeed, among the most advanced prototypes in clinical trials, there are three liposomal formulations, namely Emepepimut-S (Stimuvax®), MAGE-A3, and PRAME, all of which are indicated for non-small cell lung cancer (De Pas et al. 2012).

On the other hand, polymeric nanostructures are also gaining increasing attention in the development of therapeutic vaccines (De Temmerman et al. 2011; Silva et al. 2013a). In this sense, the knowledge accumulated from the use of poly(lactic-*co*-glycolic acid) (PLGA)-based micro- and nanoparticles in vaccination for infectious diseases has led researchers to anticipate their potential benefits in cancer immunotherapy (Csaba et al. 2009; Zolnik et al. 2010; Gonzalez-Aramundiz et al. 2012). This potential has been made visible by the presence of the first formulation of cancer vaccines based on PLGA microparticles under clinical development for the specific indication of breast cancer (intradermal or subcutaneous vaccinations of HER-2 derived p369-377 peptide, incorporated into PLGA microparticles with adjuvant sargramostim, in patients with stage III or IV HER-2-expressing cancers) (http://clinicaltrials.gov/ct2/show/NCT00005023).

Several studies have shown the possibility of incorporating tumor antigens and tumor lysates within PLGA nanoparticles. For example, following the encapsulation of peptides such as MART-1 (Ma et al. 2011) and MHC class Ia, Ib, and class II-restricted peptides (Ma et al. 2012) into PLGA nanoparticles, Ma et al. (2011) found a specific CD8+ T cell response and an antitumor response in prostate adenocarcinoma challenged mice (TRAMP-C2) after intraperitoneal immunization. PLGA nanostructures have also been engineered to provide a prolonged release of synthetic long peptides (SLPs), a new class of peptide antigens that cover the entire amino acid sequence of tumor-associated protein antigens (Silva et al. 2013b). The in vivo efficacy of this approach remains to be investigated. In other instances, PLGA nanostructures were designed for encapsulation of the whole tumor lysates obtained from immortalized cell lines and also from patients (Hanlon et al. 2011; Prasad et al. 2011, 2012). Results show that these nanoparticles were able to stimulate isolated DCs and activate a CD8+ T-cell response.

A double antigen/adjuvant encapsulation approach has been explored for specific antigens with additional adjuvants, TLRs agonists. For example, intradermal immunization of a combination of the murine melanoma antigenic peptide (p180-188) derived from the tyrosinase-related protein 2 (TRP2) with monophosphoryl lipid A (a TLR4 ligand), encapsulated into PLGA nanoparticles, significantly delayed the growth of subcutaneously inoculated B16 melanoma cells in mice in a prophylactic setting (Zhang et al. 2011). Indeed, a combined treatment with these nanosystems (intradermal injection) plus recombinant IFN-γ (intraperitoneal injection from day 1 after the tumor inoculation), for the prevention of tumor escape via MHC class I down-regulation, was highly effective for controlling tumor growth in challenged mice (B16-F10 melanoma tumor cells). Similarly, the co-encapsulation of a novel synthetic analogue of lipid A (7-acyl lipid A) with the antigenic peptide TRP2 into PLGA nanoparticles produced positive responses in mice (Hamdy et al. 2008). More precisely, following the subcutaneous administration of this novel formulation to tumor-bearing mice (B16-F10 melanoma tumor cells), the authors found increased levels of pro-inflammatory T helper 1 (Th1)-related cytokines and decreased levels of VEGF, which were all related to an enhanced therapeutic anti-tumor effect.

Acrylate polymers have also received attention as carriers for cancer vaccination. A study has made use of a polyacrylate star-polymer for the delivery of the HPV16 E7 peptide, an antigen against cervical cancer (Liu et al. 2013a). The same antigen was co-encapsulated with CpG into poly(ethylene glycol)-polylactide (PEG-PLA)-stabilized squalene oil-in-water emulsions (Chen et al. 2014; Song et al. 2014). In all cases, enhanced antigen-specific T-cell responses and the ability of the nanovaccines to stop tumor growth were shown.

Polyamino acids, such as poly-glutamic acid (PGA) and poly-L-lysine (PLL), have also been used for the encapsulation of cancer antigens. Self-assembled amphiphilic nanoparticles composed of a co-polymer of PGA and L-phenylalanine ethyl ester (Yoshikawa et al. 2008a, b) and PGA/benzalkonium chloride nanoparticles (Kurosaki et al. 2012) were both used for the encapsulation of ovalbumin. The antigen-specific CTL responses and anti-tumor effects observed by subcutaneous immunization into either mice bearing melanoma B16-OVA cells (Yoshikawa et al. 2008b) or mice challenged with intradermally (Yoshikawa et al. 2008a) or subcutaneously (Kurosaki et al. 2012) injected lymphoma E.G7-OVA cells led the authors to conclude about their potential as anticancer vaccines. This potential was subsequently illustrated for the EphA2-derived peptide (Yamaguchi et al. 2010). The results show a delayed growth of colon carcinoma MC38 cells (EphA2-positive), injected in the liver when mice were previously immunized by intraperitoneal administration of EphA2-loaded PGA-related nanoparticles (Yamaguchi et al. 2010). Interestingly, these nanoparticles have also been proposed for needle-free mucosal vaccination, in view of the antitumor efficacy observed after intranasal immunization (Matsuo et al. 2011).

A different vaccination approach that is expected to greatly benefit from the use of polymer-based nanocarriers is that based on genetic vaccines. The polyamino acid poly-L-lysine (PLL) has already been described for the association of genetic vaccines (Tang et al. 2007; Perche et al. 2011). PLL linked to OxMan and RedMan conjugates (mannan conjugated to Mucin 1 fusion protein in the oxidized form, OxMan, or in the reduced form, RedMan) were used as carriers for pDNA (Tang et al. 2007). After intradermal immunization with ovalbumin encoding pDNA associated with the complexes, mice were challenged with a subcutaneous injection of lymphoma EG7-OVA cells. The results were positive in terms of tumor growth for both OxMan-MUC1 and RedMan-MUC1 PLL/pDNA nanoparticles (OxMan-PLL-DNA and RedMan-PLL-Man conjugates). Another example refers to the preparation of mannosylated and histidylated PLL lipopolyplexes with mRNA encoding MART-1 melanoma antigen (Perche et al. 2011). Mice immunized with this formulation by intravenous injection had a prolonged survival after they were challenged with B16-F10 melanoma cells.

The polysaccharide chitosan (CS) has also been extensively studied as a biomaterial for anticancer genetic vaccination (Morille et al. 2008; Heuking et al. 2009; Chen et al. 2012; Garcia-Fuentes and Alonso 2012; Liu et al. 2013b; Buschmann et al. 2013). A CS-polyethylenimine (CS-PEI) conjugate has been reported for the transfection of DCs with a plasmid encoding gp100, a melanoma-associated antigen protein. The transfected cells were used for intradermal immunization in mice bearing melanoma orthographs (Chen et al. 2012). Unfortunately, this dendritic cell-vaccination approach did not lead to satisfactory results in terms of tumor volume and growth. Alginic acid-coated CS nanoparticles have also been described as carriers for the oral delivery of a DNA vaccine for breast cancer (Liu et al. 2013a). In this case, DNA encoding legumain, an asparaginyl endopeptidase, which is significantly overexpressed on TAMs and tumor cells, was encapsulated into the nanostructures. The results observed upon oral vaccination with this nanosystem provided evidence of the inhibition of tumor growth and the increase in survival rate in an orthotopic mouse model of breast cancer (Liu et al. 2013a). In another study, a CS derivative, 6-*O*-carboxymethyl-*N*,*N*,*N*-trimethyl-chitosan (CM-TMC), was selected to complex a model plasmid DNA expressing green fluorescent protein (GFP) and form nanostructures, which were additionally decorated with a TLR2 agonist, Pam3 Cys. The resulting nanostructures showed the capacity to stimulate human macrophages in vitro (IL-8 release) due to the TLR2 agonist present on their surface (Heuking et al. 2009). With respect to the association of peptides for vaccination, CS nanoparticles decorated with the antigen K-ras mutant peptide (K-ras+peptide) have been developed for the sensitization of DCs for pancreatic cancer vaccination (Tan et al. 2011). DCs loaded with different antigens (whole tumor cell antigen, K-ras+peptide, and K-ras+peptide-loaded nanoparticles) were used to induce CTLs, which were subsequently injected intraperitoneally to mice bearing pancreatic xenografts. In all cases, CTLs were able to kill tumor cells, and were more effective than those induced by DCs loaded with whole tumor lysates. K-ras antigen showed greater specificity for the target cells, and it was more effective when linked to the nanoparticles.

In a different formulation, a DNA vaccine encoding calreticulin together with the HPV16 E7 peptide antigen was associated with poly(methyl methacrylate) (PMMA) nanoparticles (Lou et al. 2009). The in vivo responses following intradermal administration were positive in terms of tumor growth restriction.

8.4.2 *Passive Immunotherapy*

Passive immunotherapy consists in the delivery of therapeutic monoclonal antibodies, which are designed to recognize tumor antigens expressed at the surface of cancer cells. The most well-known antitumoral monoclonal antibodies have been designed to interfere with the activity of tyrosine kinase (TK) receptors present in cancer cells, such as the epidermal growth factor receptor (EGFR) or the insulin growth factor receptor-1 (IGFR-1), as well as to TK ligands secreted by tumor cells, such as VEGF. By binding to either the extracellular domain of the receptor or the ligand itself, these monoclonal antibodies prevent receptor-ligand interactions or receptor activation and signaling, which affects tumor growth and progression (Weiner et al. 2010; Shuptrine et al. 2012; Dienstmann et al. 2012). Other antibodies have the potential to trigger tumor death through mechanisms of antibody-dependent cellular cytotoxicity; they attach to surface antigens on cancer cells, acting as a marker for the attraction of immune cells to destroy them (Alderson and Sondel 2011; Seidel et al. 2013; Shuptrine et al. 2012). Other mechanisms that have been described to account for the antitumoral activity of mAbs are complement-dependent cellular cytotoxicity, when two or more mAbs bind to a cell and the classical complement pathway is activated, along with the induction of adaptive immunity (Shuptrine et al. 2012).

The use of monoclonal antibodies (mAb) has been established as one of the most promising anticancer therapeutic strategies for both solid tumor and hematologic malignancy therapies, with more than 10 mAbs having already been approved by the US and EU regulatory agencies, as a single agent or in combination with radio- or chemotherapy (www.nccn.org) (Galluzzi et al. 2012b; Shuptrine et al. 2012; Scott et al. 2012; Dienstmann et al. 2012). However, it is broadly assumed that further improvements of this type of therapy are still needed. Unfortunately, mechanisms of resistance (tumor-associated resistance and host-associated resistance) have already been identified, and research that is already underway is aiming to circumvent such an important limitation (Villamor et al. 2003; Reslan et al. 2009).

Nanotechnology might offer the opportunity to potentially circumvent resistance and also to control the delivery of mAbs, hence increasing the efficacy of mAb-based therapies (Bhattacharyya et al. 2010; Owen et al. 2013). In fact, releasing antibodies with high spatial and temporal accuracy could result in low systemic off-target effects and improved therapeutic effects. Interestingly, despite this, only a few isolated reports have been found to describe the use of nanosystems for this purpose. Lee et al. (2009) reported the use of self-assembled micelles made of (poly{(*N*-methyldiethanolamine sebacate)-*co*-[(cholesteryl oxocarbonylamino ethyl) methyl bis(ethylene) ammonium bromide] sebacate), for the simultaneous delivery of paclitaxel and Herceptin, a monoclonal antibody that targets human epidermal growth factor receptor type 2 (HER2). Similarly, Liu and Feng (2011) proposed the association of docetaxel and Herceptin to polylactide-D-α-tocopheryl polyethylene glycol succinate (PLA-TPGS) nanoparticles, and observed a synergistic effect in cell culture. A step forward with this type of PLA-TPGS nanostructure involved the additional incorporation of iron oxide nanoparticles for hyperthermia therapy (Mi et al. 2012). These types of nanocarriers were further optimized to precisely modulate the surface density of Herceptin (Zhao et al. 2012).

Despite the few studies on the application of nanotechnology for passive immunotherapy, monoclonal antibodies have been widely used for specific targeting of nanostructures to tumor cells overexpressing the receptors, for a targeted delivery approach (Fay and Scott 2011; Altintas et al. 2012). For example, surface decorated albumin nanoparticles (Taheri et al. 2012), lipid nanocapsules (Sánchez-Moreno et al. 2013), poly(vinyl alcohol) nanocapsules (Chiang et al. 2014), polymalic acid (PMLA) nanobioconjugates (Ding et al. 2013), and P(LA-*co*-TMCC)-*g*-PEG-furan micelles (Owen et al. 2013) all incorporate monoclonal antibodies for improved interactions with cells expressing HER2. However, the contribution of the monoclonal antibodies themselves to the efficacy of the therapy has not been taken into consideration, thus disregarding any potential therapeutic effect due to the mAbs in combination with the encapsulated antitumoral drugs.

8.4.3 Targeting Immunosuppressive Cells

As indicated above, cancer therapeutic strategies have been classically directed towards tumor cells; however, in recent years, the key recognized role of immunosuppressive cells (TAMs, DCs, and MDSCs) (see Sect. 2) in tumor progression has led to the search for novel therapies specifically targeted to these cells.

In the case of TAMs and DCs, the proposed therapeutic interventions aim at interfering with the recruitment of immune cells to the tumor surroundings and with their conversion to immunosuppressive cells, and at re-educating these cells towards an immunologically active phenotype. For example, the inhibition of certain transcription factors such as c-Myc, NK-B, and Stat-3, or immunosuppressive molecules such as the B7 family, has been described for curtailing the pro-tumoral functions of TAMs and DCs and stimulating the immune system (Zou and Chen 2008; Hagemann et al. 2008; Yu et al. 2009; Pello et al. 2012). Delivery of microRNAs (miRs), such as miR-155, has also been proposed for reverting the immunosuppressive tumor microenvironment (Lindsay 2008; He et al. 2009; Ha 2011; Cubillos-Ruiz et al. 2012; Cai et al. 2012). Apart from gene therapies, certain chemotherapeutics, such as curcumin and trabectedin, have also been evaluated for their ability to re-educate and deplete TAMs (Germano et al. 2013; Zhang et al. 2013). These and other therapeutic strategies are described with more details in the previous chapter.

In the particular case of MDSCs, several therapeutic strategies have been described for targeting this cell compartment and restoring antitumor immunity. In brief, these strategies have been oriented (1) to promote MDSCs differentiation into mature APCs that do not have suppressive abilities (e.g., all-trans retinoic acid and Vitamin D), (2) to inhibit MDSCs expansion (e.g., tyrosine kinase inhibitors) and function (e.g., cyclooxygenase 2 inhibitors, phosphodiesterase 5 inhibitors, and ROS inhibitors), and (3) to deplete MDSCs (e.g., chemotherapeutic agents such as gemcitabine, curcumin, docetaxel, and 5-Fu) (Gabrilovich and Nagaraj 2009; Kao et al. 2011; Gabrilovich et al. 2012) and are extensively reviewed in the previous chapter.

Overall, these therapeutic options are viewed as novel anticancer immunotherapies, and research involving the use of nanotechnology to manipulate these cells is in a very early stage, as disclosed in the next paragraphs and detailed in Table 8.1.

8.4.3.1 Nanocarriers Targeted to TAMs

Despite the enormous potential of targeting TAMs as a novel immunotherapeutic strategy, to date there are only a few reports dealing with the development of nanocarriers specifically engineered for this purpose. The use of biomarkers expressed at the surface of TAMs as a target has been the focus of attention of nanotechnologists. To avoid recognition by peripheral macrophages, some studies have dealt with the shield of surface-decorated nanocarriers using polymers that degrade due to environmental changes in the tumor microenvironment, such as a more acidic pH (Huang et al. 2012; Zhu et al. 2013). Biomarkers aberrantly expressed at the surface of TAMs include the mannose receptor (Movahedi et al. 2012; Locke et al. 2012; Zhu et al. 2013), the galactose-type lectin (Mgl) receptor (Huang et al. 2012), the folate receptor beta (Nagai et al. 2009), scavenger receptors such as CD163 (Etzerodt et al. 2012), and legumain (Luo et al. 2006; Gomez-Cabrero et al. 2013; Zhang et al. 2013).

With the aim of preventing the nanoparticles recognition by peripheral macrophages, PLGA nanoparticles decorated with mannose ligands were shielded with PEG moieties attached via a linker that degrades at the acidic pH of the tumor microenvironment (Zhu et al. 2013). The biodistribution results obtained after intravenous administration to mice have pointed out the potential of this strategy; however, no efficacy data have been reported so far.

Similarly, galactosylated cationic dextran nanocomplexes, engineered for oligonucleotide delivery to MgI receptor expressing TAMs (Huang et al. 2012), were coated with a second material, pH-sensitive PEG-histidine-modified alginate (PHA), in order to trigger the delivery specifically into the acidic tumor microenvironment. The proposed therapy was designed to inhibit IL-10 signaling, to switch the phenotype of TAMs from M2 pro-tumoral to M1 pro-inflammatory, and simultaneously activate macrophages and dendritic cells via TLR9. The therapeutic efficacy of this approach was reported in terms of tumor growth in an experimental mice model of hepatoma after intravenous administration.

Other examples deal with the use of liposomes for targeting legumain, an asparaginyl endopeptidase that is overexpressed on TAMs. In particular, curcumin-loaded legumain-targeted liposomes (Zhang et al. 2013) have demonstrated the ability to revert M2 polarization to an M1 phenotype, which translates in a decreased tumor cell proliferation, migration, and invasion and increased apoptosis, as shown in a mice model of metastatic breast cancer after intravenous administration. Legumain-targeted liposomes for the delivery of an imidazole derivative of the synthetic oleanane triterpenoid (CDDO-Im), an inhibitor of Stat3 activation, have also been recently reported to improve the antitumor effects of a HER-2 DNA vaccine (Liao et al. 2011).

Table 8.1 Nanotherapies targeted to immunosuppressor cells

Nanocarriers	Ligand	Encapsulated molecule	Target cells	Tumor model	Observations	References
PEG-shielded mannosylated PLGA nanoparticles	Mannose	FITC	TAMs	Melanoma (B16-F10 tumor-bearing mice)	Co-localization of the nanoparticles with TAMs	Zhu et al. (2013)
PEG-histidine-modified alginate coated galactosylated cationic dextran nanocomplexes	Galactose	Oligonucleotides	TAMs	Hepatoma (hepa 1–6 tumor-bearing mice)	Accumulation of oligonucleotides in TAMs Decreased IL-10 and IL-10-RA expression Re-education of TAMs to a M1 proinflammatory phenotype Decreased tumor growth	Huang et al. (2012)
Legumain-targeted liposomes	Legumain inhibitor	Hydrazinocurcumin	TAMs	Breast cancer (4 T1 co-injected with RAW264.7 in mice)	Re-education of TAMs to a M1 proinflammatory phenotype Decreased tumor growth	Zhang et al. (2013)
Legumain-targeted liposomes	Legumain inhibitor RR-11a	Imidazole derivative (CDDO-Im)	TAMs	Breast cancer (4TO7 cells or MMTV-Neu primary tumor cells orthotopically injected in mice)	Inhibition of Stat-3 activation Decreased infiltration of TAMs Decreased tumor growth in combination with the antitumor HER2 DNA vaccine	Liao et al. (2011)
PEI/PEI-StA nanocomplexes loaded into PLGA nanoparticles	–	siRNA	DCs	–	Stat-3 down-regulation Restored ability of DCs to mediate a CTL response	Alshamsan et al. (2010)
PEI-StA nanocomplexes		siRNA	DCs	Melanoma (B16 tumor-bearing mice)	Stat-3 down-regulation Restored ability of DCs to mediate a CTL response Decreased tumor growth	Alshamsan et al. (2011)

(continued)

Table 8.1 (continued)

Nanocarriers	Ligand	Encapsulated molecule	Target cells	Tumor model	Observations	References
PEI nanocomplexes	–	siRNA	DCs	Ovarian carcinoma (ID8-Defb29/Vegf-A tumor-bearing mice)	PD-L1 downregulation Stimulation of TLR3, TLR5, and TLR7 Restored ability of DCs to mediate a CTL response Improved antitumor immune response	Cubillos-Ruiz et al. (2009)
PEI nanocomplexes	–	miRNA mimics	DCs	Ovarian carcinoma (ID8-Defb29/Vegf-A tumor-bearing mice)	Augmentation of miR-155 activity an silencing of multiple immunosuppressive mediators Restored ability of DCs to mediate a CTL response Decreased tumor growth	Cubillos-Ruiz et al. (2012)
Very small size proteoliposomes—VSSP	–	Vaccine adjuvant	MSDCs	Lymphoma/fibrosarcoma (EG.7-OVA/MCA203 tumor-bearing mice)	Modulation of the suppressive activity of tumor-derived MDSCs by inducing their differentiation Decrease in tumor growth when combined with a tumor associated antigen	Fernández et al. (2011) and Fernández et al. (2013)
Ultrasmall Pluronic-stabilized poly(propylene sulfide) nanoparticles	–	Dy649	MSDCs	Lymphoma (EG.7-OVA tumor-bearing mice)	Preferential accumulation of the nanoparticles in MDSCs in draining nodes, spleen, and tumor, after intradermal administration	Kourtis et al. (2013)

TAMs tumor associated macrophages, *DCs* dendritic cells, *MDSCs* myeloid derived suppressor cells, *PEG* polyethylene glycol, *PLGA* poly(lactic-*co*-glycolic acid), *PEI* polyethylenimine, *StA* stearic acid, *CTL* cytotoxic T lymphocyte

8.4.3.2 Nanocarriers Targeted to DCs

In the case of DCs, most of the reported works refer to the use of gene therapies. Polyethylenimine (PEI) and stearic acid-modified polyethylenimine (PEI-StA) were selected for preparation of siRNA complexes to mediate Stat3 down-regulation in DCs (Alshamsan et al. 2010). These complexes were additionally encapsulated into PLGA nanoparticles. Stat-3, one of the members of the Stat family (cytoplasmic transcription factors that are key mediators of cytokine and growth factor signaling pathways), has emerged as a negative regulator of inflammatory responses (Kortylewski et al. 2005, 2009; Kujawski et al. 2008; Lee et al. 2011). It is overexpressed in tumor cells, stromal cells, and infiltrating hematopoietic cells of many types of tumors, contributing to tumor survival, proliferation, and dissemination. Stat-3 activation in TAMs and DCs has a profound anti-inflammatory effect by preventing their complete maturation and blocking their ability to produce many pro-inflammatory cytokines such as IL-12 (Yu et al. 2009). In vitro studies have shown the efficacy of PEI-based gene nanocarriers to decrease the expression of Stat3 and restore DCs functionality (Alshamsan et al. 2010). A posterior in vivo study with PEI-StA polyplexes (intratumoral injection) aimed to assess the immunological impact, DCs activation, and therapeutic response of this approach, using cell culture models and mice bearing melanoma allografts for that purpose (Alshamsan et al. 2011).

PEI polyplexes have also been reported for the delivery of siRNA to down-regulate the programmed cell death ligand 1, PD-L1 (Cubillos-Ruiz et al. 2009). Binding of programmed death-1 (PD-1) receptor on activated T cells to its ligand PD-L1, which is expressed on DCs, is a key negative regulator of T-cell activity (Pen et al. 2014). The aim of the study was to reprogram DCs from an immunosuppressive phenotype to efficient APCs that activate tumor-reactive lymphocytes and exerted direct tumoricidal activity in aggressive ovarian carcinoma-bearing hosts (Cubillos-Ruiz et al. 2009). The authors reported that PEI-based nanoparticles were preferentially and avidly engulfed by CD11c+ DCs at tumor locations after intravenous delivery and mediated the maturation of tumor-associated DCs.

Lastly, PEI nanocomplexes were also evaluated for carrying oligonucleotide duplexes mimicking the bulged structure of endogenous pre-miRNA. After intravenous administration to a model of ovarian cancer, the authors reported a dramatic augmentation of miR-155 activity and silencing of multiple immunosuppressive mediators (Cubillos-Ruiz et al. 2012). Accordingly, tumor-infiltrating DCs were transformed from immunosuppressive to highly immunostimulatory cells capable of triggering potent antitumor responses that abrogated the progression of established ovarian cancers.

8.4.3.3 Nanocarriers Targeted to MDSCs

Despite the potential of nanotechnology to develop the therapeutic options available to target MDSCs, to the best of our knowledge, only two examples can be provided so far.

One example refers to a vaccine nanoparticulate adjuvant based on the combination of outer membrane vesicls (OMPs) from *Neisseria meningitidis* with GM3 ganglioside (very small size proteoliposomes—VSSP), which is currently under investigation in clinical trials as part of the formulation of several cancer vaccine candidates (renal carcinoma, breast cancer, prostate cancer, and cervical intraepithelial neoplasia grade III). VSSP were proved to modulate the suppressive activity of tumor-derived MDSCs in mice by inducing their differentiation to APCs (Fernández et al. 2011, 2013), and on the basis of these results, the development of a vaccine able to target these immunosuppressive cells whilst eliciting an immune response towards the associated antigen has been proposed.

A second example relates to the use of ultra-small pluronic-stabilized poly(propylene sulfide) nanoparticles, with a mean size of 30 nm, which preferentially accumulate in MDSCs in draining nodes, spleen, and tumor, after intradermal administration to immunocompetent mice bearing lymphoma and melanoma tumors (Kourtis et al. 2013). Despite the fact that therapeutic data have not yet been reported, this formulation seems very promising for the delivery of immunomodulatory therapies to MDSCs.

8.5 Conclusions

The advancements in immunology and cancer have made clear the complexity of their relationship and the necessity to tune the immune system in order to improve the effectiveness of current immunotherapies. In this review, we have shown some examples that highlight the potential of nanomedicines for the development of improved active and passive immunotherapies to treat cancer. Additionally, we focused our attention in an emerging field, the design of novel therapeutic strategies to deal with immunosuppressive cells and the tumor microenvironment. In conclusion, nanotechnology could play a critical role for generating novel immunotherapeutics in the coming years, since nanomedicines could indeed be tailored to manipulate two arms of the immune system: promoting the access of antigens and humoral and cellular responses, and manipulating the tumor microenvironment by targeting immunosuppressor cells, i.e., TAMs, DCs, and MDSCs.

Acknowledgments The authors thank the support given by the Carlos III Health Institute, Spain (Miguel Servet Program CP12/03150), EuroNanoMed 2013 (053 NICHE), and U54-CA151884.

References

Alderson KL, Sondel PM (2011) Clinical cancer therapy by NK cells via antibody-dependent cell-mediated cytotoxicity. J Biomed Biotechnol 1–7. doi:10.1155/2011/379123

Allavena P, Mantovani A (2012) Immunology in the clinic review series; focus on cancer: tumour-associated macrophages: undisputed stars of the inflammatory tumour microenvironment. Clin Exp Immunol 167:195–205. doi:10.1111/j.1365-2249.2011.04515.x

Alshamsan A, Haddadi A, Hamdy S et al (2010) STAT3 silencing in dendritic cells by siRNA polyplexes encapsulated in PLGA nanoparticles for the modulation of anticancer immune response. Mol Pharm 7:1643–1654. doi:10.1021/mp100067u

Alshamsan A, Hamdy S, Haddadi A et al (2011) STAT3 knockdown in B16 melanoma by siRNA lipopolyplexes induces bystander immune response in vitro and in vivo. Transl Oncol 4: 178–188. doi:10.1593/tlo.11100

Altintas I, Kok RJ, Schiffelers RM (2012) Targeting epidermal growth factor receptor in tumors: from conventional monoclonal antibodies via heavy chain-only antibodies to nanobodies. Eur J Pharm Sci 45:399–407. doi:10.1016/j.ejps.2011.10.015

Benencia F, Sprague L, McGinty J et al (2012) Dendritic cells the tumor microenvironment and the challenges for an effective antitumor vaccination. J Biomed Biotechnol 2012:1–15. doi:10.1155/2012/425476

Bhattacharyya S, Bhattacharya R, Curley S et al (2010) Nanoconjugation modulates the trafficking and mechanism of antibody induced receptor endocytosis. Proc Natl Acad Sci U S A 107:14541–14546. doi:10.1073/pnas.1006507107

Boghossian S, Von-Delwig A (2012) Tumour vaccines, monoclonals, proteins or whole cell therapies. J Vaccines Vaccin 3–12. doi:10.4172/2157-7560.S1-003

Buschmann MD, Merzouki A, Lavertu M et al (2013) Chitosans for delivery of nucleic acids. Adv Drug Deliv Rev 65:1234–1270. doi:10.1016/j.addr.2013.07.005

Cai X, Yin Y, Li N et al (2012) Re-polarization of tumor-associated macrophages to pro-inflammatory M1 macrophages by microRNA-155. J Mol Cell Biol 4:341–343. doi:10.1093/jmcb/mjs044

Chen Y-Z, Yao X-L, Ruan G-X et al (2012) Gene-carried chitosan-linked polyethylenimine induced high gene transfection efficiency on dendritic cells. Biotechnol Appl Biochem 59:346–352. doi:10.1002/bab.1036

Chen W-L, Liu S-J, Leng C-H et al (2014) Disintegration and cancer immunotherapy efficacy of a squalane-in-water delivery system emulsified by bioresorbable poly(ethylene glycol)-*block-polylactide*. Biomaterials 35:1686–1695. doi:10.1016/j.biomaterials.2013.11.004

Chiang C-S, Hu S-H, Liao B-J et al (2014) Enhancement of cancer therapy efficacy by trastuzumab-conjugated and pH-sensitive nanocapsules with the simultaneous encapsulation of hydrophilic and hydrophobic compounds. Nanomedicine 10:99–107. doi:10.1016/j.nano.2013.07.009

Coffelt SB, Hughes R, Lewis CE (2009) Tumor-associated macrophages: effectors of angiogenesis and tumor progression. Biochim Biophys Acta 1796:11–18. doi:10.1016/j.bbcan.2009.02.004

Colotta F, Allavena P, Sica A et al (2009) Cancer-related inflammation, the seventh hallmark of cancer: links to genetic instability. Carcinogenesis 30:1073–1081. doi:10.1093/carcin/bgp127

Correia-Pinto JF, Csaba N, Alonso MJ (2013) Vaccine delivery carriers: insights and future perspectives. Int J Pharm 440:27–38. doi:10.1016/j.ijpharm.2012.04.047

Csaba N, Garcia-Fuentes M, Alonso MJ (2009) Nanoparticles for nasal vaccination. Adv Drug Deliv Rev 61:140–157. doi:10.1016/j.addr.2008.09.005

Cubillos-Ruiz JR, Engle X, Scarlett UK et al (2009) Polyethylenimine-based siRNA nanocomplexes reprogram tumor-associated dendritic cells via TLR5 to elicit therapeutic antitumor immunity. J Clin Invest 119:2231–2244. doi:10.1172/JCI37716.a

Cubillos-Ruiz JR, Baird JR, Tesone AJ et al (2012) Reprogramming tumor-associated dendritic cells in vivo using miRNA mimetics triggers protective immunity against ovarian cancer. Cancer Res 72:1683–1693. doi:10.1158/0008-5472.CAN-11-3160

Cukierman E, Khan DR (2010) The benefits and challenges associated with the use of drug delivery systems in cancer therapy. Biochem Pharmacol 80:762–770. doi:10.1016/j.bcp.2010.04.020

De Pas T, Giovannini M, Rescigno M et al (2012) Vaccines in non-small cell lung cancer: rationale, combination strategies and update on clinical trials. Crit Rev Oncol Hematol 83:432–443. doi:10.1016/j.critrevonc.2011.12.005

De Temmerman M-L, Rejman J, Demeester J et al (2011) Particulate vaccines: on the quest for optimal delivery and immune response. Drug Discov Today 16:569–582. doi:10.1016/j.drudis.2011.04.006

Dienstmann R, Markman B, Tabernero J (2012) Application of monoclonal antibodies as cancer therapy in solid tumors. Curr Clin Pharmacol 7:137–145. doi:http://dx.doi.org/10.2174/157488412800228929

Ding H, Helguera G, Rodríguez JA et al (2013) Polymalic acid nanobioconjugate for simultaneous immunostimulation and inhibition of tumor growth in HER2/neu-positive breast cancer. J Control Release 171:322–329. doi:10.1016/j.jconrel.2013.06.001

Eifler AC, Thaxton CS (2011) Nanoparticle therapeutics: FDA approval, clinical trials, regulatory pathways, and case study. Methods Mol Biol 726:325–338. doi:10.1007/978-1-61779-052-2_21

Emens LA, Silverstein SC, Khleif S et al (2012) Toward integrative cancer immunotherapy: targeting the tumor microenvironment. J Transl Med 10:70. doi:10.1186/1479-5876-10-70

Etzerodt A, Maniecki MB, Graversen JH, Møller HJ, Torchilin VP, Moestrup SK (2012) Efficient intracellular drug-targeting of macrophages using stealth liposomes directed to the hemoglobin scavenger receptor CD163. J Control Release 160(1):72–80. doi:10.1016/j.jconrel.2012.01.034

Fay F, Scott CJ (2011) Antibody-targeted nanoparticles for cancer therapy. Immunotherapy 3:381–394. doi:10.2217/imt.11.5

Fernández A, Mesa C, Marigo I et al (2011) Inhibition of tumor-induced myeloid-derived suppressor cell function by a nanoparticulated adjuvant. J Immunol 186:264–274. doi:10.4049/jimmunol.1001465

Fernández A, Oliver L, Rábade-Chediak M et al (2013) Ganglioside containing nanoproteoliposome dampens myeloid-derived suppressor cells function and ability to cross-present tumor antigens: a new approach to recover CTL function on tumor bearing individuals. J Immunother Cancer 1:P225. doi:10.1186/2051-1426-1-S1-P225

Flemming A (2011) Nano-adjuvant: double TLR stimulation is the key. Nat Rev Drug Discov 10:258. doi:10.1038/nrd3426

Foged C, Hansen J, Agger EM (2012) License to kill: formulation requirements for optimal priming of CD8(+) CTL responses with particulate vaccine delivery systems. Eur J Pharm Sci 45:482–491. doi:10.1016/j.ejps.2011.08.016

Fricke I, Gabrilovich DI (2006) Dendritic cells and tumor microenvironment: a dangerous liaison. Immunol Invest 35:1–19. doi:10.1080/08820130600803429

Gabrilovich D, Nagaraj S (2009) Myeloid-derived suppressor cells as regulators of the immune system. Nat Rev Immunol 9:162–174. doi:10.1038/nri2506.Myeloid-derived-suppressor

Gabrilovich DI, Ostrand-Rosenberg S, Bronte V (2012) Coordinated regulation of myeloid cells by tumours. Nat Rev Immunol 12:253–268. doi:10.1038/nri3175

Galluzzi L, Senovilla L, Zitvogel L, Kroemer G (2012a) The secret ally: immunostimulation by anticancer drugs. Nat Rev Drug Discov 11:215–233. doi:10.1038/nrd3626

Galluzzi L, Vacchelli E, Fridman WH et al (2012b) Trial Watch: Monoclonal antibodies in cancer therapy. Oncoimmunology 1:28–37. doi:10.4161/onci.1.1.17938

Garcia-Fuentes M, Alonso MJ (2012) Chitosan-based drug nanocarriers: where do we stand? J Control Release 161:496–504. doi:10.1016/j.jconrel.2012.03.017

Garg AD, Nowis D, Golab J et al (2010) Immunogenic cell death, DAMPs and anticancer therapeutics: an emerging amalgamation. Biochim Biophys Acta 1805:53–71. doi:10.1016/j.bbcan.2009.08.003

Germano G, Frapolli R, Belgiovine C et al (2013) Role of macrophage targeting in the antitumor activity of trabectedin. Cancer Cell 23:249–262. doi:10.1016/j.ccr.2013.01.008

Goldinger SM, Dummer R, Baumgaertner P et al (2012) Nano-particle vaccination combined with TLR-7 and -9 ligands triggers memory and effector $CD8^{+}$ T-cell responses in melanoma patients. Eur J Immunol 42:3049–3061. doi:10.1002/eji.201142361

Gomez-Cabrero A, Wrasidlo W, Reisfeld RA (2013) IMD-0354 targets breast cancer stem cells: a novel approach for an adjuvant to chemotherapy to prevent multidrug resistance in a murine model. PLoS One 8:e73607. doi:10.1371/journal.pone.0073607

Gonzalez-Aramundiz JV, Cordeiro AS, Csaba N et al (2012) Nanovaccines: nanocarriers for antigen delivery. Biol Aujourdhui 206:249–261. doi:10.1051/jbio/2012027

Ha T-Y (2011) The role of microRNAs in regulatory T cells and in the immune response. Immune Netw 11:11–41. doi:10.4110/in.2011.11.1.11

Hagemann T, Lawrence T, McNeish I et al (2008) "Re-educating" tumor-associated macrophages by targeting NF-kappaB. J Exp Med 205:1261–1268. doi:10.1084/jem.20080108

Hamdy S, Molavi O, Ma Z et al (2008) Co-delivery of cancer-associated antigen and Toll-like receptor 4 ligand in PLGA nanoparticles induces potent CD8+ T cell-mediated anti-tumor immunity. Vaccine 26:5046–5057. doi:10.1016/j.vaccine.2008.07.035

Hamdy S, Haddadi A, Hung RW, Lavasanifar A (2011) Targeting dendritic cells with nano-particulate PLGA cancer vaccine formulations. Adv Drug Deliv Rev 63:943–955. doi:10.1016/j.addr.2011.05.02

Hanahan D, Weinberg RA (2000) The hallmarks of cancer. Cell 100:57–70

Hanahan D, Weinberg RA (2011) Hallmarks of cancer: the next generation. Cell 144:646–674. doi:10.1016/j.cell.2011.02.013

Hanlon DJ, Aldo PB, Devine L et al (2011) Enhanced stimulation of anti-ovarian cancer CD8(+) T cells by dendritic cells loaded with nanoparticle encapsulated tumor antigen. Am J Reprod Immunol 65:597–609. doi:10.1111/j.1600-0897.2010.00968.x

He M, Xu Z, Ding T et al (2009) MicroRNA-155 regulates inflammatory cytokine production in tumor-associated macrophages via targeting C/EBPbeta. Cell Mol Immunol 6:343–352. doi:10.1038/cmi.2009.45

Heuking S, Adam-Malpel S, Sublet E et al (2009) Stimulation of human macrophages (THP-1) using Toll-like receptor-2 (TLR-2) agonist decorated nanocarriers. J Drug Target 17:662–670. doi:10.1080/10611860903106034

Holback H, Yeo Y (2011) Intratumoral drug delivery with nanoparticulate carriers. Pharm Res 28:1819–1830. doi:10.1007/s11095-010-0360-y

Hrkach J, Von Hoff D, Mukkaram Ali M et al (2012) Preclinical development and clinical translation of a PSMA-targeted docetaxel nanoparticle with a differentiated pharmacological profile. Sci Transl Med 4:128ra39. doi:10.1126/scitranslmed.3003651

Huang Z, Zhang Z, Jiang Y et al (2012) Targeted delivery of oligonucleotides into tumor-associated macrophages for cancer immunotherapy. J Control Release 158:286–292. doi:10.1016/j.jconrel.2011.11.013

Jemal A, Bray F, Center MM et al (2011) Global cancer statistics. CA Cancer J Clin 61:69–90. doi:10.3322/caac.20107

Joyce JA, Pollard JW (2009) Microenvironmental regulation of metastasis. Nat Rev Cancer 9:239–252. doi:10.1038/nrc2618

Junttila MR, de Sauvage FJ (2013) Influence of tumour micro-environment heterogeneity on therapeutic response. Nature 501:346–354. doi:10.1038/nature12626

Kao J, Ko EC, Eisenstein S et al (2011) Targeting immune suppressing myeloid-derived suppressor cells in oncology. Crit Rev Oncol Hematol 77:12–19. doi:10.1016/j.critrevonc.2010.02.004

Kaplan RN, Rafii S, Lyden D (2006) Preparing the "soil": the premetastatic niche. Cancer Res 66:11089–11093. doi:10.1158/0008-5472.CAN-06-2407

Kaplan RN, Psaila B, Lyden D (2007) Niche-to-niche migration of bone-marrow-derived cells. Trends Mol Med 13:72–81. doi:10.1016/j.molmed.2006.12.003

Kepp O, Tesniere A, Schlemmer F et al (2009) Immunogenic cell death modalities and their impact on cancer treatment. Apoptosis 14:364–375. doi:10.1007/s10495-008-0303-9

Kortylewski M, Kujawski M, Wang T et al (2005) Inhibiting Stat3 signaling in the hematopoietic system elicits multicomponent antitumor immunity. Nat Med 11:1314–1321. doi:10.1038/nm1325

Kortylewski M, Xin H, Kujawski M et al (2009) Regulation of the IL-23 and IL-12 balance by Stat3 signaling in the tumor microenvironment. Cancer Cell 15:114–123. doi:10.1016/j.ccr.2008.12.018

Kourtis IC, Hirosue S, de Titta A et al (2013) Peripherally administered nanoparticles target monocytic myeloid cells, secondary lymphoid organs and tumors in mice. PLoS One 8:e61646. doi:10.1371/journal.pone.0061646

Kujawski M, Kortylewski M, Lee H et al (2008) Stat3 mediates myeloid cell – dependent tumor angiogenesis in mice. J Clin Invest 118:3367–3377. doi:10.1172/JCI35213

Kurosaki T, Kitahara T, Nakamura T et al (2012) Development of effective cancer vaccine using targeting system of antigen protein to APCs. Pharm Res 29:483–489. doi:10.1007/s11095-011-0571-x

Kvistborg P, van Buuren MM, Schumacher TN (2013) Human cancer regression antigens. Curr Opin Immunol 25:284–290. doi:10.1016/j.coi.2013.03.005
Laoui D, Van Overmeire E, Movahedi K et al (2011) Mononuclear phagocyte heterogeneity in cancer: different subsets and activation states reaching out at the tumor site. Immunobiology 216:1192–1202. doi:10.1016/j.imbio.2011.06.007
Lee ALZ, Wang Y, Cheng HY et al (2009) The co-delivery of paclitaxel and Herceptin using cationic micellar nanoparticles. Biomaterials 30:919–927. doi:10.1016/j.biomaterials.2008.10.062
Lee ALZ, Wang Y, Pervaiz S et al (2011) Synergistic anticancer effects achieved by co-delivery of TRAIL and paclitaxel using cationic polymeric micelles. Macromol Biosci 11:296–307. doi:10.1002/mabi.201000332
Leimgruber A, Berger C, Cortez-retamozo V et al (2009) Behavior of endogenous tumor-associated macrophages assessed in vivo using a functionalized nanoparticle. Neoplasia 11:459–468. doi:10.1593/neo.09356
Li X, Sloat BR, Yanasarn N, Cui Z (2011) Relationship between the size of nanoparticles and their adjuvant activity: data from a study with an improved experimental design. Eur J Pharm Biopharm 78:107–116. doi:10.1016/j.ejpb.2010.12.017
Li X, Min M, Du N et al (2013) Chitin, chitosan, and glycated chitosan regulate immune responses: the novel adjuvants for cancer vaccine. Clin Dev Immunol 2013:387023. doi:10.1155/2013/387023
Liao D, Liu Z, Wrasidlo WJ et al (2011) Targeted therapeutic remodeling of the tumor microenvironment improves an HER-2 DNA vaccine and prevents recurrence in a murine breast cancer model. Cancer Res 71:5688–5696. doi:10.1158/0008-5472.CAN-11-1264
Lindsay MA (2008) MicroRNAs and the immune response. Trends Immunol 29:343–351. doi:10.1016/j.it.2008.04.004
Liu Y, Feng S-S (2011) The synergistic effect of herceptin and docetaxel in polylactide-D-α-tocopheryl polyethylene glycol succinate (PLA-TPGS) nanoparticles. J Control Release 152(Suppl):e64–e65. doi:10.1016/j.jconrel.2011.08.125
Liu T-Y, Hussein WM, Jia Z et al (2013a) Self-adjuvanting polymer-peptide conjugates as therapeutic vaccine candidates against cervical cancer. Biomacromolecules 14:2798–2806. doi:10.1021/bm400626w
Liu Z, Lv D, Liu S et al (2013b) Alginic acid-coated chitosan nanoparticles loaded with legumain DNA vaccine: effect against breast cancer in mice. PLoS One 8:e60190. doi:10.1371/journal.pone.0060190
Locke LW, Mayo MW, Yoo AD et al (2012) PET imaging of tumor associated macrophages using mannose coated 64Cu liposomes. Biomaterials 33:7785–7793. doi:10.1016/j.biomaterials.2012.07.022
Lou P-J, Cheng W-F, Chung Y-C et al (2009) PMMA particle-mediated DNA vaccine for cervical cancer. J Biomed Mater Res A 88:849–857. doi:10.1002/jbm.a.31919
Luo Y, Zhou H, Krueger J et al (2006) Targeting tumor-associated macrophages as a novel strategy against breast cancer. J Clin Invest 116:2132–2141. doi:10.1172/JCI27648.2132
Ma W, Smith T, Bogin V et al (2011) Enhanced presentation of MHC class Ia, Ib and class II-restricted peptides encapsulated in biodegradable nanoparticles: a promising strategy for tumor immunotherapy. J Transl Med 9:34. doi:10.1186/1479-5876-9-34
Ma W, Chen M, Kaushal S et al (2012) PLGA nanoparticle-mediated delivery of tumor antigenic peptides elicits effective immune responses. Int J Nanomedicine 7:1475–1487. doi:10.2147/IJN.S29506
Mantovani A, Romero P, Palucka AK, Marincola FM (2008) Tumour immunity: effector response to tumour and role of the microenvironment. Lancet 371:771–783. doi:10.1016/S0140-6736(08)60241-X
Matsuo K, Koizumi H, Akashi M et al (2011) Intranasal immunization with poly(γ-glutamic acid) nanoparticles entrapping antigenic proteins can induce potent tumor immunity. J Control Release 152:310–316. doi:10.1016/j.jconrel.2011.03.009
Mi Y, Liu X, Zhao J et al (2012) Multimodality treatment of cancer with herceptin conjugated, thermomagnetic iron oxides and docetaxel loaded nanoparticles of biodegradable polymers. Biomaterials 33:7519–7529. doi:10.1016/j.biomaterials.2012.06.100

Montero A, Diaz-Montero C (2012) Myeloid-derived suppressor cells in cancer patients: a clinical perspective. J Immunother 35:107–115. doi:10.1097/CJI.0b013e318242169f

Morille M, Passirani C, Vonarbourg A et al (2008) Progress in developing cationic vectors for non-viral systemic gene therapy against cancer. Biomaterials 29:3477–3496. doi:10.1016/j.biomaterials.2008.04.036

Movahedi K, Schoonooghe S, Laoui D et al (2012) Nanobody-based targeting of the macrophage mannose receptor for effective in vivo imaging of tumor-associated macrophages. Cancer Res 72:4165–4177. doi:10.1158/0008-5472.CAN-11-2994

National Comprehensive Clinical Cancer Network (2013) National Comprehensive Cancer Network Guidelines

Nagai T, Tanaka M, Tsuneyoshi Y, Xu B, Michie SA, Hasui K, Hirano H, Arita K, Matsuyama T (2009) Targeting tumor-associated macrophages in an experimental glioma model with a recombinant immunotoxin to folate receptor beta. Cancer Immunol Immunother 58(10):1577–1586. doi:10.1007/s00262-009-0667-x

Nguyen-Hoai T, Baldenhofer G, Sayed Ahmed MS et al (2012) CCL21 (SLC) improves tumor protection by a DNA vaccine in a Her2/neu mouse tumor model. Cancer Gene Ther 19:69–76. doi:10.1038/cgt.2011.69

Owen SC, Patel N, Logie J et al (2013) Targeting HER2+ breast cancer cells: lysosomal accumulation of anti-HER2 antibodies is influenced by antibody binding site and conjugation to polymeric nanoparticles. J Control Release 172:395–404. doi:10.1016/j.jconrel.2013.07.011

Peinado H, Rafii S, Lyden D (2008) Inflammation joins the “niche”. Cancer Cell 14:347–349. doi:10.1016/j.ccr.2008.10.012

Peinado H, Lavotshkin S, Lyden D (2011) The secreted factors responsible for pre-metastatic niche formation: old sayings and new thoughts. Semin Cancer Biol 21:139–146. doi:10.1016/j.semcancer.2011.01.002

Pello OM, Chèvre R, Laoui D et al (2012) In vivo inhibition of c-MYC in myeloid cells impairs tumor-associated macrophage maturation and pro-tumoral activities. PLoS One 7:e45399. doi:10.1371/journal.pone.0045399

Pen JJ, Keersmaecker BD, Heirman C et al (2014) Interference with PD-L1/PD-1 co-stimulation during antigen presentation enhances the multifunctionality of antigen-specific T cells. Gene Ther 21:262–271. doi:10.1038/gt.2013.80

Perche F, Benvegnu T, Berchel M et al (2011) Enhancement of dendritic cells transfection in vivo and of vaccination against B16F10 melanoma with mannosylated histidylated lipopolyplexes loaded with tumor antigen messenger RNA. Nanomedicine 7:445–453. doi:10.1016/j.nano.2010.12.010

Plebanski M, Xiang S, Zaman M et al (2013) Nanovaccines and their mode of action. Methods 60:226–231. doi:10.1016/j.ymeth.2013.04.014

Pollard JW (2004) Tumour-educated macrophages promote tumour progression and metastasis. Nat Rev Cancer 4:71–78. doi:10.1038/nrc1256

Prasad S, Cody V, Saucier-Sawyer JK et al (2011) Polymer nanoparticles containing tumor lysates as antigen delivery vehicles for dendritic cell-based antitumor immunotherapy. Nanomedicine 7:1–10. doi:10.1016/j.nano.2010.07.002

Prasad S, Cody V, Saucier-Sawyer JK et al (2012) Optimization of stability, encapsulation, release, and cross-priming of tumor antigen-containing PLGA nanoparticles. Pharm Res 29:2565–2577. doi:10.1007/s11095-012-0787-4

Prego C, Paolicelli P, Díaz B et al (2010) Chitosan-based nanoparticles for improving immunization against hepatitis B infection. Vaccine 28:2607–2614. doi:10.1016/j.vaccine.2010.01.011

Ren Y, Cheung HW, von Maltzhan G et al (2012) Targeted tumor-penetrating siRNA nanocomplexes for credentialing the ovarian cancer oncogene ID4. Sci Transl Med 4:147ra112. doi:10.1126/scitranslmed.3003778

Reslan L, Dalle S, Dumontet C (2009) Understanding and circumventing resistance to anticancer monoclonal antibodies. MAbs 1:222–229

Rosenthal JA, Chen L, Baker JL et al (2014) Pathogen-like particles: biomimetic vaccine carriers engineered at the nanoscale. Curr Opin Biotechnol 28:51–58. doi:10.1016/j.copbio.2013.11.005

Roth L, Agemy L, Kotamraju VR et al (2012) Transtumoral targeting enabled by a novel neuropilin-binding peptide. Oncogene 31:3754–3763. doi:10.1038/onc.2011.537

Ruoslahti E (2012) Peptides as targeting elements and tissue penetration devices for nanoparticles. Adv Mater 24:3747–3756. doi:10.1002/adma.201200454

Salem AK, Weiner GJ (2009) CpG oligonucleotides as immunotherapeutic adjuvants: innovative applications and delivery strategies. Adv Drug Deliv Rev 61:193–194. doi:10.1016/j.addr.2008.12.003

Sánchez-Moreno P, Ortega-Vinuesa JL, Boulaiz H et al (2013) Synthesis and characterization of lipid immuno-nanocapsules for directed drug delivery: selective antitumor activity against HER2 positive breast-cancer cells. Biomacromolecules 14:4248–4259. doi:10.1021/bm401103t

Schlosser E, Mueller M, Fischer S et al (2008) TLR ligands and antigen need to be coencapsulated into the same biodegradable microsphere for the generation of potent cytotoxic T lymphocyte responses. Vaccine 26:1626–1637. doi:10.1016/j.vaccine.2008.01.030

Schmid MC, Varner JA (2010) Myeloid cells in the tumor microenvironment: modulation of tumor angiogenesis and tumor inflammation. J Oncol 2010:201026. doi:10.1155/2010/201026

Schmieder A, Michel J, Schönhaar K et al (2012) Differentiation and gene expression profile of tumor-associated macrophages. Semin Cancer Biol 22:289–297. doi:10.1016/j.semcancer.2012.02.002

Schroeder A, Heller DA, Winslow MM et al (2012) Treating metastatic cancer with nanotechnology. Nat Rev Cancer 12:39–50. doi:10.1038/nrc3180

Scott AM, Wolchok JD, Old LJ (2012) Antibody therapy of cancer. Nat Rev Cancer 12:278–287. doi:10.1038/nrc3236

Seidel UJE, Schlegel P, Lang P (2013) Natural killer cell mediated antibody-dependent cellular cytotoxicity in tumor immunotherapy with therapeutic antibodies. Front Immunol 4:76. doi:10.3389/fimmu.2013.00076

Seton-Rogers S (2012) Immunotherapy: combinations that work. Nat Rev Cancer 12:231. doi:10.1158/0008

Shen H, Ackerman AL, Cody V et al (2006) Enhanced and prolonged cross-presentation following endosomal escape of exogenous antigens encapsulated in biodegradable nanoparticles. Immunology 117:78–88. doi:10.1111/j.1365-2567.2005.02268.x

Shiao S, Ganesan A (2011) Immune microenvironments in solid tumors: new targets for therapy. Genes Dev 25:2559–2572. doi:10.1101/gad.169029.111.success

Shuptrine CW, Surana R, Weiner LM (2012) Monoclonal antibodies for the treatment of cancer. Semin Cancer Biol 22:3–13. doi:10.1016/j.semcancer.2011.12.009

Sica A, Schioppa T, Mantovani A, Allavena P (2006) Tumour-associated macrophages are a distinct M2 polarised population promoting tumour progression: potential targets of anti-cancer therapy. Eur J Cancer 42:717–727. doi:10.1016/j.ejca.2006.01.003

Silva JM, Videira M, Gaspar R et al (2013a) Immune system targeting by biodegradable nanoparticles for cancer vaccines. J Control Release 168:179–199. doi:10.1016/j.jconrel.2013.03.010

Silva L, Rosalia R, Sazak A et al (2013b) Optimization of encapsulation of a synthetic long peptide in PLGA nanoparticles: low-burst release is crucial for efficient CD8(+) T cell activation. Eur J Pharm Biopharm 83:338–345. doi:10.1016/j.ejpb.2012.11.006

Skak K, Kragh M, Hausman D et al (2008) Interleukin 21: combination strategies for cancer therapy. Nat Rev Drug Discov 7:231–240. doi:10.1038/nrd2482

Solinas G, Germano G, Mantovani A, Allavena P (2009) Tumor-associated macrophages (TAM) as major players of the cancer-related inflammation. J Leukoc Biol 86:1065–1073. doi:10.1189/jlb.0609385

Solinas G, Schiarea S, Liguori M et al (2010) Tumor-conditioned macrophages secrete migration-stimulating factor: a new marker for M2-polarization, influencing tumor cell motility. J Immunol 185:642–652. doi:10.4049/jimmunol.1000413

Solito S, Falisi E, Diaz-Montero CM et al (2011) A human promyelocytic-like population is responsible for the immune suppression mediated by myeloid-derived suppressor cells. Blood 118:2254–2265. doi:10.1182/blood-2010-12-325753

Song Y-C, Cheng H-Y, Leng C-H et al (2014) A novel emulsion-type adjuvant containing CpG oligodeoxynucleotides enhances CD8(+) T-cell-mediated anti-tumor immunity. J Control Release 173:158–165. doi:10.1016/j.jconrel.2013.10.027

Stewart BS, Wild CP (2014) World Cancer Report 2014. WHO. doi:9283204298
Sugahara KN, Teesalu T, Karmali PP et al (2009) Tissue-penetrating delivery of compounds and nanoparticles into tumors. Cancer Cell 16:510–520. doi:10.1016/j.ccr.2009.10.013
Taheri A, Dinarvand R, Atyabi F et al (2012) Trastuzumab decorated methotrexate-human serum albumin conjugated nanoparticles for targeted delivery to HER2 positive tumor cells. Eur J Pharm Sci 47:331–340. doi:10.1016/j.ejps.2012.06.016
Tan G, Wang Z, Zhang X et al (2011) Induction of CTLs by DCs pulsed with K-ras mutant peptide on the surface of nanoparticles in the treatment of pancreatic cancer. Oncol Rep 26:215–221. doi:10.3892/or.2011.1283
Tang CK, Lodding J, Minigo G et al (2007) Mannan-mediated gene delivery for cancer immunotherapy. Immunology 120:325–335. doi:10.1111/j.1365-2567.2006.02506.x
Vanneman M, Dranoff G (2012) Combining immunotherapy and targeted therapies in cancer treatment. Nat Rev Cancer 12:237–251. doi:10.1038/nrc3237
Vasievich EA, Huang L (2011) The suppressive tumor microenvironment: a challenge in cancer immunotherapy. Mol Pharm 8:635–641. doi:10.1021/mp1004228
Vicari AP, Treilleux I, Lebecque S (2004) Regulation of the trafficking of tumour-infiltrating dendritic cells by chemokines. Semin Cancer Biol 14:161–169. doi:10.1016/j.semcancer.2003.10.002
Vicente S, Diaz-Freitas B, Peleteiro M et al (2013a) A polymer/oil based nanovaccine as a single-dose immunization approach. PLoS One 8:e62500. doi:10.1371/journal.pone.0062500
Vicente S, Peleteiro M, Díaz-Freitas B et al (2013b) Co-delivery of viral proteins and a TLR7 agonist from polysaccharide nanocapsules: a needle-free vaccination strategy. J Control Release 172:773–781. doi:10.1016/j.jconrel.2013.09.012
Vila A, Sánchez A, Janes K et al (2004) Low molecular weight chitosan nanoparticles as new carriers for nasal vaccine delivery in mice. Eur J Pharm Biopharm 57:123–131. doi:10.1016/j.ejpb.2003.09.006
Villamor N, Montserrat E, Colomer D (2003) Mechanism of action and resistance to monoclonal antibody therapy. Semin Oncol 30:424–433
Waeckerle-Men Y, Groettrup M (2005) PLGA microspheres for improved antigen delivery to dendritic cells as cellular vaccines. Adv Drug Deliv Rev 57:475–482. doi:10.1016/j.addr.2004.09.007
Waite CL, Roth CM (2011) Binding and transport of PAMAM-RGD in a tumor spheroid model: the effect of RGD targeting ligand density. Biotechnol Bioeng 108:2999–3008. doi:10.1002/bit.23255
Wang AZ, Langer R, Farokhzad OC (2012) Nanoparticle delivery of cancer drugs. Annu Rev Med 63:185–198. doi:10.1146/annurev-med-040210-162544
Weiner LM, Surana R, Wang S (2010) Monoclonal antibodies: versatile platforms for cancer immunotherapy. Nat Rev Immunol 10:317–327. doi:10.1038/nri2744
Wels J, Kaplan RN, Rafii S, Lyden D (2008) Migratory neighbors and distant invaders: tumor-associated niche cells. Genes Dev 22:559–574. doi:10.1101/gad.1636908
Wen Z-S, Xu Y-L, Zou X-T, Xu Z-R (2011) Chitosan nanoparticles act as an adjuvant to promote both Th1 and Th2 immune responses induced by ovalbumin in mice. Mar Drugs 9:1038–1055. doi:10.3390/md9061038
Winter H, van den Engel NK, Rusan M et al (2011) Active-specific immunotherapy for non-small cell lung cancer. J Thorac Dis 3:105–114. doi:10.3978/j.issn.2072-1439.2010.12.06
Wong C, Stylianopoulos T, Cui J et al (2011) Multistage nanoparticle delivery system for deep penetration into tumor tissue. Proc Natl Acad Sci U S A 108:2426–2431. doi:10.1073/pnas.1018382108
Xiang SD, Scholzen A, Minigo G et al (2006) Pathogen recognition and development of particulate vaccines: does size matter? Methods 40:1–9. doi:10.1016/j.ymeth.2006.05.016
Yaguchi T, Sumimoto H, Kudo-Saito C et al (2011) The mechanisms of cancer immunoescape and development of overcoming strategies. Int J Hematol 93:294–300. doi:10.1007/s12185-011-0799-6
Yamaguchi S, Tatsumi T, Takehara T et al (2010) EphA2-derived peptide vaccine with amphiphilic poly(gamma-glutamic acid) nanoparticles elicits an anti-tumor effect against mouse liver tumor. Cancer Immunol Immunother 59:759–767. doi:10.1007/s00262-009-0796-2

Yoshikawa T, Okada N, Oda A et al (2008a) Development of amphiphilic gamma-PGA-nanoparticle based tumor vaccine: potential of the nanoparticulate cytosolic protein delivery carrier. Biochem Biophys Res Commun 366:408–413. doi:10.1016/j.bbrc.2007.11.153

Yoshikawa T, Okada N, Oda A et al (2008b) Nanoparticles built by self-assembly of amphiphilic gamma-PGA can deliver antigens to antigen-presenting cells with high efficiency: a new tumor-vaccine carrier for eliciting effector T cells. Vaccine 26:1303–1313. doi:10.1016/j.vaccine.2007.12.037

Yu H, Pardoll D, Jove R (2009) STATs in cancer inflammation and immunity: a leading role for STAT3. Nat Rev Cancer 9:798–809. doi:10.1038/nrc2734

Zhang Z, Tongchusak S, Mizukami Y et al (2011) Induction of anti-tumor cytotoxic T cell responses through PLGA-nanoparticle mediated antigen delivery. Biomaterials 32:3666–3678. doi:10.1016/j.biomaterials.2011.01.067

Zhang X, Tian W, Cai X et al (2013) Hydrazinocurcumin encapsuled nanoparticles "re-educate" tumor-associated macrophages and exhibit anti-tumor effects on breast cancer following STAT3 suppression. PLoS One 8:e65896. doi:10.1371/journal.pone.0065896

Zhao J, Mi Y, Liu Y, Feng S-S (2012) Quantitative control of targeting effect of anticancer drugs formulated by ligand-conjugated nanoparticles of biodegradable copolymer blend. Biomaterials 33:1948–1958. doi:10.1016/j.biomaterials.2011.11.051

Zhu S, Niu M, O'Mary H, Cui Z (2013) Targeting of tumor-associated macrophages made possible by PEG-sheddable, mannose-modified nanoparticles. Mol Pharm 10:3525–3530. doi:10.1021/mp400216r

Zolnik BS, González-Fernández A, Sadrieh N, Dobrovolskaia MA (2010) Nanoparticles and the immune system. Endocrinology 151:458–465. doi:10.1210/en.2009-1082

Zou W, Chen L (2008) Inhibitory B7-family molecules in the tumour microenvironment. Nat Rev Immunol 8:467–477. doi:10.1038/nri2326

Part IV
Gene Therapy

Chapter 9
RNAi-Based Nano-Oncologicals: Delivery and Clinical Applications

Katharina Müller and Ernst Wagner

9.1 Introduction

RNA interference (RNAi) is a natural process triggered by small double-stranded RNAs including microRNAs (miRNA, miR) and siRNA in several ways. It is a major mechanism of sequence-specific, post-transcriptional gene silencing in animals (using miRNA) and plants (using siRNA). miRNA have also been shown to be involved in the evolution of various human cancers, such as breast, colon, and lung cancer (Croce 2009). Both, loss and overexpression of miRNAs, can enhance tumor development leading to poor clinical prognosis of cancer patients. Many cases are known, where miRNAs act as tumor suppressor, when they are deleted or mutated in diverse human malignancies (Croce 2009). The first example of miRNA entanglement in human cancer was reported by Calin et al. in 2002. They showed that miR-15a and miR-16-1 are involved in chronic lymphocytic leukemia (CLL), the most common form of adult leukemia in the western world. Both miRNAs are located at chromosome 13q14, which is frequently deleted or down-regulated in more than half of B cell CLL. The loss of miR-15a and miR-16-1 leads to an upregulation in cancer genes (such as MCL1, BCL2, ETS1, or JUN) that disturb apoptosis (Calin et al. 2002, 2008; Iorio and Croce 2012). Another tumor suppressor is miRNA-200c. The loss of miRNA-200c and other members of the miRNA-200 family promotes epithelial to mesenchymal transition (EMT) in cancer cells, such as breast, endometrial, or ovarian cancer. miRNA-200c directly targets Zeb1 and Zeb2, which are transcriptional repressors of E-cadherin, a calcium-dependent cell adhesion protein (Howe et al. 2011; Gregory et al. 2008; Guilford et al. 1998).

K. Müller (✉) • E. Wagner
Pharmaceutical Biotechnology, Center for System-Based Drug Research,
Center for Nanoscience (CeNS), and Nanosystems Initiative Munich (NIM),
Ludwig-Maximilians University, Butenandtstrasse 5-13 81377, Munich, Germany
e-mail: katharina.mueller@cup.uni-muenchen.de

M.J. Alonso, M. Garcia-Fuentes (eds.), *Nano-Oncologicals: New Targeting and Delivery Approaches*, Advances in Delivery Science and Technology,
DOI 10.1007/978-3-319-08084-0_9

Inhibition of miRNA-200c was also shown to increase doxorubicin resistance by targeting TrkB and Bmi1 in BT474 and MDA-MB 436, two breast cancer cell lines (Kopp et al. 2012). Therefore restoration of miRNA-200c reduces migration, invasion, and chemoresistance in aforesaid cancer cells.

miRNAs can also act as oncogenes, when overexpression of these miRNAs causes tumor development. miR-155 is a prominent example for miRNA oncogenes, because its upregulation occurs in diffuse large B cell lymphoma (DLBCL) (Eis et al. 2005), lung cancer (Zheng et al. 2011; Kong et al. 2012), breast cancer (Iorio et al. 2005), and other forms of cancer. DLBCL patients with high levels of miR-155, which is processed from BIC, a noncoding transcript highly expressed in activated B and T cells (Tili et al. 2009), have a poor clinical prognosis (Eis et al. 2005). In lung cancer the level of miR-155 has shown to be higher in the plasma from lung cancer patients with metastasis than in those without metastasis (Zheng et al. 2011). miR-21 is overexpressed in cancer, such as breast cancer (Iorio et al. 2005), pancreas cancer (Bloomston et al. 2007), colon cancer (Asangani et al. 2008), or others. In colorectal cancer cell lines miR-21 downregulates the tumor suppressor Pdcd4. Therefore, upregulation of this miRNA leads to invasion, intravasation, and metastasis (Asangani et al. 2008).

These recently unraveled natural RNAi processes can be capitalized for development of a new class of anticancer medicines, RNAi-based nano-oncologicals. The findings listed above suggest that re-expression of lost tumor suppressor miRNAs (by introduction of pre-miRs or miRNA mimics) or repression of oncogenic miRNAs (by antisense molecules, antagomirs) could provide therapeutic benefit, which has already been proven in various studies. In addition, artificial siRNAs can be directed very specifically against any anticancer target molecule, irrespective of whether it is druggable or not druggable with conventional medicines.

The delivery of pDNA has been successfully accomplished in the last years (Lori 2011). Although siRNA and miRNA form polyplexes with oligomers through electrostatic interactions like pDNA, the delivery of these small RNAs is more challenging. To a large extent this is due to the smaller size of these RNAs. Whereas pDNA consists of several kilo bp, siRNA and mature miRNA count 21–23 bp. Therefore the electrostatic interaction with cationic carriers of RNA is not as strong as with pDNA which often leads to less stable polyplexes. Another factor influencing the stability of RNA polyplexes is the different structure of sugars. The deoxyribose in DNA is quite stable, the additional hydroxyl group in 2′-position of the ribose in RNA decreases stability (Scholz and Wagner 2012). Furthermore, the size of pDNA and siRNA or miRNA polyplexes influences the delivery. pDNA forms coiled polyplexes of 30–100 nm (Spagnou et al. 2004), whereas siRNA or miRNA forms smaller polyplexes (Dohmen et al. 2012) or even bigger ones due to aggregation (Scholz and Wagner 2012). Favorable for siRNA or miRNA delivery is their site of action. In comparison to pDNA they do not have to be transported and enter the nucleus but operate in the cytoplasm. Therefore, an important requirement for efficient siRNA or miRNA carriers is the early release of their nucleic acid cargo. However, once in the nucleus the effect of pDNA is enhanced due to amplification through transcription and translation, whereas siRNAs or miRNAs lack these mechanisms. For this reason a higher amount of siRNA or miRNA is needed for effective gene silencing compared to gene expression of pDNA (Scholz and Wagner 2012; Schwake et al. 2010).

The grand solution to be found for a broad medical translation of microRNA and siRNA is the development of effective and safe delivery systems (Wagner 2013).

9.2 Mechanism of RNAi by microRNA and siRNA

Sequence-specific, post-transcriptional gene silencing in animals (miRNA) and plants (siRNA) proceeds in significantly different fashion (Fig. 9.1). In animals, endogenous miRNAs regulate gene expression during embryonal development and cellular differentiation. miRNA genes derive from independent transcription units or from introns of protein-coding genes (Krol et al. 2010). They are mainly clustered in the human genome and miRNAs within a cluster are often related to each other (Treiber et al. 2012). miRNA genes are transcribed by RNA polymerase II (Pol II) into long primary transcripts (pri-miRNAs) (Fig. 9.1a) in the nucleus (Ma et al. 2004). Here they are processed firstly by a RNase-III-type endonuclease termed Drosha, along with the double-stranded (ds) RNA-binding protein of DiGeorge syndrome critical region gene 8 (DGCR8) into so-called miRNA precursor (pre-miRNA), a ~70 nt stem-loop structure (Meister and Tuschl 2004). Afterwards pre-miRNA is transported into the cytoplasm by exportin 5, a dsRNA-binding protein. Once in the cytoplasm, pre-miRNA is recognized by another RNase III, called Dicer. Dicer and its ds-RNA-binding protein partners, HIV-1 TAR RNA-binding protein (TRBP) and protein activator of protein kinase PKR (PACT) process pre-miRNA into ~22 nt mature miRNA (Kim and Rossi 2007). This RNA duplex possessing a 5′phosphate and ~2′nt 3′overhang, which is characteristic of a RNase III product (Bartel 2009) is then loaded into the RNA-induced silencing complex (RISC). RISC contains one member of the Argonaute protein family (Ago 1 to Ago 4). Only one of the Argonaute proteins (Ago 2) provides the RISC with endonuclease activity under special conditions (see below). The so-called antisense or guide strand of miRNA which enters the RISC is the one whose 5′end is less firmly paired (Khvorova et al. 2003). If the guide strand shows imperfect sequence homology, the passenger strand is unwinded and discarded by a bypass mechanism that necessitates helicase activity. If the guide strand has perfect sequence complementary to its passenger strand and Ago 2 is part of the RISC, the passenger strand is cleaved.

The extent of sequence complementary also influences the further procedure of RISC. In the rare case of perfect or near perfect Watson–Crick base pairing between the miRNA and the 3′untranslated region (3′UTR) of its target mRNA, this leads to direct sequence specific cleavage of the mRNA, comparable to the siRNA pathway (see below). However, the more common mechanism of miRNA induced gene silencing occurs with miRNA binding with imperfect base pairing. The first 2-7 or 2-8 nucleotides from the 5′end of the miRNA, the seed sequence, must have perfect complementarity. Mismatches at the 3′end of the miRNA are tolerated. In this case

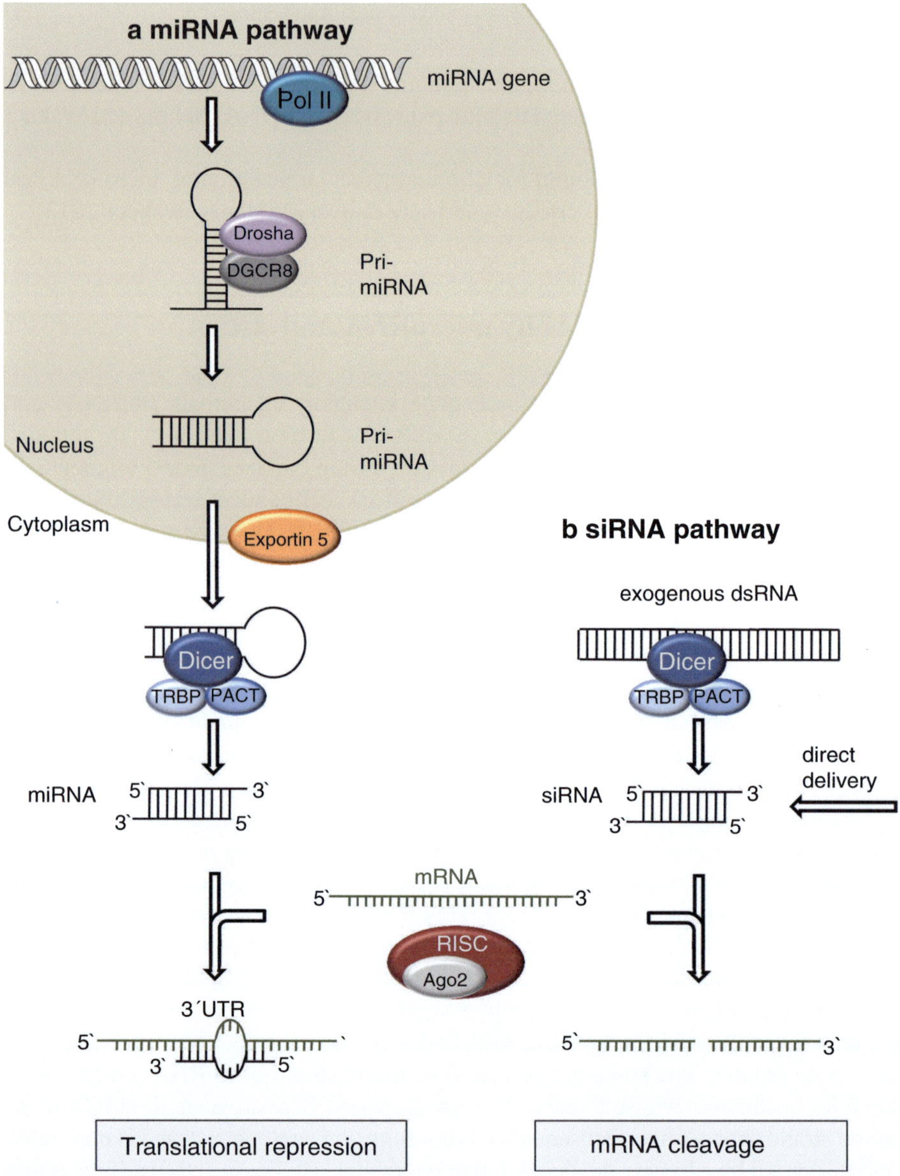

Fig. 9.1 Mechanism of RNAi in mammalian cells. (**a**) miRNA pathway. Endogenous miRNA genes are transcribed by RNA polymerase II into primary miRNA (pri-miRNA). The endonuclease Drosha and ds-RNA binding protein DGCR8 then process pri-miRNA into precursor miRNA (pre-miRNA), which is exported into cytoplasm by Exportin 5. In the cytoplasm pre-miRNA is cut by the RNase III, Dicer, together with TRBP (HIV-1 TAR RNA-binding protein) and PACT (protein activator of protein kinase PKR) into a dsRNA consisting of ~22 nt and a 5′phosphate and ~2 nt 3′overhang. The miRNA guide strand is then loaded into the RISC (RNA-induced silencing complex)- Argonaute (Ago 1 or Ago 2) complex and recognizes its targeting mRNA. In mammalian cells, the guide strand has imperfect complementary to the target mRNA, which leads to translational repression and therefore less protein. (**b**) siRNA pathway. Analogous to pre-miRNA exogenous dsRNA is processed in the cytoplasm by Dicer, TRBP, and PACT into siRNA. Alternatively exogenous siRNA can be directly delivered. siRNA/Ago 2 RISCs require perfect complementary for target mRNA cleavage

miRNA induces mRNA degradation and therefore translational repression (Kim and Rossi 2007). Because of their gene silencing capability without perfect match, one miRNA can regulate up to hundreds of different mRNAs (Krol et al. 2010). In 2005 Lim et al. already observed this effect using microarray analysis (Lim et al. 2005). Moreover, Dicer can process pre-miRNA in miRNAs altering their length in 1–2 nt. This length difference influences the miRNA seed sequences as well as guide strand loading into the RISC, thus increasing the target mRNAs for a single miRNA (Yates et al. 2013). In contrast, different miRNAs can silence a common single mRNA (Kim 2005). Therefore, reestablishment or overexpression of a single miRNA or silencing of miRNAs using antagomirs in cancer could lead to off-target effects (OTEs) that must be considered carefully.

The siRNA mechanism is also based on endogenous double-stranded RNAs, which often derive from mRNAs, transposons, viruses, or heterochromatic DNAs (Bartel 2004). These RNAs are processed by Dicer (Fig. 9.1b) but, in contrast to miRNA, request a perfect match with the target mRNA along about 20 nucleotides and therefore, at least in theory, each siRNA has only one specific mRNA target. siRNA triggered gene silencing mainly serves as an innate immune defense protecting nematodes, insects, and plants against invasive nucleic acids from pathogens (Guzman-Villanueva et al. 2012). The siRNA guide strand always has to have perfect complementary to its target mRNA and therefore leads to mRNA cleavage. The guide strand interacts with the catalytic, RNase H-like PIWI domain of Ago 2 at the 5′end and with a PIWI-Argonaute-Zwille (PAZ) domain at the 3′end (Ma et al. 2004). The targeting mRNA is cleaved between bases 10 and 11 relative to the 5′end of the siRNA guide strand (Kim and Rossi 2007). In contrast to miRNA, targeted cleavage sites can be both translated and untranslated regions of the target mRNA.

Mammalian cells do not express siRNAs or miRNAs with perfect match to mRNA targets. However artificially delivered or transfected perfectly matched exogenous dsRNA can be processed in the cytoplasm similar to pre-miRNA by Dicer into siRNA of 21–25 nucleotides in length (Elbashir et al. 2001). The siRNA duplex is then incorporated into the RISC and the passenger strand is cleaved and expelled. Sequence-specific gene silencing by such artificial RISCs is found. As short 21 bp siRNAs, when introduced into the cytoplasm, can directly intervene with the RNAi pathway without need for cleavage by Dicer, synthetic 21 bp siRNAs are widely used for research and therapeutic applications of RNAi (Guzman-Villanueva et al. 2012). The gene knockdown caused by a siRNA is often temporary. Edinger et al. (2014) found a maximum downregulation at mRNA level after 24 h and at protein level after 48 h in in vitro studies. In other studies a silencing of 3–4 weeks in nondividing liver cells was observed in vivo, indicating that the transient gene knockdown is due to dilution effects of siRNA concentration in the cytosol through cell division (Bartlett and Davis 2006). Zimmermann et al. (2006) observed silencing of apolipoprotein B for 11 days, when they applied a liposomal siRNA formulation systemically into cynomolgus monkeys. Considering these different findings, the dose regime for RNAi based nanodrugs should be calculated carefully, when transferred to clinical application (Bartlett and Davis 2006).

9.3 Delivery of siRNA and miRNA as Therapeutic Agents

9.3.1 Limitations to Direct Therapeutic Use of siRNA and miRNA

There are several barriers that restrict the direct use of miRNA or siRNA for therapeutic application. First of all, naked RNA is rapidly cleaved by ribonucleases (RNases) enzymes which are commonly occurring in a wide variety of organism. RNase interacts with the RNA backbone catalyzing its hydrolysis (Yazbeck et al. 2002). Therefore, the poor RNase resistance limits the application of non-modified RNA in vitro and in vivo. Moreover, synthetic miRNAs or siRNAs are eliminated from the bloodstream by excretion in urine or bile. Due to their small size RNA molecules are able to pass the capillaries of the kidney glomerulus easily (Kanasty et al. 2012). The biological half-life of siRNA has been reported to be 2–6 min, whereas chemically stabilized siRNA circulates for 30–50 min (Soutschek et al. 2004).

Another limiting factor concerning siRNA and miRNA is toxicity. Naked dsRNAs are able to cause OTEs. For example one siRNA can regulate numerous unintended transcripts. This emerges when siRNA acts in a miRNA-related manner with partial sequence complementary. It has been reported that many off-targets silenced by siRNA showed 3′UTR perfect complementary to the seed region of the siRNA but not throughout the entire guide strand (Jackson et al. 2006; Birmingham et al. 2006). In addition, siRNA can induce an innate immune response. This immunogenicity is divided into two groups: activation of toll-like receptors (TLR) or non-TLR-mediated immune response (Pecot et al. 2011). This immunogenicity derives from the stimulation of pattern recognition receptors (PRRs). PPRs recognize invariant molecular structures of pathogens (Medzhitov and Janeway 2002; Kanasty et al. 2012). Relating to siRNA, two classes of PPRs are affected: TLRs and cytoplasmic receptors. TLR3, TLR7, and TLR8 recognize synthetic siRNA (Rettig and Behlke 2012). TLR3 located in endosomes and cell surfaces is only activated by dsRNA, which is typically for viruses (Seth et al. 2006). After siRNA detection TLR3 activates interferon-γ (INFγ) and interleukin-12 (IL-12), causing anti-angiogenic effects (Pecot et al. 2011). TLR7 and TLR8 are stimulated by either single-stranded RNA (ssRNA) or RNA duplexes. They are expressed in endosomes and lysosomes and in the endoplasmic reticulum of plasmacytoid dendritic cells, B cells, and myeloid cells (Kanasty et al. 2012). TLR7 and TLR8 cause nuclear translocation of nuclear factor κ-light-chain-enhancer of activated B cells (NF-κB) and downstream activation of interferon-α (INFα) and inflammatory cytokines (Hornung et al. 2005; Pecot et al. 2011). siRNAs can also provoke a non-TLR-mediated immune response by activating cytoplasmic receptors. dsRNA-binding protein kinase (PKR) leads to inhibition of protein translation and interferon response after recognizing siRNA. In addition, retinoic acid-inducible gene 1 (RIG-1), another cytoplasmic PRR, causes interferon response and upregulation of other inflammatory mediators (Kanasty et al. 2012). Chemical modification and variations in siRNA design reduce the risk of OTEs and immune responses (Rettig and Behlke

2012). In correlation to siRNA, the risk of miRNAs causing toxicity is likely to be lower since physiologic gene expression networks have managed to adapt to regulatory effects of endogenous miRNAs (Kota et al. 2009). Nevertheless, side effects of miRNA-based therapies have to be examined before usage in clinical application.

Naked siRNA or miRNA in the circulation have to find their site of action, such as a specific tissue or cancer cells. Since these molecules do not possess any targeting domain and, because of their polyanionic nature, do not passively enter cells across cell membranes, they have to be delivered by conjugation or formulations with targeting ligands and shielding domains (Gallas et al. 2013). Animal cells are surrounded by the complex extracellular matrix (ECM), a dense mesh of fibrous proteins and glycosaminoglycans, which has to be overcome by the RNA molecules, as well (Zamecnik et al. 2004). Next, after siRNA and miRNA entered their target cell via endocytosis, they must escape from endosomes, avoiding entrapment and degradation in lysosomes (Shim and Kwon 2010). These extracellular and intracellular barriers have to be addressed during the development of potent delivery systems (Wagner 2013).

9.3.2 Chemical Modifications of siRNA and miRNA

Chemical changes in the backbone of siRNA and miRNA were introduced to improve stability and reduce immunogenicity of these molecules. A common strategy comprises the modification of the ribose 2′-OH group of siRNA (Fig. 9.2), which is involved in the hydrolysis mechanism of many serum RNases. The substitution of this functional group with 2′-*O*-methyl (2′OMe), 2′-deoxy (2′H), 2′-fluoro (2′F), or 2′-methoxyethyl (2′-MOE) ribonucleotides (Fig. 9.2a) has proven to increase RNase resistance without losing the ability to enter the RISC and to interact with Ago 2 (Behlke 2006; Rana 2007). Furthermore, 2′-OMe uridine or guanosine modifications have been shown to prevent stimulation of the innate immune system by avoiding activation of TLRs (Judge et al. 2005). Other established procedures are the incorporation of phosphorothioate backbone linkages at the 3′-end of the siRNA strand to improve the stability against endonucleases, or the replacement of a nonbridging phosphodiester oxygen by an isoelectronic borane (BH_3^-) moiety to increase activity and RNase resistance (Fig. 9.2b) (Bumcrot et al. 2006).

Locked nucleic acid (LNA) (Fig. 9.2c) includes a methylene bridge linking the 2′-oxygen to the 4′-carbon of the ribose ring. This modification ensures that the ribose ring is locked in the 3′-endo conformation and therefore increases nuclease resistance to RNA oligonucleotides. Moreover, LNAs are processed by the siRNA machinery, because they maintain an A-form helix geometry typical for RNA–RNA duplexes, in a normal way and reduce unintended OTEs (Elmen et al. 2005). LNAs are widely used as antagomirs because of their thermodynamic stability and therefore efficient binding to miRNA seeds. Antagomirs are single-stranded oligonucleotides with perfect complementary to a miRNA. If hybridized with its corresponding miRNA, target mRNA recognition and therefore gene regulation is hampered (Obad et al. 2011).

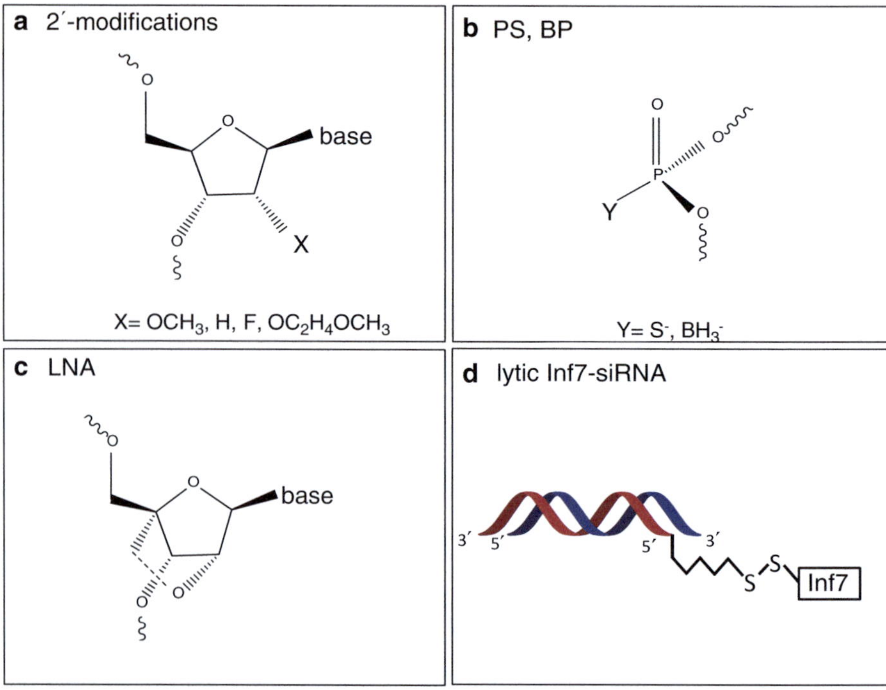

Fig. 9.2 Chemical modifications of RNA. (**a**) Stabilizing 2′-OH modifications. 2′OMe: X=OCH_3; 2′deoxy: X=H; 2′fluoro: X=F; 2′methoxyethyl (2′-MOE): X=$OCH_2H_4OCH_3$. (**b**) Stabilizing backbone modifications. Phosphorothioate linkage (PS): Y=S^-; boranophosphate (BP): Y=BH_3^-. (**c**) Locked nucleic acid (LNA); methylene bridge between 2′-oxygen and 4′-carbon of the ribose ring. (**d**) Endosomolytic Inf7-siRNA; the pH-triggered fusogenic peptide Inf7 (sequence: GLFE AIEG FIEN GWEG MIDG WYGC) is covalently linked to the 5′-end of the siRNA sense strand

Another strategy to increase the delivery of RNA molecules comprises their conjugation to small molecules or peptides. A prominent example is the attachment of lipophilic moieties to siRNA. Cholesterol was linked to the 3′-end of the siRNA sense strand, resulting in silencing of its target apolipoprotein B (apoB) in vitro and in vivo in the liver and jejunum (Soutschek et al. 2004). This is accomplished by interactions of siRNA with lipoproteins in the circulation and uptake into cell by low-density-lipoprotein-receptor (LDL-receptor). Long-chain fatty acids and bile acids conjugated to apoB-siRNA also mediate silencing of apoB in mice and hamsters (Wolfrum et al. 2007). Furthermore, improved delivery due to cholesterol conjugation has been reported for antisense oligonucleotides (Oberhauser and Wagner 1992). Krützfeldt et al. successfully silenced miR-122, frequently expressed in hepatocytes, by treating mice with a chemically modified cholesterol antagomir. In liver, levels of miR-122 and 3-hydroxy-3-methylglutaryl-CoA-reductase (Hmgcr), a miR-122 target, were decreased after tail-vain injection of miR-122 antagomir. Furthermore, due to the fact that Hmgcr is involved in endogenous cholesterol biosynthesis, cholesterol levels in plasma were significantly reduced (Krutzfeldt et al. 2005).

To improve endosomal escape Dohmen et al. conjugated Inf-7, an acidic peptide analogue of the amino terminus of the influenza virus hemagglutinin, to the 5′-end of the siRNA sense strand (Fig. 9.2d). This Inf7-siRNA showed pH-dependent lytic activity and therefore an increased gene silencing efficiency when delivered with sequence defined oligomers in vitro (Dohmen et al. 2012).

9.3.3 Various RNA Delivery Strategies

In the last one and a half decades several delivery systems for siRNA and miRNA have been developed to overcome the barriers restricting successful therapeutic application (Aigner 2006; Behlke 2006; de Fougerolles et al. 2007; Gao and Huang 2009; Whitehead et al. 2009; Burnett et al. 2011).

One strategy is the delivery of RNAi-inducing agents via viruses. Adenovirus-associated viral (AAV) based and lentiviral based vector systems are commonly used as vesicles to introduce nucleic acid into cells. They are able to transduce dividing and nondividing cells leading to stable expression of siRNAs in these cells (Morris and Rossi 2006). For instance, lentiviral vector systems based on the human U6 small nuclear promoter (U6) have been reported to induce efficient gene silencing of green fluorescent protein (GFP) in vitro and in vivo persistent for several months (Makinen et al. 2006). miR-26a has been successfully delivered into hepatocellular carcinoma (HCC) cells via an AAV vector. HCC cells having low levels of miR-26a have shown suppressed cell proliferation and activated tumor-specific apoptosis in vivo after AAV-mediated miR-26a delivery due to downregulation of cyclins D2 and E2 and induction of G1 arrest (Kota et al. 2009). Nevertheless, therapeutic nucleic acid delivery via viral vectors is fraught with some risks. Viral vector systems are recognized by the innate immune system, they are capable to insert mutagenesis (Shim and Kwon 2010) and they are restricted in size and payload.

An alternative nonviral delivery strategy is utilized by polymer therapeutics (Duncan and Vicent 2013). Due to their negative charges siRNAs and miRNAs are able to form complexes with polycationic agents (Wagner 2012). Polyethylenimine (PEI) is one of the most commonly used polymers for the delivery of nucleic acids (Bolcato-Bellemin et al. 2007). PEI is a linear or branched polymer with protonable amino groups in every third position forming complexes with siRNA and miRNA through electrostatic interactions. Its buffering capacity supports endosomal escape of nucleic acids by osmotic swelling and rupture of the endosome called proton sponge effect (Sonawane et al. 2003; Wagner 2012). Complexes consisting of low molecular weight PEI and siRNA targeting c-erbB2/neu (HER-2) receptor have shown successful delivery in vivo. HER-2 belongs to the epidermal growth factor (EGF) receptor family and is frequently overexpressed in human cancer. After intraperitoneal injection of PEI/siRNA complexes in mice with subcutaneous SKOV-3 ovarian carcinomas, tumor growth was reduced due to downregulation of HER-2 (Urban-Klein et al. 2005). PEI has also been reported to deliver miRNA. Ibrahim et al. applied complexes of PEI and miR-145 or miR-33a in mice with colon carcinoma. miR-145 delivery repressed its targets c-Myc and ERK5 and led to diminished tumor proliferation and enhanced apoptosis. Application of PEI/miR-33a complexes repressed Pim-1 and showed antitumor effects (Ibrahim et al. 2011). However, nucleic acid delivery using PEI is afflicted with certain disadvantages regarding toxicity and biocompatibility. Due to its strong positive charge and its tendency to form aggregates PEI leads to interaction with cell surfaces leading to liver necrosis, activation of lung endothelium, adhesion of aggregated platelets, and shock after systemic injection (Gunther et al. 2011).

The nature-derived polymer chitosan shows a better biocompatibility and also high efficiency in siRNA delivery (Howard et al. 2006; Gao et al. 2009). Another biopolymer based delivery strategy uses cyclodextrin-based carriers. Cyclodextrins are cyclic oligomers consisting of glucose providing an overall amphipathic structure with high biocompatibility and low toxicity (Pecot et al. 2011; Guzman-Villanueva et al. 2012). Heidel et al. used a cyclodextrin polycation system with transferrin (Tf), serving as a targeting ligand to deliver siRNA. Cyclodextrin/siRNA nanoparticles were applied systemically and in multiple doses into non-human primates. Even at high siRNA levels (27 mg siRNA/kg) no toxic effects were observed (Heidel et al. 2007). In another study applying siRNA targeting EWS-FLI1 mRNA, tumor growth was shown to be reduced in mice with Ewing's carcinoma upon intravenous delivery of Tf-targeted cyclodextrin nanoparticles (Hu-Lieskovan et al. 2005).

Cationic lipids present another class of non-viral carriers for the delivery of nucleic acids. Cationic lipids comprise a positive charged headgroup, a hydrophobic chain, and a linker connecting the polar and non-polar regions (Lv et al. 2006). They form lipoplexes with oligonucleotides via electrostatic interactions (Gao and Huang 2009; Gallas et al. 2013). These lipoplexes are characterized by high transfection efficiency, resistance to nucleases, and improved endosomal escape due to interactions with phospholipids within the endosome. However, as with PEI, the strong positive charge of cationic lipids presents a reason for their cytotoxicity and therefore limited use in vivo. They have the tendency to aggregate with anionic serum proteins, causing an immune response (Guzman-Villanueva et al. 2012). To overcome these drawbacks, surface modifications and different compositions of lipids have been investigated.

An important lipid system that has entered advanced preclinical and already clinical studies is based on the encapsulation of oligonucleotides in stable nucleic acid lipid particles (SNALPs), consisting mainly of 1,2-dilinoleyloxy-3-dimethylaminopropane (DLinDMA). siRNA targeting apolipoprotein B (ApoB), a protein essential for the cholesterol transport and metabolism, has been delivered systemically in non-human primates using SNALPs. The efficiency of this system has been demonstrated in significant reductions of ApoB, serum cholesterol, and LDL levels (Zimmermann et al. 2006).

For surface modification, liposome-polycation-hyaluronic acid (LPH) nanoparticles have been used to bring siRNA and miRNA together into lung metastasis bearing mice. Tumor-targeting single-chain antibody fragments (scFv) have been adapted to the LPH nanoparticles. miRNA-34a together with siRNAs targeting c-Myc, MDM2, and VEGF has been encapsulated with these particles. Intravenous injection resulted in downregulation of the siRNAs target genes. miRNA-34a delivered with LPH nanoparticles leads to apoptosis and reduced tumor growth in the lung (Chen et al. 2010).

Hydrogels present an interesting strategy to improve the delivery of siRNA. Krebs et al., for example, integrated siRNA in three different biodegradable hydrogels: calcium crosslinked alginate, photocrosslinked alginate, and collagen. The hydrogels released bioactive siRNA that could silence GFP in stably transfected HEK293 cells to less than 20 % (Krebs et al. 2009). Furthermore, chitosan hydrogel was used to deliver siRNA against tissue transglutaminase (TG2). The biodegradable hydrogel loaded with siRNA and docetaxel was injected intra-tumoral in melanoma or breast cancer bearing mice and reduced tumor growth significantly (Han et al. 2011). Another approach is the delivery of hydrogel nanoparticles. Dunn et al. encapsulated siRNA in particles prepared with a method known as particle replication in nonwetting templates (PRINT). siRNA was delivered as polymerizable prodrug with reducible disulfide linkages to prevent its dissociation from hydrogel nanoparticles before reaching its site of action. The nanoparticles efficiently released siRNA that silenced luciferase when transfected in HeLa/luc cells in vitro without harming cell viability (Dunn et al. 2012).

9.3.4 RNA Delivery via Sequence Defined Oligomers

As multiple different barriers have to be overcome in therapeutic nucleic acid delivery, carriers need to comprise diversified functions. Using classical polymers it is impossible to incorporate various functional domains in precise manner. As a solution, sequence defined oligomers have been designed which combine multifunctionality with chemical precision. These potent nucleic acid delivery vehicles are obtained via solid phase-supported peptide synthesis (SPPS). This method has been adapted by Hartmann et al. (2006) for the synthesis of defined polyamides starting from a fixed resin and alternate assembly of diacids and diamines. Schaffert et al. (2011a) extended the strategy to the synthesis of larger cationic oligomers using Fmoc-protected artificial polyamino acids. Different artificial oligoamino acids were generated (Fig. 9.3a) based on the idea to lend an effective microdomain from PEI: the protonatable diaminoethane motif (Fig. 9.3a bottom left). The artificial building blocks differ in the type of acid (glutaric acid, phthalic acid, succinic acid) and in the number of protonable amine repeats: triethylene tetramine (tt), tetraethylene pentamine (tp), and pentaethylene hexamine (ph). Artificial cationic building blocks such as Gtp, Stp, Ptp, or Sph could be easily merged with natural peptide sequences or lipidic domains, resulting in precise and pure carriers for nucleic acid delivery. Because of the ease of synthesis, small libraries of hundreds of carriers with different topologies (Fig. 9.3b) have been generated (Schaffert et al. 2011b; Salcher et al. 2012; Fröhlich et al. 2012; Troiber et al. 2013). The oligoamine components are only partly cationic at physiological neutral pH, but sufficiently charged for binding nucleic acids via electrostatic interactions. Upon endosomal acidification they gain additional cationic charges required for destabilizing lipid membranes and escape from endolysosomes to the cytoplasm.

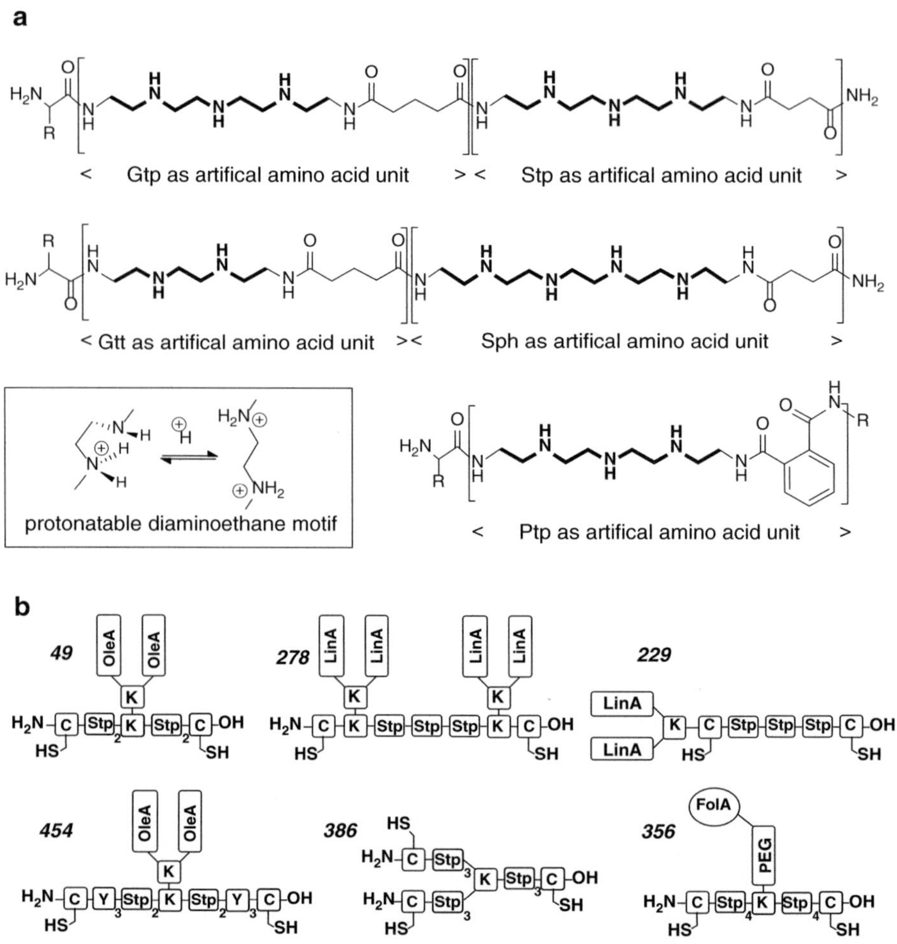

Fig. 9.3 Sequence-defined oligomers as RNA carriers. (**a**) Artificial amino acids: glutaryl tetraethylene pentamine (Gtp), succinoyl tetraethylene pentamine (Stp), glutaroyl triethylene tetramine (Gtt), succinoyl pentaethylene hexamine (Sph), and phthaloyl tetraethylene pentamine (Ptp). Endosomal protonation of the diaminoethane motif is presented in the box (*bottom left*). (**b**) Oligomers: ***49*** (T-shape), ***278*** (U-shape), ***229*** (i-shape), ***454*** (T-shape with tyrosines), ***386*** (three-arm), ***356*** (two-arm with PEG and folate targeting). C, cysteine; K, lysine; Y3, tyrosine trimer; FolA, folic acid

For stable polyplex formation using different nucleic acid cargos (such as siRNA versus pDNA), in general, different polymer requirements have been observed (Kwok and Hart 2011; Scholz and Wagner 2012). Also with sequence-defined oligomers, additional stabilization was required for siRNA (as opposed to pDNA) in forming stable polyplexes. Oligomers modified with two fatty acids (oleic acid or linoleic acid) at central (T-shape) or terminal (i-shape, U-shape) positions show enhanced stability due to hydrophobic interactions. In addition, terminal cysteines (two-arms, three-arms) which are able to form disulfide bridges after siRNA complexation stabilize polyplexes in a covalent manner. With regard to the artificial amino acids, Stp or Sph containing oligomers showed a higher transfection efficiency than Ptp or Gtt (Salcher et al. 2012; Fröhlich et al. 2012). High stability is a very important feature for polyplex application in vivo. Incorporation of tyrosine trimers increased stability in vitro, in full serum (as evaluated by FCS), and in vivo (as evaluated by NIR fluorescence bioimaging in mice) (Troiber et al. 2012, 2013; Fröhlich et al. 2012).

Modifications of oligomers with lipid moieties did not only improve the stability of polyplexes but also their endosomal escape. The lipooligomers due to their amphiphilic character cationization at endosomal pH of 5–6 are able to lyse lipid membranes in a favorable pH-specific manner (Schaffert et al. 2011b). Oleic or linoleic acid emerged as most potent fatty acids in gene silencing without significant cytotoxicity (Fröhlich et al. 2012).

Alternatively, lytic peptides can be introduced to enhance endosomal escape due to their ability to directly disrupt endosomal membranes (Kos and Wagner 2013). For example, influenza peptide-siRNA conjugates (as mentioned before) can successfully enhance endosomal escape when delivered with oligomers (Dohmen et al. 2012). A pH-responsive endosomolytic form of the peptide melittin has been previously shown to improve transfection efficiency when incorporated into polymers. The amines of melittin had to be modified with dimethylmaleic anhydride (DMMAn) to reduce extracellular toxicity. In the acidic endosome these protecting groups are cleaved and the lytic activity of melittin is restored (Meyer et al. 2008, 2009).

Polyplexes are usually positively charged because of surplus of cationic oligomer at the surface, which can lead to aggregation and unspecific interaction with blood compounds and non-target cells. Hydrophilic polyethylene glycol (PEG) is widely used for nanoparticle shielding to overcome this drawback. Furthermore, the incorporation of PEG protects polyplexes from recognition by the innate immune system, decreases cytotoxicity, and increases solubility, stability, and circulation time (Lee and Kim 2005; Meyer et al. 2008; Whitehead et al. 2009). PEG can also serve as spacer between a targeting ligand and the cationic oligomers, increasing the binding of ligand and receptor (Lee and Kim 2005; Dohmen et al. 2012).

In sequence-defined oligomers, incorporation of targeting ligands was proven as useful. For example, peptide B6 targeting the transferrin receptor, or cyclopeptide c(RGDfK) targeting $\alpha_v\beta_3$ integrin, was coupled to oligomers using PEG spacers. These targeted oligomers showed high ligand-dependent transfection efficiency for pDNA (Martin et al. 2012). To elaborate the targeting concept for siRNA delivery, Dohmen et al. developed folic acid modified oligomers, targeting the folic acid receptor overexpressed in tumor cells. The polyplexes consisting of these oligomers and Inf7-siRNA showed efficient and ligand specific gene silencing in vitro and an increased retention in KB tumor tissue due to targeting effects in vivo (Dohmen et al. 2012).

Screening oligomer efficiencies in vitro does not necessarily inform about efficiencies in the real in vivo situation (Fröhlich et al. 2012). Edinger et al. (2014) examined the in vitro and in vivo siRNA delivery and antitumoral effects of three oligomers: one T-shaped lipo-oligomer containing two terminal cysteines, four Stp units, and two oleic acids (***49***); one i-shaped lipo-oligomer containing two terminal cysteines, three Stp units, and two linoleic acids (***229***); and one branched three-arm oligomer containing terminal cysteines and nine Stp units (***386***). In vitro the oligomers showed similar gene downregulation in murine neuroblastoma cells when delivered with siRNA targeting RAS related nuclear protein (Ran). Ran is a small Ras-related GTPase protein involved in many aspects of nuclear functions, such as transport in and out of the nucleus or postmitotic nuclear assembly required for cell survival. Knockdown was found to result in preferential cell death in activated K-Ras mutant tumor cells (Morgan-Lappe et al. 2007). In vivo in the same neuroblastoma tumor model, ***386*** showed the best antitumoral efficiency reducing tumor growth. Oligomer ***49*** emerged as second best delivery vesicle in vivo, whereas ***229*** failed in vivo tumor treatment (Edinger et al. 2014). These first data demonstrate the encouraging potential of precise sequence-defined oligomers to serve as delivery vesicles for therapeutic nucleic acids.

9.4 Clinical Application of RNAi for Cancer Therapy

Several human clinical trials using RNAi or related RNA-modulating nano-oncological drugs for cancer therapy are currently carried out (see Table 9.1). Major investments in the development of these therapeutics have been made in the last years with the prospect of commercial drugs in the next 5–10 years (Haussecker 2012).

9.4.1 siRNA Therapeutics

Davis and colleagues firstly reported systemic siRNA delivery in humans for cancer treatment via targeted nanoparticles. Their formulation consisted of a cyclodextrin-containing polymer (CDP); adamantane conjugated to PEG (AD-PEG) for steric stabilization; AD-PEG with the targeting ligand transferrin binding to transferrin receptors, which are frequently upregulated in cancer cells; and siRNA targeting the M2 subunit of ribonucleotide reductase (RRM2), a crucial factor of tumor malignancy. These nanoparticles, named CALAA-01, were systemically administered to patients with solid cancers. The study was carried out on three patients receiving different doses (18, 24, and 30 mg/m^2) on days 1, 3, 8, and 10 of a 21-day cycle by a 30 min intravenous infusion. Biopsies from tumor tissues were analyzed and a reduction of mRNA and protein levels of RRM2 were observed. The RNAi was confirmed by 5′-RNA-ligand-mediated RACE PCR technique detecting a RRM2 mRNA fragment in tumor tissue of the patient receiving the highest dose (Davis 2009; Davis et al. 2010).

Atu027 is a lipid-based siRNA nanoparticle approach which has recently been tested in a phase I clinical trial. Atu027 contains siRNA targeting protein kinase N3 (PKN3). PKN3 mediates malignant cell growth (Santel et al. 2006). Atu027 has been

Table 9.1 RNAi and related nano-oncologicals in clinical evaluation

Name	Nucleic acid	Carrier type	Target	Cancer type	Phase	Status	Company
CALAA-01	siRNA	Cyclodextrins-containing polymer (CDP)	M2 subunit of ribonucleotide reductase (RRM2)	Solid cancer	I	Terminated	Calando Pharmaceuticals
Atu027	siRNA	Lipid nanoparticle	Protein kinase N3 (PKN3)	Advanced solid tumors	I	Completed	Silence Therapeutics AG
				Advanced pancreatic cancer	Ib/IIa in combination with gemcitabine	Recruiting	
ALN-VSP	Two siRNAs	Lipid nanoparticle	Vascular endothelial growth factor A (VEGF) and kinesin spindle protein (KSP)	Advanced tumors and hepatic and extrahepatic metastases	I	Completed	Alnylam Pharmaceuticals
MRX34	miR-34 mimic	Lipid nanoparticle	miRNA replacement	Primary liver cancer	I	Recruiting	Mirna Therapeutics
EZN-2968	RNA antagonist		α-subunit of Hypoxia-inducible factor-1 (HIF-1α)	Solid tumors	I	Completed	Enzon Pharmaceuticals
				Solid tumors with liver metastases	I	Completed	
OGX-427	Antisense oligonucleotide		Chaperone protein Hsp27	Metastatic castration-resistant prostate cancer (mCRPC)	I	Completed	OncoGenex Pharmaceuticals
				CRPC, non-small cell lung cancer (NSCLC), or ovarian cancer	I	Completed	
				mCRPC	II	Completed	

shown to silence PKN3 when it was systemically delivered in mice, rats, and non-human primates without stimulation of the innate immune system. In pancreatic and prostate cancer mouse models Atu027 decreased tumor growth and lymph node metastasis formation (Aleku et al. 2008). In a phase I clinical trial 24 patients suffering from advanced solid tumors were treated with Atu027. Atu027 was well tolerated, one patient showed disease stabilization and one patient showed reduction of pulmonary metastases demonstrating the potential of this siRNA delivery system. Currently participants are recruited for a clinical trial examining the effects of Atu027 with Gemcitabine in advanced and pancreatic cancer (www.clinicaltrials.gov).

A lipidic dual siRNA formulation was recently investigated in a phase I oncology trial. The formulation (ALN-VSP) consists of a lipid nanoparticle with two siRNAs, one targeting vascular endothelial growth factor A (VEGF) and the other one targeting kinesin spindle protein (KSP). RNAi specific effectiveness of ALN-VSP as well as antitumor activity has been proven previously in preclinical animal studies. In a phase I dose-escalation study 41 patients with advanced tumors and hepatic and extrahepatic metastases were intravenously treated with ALN-VSP with doses of 0.01–1.5 mg/kg every 2 weeks. ALN-VSP was generally well tolerated, and occurring adverse events were comparable to other targeted chemotherapies. Furthermore, both siRNAs and cleavage products of VEGF mRNA could be detected in hepatic and extrahepatic tumor biopsies, indicating that the antitumoral effects of ALN-VSP underlie an RNAi mechanism. Fifty percent of patients treated with doses greater than 0.7 mg/kg achieved stable disease including one patient with endometrial cancer and multiple liver metastases, who attained a major response, meaning disappearance of all target and nontarget lesions (Tabernero et al. 2013).

9.4.2 miRNA Replacement Therapy

MRX34 was developed to deliver miR-34 mimic, a synthetic double stranded miRNA of 20–25 nucleotides, which will be the first microRNA assessed in a clinical phase I study. MRX34 is a delivery system based on a mixture of different lipids, such as palmitoyl oleoyl phosphatidyl choline or cholesterol. miR-34 is frequently downregulated in many human cancer types leading to metastasis, anti-apoptosis, chemoresistance, and tumor proliferation (Bouchie 2013). Anti-tumor activity of miR-34 has been shown in non-small cell lung cancer (NSCLC) in mice. miR-34 was systemically delivered using a neutral lipid emulsion, which led to a decrease of proliferation markers, increased apoptosis of tumors, and therefore reduced tumor burden (Trang et al. 2011). The efficiency of MRX34 has been proven in preclinical studies. In a survival study MRX34 was intravenously delivered in mice with HCC. All treated animals stayed alive and appeared healthy in comparison to control groups. Patients suffering from primary liver cancer and liver metastasis from other cancers are currently recruited for a multicenter phase I clinical trial to investigate safety, pharmacokinetics, and pharmacodynamics of MRX34 (see www.clinicaltrials.gov for details).

9.4.3 *RNAi Oligonucleotide Based Therapeutics*

Two phase I clinical trials have been initiated for EZN-2968, a RNA antagonist. Chemically it is an oligonucleotide of 16 nucleotides, of which 6 are exchanged by LNA nucleotides. EZN-2968 downregulates the expression of the α-subunit of Hypoxia-inducible factor-1 (HIF-1α). Upregulation of HIF-1α signals hypoxia or activation of growth factor pathways and therefore malignant development of many cancer types. Intravenous application of EZN-2968 in mice led to dose-dependent downregulation of HIF-1α mRNA in the liver and tumor reduction (Greenberger et al. 2008). A phase I study on patients with solid tumors showed well tolerability of EZN-2968 and tumor shrinkage in two patients (of 18). Another pilot study of EZN-2986 has been performed on patients suffering from solid tumors with liver metastases and completed in September 2013, but the results are not published yet (see www.clinicaltrials.gov for details).

OGX-427 is an antisense oligonucleotide targeting the chaperone protein Hsp27. Hsp27 is upregulated in many cancer types such as breast and prostate cancer due to enhanced cellular stresses. Shiota et al. reported that Hsp27 causes EMT in prostate cancer and silencing of Hsp27 decreases migration in vitro and in PC-3 M tumor-bearing mice. They found that Hsp27 enhances EMT by mediating IL-6-dependent STAT3 phosphorylation, nuclear translocation, and STAT3 binding to the Twist promoter. In a phase I study patients with metastatic castration-resistant prostate cancer (mCRPC) received 400–1,000 mg of OGX-427, and levels of circulating tumor cells (CTC) were analyzed. OGX-427 treatment reduced CTC counts indicating the first evidence of anticancer potential of OGX-427 in the clinics (Shiota et al. 2013). In another phase I study patients with CRPC, NSCLC, or ovarian cancer were treated intravenously with OGX-427 alone or with docetaxel in order to examine adverse effects and dose for phase II trials. OGX-427 treatment led to infusion reactions but was overall well tolerated. Furthermore OGX-427 showed antitumoral effects like changes in tumor markers and CTC levels (Hotte et al. 2010). In a phase II study patients suffering from mCRPC were treated with OGX-427 plus prednisone or with prednisone alone. 82 % of patients receiving the combined therapy showed a decrease in prostate-specific antigen (PSA), and 60 % had a CTC conversion from ≥ 5 to $<5/7.5$ ml compared to 40 and 20 % respectively of patients treated with prednisone alone (Chi et al. 2012).

9.5 Conclusions

The modulation of mRNA levels using siRNA, miRNA, antagomirs, or antisense oligonucleotides offers the opportunity to treat various diseases. In many cancer types the natural upregulation of onco-miRs or loss of suppressor-miRs leads to invasion, metastasis, chemoresistance, and poor clinical prognosis for patients. Delivery of exogenous suppressor-miRs or anti onco-miR antagomirs may revert cancer aggressiveness by modulating translation of multiple target gene mRNAs. Complementary, artificial siRNAs can be designed for specific downregulation of basically any target gene that promotes tumor development. Despite the great potential of these RNAs the therapeutic application still faces considerable drawbacks, like instability in vivo, immune response, weak tissue targeting, and inefficient cellular uptake. Various synthetic delivery strategies are taken to overcome these barriers. Modern chemistry using solid-phase supported synthesis provides the possibility to produce chemically modified oligonucleotides with enhanced stability, potency, and reduced side effects. Also sequence-defined oligomers as nucleic acid carriers containing stabilizing, lytic, and targeting moieties can be designed in a pure and precise way by solid-phase supported synthesis.

While RNAi nanodrugs have already shown very encouraging clinical effects especially in liver-based diseases, up to now no RNAi-based formulations are on the market for cancer treatment. In the last 15 years, billions of dollars have been spent on the development of RNAi therapeutics indicating the significance but also the complexity of this field of research. However, several promising RNAi-based or RNAi-modulating drugs for treating human cancer are now in the pipeline of phase I and phase II clinical studies, hopefully leading to efficient and safe nano-oncologicals in the near future.

Acknowledgments Funding of related research of Katharina Müller and Ernst Wagner by DFG SFB 1032 project B4 is gratefully acknowledged.

References

Aigner A (2006) Gene silencing through RNA interference (RNAi) in vivo: strategies based on the direct application of siRNAs. J Biotechnol 124(1):12–25

Aleku M, Schulz P, Keil O, Santel A, Schaeper U, Dieckhoff B, Janke O, Endruschat J, Durieux B, Roder N, Loffler K, Lange C, Fechtner M, Mopert K, Fisch G, Dames S, Arnold W, Jochims K, Giese K, Wiedenmann B, Scholz A, Kaufmann J (2008) Atu027, a liposomal small interfering RNA formulation targeting protein kinase N3, inhibits cancer progression. Cancer Res 68(23):9788–9798. doi:10.1158/0008-5472.CAN-08-2428

Asangani IA, Rasheed SA, Nikolova DA, Leupold JH, Colburn NH, Post S, Allgayer H (2008) MicroRNA-21 (miR-21) post-transcriptionally downregulates tumor suppressor Pdcd4 and stimulates invasion, intravasation and metastasis in colorectal cancer. Oncogene 27(15): 2128–2136. doi:10.1038/sj.onc.1210856

Bartel DP (2004) MicroRNAs: genomics, biogenesis, mechanism, and function. Cell 116(2):281–297

Bartel DP (2009) MicroRNAs: target recognition and regulatory functions. Cell 136(2):215–233. doi:10.1016/j.cell.2009.01.002

Bartlett DW, Davis ME (2006) Insights into the kinetics of siRNA-mediated gene silencing from live-cell and live-animal bioluminescent imaging. Nucleic Acids Res 34(1):322–333

Behlke MA (2006) Progress towards in vivo use of siRNAs. Mol Ther 13(4):644–670

Birmingham A, Anderson EM, Reynolds A, Ilsley-Tyree D, Leake D, Fedorov Y, Baskerville S, Maksimova E, Robinson K, Karpilow J, Marshall WS, Khvorova A (2006) 3′ UTR seed matches, but not overall identity, are associated with RNAi off-targets. Nat Methods 3(3): 199–204. doi:10.1038/nmeth854

Bloomston M, Frankel WL, Petrocca F, Volinia S, Alder H, Hagan JP, Liu CG, Bhatt D, Taccioli C, Croce CM (2007) MicroRNA expression patterns to differentiate pancreatic adenocarcinoma from normal pancreas and chronic pancreatitis. JAMA 297(17):1901–1908. doi:10.1001/jama.297.17.1901

Bolcato-Bellemin AL, Bonnet ME, Creusat G, Erbacher P, Behr JP (2007) Sticky overhangs enhance siRNA-mediated gene silencing. Proc Natl Acad Sci U S A 104(41):16050–16055

Bouchie A (2013) First microRNA mimic enters clinic. Nat Biotechnol 31(7):577. doi:10.1038/nbt0713-577

Bumcrot D, Manoharan M, Koteliansky V, Sah DW (2006) RNAi therapeutics: a potential new class of pharmaceutical drugs. Nat Chem Biol 2(12):711–719

Burnett JC, Rossi JJ, Tiemann K (2011) Current progress of siRNA/shRNA therapeutics in clinical trials. Biotechnol J 6(9):1130–1146. doi:10.1002/biot.201100054

Calin GA, Dumitru CD, Shimizu M, Bichi R, Zupo S, Noch E, Aldler H, Rattan S, Keating M, Rai K, Rassenti L, Kipps T, Negrini M, Bullrich F, Croce CM (2002) Frequent deletions and down-regulation of micro- RNA genes miR15 and miR16 at 13q14 in chronic lymphocytic leukemia. Proc Natl Acad Sci U S A 99(24):15524–15529. doi:10.1073/pnas.242606799

Calin GA, Cimmino A, Fabbri M, Ferracin M, Wojcik SE, Shimizu M, Taccioli C, Zanesi N, Garzon R, Aqeilan RI, Alder H, Volinia S, Rassenti L, Liu X, Liu CG, Kipps TJ, Negrini M, Croce CM (2008) MiR-15a and miR-16-1 cluster functions in human leukemia. Proc Natl Acad Sci U S A 105(13):5166–5171. doi:10.1073/pnas.0800121105

Chen Y, Zhu X, Zhang X, Liu B, Huang L (2010) Nanoparticles modified with tumor-targeting scFv deliver siRNA and miRNA for cancer therapy. Mol Ther 18(9):1650–1656. doi:10.1038/mt.2010.136

Chi KN, Hotte SJ, Ellard S, Gingerich JR, Joshua AM, Yu EY, Gleave ME (2012) A randomized phase II study of OGX-427 plus prednisone (P) versus P alone in patients (pts) with metastatic castration resistant prostate cancer (CRPC). J Clin Oncol 30(Suppl) [abstr 4514]

Croce CM (2009) Causes and consequences of microRNA dysregulation in cancer. Nat Rev Genet 10(10):704–714. doi:10.1038/nrg2634

Davis ME (2009) The first targeted delivery of siRNA in humans via a self-assembling, cyclodextrin polymer-based nanoparticle: from concept to clinic. Mol Pharm 6(3):659–668. doi:10.1021/mp900015y

Davis ME, Zuckerman JE, Choi CH, Seligson D, Tolcher A, Alabi CA, Yen Y, Heidel JD, Ribas A (2010) Evidence of RNAi in humans from systemically administered siRNA via targeted nanoparticles. Nature 464:1067–1070. doi:10.1038/nature08956

de Fougerolles A, Vornlocher HP, Maraganore J, Lieberman J (2007) Interfering with disease: a progress report on siRNA-based therapeutics. Nat Rev Drug Discov 6 (6):443–453. doi:10.1038/nrd2310

Dohmen C, Edinger D, Frohlich T, Schreiner L, Lachelt U, Troiber C, Radler J, Hadwiger P, Vornlocher HP, Wagner E (2012) Nanosized multifunctional polyplexes for receptor-mediated SiRNA delivery. ACS Nano 6(6):5198–5208. doi:10.1021/nn300960m

Duncan R, Vicent MJ (2013) Polymer therapeutics-prospects for 21st century: the end of the beginning. Adv Drug Deliv Rev 65(1):60–70. doi:10.1016/j.addr.2012.08.012

Dunn SS, Tian S, Blake S, Wang J, Galloway AL, Murphy A, Pohlhaus PD, Rolland JP, Napier ME, DeSimone JM (2012) Reductively responsive siRNA-conjugated hydrogel nanoparticles for gene silencing. J Am Chem Soc 134(17):7423–7430. doi:10.1021/ja300174v

Edinger D, Kläger R, Troiber C, Dohmen C, Wagner E (2014) Gene silencing and antitumoral effects of Eg5 or Ran siRNA oligoaminoamide polyplexes. Drug Deliv Transl Res 4:84–95

Eis PS, Tam W, Sun L, Chadburn A, Li Z, Gomez MF, Lund E, Dahlberg JE (2005) Accumulation of miR-155 and BIC RNA in human B cell lymphomas. Proc Natl Acad Sci U S A 102(10):3627–3632. doi:10.1073/pnas.0500613102

Elbashir SM, Harborth J, Lendeckel W, Yalcin A, Weber K, Tuschl T (2001) Duplexes of 21-nucleotide RNAs mediate RNA interference in cultured mammalian cells. Nature 411(6836):494–498

Elmen J, Thonberg H, Ljungberg K, Frieden M, Westergaard M, Xu Y, Wahren B, Liang Z, Orum H, Koch T, Wahlestedt C (2005) Locked nucleic acid (LNA) mediated improvements in siRNA stability and functionality. Nucleic Acids Res 33(1):439–447. doi:10.1093/nar/gki193

Fröhlich T, Edinger D, Kläger R, Troiber C, Salcher E, Badgujar N, Martin I, Schaffert D, Cengizeroglu A, Hadwiger P, Vornlocher HP, Wagner E (2012) Structure-activity relationships of siRNA carriers based on sequence-defined oligo (ethane amino) amides. J Control Release. doi:10.1016/j.jconrel.2012.03.018. pii: S0168-3659(12)00206-4

Gallas A, Alexander C, Davies MC, Puri S, Allen S (2013) Chemistry and formulations for siRNA therapeutics. Chem Soc Rev. doi:10.1039/c3cs35520a

Gao K, Huang L (2009) Nonviral methods for siRNA delivery. Mol Pharm 6(3):651–658. doi:10.1021/mp800134q

Gao S, Dagnaes-Hansen F, Nielsen EJ, Wengel J, Besenbacher F, Howard KA, Kjems J (2009) The effect of chemical modification and nanoparticle formulation on stability and biodistribution of siRNA in mice. Mol Ther 17(7):1225–1233. doi:10.1038/mt.2009.91

Greenberger LM, Horak ID, Filpula D, Sapra P, Westergaard M, Frydenlund HF, Albaek C, Schroder H, Orum H (2008) A RNA antagonist of hypoxia-inducible factor-1alpha, EZN-2968, inhibits tumor cell growth. Mol Cancer Ther 7(11):3598–3608. doi:10.1158/1535-7163. MCT-08-0510

Gregory PA, Bert AG, Paterson EL, Barry SC, Tsykin A, Farshid G, Vadas MA, Khew-Goodall Y, Goodall GJ (2008) The miR-200 family and miR-205 regulate epithelial to mesenchymal transition by targeting ZEB1 and SIP1. Nat Cell Biol 10(5):593–601. doi:10.1038/ncb1722

Guilford P, Hopkins J, Harraway J, McLeod M, McLeod N, Harawira P, Taite H, Scoular R, Miller A, Reeve AE (1998) E-cadherin germline mutations in familial gastric cancer. Nature 392(6674):402–405. doi:10.1038/32918

Gunther M, Lipka J, Malek A, Gutsch D, Kreyling W, Aigner A (2011) Polyethylenimines for RNAi-mediated gene targeting in vivo and siRNA delivery to the lung. Eur J Pharm Biopharm 77(3):438–449. doi:10.1016/j.ejpb.2010.11.007

Guzman-Villanueva D, El-Sherbiny IM, Herrera-Ruiz D, Vlassov AV, Smyth HD (2012) Formulation approaches to short interfering RNA and MicroRNA: challenges and implications. J Pharm Sci 101(11):4046–4066. doi:10.1002/jps.23300

Han HD, Mora EM, Roh JW, Nishimura M, Lee SJ, Stone RL, Bar-Eli M, Lopez-Berestein G, Sood AK (2011) Chitosan hydrogel for localized gene silencing. Cancer Biol Ther 11(9):839–845

Hartmann L, Krause E, Antonietti M, Borner HG (2006) Solid-phase supported polymer synthesis of sequence-defined, multifunctional poly(amidoamines). Biomacromolecules 7(4):1239–1244

Haussecker D (2012) The business of RNAi therapeutics in 2012. Mol Ther Nucleic Acids 1:e8. doi:10.1038/mtna.2011.9

Heidel JD, Yu Z, Liu JY, Rele SM, Liang Y, Zeidan RK, Kornbrust DJ, Davis ME (2007) Administration in non-human primates of escalating intravenous doses of targeted nanoparticles containing ribonucleotide reductase subunit M2 siRNA. Proc Natl Acad Sci U S A 104(14):5715–5721

Hornung V, Guenthner-Biller M, Bourquin C, Ablasser A, Schlee M, Uematsu S, Noronha A, Manoharan M, Akira S, de Fougerolles A, Endres S, Hartmann G (2005) Sequence-specific potent induction of IFN-alpha by short interfering RNA in plasmacytoid dendritic cells through TLR7. Nat Med 11(3):263–270. doi:10.1038/nm1191

Hotte SJ, Yu EY, Hirte HW, Higano CS, Gleave ME, Chi KN (2010) Phase I trial of OGX-427, a 2′methoxyethyl antisense oligonucleotide (ASO), against heat shock protein 27 (Hsp27): final results. J Clin Oncol 28:15s (suppl; abstr 3077)

Howard KA, Rahbek UL, Liu X, Damgaard CK, Glud SZ, Andersen MO, Hovgaard MB, Schmitz A, Nyengaard JR, Besenbacher F, Kjems J (2006) RNA interference in vitro and in vivo using a novel chitosan/siRNA nanoparticle system. Mol Ther 14(4):476–484. doi:10.1016/j.ymthe.2006.04.010. pii: S1525-0016(06)00180-8

Howe EN, Cochrane DR, Richer JK (2011) Targets of miR-200c mediate suppression of cell motility and anoikis resistance. Breast Cancer Res 13(2):R45. doi:10.1186/bcr2867

Hu-Lieskovan S, Heidel JD, Bartlett DW, Davis ME, Triche TJ (2005) Sequence-specific knockdown of EWS-FLI1 by targeted, nonviral delivery of small interfering RNA inhibits tumor growth in a murine model of metastatic Ewing's sarcoma. Cancer Res 65(19):8984–8992

Ibrahim AF, Weirauch U, Thomas M, Grunweller A, Hartmann RK, Aigner A (2011) MicroRNA replacement therapy for miR-145 and miR-33a is efficacious in a model of colon carcinoma. Cancer Res 71(15):5214–5224. doi:10.1158/0008-5472.CAN-10-4645

Iorio MV, Croce CM (2012) MicroRNA dysregulation in cancer: diagnostics, monitoring and therapeutics. A comprehensive review. EMBO Mol Med 4(3):143–159. doi:10.1002/emmm.201100209

Iorio MV, Ferracin M, Liu CG, Veronese A, Spizzo R, Sabbioni S, Magri E, Pedriali M, Fabbri M, Campiglio M, Menard S, Palazzo JP, Rosenberg A, Musiani P, Volinia S, Nenci I, Calin GA, Querzoli P, Negrini M, Croce CM (2005) MicroRNA gene expression deregulation in human breast cancer. Cancer Res 65(16):7065–7070. doi:10.1158/0008-5472.CAN-05-1783

Jackson AL, Burchard J, Schelter J, Chau BN, Cleary M, Lim L, Linsley PS (2006) Widespread siRNA "off-target" transcript silencing mediated by seed region sequence complementarity. RNA 12(7):1179–1187. doi:10.1261/rna.25706

Judge AD, Sood V, Shaw JR, Fang D, McClintock K, MacLachlan I (2005) Sequence-dependent stimulation of the mammalian innate immune response by synthetic siRNA. Nat Biotechnol 23(4):457–462. doi:10.1038/nbt1081

Kanasty RL, Whitehead KA, Vegas AJ, Anderson DG (2012) Action and reaction: the biological response to siRNA and its delivery vehicles. Mol Ther 20(3):513–524. doi:10.1038/mt.2011.294

Khvorova A, Reynolds A, Jayasena SD (2003) Functional siRNAs and miRNAs exhibit strand bias. Cell 115(2):209–216

Kim VN (2005) MicroRNA biogenesis: coordinated cropping and dicing. Nat Rev Mol Cell Biol 6(5):376–385. doi:10.1038/nrm1644

Kim DH, Rossi JJ (2007) Strategies for silencing human disease using RNA interference. Nat Rev Genet 8(3):173–184. doi:10.1038/nrg2006

Kong YW, Ferland-McCollough D, Jackson TJ, Bushell M (2012) MicroRNAs in cancer management. Lancet Oncol 13(6):e249–e258. doi:10.1016/S1470-2045(12)70073-6

Kopp F, Oak PS, Wagner E, Roidl A (2012) miR-200c sensitizes breast cancer cells to doxorubicin treatment by decreasing TrkB and Bmi1 expression. PLoS One 7(11):e50469. doi:10.1371/journal.pone.0050469

Kos P, Wagner E (2013) Polymers for siRNA delivery: combining precision with multifunctionality. Chem Today 31(2):6–10

Kota J, Chivukula RR, O'Donnell KA, Wentzel EA, Montgomery CL, Hwang HW, Chang TC, Vivekanandan P, Torbenson M, Clark KR, Mendell JR, Mendell JT (2009) Therapeutic microRNA delivery suppresses tumorigenesis in a murine liver cancer model. Cell 137(6):1005–1017. doi:10.1016/j.cell.2009.04.021. pii: S0092-8674(09)00446-2

Krebs MD, Jeon O, Alsberg E (2009) Localized and sustained delivery of silencing RNA from macroscopic biopolymer hydrogels. J Am Chem Soc 131(26):9204–9206. doi:10.1021/ja9037615

Krol J, Loedige I, Filipowicz W (2010) The widespread regulation of microRNA biogenesis, function and decay. Nat Rev Genet 11(9):597–610. doi:10.1038/nrg2843

Krutzfeldt J, Rajewsky N, Braich R, Rajeev KG, Tuschl T, Manoharan M, Stoffel M (2005) Silencing of microRNAs in vivo with 'antagomirs'. Nature 438(7068):685–689. doi:10.1038/nature04303

Kwok A, Hart SL (2011) Comparative structural and functional studies of nanoparticle formulations for DNA and siRNA delivery. Nanomedicine 7(2):210–219. doi:10.1016/j.nano.2010.07.005. pii: S1549-9634(10)00239-X

Lee M, Kim SW (2005) Polyethylene glycol-conjugated copolymers for plasmid DNA delivery. Pharm Res 22(1):1–10

Lim LP, Lau NC, Garrett-Engele P, Grimson A, Schelter JM, Castle J, Bartel DP, Linsley PS, Johnson JM (2005) Microarray analysis shows that some microRNAs downregulate large numbers of target mRNAs. Nature 433(7027):769–773. doi:10.1038/nature03315

Lori F (2011) DermaVir: a plasmid DNA-based nanomedicine therapeutic vaccine for the treatment of HIV/AIDS. Expert Rev Vaccines 10(10):1371–1384. doi:10.1586/erv.11.118

Lv H, Zhang S, Wang B, Cui S, Yan J (2006) Toxicity of cationic lipids and cationic polymers in gene delivery. J Control Release 114(1):100–109. doi:10.1016/j.jconrel.2006.04.014

Ma JB, Ye K, Patel DJ (2004) Structural basis for overhang-specific small interfering RNA recognition by the PAZ domain. Nature 429(6989):318–322. doi:10.1038/nature02519

Makinen PI, Koponen JK, Karkkainen AM, Malm TM, Pulkkinen KH, Koistinaho J, Turunen MP, Yla-Herttuala S (2006) Stable RNA interference: comparison of U6 and H1 promoters in endothelial cells and in mouse brain. J Gene Med 8(4):433–441. doi:10.1002/jgm.860

Martin I, Dohmen C, Mas-Moruno C, Troiber C, Kos P, Schaffert D, Lachelt U, Teixido M, Gunther M, Kessler H, Giralt E, Wagner E (2012) Solid-phase-assisted synthesis of targeting peptide-PEG-oligo(ethane amino)amides for receptor-mediated gene delivery. Org Biomol Chem 10(16):3258–3268. doi:10.1039/c2ob06907e

Medzhitov R, Janeway CA Jr (2002) Decoding the patterns of self and nonself by the innate immune system. Science 296(5566):298–300. doi:10.1126/science.1068883

Meister G, Tuschl T (2004) Mechanisms of gene silencing by double-stranded RNA. Nature 431(7006):343–349

Meyer M, Philipp A, Oskuee R, Schmidt C, Wagner E (2008) Breathing life into polycations: functionalization with pH-responsive endosomolytic peptides and polyethylene glycol enables siRNA delivery. J Am Chem Soc 130(11):3272–3273

Meyer M, Dohmen C, Philipp A, Kiener D, Maiwald G, Scheu C, Ogris M, Wagner E (2009) Synthesis and biological evaluation of a bioresponsive and endosomolytic siRNA-polymer conjugate. Mol Pharm 6(3):752–762. doi:10.1021/mp9000124

Morgan-Lappe SE, Tucker LA, Huang X, Zhang Q, Sarthy AV, Zakula D, Vernetti L, Schurdak M, Wang J, Fesik SW (2007) Identification of Ras-related nuclear protein, targeting protein for Xenopus kinesin-like protein 2, and stearoyl-CoA desaturase 1 as promising cancer targets from an RNAi-based screen. Cancer Res 67(9):4390–4398

Morris KV, Rossi JJ (2006) Lentiviral-mediated delivery of siRNAs for antiviral therapy. Gene Ther 13(6):553–558. doi:10.1038/sj.gt.3302688

Obad S, dos Santos CO, Petri A, Heidenblad M, Broom O, Ruse C, Fu C, Lindow M, Stenvang J, Straarup EM, Hansen HF, Koch T, Pappin D, Hannon GJ, Kauppinen S (2011) Silencing of microRNA families by seed-targeting tiny LNAs. Nat Genet 43(4):371–378. doi:10.1038/ng.786

Oberhauser B, Wagner E (1992) Effective incorporation of 2′-*O*-methyl-oligoribonucleotides into liposomes and enhanced cell association through modification with thiocholesterol. Nucleic Acids Res 20(3):533–538

Pecot CV, Calin GA, Coleman RL, Lopez-Berestein G, Sood AK (2011) RNA interference in the clinic: challenges and future directions. Nat Rev Cancer 11(1):59–67. doi:10.1038/nrc2966

Rana TM (2007) Illuminating the silence: understanding the structure and function of small RNAs. Nat Rev Mol Cell Biol 8(1):23–36. doi:10.1038/nrm2085

Rettig GR, Behlke MA (2012) Progress toward in vivo use of siRNAs-II. Mol Ther 20(3):483–512. doi:10.1038/mt.2011.263

Salcher EE, Kos P, Frohlich T, Badgujar N, Scheible M, Wagner E (2012) Sequence-defined four-arm oligo(ethanamino)amides for pDNA and siRNA delivery: impact of building blocks on efficacy. J Control Release 164(3):380–386. doi:10.1016/j.jconrel.2012.06.023. pii: S0168-3659(12)00523-8

Santel A, Aleku M, Keil O, Endruschat J, Esche V, Fisch G, Dames S, Loffler K, Fechtner M, Arnold W, Giese K, Klippel A, Kaufmann J (2006) A novel siRNA-lipoplex technology for RNA interference in the mouse vascular endothelium. Gene Ther 13(16):1222–1234. doi:10.1038/sj.gt.3302777

Schaffert D, Badgujar N, Wagner E (2011a) Novel Fmoc-polyamino acids for solid-phase synthesis of defined polyamidoamines. Org Lett 13(7):1586–1589. doi:10.1021/ol200381z

Schaffert D, Troiber C, Salcher EE, Frohlich T, Martin I, Badgujar N, Dohmen C, Edinger D, Klager R, Maiwald G, Farkasova K, Seeber S, Jahn-Hofmann K, Hadwiger P, Wagner E (2011b) Solid-phase synthesis of sequence-defined T-, i-, and U-shape polymers for pDNA and siRNA delivery. Angew Chem Int Ed Engl 50(38):8986–8989. doi:10.1002/anie.201102165

Scholz C, Wagner E (2012) Therapeutic plasmid DNA versus siRNA delivery: common and different tasks for synthetic carriers. J Control Release 161(2):554–565. doi:10.1016/j.jconrel.2011.11.014. pii: S0168-3659(11)01044-3

Schwake G, Youssef S, Kuhr JT, Gude S, David MP, Mendoza E, Frey E, Radler JO (2010) Predictive modeling of non-viral gene transfer. Biotechnol Bioeng 105(4):805–813. doi:10.1002/bit.22604

Seth RB, Sun L, Chen ZJ (2006) Antiviral innate immunity pathways. Cell Res 16(2):141–147. doi:10.1038/sj.cr.7310019

Shim MS, Kwon YJ (2010) Efficient and targeted delivery of siRNA in vivo. FEBS J 277(23): 4814–4827. doi:10.1111/j.1742-4658.2010.07904.x

Shiota M, Bishop JL, Nip KM, Zardan A, Takeuchi A, Cordonnier T, Beraldi E, Bazov J, Fazli L, Chi K, Gleave M, Zoubeidi A (2013) Hsp27 regulates epithelial mesenchymal transition, metastasis, and circulating tumor cells in prostate cancer. Cancer Res 73(10):3109–3119. doi:10.1158/0008-5472.CAN-12-3979

Sonawane ND, Szoka FC Jr, Verkman AS (2003) Chloride accumulation and swelling in endosomes enhances DNA transfer by polyamine-DNA polyplexes. J Biol Chem 278(45): 44826–44831

Soutschek J, Akinc A, Bramlage B, Charisse K, Constien R, Donoghue M, Elbashir S, Geick A, Hadwiger P, Harborth J, John M, Kesavan V, Lavine G, Pandey RK, Racie T, Rajeev KG, Rohl I, Toudjarska I, Wang G, Wuschko S, Bumcrot D, Koteliansky V, Limmer S, Manoharan M, Vornlocher HP (2004) Therapeutic silencing of an endogenous gene by systemic administration of modified siRNAs. Nature 432(7014):173–178

Spagnou S, Miller AD, Keller M (2004) Lipidic carriers of siRNA: differences in the formulation, cellular uptake, and delivery with plasmid DNA. Biochemistry 43(42):13348–13356. doi:10.1021/bi048950a

Tabernero J, Shapiro GI, LoRusso PM, Cervantes A, Schwartz GK, Weiss GJ, Paz-Ares L, Cho DC, Infante JR, Alsina M, Gounder MM, Falzone R, Harrop J, White AC, Toudjarska I, Bumcrot D, Meyers RE, Hinkle G, Svrzikapa N, Hutabarat RM, Clausen VA, Cehelsky J, Nochur SV, Gamba-Vitalo C, Vaishnaw AK, Sah DW, Gollob JA, Burris HA 3rd (2013) First-in-humans trial of an RNA interference therapeutic targeting VEGF and KSP in cancer patients with liver involvement. Cancer Discov 3(4):406–417. doi:10.1158/2159-8290.CD-12-0429

Tili E, Croce CM, Michaille JJ (2009) miR-155: on the crosstalk between inflammation and cancer. Int Rev Immunol 28(5):264–284. doi:10.1080/08830180903093796

Trang P, Wiggins JF, Daige CL, Cho C, Omotola M, Brown D, Weidhaas JB, Bader AG, Slack FJ (2011) Systemic delivery of tumor suppressor microRNA mimics using a neutral lipid emulsion inhibits lung tumors in mice. Mol Ther 19(6):1116–1122. doi:10.1038/mt.2011.48

Treiber T, Treiber N, Meister G (2012) Regulation of microRNA biogenesis and function. Thromb Haemost 107(4):605–610. doi:10.1160/TH11-12-0836

Troiber C, Kasper JC, Milani S, Scheible M, Martin I, Schaubhut F, Kuchler S, Radler J, Simmel FC, Friess W, Wagner E (2012) Comparison of four different particle sizing methods for siRNA polyplex characterization. Eur J Pharm Biopharm. doi:10.1016/j.ejpb.2012.08.014. pii: S0939-6411(12)00320-7

Troiber C, Edinger D, Kos P, Schreiner L, Klager R, Herrmann A, Wagner E (2013) Stabilizing effect of tyrosine trimers on pDNA and siRNA polyplexes. Biomaterials 34(5):1624–1633. doi:10.1016/j.biomaterials.2012.11.021. pii: S0142-9612(12)01265-3

Urban-Klein B, Werth S, Abuharbeid S, Czubayko F, Aigner A (2005) RNAi-mediated gene-targeting through systemic application of polyethylenimine (PEI)-complexed siRNA in vivo. Gene Ther 12(5):461–466

Wagner E (2012) Polymers for siRNA delivery: inspired by viruses to be targeted, dynamic, and precise. Acc Chem Res 45(7):1005–1013. doi:10.1021/ar2002232

Wagner E (2013) Biomaterials in RNAi therapeutics: quo vadis? Biomater Sci 1:804–809. doi:10.1039/C3BM60071H

Whitehead KA, Langer R, Anderson DG (2009) Knocking down barriers: advances in siRNA delivery. Nat Rev Drug Discov 8(2):129–138. doi:10.1038/nrd2742

Wolfrum C, Shi S, Jayaprakash KN, Jayaraman M, Wang G, Pandey RK, Rajeev KG, Nakayama T, Charrise K, Ndungo EM, Zimmermann T, Koteliansky V, Manoharan M, Stoffel M (2007) Mechanisms and optimization of in vivo delivery of lipophilic siRNAs. Nat Biotechnol 25(10):1149–1157. doi:10.1038/nbt1339

Yates LA, Norbury CJ, Gilbert RJ (2013) The long and short of microRNA. Cell 153(3):516–519. doi:10.1016/j.cell.2013.04.003

Yazbeck DR, Min KL, Damha MJ (2002) Molecular requirements for degradation of a modified sense RNA strand by *Escherichia coli* ribonuclease H1. Nucleic Acids Res 30(14):3015–3025

Zamecnik J, Vargova L, Homola A, Kodet R, Sykova E (2004) Extracellular matrix glycoproteins and diffusion barriers in human astrocytic tumours. Neuropathol Appl Neurobiol 30(4): 338–350. doi:10.1046/j.0305-1846.2003.00541.x

Zheng D, Haddadin S, Wang Y, Gu LQ, Perry MC, Freter CE, Wang MX (2011) Plasma microRNAs as novel biomarkers for early detection of lung cancer. Int J Clin Exp Pathol 4(6): 575–586

Zimmermann TS, Lee AC, Akinc A, Bramlage B, Bumcrot D, Fedoruk MN, Harborth J, Heyes JA, Jeffs LB, John M, Judge AD, Lam K, McClintock K, Nechev LV, Palmer LR, Racie T, Rohl I, Seiffert S, Shanmugam S, Sood V, Soutschek J, Toudjarska I, Wheat AJ, Yaworski E, Zedalis W, Koteliansky V, Manoharan M, Vornlocher HP, MacLachlan I (2006) RNAi-mediated gene silencing in non-human primates. Nature 441(7089):111–114

Chapter 10
Nano-Platforms for Tumor-Targeted Delivery of Nucleic Acid Therapies

Arun K. Iyer, Shanthi Ganesh, and Mansoor M. Amiji

10.1 Introduction

The treatment of several types of cancer has remained elusive even after persistent efforts by researchers and clinicians to combat this deadly disease. In an insightful analysis, Hanahan and Weinberg (2011) have described the hallmark features of cancers and the multitude of deregulation and signaling pathways that are involved in the development and progression of tumors. Although very different in nature, all forms of cancer are derived from normal cells that have acquired abnormal phenotypes (Varmus and Weinberg 1993). Either by enhanced proliferation or reduced apoptosis, these neoplastic cells begin to neo-vascularize forming new blood vessels, establishing subpopulation of virulent cells, with the capability to invade and infiltrate into adjacent normal cells and tissues. Also, the genomic instability in cancers causes several types of mutations that lead to intra-tumoral heterogeneity in successive generations of tumors (Cahill et al. 1999; Loeb et al. 2003). In addition, tumor cells are smart in reprogramming themselves to cater to demanding conditions, such as an ability to survive and often thrive in hypoxic conditions and/or highly acidic low pH microenvironment of the tumors (Milane et al. 2011a; Vaupel and Mayer 2007; Dang et al. 1997; Asosingh et al. 2005; Martinez-Zaguilan et al. 1996; Gatenby and Gawlinski 2003).

A.K. Iyer
Department of Pharmaceutical Sciences, School of Pharmacy,
Bouvé College of Health Sciences, Northeastern University, Boston, MA, USA

S. Ganesh
Dicerna, Inc., 480 Arsenal Street, Building 1, Suite 120 Watertown, MA 02472, USA

Department of Pharmaceutical Sciences, School of Pharmacy, Bouvé College of Health Sciences, Northeastern University, 140 The Fenway Building, Boston, MA 02115, USA

M.M. Amiji (✉)
Department of Pharmaceutical Sciences, School of Pharmacy, Bouvé College of Health Sciences, Northeastern University, 140 The Fenway Building, Boston, MA 02115, USA
e-mail: m.amiji@neu.edu

M.J. Alonso, M. Garcia-Fuentes (eds.), *Nano-Oncologicals: New Targeting and Delivery Approaches*, Advances in Delivery Science and Technology,
DOI 10.1007/978-3-319-08084-0_10

While the micro-environmental selection pressures lead to development of multidrug resistance (MDR) (Donnenberg and Donnenberg 2005), the derivation of tumors from normal tissue poses another major barrier in the safe and effective eradication of tumors and cancerous cells. Both molecular and physical boundaries between normal and cancerous tissue are often diffused, rendering many of the current treatments highly unspecific. Also, conventional forms of chemotherapy using small molecule drugs and anticancer agents do not differentiate tumor and normal tissues and cells causing acute systemic side effects, necessitating the need for the development of novel targeted strategies and therapeutics for more effective management of cancers.

In response to the failures of conventional chemotherapies in the clinics, there has been a growing interest in devising advanced targeted delivery systems based on nano-medical technologies for early detection, diagnosis, and treatment of cancers (Printz 2012; Bakht et al. 2012; Parhi et al. 2012; Kolhe and Parikh 2012; Schroeder et al. 2012; Wang et al. 2009). Since the microenvironment and physiological properties of solid tumors are uniquely different from the (healthy) normal tissues, it is possible to design nano-sized delivery platforms that specifically transport drugs and genes to the sites of tumors (Danhier et al. 2010). Furthermore, nanoparticle systems have the ability to achieve temporal and spatial site-specific drug delivery (Couvreur and Vauthier 2006). These unique properties of nanoparticle systems not only enhance the efficacy of chemotherapy but also significantly reduce the adverse side effects associated with the free drug and allow for fewer doses of drug administration to cancer patients. In this review we will discuss the development of some such nanoparticles based delivery systems such as polymeric micelles and nanoparticles, liposomes, and multifunctional nanoparticle systems with emphasis on nucleic acid therapeutics.

10.2 Mechanisms of Tumor Selective Delivery

10.2.1 Passive and Active Targeting to Tumors

The development of tumors involves several anatomical and pathophysiological changes that could be utilized for selective tumor targeting (Iyer et al. 2006). As the tumor cells multiply and grow to a size of ~1 mm in diameter, they start to form neovasculatures (Voest 1998). In order to sustain their growth, the so-formed tumor nodules develop more complex network of blood vessels around them, a process called angiogenesis (Folkman 1990). The endothelial cells lining the tumor blood vessels are highly disorganized and defective in architecture. Moreover, the tumor blood vessels have wide gap junctions leading to "leaky" vascular architectures (Dvorak 1990; Nagy et al. 1989). Apart from the unique anatomical features described above, tumor cells secrete elevated levels of permeability mediators such as vascular endothelial growth factor (VEGF) [also known as vascular permeability factor (VPF)], bradykinin (BK), prostaglandins (PGs), matrix metalloproteinases (MMPs), nitric oxide (NO), and peroxynitrite (Iyer et al. 2006; Greish et al. 2003; Maeda et al. 2000,

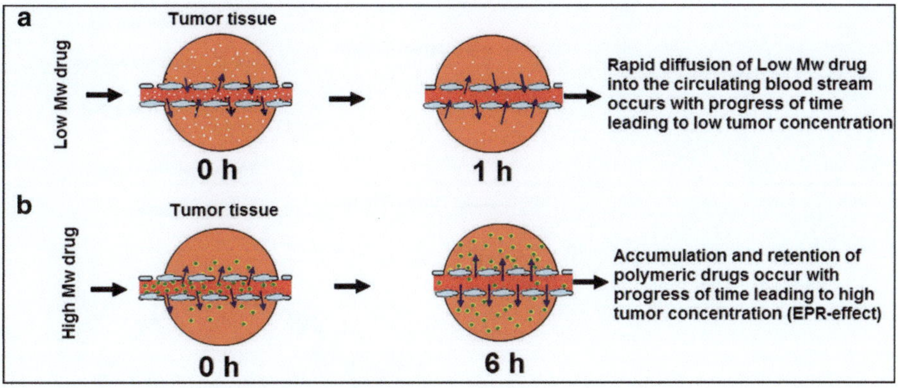

Fig. 10.1 Concept of EPR effect for tumor targeted delivery of polymeric drugs (adapted from Iyer et al. 2006 with permissions ©Elsevier B.V.)

2001). The overproduction of permeability mediators coupled with the anatomical and pathophysiological abnormalities leads to extensive accumulation of blood plasma components, macromolecules, and nanoparticles into the tumor interstitium (Iyer et al. 2006; Greish et al. 2003; Maeda 2001; Yuan et al. 1994). This phenomenon was coined the *enhanced permeability and retention* (EPR) effect, first discovered by Matsumura and Maeda more than three decades ago (Matsumura and Maeda 1986) (Fig. 10.1). In this regard, it is also important to note that polymeric drugs and macromolecules with molecular weights >40 kDa (which is above the renal excretion threshold) are able to circulate longer in the blood and show prolonged accumulation in the solid tumors (Greish et al. 2003; Maeda et al. 2001; Maeda 2001, 2012; Torchilin 2011). The EPR phenomenon was later observed for many types of polymer-drug conjugates, micelles, nanoparticles, and liposomal delivery systems (Iyer et al. 2006; Maeda et al. 2001; Duncan 2003; Jain 1987; Noguchi et al. 1998). Also, the tumor tissues were found to have impaired lymphatic clearance due to which the accumulation and retention of nanoparticles continued to occur in the tumor as long as they could circulate in the blood. Furthermore, the EPR effect was found to be more effective if the nanoparticles could escape mononuclear phagocytic systems (MPS) and show prolonged circulation half-life in the blood. In this regard, incorporation of amphipathic molecules such as poly(ethylene glycol) (PEG) on the surface of nanosystems was found very useful in facilitating MPS escape and rendering long plasma residence time, thus enhancing tumor accumulation (Pasut and Veronese 2009; Veronese and Pasut 2005). For instance, PEG-modified "stealth" liposomes encapsulated with doxorubicin could circulate for prolonged periods in the blood and exhibited improved tumor accumulation in addition to lowering the toxicities associated with free form of doxorubicin (Gabizon 1992). In general, nanosystems in the size range of 20–200 nm have been found effective in permeating and accumulating in the solid tumor tissue (Danhier et al. 2010), although there has been some indication that the hyperpermeability of tumor vasculature with wide gap junctions can facilitate particles as large as 700 nm to accumulate effectively in solid tumors (Yuan et al. 1994; Gao et al. 2012; Hobbs et al. 1998).

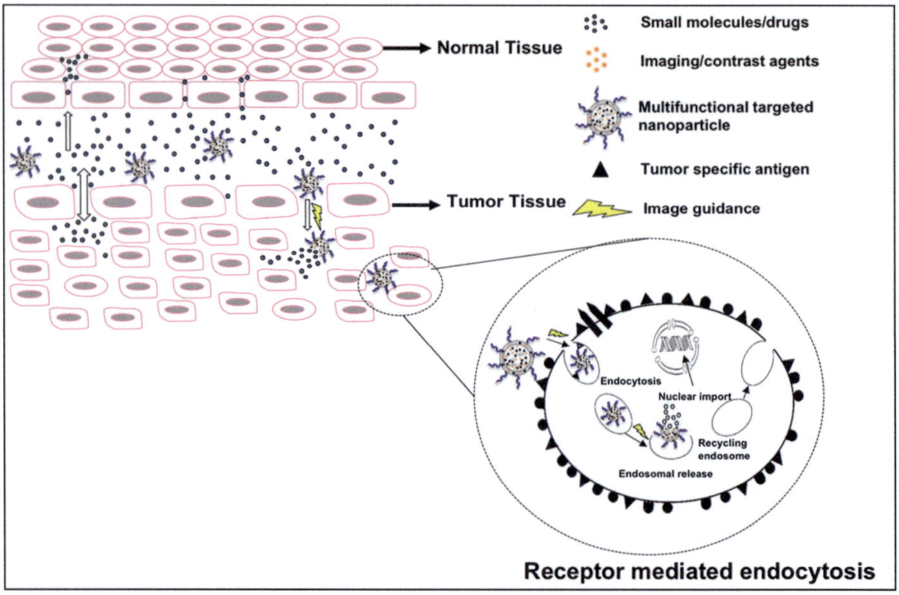

Fig. 10.2 Passive and active tumor drug targeting. The schematic shows the passive and active targeting mechanisms of multifunctional image guided nanoparticles and the difference in the vasculature of normal and tumor tissues; drugs and small molecules diffuse freely in and out of the normal and tumor blood vessels due to their small size, and thus the effective drug concentration in the tumor drops rapidly with time. Macromolecular drugs and nanoparticles can passively target tumors due to the leaky vasculature or the EPR effect; however, they cannot diffuse back into the bloodstream due to their large size and impaired lymphatic clearance, leading to enhanced tumor accumulation and retention. Targeting molecules such as antibodies or peptides present on the nanoparticles can selectively bind to cell surface receptors/antigens overexpressed by tumor cells and can be taken up by receptor-mediated endocytosis (active targeting). The image guiding molecules and contrast agents conjugated/encapsulated in the nanoparticles can be useful for targeted imaging and (noninvasive) visualization of nanoparticle accumulation/localization, as well as for mechanistic understanding of events and efficacy of drug treatment simultaneously (reprinted with permissions from Iyer et al. 2012 ©Bentham Science)

Although EPR effect provides a "first pass" for selective accumulation of nanoparticles, micelles, and liposomal formulations into the tumor interstitium, their intracellular delivery still remains challenging (Torchilin et al. 2001). It is critical for nanoparticle system to enter into the cancer cells and more importantly, release the drug/gene cargo in the right location, for effective cell killing. Intracellular delivery to specific location within the cells and organelles is essential for almost all anticancer drugs and genes. For instance, intracellular delivery of siRNAs and its release in the cytoplasm is a key for the success of RNA interference (RNAi) based gene silencing strategies (Ganesh et al. 2013a). In this regard the specificity and targeting ability of nanosystems can be remarkably improved when tumor-targeting ligands are used as part of the nano-delivery systems (Fig. 10.2) (Torchilin et al. 2001; Pasqualini et al. 1997). Such targeted delivery systems can selectively home to tumor cells that overexpress specific receptors or antigens (Marcucci and Lefoulon 2004), thereby promoting intracellular delivery (Torchilin et al. 2001; Ganesh et al. 2013a; Milane et al. 2011b; Magadala and Amiji 2008). Use of such mechanism is called "active" tumor targeting (Fig. 10.2) (Danhier et al. 2010). Thus, active targeting in effect aids in more specific "secondary" targeting after "primary" targeting based on the EPR-effect and in combination, passive and active targeting provide for increased accumulation and penetration of nanoparticles at the tumor site, thereby facilitating improved maintenance of high intracellular drug concentrations. Such systems provide several fold increased effectiveness as compared to free drug administration (Dang et al. 1997). Indeed currently, active targeting has become a widely recognized potential route for increasing therapeutic indexes in cancer treatment.

Several ligands or targeting agents can be used to surface decorate nanoparticle systems for active targeting to tumors (Lue et al. 2010; Rihova 1998; Farokhzad et al. 2006). For example, RGD peptide coupled nanoparticles can target integrin receptors ($\alpha_v\beta_5$ or $\alpha_v\beta_3$) overexpressed on vascular endothelial cells of angiogenic blood vessels and tumor cells (Ruoslahti and Pierschbacher 1987). In a recent study we functionalized poly(epsilon-caprolactone) (PCL) microparticles containing colloidal gold with RGD peptide that specifically homes to colon tumors overexpressing integrin receptors (Fig. 10.3) (Lue et al. 2010; Iftimia et al. 2012). These particles could increase the localization of fluorescent probes loaded in them for the diagnos-

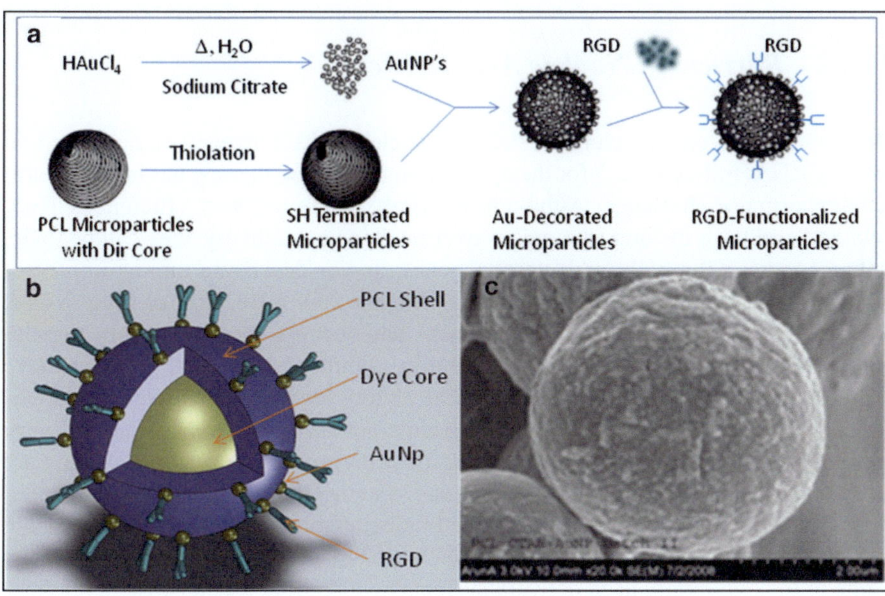

Fig. 10.3 RGD peptide functionalized poly(epsilon-caprolactone)-gold microparticle design for colon cancer screening. (**a**) Fabrication steps; (**b**) 3D design; and (**c**) scanning electron micrograph (SEM) revealing a size of ~1.5 μm for the microparticle. See Lue et al. (2010) and Iftimia et al. (2012) for details

tic imaging and screening of colon cancers. There are several such examples in the literature where a combination of active and passive targeting has been utilized for tumor targeted drug delivery (Danhier et al. 2010; Torchilin et al. 2001; Ganesh et al. 2013a; Magadala and Amiji 2008; Torchilin 2010; Keereweer et al. 2012; Kim et al. 2010; Cho et al. 2008; Chen et al. 2010a). In this chapter, we will be discussing the use of various targeted and non-targeted delivery systems that are being currently used for the effective delivery of various nucleic acid drugs.

Delivery of nucleic acids into cells to alter functions at the genetic level is a powerful approach to treat variety of genetic diseases. It can be achieved by introducing a transgene into specific cells of the patient where the endogenous gene is under expressing or by introducing a therapeutic agent such as antisense oligonucleotides (AON), small interfering RNA (siRNA), or miRNA to inhibit transcription and/or translation of an overexpressing endogenous gene or a cancer causing oncogene. Nucleic acids have also been utilized as vaccines for the treatment of several diseases, including cancer. However, there are many hurdles to overcome in developing effective nucleic acid based therapies, including cellular barriers, enzymatic degradation, and rapid clearance after administration. Successful transfer of plasmid DNA, siRNA, miRNA, and vaccines into cells therefore needs efficient carriers. The ideal carrier should be safe and ensures that the nucleic acid survives the extra-/intracellular environment and can efficiently transfer the payload to the appropriate cellular compartments.

10.2.2 Barriers to Nucleic Acid Delivery

Developing safe and efficient carriers for nucleic acid delivery (such as plasmid DNA, siRNA, miRNA, etc.) for the applications in human therapy has been and still remains a major challenge. Although the viral vectors are very efficient delivery vehicles, and have distinct advantages over nucleic acid delivery, they can be toxic, immunogenic, potentially activate oncogenes, and/or deactivate tumor suppressor genes (Li and Huang 2006; Schaffert and Wagner 2008). Because of these issues, the non-viral/synthetic nanoparticle system has become an increasingly popular alternative for nucleic acid delivery. Therefore, synthetic lipids and polymers have been developed to offer alternatives to viral vectors for nucleic acid delivery applications. These non-viral nanoparticle systems are designed and formulated very carefully to have appropriate dimensions for cell entry and endocytosis. They are also built to avoid stimulation of the immune system. Clearance of larger molecular mass lipids/polymers requires them to be biodegradable. The use of biodegradable materials would have linkages that can be cleaved inside the cell to reduce toxicity. Some of the commonly used delivery systems include lipid based carriers, polymer systems, peptides/proteins, and conjugates. Although these non-viral vehicles overcome number of issues that the viral vectors have, they still struggle to achieve the same high transfection efficiencies. Several attempts have been made over the years to improve the systems. Nucleic acid delivery involves similar processes such as condensation, protection, and endosomal release. The mechanism of release from the carrier for other nucleic acids such as siRNA, miRNA, and ODNs is however different from the plasmid DNA release. Both siRNA and ODNs need to reach the cytosol to be a part of the action, whereas the plasmid DNA requires further transport to the nucleus; thus, the siRNA requires an early release (Scholz and Wagner 2012). Plasmid DNA is condensed in polymeric or peptide based vectors to improve the ability to penetrate the nuclear membrane. More commonly, nuclear localization signals (NLS) have been exploited for the active transport of plasmids through NPC. Various genetic elements can be inserted into the expression cassette to facilitate nuclear targeting and modulate transgene expression.

Systemic delivery of RNAi therapeutics to the targeted tissue and cells has also been proven challenging and has many barriers. Intravascular barriers include engulfment of naked siRNAs by macrophages or degradation by serum nucleases (Czech et al. 2011). Extracellular nanoparticles, on the other hand, encounter degradation by immune cells and ECM. Particles that escape these barriers preferentially traffic to tumors by the EPR effect (Figs. 10.2 and 10.4). Naked siRNAs with large molecular weight (about 13 kDa) and net negative charge can neither readily enter the cells nor

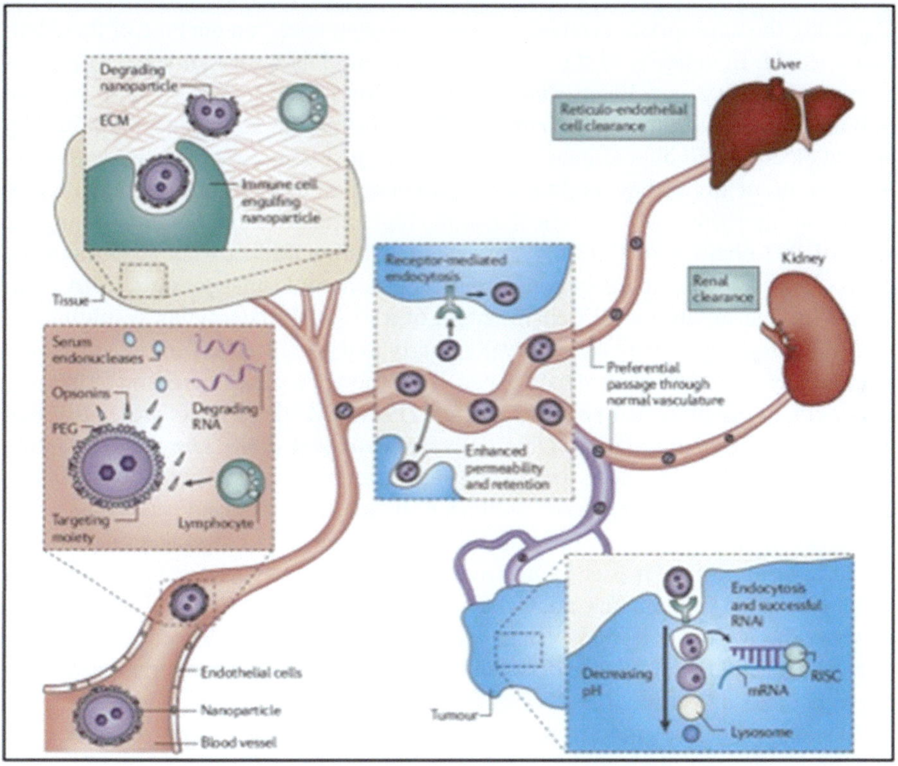

Fig. 10.4 Biological barriers of RNAi delivery (reprinted with permission from Pecot et al. 2011. *Nature Reviews Cancer*, 11: 59–67, 2011)

easily escape from the lumen of intracellular membranes. Even with direct injection into a targeted site, such as the vitreous humor of the eye, or inhalation directly into the lung, siRNAs must evade nuclease activities, penetrate the interstitial space of cell matrices, cross cell plasma membranes, escape from endosomes, and then load into the RISC complex within the cell cytoplasm to be effective (Peer and Lieberman 2011). Another problem is that double-stranded RNAs induce undesirable off-target effect, like inflammatory response. Thus, unprotected and unmodified native siRNA by far is not used for gene silencing in directed in vivo applications.

To understand this, we will illustrate appropriate non-viral delivery systems that are being currently pursued for the delivery of various nucleic acids and discuss how these different systems overcome the issues to mediate effective delivery.

10.3 Multifunctional Targeted Nano-Platforms for Nucleic Acid Delivery

10.3.1 Plasmid DNA Delivery and Gene Therapy

Generally the appropriate synthetic carrier is chosen based on the type of therapeutic molecule. In majority of the studies, the plasmid DNA was delivered to express therapeutic transgenes. Electrostatic complexation of plasmid DNA with cationic polymers, such as poly(ethyleneimine) (PEI), forms the so-called polyplexes by simple electrostatic interactions. Since these are just complexes formed with electrostatic interactions, they might not be stable enough to withstand the harsh serum conditions and also could be toxic due to the higher positive charge. Fine-tuning the ratio of PEI chain with PEG molecules can stabilize these particles. DeRouchey et al. (2006) showed that they could construct monomolecular plasmid DNA complexes by fine-tuning the PEI chain length of a PEG-PEI copolymer resulting in a high PEG grafting density and correspondingly to a good stability in salt and serum without any aggregation tendency. Apart from the approaches based on PEI based polymers, hyper-branched polyamidoamine (PAMAM) dendrimers were also used as polymer carriers for plasmid DNA delivery (Patil et al. 2008; Lee et al. 2003; Sun and Zhang 2010). To reduce the toxicity and increase the stability of dendrimers, amine modified dendrimers have been used. In addition, chitosan was also used as an attractive delivery system for plasmid DNA, which shows high biocompatibility (Lavertu et al. 2006). It was found from previous studies that the high molecular weight chitosan offers better nucleic acid complexation and extracellular stability, whereas the lower molecular weight ones allow for a more efficient intracellular release. Thus finding a balance between these effects is actually important in finding the right chitosan carrier.

In another study, a group from Hopkins has designed and synthesized a series of biocompatible and biodegradable polyphosphoramidate (PPA) gene carriers that self-assemble with DNA to form micellar nanoparticles (Jiang et al. 2013). These contain a DNA/PPA complex core and a PEG corona rendering particles more stability. To have direct access to liver parenchyma cells and to avoid the particles being scavenged by the Kupffer cells, researchers undertaking the study used an alternative injection route such as retrograde intrabiliary infusion (RII) to transport these particles (Jiang et al. 2013). By this method, they managed to overcome several key limitations of intravenous (i.v.) delivery and showed significantly higher levels of transgene expression in the liver. Authors from the study claim that the uniform size and high stability of PEG-PPA/DNA micellar nanoparticles in bile containing medium make them excellent candidate for delivering genes through RII.

Apart from plasmid DNA-polymer based particles, stealth liposomes could provide a stable and safe alternative to cationic DNA complexes for effective DNA delivery (Li and Szoka 2007). Such systems are also easily modified with targeting ligands for improved site-specific delivery. Direct complexation of plasmid DNA with cationic lipids results in self-assembly of the so-called lipoplexes. This again leads to physically unstable nanoparticles with irregular morphologies. Alternatively, the encapsulation of plasmid DNA in PEG shielded cationic liposomes seems a great improvement leading to small particle size and controlled structures of regular morphology and good stability. One such example is the stable nucleic acid lipid particles (SNALP) formulations consisting of a lipid bilayer including a mixture of cationic and neutral lipids and a PEG lipid coating (Morrissey et al. 2005). The particles were found to be around 140 nm in size, independent of the payload. Some previous work has also shown that the plasmid DNA can be encapsulated in neutral liposomes that were functionalized with targeting peptide. To improve the pharmacokinetics and inflammatory toxicity of the nanoparticles, PEG has been added to the surface of the particles.

10.3.2 Nucleic Acid Vaccines

Effective use of naked nucleic acids as vaccines was one of the most important advances in the history of vaccinology. This has been promoted as a new approach to prime specific humoral and cellular responses to protein antigens. Although the nucleic acids showed promise for use as vaccine vectors in experimental animal models, they have lacked potency in human clinical trials (Restifo et al. 2000). Recombinant viral vectors have the advantage of efficient delivery of the nucleic acid payload, but their utility is hampered by antiviral immunity and safety concerns. Injection of naked mRNA or self-amplifying RNA in vivo induces gene expression and generates immune responses. However, due to the instability of these RNA in vivo, this method resulted in limited potency. Hence, one effective way is to formulate the nucleic acid vaccines in synthetic delivery vehicles to increase the delivery efficiency and potency. Among them, nanoparticles such as lipid/polymer based and micro-/nanoparticles have emerged as effective vaccine adjuvants for cancer therapy (Bodles-Brakhop and Draghia-Akli 2008). Cationic polymers/lipids can be combined with DNA or RNA to form particulate complexes capable of gene transfer into the target cells. Upon systemic administration, a TLR directed cationic lipid/DNA complex (JVRS-100) enters dendritic cells (DCs) and macrophages where the DNA binds to and activates Toll like receptors and results in anti-tumor responses generated by the innate immune system. This vaccine adjuvant also induces strong cytotoxic T lymphocyte response. The efficacy of this vaccine has been evaluated in Phase 1 trial for the treatment of leukemia (Bolhassani et al. 2011). Cationic polymers such as PLL and PEI were most widely used for gene therapy and have been utilized for vaccine delivery as well. When they are PEG-modified and targeted, they can effectively deliver the nucleic acids to the target cells with reduced toxicity (Brown et al. 2000). Cationic polymers are generally effective by non-specific stimulation of host immune response as well as by direct electrostatic cytotoxic interaction with tumor cells. Another approach to deliver nucleic acid vaccines is using microparticle based technology to target antigen presenting cells, APCs (Ulmer et al. 2006). Microencapsulation of nucleic acids leads to enhancement of CTL responses to encoded proteins (Doria-Rose and Haigwood 2003). Biodegradable, non-antigenic PLGA microspheres are known to be a good choice for this purpose. Antigens encapsulated or complexed with the PLGA particles are capable of enhancing humoral and cellular immune responses. These nanoparticles have been shown to stimulate immune response as measured by an increase in IL2 and IFN-gamma in spleen (Morrissey et al. 2005). These PLGA polymers can offer long-term release of the contents too. It has been shown that these polymers can effectively deliver antigens or adjuvants to a desired location at predetermined rates and durations, effectively regulating the immune responses over a period of time (Bolhassani et al. 2011). It was also reported in the literature that these polymers can effectively direct the antigens to APCs by trafficking through local lymphoid tissue for uptake by DCs. Similar to microspheres, PLGA NPs are also known to effectively enhance immune responses. In addition, multifunctional nano-based dendrimers and biodegradable chitosan nanoparticles are also being evaluated for variety of similar cancer therapies (Baker Jr 2009). Overall, these nanoparticulate forms of delivery promise the advantage of enhanced tumor selectivity and longer half-lives, thereby enhancing the effectiveness of the immune responses and reduction of systemic toxicity.

Apart from delivering plasmid DNA and RNA, a group at Novartis Vaccines managed to deliver a self-amplifying RNA using clinically suitable synthetic lipid nanoparticles as a means to increase the efficiency of antigen production and immunogenicity in vivo (Geall et al. 2013). For this purpose, researchers involved in the study used a self-amplifying RNA based on an alpha virus genome. This contains the genes encoding the alpha virus RNA replication machinery but lacks the genes encoding the structural proteins. They replaced the structural genes with genes encoding antigens. They produced this in vitro by an enzymatic transcription reaction method and encapsulated that in the lipid nanoparticles. Using this system, the researchers demonstrated effective delivery of a 9 kb self-amplifying RNA and substantial increase in immunogenicity compared with unformulated RNA. The same group also explored the utility of electroporation method to deliver the self-amplifying mRNA. By this method, the investigators involved in the study demonstrated reported gene expression and immunogenicity of genes encoding HIV envelope protein that is comparable to EP delivery of plasmid DNA.

10.3.3 siRNA Delivery and Gene Silencing

As previously described, therapeutic applications of any type of nucleic acid require successful delivery to the host cell and subsequent release in the target components. Although there are considerable mechanistic similarities in the delivery of various nucleic acids, the mechanism of release from the carrier is different for plasmid DNA versus siRNA (Scholz and Wagner 2012). As pointed out before, the siRNA should be released in the cytosol to be a part of the action, thus requiring an early release from the endosome. Synthetic carriers should therefore be designed carefully by modulating chemical and physical properties. Just like in plasmid DNA delivery, most of the carriers used for siRNA delivery also include cationic liposomes, polymers, etc. siRNA should be encapsulated in these particles to prevent the enzymatic degradation by nucleases in the blood circulation and the destabilization by electrostatic interactions with serum proteins. The particles need to be accumulated at the target site by EPR effect. Targeting may be necessary to facilitate the effective receptor mediated endocytosis followed by the crucial step of endosomal release of the siRNA in the cytosol.

Several delivery systems have been described for siRNA delivery over the years. Of all carriers, currently available, the most widely validated system is the cationic lipid-based nanoparticles (LNP) (Bodles-Brakhop and Draghia-Akli 2008). These lipid based nanoparticles are known to enter the cells by endocytosis, and the theory is that certain cationic lipids mediate endosome escape by destabilizing the endosome membrane or by causing endosomal rupture via proton sponge mechanism. These lipid-based carriers are being used successfully to deliver siRNA to targets such as liver and solid tumors. Investigators at Alnylam Pharmaceuticals have advanced its RNAi therapeutics (ALN-VSP) using their LNP system into clinic for the treatment of liver cancers. ALN-VSP targets two key genes (KSP and VEGF) involved in progression of liver cancer. The Phase I study demonstrated that the ALN-VSP was well tolerated in patients at doses from 0.1 to 1.5 mg/kg. The investigating team also showed that the two siRNAs were detected in nearly all the biopsy samples collected from livers of the patients. Authors claim that these levels are pharmacologically relevant since pre-clinical studies have shown that the same siRNA levels of 1 ng/g tissue are associated with 50 % gene silencing (Landesman et al. 2010). In addition, multiple patients achieved stable disease or better including a patient with endometrial cancer metastasis to the liver achieved a complete response in this study.

The Alnylam team recently reported the clinical trial data of the first generation and second generation nanoparticle formulations, ALN-TTR01 and ALN-TTR02, for the treatment of transthyretin (TTR)—a protein that transports thyroxine and retinol—mediated amyloidosis (ATTR) in a recent issue of the *New England Journal of Medicine* (Coelho et al. 2013). In this study, the authors found that the second generation nanoparticles ALN-TTR02 showed exceptional improvement in gene silencing efficacy without any loss in tolerability and reported that they achieved 94 % reduction of serum TTR and nearly 80 % level of suppression sustained for 1 month with just a single siRNA dose. The investigating team also reported evidence showing that as little as 50 % reduction of the disease causing protein can result in disease improvement.

In addition, researchers at Alnylam and MIT have also recently developed a class of lipid like carriers that are currently 100 times more efficient at delivering siRNA than the previously described cationic liposome systems (LNP) (Akinc et al. 2009). They screened a novel combinatorial cationic lipidoid library and found a number of effective systemic deliverable vehicles for siRNA. In addition to demonstrating in vitro activity in multiple cell lines, they also demonstrated in vivo efficacy against multiple targets in multiple animal species including non-human primates.

Although these cationic liposomes/lipidoids are the most popular siRNA delivery agents so far, there are some concerns still remaining regarding their safety for therapeutic use. Excessive positive charge is thought to be associated with certain negative consequences in vivo. They are most likely to non-specifically bind with biological surfaces, serum proteins and lead to rapid clearance.

Conjugating small molecules, peptides, proteins, polymers, and aptamers to siRNA are another class of carriers and are known to improve the stability, cellular internalization, or cell specific active targeting delivery (Jeong et al. 2009; Akhtar and Benter 2007; David et al. 2010). Cholesterol conjugated siRNA, upon systemic administration, silenced the expression of an endogenous gene encoding ApoB in the liver and jejunum, and reduced the plasma apoB levels as well as the total cholesterol level, whereas the naked siRNA did not show any activity in these tissues (Anderson et al. 2004). However, it has been demonstrated that these conjugates need to be administered at very high doses to be active in mice. Surprisingly, no off-target effect or immune stimulation was reported despite the higher doses used. Alnylam recently attached monosaccharides called *N*-acetylgalactosamine (GalNAc) to the siRNA that helps to cross the membrane. The conjugated siRNA-GalNAc conjugates targeted the liver as hepatocytes express the asialoglycoprotein receptor (ASGPR) to which GalNAc binds. This endocyte cell membrane receptor is known to be triggered by the GalNAc binding event to internalize the conjugate by endocytosis and mediate drug delivery. Alnylam recently implemented this approach in its drug called ALN-AT3 for the treatment of hemophilia in clinic.

Cationic polymers also readily bind and condense large nucleic acids and serve as efficient transfection agents. Because of this cationic nature, they were also shown to stimulate endocytosis and endosome escape. These polymers are widely divided into 2 classes such as natural polymers (chitosan, atelocollagen, or cyclodextrin) and synthetic polymers (PEI, PLL, and dendrimers) (Wang et al. 2010). Cyclodextrin, a natural polymer, has been developed as a delivery vehicle to efficiently deliver siRNA to tumors (Hu-Lieskovan et al. 2005). β-cyclodextrin containing polycations has been self-assembled with adamantane conjugated with PEG5000 and transferrin ligand to target transferrin receptors on the cell surface. These nanoparticles seemed to be stable in the bloodstream. Unlike the cationic liposomes, transferrin targeted cyclodextrin polymer particles did not elicit any immune stimulatory effects in mouse models. They also demonstrated good safety profile in nonhuman primates (Heidel et al. 2007). Not only did they demonstrate good efficacy in a number of different pre-clinical models, they also showed activity and efficient gene silencing in human cancer patients in clinic recently (Davis et al. 2010).

PEI is a synthetic polymer and it has been broadly investigated for nucleic acid delivery. PEG-modified PEI with RGD ligand has been extensively studied by Schiffelers et al. for siRNA delivery to target angiogenesis (Temming et al. 2005). Since the commonly used branched PEI is generally known to cause severe toxicity, the use of higher molecular weight unmodified PEI is likely to be limited. Several groups have been successful in minimizing the toxicity by using low-molecular weight or biodegradable PEI, although this favors the plasmid delivery than the siRNA delivery in mouse models. As an alternative to PEI carriers, PLGA is another extensively studied copolymer for delivery. As compared to cationic polymers and lipids, it has the advantage of lower toxicity. Tertiary amine modified polyvinyl alcohol backbones grafted to PLGA yielded rapid degradation properties for siRNA release. Incorporation of PEI into PLGA nanoparticle further improved the loading and activity. Dendrimers are another class of synthetic, cationic macromolecules that are being used as a carrier for small molecular drugs and large biomolecules (Wang et al. 2010). Although the most advanced delivery system at present includes a cationic group for ionic interaction with siRNA (PEG for steric hindrance, an endosomolytic group for endosome escape and a targeting ligand for specific delivery), several researchers have been working to come up with a non-condensing lipid/polymer agent as an siRNA/miRNA carrier to minimize the toxicity effects coming from the cationic charged molecules.

In addition, we have developed a self-assembled hyaluronic acid (HA) based nanosystem to effectively deliver siRNA and chemotherapeutic drugs specifically to tumors (Ganesh et al. 2013a, b). HA, also called hyaluronan, is a naturally occurring polysaccharide and the only non-sulfated glycosaminoglycan that is abundant in extracellular matrix and synovial fluids of all vertebrates. This highly negatively charged anionic polymer is composed of alternating disaccharide units of D-GLUCURONIC acid and *N*-acetyl D glucosamine with β (1–4) interglycosidic linkage. CD44 has been identified as HA receptors and is known to be overexpressed in many types of cancer cells, demonstrating enhanced binding and internalization of HA. In addition to the targeting ability, the HA polymer also has several favorable properties to be an excellent delivery vehicle. It is biodegradable, biocompatible, nontoxic, non-immunogenic, and noninflammatory, which makes it even more ideal for a drug delivery application. A relatively simple chemical structure also allows HA to be further modified to create a wide range of possible drug delivery carriers.

To make use of the favorable properties of HA, we have synthesized a series of functionally variant library of self-assembling CD44 targeting HA based macrostructures by varying the carbon chain length, nitrogen content, and polyamine side chain grafting onto the HA backbone. The modification of anionic HA polymer resulted in the effective lowering/shielding of its negative charge density that could enable siRNA encapsulation. After screening a library of derivatives, we chose a PEI modified HA nanosystem for siRNA delivery and ODA modified HA system for cisplatin delivery. For in vivo applications, we also added an additional component of HA-PEG to the HA-PEI system to improve the half-life of the system in circulation. We demonstrated target specific delivery of multiple siRNAs to tumors that overexpress CD44 and specific delivery of cisplatin to the same tumors using these HA based nanosystems (Ganesh et al. 2013a, b).

10.3.4 siRNA/Drug Combination Delivery

Drug-gene combination delivery has promising potentials for treating refractory diseases such as MDR cancers. Since one of the goals of our study was to reverse the drug resistance in cisplatin resistant tumors, in our lab, we used these HA based systems to effectively deliver siRNA and cisplatin to ultimately reverse the resistance. To do this, we initially picked the resistant tumors that overexpress CD44 and identified the resistant genes that are expressed in those cells. After careful designing and screening of multiple siRNA sequences to target those resistant genes in the resistant cells, we picked the most efficacious sequences (Ganesh et al. 2013b). We also modified those siRNAs to minimize the off-target effects coming from the unmodified sequences. Using our systems described before, we managed to first down regulate the resistant genes such as survivin and bcl2 using our HA-PEI/HA-PEG system carrying the corresponding siRNAs and sensitized the resistant tumors to cisplatin. Then we delivered cisplatin using our selected HA-ODA system to enhance the cell killing of the already sensitized cells. By delivering siRNA and a chemotherapeutic drug separately in two different systems, we demonstrated synergistic effect in killing (Ganesh et al. 2013b). The tumor growth inhibition was significantly improved in the tumors that had combination treatment (~60 %) when compared to the tumors that had single agent treatment (~30 %).

Apart from our study, few other groups also work on these combination strategies using siRNA and chemotherapeutic drugs. In this regard, a group from Rutgers University managed to co-deliver siRNA and doxorubicin together using a cationic liposome system and demonstrated synergistic effect in resistant cells (Saad et al. 2008). Leaf Huang and colleagues developed two different but novel nanoparticle formulations such as cationic liposome-polycation-DNA (LPD) and anionic liposomepolycation-DNA (LPD-II) for systemic delivery of doxorubicin and therapeutic siRNA to MDR tumors to overcome the drug resistance for cancer therapy (Chen et al. 2010b). For the first time they demonstrated that they could co-deliver siRNA and a chemotherapeutic agent to drug resistant tumor using two different multifunctional delivery systems.

10.3.5 MicroRNA Delivery

Certain miRNAs are known to be down-regulated in cancer cells and they increase the translation of oncogenes and hence formation of tumors. Therefore, recovering the expression of the down-regulated miRNAs by the introduction of miRNA mimics into the cells may reduce cancer progression. On the other hand, when miRNAs are up regulated in cancer cells, they block the tumor suppressor genes and lead to tumor formation (Garzon et al. 2010; McManus 2003; Visone and Croce 2009). One of the strategies to target these overexpressed miRNA expression in cancer is the use of oligonucleotides or antagomirs to block the expression. The key challenge in achieving effective miRNA therapeutics therefore is the development of an efficient delivery system, which can specifically deliver antagomirs or miRNA mimics to target cells in living animals.

Atelocollagen, a low immunogenic cationic biomaterial, has been used in the clinic for a wide range of purposes including wound healing, etc. It has been also shown previously that this atelocollagen system can be used for successful delivery of plasmid DNA, AON, and siRNA both in vitro and in vivo (Ochiya et al. 1999). Recently this system has been utilized for delivery of tumor suppressor miRNAs down-regulated in cancer cells as a strategy to prevent cancer cell metastasis. Mouse tail-vein injection of miR-16 complexed with atelocollagen significantly inhibited the growth of prostate tumors in bones in a therapeutic bone metastasis model. In addition, the same group also intravenously injected a complex of miR-143 and atelocollagen once in 3 days for 3 weeks and significantly suppressed the lung metastasis of metastatic human osteosarcoma cell lines (Takeshita et al. 2010). miR-143 was known to be the most down-regulated miRNA in metastatic human osteosarcoma cell lines relative to the parental cell line, and miR-16 was known to be the lowest expressing one in prostate cancer cells.

Bader and his group identified the tumor suppressor miRNA, let-7, in majority of NSCLC cells and patient samples and used the concept to reintroduce this miRNA mimics to cells/tumors to reactivate cellular pathways to drive a therapeutic response. Studies have shown that the reduced let-7 was significantly associated with shortened postoperative survival and overexpression resulted in the inhibition of lung cancer cell growth (Trang et al. 2010). Using a neutral lipid delivery system, Bader et al., efficiently delivered the miRNA mimic to cells and demonstrated significant growth inhibition and proliferation. They also demonstrated lung tumor growth inhibition in sc NSCLC tumors (H460) that lack the expression of let-7 using the same delivery system (Trang et al. 2010). In addition, it was also demonstrated that let-7g works against the oncogene K-ras that has a crucial role in the proliferation of lung cancer cells, especially in the cells with K-ras mutations. Given this information, this group also demonstrated a significant tumor growth inhibition/antitumor efficacy in a genetically engineered K-ras mutated oncogenic mouse model, a model that resembles very closely with actual human lung cancer (Trang et al. 2011).

In another study, transferrin-targeted protamine-containing liposomes were developed to deliver miRNA (Liang et al. 2011). miR29b is a member of the miR-29 family, which downregulates the cellular expression of antiapoptotic Mcl-1 proteins. The formulation was designated as Tf-LPmR, has protamine to increase the delivery efficiency and transferrin ligand for receptor mediated internalization in leukemia cells (K562) that overexpress transferrin receptors. It was demonstrated that this Tfr targeted system containing miR29b liposome formulation resulted in enhanced biological effects on the suppression of the target gene's expression (efficient inhibition of Mcl-1 expression at the mRNA and protein levels) compared with the nontargeted or non-protamine-containing liposomes (Rai et al. 2011). Leaf Huang's group demonstrated efficient miRNA delivery using a tumor targeted delivery system into experimental lung metastasis of murine B16F10 melanoma. In this study researchers developed a liposome-polycation-hyaluronic acid nanoparticle system modified with a tumor targeting single chain antibody fragment (scFv) for systemic delivery (Chen et al. 2010a). Delivery of miR34a using this system demonstrated apoptosis, inhibition of survivin expression, and downregulation of MAPK pathway in B16F10 cells. Systemic delivery of miR34a using this system significantly downregulated survivin expression in the metastatic tumor and reduced tumor load in the lung.

Apart from using liposome systems, a number of groups also worked on polymer delivery system for the same purpose. One such example was using PLGA based nanoparticles. PLGA was extensively studied for its ability to deliver different therapeutic agents. These polymer-based nanoparticles were shown to escape from the endosomal compartment to the cytoplasmic compartment and release its contents over longer periods of time (Bian et al. 2013; Lu et al. 2009). These features rendered PLGA nanoparticles as a potential tool for delivering oligonucleotide/miRNA efficiently. A group from China has modified the biodegradable PLGA nanoparticles with a polyplexed PEI coating, in which the PLGA acts as the core and cationic PEI as the shell with an aim of encapsulating negatively charged nucleic acids. With the help of cationic shell, they efficiently encapsulated miR-26a, a cell cycle suppressor, and demonstrated efficient delivery to human hepatocellular carcinoma cells such as HepG2 cells (Liang et al. 2011). This significantly increased the expression levels of miR-26a and inhibited the cell cycle progression by the induction of G1 phase arrest in transfected HepG2 cells. These PLGA nanoparticles also showed a better safety profile compared to PEIs and liposomes. In another instance, Paulmurugan et al. encapsulated chemically modified antagomirs in PLGA nanoparticles and showed efficient internalization and sustained release into cells. They also demonstrated tumor regression with the delivery of antagomirs. The results showed that the PLGA used for antagomir delivery is not only efficient in crossing the cell membrane but also can maintain functional intracellular antagomir level for an extended period of time and achieve therapeutic effect in living animals.

A highly branched dendritic polymer including poly(amidoamine, PAMAM) has attracted interest as nucleic acid delivery vectors (Rahbek et al. 2010). The presence of primary amine groups on their branched surface binds nucleic acids, compacts it into polyplexes, and promotes cellular uptake in a wide variety of cultured cells. Primary amines on the surface also make it possible to conjugate suitable ligands such as transferrin for tumor or brain delivery. Overexpression of miR-21 plays a key role in majority of cancers. Downregulation of miR-21 therefore leads to repression of cell growth, increased cellular apoptosis, and cell cycle arrest, which can theoretically enhance the chemotherapeutic effect in cancer. Zhou et al. used this PAMAM dendrimer as a carrier polymer to co-deliver anti-miR21 and 5-fluorouracil to achieve delivery of miRNA to human glioblastoma cells and enhance the cytotoxicity of 5-FU. With the help of positively charged primary amine groups, the miRNA gets encapsulated easily. Because of the presence of well-defined cavities and open architecture in PAMAM, guest molecules such as 5-FU get encapsulated into the macromolecular interior through hydrophobic interactions. Through their charge-based interactions, 5-FU-PAMAM could conjugate with miR21 (Ren et al. 2010). With this system, the authors demonstrated significant improvement in cytotoxicity of 5-FU and dramatic increase in apoptotic percentage of U251 cells. They also showed that this co-delivery brought down the migration ability of tumor cells, suggesting that the co-delivery system may have an important clinical application in the treatment of miR21 overexpressing glioblastoma.

Once the role of specific miRNA in disease pathogenesis is established, selecting anti-miRNA or miRNA mimics and appropriate delivery strategies will be the next critical steps for successful therapy. It is the hope that this miRNA therapeutics with the help of an effective and safe delivery system will soon lead to effective treatments for undruggable targets.

10.4 Conclusions and Future Prospects

Gene therapy offers great opportunities for the treatment of several diseases including cancer. Designing synthetic carriers to specifically reach the target tissues and release the cargo has become a research field of increasing interest. Despite significant progress in the field over the years, improving the safety and effectiveness of nucleic acid carriers for the application in human therapy still remains a major challenge. We have discussed some of the recent developments in solving some of the delivery issues for successful nucleic acid therapies in vitro and in vivo. Although each system has its own merits and shortcomings, multifunctional nanosystems have shown tremendous potentials for nucleic acid therapy. In addition to come up with an ideal system that specifically reaches the target tissue, there are other aspects of these carriers that should also be considered. Further advancements in gene delivery research also depend greatly on the development and scale-up of nano-based materials to be used in the clinical settings (Desai 2012). The manufacture of nanoparticles requires multiple steps involving multicomponent systems. In many cases, the small-scale processes achieve reproducibility with well-characterized components, but the reproducibility and consistency remain a constant challenge for scale-up and manufacturing process of complex nanoparticle systems. The formulation process must be robust to ensure reproducibility. All the nanoparticle systems discussed in this chapter, in addition to having lipid or polymer material, also have complex components such as proteins or nucleic acids as integral part, which may be sensitive to the manufacturing process conditions. In the multi-step process, it is therefore important to carry out in-process testing for critical parameters with reliable and rapid analytical methods. Liposome systems are in the clinic and these provide a good example to illustrate the issues of scale-up, stability, and importance of critical process parameters. Differing the process methods are reported to create liposomes with multilamellar vesicles, small unilamellar vesicles, or large unilamellar vesicles with instability (Zhu et al. 2007). It is therefore very important to carefully consider and evaluate the critical manufacturing parameters during the development process. The better one can understand these components and products in early stages of development, the more likely the successful reproducible manufacturing process will be achieved. The National Cancer Institute's Alliance for Nanotechnology in Cancer established the Nanotechnology Characterization Laboratory (NCL) as a resource for robust preclinical characterization of the nanoparticle systems prior to Investigational New Drug (IND) application.

In addition, the FDA has also highlighted the importance and challenges of nanotechnology in a draft guidance document.

Some of the nanoparticles discussed in this section have utilized the concept of self-assembly, where two or more components can be mixed together under appropriate conditions at the bedside to create particles as the final product for human use (Davis 2009; Eifler and Thaxton 2011). In this case, only the individual components should be manufactured and there is no need to create a nanomedicine as a finished product, which significantly reduces the cost and the complexity of the process. But, however, despite these advantages, this method also raises several questions: if the product made at the bedside makes the same reproducible structure of nanomedicine that was made at the shelf and, in that case, if these products should be subjected to some release tests at the bedside to make sure the product is made the way it was made and characterized at the shelf. One such example is the CALA-001, the transferrin targeted cyclodextrin polymer nanoparticles, the first targeted system to reach the clinical trial to treat cancer (Davis 2009). The individual components were just mixed at the bedside before injecting into humans. However, the pharmaceutical company, which brought this to clinic, has established a quality control and quality assurance program, including a set of standard operating procedures and specifications, designed to ensure that its products are manufactured in accordance with the current Good Manufacturing Practices or cGMPs. As discussed, some of the issues need careful consideration as policies and guidelines evolve for nanoparticle products in the future.

References

Akhtar S, Benter IF (2007) Nonviral delivery of synthetic siRNAs in vivo. J Clin Invest 117:3623–3632

Akinc A, Goldberg M, Qin J, Dorkin JR, Gamba-Vitalo C, Maier M, Jayaprakash KN, Jayaraman M, Rajeev KG, Manoharan M, Koteliansky V, Rohl I, Leshchiner ES, Langer R, Anderson DG (2009) Development of lipidoid-siRNA formulations for systemic delivery to the liver. Mol Ther 17:872–879

Anderson DG, Peng W, Akinc A, Hossain N, Kohn A, Padera R, Langer R, Sawicki JA (2004) A polymer library approach to suicide gene therapy for cancer. Proc Natl Acad Sci U S A 101:16028–16033

Asosingh K, De Raeve H, de Ridder M, Storme GA, Willems A, Van Riet I, Van Camp B, Vanderkerken K (2005) Role of the hypoxic bone marrow microenvironment in 5T2MM murine myeloma tumor progression. Haematologica 90:810–817

Baker JR Jr (2009) Dendrimer-based nanoparticles for cancer therapy. Hematology Am Soc Hematol Educ Program 708–719

Bakht MK, Sadeghi M, Pourbaghi-Masouleh M, Tenreiro C (2012) Scope of nanotechnology-based radiation therapy and thermotherapy methods in cancer treatment. Curr Cancer Drug Targets 12:998–1015

Bian X, Liang S, John J, Hsiao CH, Wei X, Liang D, Xie H (2013) Development of PLGA-based itraconazole injectable nanospheres for sustained release. Int J Nanomedicine 8:4521–4531

Bodles-Brakhop AM, Draghia-Akli R (2008) DNA vaccination and gene therapy: optimization and delivery for cancer therapy. Expert Rev Vaccines 7:1085–1101

Bolhassani A, Safaiyan S, Rafati S (2011) Improvement of different vaccine delivery systems for cancer therapy. Mol Cancer 10:3
Brown MD, Schatzlein A, Brownlie A, Jack V, Wang W, Tetley L, Gray AI, Uchegbu IF (2000) Preliminary characterization of novel amino acid based polymeric vesicles as gene and drug delivery agents. Bioconjug Chem 11:880–891
Cahill DP, Kinzler KW, Vogelstein B, Lengauer C (1999) Genetic instability and Darwinian selection in tumours. Trends Cell Biol 9:M57–M60
Chen Y, Zhu X, Zhang X, Liu B, Huang L (2010a) Nanoparticles modified with tumor-targeting scFv deliver siRNA and miRNA for cancer therapy. Mol Ther 18:1650–1656
Chen Y, Bathula SR, Li J, Huang L (2010b) Multifunctional nanoparticles delivering small interfering RNA and doxorubicin overcome drug resistance in cancer. J Biol Chem 285:22639–22650
Cho K, Wang X, Nie S, Chen ZG, Shin DM (2008) Therapeutic nanoparticles for drug delivery in cancer. Clin Cancer Res 14:1310–1316
Coelho T, Adams D, Silva A, Lozeron P, Hawkins PN, Mant T, Perez J, Chiesa J, Warrington S, Tranter E, Munisamy M, Falzone R, Harrop J, Cehelsky J, Bettencourt BR, Geissler M, Butler JS, Sehgal A, Meyers RE, Chen Q, Borland T, Hutabarat RM, Clausen VA, Alvarez R, Fitzgerald K, Gamba-Vitalo C, Nochur SV, Vaishnaw AK, Sah DW, Gollob JA, Suhr OB (2013) Safety and efficacy of RNAi therapy for transthyretin amyloidosis. N Engl J Med 369:819–829
Couvreur P, Vauthier C (2006) Nanotechnology: intelligent design to treat complex disease. Pharm Res 23:1417–1450
Czech MP, Aouadi M, Tesz GJ (2011) RNAi-based therapeutic strategies for metabolic disease. Nat Rev Endocrinol 7:473–484
Dang CV, Lewis BC, Dolde C, Dang G, Shim H (1997) Oncogenes in tumor metabolism, tumorigenesis, and apoptosis. J Bioenerg Biomembr 29:345–354
Danhier F, Feron O, Preat V (2010) To exploit the tumor microenvironment: passive and active tumor targeting of nanocarriers for anti-cancer drug delivery. J Control Release 148:135–146
David S, Pitard B, Benoit JP, Passirani C (2010) Non-viral nanosystems for systemic siRNA delivery. Pharmacol Res 62:100–114
Davis ME (2009) The first targeted delivery of siRNA in humans via a self-assembling, cyclodextrin polymer-based nanoparticle: from concept to clinic. Mol Pharm 6:659–668
Davis ME, Zuckerman JE, Choi CH, Seligson D, Tolcher A, Alabi CA, Yen Y, Heidel JD, Ribas A (2010) Evidence of RNAi in humans from systemically administered siRNA via targeted nanoparticles. Nature 464:1067–1070
DeRouchey J, Walker GF, Wagner E, Radler JO (2006) Decorated rods: a "bottom-up" self-assembly of monomolecular DNA complexes. J Phys Chem B 110:4548–4554
Desai N (2012) Challenges in development of nanoparticle-based therapeutics. AAPS J 14: 282–295
Donnenberg VS, Donnenberg AD (2005) Multiple drug resistance in cancer revisited: the cancer stem cell hypothesis. J Clin Pharmacol 45:872–877
Doria-Rose NA, Haigwood NL (2003) DNA vaccine strategies: candidates for immune modulation and immunization regimens. Methods 31:207–216
Duncan R (2003) The dawning era of polymer therapeutics. Nat Rev Drug Discov 2:347–360
Dvorak HF (1990) Leaky tumor vessels: consequences for tumor stroma generation and for solid tumor therapy. Prog Clin Biol Res 354A:317–330
Eifler AC, Thaxton CS (2011) Nanoparticle therapeutics: FDA approval, clinical trials, regulatory pathways, and case study. Methods Mol Biol 726:325–338
Farokhzad OC, Karp JM, Langer R (2006) Nanoparticle-aptamer bioconjugates for cancer targeting. Expert Opin Drug Deliv 3:311–324
Folkman J (1990) What is the evidence that tumors are angiogenesis dependent? J Natl Cancer Inst 82:4–6
Gabizon AA (1992) Selective tumor localization and improved therapeutic index of anthracyclines encapsulated in long-circulating liposomes. Cancer Res 52:891–896

Ganesh S, Iyer AK, Morrissey DV, Amiji MM (2013a) Hyaluronic acid based self-assembling nanosystems for CD44 target mediated siRNA delivery to solid tumors. Biomaterials 34:3489–3502

Ganesh S, Iyer AK, Weiler J, Morrissey DV, Amiji MM (2013b) Combination of siRNA-directed gene silencing with cisplatin reverses drug resistance in human non-small cell lung cancer. Mol Ther Nucleic Acids 2:e110

Gao GH, Li Y, Lee DS (2012) Environmental pH-sensitive polymeric micelles for cancer diagnosis and targeted therapy. J Control Release 169:180–184

Garzon R, Marcucci G, Croce CM (2010) Targeting microRNAs in cancer: rationale, strategies and challenges. Nat Rev Drug Discov 9:775–789

Gatenby RA, Gawlinski ET (2003) The glycolytic phenotype in carcinogenesis and tumor invasion insights through mathematical models. Cancer Res 63:3847–3854

Geall AJ, Mandl CW, Ulmer JB (2013) RNA: the new revolution in nucleic acid vaccines. Semin Immunol 25:152–159

Greish K, Fang J, Inutsuka T, Nagamitsu A, Maeda H (2003) Macromolecular therapeutics: advantages and prospects with special emphasis on solid tumour targeting. Clin Pharmacokinet 42:1089–1105

Hanahan D, Weinberg RA (2011) Hallmarks of cancer: the next generation. Cell 144:646–674

Heidel JD, Yu Z, Liu JY, Rele SM, Liang Y, Zeidan RK, Kornbrust DJ, Davis ME (2007) Administration in non-human primates of escalating intravenous doses of targeted nanoparticles containing ribonucleotide reductase subunit M2 siRNA. Proc Natl Acad Sci U S A 104:5715–5721

Hobbs SK, Monsky WL, Yuan F, Roberts WG, Griffith L, Torchilin VP, Jain RK (1998) Regulation of transport pathways in tumor vessels: role of tumor type and microenvironment. Proc Natl Acad Sci U S A 95:4607–4612

Hu-Lieskovan S, Zhang J, Wu L, Shimada H, Schofield DE, Triche TJ (2005) EWS-FLI1 fusion protein up-regulates critical genes in neural crest development and is responsible for the observed phenotype of Ewing's family of tumors. Cancer Res 65:4633–4644

Iftimia N, Iyer AK, Hammer DX, Lue N, Mujat M, Pitman M, Ferguson RD, Amiji M (2012) Fluorescence-guided optical coherence tomography imaging for colon cancer screening: a preliminary mouse study. Biomed Opt Express 3:178–191

Iyer AK, Khaled G, Fang J, Maeda H (2006) Exploiting the enhanced permeability and retention effect for tumor targeting. Drug Discov Today 11:812–818

Iyer AK, He J, Amiji MM (2012) Image-guided nanosystems for targeted delivery in cancer therapy. Curr Med Chem 19:3230–3240

Jain RK (1987) Transport of molecules in the tumor interstitium: a review. Cancer Res 47:3039–3051

Jeong JH, Mok H, Oh YK, Park TG (2009) siRNA conjugate delivery systems. Bioconjugate Chem 20:5–14

Jiang X, Ren Y, Williford JM, Li Z, Mao HQ (2013) Liver-targeted gene delivery through retrograde intrabiliary infusion. Methods Mol Biol 948:275–284

Keereweer S, Mol IM, Kerrebijn JD, Van Driel PB, Xie B, Baatenburg de Jong RJ, Vahrmeijer AL, Lowik CW (2012) Targeting integrins and enhanced permeability and retention (EPR) effect for optical imaging of oral cancer. J Surg Oncol 105:714–718

Kim K, Kim JH, Park H, Kim YS, Park K, Nam H, Lee S, Park JH, Park RW, Kim IS, Choi K, Kim SY, Park K, Kwon IC (2010) Tumor-homing multifunctional nanoparticles for cancer theragnosis: simultaneous diagnosis, drug delivery, and therapeutic monitoring. J Control Release 146:219–227

Kolhe S, Parikh K (2012) Application of nanotechnology in cancer: a review. Int J Bioinform Res Appl 8:112–125

Landesman Y, Svrzikapa N, Cognetta A 3rd, Zhang X, Bettencourt BR, Kuchimanchi S, Dufault K, Shaikh S, Gioia M, Akinc A, Hutabarat R, Meyers R (2010) In vivo quantification of formulated and chemically modified small interfering RNA by heating-in-Triton quantitative reverse transcription polymerase chain reaction (HIT qRT-PCR). Silence 1:16

Lavertu M, Methot S, Tran-Khanh N, Buschmann MD (2006) High efficiency gene transfer using chitosan/DNA nanoparticles with specific combinations of molecular weight and degree of deacetylation. Biomaterials 27:4815–4824

Lee JH, Lim YB, Choi JS, Lee Y, Kim TI, Kim HJ, Yoon JK, Kim K, Park JS (2003) Polyplexes assembled with internally quaternized PAMAM-OH dendrimer and plasmid DNA have a neutral surface and gene delivery potency. Bioconjug Chem 14:1214–1221

Li SD, Huang L (2006) Gene therapy progress and prospects: non-viral gene therapy by systemic delivery. Gene Ther 13:1313–1319

Li W, Szoka FC Jr (2007) Lipid-based nanoparticles for nucleic acid delivery. Pharm Res 24:438–449

Liang GF, Zhu YL, Sun B, Hu FH, Tian T, Li SC, Xiao ZD (2011) PLGA-based gene delivering nanoparticle enhance suppression effect of miRNA in HePG2 cells. Nanoscale Res Lett 6:447

Loeb LA, Loeb KR, Anderson JP (2003) Multiple mutations and cancer. Proc Natl Acad Sci U S A 100:776–781

Lu JM, Wang X, Marin-Muller C, Wang H, Lin PH, Yao Q, Chen C (2009) Current advances in research and clinical applications of PLGA-based nanotechnology. Expert Rev Mol Diagn 9:325–341

Lue N, Ganta S, Hammer DX, Mujat M, Stevens AE, Harrison L, Ferguson RD, Rosen D, Amiji M, Iftimia N (2010) Preliminary evaluation of a nanotechnology-based approach for the more effective diagnosis of colon cancers. Nanomedicine (Lond) 5:1467–1479

Maeda H (2001) The enhanced permeability and retention (EPR) effect in tumor vasculature: the key role of tumor-selective macromolecular drug targeting. Adv Enzyme Regul 41:189–207

Maeda H (2012) Macromolecular therapeutics in cancer treatment: the EPR effect and beyond. J Control Release 164:138–144

Maeda H, Wu J, Sawa T, Matsumura Y, Hori K (2000) Tumor vascular permeability and the EPR effect in macromolecular therapeutics: a review. J Control Release 65:271–284

Maeda H, Sawa T, Konno T (2001) Mechanism of tumor-targeted delivery of macromolecular drugs, including the EPR effect in solid tumor and clinical overview of the prototype polymeric drug SMANCS. J Control Release 74:47–61

Magadala P, Amiji M (2008) Epidermal growth factor receptor-targeted gelatin-based engineered nanocarriers for DNA delivery and transfection in human pancreatic cancer cells. AAPS J 10:565–576

Marcucci F, Lefoulon F (2004) Active targeting with particulate drug carriers in tumor therapy: fundamentals and recent progress. Drug Discov Today 9:219–228

Martinez-Zaguilan R, Seftor EA, Seftor RE, Chu YW, Gillies RJ, Hendrix MJ (1996) Acidic pH enhances the invasive behavior of human melanoma cells. Clin Exp Metastasis 14:176–186

Matsumura Y, Maeda H (1986) A new concept for macromolecular therapeutics in cancer chemotherapy: mechanism of tumoritropic accumulation of proteins and the antitumor agent smancs. Cancer Res 46:6387–6392

McManus MT (2003) MicroRNAs and cancer. Semin Cancer Biol 13:253–258

Milane L, Duan Z, Amiji M (2011a) Role of hypoxia and glycolysis in the development of multi-drug resistance in human tumor cells and the establishment of an orthotopic multi-drug resistant tumor model in nude mice using hypoxic pre-conditioning. Cancer Cell Int 11:3

Milane L, Duan Z, Amiji M (2011b) Development of EGFR-targeted polymer blend nanocarriers for combination paclitaxel/lonidamine delivery to treat multi-drug resistance in human breast and ovarian tumor cells. Mol Pharm 8:185–203

Morrissey DV, Lockridge JA, Shaw L, Blanchard K, Jensen K, Breen W, Hartsough K, Machemer L, Radka S, Jadhav V, Vaish N, Zinnen S, Vargeese C, Bowman K, Shaffer CS, Jeffs LB, Judge A, MacLachlan I, Polisky B (2005) Potent and persistent in vivo anti-HBV activity of chemically modified siRNAs. Nat Biotechnol 23:1002–1007

Nagy JA, Brown LF, Senger DR, Lanir N, Van de Water L, Dvorak AM, Dvorak HF (1989) Pathogenesis of tumor stroma generation: a critical role for leaky blood vessels and fibrin deposition. Biochim Biophys Acta 948:305–326

Noguchi Y, Wu J, Duncan R, Strohalm J, Ulbrich K, Akaike T, Maeda H (1998) Early phase tumor accumulation of macromolecules: a great difference in clearance rate between tumor and normal tissues. Jpn J Cancer Res 89:307–314

Ochiya T, Takahama Y, Nagahara S, Sumita Y, Hisada A, Itoh H, Nagai Y, Terada M (1999) New delivery system for plasmid DNA in vivo using atelocollagen as a carrier material: the Minipellet. Nat Med 5:707–710

Parhi P, Mohanty C, Sahoo SK (2012) Nanotechnology-based combinational drug delivery: an emerging approach for cancer therapy. Drug Discov Today 17:1044–1052

Pasqualini R, Koivunen E, Ruoslahti E (1997) Alpha v integrins as receptors for tumor targeting by circulating ligands. Nat Biotechnol 15:542–546

Pasut G, Veronese FM (2009) PEGylation for improving the effectiveness of therapeutic biomolecules. Drugs Today (Barc) 45:687–695

Patil ML, Zhang M, Betigeri S, Taratula O, He H, Minko T (2008) Surface-modified and internally cationic polyamidoamine dendrimers for efficient siRNA delivery. Bioconjug Chem 19:1396–1403

Pecot CV, Calin GA, Coleman RL, Lopez-Berestein G, Sood AK (2011) RNA interference in the clinic: challenges and future directions. Nat Rev Cancer 11:59–67

Peer D, Lieberman J (2011) Special delivery: targeted therapy with small RNAs. Gene Ther 18:1127–1133

Printz C (2012) Keeping it small: nanotechnology enables scientists to target cancer in new ways. Cancer 118:3879–3880

Rahbek UL, Nielsen AF, Dong M, You Y, Chauchereau A, Oupicky D, Besenbacher F, Kjems J, Howard KA (2010) Bioresponsive hyperbranched polymers for siRNA and miRNA delivery. J Drug Target 18:812–820

Rai K, Takigawa N, Ito S, Kashihara H, Ichihara E, Yasuda T, Shimizu K, Tanimoto M, Kiura K (2011) Liposomal delivery of microRNA-7-expressing plasmid overcomes epidermal growth factor receptor tyrosine kinase inhibitor-resistance in lung cancer cells. Mol Cancer Ther 10:1720–1727

Ren Y, Kang CS, Yuan XB, Zhou X, Xu P, Han L, Wang GX, Jia Z, Zhong Y, Yu S, Sheng J, Pu PY (2010) Co-delivery of as-miR-21 and 5-FU by poly(amidoamine) dendrimer attenuates human glioma cell growth in vitro. J Biomater Sci Polym Ed 21:303–314

Restifo NP, Ying H, Hwang L, Leitner WW (2000) The promise of nucleic acid vaccines. Gene Ther 7:89–92

Rihova B (1998) Receptor-mediated targeted drug or toxin delivery. Adv Drug Deliv Rev 29:273–289

Ruoslahti E, Pierschbacher MD (1987) New perspectives in cell adhesion: RGD and integrins. Science 238:491–497

Saad M, Garbuzenko OB, Minko T (2008) Co-delivery of siRNA and an anticancer drug for treatment of multidrug-resistant cancer. Nanomedicine (Lond) 3:761–776

Schaffert D, Wagner E (2008) Gene therapy progress and prospects: synthetic polymer-based systems. Gene Ther 15:1131–1138

Scholz C, Wagner E (2012) Therapeutic plasmid DNA versus siRNA delivery: common and different tasks for synthetic carriers. J Control Release 161:554–565

Schroeder A, Heller DA, Winslow MM, Dahlman JE, Pratt GW, Langer R, Jacks T, Anderson DG (2012) Treating metastatic cancer with nanotechnology. Nat Rev Cancer 12:39–50

Sun X, Zhang N (2010) Cationic polymer optimization for efficient gene delivery. Mini Rev Med Chem 10:108–125

Takeshita F, Patrawala L, Osaki M, Takahashi RU, Yamamoto Y, Kosaka N, Kawamata M, Kelnar K, Bader AG, Brown D, Ochiya T (2010) Systemic delivery of synthetic microRNA-16 inhibits the growth of metastatic prostate tumors via downregulation of multiple cell-cycle genes. Mol Ther 18:181–187

Temming K, Schiffelers RM, Molema G, Kok RJ (2005) RGD-based strategies for selective delivery of therapeutics and imaging agents to the tumour vasculature. Drug Resist Updat 8:381–402

Torchilin VP (2010) Passive and active drug targeting: drug delivery to tumors as an example. Handb Exp Pharmacol (197):3–53

Torchilin V (2011) Tumor delivery of macromolecular drugs based on the EPR effect. Adv Drug Deliv Rev 63:131–135

Torchilin VP, Rammohan R, Weissig V, Levchenko TS (2001) TAT peptide on the surface of liposomes affords their efficient intracellular delivery even at low temperature and in the presence of metabolic inhibitors. Proc Natl Acad Sci U S A 98:8786–8791

Trang P, Medina PP, Wiggins JF, Ruffino L, Kelnar K, Omotola M, Homer R, Brown D, Bader AG, Weidhaas JB, Slack FJ (2010) Regression of murine lung tumors by the let-7 microRNA. Oncogene 29:1580–1587

Trang P, Wiggins JF, Daige CL, Cho C, Omotola M, Brown D, Weidhaas JB, Bader AG, Slack FJ (2011) Systemic delivery of tumor suppressor microRNA mimics using a neutral lipid emulsion inhibits lung tumors in mice. Mol Ther 19:1116–1122

Ulmer JB, Wahren B, Liu MA (2006) Gene-based vaccines: recent technical and clinical advances. Trends Mol Med 12:216–222

Varmus HE, Weinberg RA (1993) Genes and the biology of cancer. Scientific American Library

Vaupel P, Mayer A (2007) Hypoxia in cancer: significance and impact on clinical outcome. Cancer Metastasis Rev 26:225–239

Veronese FM, Pasut G (2005) PEGylation, successful approach to drug delivery. Drug Discov Today 10:1451–1458

Visone R, Croce CM (2009) MiRNAs and cancer. Am J Pathol 174:1131–1138

Voest EE (1998) Neovascularization: the Achilles' heel of tumors? Drug Resist Updat 1:86–87

Wang X, Wang Y, Chen ZG, Shin DM (2009) Advances of cancer therapy by nanotechnology. Cancer Res Treat 41:1–11

Wang J, Lu Z, Wientjes MG, Au JL (2010) Delivery of siRNA therapeutics: barriers and carriers. AAPS J 12:492–503

Yuan F, Salehi HA, Boucher Y, Vasthare US, Tuma RF, Jain RK (1994) Vascular permeability and microcirculation of gliomas and mammary carcinomas transplanted in rat and mouse cranial windows. Cancer Res 54:4564–4568

Zhu J, Xue J, Guo Z, Marchant RE (2007) Vesicle size and stability of biomimetic liposomes from 3′-sulfo-Lewis a (SuLea) containing glycolipids. Colloids Surf B Biointerfaces 58:242–249

Part V
Nanotheranostics

Chapter 11
Multifunctional Gold Nanocarriers for Cancer Theranostics: From Bench to Bedside and Back Again?

João Conde, Furong Tian, Pedro V. Baptista, and Jesús M. de la Fuente

11.1 Introduction

The National Cancer Institute predicts that over the next years, nanotechnology will result in important advances in early detection, molecular imaging, targeted and multifunctional therapeutics, prevention and control of cancer (National Cancer Institute 2010). Nanotechnology offers numerous tools to diagnose and treat cancer, such as new imaging agents, multifunctional devices capable of overcoming biological barriers to deliver therapeutic agents directly to cells and tissues involved in cancer growth and metastasis, and devices capable of predicting molecular changes to prevent action against precancerous cells (Baptista 2009).

Nanoparticle-based delivery systems in Theranostics (Diagnostics & Therapy) provide better penetration of therapeutic and diagnostic substances within the body

J. Conde (✉)
Instituto de Nanociencia de Aragon (INA), Universidad de Zaragoza, Zaragoza 50018, Portugal

Research Centre for Human Molecular Genetics, Science and Technology Faculty, New University of Lisbon, Caparica, Portugal

Massachusetts Institute of Technology, Harvard–MIT Biomedical Engineering Center, E25-449, 45 Carleton Street, Cambridge, MA 02139, USA
e-mail: jdconde@mit.edu

F. Tian
Focas Research Institute, Dublin Institute of Technology, Camden Row, Dublin, Ireland

P.V. Baptista
Research Centre for Human Molecular Genetics, Science and Technology Faculty, New University of Lisbon, Caparica, Portugal

J.M. de la Fuente
Instituto de Nanociencia de Aragon (INA), Universidad de Zaragoza, Zaragoza, Spain

M.J. Alonso, M. Garcia-Fuentes (eds.), *Nano-Oncologicals: New Targeting and Delivery Approaches*, Advances in Delivery Science and Technology,
DOI 10.1007/978-3-319-08084-0_11

at a reduced risk in comparison with conventional therapies (Ma et al. 2011; Praetorius and Mandal 2007). Limitations in medical practice are intimately associated with the fact that diagnostics, therapy, and therapy guidance are mostly separate from each other. It is here that theranostics unites the three stages in one single process, supporting early-stage diagnosis and treatment, overcoming some of the sensitivity and specificity of current medicines (Lammers et al. 2010, 2011; Pene et al. 2009). At the present time, there is a growing need to enhance the capability of theranostics procedures where nanoparticle-based sensors may provide for the simultaneous detection of several gene-associated conditions and nanodevices with the ability to monitor real-time drug action. These innovative multifunctional nanocarriers for cancer theranostics may allow the development of diagnostics systems such as colorimetric and immunoassays, and in therapy approaches through gene therapy, drug delivery, and tumor targeting systems (Fig. 11.1).

Nowadays the main challenge is to develop a system for molecular therapy capable of circulating in the blood stream undetected by the immune system and capable to recognize the desirable target and signal it for effective drug delivery

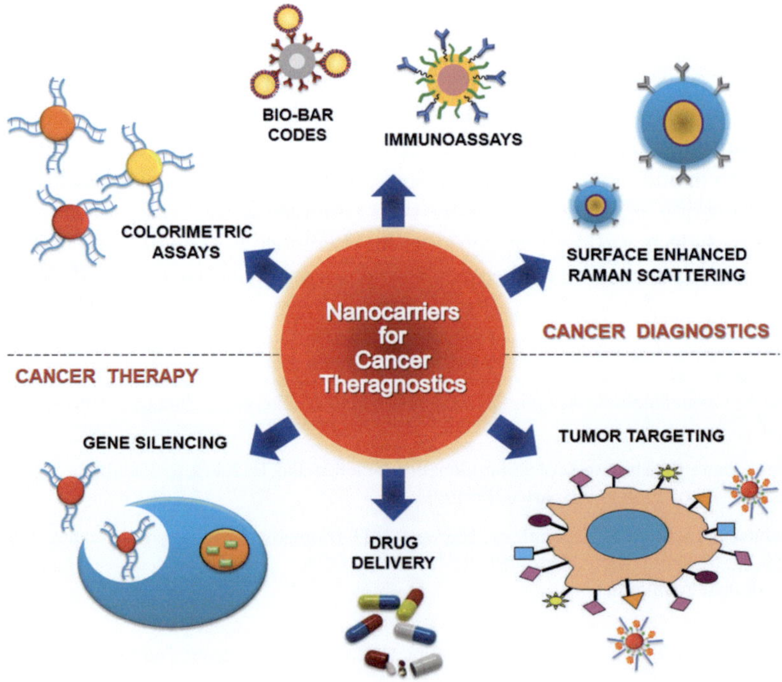

Fig. 11.1 Nanocarriers for cancer theranostics. Nanoparticles-based strategies can be used for biosensing using plasmonic nanosensors for colorimetric assays and bio-barcodes for protein detection or intense labels for immunoassays. Moreover, the use of metal surfaces to enhance the Raman scattering signal of target molecules may be used for cancer diagnostics. Engineered nanocarriers can also act as therapeutic agents via gene silencing and drug delivery systems. Some nanocarriers can be attached to specific targets for selective damage to cancer cells through tumor targeting approaches

or gene silencing. As a result, nanotechnology is playing a role in providing new types of nanotherapeutics for cancer that have the potential to provide effective therapies with minimal side effects and with high specificity (Heath and Davis 2008). The interdisciplinary and vibrant field of nanotechnology continues giving us hope of a personalized medicine as a part of cancer patient management. Gold nanoparticles (AuNPs) are one of those nanosystems that provide non-toxic carriers for drug and gene delivery applications (Nishiyama 2007; Ghosh et al. 2008a). They are versatile agents with a variety of biomedical applications including use in highly sensitive diagnostic assays (Goodman et al. 2004), thermal ablation, and radiotherapy enhancement (Hirsch et al. 2003; Hainfeld et al. 2004, 2008), as well as drug and gene delivery (Hong et al. 2006; Thomas and Klibanov 2003).

The unique characteristics of AuNPs in the nanometer range, such as high surface-to-volume ratio or size-dependent optical properties, are drastically different from those of their bulk materials and hold pledge in the clinical field for disease therapeutics (Kim 2007; Heath and Davis 2008). Nanoparticles (NPs) exhibiting these unique and broad-based optical properties, ease of synthesis and facile surface chemistry and functionalization, and appropriate size scale are generating much eagerness in clinical diagnostics and therapy. The most common applications in which gold nanocarriers have been used so far are labeling, delivering, heating, sensing, and detection (Sperling et al. 2008).

In spite of these advantages, nanoparticles show some limitations, such as their small size and large surface area can lead to particle–particle aggregation and may result in limited loading of functional components and burst release. In fact, only NPs with the appropriate size (and surface chemistry) will not be immediately recognized by our immune system and committed to removal from the organism, thus showing increased circulation times. Also, size and surface properties play an important role to avoid clearance (Sperling and Parak 2010). For example, hydrophilic nanoparticles with an effective size in the range of 10 to 100 nm are small enough to slow down activation of the mononuclear phagocyte system and are big enough to avoid renal filtration (Gil and Parak 2008). Nanoparticles with unique and broad-based optical properties, ease of synthesis and facile surface chemistry and functionalization, and appropriate size scale are generating much eagerness in biotechnology and biomedicine with particular emphasis in clinical diagnostics and therapy (Doria et al. 2012). However for the biological application of these multifunctional NPs, their functionalization with one or several biomolecules such as DNA/RNA, oligonucleotides (i.e., ssDNA/RNA, dsDNA/RNA), peptides and antibodies, fluorescent dyes, polymers, drugs, tumoral markers, enzymes, and other proteins that will introduce several functionalities and moieties is necessary. In the end the conjugation strategy is directly dependent on a numbers of factors, namely the NPs size, surface chemistry and shape, as well as the type of ligands and functional groups one desires to exploit in the functionalization (Sperling and Parak 2010).

When referring to cancer therapy, targeting and localized delivery are of utmost importance to enhance the therapeutic effect and decrease the adverse distribution to healthy organs and tissues. Multifunctional gold nanocarriers may potentiate the development of individualized cancer therapy based on the individual's biological information within the tumor (i.e., biomolecular profiling). Gold nanocarriers can be modified with multiple cell-targeting and membrane translocating peptides, loaded with DNA/RNA and used as nanovectors (Gil and Parak 2008; Peer et al. 2007).

In this chapter, we will focus on the exciting new methods and applications of AuNPs for cancer diagnosis and therapy with particular emphasis on their use in vivo and their potential to be translated into clinical settings.

11.2 Nanodiagnostics

Nanodiagnostics can be defined as the use of nano-sized materials, devices, or systems for diagnostics purposes (Doria et al. 2007). It is a promising field as more and improved techniques are becoming available for clinical diagnostics with increased sensitivity at lower costs (Baptista et al. 2005, 2006, 2008; Doria et al. 2007).

The use of the colloidal gold color change upon aggregation is the best characterized example for diagnostic systems using AuNPs. In fact, AuNPs functionalized with ssDNA capable of specifically hybridizing to a complementary target for the detection of specific nucleic acid sequences in biological samples have been extensively used (Li and Rothberg 2004; Mirkin et al. 1996; Thaxton et al. 2006; Cheng et al. 2006; Baptista et al. 2008; Doria et al. 2007; Taton et al. 2000; Qin and Yung 2007; Sato et al. 2003, 2005; Elghanian et al. 1997).

The use of thiol-linked ssDNA-modified gold nanoparticles for the colorimetric detection of gene targets represents an inexpensive and easy to perform alternative to fluorescence or radioactivity-based assays (Storhoff et al. 2004). In 1996, *Mirkin* et al. (1996) described the use of a cross-linking method that relies on the detection of single-stranded oligonucleotide targets using two different Au-nanoprobes, each of them functionalized with a DNA-oligonucleotide complementary to one half of the given target. A mixture of gold nanoparticles with surface-immobilized non-complementary DNA sequences appears red in color and has a strong absorbance at 520 nm. When a complementary DNA sequence is added, the nanoparticles are reversibly aggregated causing a red shift in the surface plasmon absorbance to 574 nm, and the solution becomes purple in color. Consequently, these AuNPs reversibly assemble by the formation of the DNA duplex, linking the particles together. These pioneer works gave the research community the idea of an easy method to sensitize oligonucleotide–AuNP conjugates, with intense and highly tunable optical properties, ease of conjugation through the gold–thiol bond, catalytic properties, and relative biocompatibility (Thaxton et al. 2006; Cao et al. 2005).

On the other hand, in 2005 Baptista et al. introduced a non-cross linking method where thiol-linked DNA–gold nanoparticles were used in a novel colorimetric method to detect the presence of specific mRNA from a total RNA extract of yeast cells (Baptista et al. 2005). The method consists in visual and/or spectrophotometric comparison of solutions before and after salt induced Au-nanoprobe aggregation (see Fig. 11.2)—the presence of a complementary target prevents aggregation and the solution remains red and has a strong absorbance at ±520 nm; non-complementary/mismatched targets do not prevent Au-nanoprobe aggregation, resulting in a visible change of color from red to blue and therefore a shift in the surface plasmon absorbance to 600–650 nm. This method has been successfully applied to detect eukaryotic gene expression without retro-transcription or PCR amplification steps (Conde et al. 2010b; Baptista et al. 2005); to distinguish fully complementary from mismatched sequences, with a single base mismatch, i.e., to detect common mutations within the β-globin gene (Doria et al. 2007); and in a fast and straightforward assay for *Mycobacterium tuberculosis* DNA detection in clinical samples (Costa et al. 2009; Baptista et al. 2006).

Other approaches are the use of AuNPs as a core/seed that can be tailored with a wide variety of surface functionalities to provide highly selective nanoprobes for diagnosis (You et al. 2007); the utilization of surface plasmon resonance (SPR) scattering imaging or SPR absorption spectroscopy generated from antibody conjugated AuNPs in molecular biosensor techniques for the diagnosis and investigation of oral epithelial living cancer cells in vivo and in vitro (El-Sayed et al. 2005a); the use of multifunctional AuNPs which incorporate both cytosolic delivery and targeting moieties on the same particle functioning as intracellular sensors to monitoring actin rearrangement in live fibroblasts (Kumar et al. 2007); and the employment of AuNPs in electrochemical based methods that can be coupled with metal deposition for signal enhancement (Castaneda et al. 2007).

Gold NPs have already proven to be one of the most important groups of nanomaterials for biosensing approaches. Highly sensitive and specific biosensors based on AuNPs have open up the possibility of creating new diagnostic platforms for disease markers, biological and infectious agents in the early-stage detection of disease and threats, especially in cancer (Conde et al. 2012b; Doria et al. 2012).

11.2.1 Gold Nanocarriers in Cancer Diagnosis

Cancer is the one of first leading causes of mortality in the modern world, with more than 10 million new cases every year (Hanahan and Weinberg 2000; Siegel et al. 2012). Some continue to argue that the search for the origin and treatment of this disease will continue through the next quarter of century, adding successive layers of complexity to an investigation that per se is very complex. In fact, advances in diagnosis and treatment of this disease that kills millions of people each year worldwide have not been as effective as for other chronic diseases, and only for some types of cancer there are effective methods of detection (Hanahan and Weinberg 2000). The high mortality rate in cancer is commonly attributed to the difficulties in detecting the disease at an early treatable stage. The main challenge is to find new and more effective diagnostic agents for the monitorization of predictive cell

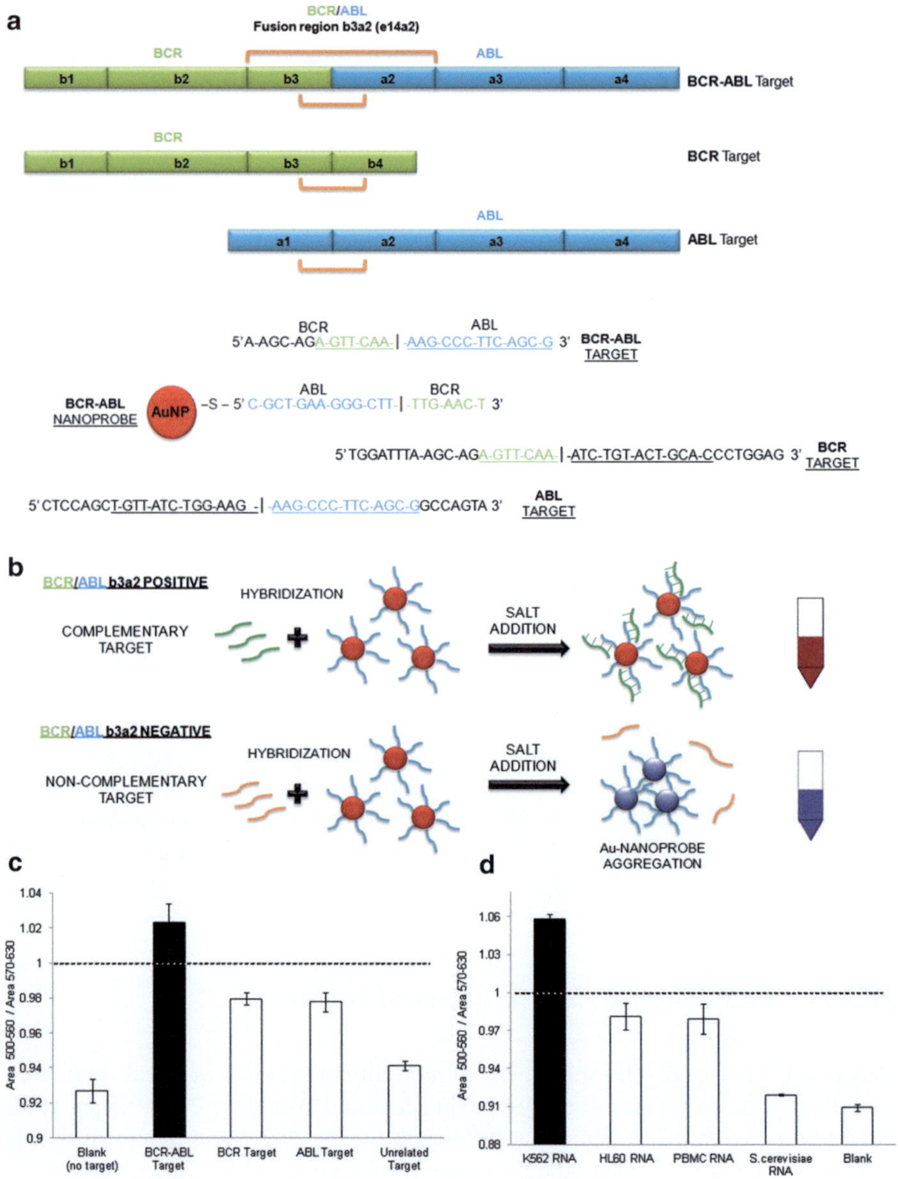

Fig. 11.2 The colorimetric assay. (**a**) Oligonucleotide probe and target sequences designed for BCR–ABL b3a2 (e14a2) junction and for BCR and ABL genes. Complementary and non-complementary target sequences were used to study the level of specific interaction between the target and the Au-nanoprobes. BCR–ABL fusion positive (100 % complementary); BCR and ABL gene sequences were used as controls (50 % non-complementary); and a completely unrelated sequence (100 % non-complementary) was used as negative control. (**b**) The assay is based on the increased stability of the Au-nanoprobes upon hybridization with the complementary target in solution, while non-hybridized Au-nanoprobes easily aggregate once the solution's ionic strength is increased. The absence of full complementarity is revealed by a change of color from *red* to *blue*

molecular changes that are involved in tumor development as the key to the efficient and ultimately triumphant treatment of cancer is early and accurate diagnosis (Etzioni et al. 2003).

It is here that nanotechnology enters the fight in the technological leap of controlling materials at nanoscale by offering a "big revolution" in new medical and healthcare diagnostic systems (Gil and Parak 2008). In fact, nanotechnology combined with biology and medicine is the most advanced technology both from an academic point-of-view and for commercial applications, producing major advances in molecular diagnostics and bioengineering (Giljohann et al. 2010; Salata 2004). AuNPs are one of those nanosystems that provide non-toxic carriers with a variety of biomedical applications including use in highly sensitive diagnostic assays (Conde et al. 2012b; Doria et al. 2012).

In reality the multiplexed marker protein assays are critical in the diagnosis of complex diseases like cancer. In fact, AuNP probes barcoded with reporter DNAs and a capture antibody have been extensively used with great promise (Stoeva et al. 2006; Son and Lee 2007). For instance, AuNP immunoassays are one of the most used nanosystems in cancer detection. Mirkin and co-workers developed an ultrasensitive method for detecting protein analytes. This facile immunoassay was used for the detection of prostate specific antigen (PSA, a valuable biomarker for prostate cancer screening) and had an almost one million-fold higher sensitivity than a conventional ELISA-based assay (Nam et al. 2003). This study was the first step to use 6 years later the AuNP bio-barcode assay probe for the detection of PSA in a clinical pilot study with 18 men who have undergone radical prostatectomy for prostate cancer. This new bio-barcode PSA assay is approximately 300 times more sensitive than commercial immunoassays (Thaxton et al. 2009).

Also, Huang et al. have reported a PSA immunoassay on a commercially available SPR biosensor. This sandwich assay with AuNPs was used for a major enhancement in sensitivity of PSA detection at clinically relevant concentrations (Huang et al. 2005). Other studies demonstrated a highly sensitive organic electrochemical transistor based immunosensor with secondary antibody-conjugated AuNPs with a low detection limit for PSA. These sensor performances were particularly improved in the lower concentration range where the detection is clinically important for the preoperative diagnosis and screening of prostate cancer (Huang et al. 2005).

Fig. 11.2 (continued) due to Au-nanoprobe aggregation which is corroborated by naked eye and UV/vis spectroscopy. BCR–ABL positive: sample in the presence of complementary target (solution remains *red*); BCR–ABL negative: sample in the presence of non-complementary target (solution turns *blue*). (**c**) Spectrophotometry relative to the detection of synthetic BCR–ABL oligonucleotide target. Oligonucleotides with BCR or ABL sequence only (showing 50 % complementarity) were used as controls and an unrelated target (showing 100 % non-complementarity to the Au-nanoprobe) as negative control. (**d**) Detection of BCR–ABL in total RNA from K562 cell line, HL-60 cell line, and human PBMC (harboring 50 % complementary targets to the nanoprobe) and *S. cerevisiae* cells (100 % non-complementary). Nanoprobe aggregation as measured by ratio of area under the curve AUC500 nm-560 nm/AUC570 nm-630 nm. The *dashed line* represents the threshold of 1 considered for discrimination between positive and negative

Additionally, AuNP probes coupled with dynamic light scattering (DLS) measurements can also be used for the development of a one-step homogeneous immunoassay for the detection of free-PSA. The light scattering intensity of nanoparticles and nanoparticle oligomers is several orders of magnitude higher than proteins, making it possible to detect nanoparticle probes in the low picomolar concentration range (Liu et al. 2008).

AuNPs scatter light intensely at or near their surface plasmon wavelength region and when coupled with DLS detection can be very useful for serum protein biomarker detection and analysis. Huo et al. reported the use of citrate-AuNP to absorb proteins from the serum and form a protein corona on the nanoparticle surface. The protein corona formation and the subsequent binding of antibody to the target proteins in the protein corona were detected by DLS. Using this simple assay, the authors discovered multiple molecular aberrations associated with prostate cancer from both mice and human blood serum samples (Huo et al. 2011, 2012).

Other authors have taken a different approach to addressing the use of AuNPs in cancer biomarker detection by employing the biological applications of antibody-conjugated AuNPs in several types of cancer, such as breast cancer (Ambrosi et al. 2010; Lu et al. 2010a), pancreatic adenocarcinoma (Eck et al. 2008), cervical cancer (Rahman et al. 2005), epithelial cancer (Yang et al. 2008), liver cancer (Lan et al. 2011), prostate cancer (Lukianova-Hleb et al. 2011), and oral cancer (El-Sayed et al. 2005a).

The mechanism of selectivity and all these immunoassay's response open up a new possibility of rapid, simple, clean, easy, economically very cheap, non-toxic, and reliable diagnosis of cancer. In fact, antibody-conjugated AuNPs are one of the most used nanosystems in cancer diagnostics, being useful in molecular biosensor techniques for the diagnosis and investigation of cancer cells in vivo and in vitro. The importance of these nanosystems can be demonstrated by a significant number of companies involved in the synthesis and applications of antibody-conjugated nanoparticles, such as Magnisense SAS, Diagnostic Biosensors, LLC, Alnis Biosciences, Inc., and Invitrogen Corp (Arruebo et al. 2009).

Direct detection of cancer cells using colorimetric assays with AuNPs has been extensively used due to their simplicity and versatility, among which those based on LSPR. LSPR is the collective oscillation of the electrons in the conduction band, which is usually in the visible region giving rise to the strong SPR absorption (Daniel and Astruc 2004). These AuNPs are commonly functionalized with biomolecules (e.g., DNA, RNA, peptides, enzymes) and capable of recognizing molecular events associated with cancer development down to femtomolar level with single base discrimination resolution (Kang et al. 2010; Li et al. 2005; Medley et al. 2008). Molecular nanodiagnostics applied to cancer may provide rapid and sensitive detection of cancer related molecular alterations, which would enable early detection even when those alterations occur only in a small percentage of cells. For instance, Conde et al. present an AuNPs based approach for the molecular recognition and quantification of the *BCR–ABL* fusion transcript, which is responsible for chronic myeloid leukemia (CML). This inexpensive and very easy to perform method allows quantification of unamplified total human RNA and specific detection of the oncogene transcript. This assay may constitute a promising tool in early diagnosis of CML and could easily be extended to further target genes with proven involvement in cancer development (Conde et al. 2010b, 2012c). The sensitivity settled by the Au-nanoprobes allows differential gene expression from 10 ng/μl of total RNA and takes less than 30 min to complete after total RNA extraction, minimizing RNA degradation (see Fig. 11.2).

Also, aptamer-conjugated AuNPs have become a powerful tool for point of care diagnostics (Mukerjee et al. 2012). Most of the common aptamer–AuNP assays are able to differentiate between different types of target and control cells based on the aptamer used in the assay, indicating the wide applicability for cancer cell detection. For instance, Liu et al. (2009) reported the use of an aptamer–nanoparticle strip biosensor for the rapid, specific, sensitive, and low-cost detection of circulating cancer cells in human blood, showing great promise for in-field and point-of-care cancer diagnosis and therapy. In another study, Medley et al. have developed a colorimetric assay for the direct detection of diseased cells and thus capable of distinguishing between cancer cells and noncancerous cells. This assay combines the selectivity and affinity of aptamers and the spectroscopic advantages of AuNPs to allow for the sensitive detection of cancer cells with both the naked eye and based on absorbance measurements (Medley et al. 2008). Another important aspect correlated with disease state in cancer patients is the presence of circulating tumor cells in the bloodstream. In order to induce optical contrast in non-pigmented cancer cells, Viator et al. (2010) attached AuNPs to a prostate cancer cell line, using optical absorption to detect such cells in a photoacoustic flowmeter designed to find circulating tumor cells in blood samples.

Another application of AuNPs in cancer is their capability to target and provide in vivo tumor detection using surface-enhanced Raman scattering (SERS). SERS has led the way in terms of use of spectroscopic methods for signal enhancement by nanostructured metal surfaces towards in vivo tracking of biomolecules trafficking (Stiles et al. 2008; Wilson and Willets 2013). SERS has been extensively used for molecular/ion detection and bioimaging applications since it minimizes photoblinking or photobleaching from conventional fluorophores, decreases signal-to-noise ratio in complex in vitro and in vivo, and usually Raman reporters are stable and yield large optical enhancements (Alvarez-Puebla and Liz-Marzan 2010, 2012a, b; Samanta et al. 2011). AuNPs covered by Raman reporters have been used for SERS to detect cancer cells in vitro and tumors in vivo (Kong et al. 2012; Qian et al. 2008b). Actually, Lin et al. (2011) described the tremendous potential of using AuNP based-SERS to obtain blood serum biochemical information for non-invasive colorectal cancer detection. Raman reporters when combined with AuNPs can elicit an optical contrast to discriminate between cancerous and normal cells and their conjugation with antibodies allowed them to map the expression of relevant biomarkers for molecular imaging (Kah et al. 2007), as well as detect and characterized circulating tumor cells. These SERS nanoparticles constitute an important tool for clinical research once they can successfully identify circulating tumor cells in the peripheral blood of 19 patients with squamous cell carcinoma of the head and neck (Wang et al. 2011).

The papers discussed earlier report the development of nanoscale devices and platforms that can be used for improved biomarker detection, such as nucleic acids (DNA or RNA) or proteins. However, the development of molecular diagnosis of cancer and at the same time the selective delivery of a specific anticancer agent by joining diagnostics and therapy (theranostics) on a single nanodevice will most definitely revolutionize the way we manage cancer (Baptista 2012).

In fact, Conde et al. recently developed a highly sensitive probe for in vivo tumor recognition with the capacity to target specific cancer biomarkers such as epidermal growth factor receptors (EGFRs) on human cancer cells and xenograft tumor models. The authors used ~90 nm AuNPs capped by a Raman reporter, encapsulated and entrapped by larger polymers and an FDA antibody–drug conjugate—Cetuximab (Erbitux®). These smart SERS gold nanoantennas present a high Raman signal both in cancer cells and in mice bearing xenograft tumors, and the Raman detection signal is accomplished simultaneously by extensive tumor growth inhibition in mice. This approach seems to be an innovative and efficient theranostics system for both tumor detection and tumor cell inhibition at the same time (Conde et al. 2014a).

Table 11.1 summarizes the latest types of AuNP-based biosensors for cancer diagnostics, according to their methodology principle. Some of the described nanosystems will most likely revolutionize our understanding of biological mechanisms and push forward the clinical practice through their integration in future diagnostics platforms. Nevertheless, very little gave successful results or went to clinical trials. Therefore new synthesis, fabrication, and characterization methods are needed for developing highly advanced AuNPs capable of use in sensitive and multiple detection methods with negligible toxicity and high sensitivity. In the future, it might be possible to apply all AuNPs properties together and evolve new chemistry for synthesis of smart materials for diagnostic applications and clinical trials.

11.3 Nanotherapy

In medical terms, a therapeutic effect is a consequence of a medical treatment of any kind, the results of which are judged to be desirable and beneficial. Conventional therapy methods in cancer involve the employment of anticancer agents that do not greatly differentiate between cancerous and normal cells (Minelli et al. 2010). Efficient in vivo targeting to heterogeneous population of cancer cells and tissue still requires better selectivity and decreased toxicity to surrounding normal cells, towards a decrease of systemic toxicity, adverse and severe side effects (Liu et al. 2007).

In another way, universally targeting cells within a tumor is not always feasible because some drugs cannot diffuse efficiently and the random nature of the approach makes it difficult to control the process and may induce multiple-drug resistance (MDR)—a situation where chemotherapy treatments fail patients owing to resistance of cancer cells towards one or more drugs (Peer et al. 2007). Consequently, nanotechnology could offer a less invasive alternative, enhancing the life expectancy and quality of life of the patient (Cuenca et al. 2006).

Table 11.1 Summary of the latest AuNP-based biosensors used in cancer diagnostics according to the type of NP, surface modification, type of cancer, and explored methodology principle

Method	Type of NP	Surface modification	Target/cells/samples	Type of cancer	Comments (reference)
Colorimetric Scanometric	40 nm spherical	dA-tailed probe applied to the strip, which contains oligo(dT)-conjugated AuNPs in dry form	Fusion genes in K562 cell line	Acute and chronic leukemia	Dry-reagent, disposable, dipstick test for molecular screening of seven chromosomal translocations associated with acute and chronic leukemia (Kalogianni et al. 2007)
	13 nm spherical	ssDNA	BCR–ABL b3a2 (e14a2) fusion transcript	Chronic myeloid leukemia	Detection and quantification of the BCR–ABL gene fusion using thiol-DNA modified AuNPs (Conde et al. 2010b)
	30 nm spherical	Thiol-terminated DNA barcodes; anti-PSA antibodies	Prostate-specific antigen (PSA) (biomarker for prostate cancer screening)	Prostate cancer	Nanoparticle based bio-barcode for PSA detection (Thaxton et al. 2009)
	15 nm spherical	Antibody anti-CA15-3–HRP (horseradish peroxidase)	CA15-3 breast cancer biomarker in human serum	Breast cancer	Enhanced AuNP based ELISA for a breast cancer biomarker detection (Ambrosi et al. 2010)
	Oval-shaped	Monoclonal anti-HER2/c-ErbB-2 antibody; S6 RNA aptamer-conjugated	SKBR-3 cells	Breast cancer	Colorimetric and highly sensitive two-photon scattering assay for highly selective and sensitive detection of breast cancer (Lu et al. 2010a)
Immuno-assays	15 nm spherical	Dithiol-PEG-COOH; F19 monoclonal antibodies	Tissues from cancerous and healthy human pancreas (patients undergoing pancreatic resection)	Pancreatic adenocarcinoma	Pancreatic adenocarcinoma (Eck et al. 2008)
	25 nm spherical	Anti-EGFR and non-specific IgG antibodies	SiHa cervical cancer cells	Cervical cancer	Optical imaging of cervical pre-cancers using AuNPs and CdSe QDs for reflectance and fluorescence imaging (Rahman et al. 2005)
	10 nm spherical	Antibody (Cetuximab)	Epidermal growth factor receptor (EGFR) in A431 cells	Liver cancer	High-performance probes based on AuNPs for detection of live cancer cell (Yang et al. 2008)
	16 nm spherical	NHS-PEG; mouse anti-human AFP antibodies (antibody-1 and antibody-2)	Liver cancer biomarker alpha-fetoprotein (AFP)	Liver cancer	One-step homogeneous detection of cancer marker using antibody–AuNP probes (Lan et al. 2011)
	60 nm spherical	PSMA (prostate specific membrane antigen); C225 (Erbitux, the antibody raised against human EGF receptor)	Living bone metastatic prostate cancer (C4-2B); human bone marrow stromal (HS-5) cells	Prostate cancer	Tunable plasmonic nanoprobes for theranostics of prostate cancer (Lukianova-Hleb et al. 2011)
	35 nm spherical	Monoclonal anti-epidermal growth factor receptor (anti-EGFR)	Nonmalignant epithelial cells (HaCaT); malignant oral epithelial cells (HOC 313 clone 8 and HSC 3)	Oral cancer	Surface plasmon resonance scattering and absorption of antibody–AuNPs in oral cancer diagnostics (El-Sayed et al. 2005a)

(continued)

Table 11.1 (continued)

Method	Type of NP	Surface modification	Target/cells/samples	Type of cancer	Comments (reference)
Surface-enhanced Raman scattering (SERS)	30 nm spherical	Mouse anti-human free PSA clone PSA-F65 and clone PSA-66	Prostate-specific antigen (PSA) in human serum	Prostate cancer	Detection of PSA with an immunoassay based on SERS and immunogold labels (Grubisha et al. 2003)
	43 nm spherical	Human serum	Human serum	Colorectal cancer	AuNP based-SERS to obtain blood serum biochemical information for non-invasive colorectal cancer detection (Lin et al. 2011)
	60 nm spherical	QSY reporter molecules; thiol-PEG-COOH; epidermal growth factor (EGF) peptide	Epidermal growth factor (EGF)	Squamous cell carcinoma of the head and neck	SERS AuNPs identified circulating tumor cells in the peripheral blood of cancer patients (Wang et al. 2011)
	35–50 nm spherical	Thiol-DNA hairpin tagged with a Raman label	BRCA1 SNPs	Breast cancer	Plasmonic nanoprobes for detection of SNPs in breast cancer BRCA1 gene (Wabuyele et al. 2010)
Electro-chemical	5 nm spherical	Glutathione (GSH); primary antibodies for human interleukin-8	Interleukin-8 (IL-8) cancer biomarker in human serum	Cancer biomarker	Ultrasensitive immunosensor based on a glutathione-protected AuNP sensor surface (Munge et al. 2011)
	4 nm spherical	Antibodies for IL-6; biotinylated secondary antibody with 16–18 horseradish peroxidase labels	Interleukin-6 (IL-6) cancer biomarker in calf serum	Cancer biomarker	Inkjet printed AuNP electrochemical arrays for immunodetection of a cancer biomarker protein (Jensen et al. 2011)
	13 nm spherical	Alkaline phosphatase; poly(styrene-co-acrylic acid); TNF-α antibody	Tumor necrosis factor α (TNF-α) in human serum	Tumor necrosis factor	Electrochemical immunosensor of tumor necrosis factor based on alkaline phosphatase functionalized NPs (Yin et al. 2011)

At the moment, it is expected that the greatest gains in therapeutic selectivity will be achieved by synergistic combinations of several multicomponent targeting strategies. Currently, it is essential to develop technology for target and delivery of multiple therapeutic agents and for the simultaneous capability of avoiding biological and biophysical barriers. For example, nanoparticles can extravasate into the tumor stroma through the fenestrations of the angiogenic vasculature, demonstrating targeting by enhanced permeation and retention. These particles are able to carry multiple antibodies, which further target them to epitopes on cancer cells, and direct antitumor action, leading to cell death. Irradiation might be used to activate the nanoparticles and set up the release of their cytotoxic action (Ferrari 2005).

Due to advances in nanobiotechnology, the potential therapeutic application of gold nanocarriers represents an attractive platform for cancer therapy and has been investigated by different coworkers and used in a broad range of applications (Cuenca et al. 2006). Table 11.2 summarizes the latest progresses and general considerations for AuNP delivery and targeting in cancer therapy, according to their methodology principle, type of incubation/exposure, and target organs. In this panorama we can see that over the last 10 years, the majority (approximately 80 %) of gold nanoformulations for gene therapy, tumor targeting, and drug delivery in cancer have been tested in cell cultures and normally targeting reporter genes, such as luciferase or GFP. In the future, it is imperative to develop new therapy vehicles and extensive testing in animal models in order to develop the next-generation nanoparticle translation into the clinics. There is only one active clinical trial reporting the use of AuNPs. This phase I trial is studying the side effects and best dose of TNF-bound colloidal gold in treating patients with advanced solid tumors (clinical trial number NCT00356980), sponsored by the National Institutes of Health Clinical Center (CC) and National Cancer Institute (NCI). As it can be seen in Fig. 11.3 there is a disproportional level between nanoparticle production and their translation into clinics.

In fact, clinical trials require the coordinated effort of interdisciplinary research groups, institutes, and pharmaceuticals. Clinical-stage programs will probably have to deal with more and more human data and financial investment, before they can be viewed as a mainstream proposition for resourceful nanotherapy vehicle developers. The current generation of nanoparticles varies widely in size, chemical composition, surface charge, tissue tropism, and sensitivity that makes difficult to translate them into the manufacturing process. The community has to learn how to deal with all the data produced so far. The rules for understanding how nanoparticles interact with different organs and organisms are starting to emerge, although most of the valuable evidences have to come from animal models.

11.3.1 Gene Therapy

We are in the dawn of a new age in gene therapy driven by nanotechnology vehicles. Although there are technical challenges associated with the therapeutic application of nanoparticles, the integration of therapy with diagnostic profiling has accelerated

Table 11.2 Summary of AuNPs used in cancer therapy according to the type of NP, surface modification, type of cancer, target cells/organs/organisms, and explored methodology principle

Method	Type of AuNP	Surface modification	Target cells/organs/organisms	Comments (reference)
Gene therapy	13 nm spherical	Thiol-ssDNA	RAW 264.7 (macrophage); HeLa cells (cervical carcinoma); NIH-3 T3 (fibroblast); MDCK cells (kidney)	In vitro intracellular gene regulation; control of protein expression in cells (Rosi et al. 2006)
	13 nm spherical	PEG-*block*-poly(2-(*N*,*N*-dimethylamino)ethyl methacrylate) copolymer (PEG-PAMA); siRNA	HuH-7 cells (hepatocarcinoma)	Smart PEGylated AuNPs for the in vitro delivery of siRNA and to induce gene silencing (Oishi et al. 2006)
	13 nm spherical	Thiol-siRNA	HeLa cells (cervical carcinoma)	Polyvalent RNA–nanoparticle conjugates for in vitro luciferase knockdown (Giljohann et al. 2009)
	14 nm spherical	Thiol-siRNA; naked siRNA; PEG-COOH; PEG-N_3; HIV-derived TAT peptide; RGD peptide; quaternary ammonium	HeLa cells (cervical carcinoma); freshwater polyp (*Hydra*); mice (C57BL/6J)	In vitro and in vivo RNAi triggering using hierarchical approach with three biological systems of increasing complexity (Conde et al. 2012a)
	15 nm spherical	Cationic polymers: PEI; charge-reversal PAH-Cit; MUA (mercaptoundecanoic acid)	HeLa cells (cervical carcinoma)	Charge-reversal functional AuNPs to deliver siRNA and plasmid DNA into cancer cells (in vitro) (Guo et al. 2010)
	14 nm spherical	Cy3 labeled hairpin-DNA	HCT-116 cells (carcinoma of colon)	Au-nanobeacons capable of intersecting both pathways of in vitro RNA interference, blocking exogenous siRNA and endogenous microRNA (Conde et al. 2013a)
	13 nm spherical	PEG-NH_2; siRNA; poly(β-amino esters) (PBAEs)	HeLa cells (cervical carcinoma)	Gold, poly(β-amino ester) nanoparticles that facilitate high levels of in vitro siRNA delivery (Lee et al. 2009)
	40 nm spherical	Protease-degradable poly-L-lysine (PLL); siRNA	MDA-MB231-luc cells (breast cancer); LNCaP-luc cells (prostate adenocarcinoma)	Multilayer siRNA coated AuNPs using siRNA and PLL as the charged polyelectrolytes for in vitro luciferase knockdown (Lee et al. 2011)
	Au-nanospheres	Folate receptor; siRNA	Nude mice bearing HeLa cervical cancer xenografts	Near-IR light-inducible NF-kappaB in vivo downregulation through folate receptor-targeted hollow Au-nanospheres carrying siRNA recognizing NF-kappaB p65 subunit (Lu et al. 2010b)
	40 nm Au-nanoshells	PEG-NH_2; TAT-lipid; Cy3-siRNA	Mouse endothelial cells	NIR laser-induced release of siRNA from the nanoshells and in vitro GFP silencing (Braun et al. 2009)
	Spherical nucleic acid NP	siRNA	HeLa cells (cervical carcinoma); SKH-1E and C57BL/6 J mice	In vitro and in vivo topical delivery of siRNA-based spherical nucleic acid AuNP conjugates for gene regulation (Zheng et al. 2012)

Tumor targeting	Spherical	pH low insertion peptide (pHLIP)	Mouse model	Nanogold–pHLIP conjugates used in vivo to target tumors (Yao et al. 2013)
	6 nm spherical	Amine-terminated generation 5 (G5) poly(amidoamine) (PAMAM) dendrimers pre-functionalized with folic acid (FA) and fluorescein isothiocyanate (FI)	KB-HFAR cells (human epithelial carcinoma)	Multifunctional dendrimer-stabilized AuNPs can specifically target cancer cells expressing high-affinity FA receptors in vitro (Shi et al. 2009)
	30 nm spherical	PEGylated trastuzumab (Herceptin)	MDA-MB-361 (breast cancer) tumors	Human epidermal growth factor receptor-2 (HER-2)-targeted AuNPs that enhance the radiation response of in vitro breast cancer cells and in vivo tumor xenografts to X-radiation (Chattopadhyay et al. 2013)
	4 nm spherical	VEGF antibody (AbVF)	CLL B cells (B-chronic lymphocytic leukemia)	AuNPs enhance in vitro apoptosis in B-chronic lymphocytic leukemia (Mukherjee et al. 2007)
	60 nm spherical	PEG-COOH; malachite green isothiocyanate (MGITC); ScFv B10 (antibody fragment for human EGFR)	Tu686 and H520 cells (EGFR-positive cancer cells); nude mouse xenografted with Tu686 cells	PEGylated SERS AuNPs for in vitro and in vivo tumor targeting and detection (Qian et al. 2008a)
	Ultra-small Au-nanoclusters	Folic acid (FA); near-infrared fluorescent dye (MPA); doxorubicin (DOX)	A549 cells (lung cancer); HepG-2 (liver cancer); MDA-MB-231 (breast cancer); HTC116 (carcinoma of colon)	*Cellular* and in vivo studies with Au-FA-MPA and Au-FA-DOX show high affinity and anti-tumor activity to different tumors (Chen et al. 2012)
	30 nm spherical	PEGylated trastuzumab (FDA-approved humanized monoclonal antibody); radiolabeled polymer	MDA-MB-361 cells (breast cancer); athymic CD-1 mice bearing MDA-MB-361 tumors	AuNPs that enhance tumor uptake and intracellular delivery (in vitro and in vivo) while reducing the systemic exposure by evaluation of the impact of targeting and route of administration on organ distribution (Chattopadhyay et al. 2012)

(continued)

Table 11.2 (continued)

Method	Type of AuNP	Surface modification	Target cells/organs/organisms	Comments (reference)
Drug delivery	2.5 nm spherical	Hydrophobic drugs: tamoxifen (TAF) and β-lapachone (LAP); Bodipy (fluorescent probe)	MCF-7 cells (breast cancer)	Entrapment of hydrophobic drugs in AuNP monolayers with efficient in vitro release into cancer cells (Kim et al. 2009a)
	30 nm spherical	Prostate-specific membrane antigen (PSMA) RNA aptamer; doxorubicin (DOX)	LNCaP cells (prostate adenocarcinoma)	A drug-loaded aptamer—AuNP bioconjugate for in vitro imaging and therapy of prostate cancer (Kim et al. 2010)
	25 nm spherical	Thiol-PEGylated; tamoxifen (TAM)	MDA-MB-361 cells (breast cancer); MCF-7 cells (breast cancer)	Tamoxifen–poly(ethylene glycol)–thiol AuNPs that enhance potency and selective delivery for in vitro breast cancer treatment (Dreaden et al. 2009)
	180 nm spherical	PEG; oxaliplatin	A549 cells (lung epithelial cancer); HCT116, HCT15, HT29, RKO cells (all for colon cancer)	AuNPs for the improved anticancer drug delivery in vitro of the active component of oxaliplatin (Brown et al. 2010)
	13 nm spherical	Cisplatin	A549 cells (lung cancer); tumor-bearing SCID mice	In vitro and in vivo antitumoral drugs conjugated to AuNPs (Comenge et al. 2012)
	Spherical	Cyclic peptide	CCRF-CEM cells (human leukemic lymphoblasts); SK-OV-3 cells (human ovarian adenocarcinoma)	Cyclic peptide-capped AuNPs for in vitro drug delivery (Nasrolahi et al. 2012)
	40 nm Au-nanospheres	NH_2-PEG-COOH; doxorubicin (DOX); cyclic peptide c(TNYL-RAW), a second-generation EphB4-binding antagonist	Mice bearing Hey tumors	In vivo photothermal chemotherapy using doxorubicin-loaded Au-nanospheres that target EphB4 receptors in tumors (You et al. 2012)
	Plasmonic vesicles assembled from 14 nm spherical AuNPs	Raman reporter; PEG; hydrophobic copolymer (PMMAVP) of methyl methacrylate (MMA) and 4-vinylpyridine (4VP)	SKBR-3 cells (breast adenocarcinoma)	Self-assembled plasmonic vesicles of SERS-encoded amphiphilic AuNPs for in vitro cancer cell targeting and traceable intracellular drug delivery (Song et al. 2012)

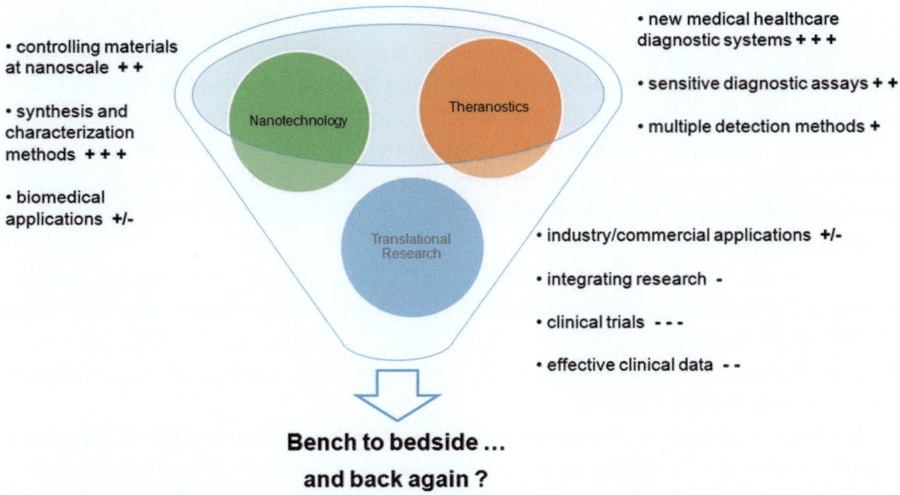

Fig. 11.3 Disproportional level between nanoparticle production and their translation into clinics. Thousands of systems were published describing different synthesis, biofunctionalization, and characterization methods that will most likely revolutionize our understanding of chemical and biological mechanisms and push forward efficient diagnostics and therapeutic platforms. Nevertheless, very few were produced to improve a bench-to-bedside approach to translational research. Outcomes like this must be followed by extensive laboratory work, which results in improved screening procedures and a new therapy of great potential, although the final product should always be part of a two-way interaction between laboratory scientists and clinicians

the pace of discovery of new nanotechnology methods. The development of a safe, efficient, specific, and nonpathogenic vehicle for gene delivery is highly attractive (Akhtar and Benter 2007; Kim et al. 2009b).

Gene therapy is receiving increasing attention and, in particular, small-interference RNA (siRNA) shows importance in novel molecular approaches in the knockdown of specific gene expression in cancerous cells. In fact, this non-viral-vector-mediated delivery of therapeutic siRNAs is highly desirable and constitutes an important challenge to gene therapy (Castanotto and Rossi 2009; Li et al. 2002; Soutschek et al. 2004).

In fact, antisense DNA (Fichou and Ferec 2006; Toub et al. 2006) and RNA interference (RNAi) via the use of small-interfering RNA (Fire et al. 1998; Baker 2010; Milhavet et al. 2003; Wall and Shi 2003) have emerged as powerful and useful tools to block gene function and for sequence-specific posttranscriptional gene silencing, playing an important role in downregulation of specific gene expression in cancer cells. Thus, one drawback of using naked siRNAs is that they show extremely short half-lives, weak protection against action by RNAses, poor chemical stability, and common dissociation from vector (Hannon and Rossi 2004). In fact, the major obstacle to clinical application is the uncertainty about how to deliver therapeutic RNAs (e.g., miRNA and/or siRNA) with maximal therapeutic impact. AuNPs have shown potential as intracellular delivery vehicles for antisense oligonucleotides (e.g., miRNA and/or siRNA) with maximal therapeutic impact. AuNPs have shown potential as intracellular delivery vehicles for antisense oligonucleotides (Rosi et al. 2006) and for therapeutic siRNA by providing protection against RNAses and ease of functionalization for selective targeting (Giljohann et al. 2009; Whitehead et al. 2009). For example, Mirkin and coworkers reported the use of polyvalent RNA–AuNP conjugates that are readily taken up by cells and that the particle bound siRNA can effectively regulate genes in the context of RNA interference (Giljohann et al. 2009).

Several other studies using engineered NPs modified with siRNA have demonstrated a cytoplasmic delivery system of siRNA and efficient gene silencing using AuNPs (Giljohann et al. 2009; Lee et al. 2008, 2009; Guo et al. 2010). However, almost all nanoconjugates using siRNA have exclusively been tested in cell cultures targeting only reporter genes.

Recently, Conde et al. provided evidence of in vitro and in vivo RNAi triggering via the synthesis of a library of novel multifunctional AuNPs, using a hierarchical approach including three biological systems of increasing complexity: in vitro cultured human cells, in vivo freshwater polyp (*Hydra vulgaris*), and in vivo mice models (Conde et al. 2012a). The authors developed effective conjugation strategies to combine, in a highly controlled way, specific biomolecules to the surface of AuNPs such as (a) biofunctional spacers: poly(ethylene glycol) (PEG) spacers used to increase solubility and biocompatibility; (b) cell penetrating peptides (CPPs) such as TAT and RGD peptides, a novel class of membrane translocating agents named CPPs that exploit more than one mechanism of endocytosis to overcome the lipophilic barrier of the cellular membranes and deliver large molecules and even small particles inside the cell for their biological actions; and (c) siRNA complementary to a master regulator gene, the protooncogene c-myc, were bond covalently (SH-siRNA) and ionically (naked/unmodified siRNA) to AuNPs (Conde et al. 2012a).

However, efforts to target siRNA-nanoparticles to organs are less advanced. Some organs need smaller and novel NPs to access different kinds of tissue. Another challenge, which also requires novel materials, is the endosomal release of siRNA, once it is transported across the cell membrane. Most of the described systems also get trapped by the lysosomes and their siRNA cargo gets compromised. Probably the endosomal escape or siRNA accessibility to form the RNA-induced silencing complex (RISC) requires substantially more circulation time.

While the AuNPs for gene therapy discussed thus far employ the load of siRNA to nanoparticles, AuNPs have also shown potential as intracellular delivery vehicles for antisense oligonucleotides (ssDNA, dsDNA) by providing protection against intracellular nucleases and ease of functionalization for selective targeting (Whitehead et al. 2009; Giljohann et al. 2009). So far, several strategies for gene delivery systems have been developed, including mixed monolayer protected AuNPs (Rana et al. 2012; Ghosh et al. 2008a), polymer–AuNPs complexes (Thomas and Klibanov 2003; Ghosh et al. 2008b), and dsDNA and ssDNA functionalized AuNPs (Conde et al. 2010a; McIntosh et al. 2001).

Recently, Conde et al. developed a new theranostic system capable of intersecting all RNA pathways: from gene specific downregulation to silencing the silencers, i.e., siRNA and miRNA pathways. The authors reported the development AuNPs functionalized with a fluorophore labeled hairpin-DNA, i.e., gold nanobeacons, capable of efficiently silencing single gene expression, exogenous siRNA, and endogenous miRNAs while yielding a quantifiable fluorescence signal directly proportional to the level of silencing (Conde et al. 2013a).

From a synthetic point of view the differences between DNA–AuNPs and siRNA–AuNPs are still unclear. From the papers reported so far the siRNA and DNA–gold nanoconjugates are almost from the same size and charge and show similar efficiencies. Nevertheless siRNA–AuNP conjugates require functionalization with thiol-PEG molecules to achieve equivalent stability to DNA–AuNPs, which may have something to do with the capacity loading of the different molecules due to singular hydrophobicity and/or hydrophilicity, molecular weight, and charge density properties between RNA and DNA (Gary et al. 2007). However, this fact needs additional clarification.

Further research into the fundamental mechanisms of in vivo gene therapy using nanodevices could unveil new dimensions of nanoparticle-mediated gene silencing that will have profound implications for understanding gene regulation, and which could also affect the development of functional genomics and therapeutic applications. One of the most important issues that are still unclear is how biocompatible AuNPs will be following intravenous injection, in particular when the ultimate destination is the cytoplasm and/or nucleus inside cells. Future in vivo work will need to cautiously consider the accurate option of chemical modifications to incorporate into the nanoparticles to avoid off-target effects. Though the nanoparticles' potential against cancer is still in need of further optimization and characterization, it is now time to start translating these promising platforms to the clinical settings towards widespread effective therapy strategies in the fight against cancer.

11.3.2 Tumor Targeting

Nanoparticles are excellent tumor-targeting vehicles because of a unique inherent property of solid tumors. Numerous tumors present with defective vasculature and poor lymphatic drainage, due to the rapid growth of solid tumors, resulting in an enhanced permeability and retention effect. This effect allows nanoparticles to accumulate specifically at the tumor site (Minelli et al. 2010; Gil and Parak 2008). Once the tumor is directly connected to the main blood circulation system, multifunctional nanocarriers can exploit several characteristics of the newly formed vasculature and efficiently target tumors. Tumor cells are supplied by blood capillaries that perfuse the cells of the tissue where nanocarriers can passively accumulate or anchor through targeting moieties to biomarkers overexpressed by tumor cells (Conde et al. 2012b).

Shi et al. developed a simple system with multifunctional amine terminated poly(amidoamine) (PAMAM) dendrimers, folic acid (FA), and fluorescein isothiocyanate functionalized in gold nanoparticles. This approach can specifically target cancer cells expressing high-affinity FA receptors in vitro (Shi et al. 2009).

Further work in tumor targeting was reported in a subcutaneous model of colon cancer, where it was demonstrated that systemically delivered AuNPs (size, approximately 33 nm) conjugated to tumor necrosis factor (TNF) accumulated in tumors (Paciotti et al. 2004). This study outlines the development of a colloidal gold nanoparticle vector that targets the delivery of TNF to a solid tumor growing in mice.

Mukherjee et al. studied B-chronic lymphocytic leukemia (CLL) which is characterized by apoptosis resistance. They found induction of significantly more apoptosis in CLL B cells by co-culture with an anti-VEGF antibody. To increase the efficacy of these agents in CLL therapy they focused on the use of AuNPs, by attaching VEGF antibody to the nanoparticle's surface. The AuNP–VEGF antibody treated cells showed significant downregulation of anti-apoptotic proteins (Mukherjee et al. 2007).

In cancer research, colloidal gold can be used to target tumors and provide detection using SERS (surface-enhanced Raman spectroscopy) in vivo. These AuNPs are surrounded with Raman reporters which provide light emission that is over 200 times brighter than quantum dots (Cai et al. 2008; Kneipp et al. 2006). It was found that the Raman reporters were stabilized when the nanoparticles were encapsulated with a thiol-modified polyethylene glycol coat and gave large optical enhancements. This allows for compatibility and circulation in vivo. When conjugated to tumor-targeting ligands, these conjugated SERS nanoparticles were able to target tumor markers such as EGFR, which is sometimes overexpressed in cells of certain cancer types (El-Sayed et al. 2005a, 2006), and then detect the location of the tumor on human cancer cells and in xenograft tumor models (Qian et al. 2008a). Qian et al. described biocompatible and nontoxic nanoparticles for in vivo tumor targeting and detection based on PEGylated gold nanoparticles and SERS. These conjugated nanoparticles were able to target tumor biomarkers such as EGFRs on human cancer cells and in xenograft tumor models (Qian et al. 2008a).

Although cancer therapies are improving, some formulations are not reaching with high efficiency tumor cells or tissues, and countless doubts remain over the efficacy of those that do. To efficiently target a cancer cell, either a circulation cell or a cell from the primary tumor or one hidden within a population of normal cells represents an exceptional challenge. In fact, there are a lot of limitations for tumor targeting as some nanocarriers can also target normal proteins which are not exclusively expressed by the cancer cell. Targeting specific cells may be completely different to target the organ. The most important aspects that the researchers need to take into account are the specificity of the nanoparticle to the target molecules, as well as toxicological and immunological effects (Schroeder et al. 2012). For example, the combination of small size nanoparticles with their special thermal, imaging, drug/gene carrier, or optical characteristics with the specific and selective recognition abilities of antibodies will definitely produce a hybrid product that shows versatility and specificity.

Actually, Conde et al. (2013b) reported the evaluation of the inflammatory response and therapeutic siRNA silencing via RGD-nanoparticles in a lung cancer mouse model. This study reported the use of siRNA/RGD gold nanoparticles capable of targeting tumor cells in two lung cancer xenograft mouse models, resulting in successful and significant *c-Myc* oncogene downregulation followed by tumor growth inhibition and prolonged survival of the animals. This delivery system can achieve translocation of siRNA duplexes directly into the tumor cell cytoplasm and accomplish successful silencing of an oncogene expression. Actually, RGD/siRNA–AuNPs can target preferentially and be taken up by tumor cells via αvβ3 integrin-receptor-mediated endocytosis with no cytotoxicity, showing that they can accumulate in tumor tissues overexpressing αvβ3 integrins and selectively deliver *c-Myc* siRNA to suppress tumor growth and angiogenesis (Conde et al. 2013b).

Therefore, multifunctional nanocarriers have the potential to join numerous therapeutic functions into a single platform, by targeting specific tumor cells, tissues, and organs.

11.3.3 Drug Delivery

The vast majority of FDA approved used drugs exhibit a short half-life in the blood stream and a high overall clearance rate. In fact, the major obstacles/limitations in drug delivery are cytoplasmic and systemic delivery of the drug, renal clearance, target site accumulation after administration, and heterogeneous vascular perfusion and diffusion. Actually, these small drug molecules usually diffuse rapidly into healthy tissues and are dispersed consistently within the body. As a consequence, just a small amount of the drug can reach the target site, which often leads to side effects. These obstacles usually occur with drugs that exhibit a narrow therapeutic index, such as anticancer biomolecules, immunosuppressive agents, as well as anti-rheumatic medicines. Poor drug delivery and accumulation at the target site frequently lead to significant limitations, such as multi-drug resistance, which leads many cancers to develop severe resistance to chemotherapy drugs (Ehdaie 2007).

Nanocarriers can be used to optimize the biodistribution of drugs to diseased organs, tissues, or cells, in order to improve and target drug delivery (Han et al. 2007b).

It is important to realize that the nanoparticle-mediated drug delivery is feasible only if the drug distribution is otherwise inadequate. These cases include drug targeting of difficult, unstable molecules (proteins, siRNA, DNA), delivery to the difficult sites (brain, retina, tumors, intracellular organelles), and drugs with serious side effects (e.g., anti-cancer agents). The performance of the nanoparticles depends on the size and surface functionalities in the particles. Also, the drug release and particle disintegration can vary depending on the system (e.g., biodegradable polymers sensitive to pH). An optimal nanodrug delivery system ensures that the active drug is available at the site of action for the correct time and duration, and their concentration should be above the minimal effective concentration (MEC) and below the minimal toxic concentration (MTC) (Han et al. 2007a; Langer 2000).

AuNPs are also being investigated as vehicles for drug delivery such as paclitaxel (Gibson et al. 2007). *Gibson* et al. describe the first example of 2 nm AuNPs covalently functionalized with the chemotherapeutic drug paclitaxel. The administrations of hydrophobic drugs require molecular encapsulation and it is found that nanosized particles are particularly efficient in evading the reticuloendothelial system. This approach gives a rare opportunity to prepare hybrid particles with a well-defined amount of drug and offers a new alternative for the design of nanosized drug-delivery systems (Kim et al. 2009a; Hwu et al. 2009; Gibson et al. 2007) (see Fig. 11.4).

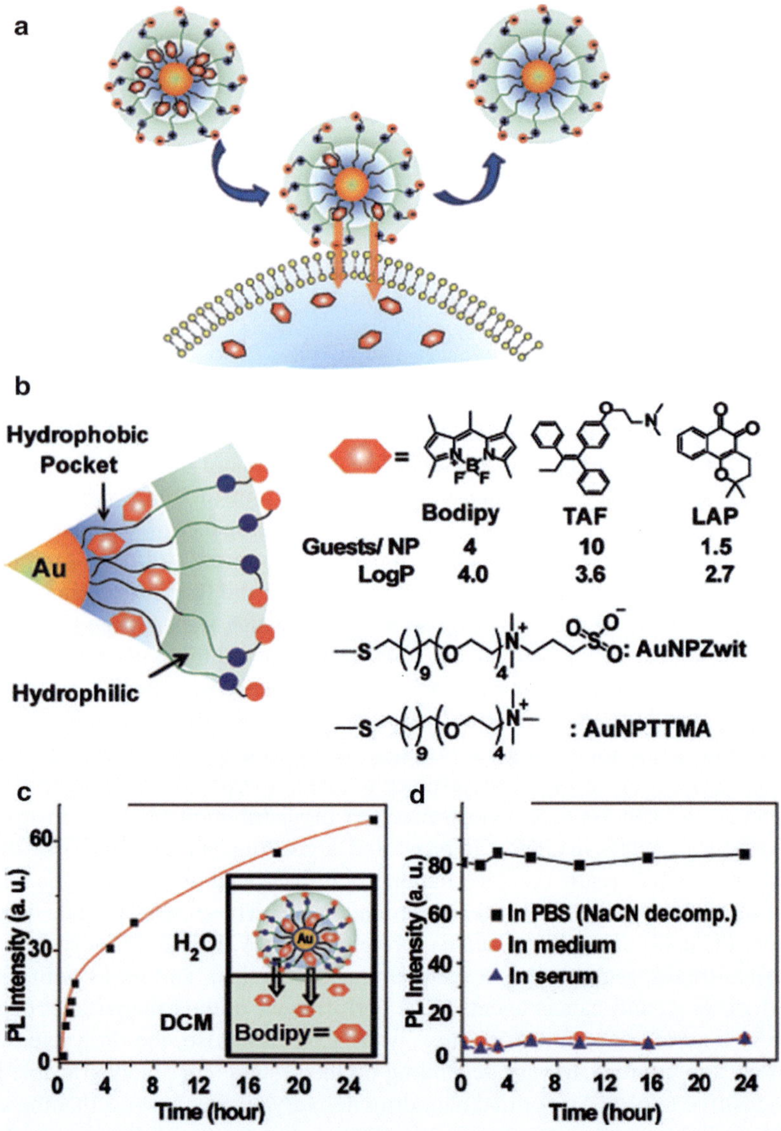

Fig. 11.4 Drug delivery. Entrapment of hydrophobic drugs in nanoparticle monolayers with efficient release into cancer cells. (**a**) Delivery of payload to cell through monolayer–membrane interactions. (**b**) Structure of particles and guest compounds: Bodipy, TAF, and LAP, the number of encapsulated guests per particle, and log *P* of the guests. (**c**) Release of Bodipy from AuNPZwit Bodipy in DCM-aqueous solution two-phase systems (λex 499 nm, λem 517 nm). (**d**) PL intensity of AuNPZwit-Bodipy in cell culture medium and 100 % serum, indicating little or no release relative to AuNPZwit-Bodipy in PBS after NaCN induced release of guest molecules (λex 499 nm, λem 510 nm). Reproduced with permission from Kim et al. (2009a). Copyright 2013, American Chemical Society

Nanotechnology has provided over the last years novel and powerful systems that may be used for the treatment of human diseases. However the majority of products, reagents, and drugs being used for the development of these nanoscale systems have to be approved by the main supervising agencies, such as the FDA and EMA (Baptista 2009).

Thus far, some limitations for the correct design and application of nanoparticles, such as pharmacokinetics, biodistribution, and side effects of the nanotherapy and safety profile of nanoparticles before and after conjugation and toxicity, need to be clarified to validate efficient clinical appliance (Baptista 2009).

11.4 NanoToxicity

The AuNPs have a proclivity in vivo and in vitro to bioaccumulate within various types of cells with a special affinity for macrophage-type cells (both histiocytes and blood phagocytic cells) and reticuloendothelial cells throughout the body. They also produce varying degrees of bioaccumulation in such tissues as lymph nodes, bone marrow, spleen, adrenals, liver, and kidneys (Lasagna-Reeves et al. 2010; Chen et al. 2009; Dobrovolskaia et al. 2008).

Research shows that nanoparticles can stimulate and/or suppress the immune responses and that their compatibility with the immune system is largely determined by their surface chemistry. In fact, the influence of size, solubility, and sur-

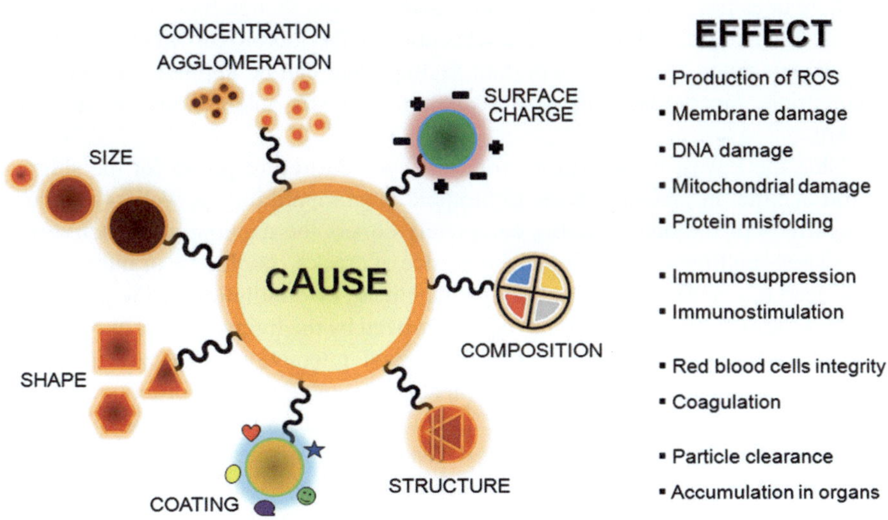

Fig. 11.5 Nanotoxicity: causes and effects. Nanoparticles biocompatibility/effects and their use in biological applications can be influenced by size, shape, solubility, composition, and surface charge and modification/chemistry

face modification on the biocompatibility of nanoparticles and their use in biological applications are well known (Dobrovolskaia and McNeil 2007) (see Fig. 11.5).

AuNPs are generally considered to be benign. However, the size similarity of AuNPs to biological matters could provide "camouflage" to cellular barriers, leading to undesired cellular entry which might be detrimental to normal cellular function (Connor et al. 2005).

Pan et al. (2007) recently conducted a systematic investigation of the size-dependent cytotoxicity of AuNPs against four cell lines. They found that AuNPs 1–2 nm in size displayed cell-type dependent cytotoxicity with high micromolar IC50s. In contrast, AuNPs 15 nm in size were nontoxic to cells at concentrations 60-fold higher than the IC50 of the smaller AuNPs. These results seemed to confirm size dependent toxicity of AuNPs (Kim et al. 2009a; Visaria et al. 2006; Paciotti et al. 2004; El-Sayed et al. 2005b; Huang et al. 2008), an inference that has hitherto been somewhat ambivalent. On the other hand, it was found that the presence of sodium citrate residues on AuNPs impaired the viability in the alveolar type-II cell lines A549 and NCIH441. Interestingly, the presence of an excess of sodium citrate on the surface of NPs not only reduced the in vitro viability of A549 and NCIH441 cell lines but also affected cellular proliferation and increased the release of lactate dehydrogenase (marker for apoptotic cell degradation) (Uboldi et al. 2009).

Although AuNPs are generally considered as highly biocompatible, previous in vitro studies have also shown that cytotoxicity of AuNPs in certain human epithelial cells was observed (Freese et al. 2012; Rothen-Rutishauser et al. 2007).

Now the most urgent questions rise up. Are the gold nanoparticles cytotoxic or biocompatible? And how can the gold nanoparticles be designed to avoid these effects?

There does not seem to have a simple answer. Even though there is not any general mechanism for making nanoparticles universally "non-toxic" to all living cells and all organisms, there are important findings that can be applied for increasing nanoparticle biocompatibility and reducing cytotoxic interactions in vivo and in vitro.

Using the lowest nanoparticle dose to get the desired response for the shortest period of time, in general, seems to promote biocompatibility as well as coating a nanoparticle of the outer coating completely covers the nanoparticle reactive surface (a non-continuous covering, the presence of cracks, roughness, or interruptions could lead to complement or antibody attachment, or dissolution of the coating by cell digestion) and cannot be removed and utilized by the living cell (Bellucci 2009).

It is essential to test nanoparticle/biological interactions experimentally and modify the nanoparticles for best biocompatibility with the cell in order to eliminate some obstacle, like the peroxidation of membrane lipids, the generation of reactive oxygen species, the acute and chronic release of pro-inflammatory factors, modification in genetic cellular function, and the possibility of nanoparticles becoming inactive/unavailable during filtration or passage through pores and fenestrations (Sun et al. 2005) due to size, inflexibility of the nanoparticle core, or protein adsorption and agglomeration (Bellucci 2009).

When interpreting nanoparticle interactions with cells and organisms, it is important to remember that living systems may appear normal and be capable of growth and function, but they may be genetically altered in subtle ways following nanoparticle exposure, which can produce serious consequences at some time in the distant future. Conversely, other cells that seem to be damaged may, in time, recover from nanoparticle exposure and function normally in the absence of the nanoparticles (Bellucci 2009).

In conclusion, the only weapon that we have to insure that these new materials are well designed and safely used is to question and test each new nanoparticle to make sure that it has been designed for safety (with maximum biocompatibility) during handling, use, and disposal. Evaluating the biocompatibility of nanomaterials is imperative. In fact, it is important to carefully characterize the biocompatibility and safety of the nanomaterials if they are to be used for medical purposes. Despite the major scientific advances made in the field of molecular and cell biology and biotechnology, the basic concepts of regulatory toxicology have hardly changed over the past decades (Dobrovolskaia and McNeil 2007). Actually, the vast majority of studies report the biocompatibility of nanomaterials only through the study of cell viability. Almost no importance is given when testing nanomaterials in the detection of genetic damages (DNA strand breaks and the formation of nuclear abnormalities), or in identifying protein markers of toxicity, or measuring the level of oxidative stress. For example, when using gene silencing technologies, the function of specific genes and proteins in toxicity pathways could be identified, once DNA-damage response (DNA repair, cell-cycle regulation, and apoptosis) encompasses gene-expression regulation at the transcriptional and post-translational levels.

In fact, Conde et al. reported a gold-nanobeacon system (Conde et al. 2013a) used for gene therapy that was extensively evaluated for the genotoxic, cytotoxic, and proteomic effects after incubation in cancer cells (Conde et al. 2014b). The exposure was evaluated by two-dimensional protein electrophoresis followed by mass spectrometry to perform a proteomic profile and MTT assay, glutathione-S-transferase assay, micronucleus test, and comet assay to assess the genotoxicity. An assessment of genome-related toxicity revealed no significant DNA damage increase, as well as no potential mutagenic or clastogenic consequences to the cell (Conde et al. 2014b).

11.5 Conclusions and Future Perspectives

Over the last decade, thousands of different gold nanocarriers were developed and published. Almost 20 % of these papers were published in 2010 alone. It is indisputable that the use of gold nanocarriers has been gaining momentum as vectors for therapy and diagnostic strategies, combining the AuNPs' ease of functionalization with numerous biomolecules, high loading capacity, and fast uptake by target cells.

Here, we have reviewed part of this exciting progress and research advances within the context of multifunctional gold nanocarriers for cancer theranostics.

Despite the significant efforts towards the use of gold nanocarriers in biologically relevant research, more in vivo studies are needed to assess the applicability of these materials as delivery agents. In fact, only a few went through feasible clinical trials. Nanoparticles have to serve as the norm rather than an exception in the future conventional cancer treatments. Future in vivo work will need to carefully consider the correct choice of chemical modifications to incorporate into the multifunctional gold nanocarriers to avoid activation off-target, side effects, and toxicity. Moreover the majority of studies on nanomaterials do not consider the final application to guide the design and functionalization of NP. Instead, the focus is predominantly on engineering materials with specific physical or chemical properties.

Although great effort has been applied to the detection of microorganisms and/or virus using gold biosensors, only a few were used for the detection of chronic diseases, such as cancer. The lack of specific and unique DNA/RNA markers and the complexity of the intricate regulation pathways have made the development of probes rather difficult. Particularly in cancer, future trends need to be focused on the detection of microRNAs and circulating DNA biomarkers or the recognition of circulating tumor cells followed by a detection step. Since most of the nanodiagnostic systems report DNA sequence analysis/detection, the great majority ignored the important genomic and transcriptomic information when detecting messenger RNA and microRNAs, which are considered to be excellent biomarkers for the early diagnosis of cancer.

To improve medicine, scientific discoveries must be translated into clinical applications. Such discoveries typically begin at "the bench" with basic/fundamental research in which scientists study disease at a molecular or cellular level and then progress to the clinical level, which is the patient's "bedside." Scientists and clinicians are increasingly aware that this bench-to-bedside approach to translational research is really a two-way interaction. However, an additional effort should appear towards the development of new clinical strategies.

Moreover, the systems discussed here are each of them unique in many aspects. Some of the nanoformulations have very well-defined structures, whereas some are highly heterogeneous. Some have a wide range of size, charge, and surface. This raises the important question about the reliability of their production protocols. The great majority of the nanosystems described here present or will present additional challenges in the scale-up of the manufacturing process. Besides, central concerns about nanoparticle's biodistribution and the proper considerations of safety for the patients. Therefore, the establishment of safe regulatory approval nanoformulations turns to be essential.

For that reason, it is imperative to learn how advances in nanosystem's capabilities are being used to identify new diagnostic and therapy tools driving the development of personalized medicine in oncology; discover how integrating cancer research and nanotechnology modeling can help patient diagnosis and treatment; recognize how to translate nanotheranostics data into an actionable clinical strategy; discuss with industry leaders how nanotheranostics is evolving and what the impact is on current research efforts; and last but not least, learn what approaches are proving fruitful in turning promising clinical data into treatment realities.

Although all studies described here provide a baseline level of data in support of the effectiveness and safety of nanomaterials, we wonder how useful the data generated will be in successfully predicting and preventing scientists from jeopardizing the safety of the future patients.

With chemists, biologists, and materials scientists working together with clinicians and engineers, but especially with "translational innovators," new solutions to crucial nanobiomedical problems will hopefully be found.

Competing Interests

The authors declare that they have no competing interests.

Authors' Contributions

JC conceptualized the manuscript and wrote the draft. FT, PVB, and JMF contributed in the draft and concept of the paper. All authors contributed in the revision process. All authors read and approved the final manuscript.

Acknowledgments The authors thank ERANET-NANOSCIERA NANOTRUCK project for financial support. JC acknowledges FCT grant (SFRH/BD/62957/2009). JMF thanks ARAID and Fondo Social Europeo for financial support. PVB thanks CIGMH/FCT/MCES (PEst-OE/SAU/UI0009/2011).

References

Akhtar S, Benter IF (2007) Nonviral delivery of synthetic siRNAs in vivo. J Clin Invest 117:3623–3632

Alvarez-Puebla RA, Liz-Marzan LM (2010) SERS-based diagnosis and biodetection. Small 6:604–610

Alvarez-Puebla RA, Liz-Marzan LM (2012a) SERS detection of small inorganic molecules and ions. Angew Chem Int Ed Engl 51:11214–11223

Alvarez-Puebla RA, Liz-Marzan LM (2012b) Traps and cages for universal SERS detection. Chem Soc Rev 41:43–51

Ambrosi A, Airo F, Merkoci A (2010) Enhanced gold nanoparticle based ELISA for a breast cancer biomarker. Anal Chem 82:1151–1156

Arruebo M, Valladares M, Gonzalez-Fernandez A (2009) Antibody-conjugated nanoparticles for biomedical applications. J Nanomater

Baker M (2010) RNA interference: from tools to therapies. Nature 464:1225

Baptista P (2009) Cancer nanotechnology – prospects for cancer diagnostics and therapy. Curr Cancer Ther Rev 5:80–88

Baptista PV (2012) Could gold nanoprobes be an important tool in cancer diagnostics? Expert Rev Mol Diagn 12:541–543

Baptista P, Doria G, Henriques D, Pereira E, Franco R (2005) Colorimetric detection of eukaryotic gene expression with DNA-derivatized gold nanoparticles. J Biotechnol 119:111–117

Baptista PV, Koziol-Montewka M, Paluch-Oles J, Doria G, Franco R (2006) Gold-nanoparticle-probe-based assay for rapid and direct detection of *Mycobacterium tuberculosis* DNA in clinical samples. Clin Chem 52:1433–1434

Baptista P, Pereira E, Eaton P, Doria G, Miranda A, Gomes I, Quaresma P, Franco R (2008) Gold nanoparticles for the development of clinical diagnosis methods. Anal Bioanal Chem 391:943–950

Bellucci S (2009) Nanoparticles and nanodevices in biological applications. Lecture Notes in Nanoscale Science and Technology edn. Springer 4. doi:978-3-540-70946-6

Braun GB, Pallaoro A, Wu G, Missirlis D, Zasadzinski JA, Tirrell M, Reich NO (2009) Laser-activated gene silencing via gold nanoshell-siRNA conjugates. ACS Nano 3:2007–2015

Brown SD, Nativo P, Smith JA, Stirling D, Edwards PR, Venugopal B, Flint DJ, Plumb JA, Graham D, Wheate NJ (2010) Gold nanoparticles for the improved anticancer drug delivery of the active component of oxaliplatin. J Am Chem Soc 132:4678–4684

Cai W, Gao T, Hong H, Sun J (2008) Applications of gold nanoparticles in cancer nanotechnology. Nanotechnol Sci Appl 1:17–32 [Ref Type: Journal (Full)]

Cao YC, Jin R, Thaxton CS, Mirkin CA (2005) A two-color-change, nanoparticle-based method for DNA detection. Talanta 67:449–455

Castaneda MT, Merkoci A, Pumera M, Alegret S (2007) Electrochemical genosensors for biomedical applications based on gold nanoparticles. Biosens Bioelectron 22:1961–1967

Castanotto D, Rossi JJ (2009) The promises and pitfalls of RNA-interference-based therapeutics. Nature 457:426–433

Chattopadhyay N, Fonge H, Cai ZL, Scollard D, Lechtman E, Done SJ, Pignol JP, Reilly RM (2012) Role of antibody-mediated tumor targeting and route of administration in nanoparticle tumor accumulation in vivo. Mol Pharm 9:2168–2179

Chattopadhyay N, Cai ZL, Kwon YL, Lechtman E, Pignol JP, Reilly RM (2013) Molecularly targeted gold nanoparticles enhance the radiation response of breast cancer cells and tumor xenografts to X-radiation. Breast Cancer Res Treat 137:81–91

Chen YS, Hung YC, Liau I, Huang GS (2009) Assessment of the in vivo toxicity of gold nanoparticles. Nanoscale Res Lett 4:858–864

Chen HY, Li SL, Li BW, Ren XY, Li SN, Mahounga DM, Cui SS, Gu YQ, Achilefu S (2012) Folate-modified gold nanoclusters as near-infrared fluorescent probes for tumor imaging and therapy. Nanoscale 4:6050–6064

Cheng MM, Cuda G, Bunimovich YL, Gaspari M, Heath JR, Hill HD, Mirkin CA, Nijdam AJ, Terracciano R, Thundat T, Ferrari M (2006) Nanotechnologies for biomolecular detection and medical diagnostics. Curr Opin Chem Biol 10:11–19

Comenge J, Sotelo C, Romero F, Gallego O, Barnadas A, Parada TGC, Dominguez F, Puntes VF (2012) Detoxifying antitumoral drugs via nanoconjugation: the case of gold nanoparticles and cisplatin. PLoS One 7:e47562

Conde J, de la Fuente JM, Baptista PV (2010a) In vitro transcription and translation inhibition via DNA functionalized gold nanoparticles. Nanotechnology 21:505101

Conde J, de la Fuente JM, Baptista PV (2010b) RNA quantification using gold nanoprobes – application to cancer diagnostics. J Nanobiotechnology 8:5

Conde J, Ambrosone A, Sanz V, Hernandez Y, Marchesano V, Tian F, Child H, Berry CC, Ibarra MR, Baptista PV, Tortiglione C, de la Fuente JM (2012a) Design of multifunctional gold nanoparticles for in vitro and in vivo gene silencing. ACS Nano 6:8316–8324

Conde J, Doria G, Baptista P (2012b) Noble metal nanoparticles applications in cancer. J Drug Deliv 2012:751075

Conde J, Doria G, de la Fuente JM, Baptista PV (2012c) RNA quantification using noble metal nanoprobes: simultaneous identification of several different mRNA targets using color multiplexing and application to cancer diagnostics. Methods Mol Biol 906:71–87

Conde J, Rosa J, de la Fuente JM, Baptista PV (2013a) Gold-nanobeacons for simultaneous gene specific silencing and intracellular tracking of the silencing events. Biomaterials 34:2516–2523

Conde J, Tian F, Hernandez Y, Bao C, Cui D, Janssen KP, Ibarra MR, Baptista PV, Stoeger T, de la Fuente JM (2013b) In vivo tumor targeting via nanoparticle-mediated therapeutic siRNA coupled to inflammatory response in lung cancer mouse models. Biomaterials 34:7744–7753

Conde J, Bao C, Cui D, Baptista PV, Tian F (2014a) Antibody–drug gold nanoantennas with Raman spectroscopic fingerprints for in vivo tumour theranostics. J Control Release 183:87–93

Conde J, Larguinho M, Cordeiro A, Raposo LR, Costa PM, Santos S, Diniz MS, Fernandes AR, Baptista PV (2014b) Gold-nanobeacons for gene therapy: evaluation of genotoxicity, cell toxicity and proteome profiling analysis. Nanotoxicology 8:521–532

Connor EE, Mwamuka J, Gole A, Murphy CJ, Wyatt MD (2005) Gold nanoparticles are taken up by human cells but do not cause acute cytotoxicity. Small 1:325–327

Costa P, Amaro A, Botelho A, Inacio J, Baptista PV (2009) Gold nanoprobes assay for identification of mycobacteria from the *Mycobacterium tuberculosis* complex. Clin Microbiol Infect 16:1464–1469

Cuenca AG, Jiang H, Hochwald SN, Delano M, Cance WG, Grobmyer SR (2006) Emerging implications of nanotechnology on cancer diagnostics and therapeutics. Cancer 107:459–466

Daniel MC, Astruc D (2004) Gold nanoparticles: assembly, supramolecular chemistry, quantum-size-related properties, and applications toward biology, catalysis, and nanotechnology. Chem Rev 104:293–346

Dobrovolskaia MA, McNeil SE (2007) Immunological properties of engineered nanomaterials. Nat Nanotechnol 2:469–478

Dobrovolskaia MA, Aggarwal P, Hall JB, McNeil SE (2008) Preclinical studies to understand nanoparticle interaction with the immune system and its potential effects on nanoparticle biodistribution. Mol Pharm 5:487–495

Doria G, Franco R, Baptista P (2007) Nanodiagnostics: fast colorimetric method for single nucleotide polymorphism/mutation detection. IET Nanobiotechnol 1:53–57

Doria G, Conde J, Veigas B, Giestas L, Almeida C, Assuncao M, Rosa J, Baptista PV (2012) Noble metal nanoparticles for biosensing applications. Sensors (Basel) 12:1657–1687

Dreaden EC, Mwakwari SC, Sodji QH, Oyelere AK, El-Sayed MA (2009) Tamoxifen-poly(ethylene glycol)-thiol gold nanoparticle conjugates: enhanced potency and selective delivery for breast cancer treatment. Bioconjug Chem 20:2247–2253

Eck W, Craig G, Sigdel A, Ritter G, Old LJ, Tang L, Brennan MF, Allen PJ, Mason MD (2008) PEGylated gold nanoparticles conjugated to monoclonal F19 antibodies as targeted labeling agents for human pancreatic carcinoma tissue. ACS Nano 2:2263–2272

Ehdaie B (2007) Application of nanotechnology in cancer research: review of progress in the National Cancer Institute's Alliance for Nanotechnology. Int J Biol Sci 3:108–110

Elghanian R, Storhoff JJ, Mucic RC, Letsinger RL, Mirkin CA (1997) Selective colorimetric detection of polynucleotides based on the distance-dependent optical properties of gold nanoparticles. Science 277:1078–1081

El-Sayed IH, Huang X, El-Sayed MA (2005a) Surface plasmon resonance scattering and absorption of anti-EGFR antibody conjugated gold nanoparticles in cancer diagnostics: applications in oral cancer. Nano Lett 5:829–834

El-Sayed ME, Hoffman AS, Stayton PS (2005b) Smart polymeric carriers for enhanced intracellular delivery of therapeutic macromolecules. Expert Opin Biol Ther 5:23–32

El-Sayed IH, Huang X, El-Sayed MA (2006) Selective laser photo-thermal therapy of epithelial carcinoma using anti-EGFR antibody conjugated gold nanoparticles. Cancer Lett 239:129–135

Etzioni R, Urban N, Ramsey S, McIntosh M, Schwartz S, Reid B, Radich J, Anderson G, Hartwell L (2003) The case for early detection. Nat Rev Cancer 3:243–252

Ferrari M (2005) Cancer nanotechnology: opportunities and challenges. Nat Rev Cancer 5:161–171

Fichou Y, Ferec C (2006) The potential of oligonucleotides for therapeutic applications. Trends Biotechnol 24:563–570

Fire A, Xu S, Montgomery MK, Kostas SA, Driver SE, Mello CC (1998) Potent and specific genetic interference by double-stranded RNA in *Caenorhabditis elegans*. Nature 391:806–811

Freese C, Uboldi C, Gibson MI, Unger RE, Weksler BB, Romero IA, Couraud PO, Kirkpatrick CJ (2012) Uptake and cytotoxicity of citrate-coated gold nanospheres: comparative studies on human endothelial and epithelial cells. Part Fibre Toxicol 9:23

Gary DJ, Puri N, Won YY (2007) Polymer-based siRNA delivery: perspectives on the fundamental and phenomenological distinctions from polymer-based DNA delivery. J Control Release 121:64–73

Ghosh P, Han G, De M, Kim CK, Rotello VM (2008a) Gold nanoparticles in delivery applications. Adv Drug Deliv Rev 60:1307–1315

Ghosh PS, Kim CK, Han G, Forbes NS, Rotello VM (2008b) Efficient gene delivery vectors by tuning the surface charge density of amino acid-functionalized gold nanoparticles. ACS Nano 2:2213–2218

Gibson JD, Khanal BP, Zubarev ER (2007) Paclitaxel-functionalized gold nanoparticles. J Am Chem Soc 129:11653–11661

Gil PR, Parak WJ (2008) Composite nanoparticles take aim at cancer. ACS Nano 2:2200–2205

Giljohann DA, Seferos DS, Prigodich AE, Patel PC, Mirkin CA (2009) Gene regulation with polyvalent siRNA-nanoparticle conjugates. J Am Chem Soc 131:2072–2073

Giljohann DA, Seferos DS, Daniel WL, Massich MD, Patel PC, Mirkin CA (2010) Gold nanoparticles for biology and medicine. Angew Chem Int Ed Engl 49:3280–3294

Goodman CM, McCusker CD, Yilmaz T, Rotello VM (2004) Toxicity of gold nanoparticles functionalized with cationic and anionic side chains. Bioconjug Chem 15:897–900

Grubisha DS, Lipert RJ, Park HY, Driskell J, Porter MD (2003) Femtomolar detection of prostate-specific antigen: an immunoassay based on surface-enhanced Raman scattering and immunogold labels. Anal Chem 75:5936–5943

Guo S, Huang Y, Jiang Q, Sun Y, Deng L, Liang Z, Du Q, Xing J, Zhao Y, Wang PC, Dong A, Liang XJ (2010) Enhanced gene delivery and siRNA silencing by gold nanoparticles coated with charge-reversal polyelectrolyte. ACS Nano 4:5505–5511

Hainfeld JF, Slatkin DN, Smilowitz HM (2004) The use of gold nanoparticles to enhance radiotherapy in mice. Phys Med Biol 49:N309–N315

Hainfeld JF, Dilmanian FA, Slatkin DN, Smilowitz HM (2008) Radiotherapy enhancement with gold nanoparticles. J Pharm Pharmacol 60:977–985

Han G, Ghosh P, Rotello VM (2007a) Functionalized gold nanoparticles for drug delivery. Nanomedicine (Lond) 2:113–123

Han G, Ghosh P, Rotello VM (2007b) Multi-functional gold nanoparticles for drug delivery. Adv Exp Med Biol 620:48–56

Hanahan D, Weinberg RA (2000) The hallmarks of cancer. Cell 100:57–70

Hannon GJ, Rossi JJ (2004) Unlocking the potential of the human genome with RNA interference. Nature 431:371–378

Heath JR, Davis ME (2008) Nanotechnology and cancer. Annu Rev Med 59:251–265

Hirsch LR, Stafford RJ, Bankson JA, Sershen SR, Rivera B, Price RE, Hazle JD, Halas NJ, West JL (2003) Nanoshell-mediated near-infrared thermal therapy of tumors under magnetic resonance guidance. Proc Natl Acad Sci U S A 100:13549–13554

Hong R, Han G, Fernandez JM, Kim BJ, Forbes NS, Rotello VM (2006) Glutathione-mediated delivery and release using monolayer protected nanoparticle carriers. J Am Chem Soc 128:1078–1079

Huang L, Reekmans G, Saerens D, Friedt JM, Frederix F, Francis L, Muyldermans S, Campitelli A, Van HC (2005) Prostate-specific antigen immunosensing based on mixed self-assembled monolayers, camel antibodies and colloidal gold enhanced sandwich assays. Biosens Bioelectron 21:483–490

Huang XL, Zhang B, Ren L, Ye SF, Sun LP, Zhang QQ, Tan MC, Chow GM (2008) In vivo toxic studies and biodistribution of near infrared sensitive Au-Au(2)S nanoparticles as potential drug delivery carriers. J Mater Sci Mater Med 19:2581–2588

Huo Q, Colon J, Cordero A, Bogdanovic J, Baker CH, Goodison S, Pensky MY (2011) A facile nanoparticle immunoassay for cancer biomarker discovery. J Nanobiotechnology 9:20

Huo Q, Litherland SA, Sullivan S, Hallquist H, Decker DA, Rivera-Ramirez I (2012) Developing a nanoparticle test for prostate cancer scoring. J Transl Med 10:44

Hwu JR, Lin YS, Josephrajan T, Hsu MH, Cheng FY, Yeh CS, Su WC, Shieh DB (2009) Targeted Paclitaxel by conjugation to iron oxide and gold nanoparticles. J Am Chem Soc 131:66–68

Jensen GC, Krause CE, Sotzing GA, Rusling JF (2011) Inkjet-printed gold nanoparticle electrochemical arrays on plastic. Application to immunodetection of a cancer biomarker protein. Phys Chem Chem Phys 13:4888–4894

Kah JC, Kho KW, Lee CG, James C, Sheppard R, Shen ZX, Soo KC, Olivo MC (2007) Early diagnosis of oral cancer based on the surface plasmon resonance of gold nanoparticles. Int J Nanomedicine 2:785–798

Kalogianni DP, Bravou V, Christopoulos TK, Ioannou PC, Zoumbos NC (2007) Dry-reagent disposable dipstick test for visual screening of seven leukemia-related chromosomal translocations. Nucleic Acids Res 35:e23
Kang JH, Asami Y, Murata M, Kitazaki H, Sadanaga N, Tokunaga E, Shiotani S, Okada S, Maehara Y, Niidome T, Hashizume M, Mori T, Katayama Y (2010) Gold nanoparticle-based colorimetric assay for cancer diagnosis. Biosens Bioelectron 25:1869–1874
Kim KY (2007) Nanotechnology platforms and physiological challenges for cancer therapeutics. Nanomedicine 3:103–110
Kim CK, Ghosh P, Pagliuca C, Zhu ZJ, Menichetti S, Rotello VM (2009a) Entrapment of hydrophobic drugs in nanoparticle monolayers with efficient release into cancer cells. J Am Chem Soc 131:1360–1361
Kim SS, Garg H, Joshi A, Manjunath N (2009b) Strategies for targeted nonviral delivery of siRNAs in vivo. Trends Mol Med 15:491–500
Kim D, Jeong YY, Jon S (2010) A drug-loaded aptamer-gold nanoparticle bioconjugate for combined CT imaging and therapy of prostate cancer. ACS Nano 4:3689–3696
Kneipp J, Kneipp H, McLaughlin M, Brown D, Kneipp K (2006) In vivo molecular probing of cellular compartments with gold nanoparticles and nanoaggregates. Nano Lett 6:2225–2231
Kong KV, Lam Z, Goh WD, Leong WK, Olivo M (2012) Metal carbonyl-gold nanoparticle conjugates for live-cell SERS imaging. Angew Chem Int Ed Engl 51:9796–9799
Kumar S, Harrison N, Richards-Kortum R, Sokolov K (2007) Plasmonic nanosensors for imaging intracellular biomarkers in live cells. Nano Lett 7:1338–1343
Lammers T, Kiessling F, Hennink WE, Storm G (2010) Nanotheranostics and image-guided drug delivery: current concepts and future directions. Mol Pharm 7:1899–1912
Lammers T, Aime S, Hennink WE, Storm G, Kiessling F (2011) Theranostic nanomedicine. Acc Chem Res 44:1029–1038
Lan T, Dong C, Huang X, Ren J (2011) Single particle technique for one-step homogeneous detection of cancer marker using gold nanoparticle probes. Analyst 136:4247–4253
Langer R (2000) Biomaterials in drug delivery and tissue engineering: one laboratory's experience. Acc Chem Res 33:94–101
Lasagna-Reeves C, Gonzalez-Romero D, Barria MA, Olmedo I, Clos A, Sadagopa Ramanujam VM, Urayama A, Vergara L, Kogan MJ, Soto C (2010) Bioaccumulation and toxicity of gold nanoparticles after repeated administration in mice. Biochem Biophys Res Commun 393:649–655
Lee SH, Bae KH, Kim SH, Lee KR, Park TG (2008) Amine-functionalized gold nanoparticles as non-cytotoxic and efficient intracellular siRNA delivery carriers. Int J Pharm 364:94–101
Lee JS, Green JJ, Love KT, Sunshine J, Langer R, Anderson DG (2009) Gold, poly(beta-amino ester) nanoparticles for small interfering RNA delivery. Nano Lett 9:2402–2406
Lee SK, Han MS, Asokan S, Tung CH (2011) Effective gene silencing by multilayered siRNA-coated gold nanoparticles. Small 7:364–370
Li H, Rothberg L (2004) Colorimetric detection of DNA sequences based on electrostatic interactions with unmodified gold nanoparticles. Proc Natl Acad Sci U S A 101:14036–14039
Li H, Li WX, Ding SW (2002) Induction and suppression of RNA silencing by an animal virus. Science 296:1319–1321
Li J, Chu X, Liu Y, Jiang JH, He Z, Zhang Z, Shen G, Yu RQ (2005) A colorimetric method for point mutation detection using high-fidelity DNA ligase. Nucleic Acids Res 33:e168
Lin D, Feng S, Pan J, Chen Y, Lin J, Chen G, Xie S, Zeng H, Chen R (2011) Colorectal cancer detection by gold nanoparticle based surface-enhanced Raman spectroscopy of blood serum and statistical analysis. Opt Express 19:13565–13577
Liu Y, Miyoshi H, Nakamura M (2007) Nanomedicine for drug delivery and imaging: a promising avenue for cancer therapy and diagnosis using targeted functional nanoparticles. Int J Cancer 120:2527–2537
Liu X, Dai Q, Austin L, Coutts J, Knowles G, Zou J, Chen H, Huo Q (2008) A one-step homogeneous immunoassay for cancer biomarker detection using gold nanoparticle probes coupled with dynamic light scattering. J Am Chem Soc 130:2780–2782
Liu G, Mao X, Phillips JA, Xu H, Tan W, Zeng L (2009) Aptamer-nanoparticle strip biosensor for sensitive detection of cancer cells. Anal Chem 81:10013–10018

Lu W, Arumugam SR, Senapati D, Singh AK, Arbneshi T, Khan SA, Yu H, Ray PC (2010a) Multifunctional oval-shaped gold-nanoparticle-based selective detection of breast cancer cells using simple colorimetric and highly sensitive two-photon scattering assay. ACS Nano 4: 1739–1749
Lu W, Zhang GD, Zhang R, Flores LG, Huang Q, Gelovani JG, Li C (2010b) Tumor site-specific silencing of NF-kappa B p65 by targeted hollow gold nanosphere-mediated photothermal transfection. Cancer Res 70:3177–3188
Lukianova-Hleb EY, Oginsky AO, Samaniego AP, Shenefelt DL, Wagner DS, Hafner JH, Farach-Carson MC, Lapotko DO (2011) Tunable plasmonic nanoprobes for theranostics of prostate cancer. Theranostics 1:3–17
Ma X, Zhao Y, Liang XJ (2011) Theranostic nanoparticles engineered for clinic and pharmaceutics. Acc Chem Res 44:1114–1122
McIntosh CM, Esposito EA III, Boal AK, Simard JM, Martin CT, Rotello VM (2001) Inhibition of DNA transcription using cationic mixed monolayer protected gold clusters. J Am Chem Soc 123:7626–7629
Medley CD, Smith JE, Tang Z, Wu Y, Bamrungsap S, Tan W (2008) Gold nanoparticle-based colorimetric assay for the direct detection of cancerous cells. Anal Chem 80:1067–1072
Milhavet O, Gary DS, Mattson MP (2003) RNA interference in biology and medicine. Pharmacol Rev 55:629–648
Minelli C, Lowe SB, Stevens MM (2010) Engineering nanocomposite materials for cancer therapy. Small 6:2336–2357
Mirkin CA, Letsinger RL, Mucic RC, Storhoff JJ (1996) A DNA-based method for rationally assembling nanoparticles into macroscopic materials. Nature 382:607–609
Mukerjee A, Ranjan AP, Vishwanatha JK (2012) Combinatorial nanoparticles for cancer diagnosis and therapy. Curr Med Chem 19:3714–3721
Mukherjee P, Bhattacharya R, Bone N, Lee YK, Patra CR, Wang S, Lu L, Secreto C, Banerjee PC, Yaszemski MJ, Kay NE, Mukhopadhyay D (2007) Potential therapeutic application of gold nanoparticles in B-chronic lymphocytic leukemia (BCLL): enhancing apoptosis. J Nanobiotechnology 5:4
Munge BS, Coffey AL, Doucette JM, Somba BK, Malhotra R, Patel V, Gutkind JS, Rusling JF (2011) Nanostructured immunosensor for attomolar detection of cancer biomarker interleukin-8 using massively labeled superparamagnetic particles. Angew Chem Int Ed Engl 50:7915–7918
Nam JM, Thaxton CS, Mirkin CA (2003) Nanoparticle-based bio-bar codes for the ultrasensitive detection of proteins. Science 301:1884–1886
Nasrolahi SA, Mandal D, Tiwari RK, Guo L, Lu W, Parang K (2012) Cyclic peptide-capped gold nanoparticles as drug delivery systems. Mol Pharm 10:500–511
National Cancer Institute (2010) http://nano.cancer.gov/. Nanotechnology
Nishiyama N (2007) Nanomedicine: nanocarriers shape up for long life. Nat Nanotechnol 2:203–204
Oishi M, Nakaogami J, Ishii T, Nagasaki Y (2006) Smart PEGylated gold nanoparticles for the cytoplasmic delivery of siRNA to induce enhanced gene silencing. Chem Lett 35:1046–1047
Paciotti GF, Myer L, Weinreich D, Goia D, Pavel N, McLaughlin RE, Tamarkin L (2004) Colloidal gold: a novel nanoparticle vector for tumor directed drug delivery. Drug Deliv 11:169–183
Pan Y, Neuss S, Leifert A, Fischler M, Wen F, Simon U, Schmid G, Brandau W, Jahnen-Dechent W (2007) Size-dependent cytotoxicity of gold nanoparticles. Small 3:1941–1949
Peer D, Karp JM, Hong S, Farokhzad OC, Margalit R, Langer R (2007) Nanocarriers as an emerging platform for cancer therapy. Nat Nanotechnol 2:751–760
Pene F, Courtine E, Cariou A, Mira JP (2009) Toward theragnostics. Crit Care Med 37:S50–S58
Praetorius NP, Mandal TK (2007) Engineered nanoparticles in cancer therapy. Recent Pat Drug Deliv Formul 1:37–51
Qian X, Peng XH, Ansari DO, Yin-Goen Q, Chen GZ, Shin DM, Yang L, Young AN, Wang MD, Nie S (2008a) In vivo tumor targeting and spectroscopic detection with surface-enhanced Raman nanoparticle tags. Nat Biotechnol 26:83–90

Qian X, Zhou X, Nie S (2008b) Surface-enhanced Raman nanoparticle beacons based on bioconjugated gold nanocrystals and long range plasmonic coupling. J Am Chem Soc 130: 14934–14935

Qin WJ, Yung LY (2007) Nanoparticle-based detection and quantification of DNA with single nucleotide polymorphism (SNP) discrimination selectivity. Nucleic Acids Res 35:e111

Rahman M, Abd-El-Barr M, Mack V, Tkaczyk T, Sokolov K, Richards-Kortum R, Descour M (2005) Optical imaging of cervical pre-cancers with structured illumination: an integrated approach. Gynecol Oncol 99:S112–S115

Rana S, Bajaj A, Mout R, Rotello VM (2012) Monolayer coated gold nanoparticles for delivery applications. Adv Drug Deliv Rev 64:200–216

Rosi NL, Giljohann DA, Thaxton CS, Lytton-Jean AK, Han MS, Mirkin CA (2006) Oligonucleotide-modified gold nanoparticles for intracellular gene regulation. Science 312:1027–1030

Rothen-Rutishauser B, Muhlfeld C, Blank F, Musso C, Gehr P (2007) Translocation of particles and inflammatory responses after exposure to fine particles and nanoparticles in an epithelial airway model. Part Fibre Toxicol 4:9

Salata O (2004) Applications of nanoparticles in biology and medicine. J Nanobiotechnology 2:3

Samanta A, Maiti KK, Soh KS, Liao XJ, Vendrell M, Dinish US, Yun SW, Bhuvaneswari R, Kim H, Rautela S, Chung JH, Olivo M, Chang YT (2011) Ultrasensitive near-infrared Raman reporters for SERS-based in vivo cancer detection. Angew Chem Int Ed Engl 50: 6089–6092

Sato K, Hosokawa K, Maeda M (2003) Rapid aggregation of gold nanoparticles induced by non-cross-linking DNA hybridization. J Am Chem Soc 125:8102–8103

Sato K, Hosokawa K, Maeda M (2005) Non-cross-linking gold nanoparticle aggregation as a detection method for single-base substitutions. Nucleic Acids Res 33:e4

Schroeder A, Heller DA, Winslow MM, Dahlman JE, Pratt GW, Langer R, Jacks T, Anderson DG (2012) Treating metastatic cancer with nanotechnology. Nat Rev Cancer 12:39–50

Shi XY, Wang SH, Van Antwerp ME, Chen XS, Baker JR (2009) Targeting and detecting cancer cells using spontaneously formed multifunctional dendrimer-stabilized gold nanoparticles. Analyst 134:1373–1379

Siegel R, Naishadham D, Jemal A (2012) Cancer statistics, 2012. CA Cancer J Clin 62:10–29

Son SJ, Lee SB (2007) A platform for ultrasensitive and selective multiplexed marker protein assay toward early-stage cancer diagnosis. Nanomedicine (Lond) 2:79–82

Song JB, Zhou JJ, Duan HW (2012) Self-assembled plasmonic vesicles of SERS-encoded amphiphilic gold nanoparticles for cancer cell targeting and traceable intracellular drug delivery. J Am Chem Soc 134:13458–13469

Soutschek J, Akinc A, Bramlage B, Charisse K, Constien R, Donoghue M, Elbashir S, Geick A, Hadwiger P, Harborth J, John M, Kesavan V, Lavine G, Pandey RK, Racie T, Rajeev KG, Rohl I, Toudjarska I, Wang G, Wuschko S, Bumcrot D, Koteliansky V, Limmer S, Manoharan M, Vornlocher HP (2004) Therapeutic silencing of an endogenous gene by systemic administration of modified siRNAs. Nature 432:173–178

Sperling RA, Parak WJ (2010) Surface modification, functionalization and bioconjugation of colloidal inorganic nanoparticles. Philos Transact A Math Phys Eng Sci 368:1333–1383

Sperling RA, Rivera GP, Zhang F, Zanella M, Parak WJ (2008) Biological applications of gold nanoparticles. Chem Soc Rev 37:1896–1908

Stiles PL, Dieringer JA, Shah NC, Van Duyne RR (2008) Surface-enhanced Raman spectroscopy. Annu Rev Anal Chem 1:601–626

Stoeva SI, Lee JS, Smith JE, Rosen ST, Mirkin CA (2006) Multiplexed detection of protein cancer markers with biobarcoded nanoparticle probes. J Am Chem Soc 128:8378–8379

Storhoff JJ, Lucas AD, Garimella V, Bao YP, Muller UR (2004) Homogeneous detection of unamplified genomic DNA sequences based on colorimetric scatter of gold nanoparticle probes. Nat Biotechnol 22:883–887

Sun X, Rossin R, Turner JL, Becker ML, Joralemon MJ, Welch MJ, Wooley KL (2005) An assessment of the effects of shell cross-linked nanoparticle size, core composition, and surface PEGylation on in vivo biodistribution. Biomacromolecules 6:2541–2554

Taton TA, Mirkin CA, Letsinger RL (2000) Scanometric DNA array detection with nanoparticle probes. Science 289:1757–1760

Thaxton CS, Georganopoulou DG, Mirkin CA (2006) Gold nanoparticle probes for the detection of nucleic acid targets. Clin Chim Acta 363:120–126

Thaxton CS, Elghanian R, Thomas AD, Stoeva SI, Lee JS, Smith ND, Schaeffer AJ, Klocker H, Horninger W, Bartsch G, Mirkin CA (2009) Nanoparticle-based bio-barcode assay redefines "undetectable" PSA and biochemical recurrence after radical prostatectomy. Proc Natl Acad Sci U S A 106:18437–18442

Thomas M, Klibanov AM (2003) Conjugation to gold nanoparticles enhances polyethylenimine's transfer of plasmid DNA into mammalian cells. Proc Natl Acad Sci U S A 100:9138–9143

Toub N, Malvy C, Fattal E, Couvreur P (2006) Innovative nanotechnologies for the delivery of oligonucleotides and siRNA. Biomed Pharmacother 60:607–620

Uboldi C, Bonacchi D, Lorenzi G, Hermanns MI, Pohl C, Baldi G, Unger RE, Kirkpatrick CJ (2009) Gold nanoparticles induce cytotoxicity in the alveolar type-II cell lines A549 and NCIH441. Part Fibre Toxicol 6:18

Viator JA, Gupta S, Goldschmidt BS, Bhattacharyyal K, Kannan R, Shukla R, Dale PS, Boote E, Katti K (2010) Gold nanoparticle mediated detection of prostate cancer cells using photoacoustic flowmetry with optical reflectance. J Biomed Nanotechnol 6:187–191

Visaria RK, Griffin RJ, Williams BW, Ebbini ES, Paciotti GF, Song CW, Bischof JC (2006) Enhancement of tumor thermal therapy using gold nanoparticle-assisted tumor necrosis factor-alpha delivery. Mol Cancer Ther 5:1014–1020

Wabuyele MB, Yan F, Vo-Dinh T (2010) Plasmonics nanoprobes: detection of single-nucleotide polymorphisms in the breast cancer BRCA1 gene. Anal Bioanal Chem 398:729–736

Wall NR, Shi Y (2003) Small RNA: can RNA interference be exploited for therapy? Lancet 362:1401–1403

Wang X, Qian X, Beitler JJ, Chen ZG, Khuri FR, Lewis MM, Shin HJ, Nie S, Shin DM (2011) Detection of circulating tumor cells in human peripheral blood using surface-enhanced Raman scattering nanoparticles. Cancer Res 71:1526–1532

Whitehead KA, Langer R, Anderson DG (2009) Knocking down barriers: advances in siRNA delivery. Nat Rev Drug Discov 8:129–138

Wilson AJ, Willets KA (2013) Surface-enhanced Raman scattering imaging using noble metal nanoparticles. Wiley Interdiscip Rev Nanomed Nanobiotechnol 5:180–189

Yang J, Eom K, Lim EK, Park J, Kang Y, Yoon DS, Na S, Koh EK, Suh JS, Huh YM, Kwon TY, Haam S (2008) In situ detection of live cancer cells by using bioprobes based on Au nanoparticles. Langmuir 24:12112–12115

Yao L, Danniels J, Moshnikova A, Kuznetsov S, Ahmed A, Engelman DM, Reshetnyak YK, Andreev OA (2013) pHLIP peptide targets nanogold particles to tumors. Proc Natl Acad Sci U S A 110:465–470

Yin ZZ, Liu Y, Jiang LP, Zhu JJ (2011) Electrochemical immunosensor of tumor necrosis factor alpha based on alkaline phosphatase functionalized nanospheres. Biosens Bioelectron 26: 1890–1894

You CC, Miranda OR, Gider B, Ghosh PS, Kim IB, Erdogan B, Krovi SA, Bunz UH, Rotello VM (2007) Detection and identification of proteins using nanoparticle-fluorescent polymer 'chemical nose' sensors. Nat Nanotechnol 2:318–323

You J, Zhang R, Xiong CY, Zhong M, Melancon M, Gupta S, Nick AM, Sood AK, Li C (2012) Effective photothermal chemotherapy using doxorubicin-loaded gold nanospheres that target EphB4 receptors in tumors. Cancer Res 72:4777–4786

Zheng D, Giljohann DA, Chen DL, Massich MD, Wang XQ, Iordanov H, Mirkin CA, Paller AS (2012) Topical delivery of siRNA-based spherical nucleic acid nanoparticle conjugates for gene regulation. Proc Natl Acad Sci U S A 109:11975–11980

Chapter 12
Microbubbles as Theranostics Agents

Tuan Pham, Carl Beigie, Yoonjee Park, and Joyce Y. Wong

12.1 Introduction

Ultrasound (US) is among the most commonly used imaging modalities in the clinic. In ultrasound imaging, high-frequency (>20-kHz) sound waves are emitted through the skin. Portions of these waves are reflected back towards the transducer when they encounter tissues, and these reflected waves are computed to generate images based on sound attenuation, backscatter, and sound speed (Massoud and Gambhir 2003). Imaging contrast can be greatly improved (Blomley et al. 2001) through the addition of agents with acoustic properties different from surrounding tissues. Among the most commonly used US contrast agents are gas-filled microbubbles.

Microbubbles are gas-filled spheres with a shell that can be composed of phospholipids, polymers, or a variety of other substances (Janib et al. 2010; Unger et al. 2004). When exposed to US, these microbubbles undergo alternating contraction and expansion, a process known as cavitation, due to compressibility of encapsulated gas when subjected to acoustic pressure (Phillips and Gardner 2004; Cosgrove 2004; Stride and Saffari 2003). These vibrations cause the microbubbles to return a greater amount of US signal to the transducer, as compared to the practically incompressible surrounding tissues. This difference in signal return accounts for the majority of contrast provided by microbubbles in conventional grayscale ultrasound imaging (Cosgrove 2006). Although there are many other methods of obtaining contrast using microbubbles [Doppler (Harvey et al. 2001) and contrast specific imaging (Phillips and Gardner 2004) among others], the physics underlying these methods are beyond the scope of this chapter.

T. Pham • C. Beigie • Y. Park, Ph.D. • J.Y. Wong, Ph.D. (✉)
Department of Biomedical Engineering, Boston University,
44 Cummington Mall, Boston, MA 02215, USA
e-mail: jywong@bu.edu

M.J. Alonso, M. Garcia-Fuentes (eds.), *Nano-Oncologicals: New Targeting and Delivery Approaches*, Advances in Delivery Science and Technology,
DOI 10.1007/978-3-319-08084-0_12

Currently the only FDA-approved use of microbubble ultrasound contrast agents is for left ventricle opacification. The albumin shelled microbubble, Albunex, was first approved in 1994. Subsequently, only two commercial microbubbles remain in clinical use: Optison™ (GE Medical Diagnostics, Princeton, NJ) and Definity® (Lantheus Medical Imaging, Billerica, NJ), which were FDA-approved in 1997 and 2001, respectively (Castle et al. 2013). While microbubbles have only been used as ultrasound contrast agents in the clinic, a significant amount of work has been devoted to demonstrating the potential role of microbubbles as a complete theranostic tool. Theranostics, or the combination of imaging and therapy, has gained increasing amounts of attention because of its proposed goal of allowing clinicians to diagnose and treat in real time with a single agent. The purpose of this chapter is to highlight strategies and uses of microbubbles in the application of theranostic cancer therapy. After a brief overview of microbubble fabrication and properties affecting their stability, the concept of microbubble use for theranostics is further developed with discussion of their potential for multi-modal imaging. In the remainder of the chapter, techniques for incorporation of drugs, genes, and proteins, and factors that affect success of these payload types are presented from key literature. The work presented here is by no means all-inclusive but merely attempts to introduce these key concepts. We also examine targeting strategies and their use in improving efficiency of microbubble accumulation at the tumor site. Finally, the chapter summarizes future work required to bring theranostic microbubble platforms to the clinic.

12.2 Microbubbles: Basic Fabrication Methods

Microbubbles can be made in a variety of ways, with the two most common methods being mechanical agitation (Fang et al. 2007; Klibanov 1999) and microfluidic flow focusing (Hashimoto et al. 2008; Ganan-Calvo and Gordillo 2001; Hettiarachchi et al. 2007). In mechanical agitation, a tip sonicator or vial shaker is used to violently mix the solution containing the shell components with a headspace of gas in order to create a microbubble emulsion. Microfluidic flow focusing uses micron sized flow channels in order to mitigate flow of immiscible fluids to emulsify one phase into another immiscible phase. In the context of microbubbles, the gas phase (microbubble core) flows into the liquid phase containing shell components (lipids, polymers, etc.). These two phases are then focused at a small opening called the orifice and pinch off to form gas-filled microbubbles (Hashimoto et al. 2008). Figure 12.1 shows a general overview of these two techniques.

There are many things to consider when choosing which method of microbubble fabrication to use. The sonication technique can produce a large number of microbubbles within minutes; however, some issues may arise from resulting broad distributions of size and shell thickness (Xu et al. 2012; Lee and Lee 2010; Parhizkar et al. 2013). In contrast, microfluidic flow focusing methods yield uniform size

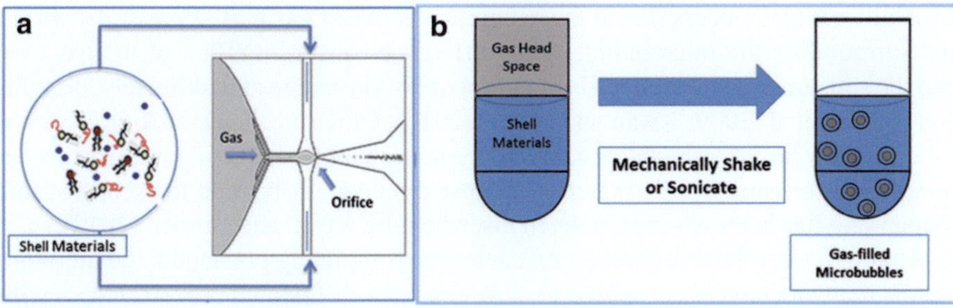

Fig. 12.1 Fabrication methods of microbubbles. (**a**) The microbubble starting materials (shell materials and gas) are flowed through a microfluidic device designed to focus the two immiscible phases at an orifice. The two colliding flows then pinch off and form microbubbles. (**b**) The shell materials are placed into a vial with a gas headspace. Then, using mechanical agitation (shaking or sonicating), the two phases are forced to partition into gas-cored microbubbles

distributions; yet different shell thicknesses and bubble sizes can still be created by modifying fabrication parameters such as flow rates or device geometries (Lee and Lee 2010; Parhizkar et al. 2013; Gunduz et al. 2012). The narrow microbubble size distributions from flow focusing have been shown to improve echogenicity (Talu et al. 2007). However, a major drawback of microfluidic-driven microbubble production is the difficulty to mass-produce these bubble populations. To address these concerns, there has been a growing literature of new methods for mass production of bubbles that include increasing microbubble production speed (Castro-Hernandez et al. 2011) and running multiple devices in parallel on a single chip (Nisisako and Torii 2008; Romanowsky et al. 2012; Kendall et al. 2012).

12.3 Microbubble Stabilization

The visualization of blood flow in left ventricle opacification is a relatively brief imaging technique and does not require long-term microbubble stability for systemic circulation. However, with the expanding number of applications for which microbubbles are being explored, long-term stability of these contrast agents has become increasingly important. This section focuses on microbubble stability against gas dissolution, coalescence, and under ultrasound.

Microbubble instability arises from a variety of physical properties. Microbubble emulsions are thermodynamically unstable and have a tendency to minimize free energy of the system by separating gas and liquid into two bulk phases. This energy minimization leads to coalescence in which microbubbles aggregate together. These patches of gathered microbubbles eventually undergo Ostwald ripening and fuse into larger microbubbles, continuing this process until gas and liquid phases are completely separated (Kwan and Borden 2012). Another form of instability is gas dissolution which occurs due to the difference between gas partial pressures within and surrounding the microbubble. This difference causes an efflux of the gas core into the surroundings, which leads to microbubble shrinkage and ultimately, dissolution (Sarkar et al. 2009; Kwan and Borden 2010). Other mechanisms for microbubble instability have been explored, including microbubble shell collapse. However, a complete discussion of microbubble collapse dynamics is beyond the scope of this chapter and has been covered in detail elsewhere by Kwan and Borden (2012).

Although microbubble dissolution cannot be completely prevented, the literature reports many techniques proposed to slow down the diffusional process. Gases with low aqueous solubility, such as perfluorocarbons, have been used as microbubble cores to decrease gas dissolution (Cui et al. 2005) and different microbubble shell compositions have been explored in an effort to increase structural integrity (Paradossi et al. 2003; Ferrara et al. 2007; Duncanson et al. 2010). Additionally, researchers have added PEG to the microbubble shell to prevent aggregation and subsequent Ostwald ripening (Duncanson et al. 2010; Park et al. 2012). By using polymer-based shell compositions, increased shelf life, suspension stability, and stability under ultrasound have been obtained. However, these microbubbles exhibit decreased echogenicity and nonlinearity when compared with soft, lipid shelled microbubbles (Sciallero and Trucco 2013).

A relatively new concept of microbubble formulation has been explored in which a lipid-shelled microbubble is made using polymerizable diacetylene lipids. Park et al. (2012) showed that by varying the amount of this polymerizable lipid, microbubble stability could be tuned. Figure 12.2 shows that the 25 % polymerizable lipid shell condition provided increased stability against microbubble dissolution (Fig. 12.2a) as well as increased stability under ultrasound (Fig. 12.2b) when compared with lower polymerizable lipid shell concentrations and commercially available microbubbles, Vevo MicroMarker (VMM).

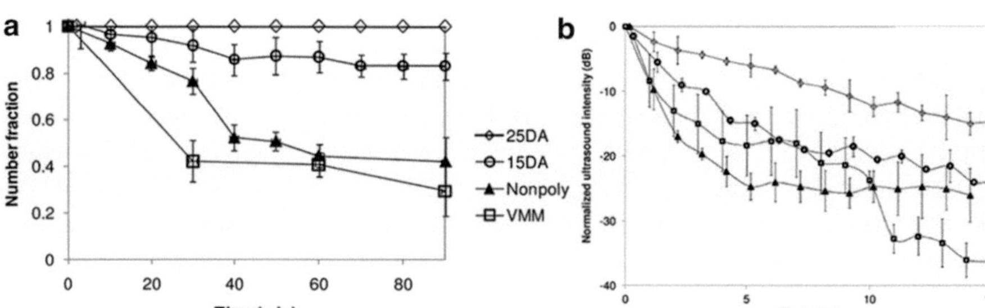

Fig. 12.2 (**a**) Normalized number of microbubbles remained with time. The stability against gas dissolution of the different shell materials is informed: (*diamond*) 25 % DA; (*circle*) 15 % DA; (*triangle*) non-polymerizable lipids; (*square*) VMM. (**b**) Ultrasound image intensity (at 7.5 MHz) vs. time for a variety of microbubble shell materials: (*diamond*) 25 % DA; (*circle*) 15 % DA; (*triangle*) nonpolymerizable lipids; (*square*) Vevo MicroMarker (VMM). The intensity in the ROI was normalized by the initial intensity and then converted to dB. The *vertical lines* represent standard error [adapted with permission from Park et al. 2012). Copyright (2012) American Chemical Society]

12.4 Theranostics

The advances made in improving microbubble stability have expanded the potential applications of microbubbles to include triggered delivery of a therapeutic payload. Many researchers envision the use of microbubbles as a theranostic particle that both provides targeted image contrast to aid in diagnostic procedures and carries a therapeutic agent to the intended target. Ideally, such theranostic particles would be injected systemically and accumulate at the desired target through use of specific targeting moieties, such as antibodies. Once at the intended target, a clinician could image and identify the location of the target using clinical US. Finally, US of a frequency specific to the microbubble, distinct from the frequency used for imaging, can be applied to burst the microbubble and release the payload. Figure 12.3 gives a simplistic overview of this concept, which remains the ultimate goal of the field.

Some early studies have shown microbubbles as a promising platform to develop this theranostic concept. Leong-Poi et al. used microbubbles loaded with plasmid DNA encoded for VEGF to induce formation of microvessels in hind limbs of rats, which in turn increased microvascular blood flow (Leong-Poi et al. 2007). Although these particular microbubbles were not used for imaging in the rats, but merely as DNA delivery vehicles, it could be reasoned that acoustic contrast enhancement of these particles would be ideal for visualization of increased microvascular blood flow (Kiessling et al. 2012). Rapoport et al. used doxorubicin-loaded nanobubbles which collected at tumor sites and coalesced into microbubbles. These in vivo-formed microbubbles were then imaged using US and destroyed to release the drug, which led to tumor growth inhibition (Rapoport et al. 2007). Although these studies and others show potential of microbubbles as theranostic tools, the field is still relatively young, and much more work is needed before true clinical translation can be achieved. While the role of microbubbles in ultrasound contrasting has already been elucidated, combinations of microbubbles with other contrasting agents allow for a more complete diagnostic tool. In the following section we review further enhancements to the imaging aspect of theranostics. Beginning from Sect. 5, we will discuss current efforts in the literature that address drug delivery aspects of theranostic function.

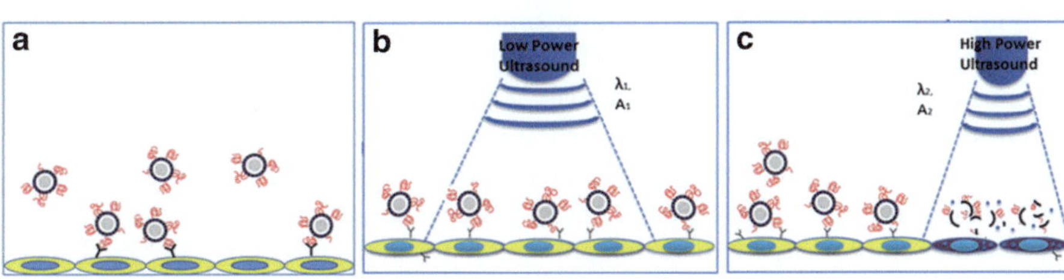

Fig. 12.3 (**a**) Drug-loaded, targeted microbubbles are injected into the bloodstream and only attach to diseased tissues that exhibit specific targets. (**b**) Using a low powered ultrasound, this diseased area is located via ultrasound for diagnostic purposes and then (**c**) switching to a higher powered ultrasound, the bubbles in the specified area will burst and locally deliver therapeutic agents to the diseased tissue

12.4.1 Combined Imaging Modalities to Enhance Diagnostic Imaging

While US is the primary mechanism for both imaging and triggering therapeutic release, others have further developed these particles to provide greater diagnostic practicality by combining them with other imaging modalities. For example, there have been microbubbles developed that are infused with contrast agents for magnetic resonance imaging (MRI) to provide a theranostic agent with both imaging capabilities. MRI is a non-invasive and widely clinically used technique that provides excellent soft tissue contrast with high morphological and spatial resolution. Superparamagnetic iron oxide nanoparticles (SPIONs) alter transverse (*T2*) relaxation times of water protons, providing dark negative signal intensity in images that can be used to visualize tumors (Lee et al. 2006). Additionally, SPIONs are non-toxic, biodegradable, and have a wide degree of versatility in biomedical applications. Current research of SPION applications includes drug delivery, bio-sensing, and molecular targeting (Laurent et al. 2008). By combining SPIONS with microbubbles, clinicians can optimize use of two imaging modalities through only one injectable contrasting agent. Using polymer-coated microbubbles, SPIONs can be incorporated into the shell of ultrasound contrast agents during synthesis steps and released upon microbubble rupture. Using a double emulsion solvent evaporation interfacial deposition (water-in-oil-in-water emulsion) process, SPIONs can be incorporated into an internal shell layer of polymer and then further encapsulated within a second shell layer all while maintaining a gas core (Yang et al. 2009). Incorporation of SPIONs into the inner shell increased MRI relaxation time as SPION concentration increased but reached a maximum in ultrasound contrast, after which additional SPIONs reduced the response of the microbubble. Thus, there exists a trade-off between MRI and ultrasound signals that must be optimized for an effective contrast agent. Using a similar oil-in-water emulsion technique, Liu et al. (2011) demonstrated significant image contrast observed in vascularized areas of a myxoid liposarcoma (MLS) ovarian carcinoma in rats upon i.v. injection of their hybrid contrast agent. Additionally, this study demonstrated a significant increase in longitudinal and transversal relaxivities after US-induced bubble destruction, which demonstrated triggerable MR imaging properties.

While MRI/US contrast combination continues to be evaluated as a future clinical technique, microbubbles are also being investigated in combination with other imaging modalities. Specifically, photoacoustic imaging can be coupled with ultrasound to provide enhanced spatial and temporal resolution (Xu and Wang 2006). Photoacoustic imaging uses nonionizing waves, such as short laser or radio-frequency pulses, of electromagnetic energy to excite ultrasound waves in biological tissues. A promising biomedical application lies in the detection of breast cancer by near-infrared light or radio-frequency–wave-induced photoacoustic imaging, which has important potential for early detection (Pramanik et al. 2008). Dove et al. demonstrated that binding of avidin-conjugated gold nanoparticles to a biotinylated microbubble shell provided stoichiometric control over nanoparticle surface density and optical absorption, while still retaining essential acoustic properties of the microbubble. Furthermore, the acoustic efficiency was 8.9-fold higher for microbubbles with conjugated nanoparticles than for nanoparticles alone, suggesting a potential for beneficial coupling between microbubbles and nanoparticles for enhanced photoacoustic response (Dove et al. 2013).

12.5 Microbubbles as a Therapeutic Delivery Device

To achieve a fully integrated theranostic particle, various techniques are employed to specifically tailor microbubble properties for payload incorporation and release. Microbubbles with a therapeutic payload have been used for a number of therapeutic strategies including delivering drugs across the blood–brain barrier (Hynynen et al. 2001; Mesiwala et al. 2002; Mesiwala and Mourad 2002), gene delivery (Martinez et al. 2002; Urban-Klein et al. 2005; Kinoshita and Hynynen 2005; Carson et al. 2012; Li et al. 2009), and targeted drug delivery (Willmann et al. 2008; Borden et al. 2008; Pochon et al. 2010; Tardy et al. 2010). As shown in Fig. 12.4, the methods to incorporate drugs into a microbubble can vary depending on the properties of the therapeutic payload.

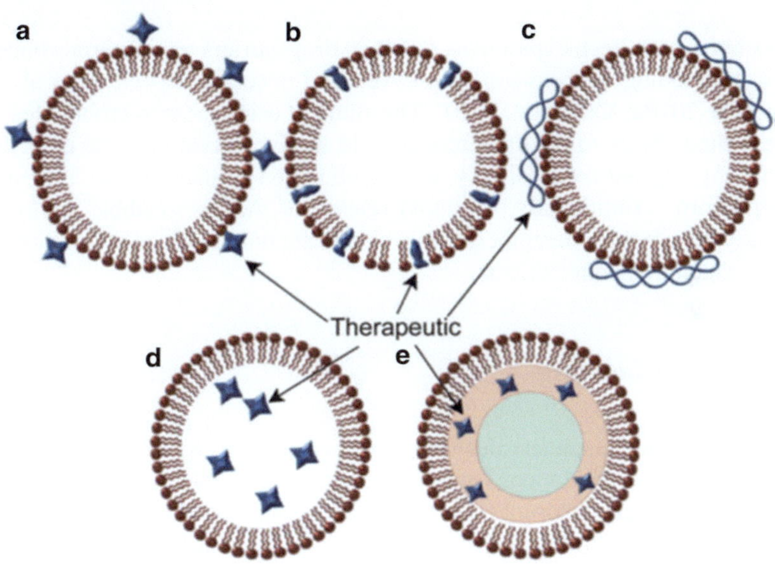

Fig. 12.4 Different ways that drugs can be incorporated into a microbubble. The drugs can be attached to the outer shell of the bubble (*a*), incorporated into the bubble's shell (*b*), bound noncovalently to the surface of the microbubbles (*c*), and entrapped within the microbubbles (*d*), or hydrophobic drugs can be incorporated into a layer of oil within the microbubble (*e*). Reprinted with permission from Elsevier (Unger et al. 2001, 2004)

Therapeutic agents can be attached to the outer shell of the microbubble either as a free drug bound to the surface or within another carrying vehicle (micelles or liposomes) that is in turn bound to the microbubble (Fig. 12.4a) (Lentacker et al. 2007). Hydrophobic drugs, such as paclitaxel and doxorubicin, can also be incorporated into the shell layer of the microbubble (Fig. 12.4b) (Cochran et al. 2011; Tinkov et al. 2010a; Kang et al. 2010). For gene therapy applications, genetic material can potentially be electrostatically bound to the microbubble surface using shell components that give an overall positive shell charge (Fig. 12.4c) (Chen et al. 2006). Therapeutic agents can also be encapsulated directly within the microbubble itself (Fig. 12.4d). For lipophilic drugs that require a larger payload than microbubble surface binding can provide, an additional oil layer can be introduced within the microbubble between the gas core and bubble shell (Fig. 12.4e) (Fang et al. 2007; Lentacker et al. 2010; Shortencarier et al. 2004). Due to numerous drug loading options and manipulations of the microbubble itself, there has been much work done in exploring possible clinical uses for these drug-loaded microbubbles. The following sections will discuss current literature in regard to payload categories of small molecule drugs, gene delivery, or protein therapeutics and application of these loading techniques.

12.5.1 Drug Therapies

Microbubbles have been explored as potential drug carriers for numerous oncological drugs including doxorubicin, paclitaxel, and docetaxel (Cochran et al. 2011; Tinkov et al. 2010a; Kang et al. 2010). The allure of using these microbubbles is due in part to their ability to facilitate reduction of in vivo toxicity (Unger et al. 1998), potential for targeted drug delivery, and combination of therapy and imaging in a single platform. Additionally, triggered release of the microbubble payload can potentially ensure that chemotherapeutic agents are only released at the tumor site, thereby reducing systemic effects.

12.5.2 Drug Loading

Depending on drug characteristics (size, hydrophobicity, charge) and microbubble properties, one method of incorporation into the microbubble may be more suitable than others. Although hydrophobic drugs can be incorporated directly to microbubble shells through hydrophobic interactions, there are many instances in which this method alone does not provide the necessary effective payload. Unger et al. (1998) circumvented this problem by solubilizing paclitaxel into soybean oil and by using the drug-loaded oil as a medial layer between the gas core of the microbubble and the lipid shell. Hettiarachchi et al. (2009) used a double emulsion microfluidic device design to create a gas-in-oil-in-water microbubble with paclitaxel dissolved in the oil phase. Each of these methods increased the payload of drug per microbubble, allowing for release of a more effective dose from the microbubble.

Hydrophilic drugs are less amenable to such methods of loading, as they generally prefer to leave hydrophobic spaces and solubilize into the surrounding water-based solution. Various approaches have been used to encapsulate the hydrophilic drug, doxorubicin. One common technique is through tethering of smaller drug carrying particles to the microbubble instead of through hydrophobic interaction between the drug and microbubble shell, as with hydrophobic drugs. Lentacker et al. loaded doxorubicin into aqueous cores of liposomes. Then, by using an avidin–biotin linkage, they tethered these drug loaded liposomes onto gas-filled microbubbles (Lentacker et al. 2010). Geers et al. (2011) performed a similar experiment, but tethered doxorubicin-loaded liposomes to microbubbles using a thiol–maleimide linkage. Once microbubbles are ruptured, these drug-loaded liposomes are released, and the drug payload is released either by diffusion or through being endocytosed by target cells. Other research has attempted fabricating microbubbles in high concentrations of drug solution to load the microbubble through use of a concentration gradient. Villa et al. (2013) used a polyvinyl alcohol (PVA) shelled microbubble with a chitosan brush coating to load doxorubicin by dialyzing microbubbles in the presence of high concentration. Once microbubbles were removed from the drug solution, the chitosan brush acted as a physical barrier to diffusion, slowing down drug leakage from the microbubble. Fabiilli et al. dissolved drugs into water and created a water-in-gas emulsion by mixing drug-doped water with gas. This gas was then used to form the core of a microbubble, creating an acoustically active microbubble, with drug loaded into the gas core of the microbubble (Fabiilli et al. 2010).

Other properties, such as charge, have also been used to incorporate drugs into microbubbles. Tinkov et al. (2010b) used anionic phospholipids to form microbubbles and electrostatically bound doxorubicin, which has a positive charge in physiological conditions, to the surface of the microbubbles. With the wide variety of possible characteristics to manipulate for drug incorporation, microbubbles offer a versatile palette for incorporation and delivery of drugs.

12.5.2.1 US-Triggered Anti-cancer Drug Delivery

Most anti-cancer drug delivery techniques use US to trigger drug release through microbubble collapse. Cochran et al. used scanning electron microscopy to compare the morphology of drug containing polymer-shelled microbubbles after fabrication (Fig. 12.5a) to the morphology after ultrasound destruction (Fig. 12.5b). The change in morphology confirmed microbubble destruction due to ultrasound. They also found that with destruction of these drug-loaded microbubbles, there was an increase in the amount of drug released (Fig. 12.5c) (Cochran et al. 2011).

Additionally, US sonoporation increases effectiveness of the chemotherapy agent by increasing drug accumulation in nearby cells. Lin et al. showed that by adding microbubbles with focused ultrasound and freely dissolved drug, the ability of doxorubicin to reduce tumors in mice was enhanced. With doxorubicin alone, tumor size was consistently larger than when treated in conjunction with microbubbles and ultrasound (Fig. 12.6a). They found that addition of US and microbubbles

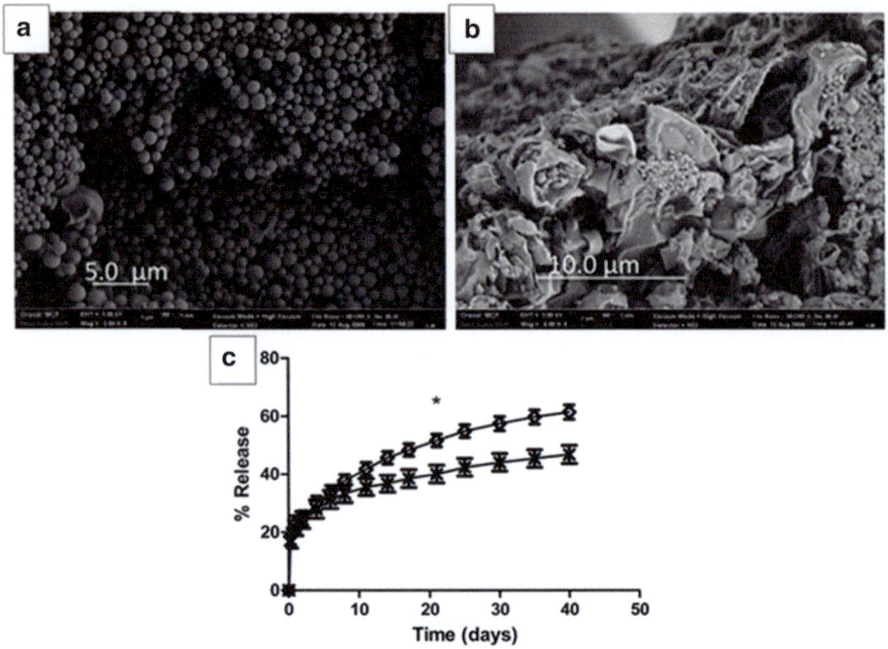

Fig. 12.5 Scanning electron images of drug loaded microbubbles before (**a**) and after (**b**) ultrasound exposure. (**c**) The *graph* shows that the insonated drug-loaded microbubbles [*diamond*] released more drugs than non-insonated microbubbles [*cross*] (*$p<0.0374$). Reprinted with permission from Elsevier (Cochran et al. 2011)

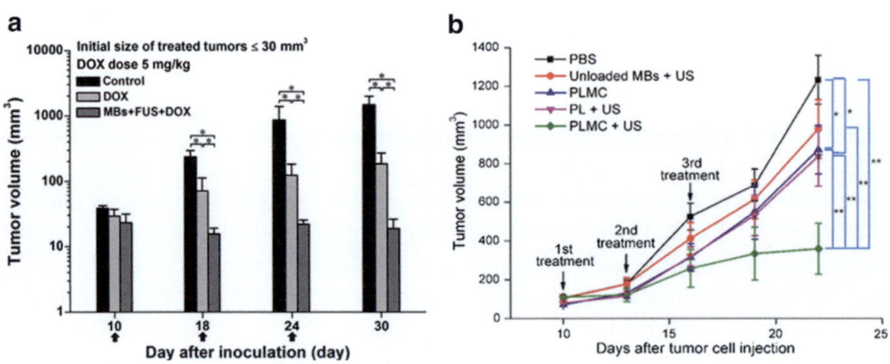

Fig. 12.6 (**a**) Effect of initial size of treated tumors on tumor growth response for treatments with 5 mg/kg of DOX. The *arrows* indicate schedule for treatments (Lin et al. 2012). (**b**) 4T1-tumor bearing mice were treated with PBS (*squares*, PBS only control), unloaded MBs (*circles*, unloaded MBs + US), PLMC (*upwards triangles*, PLMC without US), PL (*downwards triangles*, PL + US), and PLMC (*diamonds*, PLMC + US) on day 10, day 13, and day 16 after tumor cell injection. Systemic therapy with PLMC resulted in significantly greater tumor growth inhibition vs. PBS control, PLMC without US, unloaded MBs, and PL under US exposure. The results represent the mean ± SD, $n=6$. *$p<0.05$, **$p<0.01$. PTX = paclitaxel. Reprinted with permission from Elsevier (Yan et al. 2013)

led to nearly a tenfold more effective treatment over free drug alone. This indicated that the combination of ultrasound and microbubble significantly increased the amount of drug that accumulated in tumor tissues (Lin et al. 2012). Using drugs that were actually attached to the microbubble itself, Yan et al. showed that using PTX-liposome–microbubble complexes (PLMC) for breast tumor treatment reduced tumor sizes in mice. When using just PLMC, the tumor was able to grow to 872.46 ± 125.82 mm^3. However, addition of ultrasound reduced final tumor size to only 360.01 ± 131.24 mm^3 in volume, indicating that ultrasound could be used to trigger localized release of paclitaxel even when anchored to the microbubble (Yan et al. 2013). Figure 12.6b summarizes in vivo work done and shows effectiveness of these PLMCs when ultrasound is applied in addition to PLMCs alone. We will further discuss the mechanism by which sonoporation increases effectiveness of drug delivery later in the chapter.

12.5.3 Gene Therapy

Gene therapy is a promising alternative for treatment of a diverse array of cancers. Through targeting and suppression of oncogenes or insertion of apoptotic genes, gene therapy has become an intense area of research due to the possibility of avoiding severe side effects associated with traditional chemotherapy agents. However, naked nucleic acids are quickly degraded by nuclease activity or cleared from the tumor area, posing a difficult challenge to delivery (Roth and Cristiano 1997). In an attempt to improve gene transfection and prolong circulation, both viral and non-viral vectors have been evaluated for gene delivery. Viral vectors, such as retrovirus and herpes simplex virus (HSV), delivered systemically result in transduction of non-target tissues leading to adverse side effects. Additionally, viral vectors also elicit an immune response that can limit their effectiveness and prevent repeated use of the vectors (El-Aneed 2004). Non-viral vectors currently suffer from low transfection efficiency, where the amount of genetic material delivered fails to result in a functional effect (Niidome and Huang 2002). As a result of these difficulties, current gene therapy treatments remain in clinical or preclinical testing.

12.5.3.1 Gene Therapy Loading

Incorporation of oligonucleotides directly into the microbubble not only protects genetic material from degradation but also improves spatial targeting of the genetic payload through selective microbubble bursting. The most common method of incorporating oligonucleotides to the microbubble is through electrostatic interactions. Because of the negative charge of DNA, inclusion of positively charged species into or on the microbubble shell can aid in loading a nucleotide payload. In a comparison of cationic and control neutral charged microbubbles, Panje et al. (2012) showed that cationic lipid microbubbles were able to bind plasmid DNA with a

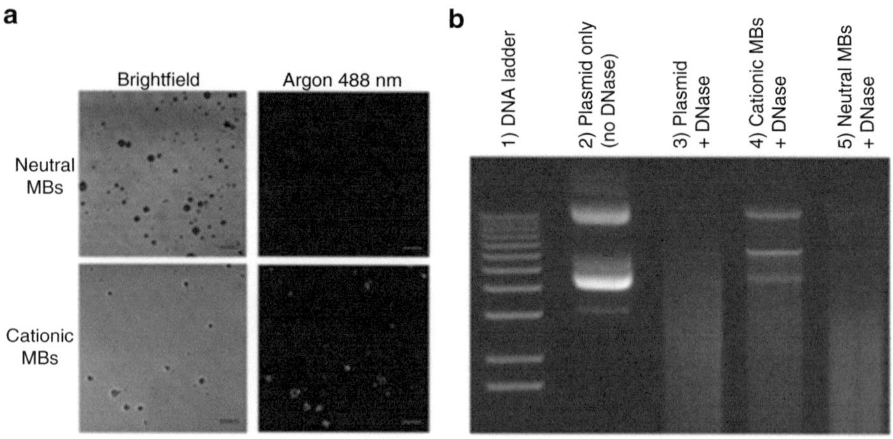

Fig. 12.7 Binding of plasmid DNA to cationic MBs and DNase protection assay. (**a**) Direct binding of YOYO-1-labeled plasmid DNA (10 μg) to cationic MBs (1 × 108) but not neutral MBs (1 × 108) was confirmed visually by confocal microscopy (1,000×, scale bar = 10 μm). (**b**) Gel electrophoresis confirmed partial protection of plasmid DNA bound to cationic MBs (lane 4) from DNase degradation. Plasmid DNA mixed with and not bound to neutral MBs (lane 5) were completely degraded after DNase treatment. Lane 2 represents control samples containing only plasmid DNA without MBs that were not exposed to DNase. Lane 3 shows complete DNase degradation of plasmid DNA not incubated with either MB type. Original figure from Panje CM, et al. Ultrasound-Mediated Gene Delivery with Cationic Versus Neutral Microbubbles: Effect of DNA and Microbubble Dose on In Vivo Transfection Efficiency. Theranostics 2012; 2(11):1078–1091. http://www.thno.org/v02p1078.htm. Reprinted with permission from Theranostics Publishing (Panje et al. 2012)

capacity that was 7.5-fold higher than neutral microbubbles. As shown in Fig. 12.7, this DNA binding by cationic microbubbles shielded DNA from DNase activity better than neutral microbubbles or free plasmid, as indicated by gel electrophoresis.

As an alternative to including cationic lipid in the microbubble shell, cationic polymers can also be used to coat the surface of the microbubble with complexed oligonucleotides. By using the layer-by-layer (LbL) assembly technique to attach multiple layers of DNA and poly-L-lysine (PLL) onto preformed lipid-coated microbubbles, Borden et al. (2007) showed that a tenfold increase in DNA loading capacity was possible. Similarly, Lentacker et al. coated albumin-shelled microbubbles with poly(allylamine hydrochloride) (PAH) and showed minimal changes in microbubble size distributions and ultrasonic function. Additionally, the cationic coating yielded increased binding of DNA to microbubbles and helped to protect bound DNA from nuclease degradation (Lentacker et al. 2006).

12.5.3.2 Gene Therapy Treatment

Two major types of gene therapies involve use of either DNA or short interfering RNA (siRNA). The delivery of DNA seeks to up-regulate expression of an immune response marker or suicide gene in tumor cells. Cells expressing HSV thymidine kinase (TK) are easily targeted and eliminated by treatment with ganciclovir.

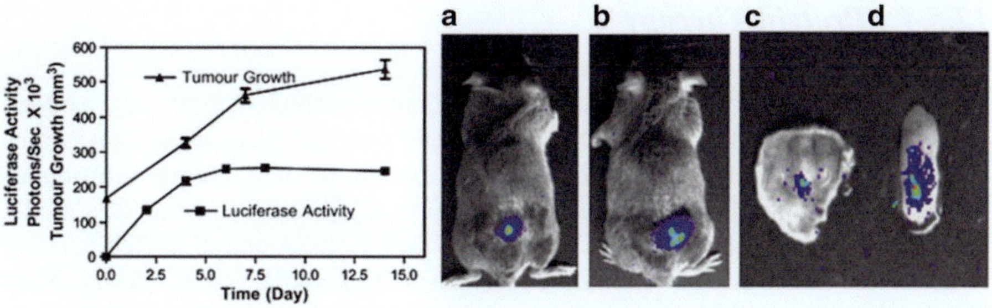

Fig. 12.8 Demonstration of prolonged luciferase activity in mouse hind leg muscle tumors after delivery of luciferase reporter, pEIP-1-Luc, following ultrasound-mediated and MB101®-assisted transfection in vivo. (**a**) The signal obtained from an animal at day 4 following transfection and (**b**) the same animal at day 20 following transfection. (**c** and **d**) The luminescent signal generated by the skin overlying the tumor and excised tumor, respectively. Reprinted with permission from Elsevier (Li et al. 2009)

Using a mouse model of squamous cell carcinoma, tumors sonoporated in the presence of microbubbles loaded with TK displayed longer doubling times ($p<0.02$), and TK-treated tumors displayed increased apoptosis ($p<0.04$) and more areas of cellular dropout ($p<0.03$) (Carson et al. 2011). Further studies additionally demonstrated effective in vivo delivery of DNA genes to cancer tumors. Using a commercially available microbubble, Sonidel MB101 microbubble (Sonidel Ltd, Ireland), Li et al. optimized cell transfection in vitro and demonstrated transfection efficiencies of up to 18 % using ultrasound. After optimization, the group used episomal plasmid-based gene with a luciferase reporter to deliver DNA in mouse hind leg muscle tumors. Luciferase activity, indicating gene expression, did not diminish during tumor growth and suggested plasmid replication during tumor expansion (Fig. 12.8) (Li et al. 2009).

Short interfering RNA (siRNA) are double stranded RNA oligonucleotides of approximately 20–25 kb in length and are complementary in sequence to an area of a gene of interest. siRNA bind to a protein called RISC (RNA-induced silencing complex) which uses RNase to cleave mRNA strands that match the complementary siRNA strand (Martinez et al. 2002). Delivery of siRNA targeted to oncogenes can be used to stop expression of key proteins and reduce tumor sizes (Urban-Klein et al. 2005). Using a GFP expressing rat C166 cell line, Kinoshita and Hynynen showed that 11 % of cells showed reduced GFP expression 48 h post-sonoporation with continuous wave ultrasound (Kinoshita and Hynynen 2005). Delivery of microbubbles loaded with epidermal growth factor receptor (EGFR) directed siRNA to murine squamous carcinoma cells in vivo resulted in three to fivefold increase in the doubling time of treated tumors compared with controls (Carson et al. 2012). No EGFR knockdown was observed in non-sonoporated tissues (i.e., heat, lungs, etc.), and there was no detectable increase of cytokines 24 h post-injection, demonstrating selectivity of the treatment to the targeted area. While the types of tumors suitable for targeting by siRNA/microbubble therapy can be limited due to US scattering by various tissues, further effectiveness of siRNA transfection may be achieved through optimization of acoustic parameters and bubble response.

12.5.4 Protein Therapy

Even though the majority of the literature using microbubble delivery of a payload pertains to drugs smaller than 2,000 Da or genetic material, there are several studies using microbubbles for protein delivery by exploiting their unique ultrasound-responsive property. Bekeredjian et al. incorporated luciferase enzyme as a model protein drug into the lipid shell of a microbubble for ultrasound-mediated delivery to the heart (Bekeredjian et al. 2005) or testes (Bekeredjian et al. 2007). They showed that protein delivery into the organ is augmented due to ultrasonic destruction via in vivo studies.

12.5.4.1 Therapeutic Protein Loading

Because of their size and availability of potential linkages, protein loading commonly involves covalent attachment between protein and the surface of microbubble. An amide bond between an amino group on the protein ligand can be formed between a carboxyl group or a maleimide on the microbubble shell with a protein thiol group to load the microbubble surface. The required moieties on the microbubble shell are generally incorporated via functionalized end groups of poly(ethylene glycol).

12.5.4.2 Current Protein Therapy Uses

As an alternative to traditional drug action, protein-conjugated microbubbles can act to modulate cell activity. For antigen delivery during vaccination, adjuvants are required to promote immunogenicity. Microbubbles coated with surface proteins can act as an antigen delivery system without the need for ultrasound application. Bioley et al. (2012) used covalent bonds between maleimide molecules on the microbubbles and thiol groups from the antigen to coat the microbubble surface. Microbubbles could have a direct effect on structural and immune cells of the lymph node, which are primary sites of adaptive immunity induction, because they have been shown to directly drain from peripheral administration sites to afferent lymph node (Sever et al. 2012).

12.6 Targeted Therapies Using Microbubbles

Microbubbles have been used in the clinic as ultrasound imaging contrast agents for echocardiography since 1997 in the United States. In combination with ultrasound, microbubbles have shown potential for molecular level imaging and delivery of drugs or genes to cells or tissue. While these advances are promising, microbubbles for molecular imaging or therapeutic purposes for the most part remain in the preclinical stage of development.

12.6.1 Targeting Methods

Tissue and cell specificity can be achieved by using passively or actively targeted microbubbles. Passive targeting refers to taking advantage of the body's intrinsic mechanism, phagocytosis, and inherent properties of the microbubble shell to use their affinity for a specific cell type or tissue. For example, albumin-shelled microbubbles were shown to bind to adherent leukocytes via cell-surface integrins or complement-mediated opsonization solely due to their shell composition (Ferrara et al. 2007; Kiessling et al. 2009; Deshpande et al. 2010). On the other hand, active targeting refers to covalent or noncovalent attachment of specific targeting moieties to the microbubble shell, to allow for binding to specific receptors. Most microbubble shell composition consists of lipid; thus, by using polymer-attached lipid, targeting moieties can be tethered at the end of the polymer brush. The polymer brush plays a role as a spacer between the ligand and monolayer shell for the ligand to bind its receptor efficiently (Kim et al. 2000). The method of linking targeting agents to the polymer brush is analogous to tethering of drug-carrying particles, which we discussed in the previous sections. For non-covalent attachment of ligand, the biotin–avidin linkage is the most widely used technique (Fig. 12.9).

Covalent attachment includes creating an amide bond between an amino group on the ligand and a carboxyl group on the microbubbles or, alternatively, using a maleimide on the microbubble shell to react with a thiol group on the ligand.

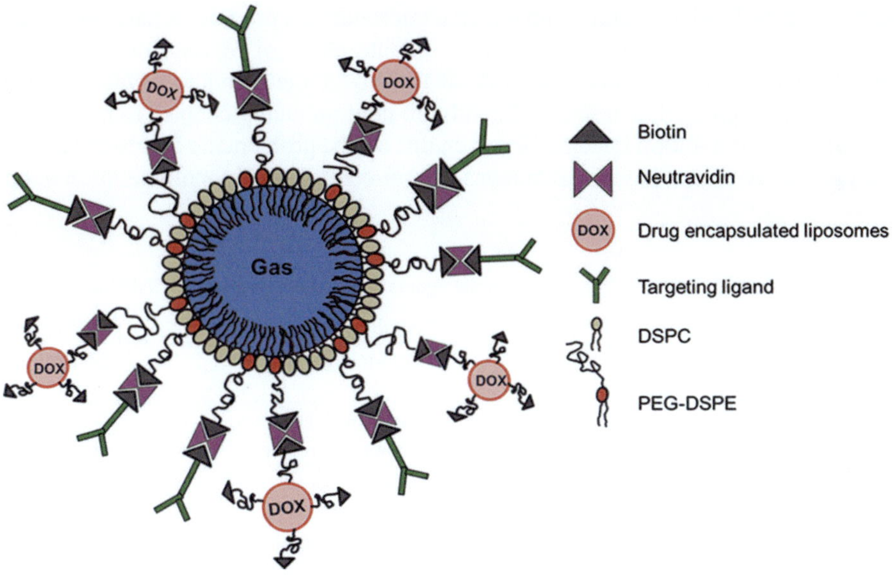

Fig. 12.9 Targeted drug-loaded microbubbles via noncovalent biotin–avidin linkage

12.6.2 Preclinical Applications of Targeted Microbubbles

There are 52 targeted ultrasound agents up to date (June 27, 2013) in the Molecular Imaging and Contrast Agent Database (MICAD) collected by the National Center for Biotechnology Information (NCBI), at the National Institutes of Health (NIH). Twenty-one of these have been developed for diagnosing and treating tumor angiogenesis, while 18 have been developed for inflammation in preclinical models. Table 12.1 summarizes targeted microbubbles that are currently involved in clinical trials. The most used molecular markers for assessment of tumor angiogenesis are vascular endothelial growth factor receptor 2 (VEGFR2) and $\alpha_v\beta_3$ integrin. Willmann et al. (2008) studied the effect of VEGFR2-targeted microbubbles on quantitative video intensity of ultrasound contrast enhanced molecular imaging in angiosarcoma and malignant glioma mice models. The average video intensity was significantly higher when using VEGFR2 targeted microbubbles vs. control immunoglobulin G2 antibody microbubbles ($p<0.001$).

Commonly used targets for detecting and treating diseases by inflammation, such as atherosclerosis or myocardial ischemia, are VCAM-1 (vascular cell adhesion molecule 1), ICAM-1 (intercellular adhesion molecule 1), E-selectin, and P-selectin. Yan et al. studied the degree of myocardial inflammation and levels of ICAM-1 protein using ultrasound molecular imaging with ICAM-1 targeted microbubbles, as well as routine non-invasive methods including electrocardiography, echocardiography, and plasma troponin I levels (Yan et al. 2011). The video intensity of ICAM-1 molecular images of the ischemic anterior wall was almost three times greater than that in the non-ischemic posterior wall during the late phase (8–24 h) of reperfusion. In contrast, routine methods yielded only weak evidence of ischemia, suggesting the ultrasound molecular imaging with targeted microbubbles provides reliable evidence of a recent myocardial ischemic event in the late phase of reperfusion. Also they investigated binding efficiency and rate of ICAM-1-targeted microbubbles in vitro (Fig. 12.8). In preclinical research, ligands are predominantly coupled to microbubbles via an avidin–biotin linkage. However, this can be problematic because of inherent immunogenicity of avidin in humans. Biotin-conjugated lipopolymer present on

Table 12.1 Typical targeted ultrasound contrast agents tested in vivo (from MICAD list)

Target	Application	PubMed ID
Vascular endothelial growth factor receptor 2 (VEGFR2)	Tumor angiogenesis	22787696
Alpha(v)beta(3) integrin	Tumor angiogenesis	21204315
Endoglin	Tumor angiogenesis	21977535
Vascular cell adhesion molecule-1 (VCAM-1)	Atherosclerosis	20641951
Intercellular adhesion molecule-1 (ICAM-1)	Atherosclerosis	20641950
P-selectin	Inflammation, tissue injury, and other endothelial responses	20641954
Matrix-metalloproteinase-2 (MMP-2)	Myocardial ischemia	22132430
Glycoprotein IIb/IIIa receptor (GPIIb/IIIa receptor)	Thrombosis	20641949

microbubbles activates the complement system in humans and mice (Borden et al. 2008). The first formulation for clinical application is BR55, that is, avidin–biotin-free soft shell microbubbles that target mouse and human VEGFR2. BR55 was found to strongly bind to the tumor endothelium of breast (Pochon et al. 2010) and prostate cancer xenografts in rodents (Tardy et al. 2010).

12.7 Mechanisms for Delivery of Therapeutic Agents Using Microbubbles and Ultrasound

As discussed previously, drug-loaded microbubbles are more effective when in combination with US. Therefore, optimization of the mechanical response for stable microbubble cavitation under sonoporation is a key factor in increasing the potential of therapeutic delivery. Traditional systemic drug delivery of small molecules and passive drug delivery particles rely on tumor enhanced permeability and retention (EPR) effect to perfuse tumor tissues (Maeda 2001). Due to increased angiogenesis and nutrient demand of tumor aggregates, high molecular weight molecules and particles are trapped in the defective vascular architecture near the tumor and remain there due to impaired lymphatic drainage/recovery system. This effect is commonly referred to as the EPR effect. However, small molecules are able to diffuse away from the tumor site and are unaffected by EPR (Maeda et al. 2000). Additionally, some tumors are as far as 100 μm from vasculature or are located in hypoxic regions, which remain difficult to reach from systemic delivery (Helmlinger et al. 1997). Stable microbubble cavitation is able to increase extravasation of small molecules from the vasculature to tumor sites. Using a flow-through tissue-mimicking phantom composed of a biocompatible hydrogel with embedded tumor cells, luciferase expression was shown to increase by 60-fold even for modest microbubble concentrations undergoing inertial cavitation (bubble collapse) when delivering a non-replicating luciferase-expressing adenovirus (Arvanitis et al. 2011). Using commercially available SonoVue, none of the insonation conditions induced cell death under ultrasound alone. However, extravasation from the tissue phantom vessel was shown to exhibit short-range vessel disruption during stable cavitation and formation of micro-channels during inertial cavitation. The findings suggest that enhanced drug delivery extravasation of anticancer agents through the use of microbubbles is possible.

Sonoporation alone has been shown to increase therapeutic retention in cells, indicating the role of acoustic energy in effecting cell permeability. In gene therapy, sonoporation of various cancer cell lines (U937, HeLa, PC-3, Meth-A, and T-24) in the presence of a reporter gene was shown to significantly increase reporter gene expression over both electrotransfection and liposomal transfection without ultrasound (Feril et al. 2006). In vivo ultrasound stimulation of prostate tumor cells implanted subcutaneously in rats induced a tenfold increase of a β-galactosidase reporter measured via histology and a 15-fold increase of β-galactosidase protein expression via ELISA (Huber and Pfisterer 2000). However, at these low frequencies (<20 kHz), prolonged ultrasound exposure has been demonstrated to result in cell death and tissue damage (Miller et al. 2002). Furthermore, these low frequencies are not in the range of clinical use ultrasound. Increasing the applied frequency to the clinical range of 1–3 MHz or using pulse-mode ultrasound reduces tissue damage but also results in lowered transfection (Hallow et al. 2006). One way to increase the efficiency of high frequency ultrasound sonoporation is through combination with microbubbles.

Furthermore, US and microbubbles have been used as tools to increase blood–brain-barrier permeability of drugs and drug delivery vehicles. Sheikov et al. showed that when US is applied in the presence of microbubbles, the acoustic energies stimulate active vesicular transport and temporarily disrupt tight junctions to allow previously size occluded drugs to pass and treat the diseased areas (Sheikov et al. 2004, 2008). Treat et al. (2012) used liposomal doxorubicin and focused US to increase permeability of the blood–brain barrier and showed a greater decrease in brain tumor size when compared with administration of free doxorubicin alone. Some groups have even used the microbubble cavitation phenomenon to accelerate lysis of thrombus and blood clots (Porter and Xie 2001; Porter et al. 1996; Culp et al. 2003).

12.8 Summary

There is a large body of literature for use of various microbubble formulations in imaging and treatment of oncological diseases. The purpose of this chapter was to introduce the reader to the basic form and function of microbubbles and to discuss current work that have laid the foundation for the use of microbubbles as contrast agents, drug delivery vehicles, and theranostic particles. The articles reviewed here only constitute a small fraction of the literature on this topic, and there is still work being done in order to move these technologies toward clinical use. More advanced studies in vivo need to be performed, and methods of optimization of acoustic parameters of the ultrasound contrast agents in vivo are being developed to aid in truly realizing the full potential of microbubbles in the clinic. Through the combination of these hybrid drug-loaded contrast agents and molecular targeting, a complete theranostic tool may be developed that would allow a clinician to potentially image, locate, and treat localized tumors with just one injection of these multipurpose particles. Furthermore, the spatial control given to clinicians over the delivery of therapeutic agents directly to the tumor site will minimize damage done to surrounding healthy tissues and has the potential to prevent side effects that currently plague chemotherapy. Using specific targeting agents and combination of multi-image modalities, microbubble platforms may one day have the capability of reaching the tumor site regardless of its location in the body. The further development of these systems could have a considerable impact on treatment and diagnosis of not only cancer but also on numerous other diseases and disorders.

Acknowledgments J. Y. W. acknowledges support from Boston University's College of Engineering Distinguished Faculty Fellowship, T. P. acknowledges support from the NIH/NIGMS T32 GM008764 and the NIH/NIAID T32 AI089673, and C. B. acknowledges support from the National Science Foundation Graduate Research Fellowship Program.

References

Arvanitis CD et al (2011) Cavitation-enhanced extravasation for drug delivery. Ultrasound Med Biol 37(11):1838–1852
Bekeredjian R et al (2005) Augmentation of cardiac protein delivery using ultrasound targeted microbubble destruction. Ultrasound Med Biol 31(5):687–691
Bekeredjian R et al (2007) Ultrasound-targeted microbubble destruction augments proteins delivery into testes. Urology 69(2):386–389
Bioley G et al (2012) Gas-filled microbubble-mediated delivery of antigen and the induction of immune responses. Biomaterials 33(25):5935–5946
Blomley MJ et al (2001) Microbubble contrast agents: a new era in ultrasound. BMJ 322(7296):1222–1225
Borden MA et al (2007) DNA and polylysine adsorption and multilayer construction onto cationic lipid-coated microbubbles. Langmuir 23(18):9401–9408
Borden MA et al (2008) A stimulus-responsive contrast agent for ultrasound molecular imaging. Biomaterials 29(5):597–606
Carson AR et al (2011) Gene therapy of carcinoma using ultrasound-targeted microbubble destruction. Ultrasound Med Biol 37(3):393–402
Carson AR et al (2012) Ultrasound-targeted microbubble destruction to deliver siRNA cancer therapy. Cancer Res 72(23):6191–6199
Castle J et al (2013) Ultrasound-mediated targeted drug delivery: recent success and remaining challenges. Am J Physiol Heart Circ Physiol 304(3):H350–H357
Castro-Hernandez E et al (2011) Microbubble generation in a co-flow device operated in a new regime. Lab Chip 11(12):2023–2029
Chen SY et al (2006) Efficient gene delivery to pancreatic islets with ultrasonic microbubble destruction technology. Proc Natl Acad Sci U S A 103(22):8469–8474
Cochran MC et al (2011) Doxorubicin and paclitaxel loaded microbubbles for ultrasound triggered drug delivery. Int J Pharm 414(1–2):161–170
Cosgrove D (2004) The advances are significant improvements in both the microbubbles used as contrast agents and in the software that allows their selective detection. Eur Radiol 14(Suppl 8): P1–P3
Cosgrove D (2006) Ultrasound contrast agents: an overview. Eur J Radiol 60(3):324–330
Cui W et al (2005) Preparation and evaluation of poly(L-lactide-*co*-glycolide) (PLGA) microbubbles as a contrast agent for myocardial contrast echocardiography. J Biomed Mater Res B Appl Biomater 73(1):171–178
Culp WC et al (2003) Microbubble-augmented ultrasound declotting of thrombosed arteriovenous dialysis grafts in dogs. J Vasc Interv Radiol 14(3):343–347
Deshpande N, Needles A, Willmann JK (2010) Molecular ultrasound imaging: current status and future directions. Clin Radiol 65(7):567–581
Dove JD, Murray TW, Borden MA (2013) Enhanced photoacoustic response with plasmonic nanoparticle-templated microbubbles. Soft Matter 9(32):7743–7750
Duncanson WJ et al (2010) Targeted binding of PEG-lipid modified polymer ultrasound contrast agents with tiered surface architecture. Biotechnol Bioeng 106(3):501–506
El-Aneed A (2004) An overview of current delivery systems in cancer gene therapy. J Control Release 94(1):1–14
Fabiilli ML et al (2010) Delivery of water-soluble drugs using acoustically triggered perfluorocarbon double emulsions. Pharm Res 27(12):2753–2765
Fang JY et al (2007) A study of the formulation design of acoustically active lipospheres as carriers for drug delivery. Eur J Pharm Biopharm 67(1):67–75
Feril LB et al (2006) Optimized ultrasound-mediated gene transfection in cancer cells. Cancer Sci 97(10):1111–1114
Ferrara K, Pollard R, Borden M (2007) Ultrasound microbubble contrast agents: fundamentals and application to gene and drug delivery. Annu Rev Biomed Eng 9:415–447

Ganan-Calvo AM, Gordillo JM (2001) Perfectly monodisperse microbubbling by capillary flow focusing. Phys Rev Lett 87(27 Pt 1):274501

Geers B et al (2011) Self-assembled liposome-loaded microbubbles: the missing link for safe and efficient ultrasound triggered drug-delivery. J Control Release 152(2):249–256

Gunduz O et al (2012) Bioinspired bubble design for particle generation. J R Soc Interface 9(67):389–395

Hallow DM et al (2006) Measurement and correlation of acoustic cavitation with cellular bioeffects. Ultrasound Med Biol 32(7):1111–1122

Harvey CJ et al (2001) Developments in ultrasound contrast media. Eur Radiol 11(4):675–689

Hashimoto M et al (2008) Formation of bubbles and droplets in parallel, coupled flow-focusing geometries. Small 4(10):1795–1805

Helmlinger G et al (1997) Interstitial pH and pO(2) gradients in solid tumors in vivo: high-resolution measurements reveal a lack of correlation. Nat Med 3(2):177–182

Hettiarachchi K et al (2007) On-chip generation of microbubbles as a practical technology for manufacturing contrast agents for ultrasonic imaging. Lab Chip 7(4):463–468

Hettiarachchi K et al (2009) Controllable microfluidic synthesis of multiphase drug-carrying lipospheres for site-targeted therapy. Biotechnol Prog 25(4):938–945

Huber PE, Pfisterer P (2000) In vitro and in vivo transfection of plasmid DNA in the Dunning prostate tumor R3327-AT1 is enhanced by focused ultrasound. Gene Ther 7(17):1516–1525

Hynynen K et al (2001) Noninvasive MR imaging-guided focal opening of the blood–brain barrier in rabbits. Radiology 220(3):640–646

Janib SM, Moses AS, MacKay JA (2010) Imaging and drug delivery using theranostic nanoparticles. Adv Drug Deliv Rev 62(11):1052–1063

Kang J et al (2010) Antitumor effect of docetaxel-loaded lipid microbubbles combined with ultrasound-targeted microbubble activation on VX2 rabbit liver tumors. J Ultrasound Med 29(1):61–70

Kendall MR et al (2012) Scaled-up production of monodisperse, dual layer microbubbles using multi-array microfluidic module for medical imaging and drug delivery. Bubble Sci Eng Technol 4(1):12–20

Kiessling F, Huppert J, Palmowski M (2009) Functional and molecular ultrasound imaging: concepts and contrast agents. Curr Med Chem 16(5):627–642

Kiessling F et al (2012) Ultrasound microbubbles for molecular diagnosis, therapy, and theranostics. J Nucl Med 53(3):345–348

Kim DH, Klibanov AL, Needham D (2000) The influence of tiered layers of surface-grafted poly(ethylene glycol) on receptor-ligand-mediated adhesion between phospholipid monolayer-stabilized microbubbles and coated class beads. Langmuir 16(6):2808–2817

Kinoshita M, Hynynen K (2005) A novel method for the intracellular delivery of siRNA using microbubble-enhanced focused ultrasound. Biochem Biophys Res Commun 335(2): 393–399

Klibanov AL (1999) Targeted delivery of gas-filled microspheres, contrast agents for ultrasound imaging. Adv Drug Deliv Rev 37(1–3):139–157

Kwan JJ, Borden MA (2010) Microbubble dissolution in a multigas environment. Langmuir 26(9):6542–6548

Kwan JJ, Borden MA (2012) Lipid monolayer collapse and microbubble stability. Adv Colloid Interface Sci 183:82–99

Laurent S et al (2008) Magnetic iron oxide nanoparticles: synthesis, stabilization, vectorization, physicochemical characterizations, and biological applications. Chem Rev 108(6):2064–2110

Lee MH, Lee D (2010) Elastic instability of polymer-shelled bubbles formed from air-in-oil-in-water compound bubbles. Soft Matter 6(18):4326–4330

Lee H et al (2006) Antibiofouling polymer-coated superparamagnetic iron oxide nanoparticles as potential magnetic resonance contrast agents for in vivo cancer imaging. J Am Chem Soc 128(22):7383–7389

Lentacker I et al (2006) Microbubbles which bind and protect DNA against nucleases. J Control Release 116(2):E73–E75

Lentacker I et al (2007) Lipoplex-loaded microbubbles for gene delivery: a Trojan horse controlled by ultrasound. Adv Funct Mate 17(12):1910–1916
Lentacker I et al (2010) Design and evaluation of doxorubicin-containing microbubbles for ultrasound-triggered doxorubicin delivery: cytotoxicity and mechanisms involved. Mol Ther 18(1):101–108
Leong-Poi H et al (2007) Therapeutic arteriogenesis by ultrasound-mediated VEGF165 plasmid gene delivery to chronically ischemic skeletal muscle. Circ Res 101(3):295–303
Li YS et al (2009) Optimising ultrasound-mediated gene transfer (sonoporation) in vitro and prolonged expression of a transgene in vivo: potential applications for gene therapy of cancer. Cancer Lett 273(1):62–69
Lin CY et al (2012) Enhancement of focused ultrasound with microbubbles on the treatments of anticancer nanodrug in mouse tumors. Nanomedicine 8(6):900–907
Liu Z et al (2011) Iron oxide nanoparticle-containing microbubble composites as contrast agents for MR and ultrasound dual-modality imaging. Biomaterials 32(26):6155–6163
Maeda H (2001) The enhanced permeability and retention (EPR) effect in tumor vasculature: the key role of tumor-selective macromolecular drug targeting. Adv Enzyme Regul 41:189–207
Maeda H et al (2000) Tumor vascular permeability and the EPR effect in macromolecular therapeutics: a review. J Control Release 65(1–2):271–284
Martinez J et al (2002) Single-stranded antisense siRNAs guide target RNA cleavage in RNAi. Cell 110(5):563–574
Massoud TF, Gambhir SS (2003) Molecular imaging in living subjects: seeing fundamental biological processes in a new light. Genes Dev 17(5):545–580
Mesiwala AH, Mourad PD (2002) Monitoring of biologic effects of focused ultrasound beams on the brain. Radiology 224(1):294–296, author reply 296–7
Mesiwala AH et al (2002) High-intensity focused ultrasound selectively disrupts the blood–brain barrier in vivo. Ultrasound Med Biol 28(3):389–400
Miller DL, Pislaru SV, Greenleaf JE (2002) Sonoporation: mechanical DNA delivery by ultrasonic cavitation. Somat Cell Mol Genet 27(1–6):115–134
Niidome T, Huang L (2002) Gene therapy progress and prospects: nonviral vectors. Gene Ther 9(24):1647–1652
Nisisako T, Torii T (2008) Microfluidic large-scale integration on a chip for mass production of monodisperse droplets and particles. Lab Chip 8(2):287–293
Panje CM et al (2012) Ultrasound-mediated gene delivery with cationic versus neutral microbubbles: effect of DNA and microbubble dose on in vivo transfection efficiency. Theranostics 2(11):1078–1091
Paradossi G et al (2003) Poly(vinyl alcohol) as versatile biomaterial for potential biomedical applications. J Mater Sci Mater Med 14(8):687–691
Parhizkar M, Edirisinghe M, Stride E (2013) Effect of operating conditions and liquid physical properties on the size of monodisperse microbubbles produced in a capillary embedded T-junction device. Microfluid Nanofluidics 14(5):797–808
Park Y et al (2012) Tunable diacetylene polymerized shell microbubbles as ultrasound contrast agents. Langmuir 28(8):3766–3772
Phillips P, Gardner E (2004) Contrast-agent detection and quantification. Eur Radiol 14(Suppl 8): P4–P10
Pochon S et al (2010) BR55: a lipopeptide-based VEGFR2-targeted ultrasound contrast agent for molecular imaging of angiogenesis. Invest Radiol 45(2):89–95
Porter TR, Xie F (2001) Ultrasound, microbubbles, and thrombolysis. Prog Cardiovasc Dis 44(2):101–110
Porter TR et al (1996) Thrombolytic enhancement with perfluorocarbon-exposed sonicated dextrose albumin microbubbles. Am Heart J 132(5):964–968
Pramanik M et al (2008) Design and evaluation of a novel breast cancer detection system combining both thermoacoustic (TA) and photoacoustic (PA) tomography. Med Phys 35(6):2218–2223
Rapoport N, Gao Z, Kennedy A (2007) Multifunctional nanoparticles for combining ultrasonic tumor imaging and targeted chemotherapy. J Natl Cancer Inst 99(14):1095–1106

Romanowsky MB et al (2012) High throughput production of single core double emulsions in a parallelized microfluidic device. Lab Chip 12(4):802–807
Roth JA, Cristiano RJ (1997) Gene therapy for cancer: what have we done and where are we going? J Natl Cancer Inst 89(1):21–39
Sarkar K, Katiyar A, Jain P (2009) Growth and dissolution of an encapsulated contrast microbubble: effects of encapsulation permeability. Ultrasound Med Biol 35(8):1385–1396
Sciallero C, Trucco A (2013) Ultrasound assessment of polymer-shelled magnetic microbubbles used as dual contrast agents. J Acoust Soc Am 133(6):El478–El484
Sever AR et al (2012) Sentinel node identification using microbubbles and contrast-enhanced ultrasonography. Clin Radiol 67(7):687–694
Sheikov N et al (2004) Cellular mechanisms of the blood–brain barrier opening induced by ultrasound in presence of microbubbles. Ultrasound Med Biol 30(7):979–989
Sheikov N et al (2008) Effect of focused ultrasound applied with an ultrasound contrast agent on the tight junctional integrity of the brain microvascular endothelium. Ultrasound Med Biol 34(7):1093–1104
Shortencarier MJ et al (2004) A method for radiation-force localized drug delivery using gas-filled lipospheres. IEEE Trans Ultrason Ferroelectr Freq Control 51(7):822–831
Stride E, Saffari N (2003) Microbubble ultrasound contrast agents: a review. Proc Inst Mech Eng H 217(6):429–447
Talu E et al (2007) Tailoring the size distribution of ultrasound contrast agents: possible method for improving sensitivity in molecular imaging. Mol Imaging 6(6):384–392
Tardy I et al (2010) Ultrasound molecular imaging of VEGFR2 in a rat prostate tumor model using BR55. Invest Radiol 45(10):573–578
Tinkov S et al (2010a) New doxorubicin-loaded phospholipid microbubbles for targeted tumor therapy: in-vivo characterization. J Control Release 148(3):368–372
Tinkov S et al (2010b) New doxorubicin-loaded phospholipid microbubbles for targeted tumor therapy: Part I – Formulation development and in-vitro characterization. J Control Release 143(1):143–150
Treat LH et al (2012) Improved anti-tumor effect of liposomal doxorubicin after targeted blood–brain barrier disruption by MRI-guided focused ultrasound in rat glioma. Ultrasound Med Biol 38(10):1716–1725
Unger EC et al (1998) Acoustically active lipospheres containing paclitaxel: a new therapeutic ultrasound contrast agent. Invest Radiol 33(12):886–892
Unger EC et al (2001) Local drug and gene delivery through microbubbles. Prog Cardiovasc Dis 44(1):45–54
Unger EC et al (2004) Therapeutic applications of lipid-coated microbubbles. Adv Drug Deliv Rev 56(9):1291–1314
Urban-Klein B et al (2005) RNAi-mediated gene-targeting through systemic application of polyethylenimine (PEI)-complexed siRNA in vivo. Gene Ther 12(5):461–466
Villa R et al (2013) Targeted doxorubicin delivery by chitosan-galactosylated modified polymer microbubbles to hepatocarcinoma cells. Colloids Surf B Biointerfaces 110:434–442
Willmann JK et al (2008) US imaging of tumor angiogenesis with microbubbles targeted to vascular endothelial growth factor receptor type 2 in mice. Radiology 246(2):508–518
Xu MH, Wang LHV (2006) Photoacoustic imaging in biomedicine. Rev Sci Instrum 77(4):1–22
Xu JH et al (2012) Controllable gas/liquid/liquid double emulsions in a dual-coaxial microfluidic device. Lab Chip 12(11):2029–2036
Yan Y et al (2011) Late-phase detection of recent myocardial ischaemia using ultrasound molecular imaging targeted to intercellular adhesion molecule-1. Cardiovasc Res 89(1):175–183
Yan F et al (2013) Paclitaxel-liposome-microbubble complexes as ultrasound-triggered therapeutic drug delivery carriers. J Control Release 166(3):246–255
Yang F et al (2009) Superparamagnetic iron oxide nanoparticle-embedded encapsulated microbubbles as dual contrast agents of magnetic resonance and ultrasound imaging. Biomaterials 30(23–24):3882–3890

Chapter 13
Multimodal Nano-Systems for Cancer Diagnosis, Imaging, and Therapy

Amit Singh, Adwait J. Oka, Purva Pandya, and Mansoor M. Amiji

13.1 Cancer and the Role of Nanotechnology

Cancer remains the most diverse and intriguing human disease due to the remarkable capability of tumor cells to adapt to their environment at genetic and physiological level. The obstacles to target cancer as a disease are often multi-layered and involve several underlying phenomenona such as ever changing genetic composition of proliferative cells, heterogeneity in cell population within a tumor with incredible level of variability, resistance to chemotherapy intervention, and presence of a complex tumor microenvironment. Concurrently, other aspects to therapeutic challenges involve our limited capability in early detection of onset of cancer, poor bioavailability of existing cancer drugs due to poor solubility, presence of host reticulo-endothelial system (RES), and limited penetration into the complex tumor milieu. The burgeoning field of nanotechnology and nanomedicine however has presented promising approaches to circumvent some of the challenges associated to cancer diagnosis, imaging, and therapy and will be discussed in greater detail in this chapter.

13.1.1 Challenges in Cancer Therapy

Neoplastic cells exhibit some properties, often referred to as *hallmark* of cancer, which includes sustained signal for cell proliferation, evasion from effect of growth/tumor suppressing molecules, resistance to cell death, acquired cell immortality,

A. Singh • A.J. Oka • P. Pandya • M.M. Amiji (✉)
Department of Pharmaceutical Sciences, School of Pharmacy, Bouvé College of Health Sciences, Northeastern University, 140 The Fenway Building, Boston, MA 02115, USA
e-mail: m.amiji@neu.edu

M.J. Alonso, M. Garcia-Fuentes (eds.), *Nano-Oncologicals: New Targeting and Delivery Approaches*, Advances in Delivery Science and Technology,
DOI 10.1007/978-3-319-08084-0_13

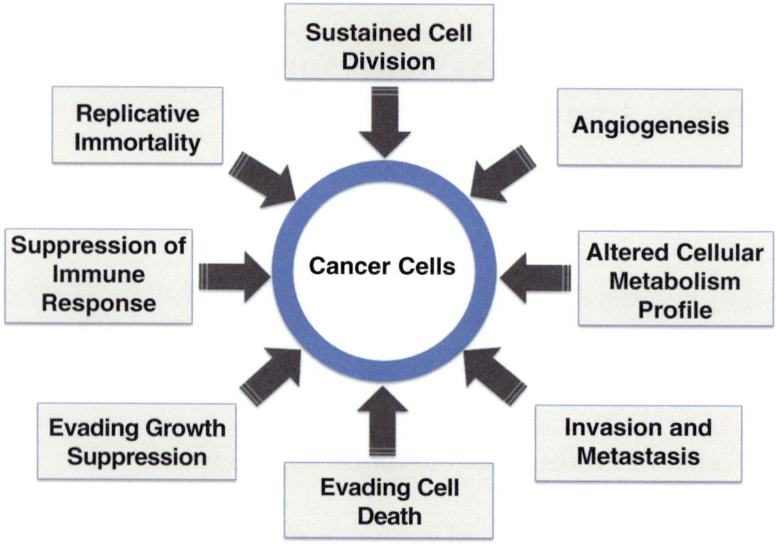

Fig. 13.1 Scheme showing the hallmark properties of cancer cells

angiogenesis and most importantly, ability to invade and metastasize to distant sites in the body (Fig. 13.1) (Hanahan and Weinberg 2011). These properties impart cancer cells the ability to thrive and grow in host body and metastasize. Molecular level analysis of cancer cells reveals complex genetic instabilities that result in alteration of physiological profile and thus aid in acquiring these hallmark properties. Phenotypic and functional heterogeneity within the cancer cells due to the acquired genetic instability and environment complexity makes it incredibly difficult to devise a universal approach to their therapy. Earlier attempts to treat cancer largely focused on chemotherapy aided by radiation therapy that demonstrated initial success in disease containment but since then has failed miserably due to onset of drug resistant. Drug resistance could be *intrinsic* where tumor cells show presence of resistance mediating factors even without being exposed to any treatment or it could be *acquired* where the presence of drug induces resistance by genetic mutations or selection of resistant minor subpopulation of cells in the tumor (Longley and Johnston 2005; Swanton 2012). The resistance in cancer tumors to the chemotherapy can be mediated by different ways such as expression of drug efflux pumps, mutation in the drug target, increased DNA repair mechanisms, activation of alternative signaling mechanisms, or evasion of cell apoptosis (Holohan et al. 2013).

ATP-binding cassette (ABC) transporter family of transmembrane proteins, including multidrug resistance-1 (MDR-1) and MDR associated protein-1 (MRP-1) among 49 other members, have been mostly implicated in regulating efflux of hydrophobic drug across the cell membrane resulting in resistance (Holohan et al. 2013). Drug inactivation, lack of drug activation, or alteration of drug target is another upstream resistance mechanism that is specific to a particular chemoagent or molecular target and contributes to resistance in cancer cells. These resistance mechanisms are regulated at molecular level by switching *on* or *off* of genes that would aid in countering the cytotoxic effect of drug or molecular target. Similarly, some drugs (such as platinum compounds) or molecular targets induce DNA damage, and the affected cells can either repair the incurred damage or undergo cell cycle arrest and subsequent cell death. Cancer cells often acquire a more sophisticated DNA damage repair mechanism, which results in an acquired resistance to such therapeutic intervention (Bouwman and Jonkers 2012). Similarly, several other downstream adaptive mechanisms confer drug resistance through apoptosis deregulation by altering the expression of anti-apoptotic genes, activation of pro-survival signaling, developing oncogenic bypass by activation of alternate pathways to negate a drug targeted pathway, or induction of epithelial to mesenchymal transition (EMT) (Holohan et al. 2013).

Tumor heterogeneity is another major contributor to limited success or complete failure of therapies that have been attempted in the past. It is often referred to as the presence of genetically *unique* subpopulation of cancer cells within the same tumor (*intra*-tumor heterogeneity) or between different tumors (*inter*-tumor heterogeneity). The origin of tumor heterogeneity is often attributed to two different lineages; clonal evolution and cancer stem cells. Rapid cell division is one of the hallmark features of the cancer cells, which along with genetic instability often results in mutant cells that could be mingled with normal cancer cells or spatially segregated. Genetic instability is the prime contributor of genotypic heterogeneity, which along with clonal selection due to environmental pressure can lead to significant variation in overall genetic, phenotypic, and functional characteristic of tumor and broadly influence the disease progression, choice of therapeutic intervention, and disease prognosis (Greaves and Maley 2012). The presence of a subpopulation of tumorigenic cancer stem cells and their non-tumorigenic progeny is further believed to add to tumor heterogeneity and influence disease growth by acquired resistance and metastasis (Shackleton et al. 2009).

The tumor microenvironment is another factor that has prominent impact on the tumor heterogeneity as well as drug resistance. Though tumor formation is largely due to neoplastic cells, these cells closely interact with other normal cells during the tumor progression and growth. Interaction of the cancer cells with the extra-cellular matrix and the bio-milieu thus formed is often referred to as tumor microenvironment. Normal fibroblasts, immune cells, and vasculature act as a barrier to tumorigenesis, but the complex interaction with cancer cells leads to their recruitment for promoting cancer. Besides, improper vasculature and poorly developed lymphatic drainage system in the tumor further leads to limited nutrient supply, hypoxic and acidic environment that causes variable bio-milieu thereby affecting drug penetration and activity within the tumor. The abundance of stromal cells, for example, has been found to have strong correlation with poor prognosis in several cancers including breast and pancreatic cancer, emphasizing the significant role that the matrix plays in the therapeutic outcome (Junttila and de Sauvage 2013). Figure 13.2 is a schematic representation of the various components of the tumor microenvironment including cancer-associated fibroblasts, inflammatory cells, blood vasculature properties, and

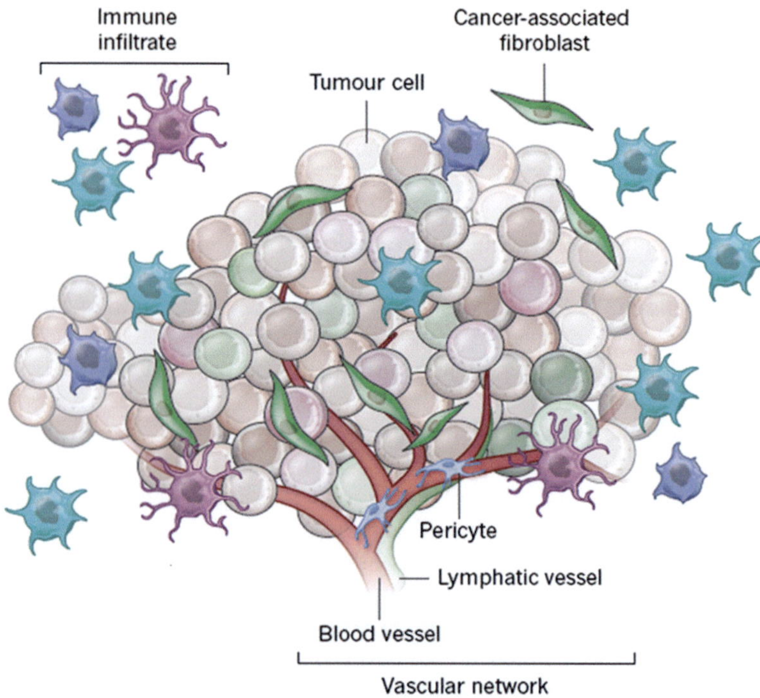

Fig. 13.2 Origins and influence of tumor heterogeneity [reprinted with permission from Junttila and de Sauvage 2013, Nature Publishing Group (NPG)]

variations in the cancer cells within the tumor. The cumulative effect of these complex phenomena within the tumor gives rise to an uphill challenge that needs to be addressed to develop an effective therapeutic strategy against treatment of cancer.

Given the tumor heterogeneity at the genetic, molecular, and functional level, the concept of *personalized* medicine is becoming extremely popular where the choice and course of therapy of an individual patient is customized based on their tumor properties. Molecular level sensitive, accurate, and rapid diagnostics therefore becomes extremely important to be able to characterize the pathways affecting tumor progression in an individual patient. The following section will discuss challenges in cancer diagnosis, need and benefits of early detection to cancer therapy, and the concept of "companion" diagnostics in cancer.

13.1.2 Challenges in Cancer Diagnostics

The poor prognosis and low overall 5-year survival rate in cancer patients undergoing therapy can be largely attributed to lack of diagnostic methods to detect the disease in its early stage of development, when the tumor growth is contained in its primary site of origin and cancer cells have not started to metastasize.

It is a well-acknowledged fact that the chance of successful treatment of a tumor is extremely high when it is localized in its primary stage of development. In a clinical setting though, the disease often is in its metastatic stage by the time diagnosis is made and therapy is initiated and thus the therapy is unable to reverse the course of disease progression. The debilitating outcome of late diagnosis can be visualized by the fact that in ovarian cancer patients, disease diagnosed in metastatic stage has a mere 40 % 5-year survival, while if the disease is diagnosed in early stage of development, the 5-year survival rate goes to nearly 90 % (Rosenthal and Jacobs 1998). Early detection of biological signatures for cancer onset has therefore become an area of extreme importance for researchers and clinicians alike. Conventional surgical procedure such as biopsy still remains the state-of-art approach since the histochemical and pathological studies give an accurate account of the disease. However, such methods are extremely laborious, time consuming and require expertise for successful execution. Recent endeavors have led to what can be called as "liquid biopsy" where patient's blood is analyzed for circulating tumor cells (CTCs) (Danila et al. 2011) or nucleic acid (Anker et al. 1999; Mitchell et al. 2008) that are shed into the circulation from the primary tumor.

CTCs specially have been actively pursued to this end because their accurate enumeration in the peripheral blood circulation strikingly correlates to the development stage of cancer (Cohen et al. 2008). Efficient capture and enumeration of CTCs however remains a daunting challenge due to their extremely low population in the blood. Typically, the CTC density is in the order of 1 cell every 10^5–10^7 blood cells and thus needs a very precise and specific capturing strategy. The problem is further compounded by the presence of circulating tumor microemboli (CTMs), which essentially are a group of CTCs that circulate in the blood. Due to the associated size discrepancy between CTCs and CTMs, different strategies might be needed to be able to capture and enumerate them independently to get a precise account of tumor burden. The isolation of CTCs and CTMs from blood has been achieved based on their physical properties (size, density, charge, etc.), use of surface biomarkers for positive selection [e.g., epithelial cell adhesion molecule (EpCAM)] or negative markers to remove CD45+ blood cells thereby enriching CTC population. However, significant advances are wanted in the existing technologies to translate them from the preclinical settings to the bedside. Advent of nanotechnology on the other hand has provided novel avenues to exploit materials properties for cancer therapy and detection, which have been discussed in detail with examples in the further sections of the chapter.

13.1.3 Nanotechnology in Cancer Diagnosis and Therapy

Besides the anatomical barriers within a tumor and lack of methodologies for early detection of onset of cancer, there are several other physical and physiological constraints that limit the efficiency of cancer detection and therapy. Conventional therapy for cancer has classically focused on surgical removal of tumor mass, an invasive method to reduce tumor burden, followed by application of anti-cancer drugs in combination with radiation therapy. Anti-cancer drugs are low molecular weight compounds that interfere and inhibit key physiological processes that are essential for cell survival. Such drugs however suffer from several limitations such as non-specificity, poor half-life in blood, and dose-dependent toxicity. In the absence of any targeting ability, these drugs are systemically circulated to every organ in the body and show an indiscriminating cytotoxic effect to cancerous and healthy cells alike. The use of imaging modalities such as radiolabelled molecules or contrast agents also show similar non-specific distribution pattern causing low signal-to-noise ratio and thus limiting the applicability of the imaging agents.

Fortunately, the unique features of tumor microenvironment that pose several challenges in cancer therapy also provide with a rescue option to address many of these challenges. One of the characteristic features of tumor is its highly permeable vasculature. More precisely, the blood vessels in tumor are characterized by highly proliferating endothelium, low concentration of pericyte cells, and abnormal basement membrane, rendering them *leaky* with very high permeability to macromolecules circulating in blood (Dvorak et al. 1988), and studies have suggested that their permeability cutoff could be as high as 400 nm (Yuan et al. 1995). This provides a unique opportunity to exploit the riches of nanotechnology to address some of the predicaments associated with cancer diagnostics and therapy. Careful design of nanoparticles of right size and characteristics would enable them to extravasate at the tumor site only and thus hold huge potential as delivery systems. Advancement in materials chemistry along with the development of sophisticated characterization instrumentation has aided in preparation of nanoparticle systems with controlled surface properties suitable for delivery application. Poly(ethylene glycol) (PEG) modified nanoparticles of right size tend to escape recognition by RES effectively, evade retention in perfused organs such as liver or kidney but show excellent accumulation at the site of tumor due to a phenomenon called *enhanced permeability and retention* (EPR) effect (Matsumura and Maeda 1986). Accumulation and retention of inert PEG modified nanoparticles in the tumor owing to EPR effect and lack of a lymphatic drainage system is often termed as *passive targeting*.

Alternatively, functionalization of the nanoparticle surface with a ligand that targets and binds to a specific receptor on the surface of cell of interest is called *active targeting*. Figure 13.3 is a schematic representation showing the passive and active targeting mechanisms of a drug delivery vehicle (Peer et al. 2007). As described before, cancer cells undergo rapid cell division, which requires enormous supply of nutrients and growth factors. These cells thus over-express several receptors on their surface to import supplies inside to meet the demand, and such receptors have been extensively exploited to direct nano-delivery system to their target cell of interest. Targeted delivery vehicles increase selectivity of cellular recognition and enhance the uptake of the payload by receptor-mediated endocytosis. Delivery vehicle relies on non-specific interaction with the target cells in the absence of a targeting ligand, which reduces the rate of internalization especially in the presence of protective coating of PEG. On the one hand, application of PEG benefits in evading the RES, thereby giving long circulating characteristics to the nanoparticles; but it also prevents interaction of the delivery system with cancer cells after extravasation into the

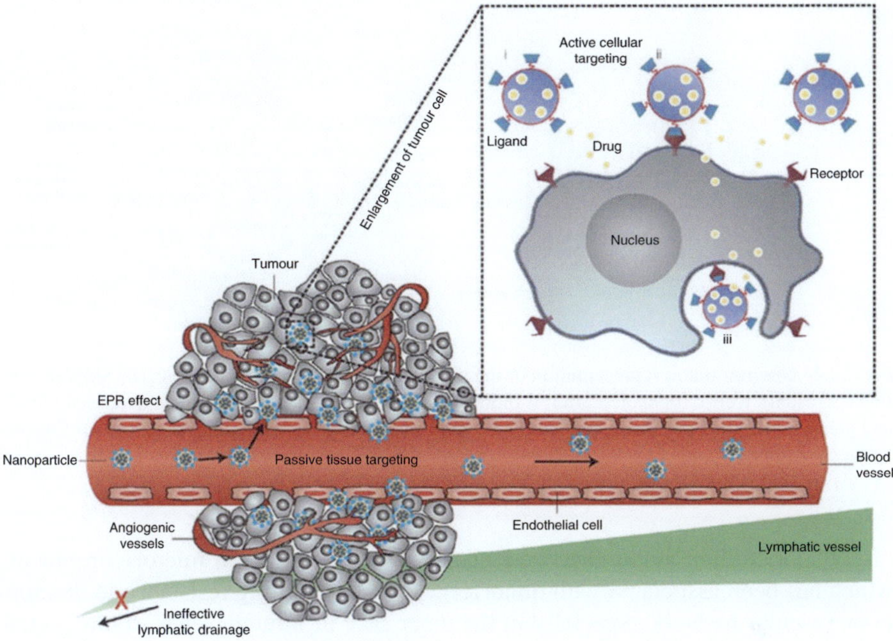

Fig. 13.3 Schematic representation of different mechanisms by which nanocarriers can deliver drugs to tumors [reprinted with permission from Peer et al. 2007 Nature Publishing Group (NPG)]

tumor environment. The presence of the targeting ligand also reduces off-target accumulation of the delivery vehicle and thus minimizes the non-specific drug associated toxicity. Targeting ligands therefore have become an essential component in designing a delivery system, and antibodies (Ab), antibody fragments, proteins, peptides, aptamers, and small molecules are some of the popular choices that have been explored. A detailed account of different types of targeting ligands used in cancer therapeutics is out of scope of this chapter, but several examples have been included in the chapter and have been reviewed comprehensively elsewhere (Wang and Thanou 2010; Brannon-Peppas and Blanchette 2004; Byrne et al. 2008).

A high tumor interstitial fluid pressure (IFP) is another feature that in general prevents the penetration and transport of the drug into the tumor interstitium (Jain 1987). The IFP in the core of the tumor is much higher and decreases outward creating a gradient that resists the drug inflow leading to a decreased transport of drug from the circulation into the distant site within the tumor. For an effective therapeutic outcome, it is important that the drug is able to penetrate to the core of the tumor, which becomes unlikely in the presence of high IFP. Nanoparticle drug delivery systems on the other hand are much larger in size compared to free drug molecules, and thus their transport against the pressure gradient is less affected. Besides, the fluid pressure within the tumor vasculature itself is one to two orders of magnitude higher than in normal tissues, resulting in increased extravasation of the nanocarrier at the site of the tumor.

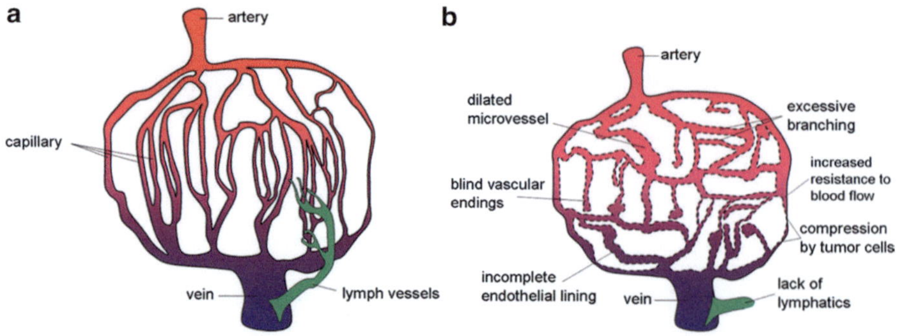

Fig. 13.4 Diagrammatic representation of the vascular system. (**a**) Normal tissue. (**b**) Solid tumor. *Red* represents well-oxygenated arterial blood, *blue* represents poorly oxygenated venous blood, and *green* represents lymphatic vessels (reprinted with permission from Tredan et al. 2007 Oxford University Press)

Hypoxic milieu is another trademark property of tumor microenvironment, which has been associated with tumorigenesis and imparting resistance to therapy. Poor vascular network especially in the inner core of a tumor leads to low partial pressure of oxygen that in turn has several metabolic implications such as prominence of glycolytic pathways leading to acidic pH, reducing environment and a gradient of cell proliferative property within the tumor (Tredan et al. 2007). Figure 13.4 gives a schematic representation of the differences in the tumor vasculature in comparison to that in a normal tissue. Tumor vasculature differs from normal blood vasculature in morphologically as well as in surface marker expression. Morphologically, blood vessels in tumor are dilated, tortuous with a highly nonuniform endothelial lining and fewer pericyte cells, making them leaky in nature. Besides, they have distinct molecular markers that aid in initiation of angiogenesis, a key early event in development of tumor. An elaborate account of tumor vasculature properties has been reviewed and is highly recommended to the readers (Ruoslahti 2002). These physiological variations can have debilitating effect on drug and imaging agent stability and penetration within the tumor. Nanotechnology however has been used to design delivery vectors that can exploit these impeding factors for therapeutic and imaging benefits. A precise control on physical and chemical properties have recently led to a new generation of stimuli-responsive nanodelivery vehicles, which can deliver their payload in response to these internal stimuli within the tumor (Ganta et al. 2008). Redox or pH responsive nanoparticles, for example, can protect the payload from various anatomical and physiological barriers within the patient's body until they reach the tumor site where they release their cargo in response to the reducing and acidic environment, respectively. The following sections of the book chapter will discuss some of these approaches in detail with examples to stress on the advantages of nanodelivery systems in diagnosis, imaging, and therapy of cancer.

13.2 Approaches to Cancer Diagnosis and Imaging

Imaging is a mainstay in cancer management as it finds application in almost every aspect of cancer care right from the early forecast and screening through detection and staging to therapy design, guidance, and monitoring up to follow-up care (Fass 2008). Imaging has carved a special niche for itself in cancer management because of several advantages over the conventional methods of diagnosis and therapy monitoring. Based on the methodology employed, imaging in cancer can be subdivided into two distinct modalities, namely optical and non-optical imaging. These imaging modalities utilize different wavelength of the electromagnetic spectrum, which have been identified in Fig. 13.5.

13.2.1 Optical Imaging Modalities

Optical imaging is based on interaction of light with biological tissues leading to photophysical and photochemical process like absorption, emission, scattering, and others. These interactions provide a basis for understanding the cellular, molecular, and morphological information about the tissue of interest (Solomon et al. 2011; Ntziachristos 2006). The benefits optical imaging has over other imaging modalities include superior sensitivity, absence of extremely sophisticated imaging hardware, compactness and portability, low level of radiation, and most importantly simultaneous multi-biomarker detection (Ntziachristos 2006; Pu et al. 2011). The commonly used optical imaging methods in cancer include bioluminescence imaging (BLI), fluorescence imaging (FI), optical coherence tomography (OCT), photoacoustic imaging (PAI), and Raman spectroscopy.

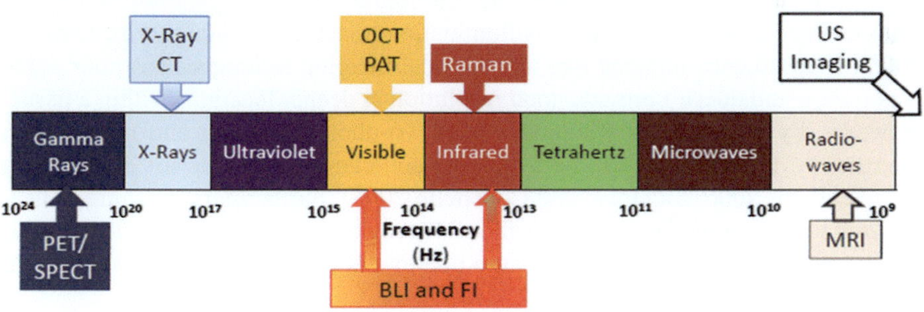

Fig. 13.5 EMR frequency spectrum for different imaging modalities. Adapted from Fass (2008). *Abbreviation*: *PET* positron emission tomography, *SPECT* single photon emission computed tomography, *X-Ray* CT- X-ray computed tomography, *OCT* optical coherence tomography, *PAT* photoacoustic tomography, *BLI* bioluminescence imaging, *FI* fluorescence imaging, *MRI* magnetic resonance imaging, *US Imaging* ultrasound imaging

13.2.1.1 Bioluminescence Imaging

BLI is an imaging technique commonly used in monitoring cell proliferation, metastasis, and therapy response and regression associated with cancer (Soling and Rainov 2003). BLI is based on the phenomenon of bioluminescence; that is light emission by living organisms. It is based on a chemical reaction involving enzyme luciferase and its substrate luciferin. Mammals do not have the gene encoding luciferase and thus genetic engineering is required for expression of this enzyme. On administration of luciferin, the enzyme-catalyzed reaction occurs associated with photon emission, which can be detected (Zabala et al. 2009).

BLI has several advantages like affordability, high sensitivity, extremely minute detection capability, and short imaging time. Also, since this technique is devoid of excitation by external light source, the background noise tends to be negligible (Zabala et al. 2009; Ji et al. 2009; Maes et al. 2009). It however suffers from drawbacks like low spatial resolution of 3–5 mm due to restricted penetration and light scattering, lack of applicability in humans and large animals, and failure to provide morphological details about tumor (Soling and Rainov 2003; Ji et al. 2009). However, the use of proteins in infra-red spectrum can solve the problem with respect to scattering and penetration (Soling and Rainov 2003).

13.2.1.2 Fluorescence Imaging

Fluorescence imaging is another diagnostic technique that finds application in cancer care. Fluorescence imaging is a very simple technique based on the principle of fluorescence, i.e., emission of light of higher wavelength occurring nanoseconds after light absorption by a fluorescent molecule (either endogenous or exogenous). Using filters to reject the exciting wavelength, only the emitted light is detected (Lichtman and Conchello 2005). In vivo fluorescence imaging has many subtypes like intravital confocal, two-photon and multiphoton microscopy, epi-illumination (fluorescence reflectance), and transillumination imaging techniques (Ntziachristos 2006). Fluorescence imaging like most optical imaging techniques has poor depth penetration and doesn't provide good resolution at deeper locations within a tissue. However, by using newer techniques like two-photon imaging and multiphoton microscopy, this problem can be overcome and can also provide for 3D imaging (Ntziachristos 2006; Lichtman and Conchello 2005). Furthermore, imaging in near infrared wavelengths (700–900 nm) allows for higher penetration and better resolution as well as high signal-to-noise ratio due to a decrease in autofluorescence (Pu et al. 2011; Kovar et al. 2007).

13.2.1.3 Optical Coherence Tomography

OCT is an essentially ultrasound imaging with the exception that light is used instead of sound (Fujimoto et al. 2000). It is based on detection and measurement of interference pattern generated from constructive or destructive interference between backscattered/back-reflected light and a reference light (Hu et al. 2006; Huang et al. 1991). The advantage of OCT over many other techniques used in cancer imaging is that it is a label-/probe-free imaging modality. Also, OCT provides a resolution of 1–20 μm. The components of OCT can also be miniaturized, thus facilitating their use within the body as well (Osiac et al. 2011; Vakoc et al. 2012). Also using modification like Doppler OCT, both structural and functional details can be elucidated (Osiac et al. 2011). Additionally in situ real-time 3D imaging can also be accomplished (John et al. 2013). Compared to non-optical imaging methods, which don't impose limitations with respect to depth, the depth of imaging for OCT (2–3 mm) is less. However, the depth penetration is better than other optical techniques (Osiac et al. 2011).

13.2.1.4 Photoacoustic Imaging

Photoacoustic imaging/photoacoustic tomography (PAT) is another technique useful in cancer detection, diagnosis, and treatment monitoring (Mallidi et al. 2011). It is based on the combination of optical and ultrasound procedures and utilizes the photoacoustic effect described by Bell (1880) wherein light absorption by tissue produces local heating which causes thermo-elastic expansion and increase in pressure. This pressure creates an ultrasonic or acoustic wave that is perceived by transducers which convert them into an electric signal (Wang 2009). PAI provides high quality imaging up to depths of 50 mm because of the combination of enhanced contrast of optical methods and high resolution associated with ultrasound technology (Fass 2008; Li et al. 2011a). A modified form of PAI, called spectroscopic photoacoustic tomography (SPAT), provides both morphological and quantitative data about multiple chromophores in a biological tissue under study (Rajian et al. 2009).

13.2.1.5 Raman Spectroscopy

Raman spectroscopy is a quantitative spectral technique, which provides chemical and morphological information of the sample. It is a vibrational spectroscopy technique based on the principle of inelastic scattering where a molecule absorbs a photon, gets excited to higher energy state, and on falling back to a different vibrational state emits photons with a different frequency. If the vibrational state is lower or higher than initial, the frequency is either higher or lower, respectively (Mahadevan 2003; Raman 1928). There are various modifications of Raman scattering that have been used in cancer imaging like Raman microspectroscopy, surface enhanced Raman scattering (SERS) spectroscopy, surface enhanced resonance Raman spectroscopy (SERRS), coherent anti-Raman scattering (CARS) spectroscopy, etc. (Surmacki et al. 2013).

Raman imaging had the advantage that it can detect multiple biological events, thereby providing multiplexing capability. Also, the technique is highly specific as each biomolecule has its own set of vibrational energy levels, and thus its signature fingerprint can be detected via Raman spectroscopy, thus providing tissue or disease specific molecular variations (Hanlon et al. 2000; Huang et al. 2003). This also enables real-time monitoring of molecular composition of a tissue (Zhang et al. 2010).

Furthermore, it also provides high spatial resolution, specificity, and sensitivity which are at par with confocal fluorescence imaging (Surmacki et al. 2013). The disadvantage of this technique is the poor penetration that is associated with other optical imaging modalities (Zhang et al. 2010). Also, the magnitude of Raman scattering achieved is very small which may be an obstacle to its usage (Zhang et al. 2010). However, this can be overcome by using metal nanoparticles based SERS phenomenon wherein scattering is increased up to a factor of 10^{15} (Surmacki et al. 2013; Zhang et al. 2010). Despite several advantages and variety of modalities available, optical imaging lacks in achievable depths of imaging as well as spatial resolution, which tend to hinder their usage in tumors located deep within the body.

13.2.2 Non-Optical Imaging Modalities

The non-optical imaging techniques can overcome the penetration problem that is often associated with the optical imaging modalities. Non-optical modalities include the traditional magnetic resonance imaging (MRI), X-ray computed tomography (CT), and ultrasound and the newer nuclear techniques such as positron emission tomography (PET) and single photon emission computed tomography (SPECT). MRI and CT are useful for anatomical imaging providing information about anatomy. SPECT, PET, and contrast enhanced CT or MRI offer imaging based functional details (Jokerst and Gambhir 2011). The disadvantage of these techniques however is the high cost, complex instrumentation and lack of capacity to handle large studies (Hillman et al. 2011).

13.2.2.1 Magnetic Resonance Imaging

MRI also sometimes called as nuclear magnetic resonance imaging (NMRI) or magnetic resonance tomography (MRT) is one of the most popular non-optical imaging modality used for cancer detection, staging, therapy/biopsy guidance, and monitoring (Fass 2008). MRI is based on the principle of nuclear magnetic resonance wherein application of magnetic field orients protons of hydrogen atoms in water or organic compounds in the direction of the applied field. On application of radiofrequency pulse, these alignments are distorted and after some time realigned producing an electromagnetic flux. The rate of relaxation of these protons in comparison to the baseline level is detected to generate MR signal. Relaxation rate depends on the physiochemical environment in which the proton is giving spin-lattice (T1) and spin-spin (T2) relaxation rates, which are used for contrast and image generation (Hendee and Morgan 1984; Massoud and Gambhir 2003). Various modifications of MRI commonly used for imaging cancer include diffusion weighted MRI (DW-MRI), dynamic contrast enhanced MRI (DCE-MRI), MR elastography (MRE), MR spectroscopy, etc. (Fass 2008).

The advantages of MRI over nuclear techniques and optical methods are its high penetration capacity allowing deep tissue tumor imaging (Fernandez-Fernandez et al. 2011). Using some of the aforementioned modifications of MRI, information about cellularity, perfusion, and staging is also possible (Fass 2008; Singh et al. 2012; Kozlowski et al. 2006). However, MRI has the disadvantage of sensitivity as a high concentration of the contrast or imaging agent is required to be localized in the tumor to get any useful information (Jokerst and Gambhir 2011; Singh et al. 2012). Besides, MRI cannot be used for real-time imaging applications (Jokerst and Gambhir 2011).

13.2.2.2 Computed Tomography

X-ray computed tomography or computed axial tomography (CAT) is another non-optical and anatomical imaging modality capable of providing 3D images (Jokerst and Gambhir 2011). CT is based on the differential absorption of X-ray by different tissues leading to contrast generation. Conventional scanners have an X-ray source and a detector that move axially around the subject, and using computer software reconstructs various slices to form an image. Newer scanners provide spiral or helical scanning and multi-slice scanning which allow scanning larger volume in relatively shorter time (Massoud and Gambhir 2003; Goldman 2007). Like MRI, CT is capable of imaging tumors located deep inside tissues (Fernandez-Fernandez et al. 2011). Also, the spatial resolution of CT is higher than MRI, Nuclear or Optical imaging techniques (Massoud and Gambhir 2003; Singh et al. 2012). The major disadvantage of CT is that it is an anatomical/structural imaging technique providing information about tumor dimensions and location, but no functional information is presented. Also, soft tissue contrast is poor which requires the use of iodine based contrast agents that show nonspecific localization and are also cleared faster resulting in shorter imaging time (Massoud and Gambhir 2003; Popovtzer et al. 2008).

13.2.2.3 Ultrasound

Ultrasound based imaging is an additional prospective imaging modality for the detection and characterization of tumors, therapy guidance, and monitoring (Kaneko and Willmann 2012). Ultrasound as an imaging technique is based on detection of back-reflected ultrasonic waves/echo. The amount of reflected echo is based on a property of tissue called as acoustic impedance. The acoustic impedance varies from tissue to tissue and depends on the product of tissue density and US wave velocity within the tissue. The contrast in US is obtained when the echo signals are obtained from interface between two tissues having different degree of acoustic impedance. The reflected echo signals attained from successive coplanar pulses are used for image generation. Various modifications of ultrasound imaging such as ultrasound biomicroscopy or power Doppler show potential in assessing tumor growth and differentiation, angiogenesis, and vascularity (Liang and Blomley 2003).

Ultrasound presents the advantages of being a low cost, readily available, and portable imaging modality. Additionally, since it does not use any ionizing radiation, it is considered to be safer. Besides, it also provides real-time imaging capability (Massoud and Gambhir 2003; Liang and Blomley 2003). However, the disadvantage of ultrasound is that its spatial resolution is poor. Furthermore, it cannot be used for imaging through bone, thereby proving a hindrance in detection of bone and brain cancers (O'Farrell et al. 2013). Also, ultrasound gives only anatomical/morphological information (Massoud and Gambhir 2003). Recent advancements such as use of Doppler and microbubble based ultrasound imaging however have shown capability in obtaining functional details of the tumor as well (Massoud and Gambhir 2003).

13.2.2.4 Positron Emission Tomography

PET is a non-optical imaging technique under the umbrella of radionuclide/nuclear imaging. It is based on the detection of high energy gamma rays emitted by a positron emitting isotope, like ^{15}O, ^{13}N, ^{11}C, and ^{18}F, incorporated in a natural bioactive molecule that is introduced into subject's body. The positron emitted during isotope decay collides with adjacent electrons leading to annihilation and production of 2 gamma rays at about 180° angle to each other, which are detected. Many such events are detected simultaneously, and using computer based reconstruction techniques similar to CT and SPECT, 3D images are created (Gambhir 2002; Brindle 2008).

PET is an extremely sensitive technique compared to other imaging modalities and also provides cellular and molecular details about a process. PET can be used in assessing the tumor metabolism, proliferation, and drug bio-distribution, which can be useful in therapy monitoring (Massoud and Gambhir 2003; O'Farrell et al. 2013; Gambhir 2002; Avril and Propper 2007). The major disadvantage associated with PET and other nuclear based imaging techniques is the small half-life of the radioactive imaging agents and the expensive and complex instruments like cyclotrons required for their generation. Also, PET cannot be used for imaging multiple probes and thus cannot monitor multiple processes. Furthermore, the radiation associated with such techniques leads to safety issues, and the resolution is also poor compared to other non-optical techniques (Massoud and Gambhir 2003; Fernandez-Fernandez et al. 2011).

13.2.2.5 Single Photon Emission Computed Tomography

SPECT is another nuclear/radionuclide imaging modality used in cancer imaging. It is based on the detection of low energy gamma ray emitted by a gamma-emitting isotope (^{99m}Tc, ^{111}In, or ^{123}I) that is injected into the patient prior to imaging. Special gamma cameras are used which rotate around the subject to get images from various angles. These are then integrated and reconstructed using computer programs to give a 3D tomographic image (O'Farrell et al. 2013; Brindle 2008). The major advantage with SPECT is that it allows concurrent monitoring of multiple imaging probes and thereby studying multiple events. Even though SPECT is ten times less sensitive compared to PET, it has high sensitivity compared to optical and other non-optical imaging techniques (Massoud and Gambhir 2003; O'Farrell et al. 2013; Gambhir 2002). The collimator design in SPECT leads to choice between spatial resolution and sensitivity. However, disadvantages associated with use of nuclear probes, as indicated under PET, apply to this technique as well (Massoud and Gambhir 2003; Fernandez-Fernandez et al. 2011).

Different imaging modalities offer different set of advantages and disadvantages, which has been summarized in Table 13.1. A universal diagnostic or imaging mechanism for all types of cancer would be hard to imagine currently. Therefore, choice of imaging/diagnostic agent and detection modality is primarily governed by the location and microenvironment of the tumor. While a good diagnostic probe can achieve sensitive level of detection owing to its own property, it is incapable of circumventing the physiological and anatomical barriers to achieve a favorable biodistribution and higher accumulation at its site of action. Appropriate choice of drug delivery vehicle therefore becomes critical. The following section of the chapter will discuss some of the commonly used delivery systems for theranostic applications.

13.3 Nano-Scale Delivery Systems

Conventional methods of drug administration suffer from several drawbacks such as poor drug bioavailability, non-specific accumulation, requirement of higher dose to achieve desired therapeutic efficiency, and thus the drug associated toxicities. Drug delivery systems often are colloidal systems, mostly synthesized using *generally regarded as safe* (GRAS) material and appropriate surface characteristics to facilitate longer circulation in the body, improve bioavailability, promote favorable biodistribution, and therefore reduce the required dose of the drug. Polymer, lipid, and inorganic delivery systems are most commonly explored for such applications and thus have been introduced in detail in the following section.

13.3.1 Polymeric Delivery Systems

Polymeric nanoparticles can be formulated by either conjugating multiple functional units to a soluble macromolecule or by a co-polymer self-assembly mechanism. The polymeric nanoparticle core can be loaded with the therapeutic agent and the imaging agent. The release mechanism of the drug and the imaging agent from the nanoparticle can be either of the following: (1) surface/bulk erosion, (2) diffusion through matrix, or (3) stimulation by the local environment (Peer et al. 2007). The polymer conjugates for theranostics essentially consist of the following four components:

Table 13.1 Characteristics of various imaging modalities

Imaging technique	EMR spectrum	Spatial resolution	Imaging depth	Sensitivity	Imaging time	Nature of molecular probe	Quantity of probe used	Scalability to human imaging	Cost
BLI	Visible light	3–5 mm	1–2 cm	NWC, possibly 10^{-15}–10^{-17} mol/L	Seconds to minutes	Activatable indirect	μg–mg	Yes but limited	\$\$
FI	Visible light or NIR	2–3 mm	<1 cm	NWC, likely 10^{-9}–10^{-12} mol/L	Seconds to minutes	Activatable, direct or indirect	μg–mg	Yes but limited	\$–\$\$
OCT	Visible or NIR	1–20 μm	2–3 mm	NA	Seconds	NA	NA	Yes	\$\$
PAI	Visible or NIR light and sound	5 μm–1 mm	50 mm–cm	NA	Seconds	NA	NA	Yes	\$\$
MRI	Radio-waves	25–100 μm	No limit	10^{-3}–10^{-5} mol/L	Minutes to hours	Activatable, direct or indirect	μg–mg	Yes	\$\$\$\$
CT	X-rays	50–200 μm	No limit	NWC	Minutes	May be possible (see text)	NA	Yes	\$\$
Ultrasound	High-frequency sound	50–500 μm	mm–cm	NWC	Seconds to minutes	Limited activatable, direct	μg–mg	Yes	\$\$
PET	High energy γ rays	1–2 mm	No limit	10^{-11}–10^{-12} mol/L	Seconds to minutes	Radiolabeled, direct or indirect	ng	Yes	\$\$\$\$
SPECT	Low energy γ rays	1–2 mm	No limit	10^{-10}–10^{-11} mol/L	Minutes	Radiolabeled direct or indirect	ng	Yes	\$\$\$

Abbreviation: *NA* not applicable, *NWC* not well characterized, *\$* to *\$\$\$\$* inexpensive to very expensive

(1) therapeutic component (drug carrier, gene carrier, hyperthermia/photodynamic therapy); (2) polymeric coating for stability, biocompatibility, and solubility; (3) functional group for drug targeting (antibodies); and (4) detection agent for imaging (MRI/fluorescence/radionuclide/photoluminescent agent, microbubbles).

There are numerous advantages of polymer-based image guided delivery apart from the combined therapy and diagnosis. It helps in analysis of the drug distribution at the site of action, visualization of the biodistribution, prediction of the drug response, quantification of the drug release, triggered drug release (in some hyperthermia directed systems), non-invasive assessment of the target site accumulation of the drug (optical imaging system), control of the drug release profile (Lammers et al. 2010), increased stability, and decreased immunogenicity when the polymer cores are shielded by stealth materials such as polyethylene glycol (Luk et al. 2012). The following section discusses some recent development in the polymer-based theranostic systems.

13.3.1.1 Polymeric Delivery for MR Imaging

Polymeric delivery systems can be conjugated with magnetic nanoparticles (MNP) of iron, nickel, cobalt, and other oxides because they are nanoscale, interact with external magnetic field which can be further exploited for use in MRI, promotion of hyperthermia treatment, drug delivery, and bioseparation processes (Krasia-Christoforou and Georgiou 2013). MNP have the unique property of responding to an external magnetic field, and therefore the delivery system can be directed to the specific site of interest (Medeiros et al. 2011). The efficacy of these particles depends on the morphology, size, charge, and surface functionalization. The pharmacokinetics, biodistribution, clearance, and toxicity are important parameters to be taken into consideration while formulating such a system. As reported in the literature, there has been a lot of focus on developing polymeric systems with two main categories of nanoparticles: (1) gadolinium (Gd^{3+}) based nanoparticles or (2) superparamagnetic iron oxide nanoparticles (SPIONs) (Krasia-Christoforou and Georgiou 2013).

Gadolinium compounds have been used for the formulation of polymeric nanoparticles based delivery system for visualization of accumulation of the drug at the site of tumor and study of other pharmacokinetic parameters. Polymeric nanoparticle system (gadolinium conjugated PGA system) was tested in MDA-MB-231 human breast cancer xenograft bearing mice. Strong MR signaling was observed in tumor periphery. The study revealed importance of the size of the nanoparticles used since very small nanoparticles tend to get cleared in circulation (Ye et al. 2006). Various other T1 contrast agents like DTPA-Gd^{3+} have been used because T1 type contrast agents offer favorable positive imaging contrast enhancement and are more widely applied clinically, though these areas have not been explored much. The application of Gd^{3+}-chelates has also been seriously impeded because of acute nephrotoxicity especially in patients with history of renal-related diseases.

13.3.1.2 Polymeric Delivery for PET Imaging

Addition of PET imaging agents (^{24}Cu and ^{18}F) to polymer-based nanocarrier system can be developed as effective nanotheranostic system for non-invasive visualization and angiogenic biomarker quantification (Welch et al. 2009). Angiogenesis (formation and differentiation of blood vessels) plays a major role in tumor growth and metastasis. Integrins (cell adhesion molecules) are involved in multiple steps of angiogenesis and metastasis (Chen et al. 2004a). Study of integrins ($\alpha_v\beta_3$ and $\alpha_v\beta_5$) has revealed that they are highly expressed in the activated epithelial cells and tumor cells but are expressed in lower levels in normal cells. This overexpression can be antagonized by RGD (Arg-Gly-Asp) peptide, which can be labeled with different radionuclides. ^{18}F labeling of RGD peptides for microPET imaging in solid tumors showed high tumor-to-background contrast. Even though they showed promising results, there were several drawbacks with this system. The procedure of labeling was tedious and relatively inefficient, the tumor retention time was short, and the uptake was modest. A ^{64}Cu labeled RGD peptide and microPET imaging of MDA-MB-435 breast cancer tumor xenograft was described, but this tracer as well did not show promising results. The tumor washout time was rapid and substantial accumulation was observed in the liver and kidneys (Chen et al. 2004b). Further efforts by the same group lead to development of Cu24 labeled PEG-modified RGD peptide for PET imaging of expression of tumor α_v integrin. The in vivo experiments suggested that DOTA-PEG-RGD was suitable ligand for labeling radiometals and could form stable complexes with DOTA moiety for therapeutics and imaging (Chen et al. 2004a).

13.3.1.3 Polymeric Delivery for Fluorescence Imaging

Fluorescence imaging can be integrated with a polymeric delivery system allowing for the image guided delivery and monitoring of the various parameters such as pharmacokinetics, pharmacodynamics, and drug tumor accumulation (Luk et al. 2012). Fluorescence imaging has many advantages like good spatial resolution in UV-NIR region; sensitivity nearly same as observed with radioisotopes (SPECT or PET) and low-cost technique. Though it is a more robust technique, one of the major weaknesses includes limited tissue penetration as well as noise and background generated due to the tissue scattering of the photons in the visible region and tissue autofluorescence. The use of near IR (NIR) light helps overcome the issue of tissue autofluorescence and thus helps in deeper tissue penetration with reduced scattering (Krasia-Christoforou and Georgiou 2013).

NIR fluorescent dyes can be combined with a therapeutic agent in a polymeric drug delivery system to develop a multifunctional polymeric nanoparticle system consisting of PEG-polycaprolactone (PCL) diblock co-polymer. The nanoparticles when loaded with NIR dye (IR-780) has shown to serve as an imaging modality as well as photothermal treatment (PTT). These particles can also be labeled with ^{188}Re for micro-SPECT guided imaging. NIR irradiation when used in such a system has been shown to cause thermal destruction of the HCT 116 colorectal carcinoma cells (Peng et al. 2011).

Quantum dots or luminescent semiconductors are passive bio-probes/labels for imaging and cellular studies. Quantum dots possess unique optical properties like high absorbency, narrow and tunable emission bands, and high stability to photobleaching. These nanoparticles have also been used for fluorescent applications. They can act as a nanocargo for therapeutic components along with a diagnostic tool (Krasia-Christoforou and Georgiou 2013). In a study by Yuan et al., zinc oxide quantum dots were loaded with doxorubicin (anti-cancer drug). A biodegradable polymer (chitosan) was used for tumor-targeted delivery, and it even imparts stability to the QDs because of the hydrophilicity and cationic charge on chitosan. Nearly 75 % drug loading was achieved with drug release being rapid initially followed by controlled release (Yuan et al. 2010).

13.3.2 Lipid-Based Delivery Systems

Lipid based delivery system could be of several types such as liposomes, nanoemulsions, solid lipid nanoparticles, etc., but liposomes are the most vastly explored among these and thus will be the prime focus here. Liposomes are vesicles consisting of concentric lipid bilayers (made up of either natural or synthetic phospholipids) called lamella with an inner aqueous core (Al-Jamal and Kostarelos 2011). Liposomes have been used as a drug delivery system for a wide range of therapeutics. The net chemical and physical properties of the liposomes such as permeability, charge density, and steric hindrance depend on the constituent phospholipids. The process of liposome formation is constitutively spontaneous which can be attributed to the amphiphilic phospholipids which self-associate into bilayers. Liposomal delivery system has tremendous advantages like biocompatibility, biodegradability, low toxicity, capability of surface modification, and small size (micro-nano) (Luk et al. 2012; Al-Jamal and Kostarelos 2011).

Liposome delivery system is an effective delivery system capable of encapsulating either a water-soluble active drug in the hydrophilic core or water-insoluble drug in the hydrophobic core. It imparts targeted and controlled delivery and thus has fewer side effects and higher efficacy compared to conventional delivery systems. Solid tumors are a potential target for liposomal delivery system due to the leaky vasculature of cancer cells (EPR effect) and the defective lymphatic drainage, allowing liposomes to accumulate at the target site.

Nanoparticles such as iron oxide MNP, quantum dots (QDs), and gold nanoparticles are often difficult to deliver in vivo because of the hydrophobic surface and poor colloidal stability. Trapping these nanoparticles within the hydrophobic core or linking them covalently to the surface of the therapeutic liposomes has shown promising results as a diagnostic and therapeutic tool with anticancer agents encapsulated in the core or embedded on the lipophilic layer. Molecular probes can also be conjugated to enhance the targeting. Such theranostic particles circulate for longer period in the blood and provide better prognosis and monitoring of the disease (Al-Jamal and Kostarelos 2011).

There are various examples where liposomes with varied properties have been incorporated with therapeutic agents as well as imaging probes.

13.3.2.1 Liposomes for MR Imaging

MRI has long been used as an imaging technique for visualization of internal body tissues. Contrast agents like Gd^{3+}, superparamagnetic iron oxide nanoparticles, and manganese-based compounds have been studied for their ability for high contrast in T2-weighted MRI. Exploiting these properties of compounds along with loading of active pharmaceutical ingredient in a delivery system is a useful approach for developing advanced nanotheranostics with non-invasive MRI (Luk et al. 2012).

Liposomes have been successfully used as a targeted delivery system for a range of therapy. Loading these liposomes with contrast agents and drug will serve as a system for MRI imaging along with drug delivery. Studies on in vivo models such as Kaposi's sarcoma expressing neural cell adhesion molecule (Grange et al. 2010) and lung carcinoma and pancreatic cancer xenograft (Erten et al. 2010) have been carried out using gadolinium salts as the MR imaging agent in combination with anti-cancer drugs. Iron oxide has also been used as MR imaging agent in a liposomal system. Loading of liposomes with magnetic fluid has shown to be the best T2 contrast agent. The higher the loading of the liposomal system with iron oxide, the better the contrast enhancement (Martina et al. 2005). Iron oxide nanoparticles are not only used as contrast agents for visualization of the tumor but also aid in hyperthermia-induced drug release (Tai et al. 2009).

Temperature triggered release of the drug from liposomes has been recently exploited to ensure time dependent targeted release of the drug. Multifunctional liposomes made using thermosensitive polymer poly(2-ethoxy(ethoxyethyl)vinyl ether) (Kono et al. 2011) loaded with doxorubicin and Gd^{3+} residues release the doxorubicin above 40 °C and show excellent relaxivity for T1 weighted MRI signal in tumor. The liposome size and tumor size determine the amount of accumulation of the liposomes in the tumor and its efficacy.

13.3.2.2 Liposome-Radionuclide Imaging

Nuclear medicine scan is an ultra-sensitive test, which is used for providing information about structure and function of specific parts of the body. It is widely used for diagnosis and monitoring of the cancerous sites, tumor localization, and staging. Radiolabeled liposomes have shown good accumulation characteristics in tumor, infection, and inflammation in in vivo studies.

Radionuclides have long been used for chemo-radionuclide therapy. Combining radionuclide agents with therapeutic agents has shown to increase the efficacy along with imaging. Liposomes have been encapsulated with modified radionuclides in order to achieve high labeling efficiency (Luk et al. 2012). Radionuclide agents like rhenium-188 and rhenium-186 have been used for labeling of liposomes and have shown to be stable in serum, in vivo models and could be used for radionuclide therapy (Bao et al. 2003). Many other radionuclides such as technetium-99, indium-111, and I-131 have been explored till now to formulate liposome based theranostic particles (Osborne et al. 1979; Harrington et al. 2001; Hoefnagel 1998).

PET radionuclides have also shown potential as theranostic agent when formulated with liposomes. In a study conducted with human colon carcinoma (HT29) tumor cells, ^{64}Cu loaded liposomal formulation was found to accumulate in tumor cells by several orders of higher magnitude than that of free ^{64}Cu. Also, the PEG-modified liposomal formulation remained in the blood circulation for a long period of time. Conjugating ligands to the surface of the liposomes can further improve the imaging capability of the formulation (Petersen et al. 2011).

13.3.2.3 Liposome-Quantum Dot Hybrid System

Quantum dots are fluorescent nanoparticles that are 10–20 times brighter than the conventional organic dyes used for fluorescence measurement. They have greater photostability and have a great potential for optical imaging application. Since they are prepared in the organic phase, QDs are hydrophobic in nature and thus possess the problem of water insolubility and bio-incompatibility (Al-Jamal and Kostarelos 2011). Functionalization of the QD surface is one approach of imparting polarity to these particles. But such surface modifications have led to the decrease in the fluorescence properties and decreased photostability of the particles. In a recent study, efforts have been made to encapsulate the QDs in the lipid bilayer, hence retaining its fluorescence properties. The hydrophobic QDs were entrapped in the lipid bilayer and the inner aqueous layer was loaded with the hydrophilic drug doxorubicin. Two types of lipid bilayer were used, rigid phase (DSPC:Chol:DSPE-PEG2000) and fluid phase of egg PC (EPC:Chol:DSP-PEG2000), to incorporate the drug loaded-liposomes-QD hybrid. Doxorubicin was loaded into the L-QD hybrid with 97 % loading efficiency. The fluorescence spectrum of the doxorubicin and the green emitting QD showed the presence of DOX and QD in a single vesicle system. This study showed that the effect of QD on the vesicle integrity was dependent on the lipid component (Tian et al. 2011).

In another study by the same group, functionalized QDs were incorporated into the hydrophilic core of the liposomes, while the hydrophobic therapeutic agent was loaded in the lipid bilayer. Such a system showed increased tumor accumulation, enhanced cellular uptake, and increased retention of the QDs due to its polymer coating (Al-Jamal et al. 2008, 2009).

13.3.3 Inorganic Delivery System

Inorganic nanoparticles have received attention recently in the cancer nanotechnology field owing to their optical properties (gold spherical nanoparticles, nanoshells, nanorods, nanowires), magnetic properties (MNP and super paramagnetic nanoparticles), size, and mechanical properties. Fluorescent quantum dots have been used for bio-sensing but are being currently reviewed for their use as a diagnostic model for cancer. MNP have been used for bio-sensing, MR imaging, cell labeling, and drug delivery. The SPIONs have been shown to be useful for hyperthermia treatment of tumor. Gold nanoparticles (nanorods, nanowires, nanoshells) have been exploited for their optical properties and are important in imaging, as drug carriers, and thermotherapy (Lim et al. 2013).

Inorganic nanoparticles used for theranostic purposes can be broadly classified into two categories: solid or porous nanoparticles. Solid consist of gold nanoparticles, which have been largely used owing to their small size, low toxicity, and absorbance in near-infrared region. The drug loaded on to these particles is released mainly by one of the following interactions: change in pH, heat, light, and glutathione. Loading of the drug can be achieved by conjugation, lipophilic interaction, and charge interaction. In order to facilitate the attachment of the drug on the inorganic nanoparticle, thiol groups have been used (largely for gold nanoparticles). Cysteine and histidine moieties have also been used for peptide delivery (Lim et al. 2013).

Porous nanoparticles are used for delivery of drugs, therapeutic moieties, genes, and antibodies. These are usually silica based porous material and enable precise drug loading due to the uniform pore size and large surface area. There are numerous properties on which the drug loading and release of the drug depends, they are pore topography, surface properties, porosity, kinetics of release, pore connectivity, and geometry. These nanoparticles have been used to deliver drugs such as doxorubicin and cisplatin (Lim et al. 2013).

Though these nanoparticles can perform all the functions separately, it is desired to have multiple functionality in order to reduce the dosing and for better disease management. Magnetic nanoparticles (MNP) have been developed with dual functional properties of delivery and diagnosis. In a recent study, combination of drugs (doxorubicin and paclitaxel) was used as the therapeutic agent with a MNP system with oleic acid capping. The hydrophobic drug was partitioned in the oleic acid layer of the MNP, and Pluronic® F-127 (a triblock co-polymer) was used to make the particles hydrophilic. The loading efficiency of the combination drug was higher than individual drugs and the release extended for up to 3 weeks. In the presence of an applied external magnetic field, the T2 relaxivity of the MNPs was higher. These particles had prolonged circulation time in mice thus providing longer time for drug delivery and imaging. These nanoparticles could thus be used for loading hydrophobic drugs with high efficacy and simultaneous MR imaging and real-time monitoring of the drug distribution and disease progression (Jain et al. 2008). Superparamagnetic nanoparticles have been lately used as MRI contrast agents and evaluated by T2 relaxivity measurement. These multifunctional and water-soluble SPIONs have demonstrated promising properties for targeted anticancer delivery and PET imaging when conjugated with cRGD and doxorubicin (Yang et al. 2011).

Gold nanoparticles have been used widely for various therapeutic as well as imaging purposes for treatment of various diseases. Functionalization of gold nanoparticle surface with peptides, PEG, and antibodies has been exploited for targeted delivery of chemotherapeutic agents. Gold nanoparticles have the unique physical and optical properties, which are used to visualize and eliminate cancer cells. Gold nanoshell based theranostic agents have the ability to enhance photoluminescence of fluorophores and MRI agents (Alexis et al. 2010). Gold nanoparticles surface was functionalized with PEG, biotin, paclitaxel, and rhodamine B which are linked by beta cyclodextrin and have been tested in cancer cell lines (HeLa, A549, MG63) against a non-cancerous cell line (NIH-3T3) using fluorescence-activated cell sorter, cell viability analyses, and CLSM. These nanoparticles have shown to play a significant role in cancer diagnosis as well as imaging (Heo et al. 2012).

Mesoporous silica nanoparticles (MSNs) have recently been developed and studied as potential theranostic platform because of their tunable size, diameter, and good biocompatibility. MSNs can also be functionalized with other inorganic moieties called MSNs-hybrid conjugates. Gold nanorods have been coated with silica matrix and have shown to be more resistant to signal degradation than uncoated gold nanorods and have higher contrast imaging (Chen et al. 2010). In order to impart magnetic properties, add targeted payload and magnetically triggered release of the drug, MSN encapsulation of iron oxide has been employed. Such nanoparticles can also provide MR imaging (Chen et al. 2013a).

13.4 Illustrative Examples of Theranostic Systems

13.4.1 Polymeric Theranostic Systems

The synthesis of star block copolymer was reported as integrated platform for cancer therapeutics and MR imaging. These polymers were structurally stable unimolecular micelles in aqueous solutions with a hydrophobic polyester core; folic acid as targeting moiety for specific recognition of folate receptor on cell surface and MRI contrast agent (Gd-DOTA) were anchored on the outer corona. Paclitaxel was encapsulated in the inner core of the micelles and controlled release was achieved. In vitro data showed enhanced T1 relaxivity and in vivo MR imaging demonstrated positive contrast enhancement and extended blood circulation duration (Li et al. 2011b). Mahmoudi and coworkers synthesized polymer coated colloidal nanocrystal clusters (CNCs) with a magnetic core of magnetite (SPIONs) using cross-linked PEG-*co*-fumarate as the polymeric coating. The CNCs were loaded with anti-estrogen tamoxifen citrate and doxorubicin. The drug release studies showed a controlled release of the therapeutic component due to the presence of cross-linked polymer. The in vitro studies confirmed the efficiency of the system as MRI contrast agent with relatively high values of the ^{1}H-NMR transverse relaxivity r_2 (Amiri et al. 2011).

Pluronic® F127 micelles were synthesized for use in controlled release delivery system and MR imaging. Doxorubicin and SPIONs were encapsulated, characterized by TEM, and dose dependent darkening of MR images was observed. Such multifunctional nano-carriers were shown to be effective for cancer therapy and imaging (Lai et al. 2010). Zinc oxide QDs were synthesized by chemical hydrolysis and fabricated as water-dispersed ZnO-QD-chitosan folate carrier loaded with doxorubicin. Nearly 75 % drug-loading was observed and the stability was enhanced due to chitosan. Initial rapid drug release was observed followed by controlled release (Yuan et al. 2010).

In a recent study by Guthi, et al., maleimide terminated multifunctional polymeric micelles (MFMs) were functionalized with lung cancer targeting peptide

(LCP) to facilitate targeting and specific uptake in lung cancer cells. The system was loaded with doxorubicin (therapeutic component) and SPIONs for MR imaging. LCP encoded MFM showed high affinity for $\alpha_v\beta_6$ integrin in the H2009 lung cancer cells. Two controls were used; a scrambled peptide was conjugated with the MFM which showed low cell targeting and another cell line was used H460 which was $\alpha_v\beta_6$ integrin negative. Decreased targeting and uptake was observed with SP-encoded MFM control and H460 cell line. Nearly threefold uptake was observed with LCP functionalized MFM which was confirmed by confocal laser scanning microscopy (CLSM) (Fig. 13.6). The MFM system showed successful doxorubicin

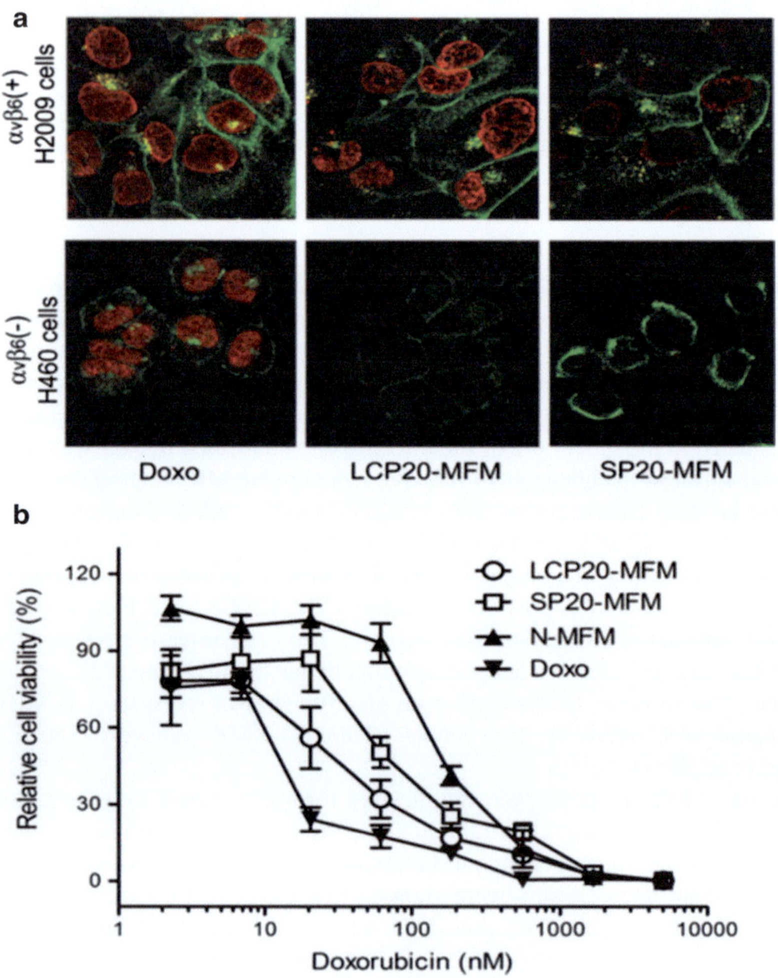

Fig. 13.6 (**a**) Confocal laser scanning microscopy of doxorubicin fluorescence in Rv-6-expressing H2009 and Rv-6-negative H460 cells treated with free doxorubicin, LCP20-MFM, and SP20-MFM. (**b**) Growth inhibition assay of H2009 cells after treatment with free doxorubicin, LCP20-MFM, SP20-MFM, and N-MFM (reprinted with permission from Guthi et al. 2009. Copyright 2010 American Chemical Society)

release and accumulation in the nucleus of the cancer cells. SPIONs were clustered inside the micellar core. The SPION release showed an increase in the T2 relaxivity values. The MR images obtained with this system showed a clear contrast difference between H2009 cells incubated with LCP encoded MFM over the SP-encoded MFM (Guthi et al. 2009).

Liu et al. report novel multifunctional pH-disintegrable micellar nanoparticles of star co-polymers containing β cyclodextrin cores (DOTA-Gd) 7-CD-(PHPMA-FA-DOX). These nanoparticles were successfully used for folate receptor targeted cancer cell delivery and MR imaging. Doxorubicin was conjugated on PHPMA arms per star molecule. This rendered the initial hydrophilic nanoparticles amphiphilic, thus initiating the self-assembly of the nanoparticles. The doxorubicin release in vitro was pH dependent. The drug release profile showed the release of doxorubicin at an acidic pH, which was attributed to the cleavage of the covalently attached doxorubicin to the PHPMA arms from the star copolymer scaffold. T1 relaxivity was observed in in vitro MR imaging for micellar nanoparticle. In vivo MR imaging showed high accumulation of nanoparticles in the kidney (Liu et al. 2012).

13.4.2 Lipid-Based Theranostic Systems

Thermosensitive liposome formulations have been used for non-invasive in vivo imaging and drug delivery. A formulation consisting of DPPC and Brij®78 was co-encapsulated with Gd^{3+}-DTPA (MRI probe) and doxorubicin. The temperature release profile of doxorubicin was 100 % release at 40–42 °C in 3 min. The DOX uptake in tumor bearing tissues was correlated with T1 response (Tagami et al. 2011). In a study by Lowery et al., tumor targeting was achieved by use of the peptide HVGGSSV, which was isolated from in vivo screening of phage-displayed peptide library. Targeting liposomes were loaded with doxorubicin and labeled with Alexa Fluor 750 for NIR imaging. Such targeting delivery system showed high anti-tumor efficacy and low systemic toxicity. The in vivo results demonstrated HVGGSSV peptide mediated image guided drug delivery in tumors (Lowery et al. 2011).

Liposomes incorporating quantum dots for imaging and docetaxel as the therapeutic agent were tested for efficacy and targeting in MCF-7 breast cancer cells. The liposomes were coated with D-alpha-tocopheryl polyethylene glycol 1000 succinate mono-ester (TPGS) and subsequently functionalized with folic acid to target the folate receptor on cell surface. The internalization of the liposomes in the cells was visualized by CLSM. The multifunctional, targeted liposomes were found to show greater efficacy over the non-targeted liposomes (Muthu et al. 2012). Grange et al. showed the combined delivery and MRI of doxorubicin containing liposomes in an in vivo model of Kaposi's sarcoma expressing neural cell adhesion molecule (NCAM). NCAM receptor was targeted by using NCAM binding peptide (C3d) -coated PEF liposomes loaded with doxorubicin and a lipophilic gadolinium-DOTA-monoamide (DOTAMA) derivative (Grange et al. 2010).

In another example, multifunctional dextran core based stealth liposomal nanoparticles were prepared encapsulating doxorubicin, iron oxide for MR imaging, and boron-dipyrromethene (BODIPY) for fluorescence. Polyethylene glycol was used for surface decoration, which increased the circulation time of the particles in the blood. A rodent dorsal window chamber was developed which facilitated the MR imaging and non-invasive optical imaging. This model was extended to the nude mice bearing lung carcinoma and MIA PaCa-2 pancreatic cancer xenograft. Enhanced contrast was observed in the blood vessels due to accumulation of the iron oxide nanoparticles in these blood vessels (Erten et al. 2010).

Multifunctional PEG-modified liposomes with both thermosensitive poly[2-(2-ethoxy)ethoxyethyl vinyl ether-*block*-octadecyl vinyl ether (EOEOVE-b*lock*-ODVE)] with a lower critical solution temperature around 40 °C and polyamidoamine (PAMAM) G3 dendron-based lipids having eight Gd^{3+} chelate residues (G3-DL-DOTA-Gd) were loaded with doxorubicin and evaluated for the accumulation of liposome at the tumor site and heat-triggered drug release of the drug. Liposomes of two different sizes, 110 and 48 nm, were administered intravenously to Balb/c nude mouse having colon 26 tumors via tail vein, and the accumulation of the liposomes at the tumor was monitored using MRI. The images revealed that the liposomes were not uniformly distributed in the tumor site and tumor tissue composition; tissue anatomy and vascularization had a role to play (Fig. 13.7a). The tumor accumulation efficiency studies revealed that irrespective of the liposome diameter, the liposomes accumulated at the tumor in a similarly time-dependent manner (Fig. 13.7b). However, the signal intensity values at the tumor sites at the same time point were different between the two mice, and the mice injected with the larger liposomes exhibited stronger signal intensity, thus suggesting that liposomes with 110 nm diameter accumulated at the tumor more efficiently than those with 48 nm diameter (Kono et al. 2011).

The multifunctional PEG-modified liposomes exhibited a considerable release of encapsulated doxorubicin after mild heating. The liposomes encapsulating doxorubicin injected intravenously into tumor-bearing mice have shown to suppress the tumor growth only when the tumor site was heated to 44 °C for 10 min at 8 h after injection (Fig. 13.8) (Kono et al. 2011).

13.4.3 Inorganic Theranostic Systems

OCT was used for monitoring effectiveness of photo-ablation therapy using gold-silica nanoshells that were exposed to NIR light. This therapy showed complete tumor regression within 12 days posttreatment in all but two mice, and additionally a 56 % contrast enhancement was achieved using the nanoshell system (Gobin et al. 2007). Recently, yolk-shell nanoparticles consisting of iron-platinum (FePt) nanoparticle yolk and iron oxide (Fe_2O_3) shell have shown promise as an anticancer cum MRI contrast agent (Gao et al. 2008). Two-photon induced luminescence and dark-field imaging have also been used to image gold nanoshells and monitor their photo-ablation potential thus serving as a beneficial prospective theranostic agent (Loo et al. 2005; Park et al. 2008).

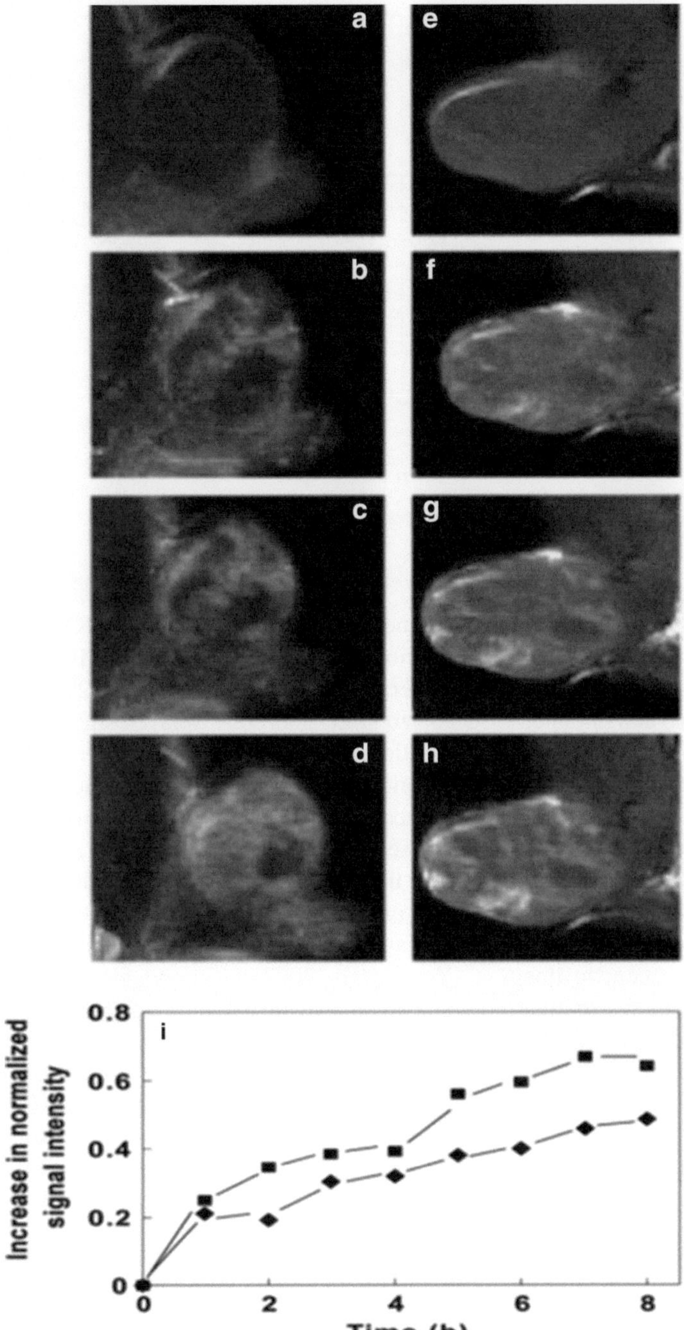

Fig. 13.7 Magnetic resonance images of tumors before (**a**, **e**) and 1 h (**b**, **f**), 3 h (**c**, **g**), and 8 h (**d**, **h**) after administration of liposomes. **a**–**d** (liposomes diameter of 110 nm) and (**e**–**h**) (liposome diameter 48 nm). I—plots of increase in the normalized signal intensity for tumor of BALB/c nude mouse injected with liposome of 110 nm (*squares*) or 48 nm (*diamonds*) as a function of time (reprinted with permission from Kono et al. 2011. Elsevier 2011)

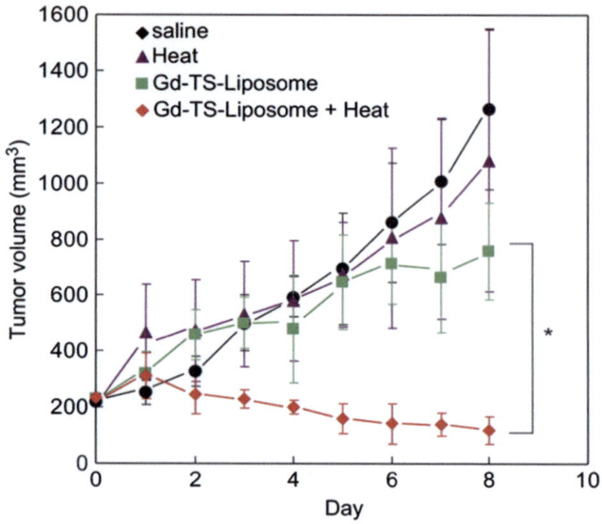

Fig. 13.8 Tumor volume measurement after doxorubicin dose of 6 mg/kg body weight. Saline treated and heat (alone) treated act as controls. Considerable decrease in the tumor volume observed with liposomes treated under heat (reprinted with permission from Kono et al. 2011. Elsevier 2011)

Oleic acid-coated hollow manganese oxide nanoparticles (HMON) were shown to efficiently encapsulate doxorubicin and also be functional as MR imaging agent and thus showing potential as a theranostic system (Shin et al. 2009). In another study amino and thiolated PEG coated gold nanorods displayed significant theranostic potential with enhanced circulation time, photothermal tumor destruction and provided for imaging by X-ray computed tomography technique (von Maltzahn et al. 2009). Plasma-magnetic nanoparticle with an iron oxide core and gold shell conjugated to EGFR antibody was used in an in vitro cancer model and showed good photothermal destruction capability that was monitored using both MR and optical (confocal reflectance and dark-field reflectance) imaging (Larson et al. 2007). Gold nanocages have been shown to tremendously increase contrasts in rat cerebral cortex using photoacoustic imaging (Yang et al. 2007a). Combination of this imaging modality with photo-ablation therapy can provide a potential theranostic system.

Cyclic RGD peptide (cRGDyK) conjugated fluorescent MSNs encapsulating doxorubicin exhibited good targeting, drug delivery, and fluorescence imaging potential (Chen et al. 2013b). Gold-silica core/shell nanoparticles with PEG conjugated EGFR antibody showed 35-fold higher Raman signal, using SERS as an imaging modality, in EGFR positive A431 tumors. This coupled with the photo-ablation capability of gold nanoparticles can serve as a prospective theranostic system (Jokerst et al. 2011). ^{64}Cu-labelled, cRGDfc peptide functionalized PEG-modified superparamagnetic iron oxide nanoparticles (SPIONS) conjugating doxorubicin by a pH-sensitive linkage demonstrated higher cellular uptake and cytotoxicity in an in vitro model. Additionally, this system showed good targeting via significant tumor accumulation as imagined by quantitative PET. Furthermore, the system was also capable of MR based imaging providing for dual mode imaging capability (Yang et al. 2011). Thiolated PEG coated gold-silica nanoshells irradiated with NIR light and imaged using magnetic resonance thermal imaging (MRTI) demonstrated localized tumor cell death in a mouse model bearing subcutaneous tumor (Hirsch et al. 2003).

13.4.4 Hybrid Theranostic Systems

PEG-PCL micelles labelled with Rhenium 188 (^{188}Re) radionuclide and incorporating a cyanine type photosensitizing NIR dye IR-780 thermally destroyed colon cancer both in in vitro and in vivo models through decreased cellular proliferation and enhanced apoptosis. Additionally, the NIR dye allowed for fluorescence imaging and the radionuclide allowed for real-time micro-SPECT based imaging of accumulation and distribution of the photosensitizer in the tumor and other tissues (Peng et al. 2011). Nucleolin targeted PEG-modified and amine functionalized polyacrylamide nanoparticles, encapsulating iron oxide nanoparticles as imaging agent and photofrin as a photosensitizing therapeutic agent, were found to be useful in specific targeting, MRI, and PDT based treatment of brain tumor (Reddy et al. 2006).

Another example of a theranostic nanoconstruct is carboxyl coated Zn/CdSe shell/core quantum dot conjugated to PEG functionalized liposomes having Anti-HER2 antibody moiety for targeting. The system showed potent in vitro cytotoxicity in cells overexpressing HER2/neu and simultaneously showed targeting via tumor localization and fluorescence imaging in a nude mice model of subcutaneously implanted HER2/neu overexpressing xenograft (Weng et al. 2008). HER2 antibody-conjugated multifunctional magneto-polymeric nanohybrids (HER-MMNPNs) were formulated and studied for their use as a system capable of carrying both chemotherapeutic and MRI imaging payload. The MMNPNs prepared from oppositely charged block polymer of PEG and polyaminoacids and encapsulating magnetic manganese iron oxide ($MnFe_2O_4$) nanoparticles and doxorubicin showed good in vivo and in vitro anti-tumor and MRI imaging capabilities (Yang et al. 2007b; Koide et al. 2006). Doxorubicin encapsulated thermally crosslinked SPIONS (or TCL-SPIONS) with anti-biofouling polymethacrylate coat were used as theranostic agent. They demonstrated substantial antitumor activity in a mice Lewis lung carcinoma model using MRI as an imaging technique (Yu et al. 2008). A multifunctional system consisting of iron oxide nanoparticles as MR imaging and hypothermic agent, quantum dots as a fluorescent imaging agent, paclitaxel as a therapeutic agent, and anti-prostate specific membrane antigen (anti-PMSA) as a targeting moiety was developed and was shown to have good targeting, imaging, and localized treatment functionality (Cho et al. 2010). HER-2 conjugated near-IR QD loaded micelles showed 77 % reduction in tumor volume in MDA-MB 231 tumor bearing mice as monitored using fluorescence imaging (Nurunnabi et al. 2010).

Acoustic phase change droplets incorporating doxorubicin and conjugated to CCRF-CEM cell targeting sgc8c aptamers were tested in vitro as ultrasound based theranostics. High intensity focused ultrasound (HIFU) was used to initiate the acoustic droplet vaporization (ADV) which showed 56.8 decrease in cell viability via mechanical (inertial cavitation) and chemical (doxorubicin based) destruction of cells. B-mode ultrasound imaging was used to determine the contrast enhancement of these vaporized bubbles (Wang et al. 2012). Another potential example of a not yet tested hybrid theranostic system is the dendriworms or dendrimer conjugated magneto-fluorescent nanoworms. It was shown that these delivery systems encapsulating siRNA against EGFR show 70 % gene silencing in vitro and significant down-regulation of EGFR in a transgenic mouse model of glioblastoma. With the crosslinked iron oxide MNP as a component, in vivo imaging using MRI was possible (Agrawal et al. 2009).

Doxorubicin and Gd^{3+}-DOTAMA that incorporated stealth liposomes exposing neural cell adhesion molecule targeted C3d peptide were used to assess the theranostic value in an experimentally induced mouse model of Kaposi's sarcoma. The system showed good therapeutic value by achieving tumor regression as visualized using gadolinium-based MRI (Grange et al. 2010). A nano-carrier system consisting of bovine serum albumin-PLGA conjugated cobalt ferrite nanoparticles can be useful as a prospective theranostic agent. The system provided clear in vitro evidence for hyperthermia-based therapy with MR as an imaging modality (Comes Franchini et al. 2010). Some of the examples of theranostic systems in preclinical or clinical trials have been summarized in Table 13.2.

13.5 Conclusions and Future Outlook

Drug delivery systems have come a long way since their inception and have shown tremendous benefits over conventional method of free drug administration in overcoming some of the key challenges in cancer therapeutics including multidrug resistance (MDR). Various types of delivery systems have been explored to improve the drug solubility, enhance retention and systemic circulation, evade recognition by immune components, and increase the bioavailability through controlled and sustained drug release kinetics, favorable biodistribution profile, and reduced drug-related toxicity. Presently, we are capable of delivering a variety of potential drug candidates including highly labile molecules such as genes, siRNA, miRNA, mRNA and proteins largely due to our thorough understanding of underlying science of drug delivery systems. This allows us to devise a multipronged approach to treat cancer by using nucleic acid therapy by either augmenting anti-cancer effect or suppress cancer-promoting factors in combination with chemotherapy. Such multifaceted attack is the key to control the disease growth and improve life expectancy of the patients.

Recently, combinatorial design has been developed to synthesize novel materials for designing drug delivery systems that allow encapsulation and controlled release of drugs with varying physical properties (Abeylath et al. 2011). Such endeavors have inherent versatility such that they give us the capability to predict the promising delivery system for a given drug, thereby reducing the time taken for formulated drug to move from preclinical to clinical stage and eventually to bedside. Similarly,

Table 13.2 Illustrative examples of theranostic systems in preclinical and clinical testing

	NP type	NP size (nm)	Therapeutic agent	Diagnostic agent	Disease state	Target	References
Preclinical	Iron oxide	10	Anti-EGFRIgG	Iron oxide	Brain CA	EGFR	Ji et al. (2009)
	Silica	100–200	Paclitaxel	Iron oxide	Many	FolicAcid	Kovar et al. (2007)
	Gold nanorod	10×40	Heat	Thermal/CT	Many	EPR	Fujimoto et al. (2000)
	QD	30	Doxorubicin	QD	Prostate CA	PSMA	Rajian et al. (2009)
Clinical	Gold nanoshell (aurolase)	150	Nanoshell (photothermal)	Nanoshell (MR and optical)	Head and neck CA	EPR	NanoSpectra (NCT00848042)
	Iron oxide	120–180	Injected cell	Iron oxide (endorem)	Healthy volunteers	None	University of Edinburgh (NCT00972946)
	Gold	27	Tumor necrosis factor	Gold NP	Solid tumors	EPR	NCI/Cytimmune (NCT00356980)

Reprinted with permission from Jokerst and Gambhir (2011). American Chemical Society (ACS) Publication. *Abbreviation*: *EGFR* epidermal growth factor receptor, *PSMA* prostate specific membrane antigen, *MSKCC* Memorial Sloan Kettering Cancer Center opportunity exist for the development of externally or internally/self-controlled stimuli-responsive nanocarriers, where the drug release is tailored by application of an external stimulus or according to the factors influenced by disease, respectively. While these systems are still in their exploratory stage of development, they certainly hold tremendous promise as the delivery vehicles of the future. In this regard, theranostic delivery systems are recent addition where the delivery cargo is loaded with diagnostic/imaging capability along with therapeutic agent to simultaneously detect/visualize cancer growth and disease prognosis upon treatment.

Great strides have been made in developing the science for imaging and diagnostics of cancer as well as delivery of chemotherapy and biologics for therapy, but the technology to exploit the cumulative benefit of the two fields is still in its infancy. In true sense, this field has developed largely in the past decade and thus much is wanted before we can realize the true application of theranostic systems. Besides, the majority of endeavors made have largely focused on the improvement of imaging of the disease to aid in following disease prognosis after initiation of therapeutic intervention, but there are no technologies that could enable detection of onset of cancer when the tumor growth is localized and dissemination of cells to other organs has not started. Early detection of cancer significantly improves the possibility of reversing the outcome of the disease is a well-established and acknowledged fact, but methods to achieve this are still elusive. Development of methods to detect tumor when it does not show the characteristic features of a tumor could only be achieved by a thorough understanding of tumor biology, signature molecular events associated, and devising strategies to find the lacunae in the otherwise perfectly disguised physiology. Even though much is known about tumor physiology and microenvironment, there are several missing links that should be connected, such as the exact role of cancer stem cells in tumor progression and metastasis. This information could hold the key to several grave problems associated with the disease such as MDR and metastasis.

Tumor cells are actively dividing and thus are in constant need of oxygen and nutrient supply to meet their growth/division rate. Angiogenesis is thus one of the hallmark properties of tumor microenvironment and signifies that tumor cells have a need to be in constant access of systemic circulation. Tumor cells shed a significant amount of their signature molecules such as proteins and nucleic acid in the blood in free form or in the form of microvesicles or exosomes, which have abundance of genetic and functional information about the tumor that could be used as a possible source of biomarkers for early cancer detection (D'Souza-Schorey and Clancy 2012). These microvesicles and exosomes have already been explored for their potential in developing biomarker for early cancer detection and analysis of their content to design *personalized therapy* for the patient. However, these vesicles show considerable heterogeneity in their morphology, properties, and surface characteristics, and thus their isolation from complex environment like blood is an extremely challenging and daunting task. Existing methods largely rely on their purification from the contaminating blood cells owing to the difference in size of these vesicles/exosomes; thus, a more robust method is required to be able to fully capture, decipher, and apply the information they carry. Potential theranostic nanocarriers have already moved to different stages of preclinical and clinical development, but much work is still needed before these multi-modal therapeutic delivery vehicles can be used to their full potential for cancer therapeutics.

References

Abeylath SC, Ganta S, Iyer AK, Amiji M (2011) Combinatorial-designed multifunctional polymeric nanosystems for tumor-targeted therapeutic delivery. Acc Chem Res 44:1009–1017

Agrawal A, Min DH, Singh N, Zhu H, Birjiniuk A, von Maltzahn G, Harris TJ, Xing D, Woolfenden SD, Sharp PA, Charest A, Bhatia S (2009) Functional delivery of siRNA in mice using dendriworms. ACS Nano 3:2495–2504

Alexis F, Pridgen EM, Langer R, Farokhzad OC (2010) Nanoparticle technologies for cancer therapy. Handb Exp Pharmacol (197):55–86

Al-Jamal WT, Kostarelos K (2011) Liposomes: from a clinically established drug delivery system to a nanoparticle platform for theranostic nanomedicine. Acc Chem Res 44:1094–1104

Al-Jamal WT, Al-Jamal KT, Bomans PH, Frederik PM, Kostarelos K (2008) Functionalized-quantum-dot-liposome hybrids as multimodal nanoparticles for cancer. Small 4:1406–1415

Al-Jamal WT, Al-Jamal KT, Tian B, Cakebread A, Halket JM, Kostarelos K (2009) Tumor targeting of functionalized quantum dot-liposome hybrids by intravenous administration. Mol Pharm 6:520–530

Amiri H, Mahmoudi M, Lascialfari A (2011) Superparamagnetic colloidal nanocrystal clusters coated with polyethylene glycol fumarate: a possible novel theranostic agent. Nanoscale 3:1022–1030

Anker P, Mulcahy H, Chen XQ, Stroun M (1999) Detection of circulating tumour DNA in the blood (plasma/serum) of cancer patients. Cancer Metastasis Rev 18:65–73

Avril N, Propper D (2007) Functional PET imaging in cancer drug development. Future Oncol 3:215–228

Bao A, Goins B, Klipper R, Negrete G, Phillips WT (2003) 186Re-liposome labeling using 186Re-SNS/S complexes: in vitro stability, imaging, and biodistribution in rats. J Nucl Med 44:1992–1999

Bell AH (1880) Upon the production of sound by radiant energy. Am J Sci 20:305–324

Bouwman P, Jonkers J (2012) The effects of deregulated DNA damage signalling on cancer chemotherapy response and resistance. Nat Rev Cancer 12:587–598

Brannon-Peppas L, Blanchette JO (2004) Nanoparticle and targeted systems for cancer therapy. Adv Drug Deliv Rev 56:1649–1659

Brindle K (2008) New approaches for imaging tumour responses to treatment. Nat Rev Cancer 8:94–107

Byrne JD, Betancourt T, Brannon-Peppas L (2008) Active targeting schemes for nanoparticle systems in cancer therapeutics. Adv Drug Deliv Rev 60:1615–1626

Chen X, Hou Y, Tohme M, Park R, Khankaldyyan V, Gonzales-Gomez I, Bading JR, Laug WE, Conti PS (2004a) Pegylated Arg-Gly-Asp peptide: 64Cu labeling and PET imaging of brain tumor alphavbeta3-integrin expression. J Nucl Med 45:1776–1783

Chen X, Park R, Tohme M, Shahinian AH, Bading JR, Conti PS (2004b) MicroPET and autoradiographic imaging of breast cancer alpha v-integrin expression using 18F- and 64Cu-labeled RGD peptide. Bioconjug Chem 15:41–49

Chen LC, Wei CW, Souris JS, Cheng SH, Chen CT, Yang CS, Li PC, Lo LW (2010) Enhanced photoacoustic stability of gold nanorods by silica matrix confinement. J Biomed Opt 15:016010

Chen N-T, Cheng S-H, Souris JS, Chen C-T, Mou C-Y, Lo L-W (2013a) Theranostic applications of mesoporous silica nanoparticles and their organic/inorganic hybrids. J Mater Chem B 1:3128–3135

Chen H, Zhen Z, Tang W, Todd T, Chuang YJ, Wang L, Pan Z, Xie J (2013b) Label-free luminescent mesoporous silica nanoparticles for imaging and drug delivery. Theranostics 3:650–657

Cho HS, Dong Z, Pauletti GM, Zhang J, Xu H, Gu H, Wang L, Ewing RC, Huth C, Wang F, Shi D (2010) Fluorescent, superparamagnetic nanospheres for drug storage, targeting, and imaging: a multifunctional nanocarrier system for cancer diagnosis and treatment. ACS Nano 4:5398–5404

Cohen SJ, Punt CJ, Iannotti N, Saidman BH, Sabbath KD, Gabrail NY, Picus J, Morse M, Mitchell E, Miller MC, Doyle GV, Tissing H, Terstappen LW, Meropol NJ (2008) Relationship of circulating tumor cells to tumor response, progression-free survival, and overall survival in patients with metastatic colorectal cancer. J Clin Oncol 26:3213–3221

Comes Franchini M, Baldi G, Bonacchi D, Gentili D, Giudetti G, Lascialfari A, Corti M, Marmorato P, Ponti J, Micotti E, Guerrini U, Sironi L, Gelosa P, Ravagli C, Ricci A (2010) Bovine serum albumin-based magnetic nanocarrier for MRI diagnosis and hyperthermic therapy: a potential theranostic approach against cancer. Small 6:366–370

D'Souza-Schorey C, Clancy JW (2012) Tumor-derived microvesicles: shedding light on novel microenvironment modulators and prospective cancer biomarkers. Genes Dev 26:1287–1299

Danila DC, Fleisher M, Scher HI (2011) Circulating tumor cells as biomarkers in prostate cancer. Clin Cancer Res 17:3903–3912

Dvorak HF, Nagy JA, Dvorak JT, Dvorak AM (1988) Identification and characterization of the blood vessels of solid tumors that are leaky to circulating macromolecules. Am J Pathol 133:95–109

Erten A, Wrasidlo W, Scadeng M, Esener S, Hoffman RM, Bouvet M, Makale M (2010) Magnetic resonance and fluorescence imaging of doxorubicin-loaded nanoparticles using a novel in vivo model. Nanomedicine 6:797–807

Fass L (2008) Imaging and cancer: a review. Mol Oncol 2:115–152

Fernandez-Fernandez A, Manchanda R, McGoron AJ (2011) Theranostic applications of nanomaterials in cancer: drug delivery, image-guided therapy, and multifunctional platforms. Appl Biochem Biotechnol 165:1628–1651

Fujimoto JG, Pitris C, Boppart SA, Brezinski ME (2000) Optical coherence tomography: an emerging technology for biomedical imaging and optical biopsy. Neoplasia 2:9–25

Gambhir SS (2002) Molecular imaging of cancer with positron emission tomography. Nat Rev Cancer 2:683–693

Ganta S, Devalapally H, Shahiwala A, Amiji M (2008) A review of stimuli-responsive nanocarriers for drug and gene delivery. J Control Release 126:187–204

Gao J, Liang G, Cheung JS, Pan Y, Kuang Y, Zhao F, Zhang B, Zhang X, Wu EX, Xu B (2008) Multifunctional yolk-shell nanoparticles: a potential MRI contrast and anticancer agent. J Am Chem Soc 130:11828–11833

Gobin AM, Lee MH, Halas NJ, James WD, Drezek RA, West JL (2007) Near-infrared resonant nanoshells for combined optical imaging and photothermal cancer therapy. Nano Lett 7:1929–1934

Goldman LW (2007) Principles of CT and CT technology. J Nucl Med Technol 35:115–128, quiz 129–130

Grange C, Geninatti-Crich S, Esposito G, Alberti D, Tei L, Bussolati B, Aime S, Camussi G (2010) Combined delivery and magnetic resonance imaging of neural cell adhesion molecule-targeted doxorubicin-containing liposomes in experimentally induced Kaposi's sarcoma. Cancer Res 70:2180–2190

Greaves M, Maley CC (2012) Clonal evolution in cancer. Nature 481:306–313

Guthi JS, Yang S-G, Huang G, Li S, Khemtong C, Kessinger CW, Peyton M, Minna JD, Brown KC, Gao J (2009) MRI-visible micellar nanomedicine for targeted drug delivery to lung cancer cells. Mol Pharm 7:32–40

Hanahan D, Weinberg RA (2011) Hallmarks of cancer: the next generation. Cell 144:646–674

Hanlon EB, Manoharan R, Koo TW, Shafer KE, Motz JT, Fitzmaurice M, Kramer JR, Itzkan I, Dasari RR, Feld MS (2000) Prospects for in vivo Raman spectroscopy. Phys Med Biol 45:R1–R59

Harrington KJ, Mohammadtaghi S, Uster PS, Glass D, Peters AM, Vile RG, Stewart JS (2001) Effective targeting of solid tumors in patients with locally advanced cancers by radiolabeled pegylated liposomes. Clin Cancer Res 7:243–254

Hendee WR, Morgan CJ (1984) Magnetic resonance imaging. Part I–physical principles. West J Med 141:491–500

Heo DN, Yang DH, Moon HJ, Lee JB, Bae MS, Lee SC, Lee WJ, Sun IC, Kwon IK (2012) Gold nanoparticles surface-functionalized with paclitaxel drug and biotin receptor as theranostic agents for cancer therapy. Biomaterials 33:856–866

Hillman EM, Amoozegar CB, Wang T, McCaslin AF, Bouchard MB, Mansfield J, Levenson RM (2011) In vivo optical imaging and dynamic contrast methods for biomedical research. Philos Trans A Math Phys Eng Sci 369:4620–4643

Hirsch LR, Stafford RJ, Bankson JA, Sershen SR, Rivera B, Price RE, Hazle JD, Halas NJ, West JL (2003) Nanoshell-mediated near-infrared thermal therapy of tumors under magnetic resonance guidance. Proc Natl Acad Sci U S A 100:13549–13554

Hoefnagel CA (1998) Radionuclide cancer therapy. Ann Nucl Med 12:61–70

Holohan C, Van Schaeybroeck S, Longley DB, Johnston PG (2013) Cancer drug resistance: an evolving paradigm. Nat Rev Cancer 13:714–726

Hu M, Chen J, Li ZY, Au L, Hartland GV, Li X, Marquez M, Xia Y (2006) Gold nanostructures: engineering their plasmonic properties for biomedical applications. Chem Soc Rev 35:1084–1094

Huang D, Swanson EA, Lin CP, Schuman JS, Stinson WG, Chang W, Hee MR, Flotte T, Gregory K, Puliafito CA et al (1991) Optical coherence tomography. Science 254:1178–1181

Huang Z, McWilliams A, Lui H, McLean DI, Lam S, Zeng H (2003) Near-infrared Raman spectroscopy for optical diagnosis of lung cancer. Int J Cancer 107:1047–1052

Iyer AK, Singh A, Ganta S, Amiji MM (2013) Role of integrated cancer nanomedicine in overcoming drug resistance. Adv Drug Deliv Rev 65:1784–1802

Jain RK (1987) Transport of molecules in the tumor interstitium: a review. Cancer Res 47:3039–3051

Jain TK, Richey J, Strand M, Leslie-Pelecky DL, Flask CA, Labhasetwar V (2008) Magnetic nanoparticles with dual functional properties: drug delivery and magnetic resonance imaging. Biomaterials 29:4012–4021

Ji X, Cheng L, Wei F, Li H, Wang M, Tian Y, Chen X, Wang Y, Wolf F, Li C, Huang Q (2009) Noninvasive visualization of retinoblastoma growth and metastasis via bioluminescence imaging. Invest Ophthalmol Vis Sci 50:5544–5551

John R, Adie SG, Chaney EJ, Marjanovic M, Tangella KV, Boppart SA (2013) Three-dimensional optical coherence tomography for optical biopsy of lymph nodes and assessment of metastatic disease. Ann Surg Oncol 20:3685–3693

Jokerst JV, Gambhir SS (2011) Molecular imaging with theranostic nanoparticles. Acc Chem Res 44:1050–1060

Jokerst JV, Miao Z, Zavaleta C, Cheng Z, Gambhir SS (2011) Affibody-functionalized gold-silica nanoparticles for Raman molecular imaging of the epidermal growth factor receptor. Small 7:625–633

Junttila MR, de Sauvage FJ (2013) Influence of tumour micro-environment heterogeneity on therapeutic response. Nature 501:346–354

Kaneko OF, Willmann JK (2012) Ultrasound for molecular imaging and therapy in cancer. Quant Imaging Med Surg 2:87–97

Koide A, Kishimura A, Osada K, Jang WD, Yamasaki Y, Kataoka K (2006) Semipermeable polymer vesicle (PICsome) self-assembled in aqueous medium from a pair of oppositely charged block copolymers: physiologically stable micro-/nanocontainers of water-soluble macromolecules. J Am Chem Soc 128:5988–5989

Kono K, Nakashima S, Kokuryo D, Aoki I, Shimomoto H, Aoshima S, Maruyama K, Yuba E, Kojima C, Harada A, Ishizaka Y (2011) Multi-functional liposomes having temperature-triggered release and magnetic resonance imaging for tumor-specific chemotherapy. Biomaterials 32:1387–1395

Kovar JL, Simpson MA, Schutz-Geschwender A, Olive DM (2007) A systematic approach to the development of fluorescent contrast agents for optical imaging of mouse cancer models. Anal Biochem 367:1–12

Kozlowski P, Chang SD, Jones EC, Berean KW, Chen H, Goldenberg SL (2006) Combined diffusion-weighted and dynamic contrast-enhanced MRI for prostate cancer diagnosis—correlation with biopsy and histopathology. J Magn Reson Imaging 24:108–113

Krasia-Christoforou T, Georgiou TK (2013) Polymeric theranostics: using polymer-based systems for simultaneous imaging and therapy. J Mater Chem B 1:3002–3025

Lai J-R, Chang Y-W, Yen H-C, Yuan N-Y, Liao M-Y, Hsu C-Y, Tsai J-L, Lai P-S (2010) Multifunctional doxorubicin/superparamagnetic iron oxide-encapsulated Pluronic F127 micelles used for chemotherapy/magnetic resonance imaging. J Appl Phys 107:09B318

Lammers T, Kiessling F, Hennink WE, Storm G (2010) Nanotheranostics and image-guided drug delivery: current concepts and future directions. Mol Pharm 7:1899–1912

Larson TA, Bankson J, Aaron J, Sokolov K (2007) Hybrid plasmonic magnetic nanoparticles as molecular specific agents for MRI/optical imaging and photothermal therapy of cancer cells. Nanotechnology 18:325101

Li W, Brown PK, Wang LV, Xia Y (2011a) Gold nanocages as contrast agents for photoacoustic imaging. Contrast Media Mol Imaging 6:370–377

Li X, Qian Y, Liu T, Hu X, Zhang G, You Y, Liu S (2011b) Amphiphilic multiarm star block copolymer-based multifunctional unimolecular micelles for cancer targeted drug delivery and MR imaging. Biomaterials 32:6595–6605

Liang HD, Blomley MJK (2003) The role of ultrasound in molecular imaging. Br J Radiol 76:S140–S150

Lichtman JW, Conchello JA (2005) Fluorescence microscopy. Nat Methods 2:910–919

Lim E-K, Jang E, Lee K, Haam S, Huh Y-M (2013) Delivery of cancer therapeutics using nanotechnology. Pharmaceutics 5:294–317

Liu T, Li X, Qian Y, Hu X, Liu S (2012) Multifunctional pH-disintegrable micellar nanoparticles of asymmetrically functionalized beta-cyclodextrin-based star copolymer covalently conjugated with doxorubicin and DOTA-Gd moieties. Biomaterials 33:2521–2531

Longley DB, Johnston PG (2005) Molecular mechanisms of drug resistance. J Pathol 205:275–292

Loo C, Lowery A, Halas N, West J, Drezek R (2005) Immunotargeted nanoshells for integrated cancer imaging and therapy. Nano Lett 5:709–711

Lowery A, Onishko H, Hallahan DE, Han Z (2011) Tumor-targeted delivery of liposome-encapsulated doxorubicin by use of a peptide that selectively binds to irradiated tumors. J Control Release 150:117–124

Luk BT, Fang RH, Zhang L (2012) Lipid- and polymer-based nanostructures for cancer theranostics. Theranostics 2:1117–1126

Maes W, Deroose C, Reumers V, Krylyshkina O, Gijsbers R, Baekelandt V, Ceuppens J, Debyser Z, Van Gool SW (2009) In vivo bioluminescence imaging in an experimental mouse model for dendritic cell based immunotherapy against malignant glioma. J Neurooncol 91:127–139

Mahadevan A (2003) Biomedical Photonics Handbook 30:31–30:27

Mallidi S, Luke GP, Emelianov S (2011) Photoacoustic imaging in cancer detection, diagnosis, and treatment guidance. Trends Biotechnol 29:213–221

Martina MS, Fortin JP, Menager C, Clement O, Barratt G, Grabielle-Madelmont C, Gazeau F, Cabuil V, Lesieur S (2005) Generation of superparamagnetic liposomes revealed as highly efficient MRI contrast agents for in vivo imaging. J Am Chem Soc 127:10676–10685

Massoud TF, Gambhir SS (2003) Molecular imaging in living subjects: seeing fundamental biological processes in a new light. Genes Dev 17:545–580

Matsumura Y, Maeda H (1986) A new concept for macromolecular therapeutics in cancer chemotherapy: mechanism of tumoritropic accumulation of proteins and the antitumor agent smancs. Cancer Res 46:6387–6392

Medeiros SF, Santos AM, Fessi H, Elaissari A (2011) Stimuli-responsive magnetic particles for biomedical applications. Int J Pharm 403:139–161

Mitchell PS, Parkin RK, Kroh EM, Fritz BR, Wyman SK, Pogosova-Agadjanyan EL, Peterson A, Noteboom J, O'Briant KC, Allen A, Lin DW, Urban N, Drescher CW, Knudsen BS, Stirewalt DL, Gentleman R, Vessella RL, Nelson PS, Martin DB, Tewari M (2008) Circulating microRNAs as stable blood-based markers for cancer detection. Proc Natl Acad Sci U S A 105:10513–10518

Muthu MS, Kulkarni SA, Raju A, Feng SS (2012) Theranostic liposomes of TPGS coating for targeted co-delivery of docetaxel and quantum dots. Biomaterials 33:3494–3501

Ntziachristos V (2006) Fluorescence molecular imaging. Annu Rev Biomed Eng 8:1–33

Nurunnabi M, Cho KJ, Choi JS, Huh KM, Lee YK (2010) Targeted near-IR QDs-loaded micelles for cancer therapy and imaging. Biomaterials 31:5436–5444

O'Farrell AC, Shnyder SD, Marston G, Coletta PL, Gill JH (2013) Non-invasive molecular imaging for preclinical cancer therapeutic development. Br J Pharmacol 169:719–735

Osborne MP, Richardson VJ, Jeyasingh K, Ryman BE (1979) Radionuclide-labelled liposomes—a new lymph node imaging agent. Int J Nucl Med Biol 6:75–83

Osiac E, Saftoiu A, Gheonea DI, Mandrila I, Angelescu R (2011) Optical coherence tomography and Doppler optical coherence tomography in the gastrointestinal tract. World J Gastroenterol 17:15–20

Park J, Estrada A, Sharp K, Sang K, Schwartz JA, Smith DK, Coleman C, Payne JD, Korgel BA, Dunn AK, Tunnell JW (2008) Two-photon-induced photoluminescence imaging of tumors using near-infrared excited gold nanoshells. Opt Express 16:1590–1599

Peer D, Karp JM, Hong S, Farokhzad OC, Margalit R, Langer R (2007) Nanocarriers as an emerging platform for cancer therapy. Nat Nanotechnol 2:751–760

Peng CL, Shih YH, Lee PC, Hsieh TM, Luo TY, Shieh MJ (2011) Multimodal image-guided photothermal therapy mediated by 188Re-labeled micelles containing a cyanine-type photosensitizer. ACS Nano 5:5594–5607

Petersen AL, Binderup T, Rasmussen P, Henriksen JR, Elema DR, Kjaer A, Andresen TL (2011) 64Cu loaded liposomes as positron emission tomography imaging agents. Biomaterials 32:2334–2341

Popovtzer R, Agrawal A, Kotov NA, Popovtzer A, Balter J, Carey TE, Kopelman R (2008) Targeted gold nanoparticles enable molecular CT imaging of cancer. Nano Lett 8:4593–4596

Pu Y, Wang WB, Xu M, Tang GC, Budansky Y, Sharanov M, Achilefu S, Eastham JA, Alfano RR (2011) Near infrared photonic finger imager for prostate cancer screening. Technol Cancer Res Treat 10:507–517

Rajian JR, Carson PL, Wang X (2009) Quantitative photoacoustic measurement of tissue optical absorption spectrum aided by an optical contrast agent. Opt Express 17:4879–4889

Raman CV (1928) New type of secondary radiation. Nature 121:501–502

Reddy GR, Bhojani MS, McConville P, Moody J, Moffat BA, Hall DE, Kim G, Koo YE, Woolliscroft MJ, Sugai JV, Johnson TD, Philbert MA, Kopelman R, Rehemtulla A, Ross BD (2006) Vascular targeted nanoparticles for imaging and treatment of brain tumors. Clin Cancer Res 12:6677–6686

Rosenthal A, Jacobs I (1998) Ovarian cancer screening. Semin Oncol 25:315–325

Ruoslahti E (2002) Specialization of tumour vasculature. Nat Rev Cancer 2:83–90

Shackleton M, Quintana E, Fearon ER, Morrison SJ (2009) Heterogeneity in cancer: cancer stem cells versus clonal evolution. Cell 138:822–829

Shin J, Anisur RM, Ko MK, Im GH, Lee JH, Lee IS (2009) Hollow manganese oxide nanoparticles as multifunctional agents for magnetic resonance imaging and drug delivery. Angew Chem Int Ed Engl 48:321–324

Singh A, Chernenko T, Amiji M (2012) Theranostic applications of plasmonic nanosystems. In: Functional nanoparticles for bioanalysis, nanomedicine, and bioelectronic devices, vol 2. American Chemical Society, pp 383–413

Soling A, Rainov NG (2003) Bioluminescence imaging in vivo – application to cancer research. Expert Opin Biol Ther 3:1163–1172

Solomon M, Liu Y, Berezin MY, Achilefu S (2011) Optical imaging in cancer research: basic principles, tumor detection, and therapeutic monitoring. Med Princ Pract 20:397–415

Surmacki J, Musial J, Kordek R, Abramczyk H (2013) Raman imaging at biological interfaces: applications in breast cancer diagnosis. Mol Cancer 12:48
Swanton C (2012) Intratumor heterogeneity: evolution through space and time. Cancer Res 72:4875–4882
Tagami T, Foltz WD, Ernsting MJ, Lee CM, Tannock IF, May JP, Li SD (2011) MRI monitoring of intratumoral drug delivery and prediction of the therapeutic effect with a multifunctional thermosensitive liposome. Biomaterials 32:6570–6578
Tai LA, Tsai PJ, Wang YC, Wang YJ, Lo LW, Yang CS (2009) Thermosensitive liposomes entrapping iron oxide nanoparticles for controllable drug release. Nanotechnology 20:135101
Tian B, Al-Jamal WT, Al-Jamal KT, Kostarelos K (2011) Doxorubicin-loaded lipid-quantum dot hybrids: surface topography and release properties. Int J Pharm 416:443–447
Tredan O, Galmarini CM, Patel K, Tannock IF (2007) Drug resistance and the solid tumor microenvironment. J Natl Cancer Inst 99:1441–1454
Vakoc BJ, Fukumura D, Jain RK, Bouma BE (2012) Cancer imaging by optical coherence tomography: preclinical progress and clinical potential. Nat Rev Cancer 12:363–368
von Maltzahn G, Park JH, Agrawal A, Bandaru NK, Das SK, Sailor MJ, Bhatia SN (2009) Computationally guided photothermal tumor therapy using long-circulating gold nanorod antennas. Cancer Res 69:3892–3900
Wang LV (2009) Multiscale photoacoustic microscopy and computed tomography. Nat Photonics 3:503–509
Wang M, Thanou M (2010) Targeting nanoparticles to cancer. Pharmacol Res 62:90–99
Wang CH, Kang ST, Lee YH, Luo YL, Huang YF, Yeh CK (2012) Aptamer-conjugated and drug-loaded acoustic droplets for ultrasound theranosis. Biomaterials 33:1939–1947
Welch MJ, Hawker CJ, Wooley KL (2009) The advantages of nanoparticles for PET. J Nucl Med 50:1743–1746
Weng KC, Noble CO, Papahadjopoulos-Sternberg B, Chen FF, Drummond DC, Kirpotin DB, Wang D, Hom YK, Hann B, Park JW (2008) Targeted tumor cell internalization and imaging of multifunctional quantum dot-conjugated immunoliposomes in vitro and in vivo. Nano Lett 8:2851–2857
Yang X, Skrabalak SE, Li ZY, Xia Y, Wang LV (2007a) Photoacoustic tomography of a rat cerebral cortex in vivo with au nanocages as an optical contrast agent. Nano Lett 7:3798–3802
Yang J, Lee CH, Ko HJ, Suh JS, Yoon HG, Lee K, Huh YM, Haam S (2007b) Multifunctional magneto-polymeric nanohybrids for targeted detection and synergistic therapeutic effects on breast cancer. Angew Chem Int Ed Engl 46:8836–8839
Yang X, Hong H, Grailer JJ, Rowland IJ, Javadi A, Hurley SA, Xiao Y, Yang Y, Zhang Y, Nickles RJ, Cai W, Steeber DA, Gong S (2011) cRGD-functionalized, DOX-conjugated, and (6)(4) Cu-labeled superparamagnetic iron oxide nanoparticles for targeted anticancer drug delivery and PET/MR imaging. Biomaterials 32:4151–4160
Ye F, Ke T, Jeong EK, Wang X, Sun Y, Johnson M, Lu ZR (2006) Noninvasive visualization of in vivo drug delivery of poly(L-glutamic acid) using contrast-enhanced MRI. Mol Pharm 3:507–515
Yu MK, Jeong YY, Park J, Park S, Kim JW, Min JJ, Kim K, Jon S (2008) Drug-loaded superparamagnetic iron oxide nanoparticles for combined cancer imaging and therapy in vivo. Angew Chem Int Ed Engl 47:5362–5365
Yuan F, Dellian M, Fukumura D, Leunig M, Berk DA, Torchilin VP, Jain RK (1995) Vascular permeability in a human tumor xenograft: molecular size dependence and cutoff size. Cancer Res 55:3752–3756
Yuan Q, Hein S, Misra RD (2010) New generation of chitosan-encapsulated ZnO quantum dots loaded with drug: synthesis, characterization and in vitro drug delivery response. Acta Biomater 6:2732–2739
Zabala M, Alzuguren P, Benavides C, Crettaz J, Gonzalez-Aseguinolaza G, Ortiz de Solorzano C, Gonzalez-Aparicio M, Kramer MG, Prieto J, Hernandez-Alcoceba R (2009) Evaluation of bioluminescent imaging for noninvasive monitoring of colorectal cancer progression in the liver and its response to immunogene therapy. Mol Cancer 8:2
Zhang Y, Hong H, Cai W (2010) Imaging with Raman spectroscopy. Curr Pharm Biotechnol 11:654–661

Part VI
Translational Aspects of Nano-Oncologicals

Chapter 14
Nanotoxicology: Towards Safety by Design

Fernando Torres Andón and Bengt Fadeel

14.1 Introduction

The development of new anti-cancer drugs remains an expensive and inefficient process, mainly due to non-specific effects and the inherent toxicity of the drugs. The rise of molecular biology and systems biology has led to major progress in our understanding of cancer cell biology. A large number of novel anti-cancer drugs have been developed and many are now implemented into routine clinical practice (Trotta et al. 2011). Some of these drugs have yielded improvements in overall survival for all patients with a given condition, whereas others have only shown benefit in smaller groups of patients with known molecular aberrations. However, less than 5 % of the drugs that reach Phase I gain a marketing authorization and only 1 in 10,000 pre-clinical compounds ever reach the market (Moreno and Pearson 2013). The high attrition rate in the drug discovery and development process has been defined by some as the "Valley of Death" in anti-cancer drug development (Adams 2012). While non-scientific reasons for this failure include lack or resources, wrong incentives, aggressive pricing strategies, or adverse regulatory environments; the scientific reasons or obstacles include the tumor microenvironment, cross-talk and negative feedback loops, development of resistance, exposure time, drug delivery, and the choice of pre-clinical models (Moreno and Pearson 2013). It is in this context that the development of novel nano-oncologicals must take place, providing original and innovative solutions to tackle each of these issues. Notably, concerns

F.T. Andón • B. Fadeel (✉)
Division of Molecular Toxicology, Institute of Environmental Medicine, Karolinska Institutet, Nobels väg 13, 17177 Stockholm, Sweden
e-mail: bengt.fadeel@ki.se

M.J. Alonso, M. Garcia-Fuentes (eds.), *Nano-Oncologicals: New Targeting and Delivery Approaches*, Advances in Delivery Science and Technology,
DOI 10.1007/978-3-319-08084-0_14

with toxicity account for approximately one-third of the cases of attrition. The aim of the current chapter is thus to tackle the question of safety and toxicological assessment in nanomedicine. Emphasis is placed on the interactions of nanoparticles with the immune system and on the role of the material-intrinsic "identity" versus biological "identity" of nanomaterials for biological outcomes. New and emerging approaches in nanotoxicological research are also highlighted.

14.1.1 Understanding the Pharmacological/Toxicological Profile of Nanomedicines

The pharmacological/toxicological profile of a pre-clinical nanomedicine can be classified into primary or secondary effects. While primary effects relate to the action of the compound at its intended target (or targets), secondary effects are due to interactions with targets other than the primary target (i.e., off-target interactions) (Bowes et al. 2012). The *off-target toxicity* is often the cause of adverse drug reactions (ADRs). Thus, in addition to the development of novel drug delivery and targeting strategies with higher selectivity for the desired target, one should aim to design nanomedicines that are effective at lower doses in order to minimize secondary toxicities. Another challenge when using anti-cancer drugs, known as *on-target toxicity*, is to separate the beneficial effects of primary target inhibition in tumors from potential adverse effects that may occur as a result of inhibition of the same target in normal tissues. In this case, nanomedicines offer novel opportunities to control the balance between efficacious and toxic doses, as well as, the inter-tissue biodistribution of the drug. Not less important are other toxicological issues mainly arising from the interaction of new nanomedicines with the immune system. The immune system is a specialized network of biological structures, cells, tissues, and organs that work together to defend our body against foreign intrusion (Fadeel 2012); it is not surprising that engineered nanomaterials are frequently sensed as being "foreign" and dangerous, leading to the triggering of immunological responses. These immunotoxicological issues, of crucial importance for the development of novel nanomedicines, are closely interlinked and can be summarized as: bioactivation, hypersensitivity and immune responses, and idiosyncratic reactions. Bioactivation could occur when the original drug is converted to reactive products (often referred to as reactive metabolites) which in turn may cause toxicity. Importantly, as novel nanostructures are prone to interact with proteins (see Sect. 3), this toxicity could arise through the modification of proteins with their consequent loss of function or induction of immune responses. The hypersensitivity and immune responses induced by some drug candidates (or their metabolites) could arise from the non-desirable interaction of drugs with immune-competent cells and/or proteins in the body, such as the complement system. Finally, the idiosyncratic reactions are rare events ($1/10^4$ individuals) which are still poorly understood and difficult to predict using current in vitro or in vivo models. Personalized medicine holds great promise in terms of avoiding these idiosyncratic reactions, which are commonly attributed to interindividual differences in the metabolism capacity and/or immune system sensitivity. Thus, drug discovery strategies must be focused on gaining a better understanding of the mechanisms of action of the novel drugs, and of the drug carriers, as well as, to achieve an appropriate balance between their efficacy and potential adverse effects. This should be done as early as possible in the drug development process. This is true also in nanomedicine (Fadeel, 2013). In the following sections, some lessons learned from nanotoxicogical studies performed in the past decade are summarized.

14.1.2 Nanotoxicology: Understanding How Nano-Scale Materials Adversely Affect Human Health

"Nanotoxicology" may be defined as the study of the adverse effects of engineered nanomaterials on living organisms and the ecosystems, including the prevention and amelioration of such adverse effects. Thus, the importance of nanotoxicology for nanomedicine is obvious, as is that of toxicology for medicine in general. In this chapter, we seek to highlight current concepts in nanotoxicological research, including the role of the "synthetic identity" of nanomaterials which is determined by the physico-chemical properties of the nanomaterial itself, and the "biological identity" which is defined by the adsorption of biomolecules onto the surface of the nanomaterial, as the nanomaterial encounters a particular biological fluid (Fadeel et al. 2013). Classical biological/toxicological endpoints (for instance, cell death, or cytokine release, as a marker of inflammation) should be evaluated also for nanomaterials; however, we emphasize the importance of not only considering whether cells are dead or alive but, more importantly, to study the molecular mechanisms underlying the cytotoxic and/or inflammatory effects of nanomaterials. We suggest that insights regarding the mode of action of different classes of nanomaterials will lead to more refined approaches for the mitigation of ADRs elicited by nanomedicines. Emerging approaches, such as high-throughput screening (HTS) and structure–activity modeling for in silico prediction of toxicological effects, may facilitate this endeavor.

14.2 Synthetic Identity: The Relevance of Material-Intrinsic Properties of Nanomaterials

Most formal definitions of engineered nanomaterials evolve around the manipulation of phenomena and materials at dimensions roughly between 1 and 100 nanometers (nm). Nevertheless, alternative ways of defining nanotechnology focus on the novel properties that arise, owing to the specific smallness of the particles or materials independent of an exact specification of size (Bleeker et al. 2013). Some classifications of nanomaterials could also be suggested regarding their origin, nature, or potential risks. Bleeker et al. (2013) have recently suggested the need to separate (1) natural, incidental, and manufactured, (2) soluble versus non-soluble, and (3) hard versus soft nanomaterials. Soft nanomaterials refer to nano-sized materials, such as micelles, emulsions, and liposomes. It is important to define exposure along with the hazard to understand if a nano-specific risk should be evaluated. Situations may be defined in which nano-related risks are negligible (i.e., oral administration of soft nanomaterials which fall apart in the gastrointestinal tract). However, if biological processes can occur within these soft nanomaterials, as may be the case in products from synthetic biology that can also be nano-sized, e.g., nano-sized "protein factories" which involve packaging the molecular machinery for protein synthesis into a nanoparticle (Schroeder et al. 2010), such conclusions should not be drawn too quickly. Nano-specific risk may also be limited when nanomaterials are quickly dissolved in physiologically or environmentally relevant media into a non-nanomaterial; however, particle dissolution may in some cases give rise to ionic species that are toxic, as shown for certain metal and metal oxide nanoparticles. It is important to realize that the nanotechnology of today, which deals largely with the study and manipulation of individual classes of nanoparticles or nanostructured surfaces, is rapidly evolving into the study of much more complex nanosystems or nanomachines analogous to endogenous "nanomachines" such as the ribosome inside the cells (Shvedova et al. 2010). We need to ensure that the battery of test methods used to assess toxicity remains valid for such complex structures.

The physico-chemical properties of engineered nanomaterials can be manipulated in numerous ways to acquire high stability, high carrier capacity, and compatibility with different administration routes. These characteristics make nanomaterials highly attractive in diagnosis and therapy of cancer. However, the same properties also have important implications for toxicity. Systematic investigations to correlate the toxic effects of nanomaterials to their physico-chemical properties are thus needed. A comprehensive understanding of the "synthetic identity" of a nanomaterial, defined by the intrinsic properties of the nanomaterial, is essential for the efficient design of safe nanomedicines (Fadeel et al. 2013).

14.2.1 Linking Toxicity to the Synthetic Identity of Nanomaterials: Size, Shape, Surface Charge, Etc.

The size of engineered nanomaterials determines their cellular uptake and biodistribution in tissues. Moreover, a higher intracellular dose of nanoparticles is often correlated with a higher toxicity. Andar et al. (2014) performed a comprehensive study of liposomes with different sizes revealing that higher uptake was correlated with smaller particles. Furthermore, these investigations revealed different uptake mechanisms. Others have demonstrated that nanoparticles smaller than 200 nm are internalized via clathrin-coated pits reaching endosomal/lysosomal compartments, whereas 500 nm particles are internalized via caveolae-mediated pathway and delivery to lysosomes was no longer apparent (Rejman et al. 2004). Extensive in vitro studies of size-dependent uptake have been performed for gold (Chithrani 2010) and silica nanoparticles (Napierska et al. 2009). Notably, the size dependence of nanoparticle uptake and cytotoxicity was studied using colloidally stable and non-agglomerated gold nanoparticles (Chithrani 2010; Chithrani and Chan 2007; Chithrani et al. 2006). These examples clearly demonstrated that size can affect the route and extent of uptake of nanomedicines by target cells. However, there are also studies demonstrating that cellular uptake and direct contact with cells is not mandatory for cytotoxicity to occur: cobalt–chromium nanoparticles were shown to damage human fibroblasts across an intact cellular barrier without having to cross the barrier. The outcome, which includes DNA damage without significant cell death, is different from that observed in cells subjected to direct exposure to nanoparticles (Bhabra et al. 2009). Besides the relevance of size for the uptake of nanomaterials and consequent cytotoxic effects, some studies have shown that size can influence the mechanism of cell death. Hence, Pan et al. (2007) showed that 1.4 nm gold particles caused necrosis, while 1.2 nm particles predominantly induced apoptosis. In contrast, the 15 nm particles were non-cytotoxic. Furthermore, Kim et al. (2012b) reported that small (10 nm) silver nanoparticles had a greater ability to induce apoptosis than other-sized silver nanoparticles (50 and 100 nm). Overall, size matters for the interaction of nanoparticles with cells, and impacts on cellular uptake, and (mode of) toxicity.

Nanoparticle shape can also determine the biological fate of nanomedicines in the body and in the cell. In particular, phagocytosis of foreign materials such as engineered nanomaterials can be influenced by their aspect ratios. Particles of ellipsoid shape are more readily engulfed by macrophages than are spherical particles (Sharma et al. 2010). In contrast, nanomaterials with dramatically high aspect ratios (i.e., typical "needle-like"-shaped, multi-walled carbon nanotubes) may resist uptake by macrophages but could cause cellular damage through piercing of the plasma membrane, as shown for the mesothelial cells (Nagai et al. 2011). Similarly, the alteration of a relatively inert material, such as TiO_2, into a fibrous structure results in a toxic nanomaterial that provokes inflammatory responses in alveolar macrophages (Hamilton et al. 2009). In addition, to the overall shape of the nanoparticles, the smoothness/roughness of the particle's surface also affects the opsonization of the particle and its subsequent uptake by phagocytes. These results indicate that materials altered into shapes that are difficult to process by phagocytic cells can result in toxicity by lysosomal disruption (Hamilton et al. 2009). Investigations by Schaeublin et al. (2012) in which keratinocytes were exposed to gold nanoparticles with different aspect ratios demonstrated that the gold nanospheres were non-toxic, whereas the gold nanorods induced apoptosis. Notably, both nanoparticles formed agglomerates in cell culture medium, but the spherical particles had a large fractal dimension (i.e., tightly bound and densely packed), while the nanorod agglomerates had a small fractal dimension (i.e., loosely bound). Similarly, the in vivo toxicity of SWCNTs has also been demonstrated to be highly dependent on their agglomeration; fibers may sometimes behave as agglomerated particles and not as ideal "fibers" (Mutlu et al. 2010; Murray et al. 2012). Importantly, the agglomeration of nanoparticles should be studied in relevant physiological media, depending on the anticipated portal of entry into the body (see Sect. 14.3).

Surface properties, such as the charge (zeta potential) and hydrophobicity, directly affect the interaction of engineered nanomaterials with biological surfaces, cell membranes, and proteins (Zhu et al. 2013). In general, owing to the overall net negative charge of cellular surfaces, positively charged nanoparticles are incorporated faster by cells than negatively charged ones, leading to high rates of non-specific internalization and a shorter half-life in the circulation. The higher toxicity of positively charged nanoparticles is generally correlated to their enhanced cellular uptake (Oh et al. 2010). Schaeublin et al. (2011) demonstrated the effects of surface charge on the modality of cell death induced by a particular nanomaterial. Thus, charged gold nanoparticles induced cell death through apoptosis, whereas neutral gold nanoparticles triggered necrosis in a human keratinocyte cell line (Schaeublin et al. 2011). To elucidate surface charge-dependent toxicity, nanoparticles with different surface charge but with other physico-chemical parameters constant are required, but this is often experimentally complicated to achieve (Fadeel et al. 2013). Walkey and Chan (2012) recently described the effects of the bio-corona on the surface charge of nanomaterials. They suggested that the protein corona tends to give nanomaterials a zeta potential of about −10 to −20 mV irrespective of the nanomaterial chemistry; this "normalization" of zeta potentials is related to the fact that most plasma proteins carry a net negative charge at physiological pH (Walkey and Chan 2012). Overall, the physico-chemical properties of nanoparticles are strongly entangled and are difficult to control independently. In fact, it is nontrivial to change only one physico-chemical parameter, without affecting others (Fadeel et al. 2013). For instance, the presence of strong electrostatic barrier created by charge repulsion may even trump other factors affecting toxicity, such as size or shape (El Badawy et al. 2011). Perhaps, the most important physico-chemical parameter is colloidal stability. In other words, unless nanoparticles are very well dispersed, any statement about size- or shape-dependent uptake or cytotoxicity is not sound, as the cell would "see" and interact with the agglomerates and not with the individual nanoparticles. Size, shape, and charge are also altered when the nanoparticles interact with the biological environment. In addition, dissolution of metal and metal oxide nanoparticles in acidic cellular compartments (lysosomes) underlies the Trojan horse-type mechanism of particle toxicity (Cronholm et al. 2013). This mechanism has been attributed to oxides of zinc, iron, manganese, cobalt, and copper nanoparticles which drive toxicity through releasing of toxic ions into cells (Stark 2011). Cho et al. (2012) correlated the toxicity of 15 different metal/metal oxide nanoparticles with one of two physico-chemical parameters: zeta potential under acid conditions for low-solubility nanoparticles and solubility (degree of dissolution) for high-solubility nanoparticles. The authors found that inflammogenicity induced by high-solubility nanoparticles depends not on the nanomaterials itself but on the ions that are produced during their dissolution inside the acidic phagolysosomes of the cells (Cho et al. 2012). A final specific characteristic of nanomaterials is their surface-to-volume ratio, which is unusually high when compared with the bulk material. This is particularly true for porous nanoparticles (Maurer-Jones et al. 2010). A significant increase in exposed functional groups on the material may lead to increased surface reactivity, further increasing the potential for interactions with biological structures such as DNA, proteins, and cell membranes (Xia et al. 2009). The interaction of molecular oxygen and electron donor or acceptor groups on the surface of nanomaterials can result in the generation of reactive oxygen species (ROS) (Nel et al. 2006). Deleterious oxidative stress effects induced by nanoparticles in vivo are commonly related with the persistence or accumulation of the nanomaterials over extended periods of time. Thus, organs such as the liver and spleen, where nanoparticles are sequestered by phagocytic cells following their introduction into the bloodstream, are the main targets of oxidative stress.

14.2.2 Linking Biodistribution to the Synthetic Identity of Nanomaterials/Nanomedicines

The biodistribution of nanomedicines is crucial for their efficacy and depends on the route of administration (pulmonary, gastro-intestinal, intravenous, intraperitoneal, and dermal) and on interactions with the following key organs: the lungs, liver, spleen, and kidney. Small nanomedicines may be removed from the blood by renal clearance (<5 nm) or rapid liver uptake (10–20 nm), whereas larger nanostructures are filtered in the sinusoidal spleen (>200 nm) or are recognized and cleared by the reticuloendothelial system (RES). Therefore, nanomedicines between 20 and 200 nm can remain in the circulation for an extended period of time (Wang et al. 2013a). Shape also matters. Huang et al. (2011) observed that short rod-shaped mesoporous silica nanoparticles are trapped in the liver, whereas long rod-shaped nanoparticles with the same composition are distributed in the spleen. Notably, using highly stable, polymer micelle assemblies designated as filomicelles, Geng et al. (2007) were able to show that filomicelles persisted in the circulation up to 1 week after intravenous injection which is about ten times longer than their spherical counterparts. The surface chemistry of nanoparticles also affects their biodistribution. Xiao et al. (2011) demonstrated that micellar nanoparticles with highly positive or highly negative surfaces undergo liver uptake, which was low for slightly negatively charged nanoparticles.

Several researchers have developed new nanomedicines for pulmonary delivery (Mansour et al. 2009; Zarogoulidis et al. 2012). Indeed, the lung is an attractive target for drug delivery due to noninvasive administration via inhalation aerosols, avoidance of first-pass metabolism, direct delivery to the site of action for the treatment of respiratory diseases, and the availability of a huge surface area for local drug action and systemic absorption of drug. In this regard, lessons can be learned about the interaction of engineered nanomaterials with the respiratory system if we look to accidental pulmonary exposure to nanoparticles, a field of great importance in occupational medicine/toxicology (Kendall and Holgate 2012; Bhattacharya et al. 2013). Several studies have suggested that alveolar macrophage-mediated clearance of nanoparticles smaller than 100 nm is quite inefficient (Geiser 2010). Hence, nanomaterials are likely to translocate beyond the epithelial barrier into the interstitium, where they might accumulate for a long time due to their resistance to phagocytic uptake (Wang et al. 2013a). In a key study, Choi et al. (2010) determined that nanoparticles with a hydrodynamic diameter less than 34 nm with non-cationic surface charge translocated rapidly from the lungs to regional lymph nodes in rats following intratracheal instillation. Furthermore, nanoparticles with a hydrodynamic diameter less than 6 nm were found to traffic rapidly from the lungs to lymph nodes and the bloodstream, ultimately being cleared from the body through the kidneys (Choi et al. 2010). Deshmukh et al. (2012) demonstrated that big particles or aggregates of particles >10 μm become passively entrapped within the capillaries of the lung. However, upon degradation, the same particles could reach >3 μm, being only transiently entrapped in the lung and subsequently being translocated to the liver. Particles in the range of 3–6 μm accumulated in the liver and spleen (Deshmukh et al. 2012). Schädlich et al. (2011) investigated the influence of the size of biodegradable PEG-PLA nanoparticles and found that nanoparticles of 111 and 141 nm accumulated in human xenograft tumor tissue, while slightly bigger nanoparticles (166 nm) were rapidly eliminated by the liver. Thus, differences in biodistribution may occur because of small alterations in nanoparticle size. These findings suggest strategies for the rational design of nanoparticles for pulmonary drug delivery. The bio-corona also needs to be considered, as discussed below.

Oral administration is the most widely used and accepted route of drug delivery in the adult population, taking advantage of the highly absorptive capacity of the gut to deliver drugs to the systemic circulation (Roger et al. 2010). The development of novel nanocarriers will be key to increasing the range of formulations for oral delivery in terms of improving drug solubility, drug stability, and bioavailability (Herrero et al. 2012). Nanomedicines for oral administration have been optimized for efficient transit under varying pH, pressure, and enzyme-catalyzed degradation (Laroui et al. 2011). In principle, nanotechnology should enable the design of the optimal drug-carrier: nanoparticles are small in size with correspondingly large surface area-to-volume ratio; engineered nanomaterials have good stability, robustness, and a long shelf-life compared with molecular carriers; drugs can be loaded into nanostructures at high concentrations, and nanomedicines may evade normal digestive processes and thus efficiently deliver drugs to specific sites; the kinetics of drug release can be modulated; and chemically tailorable surfaces may be modified with ligands to affect site-specific drug delivery. Needless to say, all the components of the nanocarrier must be biocompatible, following oral administration. Indeed, some studies have provided evidence of toxicity of metal nanoparticles (e.g., silver and gold) given through the oral route at relatively high doses in mice or rats (Park et al. 2010a; Kim et al. 2010). The highly acidic environment encountered in the stomach has been proposed to play a major role in the toxicity of copper nanoparticles by triggering release of ions (Meng et al. 2007). This and other studies highlight the potential influence of the pH shifts encountered during gastro-intestinal tract transit in both stability and dispersal/aggregation of engineered nanomaterials, potentially influencing local toxicity as well as their ability to penetrate the gastro-intestinal barriers. The possibility that nanomaterials per se could induce inflammatory responses in the gastro-intestinal tract should be evaluated and also the effect of nanomaterials on the normal microflora which is known to play a key role in gastro-intestinal tract development and regulation of inflammation (Stecher and Hardt 2011) and which has been shown more recently to play a crucial role in tuning the responses to anti-cancer therapy (Iida et al. 2013; Viaud et al. 2013). The interaction of nanomaterials with biological barriers present along the gastro-intestinal tract could lead to changes in the permeability of these barriers that allow for their even wider biodistribution with attendant toxicity (Pietroiusti et al. 2013). In fact, oral exposure to polystyrene nanoparticles was reported to influence iron uptake using an in vivo chicken intestinal loop model, and high doses of the model nanoparticle cause remodeling of the intestinal villi (Mahler et al. 2012). Schleh et al. (2012) demonstrated that size and surface charge of gold nanoparticles determine absorption across the intestinal barriers and accumulation in secondary target organs after oral administration in a rat model. While polystyrene nanoparticles are considered to be rather inert, a recent study suggested that they can induce widespread apoptosis in Caco-2 layers that is propagated to bystander cells via the generation of hydrogen peroxide (Thubagere and Reinhard 2010). Largely unexplored is the matter of the new "biological identity" of nanomaterials travelling along the gastro-intestinal tract, as a result of their interactions with biomolecules in the mucus (Fig. 14.1).

The skin is the main barrier that protects our body from the external environment. A large number of studies demonstrate that insoluble TiO_2 or ZnO nanoparticles, commonly present in sunscreens, are not able to penetrate into or through human skin. On the other hand, skin permeability to nanoparticles may increase 4 to 100 times in atopic subjects with damaged skin (Filon et al. 2011; Larese et al. 2009; Larese Filon et al. 2013). Monteiro-Riviere et al. (2011) demonstrated that UVB-damaged skin slightly enhanced penetration of nanoparticles in sunscreen formulations, but no transdermal absorption was detected.

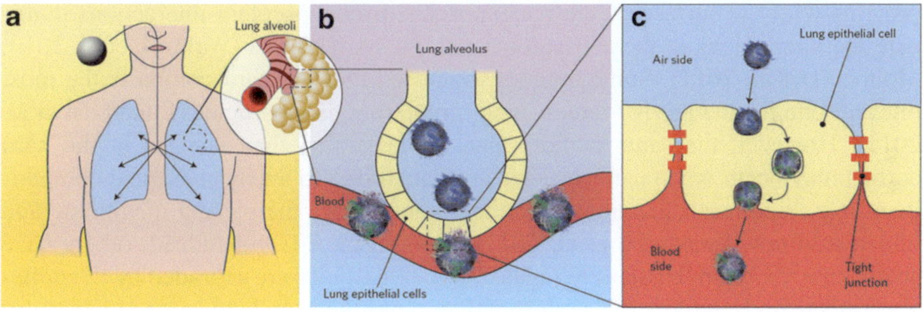

Fig. 14.1 Evolution of the bio-corona during the "fantastic voyage" of nanoparticles through the body. (**a**) Inhaled nanoparticles in the alveoli of the lungs. (**b**) An original bio-corona of proteins and lipids forms when the nanoparticle comes in contact with the lung fluids in the alveoli. On subsequent (size-dependent) translocation across the lung barrier, nanoparticles reach the bloodstream where some biomolecules from the original bio-corona may be displaced by different biomolecules in this new compartment. (**c**) A nanoparticle crossing the epithelial cells of the lung barrier is further transported through different membrane compartments and is finally exported out from the cell. Partial displacement of the original bio-corona by intracellular biomolecules could potentially occur at any point along the pathway, or once the nanoparticle is exposed to biomolecules in the blood [reprinted from Monopoli MP, Åberg C, Salvati A, Dawson KA, Biomolecular coronas provide the biological identity of nanosized materials. Nat Nanotechnol. 2012 Dec;7(12):779–86, with permission from Nature Publishing Group]

14.3 Biological Identity: Role of the Bio-Corona for the Fate of Nanomaterials in Living Systems

When nanomaterials are presented to living systems they interact with biological molecules (proteins, lipids, polysaccharides, etc.) conferring a "biological identity" to the engineered nanostructures (Walkey and Chan 2012; Fadeel et al. 2013). The bio-corona is the dynamic layer of biomolecules that adsorbs to nanoparticle surfaces immediately upon their contact with living systems and is presumed to be what organisms or cells "see" and interact with (Cedervall et al. 2007b). Thus, as suggested in a recent review (Fadeel et al. 2013), it is the combination of material-intrinsic properties (the "synthetic identity") and context-dependent properties determined, in part, by the bio-corona of a given biological compartment (the "biological identity") that will determine the interactions of engineered nanomaterials with cells and tissues and subsequent outcomes. Nanomedicines are designed to interact with biological systems at the nano-scale. Thus, in order to improve biosafety of novel nanomedicines, we must understand the bio-nano interface between these man-made nanostructures and cellular nanostructures or "nanomachines" (Shvedova et al. 2010; Yang et al. 2013b). The characteristic physico-chemical properties of engineered nanomaterials allow them to act as a scaffold for biomolecules. By far, the most studied component of the corona is the protein composition, but lipids, sugars, and other species could also play a role (Kapralov et al. 2012). Furthermore, it is important to differentiate between a "hard" (long-lived) corona of slowly exchanging proteins and an outer "soft" corona collection of (weakly interacting and rapidly exchanging) proteins (Walczyk et al. 2010). It is generally accepted that while the hard corona is formed because of the direct interaction of proteins with the surface of the nanoparticles, protein–protein interactions dominate the interactions of the soft corona with the hard corona (Walczyk et al. 2010; Monopoli et al. 2011). Some authors suggested that the hard corona is the most relevant one for the in vivo fate of long-circulating nanoparticles (Lundqvist et al. 2011). However, the same authors also demonstrated that changes in the hard corona may occur when nanoparticles are transferred to a new biological compartment, e.g., upon translocation of nanoparticles across the plasma membrane and into lysosomes (Lundqvist et al. 2011; Wang et al. 2013b). Indeed, others have shown that the biomolecules conjugated to nanoparticles are degraded by the protease cathepsin L within the endosomal compartment following endocytosis of nanoparticles; this has important implications for intracellular applications of nanomedicines (Sée et al. 2009). Thus, modification or even complete removal of the bio-corona may be expected when a nanomedicine interacts with different cells and barriers which must be surpassed to reach the target organ (Fig. 14.1). Overall, it is important to note that interactions between nanoparticles and biomolecules, and the formation of the bio-nano interface, have consequences for both the nanoparticle itself and potentially also for the proteins and other biomolecules contained in the bio-corona. The bio-corona cannot be ignored in the design of nanomedicines, especially when nanomedicines are functionalized with targeting ligands: the interplay between the "engineered" corona and the adsorbed bio-corona that is formed in situ must be understood. Importantly, Salvati et al. (2013) recently showed, in a study using transferrin-functionalized nanoparticles, that the adsorption of serum proteins obscures the targeting ligand grafted to the nanoparticles, thereby preventing their targeted uptake. In contrast, recent data from our laboratory suggest that specific (i.e., receptor-mediated) uptake of folic acid-functionalized nanoparticles by cancer cells occurs only in the presence, but not in the absence, of serum (Krais et al. 2014). These studies were conducted using in vitro models, and further studies to address the impact of the bio-corona on the in vivo targeting of nanomedicines are warranted.

Walkey and Chan (2012) provided a compilation of 26 published studies on the plasma-derived protein corona, concluding that there is no one "universal" plasma protein corona for all nanomaterials and that the relative densities of the adsorbed proteins do not, in general, correlate with their relative abundances in plasma (in other words, there is a degree of specificity). The size ratio between nanoparticle and proteins plays a vital role in determining nanomaterial–biomolecule interactions (Cedervall et al. 2007a). Deng et al. (2012) studied the role of gold nanoparticle size on binding to fibrinogen and noted that small changes in nanoparticle size (from 8 nm to 10–12 nm to 15 nm) resulted in significant differences in how the protein and nanomaterials interacted. Liu et al. (2013) recently reported that when nanoparticles and proteins are similar in size, the adsorption concept (i.e., protein corona concept) cannot be applied, and a heteroaggregation model is more appropriate. Recent studies from our laboratory show that superparamagnetic iron oxide nanoparticles (SPIONs) with different surface coating display distinct plasma protein corona compositions (Vogt et al. unpublished observations). Interestingly, using similar nanoparticles (SPIONs), Simberg et al. (2009) advocated that the nanoparticle surface could be available for recognition by cells despite the formation of a bio-corona. The authors found both the dextran coat and the iron oxide core of dextran-coated SPIONs remained accessible to specific probes after their incubation in plasma. This observation is especially relevant for the preparation of nanomedicines using targeting ligands which may thus still be accessible.

When administering nano-oncologicals to a patient, the specific route of administration will determine the composition of the initial bio-corona, and this bio-corona will be crucial for the biodistribution of the nanoparticles. The majority of bio-corona studies have been performed with plasma proteins (Dufort et al. 2012; Tenzer et al. 2013), certainly relevant for administration of nanomedicines into the bloodstream. The bio-corona formed in other compartments also merits consideration. The importance of the initial bio-corona has been demonstrated studying the interaction of magnetite nanoparticles (110–180 nm in diameter), coated with different polymers (starch, carboxymethyldextran, chitosan, poly-maleic-oleic acid, phosphatidylcholine), with alveolar macrophages (Ruge et al. 2012). In this study, cellular binding and uptake of the nanoparticles by alveolar macrophages was increased for nanoparticles treated with surfactant protein A, whereas albumin, the prevailing protein in plasma, led to a significant decrease. Kapralov et al. (2012) reported on the in vivo formation of a lipid–protein corona on the surface of SWCNTs following administration by pharyngeal aspiration in mice. The bio-corona was identical to lung surfactant, and subsequent in vitro studies demonstrated a role for the surfactant corona of lipids and proteins in macrophage uptake of SWCNTs. Plasma protein adsorption to MWCNTs was influenced by prior adsorption of pulmonary surfactant lipids (Gasser et al. 2010). These studies provide evidence that, after inhalation of nanoparticles, a different bio-corona and consequently different biological behavior may result compared to direct administration to the bloodstream.

Biomolecules forming the bio-corona strongly modify the cellular uptake of nanoparticles. The question is whether the bio-corona is capable of mediating specific cellular uptake, or whether the presence of a corona of proteins, irrespective of their identity, suffices to promote (or prevent) cellular uptake. Lunov et al. (2011a) evaluated the effect of serum proteins on cellular uptake of 50 and 100 nm fluorescently labelled carboxyl- or amine-modified polystyrene, using monocytic THP-1 cells. They demonstrated that the amount of internalized nanoparticles, the uptake kinetics, and its mechanism were critically dependent on particle opsonization by serum proteins. Similar results were obtained by Lesniak et al. (2012) who reported that silica nanoparticles incubated with A549 cells in the absence of serum have a stronger adhesion to the cell membrane and higher internalization efficiency when compared with nanoparticles with a preformed surface corona. Ehrenberg et al. (2009) reported that the capacity of polystyrene nanoparticle surfaces to adsorb protein is indicative of their tendency to associate with cells. However, removal of the most abundant proteins from the cell culture medium did not result in a differential association with endothelial cells, suggesting that the cellular uptake is not dependent on the identity of adsorbed proteins. Finally, Caracciolo et al. (2013) provided evidence that nanoparticles can acquire a selective targeting capability through the adsorption of a protein corona. The authors demonstrated that lipid/DNA nanoparticles, upon interaction with human plasma, spontaneously became coated with vitronectin which, in turn, promoted efficient uptake by cancer cells.

14.4 The Interplay of Engineered Nanomaterials with the Immune System

The interaction of any foreign agent (i.e., new drugs or novel nanostructures) with the immune system can lead to immunosuppression, to immune activation, or to immune deviation. Ideally, engineered nanomaterials can be designed to either target or avoid interactions with the immune system (Zolnik et al. 2010). The interaction between a nanomedicine and the immune system may be desirable, such as the case of vaccines or immunomodulatory cancer therapeutics, as reviewed in other chapters in this volume, and for other inflammatory and autoimmune disorders. This section will focus on undesirable interactions between nanomedicines and the immune system, and strategies to avoid them (Fig. 14.2). Based on the available pre-clinical safety data, nanoparticles are not intrinsically more immunotoxic than "conventional" drugs (Dobrovolskaia and McNeil 2007). However, care should be

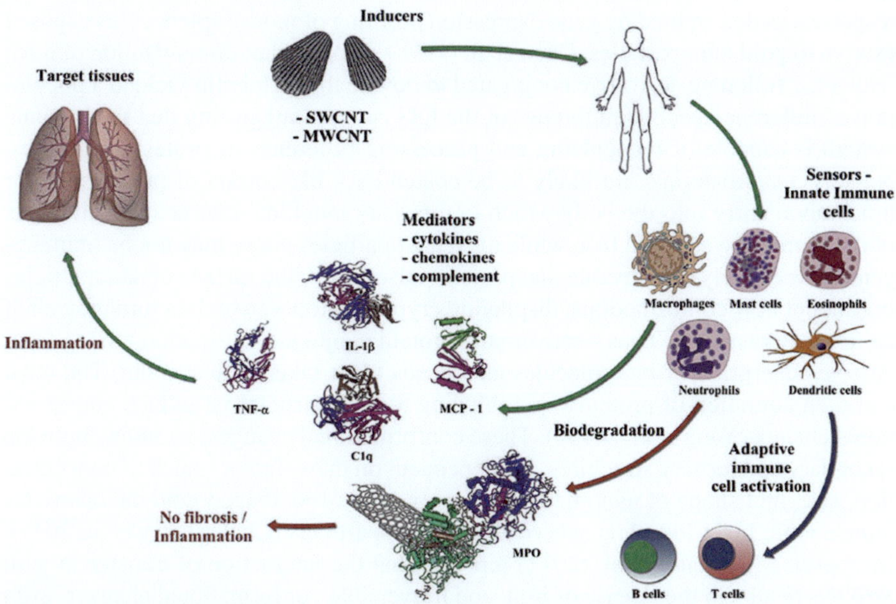

Fig. 14.2 Reciprocal interactions between carbon nanotubes and the immune system. Schematic representation of an inflammatory response. Exposure to the so-called inducers (single- and multi-walled carbon nanotubes, CNTs) initiates a multi-dimensional response from the *sensors* (i.e., innate immune cells). These cells produce soluble *mediators* (cytokines, chemokines, and complement factors) leading to inflammation in target tissues, exemplified here with the lungs. Macrophages are professional phagocytes and the first line of immune defense against foreign intrusion, including engineered nanomaterials. Recent studies have shown that neutrophils and eosinophils, key players of the innate immune system, are capable of biodegradation of CNTs through expression of myeloperoxidase (MPO) and eosinophil peroxidase (EPO), respectively, thus limiting the pathological changes in the lungs. Dendritic cells (DCs) constitute the major bridge between innate and adaptive immune responses, and their activation leads to the stimulation of B and T cells. CNTs administered through pharyngeal aspiration can interact directly with DCs, thereby leading to systemic immune suppression [reprinted from Bhattacharya K, Andón FT, El-Sayed R, Fadeel B, Mechanisms of carbon nanotube-induced toxicity: Focus on pulmonary inflammation. Adv Drug Deliv Rev. 2013 Dec;65(15):2087–97, with permission from Elsevier]

taken to ensure that suitable test methods are used for immunosafety assessment of nanomaterials (Oostingh et al. 2011). Furthermore, reformulation of conventional drugs to include nanotechnology-derived carriers often leads to reduced systemic immunotoxicity compared with that of unmodified formulations. For example, nanoparticles can be designed to evade the immune recognition by grafting of poly(ethylene glycol) (PEG) or other types of polymers onto their surface (Moghimi 2002). On the other hand, while such polymers may shield nanoparticles from recognition by the immune system, there are data suggesting the formation of PEG-specific antibodies after administration of PEG-coated liposomes, resulting in accelerated clearance (Ishida et al. 2007; Wang et al. 2007). Moyano et al. (2012) provided recent evidence that nanoparticle hydrophobicity dictates immune responses as determined by gene expression profiling of mouse splenocytes exposed ex vivo to gold nanoparticles. Chen et al. (1998) reported that immunization of mice with a C_{60} fullerene derivative conjugated to bovine thyroglobulin yielded a population of fullerene-specific antibodies of the IgG isotype, suggesting that the immune system is capable of recognizing and processing fullerenes as protein conjugates. Notably, nanomaterials are likely to be coated by a bio-corona of proteins and/or lipids upon entry into the body which in turn may modulate interactions with cells of the immune system. In fact, while most nanoparticles per se may not be immunogenic, it is entirely conceivable that proteins adsorbed to the surface of nanoparticles may adopt new conformations, displaying cryptic epitopes, which in turn may elicit an immune response. Thus, inadvertent "protein conjugate" formation arising from passive adsorption of biomolecules also needs to be taken into account. The most common alteration of proteins upon binding to nanoparticles is, in fact, conformational change (Yang et al. 2013b). These conformational changes, including both the secondary and tertiary structures, are dependent on many factors, such as concentration, size, and shape of nanomaterials. For example, rod-shaped gold nanoparticles induce more helix loss than spherical gold nanoparticles (Chakraborty et al. 2011). Moreover, Mahmoudi et al. (2011) reported that the interaction of transferrin with SPIONs results in the release of iron, and irreversible conformational changes, even after removal of the magnetic nanoparticles. In synopsis, both the "synthetic" and "biological" identities of nanomaterials (discussed above) should be taken into account where immunogenicity of nanomedicines is concerned.

The complement system is a group of proteins present in body fluids, which interact to identify and opsonize non-self, altered-self entities and also synthetic materials, for phagocytosis by cells of the innate immune system. The complement system plays a major role in innate immune defenses against microorganisms, but exaggerated activation of complement can lead to severe tissue injury. Therefore, the propensity for complement activation needs to be taken into account for any drug delivery system that comes into contact with the blood, including nano-scale carriers. Indeed, the binding of complement to nanomaterial surfaces represents a special case of undesirable bio-corona formation which may induce clinically significant adverse reactions in susceptible individuals. The interaction between nanomaterials and the complement system is complex and regulated by several factors including size, morphology, and surface characteristics (Moghimi et al. 2011). Interestingly, nanomaterials with positive or negative surface charge appear to be more efficient activators of the complement system compared to their neutral counterparts (Dobrovolskaia et al. 2009). The need to minimize complement activation is particularly clear for nanomedicine, where delivery of large amounts of nanoparticles into the blood of patients poses a clear risk of adverse reactions mediated by complement. It was, indeed, shown that hypersensitivity reactions towards Doxil® (a doxorubicin-formulation encapsulated in liposomes for use in cancer chemotherapy) were due to complement activation (Chanan-Khan et al. 2003). Furthermore, more recent data suggest that activation of the complement system at tumor sites stimulates tumor-associated immune cells and promotes their conversion into a tumor-supportive phenotype, thereby stimulating cancer progression

(Markiewski et al. 2008; Markiewski and Lambris 2009). This type of response may impact the therapeutic efficacy of nanoparticle formulations intended for cancer diagnosis or therapy. On the other hand, activation of the complement by nanomedicines can be desirable in some cases. For example, if nanostructures are intended for subcutaneous or intradermal administration, activation of the complement can benefit vaccine efficacy (Markiewski and Lambris 2009; Moghimi and Andresen 2009). Strategies to modify the interaction between nanomedicines and the complement system have been suggested by several researchers. Appropriate coating has shown to be effective in some cases for eliminating unwanted complement activation properties of nanoparticles, as demonstrated using lipid nanocapsules coated with high density of PEG (Vonarbourg et al. 2006). In addition, Moghimi and coworkers demonstrated that the complement activation can be prevented modifying the density of the polymers (Al-Hanbali et al. 2006; Hamad et al. 2010). In a recent report, Hamad et al. (2010) studied the structure–activity relationship pertaining to surface-immobilized polyethylene oxide (PEO) of various configurations on polystyrene nanoparticles and the initiation of the complement cascade and found that alteration of copolymer architecture on nanospheres from "mushroom" to "brush" configuration not only switched complement activation from the C1q-dependent classical pathway to the so-called lectin pathway but also reduced the level of generated complement activation products. These findings provide a rational basis for the intelligent design of immunologically safer and targetable nanosystems for use in the clinic.

14.4.1 Engineered Nanomaterials: Immune Activation Versus Immunosuppression

Most, if not all, nanoparticles are internalized by cells, including macrophages, through an active (endocytic) mechanism (Witasp et al. 2009; Kunzmann et al. 2011b). However, it is important to note that macrophage-like cell lines, commonly used as models to study nanomaterials, may not behave in the same manner as primary macrophages (Lunov et al. 2011a). Numerous studies have reported on the responses of immune cells towards nanomaterials and on the importance of the physico-chemical properties of nanomaterials for cellular uptake and/or for subsequent functional responses (Kunzmann et al. 2011a). Dendritic cells (DCs) are important antigen presenting cells of the immune system. Manolova et al. (2008) reported that nanoparticles target distinct DC populations in vivo in a size-dependent manner. In addition, Rettig et al. (2010) have provided evidence for a new dimension in "danger" signalling insofar as they were able to show how size quantitatively affects innate immune responses. Nanosized particles made from single-stranded RNA (ssRNA) mixed with protamine induced production of interferon-α, whereas microparticles mainly induced production of tumor necrosis factor-α (TNF-α) in human immune cells. The authors found that nanoparticles but not micro-sized particles were selectively phagocytosed by plasmacytoid DCs, which produce interferon-α (Rettig et al. 2010).

Thus, at the same time as sensing pathogen-associated molecular patterns, such as ssRNA, the immune system apparently distinguishes the size of the associated structure in such a way as to trigger anti-viral (interferon-α) or antibacterial/antifungal (TNF-α) immune responses. Shape is also important for cellular (macrophage) recognition. Particles possessing the longest dimension in the range of 2–3 μm (i.e., the typical size of bacteria) thus exhibit highest attachment to macrophages (Champion and Mitragotri 2006). Understanding how cells are capable of sensing shape or aspect ratio of nanoparticles (Meng et al. 2011) could lead to the development of more efficient drug delivery systems.

Hypersensitivity reactions in patients should be avoided, and this can be achieved in some cases, through reformulation of the drug. For example, Abraxane® is the result of reformulated paclitaxel-bound albumin nanoparticles, which did not exhibit any allergic reaction, whereas the first generation formulation of paclitaxel in the non-ionic surfactant Cremophor EL caused allergic reactions, often requiring premedication with a histamine blocker and steroids (Hawkins et al. 2008). Allergies are associated with an unbalanced Th1 versus Th2 immune response, with increased levels of cytokines, IL-4, IL-5, and IL-13, and immunoglobulin E (IgE) (Syed et al. 2013). Non-functionalized carbon nanotubes (CNTs) were shown in several pre-clinical studies to potentiate allergic responses (Ryman-Rasmussen et al. 2009; Nygaard et al. 2009; Inoue et al. 2009). Nanosized TiO_2 also increased allergic lung inflammation, Th2 cytokine production, and IgE levels in sensitized mice (Larsen et al. 2010). In contrast, silver nanoparticles have demonstrated anti-allergic effects in mice (Park et al. 2010b), and calcium carbonate or calcium phosphate nanoparticles were reported to reduce nickel allergy by binding nickel ions (Vemula et al. 2011). However, nickel allergy is based on completely different mechanisms than most common allergies, and neither Th2 cytokines nor IgE antibodies are involved. Nonetheless, these studies exemplify the "good" and the "bad" effects of nanomaterials on immune responses (allergic responses) and imply that one cannot draw conclusions about "nanomaterials," without defining the nanomaterial and the model system. One should also consider that functionalized nanomaterials may not behave as their pristine counterparts: while studies of "raw" nanomaterials are highly relevant from an occupational medicine perspective, such studies may not be directly applicable to the clinical context, where purified and functionalized nanomaterials are deployed. On the other hand, as pointed out by Cheng et al. (2012), additional functionality means additional synthetic steps and costs, more convoluted behavior in vivo, and also greater regulatory hurdles.

Nanomedicines could also induce immunosuppression, which could be either inadvertent or desirable. On the one hand, immunosuppression may lower the body's defense against infection and cancerous cells; on the other hand, it may enhance the therapeutic benefits of treatments for allergies and autoimmune diseases and prevent rejection of transplanted organs. The most obvious cause of immunosuppression is mediated by nanomaterials which induce cell death of immune-competent cells. However, other mechanisms may also come into play. Inhalation of MWCNTs was shown to suppress spleen cell function in mice, and this involved activation of cyclooxygenase enzymes in the spleen in response to a signal from the lungs, possibly TGF-β (Mitchell et al. 2009). Tkach et al. (2011) showed that pharyngeal aspiration of SWCNTs caused modified systemic immunity in mice as documented by decreased proliferation of splenic T cells, and evidence was provided for a direct effect of carbon nanotubes on DCs. Induction of immune tolerance by nanoparticles can be considered a form of desirable immunosuppression. For example, allergen-loaded nanoparticles have been reported as effective suppressors of type I and type II allergies to environmental and food allergens (Gómez et al. 2008; Schöll et al. 2004). In another study, the T cell-mediated immune response was down-regulated by simultaneous blocking of lymphocyte function-associated antigen-1 (LFA-1) and intracellular adhesion molecule-1 (ICAM-1), accomplished by using two nanoparticles targeting DCs and T cells, respectively (Chittasupho et al. 2011). To our knowledge, there are currently no studies linking autoimmune diseases to exposure to nanoparticles. However, it is possible that nanoparticles may act as scaffolds for autoantigens, leading to inadvertent immune responses, although this remains to be proven. On the other hand, a few studies have proposed the use of nanoparticles to treat autoimmune diseases (Zolnik et al. 2010). For example, delivery of IL-10-encoding DNA by nanoparticles has been shown to be successful in the suppression of autoimmune diabetes in an animal model (Basarkar and Singh 2009).

14.4.2 Inflammasome Activation: Engineered Nanomaterials as "Danger" Signals

Toll-like receptors (TLRs) are the so-called pattern recognition receptors (PRRs) expressed by cells at the front line of host defense, e.g., macrophages, DCs, and epithelial cells (Fadeel 2012). PRRs enable these cells to detect and respond to the presence of danger- and pathogen-associated molecular patterns (DAMPs and PAMPs, respectively) (Lamkanfi and Dixit 2012). Activation of the inflammasome complex in the cytoplasm of phagocytic cells occurs via engagement of TLRs, leading to subsequent assembly of the NLRP3 (NLR-related protein 3)-containing inflammasome complex and activation of caspase-1 with processing and secretion of the pro-inflammatory cytokines, IL-1β and IL-18 (Dagenais et al. 2012). NLRP3 is also activated in response to host-derived particulate matter precipitates such as uric acid and cholesterol crystals, and studies in recent years have shown that exogenous structures including asbestos fibers and crystalline silica also activate the inflammasome (Sun et al. 2013). In addition, NALP3 is the molecular target of the immunostimulatory activity of aluminum hydroxide (alum), the most commonly used vaccine adjuvant (Eisenbarth et al. 2008). Lunov et al. (2011b) showed that amino-functionalized polystyrene nanoparticles, but not carboxyl- or non-functionalized particles, trigger NLRP3 inflammasome activation and release of IL-1β in human macrophages. Tschopp and co-workers reported that nano-TiO_2 and nano-SiO_2, but not nano-ZnO, activate the NLRP3 inflammasome and cause pulmonary inflammation in mice with release of IL-1α and IL-1β (Yazdi et al. 2010).

Recent studies have shown that "needle-like" MWCNTs can activate the NLRP3 inflammasome in LPS-primed human macrophages (Palomäki et al. 2011). In another recent study, Yang et al. (2013a) demonstrated that spherical carbon nano-onions of 6 nm in size also can promote NLRP3 inflammasome activation in synergy with TLR ligands. Importantly, this was shown to be highly dependent on the physico-chemical characteristics of the nanoparticles as chemical surface functionalization significantly attenuated the inflammatory properties. Exposure of macrophages to carbon black nanoparticles of 20 nm in size was shown to result in caspase-1-dependent, inflammasome-mediated cell death (Reisetter et al. 2011). However, hollow carbon spheres of about 170 nm in diameter trigger inflammasome-dependent secretion of IL-1β in primary human macrophages independently of the induction of cell death (Andón et al. unpublished observations). Taken together, these studies point towards similarities in terms of immune sensing of nanomaterials, environmental agents such as asbestos fibers, and microorganisms. These studies also underscore that fiber-like dimensions are not necessarily a requirement for activation of the inflammasome. It is known, however, that inflammasome activation in vitro requires signaling both via TLRs and NLRP3. Demento et al. (2009) prepared nanomaterials endowed with these two signals for optimization of vaccine design. The authors incorporated LPS onto the surface of PLGA nanoparticles loaded with antigen. The LPS-modified particles were preferentially internalized by DCs, and the nanoparticle system elicited potent humoral and cellular immunity in mice. Macrophages pulsed with LPS-modified nanoparticles resulted in production of IL-1β, consistent with inflammasome activation (Demento et al. 2009). Thus, inflammasome activation by nanoparticles may potentially be exploited for therapeutic benefit.

14.4.3 Elimination of Engineered Nanomaterials Through Enzymatic Degradation

One of the maxims for the development of novel drug delivery systems is that they should be biodegradable. Non-degradable nanomaterials can accumulate in cells and tissues where they may exert detrimental effects. The typical example are carbon nanotubes; it has been shown that intravenously injected, pristine (non-functionalized) SWCNTs are highly enriched in the liver, lungs, and spleen in mice and remain in the body over an extended period of time (Yang et al. 2007). Importantly, we and others have reported on the ability of the immune system to degrade these nanomaterials (Fig. 14.2). Hence, myeloperoxidase (MPO) expressed in primary human neutrophils was found to mediate biodegradation of oxidized SWCNTs (Kagan et al. 2010). MPO is a major protein constituent of the azurophilic granules of neutrophils. It catalyzes the hydrogen peroxide-mediated oxidation of halide ions to hypohalous acids, especially HOCl. These ROS can participate in a variety of secondary reactions, leading to modifications of amino acids and many types of biological macromolecules. In the same study, neutrophils were shown to be more proficient at biodegradation of SWCNT than macrophages, presumably due to a higher level of expression of MPO (Kagan et al. 2010). SWCNT degradation was more efficient when SWCNT were coated with a "corona" of immunoglobulins, suggesting that cellular internalization of SWCNT by neutrophils via Fc receptors facilitates their degradation (Kagan et al. 2010). However, extracellular biodegradation following the degranulation of activated neutrophils cannot be excluded. In fact, our recent studies have shown that neutrophil extracellular traps (NETs) generated by ex vivo activated human neutrophils can "capture" SWCNTs and that SWCNTs undergo acellular, MPO-dependent biodegradation in NETs (Farrera et al. 2014). Additionally, we have demonstrated that eosinophil peroxidase (EPO), the major oxidant-producing enzyme in eosinophils, degrades oxidized SWCNTs following ex vivo degranulation of primary murine eosinophils (Andón et al. 2013). Furthermore, in vivo biodegradation was demonstrated in a mouse model of pharyngeal aspiration of SWCNTs (Shvedova et al. 2012). Notably, clearance of SWCNTs from the lungs of MPO-deficient mice was markedly less effective, whereas the inflammatory/fibrotic response was more pronounced as compared to wild-type mice. Thus, the immune system can apparently utilize the same enzymatic pathways for degradation of microorganisms and nanomaterials. Strategies to control the timing and the location of this biodegradation will be required in order to harness the degradative capacity of the immune system in nanomedicine.

Studies conducted in vitro and in vivo using polymeric nanocarriers have demonstrated oxidative, hydrolytic, and enzymatic mechanisms for biodegradation of these nanostructures (Kumari et al. 2010; De Gracia Lux et al. 2012). Furthermore, SPIONs undergo a particular mechanism of biodegradation, being processed by cells as part of their physiological iron metabolism (Weissleder et al. 1989). Liposomes are frequently engineered to be biodegraded by secretory phospholipase A2, a lipid hydrolyzing enzyme that is significantly up-regulated in the extracellular microenvironment of tumors (Arias 2011). Finally, Park et al. (2009) synthesized luminescent porous silicon nanoparticles and noted that these materials "self-destructed" into renally cleared components in mice with no evidence of toxicity.

14.5 Emerging Approaches in Nanotoxicology: Towards Predictive Nanotoxicology

To date, much of the nanotoxicology literature has focused on describing the effects of engineered nanomaterials using traditional in vitro and/or in vivo (animal) models. The overwhelming majority of these studies fail to provide information on the underlying mechanisms of toxicity or mode of cell death (Andón and Fadeel 2013). Nel et al. (2013a) provided an excellent summary of potentially useful mechanistic injury pathways which could be employed to develop *predictive* toxicological paradigms (Fig. 14.3). The comprehensive study of the mode of action of nanomaterials will allow for more refined approaches for the mitigation of adverse effects of these materials. A promising way forward is to use *alternative test strategies*, which reduce reliance on animal testing through the use of validated in vitro assays along

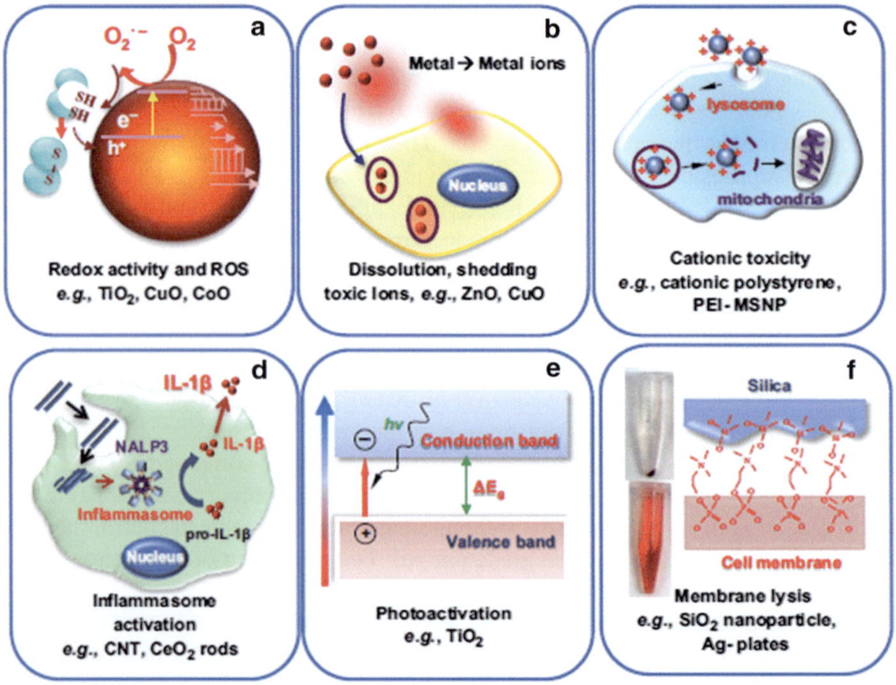

Fig. 14.3 Predictive toxicological approaches based on pathways of toxicity. The schematic diagram shows selected pathways of toxicity (POTs) that can be used for in vitro screening, but also have a role in the pathophysiology of disease. The POTs are carefully chosen to allow in vitro screening to be predictive of in vivo outcomes, thereby enabling large numbers of engineered nanomaterials to be simultaneously assayed, ranked, and subjected to further in vivo investigation. (**a**) Induction of oxidative stress by redox active nanomaterials capable of reactive oxygen species (ROS) generation by the material itself or as a consequence of interactions at the nano–bio interface. (**b**) Material dissolution or shedding of toxic metal ions by metal and metal oxide nanoparticles. (**c**) Cationic injury to the surface membrane or the lysosome. (**d**) Activation of the inflammasome by high aspect ratio materials, such as carbon nanotubes, which also induce pulmonary fibrosis. (**e**) Photoactivation and generation of electron hole pairs, leading to ROS production. (**f**) Membrane lysis by reactive surface chemistry of nanomaterials [Reprinted from Nel AE, Implementation of alternative test strategies for the safety assessment of engineered nanomaterials. J Intern Med. 2013 Dec;274(6):561–77, with permission from John Wiley and Sons]

with computational modelling approaches to predict structure–activity relationships (Nel et al. 2013b). HTS approaches, based on assays that capture specific mechanistic injury pathways, will enable the rapid testing of vast numbers of new nanomaterials. These developments are in line with the emergence of the so-called twenty-first-century paradigm in toxicology, with the aim to employ high-throughput and computational toxicology approaches to yield data predictive of results from animal toxicity studies, which will allow prioritization of chemicals for further testing and assist in prediction of risk to humans (Collins et al. 2008).

HTS is a method for scientific experimentation that comprises the screening of large chemical libraries for activity against biological targets via the use of automation, miniaturized assays, and large-scale data analysis (Mayr and Bojanic 2009). The majority of assays, biochemical or cell-based, can be adapted to a 384-well format, and this plate format has been established as the (current) format of choice among pharma and biotech companies (using standard volumes of 5 μL per well) (Mayr and Bojanic 2009). A model for how high-throughput methods might be used to assess adverse effects of nanomaterials is ToxCast (www.epa.gov/ncct/toxcast), a US Environmental Protection Agency (EPA) initiative to accelerate toxicity testing of industrial chemicals (Dix et al. 2007). In Phases I and II, researchers examined 976 different chemicals (including failed pharmaceuticals, alternative plasticizers, food additives, and pesticides) across 331 cell-free enzymatic and ligand binding HTS assays. Half-maximal activity concentrations (AC_{50}) were identified for 729 chemicals in 256 assays (Sipes et al. 2013). Interestingly, novel findings for previously unreported chemical–target combinations often clustered with known chemical–target interactions. The biochemical HTS screen offers preliminary evidence for chemical targets in a cell or tissue that, when combined with information from the literature or targeted studies, indicates potential pathways of toxicity. In recent years, similar approaches have been applied for toxicity assessment of engineered nanomaterials (Feliu and Fadeel 2010). Shaw et al. (2008) reported on the assessment of nanoparticle effects using multiple cell types and multiple in vitro assays. The cell types were selected to reflect a range of tissues relevant for evaluation of intravascularly administered (nanomedicine) agents: vascular cells (endothelial and smooth muscle cells), monocytes, and hepatocytes. Hierarchical clustering of the data identified nanomaterials with similar patterns of biologic activity across a broad sampling of cellular contexts, yielding robust structure–activity relationships for the nanomaterials tested. Furthermore, a subset of nanoparticles were tested in mice, and nanoparticles with similar activity profiles in vitro exerted similar effects in vivo, using monocyte number as the endpoint. These data suggested a strategy of multi-pronged characterization of nanomaterials in vitro that can inform the design of novel nanomaterials and guide studies of in vivo activity (Shaw et al. 2008). High-content in vitro assays combined with genome-wide expression analysis of exposed cells has also been applied to assess the toxicity of poly (ethylene glycol)-coated versus non-coated quantum dots (Zhang et al. 2006). Similarly, Jan et al. (2008) utilized high-content screening for "fingerprinting" of nanomaterials using cancer cell lines of neuronal and hepatic origin. George et al. (2010) reported a rapid cytotoxicity screen for metal oxide nanoparticles that exploits high-content screening methods. They developed a fluorescence assay that simultaneously measured four different responses in cells to oxidative stress caused by ZnO, CeO_2, and TiO_2 nanoparticles. Using the information, the authors were able to reduce the cytotoxicity of ZnO nanoparticles by decreasing ZnO dissolution through Fe doping (George et al. 2010). These examples suggest that high-content and HTS is a powerful tool for the prediction of possible hazards of nanomaterials. However, the fact that nanomaterials may interfere with commonly used in vitro assays needs to be taken into account (Lewinski et al. 2008). Standardized and validated cytotoxicity tests of nanomaterials, including assays based on new test principles using label-free detection, are therefore needed. Furthermore, the development of novel microdevices with which to reproduce complex physiological conditions (Huh et al. 2010, 2012; Kim et al. 2012a) may supplement and ultimately replace costly animal experiments, and the adaptation of such "organ-on-a-chip" devices to HTS would greatly facilitate testing of chemicals, drugs, and nanomaterials.

In addition to experimental approaches for hazard assessment of nanomaterials, there is a need for in silico methods with which to develop structure–activity relationships. Indeed, this paradigm allows for the *prediction* of toxicological effects induced by chemicals on the basis of their structural similarity with other chemicals for which toxicological endpoints have been previously measured. These structure–toxicity relationships can be quantitative or qualitative in nature, and they can predict toxicological effects directly from the physico-chemical properties of the entities/chemicals of interest. Therefore, this approach can aid in prioritizing resources in toxicological investigations while reducing the ethical and monetary costs that are related to animal testing. QSAR (quantitative structure–activity relationship) models have been successfully applied in the pharmaceutical field and in regulatory toxicology. However, the extraction of meaningful relationships between nanomaterials and toxicological properties to yield QNTR (quantitative nanostructure–toxicity relationship) models requires specific mathematical and statistical techniques (Winkler et al. 2013; Le et al. 2012). Although QSAR approaches have only recently been used to predict biological effects of nanomaterials, some encouraging initial results have already shown. Puzyn et al. (2011) reported on the cytotoxicity of 17 different types of metal oxide nanoparticles to *Escherichia coli*, and Epa et al. (2012) generated quantitative, predictive, and informative models that describe QSAR relationships for cellular uptake and apoptosis induced by metal oxide nanoparticles in several types of cells. Fourches et al. (2010) studied the cellular uptake of 109 different nanoparticles with similar core but diverse surface modifiers. Interestingly, the authors noted that the cellular uptake of nanoparticles possessing the same metal core but different organic molecules on their surface can be predicted by taking into account the chemical structure of the coating molecules (Fourches et al. 2010). Therefore, the structural determinants of the biological behavior of nanoparticles can be found both at the core of nanoparticles and at their surface. Indeed, the bio-corona is a key determinant of the "identity" of a nanomaterial and should be taken into account when modelling the behavior of nanomaterials in biological systems (Xia et al. 2010).

14.5.1 Systems Toxicology: Probing Nanomaterial Interactions with Living Systems

The introduction of engineered nanomaterials into a living system can affect a whole series of interrelated processes, simply by altering the behavior of one or two key genes, proteins, or metabolites, which in turn affects a network of interactions resulting in a biological (toxicological) effect. The emerging field of *systems biology* consists of the integration of large amounts of data obtained from high-throughput biological studies using analytical and computational modeling with the aim to understand how complex biological systems function (Pujol et al. 2010). The use of global omics approaches could enable the identification of signalling pathways governing the responses of biological systems to nanomaterials, including nano-oncologicals. Indeed, it is noted that omics approaches may provide a tool to dissect the bio-corona on the surface of nanoparticles, as shown in recent investigations of the lipid and protein composition of various nanoparticle coronas (Tenzer et al. 2011; Sund et al. 2011; Kapralov et al. 2012), as well as to define tissue responses upon in vivo exposure to nanomaterials, as illustrated in recent mass spectrometry (MS)-based lipidomics (Tyurina et al. 2011) and proteomics studies (Teeguarden et al. 2011) of pulmonary responses to SWCNTs. Additionally, the use of gene expression profiling, or transcriptomics approaches, to address the toxicity of nanomaterials has increased in recent years. Foldbjerg et al. (2012) performed global gene expression profiling of lung cells (A549) exposed to silver nanoparticles versus silver ions and found that although the transcriptional response to Ag^+ ions was highly related to the response caused by silver nanoparticles, the silver nanoparticles affected cells in a more complex manner. Tsai et al. (2011) combined proteomics approaches with transcriptomics analysis of gene expression to understand the cytotoxic effects of gold nanoparticles on a human leukemia cell line (K562). Ingenuity pathway analysis (IPA) revealed that unfolded protein-associated endoplasmic reticulum (ER) stress was the predominant event responsible for cell death (Tsai et al. 2011). Kodali et al. (2013) were able to show that SPIONs caused extensive transcriptional reprogramming of murine bone-marrow derived macrophages in the absence of acute cytotoxic or pro-inflammatory effects. Notably, exposure to nanoparticles caused diminished phagocytic capacity of macrophages towards bacteria. The authors concluded that biological effects of nanomaterials may be indirectly manifested only after challenging normal (immune) cell function. We and others (Tuomela et al. 2013) performed global gene expression profiling to identify transcriptional responses underlying the cytotoxicity caused by ZnO nanoparticles using primary human monocyte-derived macrophages (HMDM) and monocyte-derived dendritic cells (MDDC) as model systems. Gene Ontology (GO) analysis revealed that the top biological processes that were perturbed in HMDM were involved in regulation of cell death and growth. The study also revealed that the upregulation of metallothionein genes in response to cytotoxic concentrations of ZnO nanoparticles is a common gene "signature" that is shared across the different cell types, and metallothionein upregulation could potentially be used as a biomarker of exposure to this class of nanoparticles (Tuomela et al. 2013). In a more recent study, we performed next generation sequencing (RNA Seq) to determine the transcriptional response in primary human bronchial epithelial cells exposed to nanoparticles. GO enrichment analysis revealed significant downregulation of cell cycle-related genes at doses of nanoparticles that were not acutely cytotoxic, and subsequent cell-based assays confirmed that non-cytotoxic doses of the nanoparticles in question triggered cell cycle arrest (Feliu et al. unpublished observations). Our team has also developed a publically available transcriptomics database, designated NanoMiner, containing experimental results from microarray studies of cells exposed to various kinds of nanomaterials (Kong et al. 2013) (http://nanominer.cs.tut.fi/). In a recent in vivo study, Snyder-Talkington et al. (2013) used computational modelling approaches to determine biological processes strongly associated with lung inflammation and fibrosis in mice exposed to MWCNTs. Finally, metabolomics approaches are also beginning to gain ground in nanotoxicological research (Bu et al. 2010; Feng et al. 2010; Huang et al. 2012). Metabolomics is concerned with the assessment of endogenous metabolites within a biological system and the study of their variations upon different stimuli (Schnackenberg et al. 2012). In a recent systems biology study of magnetic nanoparticles, Shim et al. (2012) reported that the combination of gene expression and metabolic profiling, using gas chromatography–mass spectrometry (GC–MS), may enable a more detailed and sensitive toxicological evaluation of nanoparticles. However, the impact of nanoparticles on gene expression and metabolism was only evaluated at one time-point (12 h) and the kinetics of the response, and long-term effects, may also be of interest. Taken together, omics-based systems biology approaches offer an extraordinary opportunity to gain insights into the mechanisms of toxicity of novel nanomaterials. However, such studies should be carefully designed in order to meet the needs of the data analysis; needless to say, the nanomaterials should be subjected to physico-chemical characterization, and reference or benchmark materials should be considered in the experimental design.

14.6 Bridging Nanotoxicology and Nanomedicine: Towards the Design of Better Nanomedicines

Nanomedicines must be safe and effective, and both properties are closely connected in the design of novel nanostructures (Fig. 14.4). An effective nanocarrier should retain the drug very tightly, ideally without any release, during the transport from the portal of administration to the site of action; in addition, the same nanostructure must be able to efficiently release the drug once reaching the target to exert its pharmaceutical action (Sun et al. 2012). A safe nanomedicine, as well as any other medicine, must possess a rigorously determined chemical nature and stability, including aggregation and degradation, which can be evaluated at all the stages of its life cycle (i.e., manufacture, formulation, storage, administration, etc.) (Duncan and Gaspar 2011; Nyström and Fadeel 2012). Once in the body, nanomedicines interact with biological fluids, biological barriers, and defensive systems (including the immune system), and encounter biodegradative and excretory strategies developed by our organism. The key for the design of safer nanomedicines relies on our ability to study, predict, and manipulate the interaction between engineered nanostructures and biological systems. As we have discussed in this chapter, and as reviewed elsewhere (Monopoli et al. 2012; Fadeel et al. 2013; Lynch et al. 2014), there is growing evidence that it is the "biological identity" of nanomaterials which governs their interaction with cells and the subsequent adverse or beneficial outcomes. Yang et al.

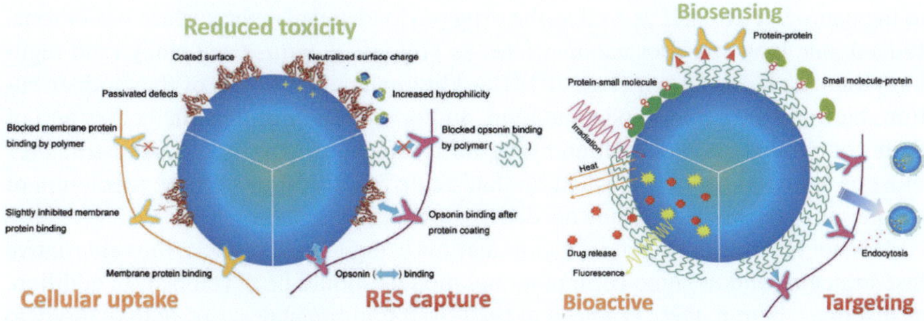

Fig. 14.4 Strategies to improve the biosafety and biomedical applications of nanomaterials. Safety and effectiveness of nanomedicines are closely connected and rely on the interplay of the synthetic and biological "identities" of the nanomaterials. (**a**) Strategies to reduce the toxicity and to modify the cellular uptake and/or reticuloendothelial system (RES) capture are schematically depicted. (**b**) Biomedical applications of nanomaterials, including targeted drug delivery and imaging, are enabled through careful design of the nanoparticle surface, to dictate nanoparticle–protein interactions [reprinted from Yang ST, Liu Y, Wang YW, Cao A, Biosafety and bioapplication of nanomaterials by designing protein-nanoparticle interactions. Small. 2013 May 27;9(9–10):1635–53, with permission from John Wiley and Sons]

(2013b) concluded that the manipulation of the bio-nano-interface is still in its infancy but found that some important lessons have been learned. Nanoparticles may thus be considered as a scaffold which has to be properly functionalized. This nanoscaffold should provide a surface with well-defined positions where functional groups and proteins can be organized (Fig. 14.4) (Yang et al. 2013b). Controlling the bio-corona through the deliberate design of binding sites for proteins may be possible, and some studies suggest that the bio-corona that is formed under physiological conditions could be exploited for targeting and/or drug delivery (Kah et al. 2012; Caracciolo et al. 2013; Cifuentes-Rius et al. 2013). The bio-nano-interface constitutes a "bridge," a common link between nanotoxicology and nanomedicine research.

14.6.1 The Future of Nanotoxicology: Concluding Remarks and Perspectives

Nanotoxicology is an emerging multi-disciplinary field involving material scientists, physicians, and toxicologists. Large amounts of data have been obtained in recent years regarding the interactions between engineered nanomaterials and biological systems. Unquestionably, these data represent a key contribution of nanotoxicology to nanomedicine. However, major challenges remain in nanotoxicological research. A comprehensive understanding of both the synthetic and biological "identity" of engineered nanomaterials is needed. During design and preclinical development of nanomedicines, not least in the specific case of nano-oncologicals, the potential toxicity of all the components must be carefully considered in respect to the nano-carrier itself as well as the drug payload and also the surface bio-corona. Indeed, one may consider nanoparticles as *biological entities* and study (and regulate) them as such (Lynch et al. 2014). Additionally, monitoring of the biodistribution, biotransformation, biodegradation, and/or excretion of nanomedicines and of their components is required, and long-term effects of exposure to nanostructures must be assessed; strategies to extrapolate acute in vitro results for the prediction of chronic in vivo effects are thus needed.

It is not feasible to screen all nanomaterials using animal models, thus alternative test strategies and in silico computational models should be developed. In addition, as discussed herein, there is a need to shift from traditional descriptive toxicology to a *predictive toxicology* (Nel et al. 2013b). An improved understanding of the complexities of the nanomaterial-induced perturbation of different cell death and cell signalling pathways may allow for a better prediction of their effects and facilitate the development of nanomedicines that are safe by design (Andón and Fadeel 2013). In addition, a thorough characterization of the effects of engineered nanomaterials on the immune system, a sentinel system designed to fend against foreign intrusion, may be indicative of the biocompatibility of these materials.

Finally, we believe that nanomaterial classification—based on physico-chemical, molecular, and physiological profiles—can and should be developed. Systems toxicological approaches are beginning to emerge and may allow for establishment of "fingerprints" of (classes of) nanomaterials, based on a detailed understanding of the synthetic and biological "identities" of nanomaterials, their interplay with living systems.

Acknowledgments This work is supported by the Swedish Research Council, Swedish Research Council for Environment, Agricultural Sciences, and Spatial Planning (FORMAS), and the European Commission (COST action MODENA, FP7-NANOMMUNE, FP7-MARINA, FP7-NANOREG, FP7-SUN, FP7-NANOSOLUTIONS, FP7-eNANOMAPPER, and the FET Flagship Project, GRAPHENE). FTA is the recipient of a postdoctoral fellowship from the Galician Government, Spain (Resolution 21, March 2013).

References

Adams DJ (2012) The Valley of Death in anticancer drug development: a reassessment. Trends Pharmacol Sci 33:173–180

Al-Hanbali O, Rutt KJ, Sarker DK et al (2006) Concentration dependent structural ordering of poloxamine 908 on polystyrene nanoparticles and their modulatory role on complement consumption. J Nanosci Nanotechnol 6:3126–3133

Andar AU, Hood RR, Vreeland WN et al (2014) Microfluidic preparation of liposomes to determine particle size influence on cellular uptake mechanisms. Pharm Res 31:401–413

Andón FT, Fadeel B (2013) Programmed cell death: molecular mechanisms and implications for safety assessment of nanomaterials. Acc Chem Res 46:733–742

Andón FT, Kapralov AA, Yanamala N et al (2013) Biodegradation of single-walled carbon nanotubes by eosinophil peroxidase. Small 9(2721–2729):2720. doi:10.1002/smll.201202508

Arias JL (2011) Drug targeting strategies in cancer treatment: an overview. Mini Rev Med Chem 11:1–17

Basarkar A, Singh J (2009) Poly (lactide-*co*-glycolide)-polymethacrylate nanoparticles for intramuscular delivery of plasmid encoding interleukin-10 to prevent autoimmune diabetes in mice. Pharm Res 26:72–81

Bhabra G, Sood A, Fisher B et al (2009) Nanoparticles can cause DNA damage across a cellular barrier. Nat Nanotechnol 4:876–883

Bhattacharya K, Andón FT, El-Sayed R, Fadeel B (2013) Mechanisms of carbon nanotube-induced toxicity: focus on pulmonary inflammation. Adv Drug Deliv Rev 65:2087–2097

Bleeker EAJ, de Jong WH, Geertsma RE et al (2013) Considerations on the EU definition of a nanomaterial: science to support policy making. Regul Toxicol Pharmacol 65:119–125

Bowes J, Brown AJ, Hamon J et al (2012) Reducing safety-related drug attrition: the use of in vitro pharmacological profiling. Nat Rev Drug Discov 11:909–922

Bu Q, Yan G, Deng P et al (2010) NMR-based metabonomic study of the sub-acute toxicity of titanium dioxide nanoparticles in rats after oral administration. Nanotechnology 21:125105

Caracciolo G, Cardarelli F, Pozzi D et al (2013) Selective targeting capability acquired with a protein corona adsorbed on the surface of 1,2-dioleoyl-3-trimethylammonium propane/DNA nanoparticles. ACS Appl Mater Interfaces 5:13171–13179

Cedervall T, Lynch I, Foy M et al (2007a) Detailed identification of plasma proteins adsorbed on copolymer nanoparticles. Angew Chem Int Ed Engl 46:5754–5756

Cedervall T, Lynch I, Lindman S et al (2007b) Understanding the nanoparticle-protein corona using methods to quantify exchange rates and affinities of proteins for nanoparticles. Proc Natl Acad Sci U S A 104:2050–2055

Chakraborty S, Joshi P, Shanker V et al (2011) Contrasting effect of gold nanoparticles and nanorods with different surface modifications on the structure and activity of bovine serum albumin. Langmuir 27:7722–7731

Champion JA, Mitragotri S (2006) Role of target geometry in phagocytosis. Proc Natl Acad Sci U S A 103:4930–4934

Chanan-Khan A, Szebeni J, Savay S et al (2003) Complement activation following first exposure to pegylated liposomal doxorubicin (Doxil): possible role in hypersensitivity reactions. Ann Oncol 14:1430–1437

Chen BX, Wilson SR, Das M, Coughlin DJ, Erlanger BF (1998) Antigenicity of fullerenes: antibodies specific for fullerenes and their characteristics. Proc Natl Acad Sci U S A 95:10809–10813

Cheng Z, Al Zaki A, Hui JZ et al (2012) Multifunctional nanoparticles: cost versus benefit of adding targeting and imaging capabilities. Science 338:903–910

Chithrani DB (2010) Intracellular uptake, transport, and processing of gold nanostructures. Mol Membr Biol 27:299–311

Chithrani BD, Chan WCW (2007) Elucidating the mechanism of cellular uptake and removal of protein-coated gold nanoparticles of different sizes and shapes. Nano Lett 7:1542–1550

Chithrani BD, Ghazani AA, Chan WCW (2006) Determining the size and shape dependence of gold nanoparticle uptake into mammalian cells. Nano Lett 6:662–668

Chittasupho C, Shannon L, Siahaan TJ et al (2011) Nanoparticles targeting dendritic cell surface molecules effectively block T cell conjugation and shift response. ACS Nano 5:1693–1702

Cho W-S, Duffin R, Thielbeer F et al (2012) Zeta potential and solubility to toxic ions as mechanisms of lung inflammation caused by metal/metal oxide nanoparticles. Toxicol Sci 126:469–477

Choi HS, Ashitate Y, Lee JH et al (2010) Rapid translocation of nanoparticles from the lung airspaces to the body. Nat Biotechnol 28:1300–1303

Cifuentes-Rius A, de Puig H, Kah JC et al (2013) Optimizing the properties of the protein corona surrounding nanoparticles for tuning payload release. ACS Nano 7:10066–10074

Collins FS, Gray GM, Bucher JR (2008) Toxicology. Transforming environmental health protection. Science 319:906–907

Cronholm P, Karlsson HL, Hedberg J et al (2013) Intracellular uptake and toxicity of Ag and CuO nanoparticles: a comparison between nanoparticles and their corresponding metal ions. Small 9:970–982

Dagenais M, Skeldon A, Saleh M (2012) The inflammasome: in memory of Dr. Jurg Tschopp. Cell Death Differ 19:5–12

De Gracia LC, Joshi-Barr S, Nguyen T et al (2012) Biocompatible polymeric nanoparticles degrade and release cargo in response to biologically relevant levels of hydrogen peroxide. J Am Chem Soc 134:15758–15764
Demento SL, Eisenbarth SC, Foellmer HG et al (2009) Inflammasome-activating nanoparticles as modular systems for optimizing vaccine efficacy. Vaccine 27:3013–3021
Deng ZJ, Liang M, Toth I et al (2012) Molecular interaction of poly(acrylic acid) gold nanoparticles with human fibrinogen. ACS Nano 6:8962–8969
Deshmukh M, Kutscher HL, Gao D et al (2012) Biodistribution and renal clearance of biocompatible lung targeted poly(ethylene glycol) (PEG) nanogel aggregates. J Control Release 164:65–73
Dix DJ, Houck KA, Martin MT et al (2007) The ToxCast program for prioritizing toxicity testing of environmental chemicals. Toxicol Sci 95:5–12
Dobrovolskaia MA, McNeil SE (2007) Immunological properties of engineered nanomaterials. Nat Nanotechnol 2:469–478
Dobrovolskaia MA, Patri AK, Zheng J et al (2009) Interaction of colloidal gold nanoparticles with human blood: effects on particle size and analysis of plasma protein binding profiles. Nanomedicine 5:106–117
Dufort S, Sancey L, Coll J-L (2012) Physico-chemical parameters that govern nanoparticles fate also dictate rules for their molecular evolution. Adv Drug Deliv Rev 64:179–189
Duncan R, Gaspar R (2011) Nanomedicine(s) under the microscope. Mol Pharm 8:2101–2141
Ehrenberg MS, Friedman AE, Finkelstein JN et al (2009) The influence of protein adsorption on nanoparticle association with cultured endothelial cells. Biomaterials 30:603–610
Eisenbarth SC, Colegio OR, O'Connor W et al (2008) Crucial role for the Nalp3 inflammasome in the immunostimulatory properties of aluminium adjuvants. Nature 453:1122–1126
El Badawy AM, Silva RG, Morris B et al (2011) Surface charge-dependent toxicity of silver nanoparticles. Environ Sci Technol 45:283–287
Epa VC, Burden FR, Tassa C et al (2012) Modeling biological activities of nanoparticles. Nano Lett 12:5808–5812
Fadeel B (2012) Clear and present danger? Engineered nanoparticles and the immune system. Swiss Med Wkly 142:w13609
Fadeel B (2013) Nanosafety: towards safer design of nanomedicines. J Intern Med 2274:578–580
Fadeel B, Feliu N, Vogt C et al (2013) Bridge over troubled waters: understanding the synthetic and biological identities of engineered nanomaterials. Wiley Interdiscip Rev Nanomed Nanobiotechnol 5:111–129
Farrera C, Bhattacharya K, Lazzaretto B, Andón FT, Hultenby K, Kotchey GP, Star A, Fadeel B (2014) Extracellular entrapment and degradation of single-walled carbon nanotubes. Nanoscale 21:6974–6983. doi:10.1039/c3nr06047k
Feliu N, Fadeel B (2010) Nanotoxicology: no small matter. Nanoscale 2:2514–2520
Feng J, Liu H, Zhang L et al (2010) An insight into the metabolic responses of ultra-small superparamagnetic particles of iron oxide using metabonomic analysis of biofluids. Nanotechnology 21:395101
Filon FL, Crosera M, Adami G et al (2011) Human skin penetration of gold nanoparticles through intact and damaged skin. Nanotoxicology 5:493–501
Foldbjerg R, Irving ES, Hayashi Y et al (2012) Global gene expression profiling of human lung epithelial cells after exposure to nanosilver. Toxicol Sci 130:145–157
Fourches D, Pu D, Tassa C et al (2010) Quantitative nanostructure – activity relationship modeling. ACS Nano 4:5703–5712
Gasser M, Rothen-Rutishauser B, Krug HF et al (2010) The adsorption of biomolecules to multi-walled carbon nanotubes is influenced by both pulmonary surfactant lipids and surface chemistry. J Nanobiotechnology 8:31
Geiser M (2010) Update on macrophage clearance of inhaled micro- and nanoparticles. J Aerosol Med Pulm Drug Deliv 23:207–217
Geng Y, Dalhaimer P, Cai S et al (2007) Shape effects of filaments versus spherical particles in flow and drug delivery. Nat Nanotechnol 2:249–255

George S, Pokhrel S, Xia T et al (2010) Use of a rapid cytotoxicity screening approach to engineer a safer zinc oxide nanoparticle through iron doping. ACS Nano 4:15–29
Gómez S, Gamazo C, San Roman B et al (2008) Allergen immunotherapy with nanoparticles containing lipopolysaccharide from Brucella ovis. Eur J Pharm Biopharm 70:711–717
Hamad I, Al-Hanbali O, Hunter AC et al (2010) Distinct polymer architecture mediates switching of complement activation pathways at the nanosphere-serum interface: implications for stealth nanoparticle engineering. ACS Nano 4:6629–6638
Hamilton RF, Wu N, Porter D et al (2009) Particle length-dependent titanium dioxide nanomaterials toxicity and bioactivity. Part Fibre Toxicol 6:35
Hawkins MJ, Soon-Shiong P, Desai N (2008) Protein nanoparticles as drug carriers in clinical medicine. Adv Drug Deliv Rev 60:876–885
Herrero EP, Alonso MJ, Csaba N (2012) Polymer-based oral peptide nanomedicines. Ther Deliv 3:657–668
Huang X, Li L, Liu T et al (2011) The shape effect of mesoporous silica nanoparticles on biodistribution, clearance, and biocompatibility in vivo. ACS Nano 5:5390–5399
Huang S-M, Zuo X, Li JJ et al (2012) Metabolomics studies show dose-dependent toxicity induced by SiO(2) nanoparticles in MRC-5 human fetal lung fibroblasts. Adv Healthc Mater 1:779–784
Huh D, Matthews BD, Mammoto A et al (2010) Reconstituting organ-level lung functions on a chip. Science 328:1662–1668
Huh D, Leslie DC, Matthews BD et al (2012) A human disease model of drug toxicity–induced pulmonary edema in a lung-on-a-chip microdevice. Sci Transl Med 4:159ra147
Iida N, Dzutsev A, Stewart CA et al (2013) Commensal bacteria control cancer response to therapy by modulating the tumor microenvironment. Science 342:967–970
Inoue K, Koike E, Yanagisawa R et al (2009) Effects of multi-walled carbon nanotubes on a murine allergic airway inflammation model. Toxicol Appl Pharmacol 237:306–316
Ishida T, Wang X, Shimizu T et al (2007) PEGylated liposomes elicit an anti-PEG IgM response in a T cell-independent manner. J Control Release 122:349–355
Jan E, Byrne SJ, Cuddihy M et al (2008) High-content screening as a universal tool for fingerprinting of cytotoxicity of nanoparticles. ACS Nano 2:928–938
Kagan VE, Konduru NV, Feng W et al (2010) Carbon nanotubes degraded by neutrophil myeloperoxidase induce less pulmonary inflammation. Nat Nanotechnol 5:354–359
Kah JC, Chen J, Zubieta A, Hamad-Schifferli K (2012) Exploiting the protein corona around gold nanorods for loading and triggered release. ACS Nano 6:6730–6740
Kapralov AA, Feng WH, Amoscato AA et al (2012) Adsorption of surfactant lipids by single-walled carbon nanotubes in mouse lung upon pharyngeal aspiration. ACS Nano 6:4147–4156
Kendall M, Holgate S (2012) Health impact and toxicological effects of nanomaterials in the lung. Respirology 17:743–758
Kim YS, Song MY, Park JD et al (2010) Subchronic oral toxicity of silver nanoparticles. Part Fibre Toxicol 7:20
Kim HJ, Huh D, Hamilton G, Ingber DE (2012a) Human gut-on-a-chip inhabited by microbial flora that experiences intestinal peristalsis-like motions and flow. Lab Chip 12:2165–2174
Kim T-H, Kim M, Park H-S et al (2012b) Size-dependent cellular toxicity of silver nanoparticles. J Biomed Mater Res A 100:1033–1043
Kodali V, Littke MH, Tilton SC et al (2013) Dysregulation of macrophage activation profiles by engineered nanoparticles. ACS Nano 7:6997–7010
Kong L, Tuomela S, Hahne L et al (2013) NanoMiner – integrative human transcriptomics data resource for nanoparticle research. PLoS One 8:e68414
Krais A, Wortmann L, Hermanns L, Feliu N, Vahter M, Stucky S, Mathur S, Fadeel B (2014) Targeted uptake of folic acid-functionalized iron oxide nanoparticles by ovarian cancer cells in the presence but not in the absence of serum. Nanomedicine. pii:S1549–9634(14)00030-6. doi:10.1016/j.nano.2014.01.006 [Epub ahead of print]
Kumari A, Yadav SK, Yadav SC (2010) Biodegradable polymeric nanoparticles based drug delivery systems. Colloids Surf B Biointerfaces 75:1–18

Kunzmann A, Andersson B, Thurnherr T et al (2011a) Toxicology of engineered nanomaterials: focus on biocompatibility, biodistribution and biodegradation. Biochim Biophys Acta 1810:361–373
Kunzmann A, Andersson B, Vogt C et al (2011b) Efficient internalization of silica-coated iron oxide nanoparticles of different sizes by primary human macrophages and dendritic cells. Toxicol Appl Pharmacol 253:81–93
Lamkanfi M, Dixit VM (2012) Inflammasomes and their roles in health and disease. Annu Rev Cell Dev Biol 28:137–161
Larese Filon F, Crosera M, Timeus E et al (2013) Human skin penetration of cobalt nanoparticles through intact and damaged skin. Toxicol In Vitro 27:121–127
Larese FF, D'Agostin F, Crosera M et al (2009) Human skin penetration of silver nanoparticles through intact and damaged skin. Toxicology 255:33–37
Laroui H, Wilson DS, Dalmasso G et al (2011) Nanomedicine in GI. Am J Physiol Gastrointest Liver Physiol 300:G371–G383
Larsen ST, Roursgaard M, Jensen KA, Nielsen GD (2010) Nano titanium dioxide particles promote allergic sensitization and lung inflammation in mice. Basic Clin Pharmacol Toxicol 106:114–117
Le T, Epa VC, Burden FR, Winkler DA (2012) Quantitative structure–property relationship modeling of diverse materials properties. Chem Rev 112:2889–2919
Lesniak A, Fenaroli F, Monopoli MP et al (2012) Effects of the presence or absence of a protein corona on silica nanoparticle uptake and impact on cells. ACS Nano 6:5845–5857
Lewinski N, Colvin V, Drezek R (2008) Cytotoxicity of nanoparticles. Small 4:26–49
Liu W, Rose J, Plantevin S, Auffan M, Bottero JY, Vidaud C (2013) Protein corona formation for nanomaterials and proteins of a similar size: hard or soft corona? Nanoscale 5:1658–1668
Lundqvist M, Stigler J, Cedervall T et al (2011) The evolution of the protein corona around nanoparticles: a test study. ACS Nano 5:7503–7509
Lunov O, Syrovets T, Loos C et al (2011a) Differential uptake of functionalized polystyrene nanoparticles by human macrophages and a monocytic cell line. ACS Nano 5:1657–1669
Lunov O, Syrovets T, Loos C et al (2011b) Amino-functionalized polystyrene nanoparticles activate the NLRP3 inflammasome in human macrophages. ACS Nano 5:9648–9657
Lynch I, Ahluwalia A, Boraschi D, Byrne HJ, Fadeel B, Gehr P, Gutleb AC, Kendall M, Papadopoulos MG (2013) The bio-nano-interface in predicting nanoparticle fate and behaviour in living organisms: towards grouping and categorising nanomaterials and ensuring nanosafety by design. BioNanoMaterials 14:195–216
Mahler GJ, Esch MB, Tako E et al (2012) Oral exposure to polystyrene nanoparticles affects iron absorption. Nat Nanotechnol 7:264–271
Mahmoudi M, Shokrgozar MA, Sardari S et al (2011) Irreversible changes in protein conformation due to interaction with superparamagnetic iron oxide nanoparticles. Nanoscale 3:1127–1138
Manolova V, Flace A, Bauer M et al (2008) Nanoparticles target distinct dendritic cell populations according to their size. Eur J Immunol 38:1404–1413
Mansour HM, Rhee Y-S, Wu X (2009) Nanomedicine in pulmonary delivery. Int J Nanomedicine 4:299–319
Markiewski MM, Lambris JD (2009) Is complement good or bad for cancer patients? A new perspective on an old dilemma. Trends Immunol 30:286–292
Markiewski MM, DeAngelis RA, Benencia F et al (2008) Modulation of the antitumor immune response by complement. Nat Immunol 9:1225–1235
Maurer-Jones MA, Lin Y-S, Haynes CL (2010) Functional assessment of metal oxide nanoparticle toxicity in immune cells. ACS Nano 4:3363–3373
Mayr LM, Bojanic D (2009) Novel trends in high-throughput screening. Curr Opin Pharmacol 9:580–588
Meng H, Chen Z, Xing G et al (2007) Ultrahigh reactivity provokes nanotoxicity: explanation of oral toxicity of nano-copper particles. Toxicol Lett 175:102–110
Meng H, Yang S, Li Z et al (2011) Aspect ratio determines the quantity of mesoporous silica nanoparticle uptake by a small GTPase-dependent macropinocytosis mechanism. ACS Nano 5:4434–4447

Mitchell LA, Lauer FT, Burchiel SW, McDonald JD (2009) Mechanisms for how inhaled multiwalled carbon nanotubes suppress systemic immune function in mice. Nat Nanotechnol 4:451–456
Moghimi SM (2002) Chemical camouflage of nanospheres with a poorly reactive surface: towards development of stealth and target-specific nanocarriers. Biochim Biophys Acta 1590:131–139
Moghimi SM, Andresen TL (2009) Complement-mediated tumour growth: implications for cancer nanotechnology and nanomedicines. Mol Immunol 46:1571–1572
Moghimi SM, Andersen AJ, Ahmadvand D et al (2011) Material properties in complement activation. Adv Drug Deliv Rev 63:1000–1007
Monopoli MP, Walczyk D, Campbell A et al (2011) Physical-chemical aspects of protein corona: relevance to in vitro and in vivo biological impacts of nanoparticles. J Am Chem Soc 133:2525–2534
Monopoli MP, Åberg C, Salvati A, Dawson KA (2012) Biomolecular coronas provide the biological identity of nanosized materials. Nat Nanotechnol 7:779–786
Monteiro-Riviere NA, Wiench K, Landsiedel R et al (2011) Safety evaluation of sunscreen formulations containing titanium dioxide and zinc oxide nanoparticles in UVB sunburned skin: an in vitro and in vivo study. Toxicol Sci 123:264–280
Moreno L, Pearson ADJ (2013) How can attrition rates be reduced in cancer drug discovery? Expert Opin Drug Discov 8:363–368
Moyano DF, Goldsmith M, Solfiell DJ et al (2012) Nanoparticle hydrophobicity dictates immune response. J Am Chem Soc 134:3965–3967
Murray AR, Kisin ER, Tkach AV, Yanamala N, Mercer R, Young SH, Fadeel B, Kagan VE, Shvedova AA (2012) Factoring-in agglomeration of carbon nanotubes and nanofibers for better prediction of their toxicity versus asbestos. Part Fibre Toxicol 10:9–10
Mutlu GM, Budinger GR, Green AA et al (2010) Biocompatible nanoscale dispersion of single-walled carbon nanotubes minimizes in vivo pulmonary toxicity. Nano Lett 10: 1664–1670
Nagai H, Okazaki Y, Chew SH, Misawa N, Yamashita Y et al (2011) Diameter and rigidity of multiwalled carbon nanotubes are critical factors in mesothelial injury and carcinogenesis. Proc Natl Acad Sci U S A 108:E1330–8
Napierska D, Thomassen LCJ, Rabolli V et al (2009) Size-dependent cytotoxicity of monodisperse silica nanoparticles in human endothelial cells. Small 5:846–853
Nel A, Xia T, Mädler L, Li N (2006) Toxic potential of materials at the nanolevel. Science 311:622–627
Nel A, Xia T, Meng H et al (2013a) Nanomaterial toxicity testing in the 21st century: use of a predictive toxicological approach and high-throughput screening. Acc Chem Res 46:607–621
Nel AE, Nasser E, Godwin H et al (2013b) A multi-stakeholder perspective on the use of alternative test strategies for nanomaterial safety assessment. ACS Nano 7:6422–6433
Nygaard UC, Hansen JS, Samuelsen M et al (2009) Single-walled and multi-walled carbon nanotubes promote allergic immune responses in mice. Toxicol Sci 109:113–123
Nyström AM, Fadeel B (2012) Safety assessment of nanomaterials: implications for nanomedicine. J Control Release 161:403–408
Oh WK, Kim S, Choi M, Kim C, Jeong YS, Cho BR, Hahn JS, Jang J (2010) Cellular uptake, cytotoxicity, and innate immune response of silica–titania hollow nanoparticles based on size and surface functionality. ACS Nano 4:5301–5313
Oostingh GJ, Casals E, Italiani P et al (2011) Problems and challenges in the development and validation of human cell-based assays to determine nanoparticle-induced immunomodulatory effects. Part Fibre Toxicol 8:8
Palomäki J, Välimäki E, Sund J et al (2011) Long, needle-like carbon nanotubes and asbestos activate the NLRP3 inflammasome through a similar mechanism. ACS Nano 5(9):6861–6870
Pan Y, Neuss S, Leifert A et al (2007) Size-dependent cytotoxicity of gold nanoparticles. Small 3:1941–1949
Park JH, Gu L, von Maltzahn G et al (2009) Biodegradable luminescent porous silicon nanoparticles for in vivo applications. Nat Mater 8:331–336

Park E-J, Bae E, Yi J et al (2010a) Repeated-dose toxicity and inflammatory responses in mice by oral administration of silver nanoparticles. Environ Toxicol Pharmacol 30:162–168

Park HS, Kim KH, Jang S et al (2010b) Attenuation of allergic airway inflammation and hyperresponsiveness in a murine model of asthma by silver nanoparticles. Int J Nanomedicine 5:505–515

Pietroiusti A, Campagnolo L, Fadeel B (2013) Interactions of engineered nanoparticles with organs protected by internal biological barriers. Small 9:1557–1572

Pujol A, Mosca R, Farrés J, Aloy P (2010) Unveiling the role of network and systems biology in drug discovery. Trends Pharmacol Sci 31:115–123

Puzyn T, Rasulev B, Gajewicz A et al (2011) Using nano-QSAR to predict the cytotoxicity of metal oxide nanoparticles. Nat Nanotechnol 6:175–178

Reisetter AC, Stebounova LV, Baltrusaitis J et al (2011) Induction of inflammasome dependent pyroptosis by carbon black nanoparticles. J Biol Chem 286:21844–21852

Rejman J, Oberle V, Zuhorn IS, Hoekstra D (2004) Size-dependent internalization of particles via the pathways of clathrin- and caveolae-mediated endocytosis. Biochem J 377:159–169

Rettig L, Haen SP, Bittermann AG et al (2010) Particle size and activation threshold: a new dimension of danger signaling. Blood 115:4533–4541

Roger E, Lagarce F, Garcion E, Benoit J-P (2010) Biopharmaceutical parameters to consider in order to alter the fate of nanocarriers after oral delivery. Nanomedicine (Lond) 5:287–306

Ruge CA, Schaefer UF, Herrmann J et al (2012) The interplay of lung surfactant proteins and lipids assimilates the macrophage clearance of nanoparticles. PLoS One 7:e40775

Ryman-Rasmussen JP, Tewksbury EW, Moss OR et al (2009) Inhaled multiwalled carbon nanotubes potentiate airway fibrosis in murine allergic asthma. Am J Respir Cell Mol Biol 40:349–358

Salvati A, Pitek AS, Monopoli MP et al (2013) Transferrin-functionalized nanoparticles lose their targeting capabilities when a biomolecule corona adsorbs on the surface. Nat Nanotechnol 8:137–143

Schädlich A, Caysa H, Mueller T et al (2011) Tumor accumulation of NIR fluorescent PEG-PLA nanoparticles: impact of particle size and human xenograft tumor model. ACS Nano 5: 8710–8720

Schaeublin NM, Braydich-Stolle LK, Schrand AM et al (2011) Surface charge of gold nanoparticles mediates mechanism of toxicity. Nanoscale 3:410–420

Schaeublin NM, Braydich-Stolle LK, Maurer EI, Park K, MacCuspie RI et al (2012) Does shape matter? Bioeffects of gold nanomaterials in a human skin cell model. Langmuir 28:3248–3258

Schleh C, Semmler-Behnke M, Lipka J et al (2012) Size and surface charge of gold nanoparticles determine absorption across intestinal barriers and accumulation in secondary target organs after oral administration. Nanotoxicology 6:36–46

Schnackenberg LK, Sun J, Beger RD (2012) Metabolomics techniques in nanotoxicology studies. Methods Mol Biol 926:141–156

Schöll I, Weissenböck A, Förster-Waldl E et al (2004) Allergen-loaded biodegradable poly (D,L-lactic-*co*-glycolic) acid nanoparticles down-regulate an ongoing Th2 response in the BALB/c mouse model. Clin Exp Allergy 34:315–321

Schroeder A, Levins CG, Cortez C et al (2010) Lipid-based nanotherapeutics for siRNA delivery. J Intern Med 267:9–21

Sée V, Free P, Cesbron Y et al (2009) Cathepsin L digestion of nanobioconjugates upon endocytosis. ACS Nano 3:2461–2468

Sharma G, Valenta DT, Altman Y, Harvey S, Xie H et al (2010) Polymer particle shape independently influences binding and internalization by macrophages. J Control Release 147:408–412

Shaw SY, Westly EC, Pittet MJ et al (2008) Perturbational profiling of nanomaterial biologic activity. Proc Natl Acad Sci U S A 105:7387–7392

Shim W, Paik MJ, Nguyen DT et al (2012) Analysis of changes in gene expression and metabolic profiles induced by silica-coated magnetic nanoparticles. ACS Nano 6:7665–7680

Shvedova AA, Kagan VE, Fadeel B (2010) Close encounters of the small kind: adverse effects of man-made materials interfacing with the nano-cosmos of biological systems. Annu Rev Pharmacol Toxicol 50:63–88

Shvedova AA, Kapralov AA, Feng WH et al (2012) Impaired clearance and enhanced pulmonary inflammatory/fibrotic response to carbon nanotubes in myeloperoxidase-deficient mice. PLoS One 7:e30923

Simberg D, Park J-H, Karmali PP et al (2009) Differential proteomics analysis of the surface heterogeneity of dextran iron oxide nanoparticles and the implications for their in vivo clearance. Biomaterials 30:3926–3933

Sipes NS, Martin MT, Kothiya P et al (2013) Profiling 976 ToxCast chemicals across 331 enzymatic and receptor signaling assays. Chem Res Toxicol 26:878–895

Snyder-Talkington BN, Dymacek J, Porter DW et al (2013) System-based identification of toxicity pathways associated with multi-walled carbon nanotube-induced pathological responses. Toxicol Appl Pharmacol 272:476–489

Stark WJ (2011) Nanoparticles in biological systems. Angew Chem Int Ed Engl 50:1242–1258

Stecher B, Hardt W-D (2011) Mechanisms controlling pathogen colonization of the gut. Curr Opin Microbiol 14:82–91

Sun Q, Radosz M, Shen Y (2012) Challenges in design of translational nanocarriers. J Control Release 164:156–169

Sun B, Wang X, Ji Z et al (2013) NLRP3 inflammasome activation induced by engineered nanomaterials. Small 9:1595–1607

Sund J, Alenius H, Vippola M, Savolainen K, Puustinen A (2011) Proteomic characterization of engineered nanomaterial-protein interactions in relation to surface reactivity. ACS Nano 5:4300–4309

Syed S, Zubair A, Frieri M (2013) Immune response to nanomaterials: implications for medicine and literature review. Curr Allergy Asthma Rep 13:50–57

Teeguarden JG, Webb-Robertson BJ, Waters KM et al (2011) Comparative proteomics and pulmonary toxicity of instilled single-walled carbon nanotubes, crocidolite asbestos, and ultrafine carbon black in mice. Toxicol Sci 120:123–135

Tenzer S, Docter D, Rosfa S et al (2011) Nanoparticle size is a critical physicochemical determinant of the human blood plasma corona: a comprehensive quantitative proteomic analysis. ACS Nano 5:7155–7167

Tenzer S, Docter D, Kuharev J et al (2013) Rapid formation of plasma protein corona critically affects nanoparticle pathophysiology. Nat Nanotechnol 8:772–781

Thubagere A, Reinhard BM (2010) Nanoparticle-induced apoptosis propagates through hydrogen-peroxide-mediated bystander killing: insights from a human intestinal epithelium in vitro model. ACS Nano 4:3611–3622

Tkach AV, Shurin GV, Shurin MR et al (2011) Direct effects of carbon nanotubes on dendritic cells induce immune suppression upon pulmonary exposure. ACS Nano 5:5755–5762

Trotta F, Leufkens HGM, Schellens JHM et al (2011) Evaluation of oncology drugs at the European Medicines Agency and US Food and Drug Administration: when differences have an impact on clinical practice. J Clin Oncol 29:2266–2272

Tsai Y-Y, Huang Y-H, Chao Y-L et al (2011) Identification of the nanogold particle-induced endoplasmic reticulum stress by omic techniques and systems biology analysis. ACS Nano 5:9354–9369

Tuomela S, Autio R, Buerki-Thurnherr T et al (2013) Gene expression profiling of immune-competent human cells exposed to engineered zinc oxide or titanium dioxide nanoparticles. PLoS One 8:e68415

Tyurina YY, Kisin ER, Murray A et al (2011) Global phospholipidomics analysis reveals selective pulmonary peroxidation profiles upon inhalation of single-walled carbon nanotubes. ACS Nano 5:7342–7353

Vemula PK, Anderson RR, Karp JM (2011) Nanoparticles reduce nickel allergy by capturing metal ions. Nat Nanotechnol 6:291–295

Viaud S, Saccheri F, Mignot G et al (2013) The intestinal microbiota modulates the anticancer immune effects of cyclophosphamide. Science 342:971–976

Vonarbourg A, Passirani C, Saulnier P et al (2006) Evaluation of pegylated lipid nanocapsules versus complement system activation and macrophage uptake. J Biomed Mater Res A 78:620–628

Walczyk D, Bombelli FB, Monopoli MP et al (2010) What the cell "sees" in bionanoscience. J Am Chem Soc 132:5761–5768
Walkey CD, Chan WCW (2012) Understanding and controlling the interaction of nanomaterials with proteins in a physiological environment. Chem Soc Rev 41:2780–2799
Wang X, Ishida T, Kiwada H (2007) Anti-PEG IgM elicited by injection of liposomes is involved in the enhanced blood clearance of a subsequent dose of PEGylated liposomes. J Control Release 119:236–244
Wang B, He X, Zhang Z et al (2013a) Metabolism of nanomaterials in vivo: blood circulation and organ clearance. Acc Chem Res 46:761–769
Wang F, Yu L, Monopoli MP, Sandin P et al (2013b) The biomolecular corona is retained during nanoparticle uptake and protects the cells from the damage induced by cationic nanoparticles until degraded in the lysosomes. Nanomedicine 9:1159–1168
Weissleder R, Stark DD, Engelstad BL et al (1989) Superparamagnetic iron oxide: pharmacokinetics and toxicity. AJR Am J Roentgenol 152:167–173
Winkler DA, Mombelli E, Pietroiusti A et al (2013) Applying quantitative structure–activity relationship approaches to nanotoxicology: current status and future potential. Toxicology 313:15–23
Witasp E, Kupferschmidt N, Bengtsson L et al (2009) Efficient internalization of mesoporous silica particles of different sizes by primary human macrophages without impairment of macrophage clearance of apoptotic or antibody-opsonized target cells. Toxicol Appl Pharmacol 239:306–319
Xia T, Li N, Nel AE (2009) Potential health impact of nanoparticles. Annu Rev Public Health 30:137–150
Xia X-R, Monteiro-Riviere NA, Riviere JE (2010) An index for characterization of nanomaterials in biological systems. Nat Nanotechnol 5:671–675
Xiao K, Li Y, Luo J et al (2011) The effect of surface charge on in vivo biodistribution of PEG-oligocholic acid based micellar nanoparticles. Biomaterials 32:3435–3446
Yang S, Guo W, Lin Y et al (2007) Biodistribution of pristine single-walled carbon nanotubes in vivo. J Phys Chem C 111:17761–17764
Yang M, Flavin K, Kopf I et al (2013a) Functionalization of carbon nanoparticles modulates inflammatory cell recruitment and NLRP3 inflammasome activation. Small 9:4194–4206
Yang S-T, Liu Y, Wang Y-W, Cao A (2013b) Biosafety and bioapplication of nanomaterials by designing protein-nanoparticle interactions. Small 9:1635–1653
Yazdi AS, Guarda G, Riteau N et al (2010) Nanoparticles activate the NLR pyrin domain containing 3 (Nlrp3) inflammasome and cause pulmonary inflammation through release of IL-1α and IL-1β. Proc Natl Acad Sci U S A 107:19449–19454
Zarogoulidis P, Chatzaki E, Porpodis K et al (2012) Inhaled chemotherapy in lung cancer: future concept of nanomedicine. Int J Nanomedicine 7:1551–1572
Zhang T, Stilwell JL, Gerion D et al (2006) Cellular effect of high doses of silica-coated quantum dot profiled with high throughput gene expression analysis and high content cellomics measurements. Nano Lett 6:800–808
Zhu M, Nie G, Meng H et al (2013) Physicochemical properties determine nanomaterial cellular uptake, transport, and fate. Acc Chem Res 46:622–631
Zolnik BS, González-Fernández A, Sadrieh N, Dobrovolskaia MA (2010) Nanoparticles and the immune system. Endocrinology 151:458–465

Chapter 15
Regulatory Aspects of Oncologicals: Nanosystems Main Challenges

Rogério S. Gaspar, Helena F. Florindo, Liana C. Silva, Mafalda A. Videira, M. Luísa Corvo, Bárbara F. Martins, and Beatriz Silva-Lima

15.1 General Regulatory Requirements for Toxicity Assessment

The developments related to critical issues in health sciences observed during the last decades as well as the successful integration of multidisciplinary technologies have led to innovative and new nanopharmaceutical products both for disease diagnosis, prevention and treatment. Among them, drug delivery systems as polymer-conjugates, liposomes and nanoparticles (NPs) have been shown to overcome important limitations of conventional treatments and thus have given rise to opportunities including very attractive and effective health technologies (Duncan and Gaspar 2011). Both increased funding and research activities, undertaken either by industry or in academia and better exploring the development of those promising nanomaterials to be used in medical applications, are mainly justified by: (1) the versatility of manufacturing process and materials (i.e., the modification of their surface to overcome formulation stability concerns and/or to include target specific molecules to cell membranes or even intracellular organelles); (2) their size scale, making it possible to overcome some important physiological barriers, and thus attaining release of the therapeutic drug at the desired target; and (3) the ability to entrap considerable amounts of pharmaceutical active molecules of different natures protecting them from aggressive environment, as well as targeting specific cells, not only leading to higher therapeutic levels with lower doses but also to a significant reduction of pharmacological side effects by avoiding non-specific release (Wang et al. 2013; Duncan and Gaspar 2011; Moghimi et al. 2012; Jung et al. 2011).

R.S. Gaspar (✉) • H.F. Florindo • L.C. Silva • M.A. Videira • M.L. Corvo
B.F. Martins • B. Silva-Lima
Faculdade de Farmácia, Universidade de Lisboa, Av. Prof. Gama Pinto, 1649-003 Lisboa, Portugal
e-mail: rgaspar@ff.ul.pt

M.J. Alonso, M. Garcia-Fuentes (eds.), *Nano-Oncologicals: New Targeting and Delivery Approaches*, Advances in Delivery Science and Technology,
DOI 10.1007/978-3-319-08084-0_15

There is a considerable number of nanoproducts approved for biomedical applications, but the absence of specific universal protocols to be used during their preclinical characterisation and development has led to the use of similar adapted strategies following those used in the development protocols of "conventional" pharmaceutical products, including in the evaluation of their safety/toxicity and overall immune compatibility (Dobrovolskaia and McNeil 2013; Ehmann et al. 2013; Schütz et al. 2013; Jung et al. 2011; Gaspar and Duncan 2009). While conventional preclinical tests and requirements might have been appropriate for the safety assessment of new medicinal products including both a new molecule and the nano-system, this might need to be adapted in the case of nanoformulations with already marketed active substances. In these cases a dedicated evaluation of the impact of the nanosized-formulation on the safety attributes of the active substance is needed. The reasoning behind the concept of the specific use for nanoformulation for the already marketed active substance needs always a justification and a comprehensive characterisation of the consequences that could derive in terms of both efficacy and safety. This means that additional pharmacology (proof of concept) and toxicity studies (mostly bridging with those already available with the active substance itself) may or should be needed. Also, the carrier activity and safety will need a full characterisation. One of the major hurdles underlying nanomedicines-related regulation is the lack of consensus on the different procedures followed by main regulatory authorities as United States (US) Food and Drug Administration (FDA) and European Medicines Agency (EMA) (Gaspar 2010; Wagner et al. 2006). European centralised procedures are being developed since 1995, with the establishment of EMA and, besides the traditional major influence of FDA, there are different positions assumed by the distinct government organisations worldwide, which have determined regulatory constraints to pharmaceutical companies during the preparation of specific Marketing Authorization Applications (MAA in Europe) or New Drug Application (NDA), for medicinal products commercialisation in the USA (Gaspar 2010). Nevertheless, regulators and industry from the USA, Europe and Japan have been brought together in the 1990s in order to develop global regulatory approaches through the International Conference on Harmonisation (ICH), increasing harmonisation of quality, safety and efficacy guidelines. An important feature presented by the European regulatory system is the opportunity that applicants have to get "scientific advice" from regulators even at early stages of the research and development cycle of medicinal products (Gaspar 2010). This practice contributes to the harmonised development of innovative pharmaceutical products, decreasing the impact of obstacles faced either by academia or global pharmaceutical companies. As we will detail later in this chapter, in the last few years the EMA has created a working group to address the specific quality, safety and efficacy aspects of nanopharmaceuticals and prepare "orientation documents" (reflection papers). While not being strict guidelines, they refer to main aspects to be considered by applicants in the development of their products. Since 2009–2010 a dialogue has been established between EMA in the EU and the FDA in the USA together with the PDMA/MHLW in Japan, which constitutes a step forward on the attempt to reach common positions in the area of nanomedicines development for those regions of the world. However, at the ICH level, nanopharmaceuticals have not yet been included as a topic.

On the other hand, as for any other innovative products, in order to pursue the commercialisation of new nanopharmaceuticals and nano-imaging products, it will be also fundamental to perform pharmacoeconomic studies to clearly show the economic and societal added value of these new products in comparison with those already available for patients. Potential increase in QALYs (quality-adjusted life expectancy years) or future consecutive hospitalisation costs are important indicators to always consider in future developments of new and innovative technologies.

As a result, having in consideration safety concerns and to prevent failures during future clinical studies, major pharmaceutical companies have also been focused on studies for early "proof of concept" intended to explore the detailed mechanisms underlying nanoproducts' both efficacy and safety, as well as the influence of pharmacogenetic variability on their clinical outcome (Ehmann et al. 2013; Schütz et al. 2013).

The preclinical study of polymer therapeutics, as well as the regulatory issues related to their design and development, has been reviewed by Gaspar and Duncan (2009). Similarly to what have been described for that class of nanopharmaceuticals, the clinical use of these complex nanosystems is strongly dependent on extensive evaluation, characterisation and understanding of key phenomena, e.g. their ability to overcome physiological barriers, including cell membranes, clarifying their uptake mechanisms, intracellular pharmacokinetics, including their effect on cell viability and functions, and their clearance from cells, tissues or organs.

Having in consideration nanopharmaceutical-based products nanometric size range and specific physicochemical properties, which are known to determine their ability to adsorb blood proteins and interact with immune cells, the studies performed during their preclinical evaluation must explore their biocompatibility and immunotoxicity (Table 15.1).

In the literature a considerable number of studies are available regarding some of these issues and also the complement activation induced by several nanopharmaceutical systems, as well as their immunological effect in vivo (Chapman et al. 2013; Dobrovolskaia and McNeil 2013; Ehmann et al. 2013; Moghimi et al. 2012; Moghimi and Hamad 2008; Szebeni and Moghimi 2009; Moghimi and Andresen 2009). The intrinsic immunogenicity of nanotechnology-based formulations is mainly dictated by their size, morphology, surface charge, targeting moieties and hydrophobicity (Table 15.1), which will also determine their interaction with plasma proteins, adjuvant properties, toxicity to immune cells and internalisation/phagocytic uptake by phagocytic cells and thus their ability to further stimulate immune cells (Dobrovolskaia and McNeil 2013; Dobrovolskaia et al. 2008, 2009a, b; McLeland et al. 2009). As a result, these properties can be manipulated to potentiate their adjuvant properties, i.e. to be used to achieve more effective vaccines. However, these nanoproducts have a wide range of applications besides vaccination purposes (prevention and treatment) and in several clinical situations, the effect of the particulate systems has to be controlled and fully known in order to prevent unexpected adverse effects, through

Table 15.1 Selection of parameters used to evaluate quality and safety of nanopharmaceuticals for biomedical use (adapted from Gaspar 2010; Dobrovolskaia and McNeil 2007)

Parameters	Goal	Product property
Physicochemical characteristics	To evaluate their average size but also particle size distribution and aggregation Particle surface charge Morphology Crystallinity Rigidity/deformability Chemistry and molecular architecture	Intermediate product quality
Manufacturing process	Control of critical points of production	Product quality
Microbial contamination	Sterility test	Contamination tests to evaluate product quality and safety
Endotoxin levels	To assess the presence of pyrogens Nitrous oxide quantification (indirect evaluation)	
Viral/mycoplasma levels	Sterility test	
Cytokine production	Identification and quantification of induced cytokines	Immunogenicity (immunosuppression or immunostimulation)/safety
Natural killer cells cytotoxicity	Effect on NK cell main functions (recognise and destroy)	
Macrophages uptake	To evaluate if nanoproducts are phagocytosed	
Complement activation	Effect on complement cascade	
Plasma protein binding	Particle charge, hydrophobicity effect	
Leucocytes proliferation	Influence on leucocyte responses	
In vivo immune response	NK cells' function Cytokine production Immunoglobulins T (Th1/Th2) and B cells proliferation	
Coagulation time	Effect on coagulation factors	Interaction with blood/ biocompatibility/safety
Platelet aggregation	Influence on coagulation cascade	
Haemolysis	Effect on red blood cells	
In vivo single dose toxicity study	Toxicity to different organs/cells, namely immune cells	Toxicity and prediction of additional toxicity tests/safety
Biodistribution studies	Pharmacokinetics, pharmacodynamics, metabolism, clearance	Product safety

their immune reactivity with potential for development of strong autoimmune responses (Dobrovolskaia et al. 2008; McLeland et al. 2009). To appropriately assess the toxicity of those nanoproducts, both the dose, administration route, type of targeted disease environment and respective therapeutic index have also to be taken into consideration along development and whenever looking at different options.

In general, most nano-based products have been known as safe, biocompatible and have shown to be able to even decrease the conventional drug toxicity. In fact, carriers keeping drugs entrapped can avoid blood and normal tissue cells toxicity, by releasing them only at desired target tissues/cells. As a result, it would normally allow the use of considerable lower doses of active pharmaceutical ingredients (API), increasing their therapeutic index and improving their safety profile (Moghimi et al. 2012). Even though, new classes of proposed nanopharmaceuticals, as dendrimers (Tekade et al. 2008; Ruiz et al. 2014), carbon nanotubes (Bianco et al. 2005a, b; Wang et al. 2013) or quantum dots (Reiss et al. 2009; Cai and Chen 2008), have given rise to additional concerns related to the use of those novel materials at human level, namely to their potential immunological deleterious effects, and will probably be absent from any clinical use at least for many years.

For the novel nanotechnology-based platforms, besides the strong demand to foster these nanomedicines in order to address unmet clinical needs, their production and their intrinsic and particular properties raised important challenges, for industry and government agencies (Wagner et al. 2006; Gaspar 2010). In fact, even if several studies have already been performed to understand the biological behaviour and pharmacokinetics (body and cell pharmacokinetics) of most innovative systems, no global trends were yet defined due to the limited data regarding the full implications of their characterisation. Also a number of issues arise due to the need of adapting to new methods of manufacture. This may also be attributed to the limited knowledge regarding the effect of small changes on manufacturing process on nanoparticles properties, as well as the limited knowledge on the modification of physicochemical properties presented by some nanosystems under specific physiological conditions. In fact, their heterogeneity might lead to minor but eventually important batch-to-batch variability, and in a number of cases it's difficult to establish an effective comparison between their behaviour in vitro verified during the research and the one showed at later development phases (Dobrovolskaia and McNeil 2013). One of the main hurdles is related to the biodegradability, release profile and characterisation of surface coating thickness, stability and even composition after in vivo administration.

These aspects can be determinant factors for the resultant pharmacokinetics and thus having also a potential pharmacodynamic impact in nanodelivery systems (Duncan and Gaspar 2011; Ehmann et al. 2013; Moghimi et al. 2012; Torchilin 2009). Most of these limitations could be partially overcome through a clearer definition of structure-activity relationships for those multivalent nanoproducts. This would better characterise their structure and the dynamic changes undergone by those nanomedicines in different bioenvironments, rather than in buffers or even water commonly used during in vitro studies (Duncan and Gaspar 2011). Some of these materials change their main physicochemical properties and even conformation depending on buffer composition, concentration and ionic strength and therefore not translating exactly what happens once administered in vivo (Dobrovolskaia and McNeil 2013).

Moreover, it is fundamental to understand their fate depending on route of administration and anatomo-physiological specific conditions defined by stage of the disease (Gaspar 2010). Table 15.1 summarises a set of main parameters that could be addressed in order to improve the data available for regulators for better evaluation of the quality, efficacy and safety of current and future nanopharmaceuticals.

The main issues related to the difficulties underlying the regulation of these complex innovative systems are thus related to the their main properties/characteristics, which can be easily modified by rather small changes in the raw materials properties, as well as by small changes in their manufacturing process. Even if those small changes seem to have a limited impact on their structures, they can significantly modify their biological properties and thus biodistribution patterns. It is highly desirable to develop quality control assays and robust methods able to effectively detect/characterise those small modifications but also to establish a relation between those physicochemical modifications and consequent effect on their biocompatibility, biological properties and therapeutic effect. The integrated knowledge regarding the physicochemical properties of nanosystems and their properties could also guide their formulation according to the desired effect (preventive, therapeutic and/or imaging), having in consideration the intended route of administration, clarifying their biodistribution, metabolism, clearance, cytotoxicity and blood biocompatibility, but also cellular uptake mechanisms and intracellular pharmacokinetics (Chapman et al. 2013; Gaspar and Duncan 2009; Dobrovolskaia and McNeil 2013; Moghimi et al. 2012).

Some studies performed with different classes of nanopharmaceuticals (polymer conjugates, liposomes, PEGylated-liposomal formulations and polymeric nanoparticles), and used to deliver a common API under mainly the same pathological conditions, have shown throughout the pharmacokinetic data the impact and main implications for carrier surface characteristics, size and manufacturing process on their biological behaviour (Hillaireau and Couvreur 2009). The pharmacokinetic and pharmacodynamic characterisation is especially relevant for nanocarriers to be applied in oncology, due to their ability to "selectively" accumulate at tumour sites (enhanced permeability and retention effect or EPR) closely related to the characteristic architecture of tumour endothelial vessels (Matsumura and Maeda 1986; Farokhzad and Langer 2009).

As mentioned above, even if the variability of raw materials can also have a critical influence on the physiological effect of these nanomedicines (Duncan and Gaspar 2011), several examples have already been demonstrated to be important for the control of raw materials, an issue fundamental but not sufficient enough to attain their successful translation into their clinical use (Wagner et al. 2006).

A number of regulatory agencies and harmonisation organisations are also focused on the development of globally harmonised assays, in order to address nanopharmaceutical's main aspects that are known to have a profound effect on both their in vivo safety and efficacy. The main hurdles associated to adequate assays are related to their ability to detect low concentrations of nanocarriers, to distinguish between nanocarrier itself and aggregates or even to detect differences between their intact and metabolised forms (Kwong et al. 2009; Nel et al. 2009). It is necessary to develop methods able to be easily reproduced and with adequate sensitivity. In order to address these limitations, alternative imaging, fluorescence or array cellular imaging techniques have been explored (Chapman et al. 2013; Choi and Frangioni 2010; Smith et al. 2009).

Major questions associated to manufacturing process of nanopharmaceuticals face current pharmaceutical innovation at production unit operations related to their scale-up potential, which is mostly impaired, by the variety of physical properties of new materials (Gaspar 2010). The identification and control of all critical points of each manufacturing process is thus an important factor to have in mind, by the implementation of adequate product experimental design, but also "quality-by-design" concepts, or improved methods like through process analytical technologies (PAT). These strategies will ensure an overall on-line/at-line quality assessment system by anticipating and considering the most probable problems that may occur and establishing the necessary automated procedures to get them resolved in line (Gaspar 2010). Consequently, those new concepts have already led to the introduction and implementation of new pharmaceutical development regulations (ICH Q8, Q9 and Q10) (Guideline 2009; Guideline 2008; Guideline 2005) presenting major implications for conventional medicinal products but now also for future nanopharmaceuticals development and manufacturing processes. Major difficulties to be faced upon are mainly related to the nature of the data that should be provided, before/during the product life cycle management, especially before and after MAA, including the need for in vivo animal or clinical studies, in order to fully characterise the impact of manufacturing changes.

The active moiety of these nanotechnology-based systems will define the type of specifications that should be analysed within the regulatory context. For example, in the presence of biological entities, as proteins, peptides, antibodies or fractions of those, the new product will have to follow the regulations defined not only for new chemical entities (NCEs) but also for biological medicinal products. More complex situations have arisen with the development of biosimilar products and thus with the regulatory specifications that should be explored during the evaluation of quality, safety and efficacy of those nanoproducts (Gaspar 2010; Ehmann et al. 2013). For example, when a biomolecule already marketed as a parenteral solution is entrapped within a delivery system, the resultant nanopharmaceutical will be evaluated under the regulatory requirements established also for biosimilar products, provided a different manufacturer is involved, and relevant factors have to be taken into consideration for comparison purposes addressing issues also related to the non-entrapped molecule.

The major differences between regulatory requirements imposed by different agencies will be emphasised during the assessment of bio(nano)similars, due to differences in data protection and patent duration of nanopharmaceuticals imposed by different regulators (Bawa et al. 2008).

15.2 Translational Aspects to Be Considered for Nanomedicines in Oncology

Over the last decades, major advances have been made in the field of nanomedicines, and several delivery systems that provide a controlled, sustained and local delivery of the therapeutic entity have reached clinical stage and showed promising results mostly in cancer therapy (Duncan and Gaspar 2011). Nevertheless, a rational development of novel biocompatible and cost-effective delivery systems able to target and deliver their payload to specific tissues, cells and/or intracellular targets is still desirable. One of the major challenges inherent to the development of delivery systems is to provide the adequate characteristics and physicochemical properties that enable them to effectively cross biological barriers and reach their targets. This is important not only to overcome systemic barriers and deliver the therapeutic entity to the cells of interest but also to efficiently target subcellular locations.

The design of delivery systems targeted to cancer cells commonly takes advantage also of the EPR effect of the tumour microenvironment (Matsumura and Maeda 1986; Torchilin 2011). The distinct tumour angiogenesis, characterised by an elevated vascular permeability, is therefore often used as an enabling strategy in order to allow passive targeting of the tumour cells (Maeda et al. 1985; Segal and Satchi-Fainaro 2009). Manipulation of the composition and surface chemistry of the delivery systems and of their physicochemical properties, such as the size, surface charge, shape and curvature, provides the modalities to develop stealthy and stable systems with prolonged circulation time in the bloodstream (Li and Huang 2008). Each of these parameters individually affects the ability of the nanosystem to overcome the systemic barrier and reach the tumour environment (Li and Huang 2010; Liu et al. 2012) being therefore difficult to anticipate the optimal characteristics for a given delivery system. Nevertheless, it is widely accepted that nanosystems with sizes ranging from 100 to 200 nm, with surface charge close to neutrality and decorated with PEG, are more promising vehicles as recently reviewed (Kong et al. 2013). This might have implications on the design and experimental setting for proof-of-concept studies to be submitted to regulators and also on the prediction of potential clinical restrictions to be considered in the design of adequately designed clinical trials.

Tumour-specific delivery or active targeting has become a popular strategy to target and deliver the nanosystems and their payload to the cell of interest (Duncan and Gaspar 2011; Muro 2012). In cancer therapy, this can be accomplished by decorating the delivery systems with ligands to receptors that are commonly overexpressed in cancer cells, such as transferrin (Davis et al. 2010), folate (Gupta et al. 2008; Leamon and Reddy 2004; Low and Kularatne 2009; Yang et al. 2013) or epidermal growth factor (Krop et al. 2010). Even though this strategy relies on specific ligand-receptor interactions and has shown promising in vitro results, limited improvement on systemic delivery has been achieved in the in vivo animal models mainly because a close interaction between the delivery system and the tumour cells is mandatory for ligand-receptor recognition (Duncan and Gaspar 2011; Kong et al. 2013). A major challenge will be thus to demonstrate their relative benefits under GLP-regulated environment with purposely designed methods, able to validate the concept.

Besides reaching cancer cells, intracellular targeting is often desirable in cancer therapy, since many of the targets are localised in subcellular compartments. Therefore, a better understanding of the endocytic, recycling and secretory pathways, as well as of the trafficking and sorting mechanisms, is decisive towards the development of improved nanomedicines, particularly in cancer therapy, where changes in those cellular processes are often observed (Roepstorff et al. 2008; Grandal and Madshus 2008). Even though grafting of a ligand in the delivery system might not improve its systemic delivery, it assures that, upon interaction with tumour cells, the nanocarriers will preferentially be internalised by these cells and primarily targeted to the endosomes through receptor-mediated endocytosis (Rajendran et al. 2010; Ritchie et al. 2013).

Cells utilise a variety of endocytic pathways to internalise their cargo (reviewed in Doherty and McMahon 2009). Despite the extensive cross-talk between the different uptake and trafficking routes, each of these pathways employs specific mechanisms for the internalisation and sorting of their cargo to its final intracellular destination (Doherty and McMahon 2009; Mercer and Helenius 2012; Parton and del Pozo 2013). That can be exploited through the design of delivery systems with physicochemical properties and biospecific targeting moieties that target them to an alternative and specific route of internalisation and subsequent desired intracellular trafficking (Paillard et al. 2010; Richardson et al. 2010; Rajendran et al. 2010). As for the systemic delivery, these issues include changes in the nanocarrier size, surface charge, ligands/targeting moieties (Kong et al. 2013). Several strategies that improve the subcellular delivery of the nanocarrier and its payload have been identified and reviewed (Rajendran et al. 2010). These include, among others, grafting of ligands that are internalised via receptor-mediated endocytosis and thus targeted to the endosomes (as discussed above) (Rajendran et al. 2010; Duncan and Gaspar 2011; Ritchie et al. 2013; Kong et al. 2013), adding of peptidic sequences (Li et al. 2004; Leopold and Crystal 2007; Mastrobattista et al. 2007), pH-sensitive polymers (Yessine et al. 2003; Yessine and Leroux 2004; Richardson et al. 2010; Cheng et al. 2013) or fusogenic lipids (Boomer et al. 2009) thus enhancing the fusogenic ability of the nanocarrier in a pH-sensitive manner and mediate its release or of its content from the endolysosomal compartments to the cytosol (Rajendran et al. 2010; Duncan and Gaspar 2011; Kong et al. 2013).

Meanwhile, other approaches need to be referred. Exosomes have gained increasingly attention in recent years. These extracellular membrane vesicles have an endosomal origin and upon release from cells transfer their cargo to—and elicit a number of cellular alterations in—the recipient cells (reviewed in Ge et al. 2012; Bobrie and Théry 2013). Their cargo includes proteins, bioactive lipids and nucleic acids and, in addition, their lipid membrane composition differs from the plasma membrane (Record et al. 2014). Together, this contributes to their role in the modulation of the homeostasis of the recipient cells, being therefore important elements in intercellular communication and in the activation of immune response (Bobrie and Théry 2013). Cancer cells exploit these vesicular structures to manoeuvre the immune system and to promote angiogenesis, cancer progression and metastisation (Camussi et al. 2010; Taylor and Gercel-Taylor 2011; Filipazzi et al. 2012; Bobrie and Théry 2013). Accordingly, strategies that envisage targeting of therapeutic entities to—and/or decrease the formation of—tumour exosomes are becoming attractive tools to modulate tumour environment and elicit a stronger tumour-suppressive immune response (EL Andaloussi et al. 2013).

Major issues will also arise from the need to develop adequately designed animal models, avoiding the hurdles of xenograft models but also allowing for identification of specific nanomedicinal biomarkers with potential use in appropriate clinical setting.

15.3 Nanomedicinal Oncological Products Intended for Clinical Practice

There are different classes of nanomaterials developed for drug delivery, therapeutics and/or as imaging tools: lipid-based systems (liposomes, solid lipid nanoparticles, lipopoliplexes), polymer-based (nanogels, polymeric particles, core-shell systems, vesicles, micelles, polymer therapeutics, dendrimers), biotechnology-based systems (transporter peptides, monoclonal antibodies, inhibitory kinases, siRNA, miRNA) and other nanosystems (metal and ceramic nanoparticles, carbon nanotubes, quantum dots or nanocrystals).

A considerable number of those nanotechnology-based products are in the market, and many others are in pharmaceutical pipelines, at different stages of development. Marketed nanomedicines and nano-imaging agents have recently been reviewed (Duncan and Gaspar 2011; Wang et al. 2013; Schütz et al. 2013), providing extensive information regarding some of those products in clinical development.

Besides the difficulties faced by researchers, pharmaceutical industries and regulators regarding the development of nano-based products, some of the already approved systems are usually used as case studies and thus have been in the central discussion regarding the regulation of similar pharmaceutical products.

Within nanomedicinal oncological products in clinical practice are the liposomal formulations. Liposomes are concentric phospholipidic bilayers that can encapsulate hydrophilic compounds in the internal aqueous space and hydrophobic compounds in the lipidic matrix (Allen and Cullis 2013). **Ambisome**® (amphotericin B-loaded liposomes) was commercially approved in Europe since the late 1980s, being in fact the first nanopharmaceutical product (as such) to receive marketing authorisation for systemic fungal infections mainly in neutropenic oncology patients (Moen et al. 2009; Wang et al. 2013; Puri et al. 2009). Since then more than 40 products have been already marketed worldwide, being more than 40 % of those protein–polymer conjugates and liposomal based formulations (Puri et al. 2009; Schütz et al. 2013). In addition, more than 20 % of nanoparticulate systems currently in the marked are indicated for cancer treatment, being developed by taking advantage of their ability for selective targeting, to promote higher cellular uptake of drugs, as well as through the entrapment of two different molecules within a single carrier. These important features have been used to improve the pharmacokinetics and bioavailability of cytotoxic drugs (e.g. paclitaxel) and/or decrease their adverse side effects (doxorubicin) that normally impaired their application in clinical use (Wang et al. 2013).

For example, doxorubicin-loaded PEGylated liposomes were the first approved liposomal formulation for treatment of cancer. In addition, those were the first PEGylated formulations that were granted approval by the USA and the European Union in 1994/1995, being called **Doxil**® and **Caelyx**®, respectively. These formulations consisted of a long circulating liposomal formulation with a PEG2000 coating (hydrogenated soy phosphatidyl choline, cholesterol, MPEG2000-distearoyl phosphatidylethanolamine, 56:39:5 molar ratio) and with Doxorubicin hydrochloride encapsulated/entrapped in the internal aqueous space. It was first authorised for the treatment of Kaposi sarcoma in HIV patients, being further approved for ovarian cancer treatment in 1999, as a monotherapy treatment for patients with metastatic breast cancer in 2003 (for use in 69 countries) and multiple myeloma in 2007 (Wang et al. 2013; Plosker 2008; Paliwal et al. 2011). In the European Union, **Caelyx/Doxil®** has reached routine clinical use to treat breast cancer in patients at risk of heart problems.

Myocet® is the non-PEGylated liposomal-loaded doxorubicin product that was later approved in Canada and Europe through a centralised procedure for the treatment of metastatic breast cancer in combination with cyclophosphamide (Wang et al. 2013; Schütz et al. 2013). The liposomal formulation consists of egg phosphatidylcholine and cholesterol (molar ration of 55:45).

Two main differences are very important that distinguished the two Doxorubicin formulations. First, the form in which Doxorubicin is encapsulated in the liposomes: as a sulphate salt (drug precipitates as sulphate salt forming striated gel) and a citrate fibre (drug forms organised fibre bundles with citrate) for **Caelyx/Doxil®** and **Myocet®**, respectively. Secondly, the presence of PEG coating in **Caelyx/Doxil®** confers it long circulating properties, in contrast with the non-PEGylated formulation. This difference leads to different relative rates of Mononuclear Phagocyte System (MPS) clearance with **Caelyx/Doxil®** avoiding the system and **Myocet®** being uptaken. In 2012, the FDA has approved a generic version of **Doxil®** (doxorubicin hydrochloride liposome injection) for the treatment of ovarian cancer (progressed or recurred disease after platinum-based chemotherapy) and for AIDS-related Kaposi's sarcoma also with failure or intolerance to other systemic chemotherapy (Allen and Cullis 2013). This was done through controversy on criteria and relevance of shortage regarding **Doxil®** availability in the market, after a number of GMP issues.

The other Doxorubicin liposomal formulation, **Lipodox®**, is also approved since 1998 in Taiwan (Kan 2007). This PEGylated liposomal doxorubicin consisted of Distearoyl Phosphatidylcholine, cholesterol, MPEG2000-distearoyl phosphatidylethanolamine (56:39:5 molar ratio), and the therapeutic indications are for Kaposi sarcoma, breast and ovarian cancer. **Lipodox®** is considered a formulation with high similarity to **Doxil®** (same active ingredient, dosage, route of administration) but not approved by FDA and not considered as a "generic" of **Doxil®**. However, in 2012 FDA has approved discretion for the imports of **Lipodox®** during a critical shortage of **Doxil®** (Allen and Cullis 2013).

DaunoXome® is a liposome-encapsulated form of daunorubicin with the liposomes consisting of small unilamellar vesicles (mean diameter size 45 nm) composed of a 2:1 molar ratio of distearoyl phosphatidylcholine and cholesterol. This formulation has been approved since 1996 in Europe and in the USA for the treatment of Kaposi's sarcoma (Minotti et al. 2004).

Marqibo®, an orphan drug, was approved by FDA in 2012 under agency's accelerated approval for patients at second or more relapsed acute lymphoblastic leukaemia or who had already followed two or more regimens of anti-leukaemia therapy. It is a liposomal formulation of vincristine sulphate with a lipid composition of cholesterol and egg sphingomyelin (45:55 molar ratio) to be administered intravenously (Rodriguez et al. 2009).

On the other hand, albumin-based nanoparticles (**Abraxane**®) were the first non-inorganic nanoparticles that was granted permission to be commercialised, initially for breast cancer treatment in 2005 in the USA and 2008 by EMA (Schütz et al. 2013; Gradishar 2006). In addition, FDA has approved its use against non-small cell lung cancer in October 2012.

Besides those approved **liposomal formulations**, the driving force of research is leading in two ways: the search of new therapeutic indications for the approved formulations (different doses, treatment schedule, etc.) or the development of new formulations with different drugs and/or different liposomes (lipid composition, size, targeting agents, triggered action, multifunctional, etc.) or the combination therapy.

Within these approaches, several liposomal formulations reached different stages of clinical trials. In 2013, more than 450 **clinical trials** are referred by US National Institutes of Health under liposomal treatment for cancer therapy. Some selected examples of different liposomal formulations either by new drugs or new therapeutic approaches were as follows:

ThermoDox (Celsion Corp) is a liposomal doxorubicin composed of DPPC, mono stearoyl PC and PEG2000-DSPE (90:10:4 molar ratio). This formulation has the particularity of being temperature-triggered (41–42 °C) release. These long circulating liposomes will release their content inside the tumour upon heating locally at low hyperthermia. The currently explored potential therapeutic indications are along primary liver cancer (hepatocellular carcinoma) not eligible for resection surgery. Currently it is in phase III clinical trial (Poon and Borys 2011; Staruch et al. 2011).

SPI-077 (Alza) is a liposomal cisplatin formulation composed of hydrogenated soy phosphatidylcholine, cholesterol and methoxy polyethylene glycol distearoyl phosphatidylethanolamine (mPEG-DSPE) (51:44:5 molar ratio). Liposomal encapsulation of cisplatin was developed to deliver cisplatin to tumours with increased selectivity while decreasing the exposure of cisplatin to normal tissue by the enhanced permeation and retention process. The proposed therapeutic indications are directed to solid tumours. Currently it attained the stage of phase II clinical trial (Seetharamu et al. 2010; White et al. 2006; Vail et al. 2002).

Lipoplatin™ (Regulon) is a liposomal cisplatin formulation composed of DPPG, soy PC, cholesterol and PEG2000-DSPE. The main difference with previous described cisplatin liposomal formulation is based on long circulating properties of this system. The main therapeutic indication currently pursued is non-small cell lung cancer, already in phase III clinical trial (Boulikas 2009).

Endo-Tag®-1 (Medigene) is a cationic liposomal paclitaxel composed of DOTAP, DOPC and paclitaxel (50:47:3 molar ratio). The combination with neutral and positive lipids of the established cytostatic drug paclitaxel allows EndoTAG®-1 to interact with negatively charged endothelial cells, which are especially required for the growth of tumour blood vessels; given to this formulation anti-angiogenesis properties. The main therapeutic indication is pancreatic cancer and triple negative breast cancer. Currently it is in a phase II clinical trial for both indications (Fasol et al. 2012).

ALN-VSP (Alnylam) aims at RNAi targeting for liver cancer, and the research program is currently partnered with Ascletis for HCC in China (including Hong

Kong, Macau and Taiwan). The main therapeutic indication is liver cancer and liver metastasis. Currently it is in a phase II clinical trial for both indications.

Several **nanoparticle-based formulations** are under clinical evaluation for a wide range of cancer treatments (Schütz et al. 2013; Wang et al. 2013), as **Livatag** at phase III (doxorubicin-loaded polymeric NP from BioAlliance Pharma SA), **Aurimmune** at phase II (TNF-bound colloidal gold NP from CytImmune Sciences) and **camptothecin-loaded polymeric nanoparticles** also at phase II (Cerulean Pharma Inc./Calando and Cell Therapeutics). **Docetaxel-labelled nanoparticles** are still at initial stages of their clinical development.

Polymer therapeutics is an important class of nanopharmaceutical products within nanomedicine field, including protein/peptide-polymer conjugates (e.g. pegylated interferons, cytokines, monoclonal antibodies, erythropoietin) and drug-polymer conjugates (Duncan 2006). Paclitaxel conjugation with polyglutamic acid (**Xyotax**, **Taxoprexin**) leads to a very promising polymer therapeutic formulation, but severe adverse effects have interrupted their evaluation by EMA for its use in the treatment of non-small cell lung cancer (NSCLC), being then proposed for ovarian cancer treatment in Phase III (Peer et al. 2007; Vicent et al. 2009). HPMA conjugates with doxorubicin for liver targeting, camptothecin-polymer conjugates and platinum-polymer conjugates are currently still under clinical evaluation (Seymour et al. 2002).

The first micellar system for paclitaxel, Taxol®, was launched in the USA in 1992 and has since been one of the most used anticancer agents. Due to some problems arising from surfactant toxicity, a demand for other amphiphilic molecules is required in order to allow the administration of poorly soluble drugs (Zhang et al. 2005). Nowadays, amphiphilic materials composed of block copolymers (generally di- and triblock-copolymers) or polymers conjugated with lipids or phospholipids are largely used (Torchilin 2007). Anticancer agents-loaded micelles, namely including paclitaxel, doxorubicin, camptothecin or cisplatin derivatives, are currently in phase II clinical trials showing their potential as drug carriers (Schütz et al. 2013; Ruiz et al. 2014). With the extensive number of PEG-block-copolymers being developed, it is expected an increased industry desire to foster their translation to clinic. Again, the most challenging issue is that regulation considers polymeric micelles as an exception thus applying a case-by-case policy. Nevertheless, since current nanopharmaceuticals development relies heavily on surface functionalisation, the most widely studied block copolymers, such as poly(ethylene glycol)-poly(propylene oxide)-poly(ethylene glycol) (PEG-PPO-PEG), poly(E-caprolactone)-*b*-chitooligosaccharide-*b*-poly(ethylene glycol) (PCL-b-COS-*b*-PEG), poly(ethylene glycol)-phosphatidylethanolamine (PEG-PE), polylactide-poly(ethylene glycol) (PLA-PEG), will allow the binding of a variety of targeting moieties potentiating both combined and selective therapies (Ruiz et al. 2014; Torchilin 2009).

Crucial to the unlocking of the therapeutic potential of some of the already developed platforms is how to obtain the first approval. A significant opportunity to enter in clinical trials is consequently open for those systems under investigation derived from the same family, whether they are adaptations or previous products or novel nanotherapeutics. Examples of obvious nanopharmaceutical candidates to reach phase III clinical trials are systems that use albumin-nanoparticles technology platform like the already approved Abraxane® (Abraxis Biosciences/Celgene). In fact, we can find in this pipeline, the ABI 008 Nab-docetaxel in a clinical phase III and ABI-009, ABI-010 and ABI-011 in phase I clinical trials (Schütz et al. 2013).

Pharmaceutical companies that intended to drive growth based on **innovative platforms** able to target multiple stage diseases know that preclinical proof-of-concept tools remain critically important. Shaping this needs the input from different key players in the pharmaceutical development chain: academia, patients, policy makers, pharmaceutical industry. Moreover, as stressed out previously, nanomedicines are drug delivery systems beyond the 100 nm (Gaspar and Duncan 2009). Thus, nanotherapeutics target-specific interaction with cell components need that they are capable of passing cell membrane and move through the cell cytosol to exert their biological activity.

Specific aspects can be identified from modification of particles surface and from inherent difficulties in the delivery of nucleic acids.

Several approaches have been explored to modify particle surface in order to modulate their in vivo biodistribution profile and thus to optimise the resulting therapeutic outcome. However, the interaction of these particulate systems with plasma proteins is strongly regulated by their surface properties and thus, proposed modifications have to be fully addressed and characterised at preclinical level in order to attest their safety before moving to clinical studies (Dobrovolskaia and McNeil 2007). Accordingly, it is fundamental not only to characterise their distribution at whole body level but also to know and establish a relation between surface characteristics and particle interaction with plasma membranes and further uptake and intracellular trafficking. These properties have a major impact not only in their biodistribution but also inherent metabolic profile and thus have shown to have major implication in nanopharmaceuticals safety. pH-sensitive systems (Torchilin 2009), light-activated materials (Reddy et al. 2006; Wood et al. 2012), magnetic-based nanostructures, temperature-sensitive systems (Ponce et al. 2007) and the grafting of molecules for specifically targeting organelles within the cytosol are some of the strategies used to manipulate their pharmacokinetics and pharmacodynamic profiles (Ponce et al. 2007; Reddy et al. 2006; Iyer et al. 2013; Wood et al. 2012).

This feature, while exciting for the next generation of nanotherapeutics, is also a challenge for the regulatory authorities and is also responsible for the lengthy development as previously mentioned. Moreover, combinatory systems could perhaps improve therapeutic outcomes, meaning that clear harmonised paths to the clinic must be established to achieve further regulatory acceptance (Ruenraroengsak et al. 2010).

Concerning nucleic acid delivery towards cancer treatment, work previously done showed that the effective regulation of tumour unregulated protein expression is definitely a powerful tool, since theoretically any gene could be elected as a RNAi target, opening up the possibility for biomedical research to truly solve unmet clinical challenges (Videira et al. 2013; Pecot et al. 2011; Bader et al. 2011; Oh and Park 2009). More impactful, different tumour stages, from cell differentiation, progression or resistance to tumour-host interaction, could be targeted eliciting a new class of RNAs-based therapeutics (Croce 2009).

To emphasise the broader impact of the siRNA-based technology, (Reischl and Zimmer 2009), divided it in four major anti-mRNA strategies: (1) single-stranded oligodeoxyribonucleotides (ODNs) that inhibit the translation of a specific gene by hybridisation with the corresponding mRNA through the Watson-Crick binding; (2) the anti-mRNA strategy involving ribozymes, able to cleave single-stranded regions in mRNA through trans-esterification or hydrolysis reactions; (3) microRNAs with 20–24 nucleotides (nt) that target mRNAs post-transcriptionally interacting with partially mismatched sequences in the 3′-untranslated (UTR) regions; and (4) the siRNA strategy (Reischl and Zimmer 2009).

The demand for therapeutic solutions prompted an extensively investigation in the nucleic acid delivery with a gradual shift towards the nanocarrier-based strategies (Tokatlian and Segura 2010). Different synthetic siRNA carriers have been engineered using current nanotechnologies in order to achieve carrier-mediated transport, recognition of the cell surface, cell internalisation and cytosolic delivery of siRNA in vitro and in vivo while providing modulation of the innate immune system and protection from premature serum nucleases degradation and elimination from blood stream (Grimm 2009; Singh and Hajeri 2009).

Despite the initial expectations with the viral-based vectors, induced immunogenicity and other safety issues have prompted the nanotechnology development focus on engineered nanocarriers. In fact, currently RNAi-based nanomedicines under preclinical and clinical evaluation are mainly based on lipid or polymeric materials (Videira et al. 2013).

Although engineering siRNA-nanocarriers could render powerful cell modulators as can be proved by the enormous amount of recent published literature, translation to the clinic involves a harmonisation on regulatory requirements for safety and efficacy. A truly global framework for physical characterisation of nanomedicines and biological compatibility studies is urgently needed. A large number of preclinical studies have presented favourable outcomes for dramatic diseases after in vivo siRNA delivery and further gene silencing critical but have not reached the stage of clinical trials. Only a few nanoparticulate formulations are now being evaluated in clinical trials. CALAA-01, a cyclodextrin-containing polymer system for encapsulation of an siRNA against ribonucleotide reductase subunit 2 (RRM2), is the first system engaged in a phase I clinical trial. The formulation modified with polyethylene glycol (PEG) and the human targeting ligand transferrin (Tf) was designed to inhibit and/or reduce tumour growth (Davis 2009). TKM-PLK1 and ALN-PCL are examples of stable nucleic acid lipid particles (SNALPs)-based siRNA systems, also under phase I clinical trial (Simon 2012). The focus in this strategy, as with other nanomedicines, is the development of common tools and standardised analysis protocols for critical parameters evaluation. In fact, market growth is challenged by lengthy development procedures. To accept the preclinical proof-of-concept and move into the early clinical research stage, a particular emphasis should be given to regulatory harmonisation concerning universal characterisation frameworks, and therefore improvements are needed from current "systems pharmacology" approaches.

New innovative materials can also be a way to move forward in a more comprehensive manner, opening new challenges to product innovation and development.

Sphingolipids (SL) have emerged as important modulators in cancer progression and therefore alternative therapies based both on the development of SL-based-nanotechnologies and on the targeting of SL metabolism to enhance cancer cell death are currently being explored (Hawkins et al. 2008; Morad and Cabot 2013). The principle underlying these strategies relies on the ability of ceramide, the backbone of all SL, in enhancing tumour cell death. Accordingly, nanocarriers containing ceramide in its composition have been extensively exploited (Hawkins et al. 2008). Several different ceramide delivery systems have been developed, including polymer-based (van Vlerken et al. 2007; Devalapally et al. 2007, 2008; Barth et al. 2010b) and lipid-based nanocarriers (Barth et al. 2010a; van Lummel et al. 2011). Focus has been given to the development of combinatorial strategies based on the use of ceramide nanosystems as a platform to deliver either chemotherapeutic agents (Tran et al. 2008; Jiang et al. 2011) or modulators of the SL metabolism (Barth et al. 2010b; Jiang et al. 2011; Dickson et al. 2011). Both the in vitro and the in vivo studies showed a synergistic enhancement of cancer cell death in this combinatorial nanomedicine approach. Ceramide-based nanotherapies have also been engineered in order to specifically target cancer cells (Pastorino et al. 2006; Adrian et al. 2011; Koshkaryev et al. 2012) and even subcellular compartments (Boddapati et al. 2008) by taking advantage of the strategies discussed in the previous section. Despite the promising in vitro and in vivo results, none of the SL-based-nanocarriers has, so far, reached the clinic.

Besides cancer therapeutics, nano-based products have been extensively developed as imaging tools, the second major application of nanoproducts already available in the market.

Oncology imaging plays an important role in disease detection, prognosis and treatment planning (Akin et al. 2012). Major medical imaging tools include X-ray computed tomography (CT), magnetic resonance imaging (MRI), ultrasound imaging, positron emission tomography (PET), single-photon emission CT (SPECT), optical imaging and photoacoustic imaging. Most of these techniques require the use of imaging agents like labels, probes or contrast agents. Some imaging agents for cancer imaging are in nanoparticles; a table with those in clinical use or in clinical trials is provided in Thakor and Gambhir (2013).

A number of relevant developments are the combined imaging equipment for clinical and preclinical diagnostics such as the combination of CT and MRI and of PET and MRI and more recently the combination of SPECT and MRI. Consequently a diversity of developments on bimodal and multimodal imaging agents is in progress. Although, combining more than one type of imaging agent in the same nanoparticle for multimodal detectability is a challenge with impact on imaging and molecular diagnostics (Kievit and Zhang 2011; Yu et al. 2012; Gautier et al. 2013).

Furthermore, nanoplatforms for therapeutics and diagnostics (theranostics) in oncology have attracted much interest in recent years. These nanoplatforms are based on combining imaging agents and medicines in the same nanostructure, i.e. engineering nanoparticles to integrate simultaneously several functions of clinical relevance such as (1) the delivery of imaging agents for one or several clinical imaging tools, like magnetic resonance imaging or radionuclide imaging or fluorescence imaging; (2) the co-delivery of specific therapeutic agents such as a drug or a gene; and also (3) to allow the functionalisation of the external shell surface for passive targeting or active targeting or for other functionalities.

The scientific interest in this area is evidenced by the exponential increase in the number of scientific papers (Xie and Jon 2012). Other imaging technologies include MRI and the potential for theranostic approaches.

Superparamagnetic iron oxide nanoparticles (SPIONs) were approved as vascular negative contrast agents for improving cancer imaging using MRI in clinic. A significant amount of research work used SPIONs coated with biocompatible moieties and acting as a platform for attach targeting tags and/or drugs and/or additional imaging agents, making them multimodal theranostic platforms (Yu et al. 2012; Wahajuddin 2012; Hervé et al. 2008; Santhosh and Ulrih 2013).

Furthermore SPIONs, having a size comparable to functional biomolecules, are suitable for being loaded into a diversity of drug delivery systems benefiting from the extensive knowledge of biodistribution, pharmacokinetics and specific targeting already studied for the nanomedicinal oncological products already in the market or clinical phases (Duncan and Gaspar 2011; Duncan and Richardson 2012). The consequent benefits are both the modulation of the imaging agent to the target site in analogy with the therapeutics, limiting its exposure to healthy tissue, and the accumulation into tumours where these drug delivery systems were designed for, allowing an improved image acquisition by MRI (Cohen and Shoushan 2013; Kim et al. 2013; Lv et al. 2013; Gautier et al. 2013; Kocbek et al. 2013; Gultepe et al. 2010).

Clinical trials evaluating the utility of magnetic nanoparticles as therapeutics and diagnostic agents in cancer and other diseases have been reported (Singh and Sahoo 2013).

Most of the published theranostic platforms are for oncology applications in general including magnetic nanoparticles (Cohen and Shoushan 2013). Relevant technological advances were attained either in preparation, functionalisation or concerning the design of strategies for clinical applications. Their success as theranostics agents in the clinical use (Cohen and Shoushan 2013) depends on some crucial issues for clinical translation (Duncan and Gaspar 2011), and an effective collaboration between chemists, biologists, material scientists, engineers and clinicians can lead to promising application in several clinical situations or management of patients (Cohen and Shoushan 2013).

Beyond the efficacy, safety is also an absolute prerequisite for biological imaging probes, such as magnetic-based nanoparticles (currently at phase I) or the highly versatile quantum dots-based systems (QDs) (Gandon et al. 1991; Gao et al. 2007). In order to explore their ability to be used as multiplexed bioassays, medical diagnostics or theranostics, large-scale toxicity data is needed, especially in living animals. Although different surface-coating technologies have emerged for these inorganic nanoparticles, only the implementation of high-throughput technologies could support its clinical application. The setting up of well-defined analysis and standardisation protocols will certainly contribute to have new insights in understanding nanopharmaceutical complex systems.

15.4 Regulatory Environment Overview: Changing Times

The regulation of medicinal products has been in a changing and evolutive path since ICH started in the early 1990s (Fig. 15.1). Meanwhile also the regulatory environment around the development of nanopharmaceuticals for oncology clinical use has been under increased challenge.

First, the evolution from clinical oncology development has demanded more and more proof of superiority in clinical efficacy from new innovative approaches. Demonstration of non-inferiority is no longer acceptable under current clinical standards for most applications in oncology. Also the need to demonstrate increased advantage in pharmacoeconomic analysis has been a constant and increased pressure on the development of new and innovative medicinal products. Moreover the new perspectives in healthcare management, incorporating health technologies assessment (HTA), also bring into play the need to integrate different medical interventions complementary to the therapeutic classical oncology setting (Main et al. 2006).

Second, the major advances in biopharmaceuticals in the oncology area also have demonstrated their potential for superiority regarding some of the nano-based approaches (most of which were based on molecules that were innovative in the late 1970s or early 1980s but no longer today, like doxorubicin) but also their limitations when regarding evidence for critical "hard" clinical efficacy end points in clinical trials, like survival rates over 5 years.

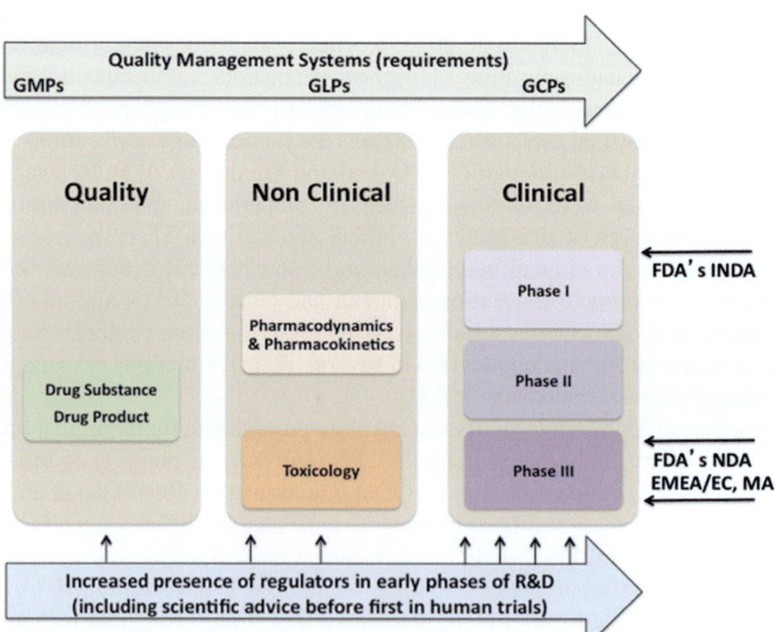

Fig. 15.1 Overview of regulatory presence on the research and development life cycle of any medicinal product

Another critical issue is the discussion on nanosimilars or also the debate over similar formulations within non-biological complex drugs (NBCD), brought to the forefront a number of critical issues in specific formulations (iron oxide nanoparticles, liposomes, polymeric micelles) and now under increased regulatory attention.

It is surprising for some that only the discussion about similar medicinal products triggered a regulatory discussion on innovative drug delivery systems. But for those more used to the practice of regulatory agencies, this will not come as a surprise. Medicine's regulators tend to move in the direction of major regulatory changes only when major problems arise on the dispute between the reliability of products or when science identifies major concerns about potential implications on quality, safety and efficacy of medicinal products. On the other hand, as referred above, it can be considered that most of the regulatory guidance in place has demonstrated to be sufficient to address the development of a new medicinal product which includes simultaneously a new molecular entity and a (innovative) nanosystem, as all those will have to be studied, characterised and justified in their proof of concept, quality, safety and efficacy. Therefore, it makes sense that the major concerns triggering the need for regulatory reaction have occurred with the innovative nano-sized systems incorporating already known and well-characterised molecules. Indeed, contrariwise to most situations where a new formulation of a well-known active substance would not trigger safety relevant concerns, regulatory authorities recognised that, associated to the claimed and intended activity changes in the activity profile, also safety changes could occur, due either to the nanosystems themselves, or/and to the associated changes in target access, cellular, tissue, organ and body distribution, immune reactions.

The burst of generic submissions of MAA (marketing authorisation applications) in the middle of the last decade prompted a European response to a problem that (basically) arises from a wrong classification of medicinal products based on iron-oxide nanoparticles, identified since the 1960s, when they were introduced in the market as "injectable aqueous solutions of iron". This was considered to be of major importance in cancer therapy, due to their use in severe anaemia patients, with significant impact in an oncology setting. The reaction of European regulators to the event of a huge number of submissions of national MAAs related to the claimed similarity to those old formulations prompted a regulatory statement that dominated and initiated the debate over new regulation for (nano)similar medicinal products.

In parallel the ITF (innovation task force) at the European Medicines Agency had already started to assemble data and in 2009 started a number of activities from the ad hoc expert group in nanomedicines to the first global conference on the topic mobilising also regulators and stakeholders from the USA and Japan among others. In parallel different discussions have been triggered, from the European Union discussion on nanomaterials classification to the "similarity" discussion, namely from a TI Pharma-based international expert group in the Netherlands (NBCDs or non-biological complex drugs expert group) composed of a number of experts originated from academia and industry. FDA and Japanese authorities, in parallel with Canada, Switzerland, Australia and others, also developed their own initiatives, some of them coordinated with the European Medicines Agency.

The major outputs from this burst of new regulatory initiatives were the guidance documents from the European Medicines Agency related to iron-oxide nanoparticles (Reflection Paper on Non-Clinical Studies for Generic Nanoparticle Iron Medicinal Product Applications- Ema/Chmp/Swp/100094/2011 2011), liposomal similar products (Reflection Paper on the Data Requirements for Intravenous Liposomal Products Developed with Reference to an Innovator Liposomal Product: Ema/Chmp/Swp/100094/2011 2013) and polymeric micelles (Joint MHLW/EMA Reflection Paper on the Development of Block Copolymer Micelle Medicinal Products:Ema/Chmp/13099/2013 2013). The latter one can be considered a "revolutionary" document in two ways: it considers a technology being developed before it comes to market; results from the integrative collaboration of EMA and the Japanese authorities being a first example of a "global" nanomedicines regulatory guidance document. An integration of the concepts behind those documents has been recently published (Ehmann et al. 2013).

Also relevant to the current trends are the documents from the FDA on liposomal products and the scientific discussions about regulatory implications in NBCDs. The latest took place not only at the NBCD expert group but also recently in open-fora at FIP (International Pharmaceutical Federation), EUFEPS (European Federation of Pharmaceutical Sciences), AAPS (American Association of Pharmaceutical Scientists) and NYAS (New York Academy of Sciences) (Holloway et al. 2012). Two relevant papers in NBCDs were also recently published (Crommelin et al. 2014; Schellekens et al. 2014).

An increased concern both at European and US level has been the question of the harmonisation of methodology essential to characterise the quality requirements. The Nanotechnology Characterization Laboratory (NCL) at the National Cancer Institute in the USA has provided a major contribution to this matter assembling data with several innovative platforms for the development of nanomedicines in oncology. Their role as a platform for support of innovation and helping SMEs and start-ups to move upwards to the clinical stage has been considered of major relevance. The data assembled over the years by the NCL have been inspiring for a number of critical issues in nanomedicines regulation that will be present in the coming years. Mostly, the needs to better understand the implications of subtle changes in surface modification and their impact on the biological interaction of nanosystems in specific applications (McNeil 2011).

A major focus in the coming years will also be played by the need to better understand the pathophysiology of tumours at specific stages. Probably the fact that the worst clinical conditions are selected first in human trials triggered a lower success rate than expected, due to the fact that huge tumours, with considerable tissue necrosis and hypoxia, will cause difficulty on the impact of EPR. Better selection of patients that can benefit largely from therapeutic strategies using the EPR and/or specifically targeted formulations will need companion diagnostics including imaging agents to better characterise the relevant patient (sub)populations.

The complexity of integrated platforms of different technologies, from therapeutics to imaging, including cancer therapeutic vaccines and multiplex medical devices, will foster new and critical regulatory considerations, along the theranostics and combination product pathways.

A critical role will be played by regulatory authorities through their scientific advise procedures, and an increased pressure will be present to have more cooperative work between different regional regulatory bodies (EMA, FDA, Japanese) major academic and industry stakeholders.

References

Adrian JE, Wolf A, Steinbach A, Rössler J, Süss R (2011) Targeted delivery to neuroblastoma of novel siRNA-anti-GD2-liposomes prepared by dual asymmetric centrifugation and sterol-based post-insertion method. Pharm Res 28(9):2261–2272. doi:10.1007/s11095-011-0457-y

Akin O, Brennan SB, David Dershaw D, Ginsberg MS, Gollub MJ, Schöder H, Panicek DM, Hricak H (2012) Advances in oncologic imaging: update on 5 common cancers. Cancer J Clin 62(6):364–393. doi:10.3322/caac.21156

Allen TM, Cullis PR (2013) Liposomal drug delivery systems: from concept to clinical applications. Adv Drug Deliv Rev 65(1):36–48. doi:10.1016/j.addr.2012.09.037

Bader AG, Brown D, Stoudemire J, Lammers P (2011) Developing therapeutic microRNAs for cancer. Gene Ther 18(12):1121–1126

Barth BM, Sharma R, Altinoğlu EI, Morgan TT, Shanmugavelandy SS, Kaiser JM, McGovern C et al (2010a) Bioconjugation of calcium phosphosilicate composite nanoparticles for selective targeting of human breast and pancreatic cancers in vivo. ACS Nano 4(3):1279–1287. doi:10.1021/nn901297q

Barth BM, Gustafson SJ, Young MM, Fox TE, Shanmugavelandy SS, Kaiser JM, Cabot MC, Kester M, Kuhn TB (2010b) Inhibition of NADPH oxidase by glucosylceramide confers chemoresistance. Cancer Biol Ther 10(11):1126–1136. doi:10.4161/cbt.10.11.13438

Bawa R, Melethil S, Simmons WJ, Harris D (2008) Nanopharmaceuticals: patenting issues and FDA regulatory challenges. The SciTech Lawyer 5(2):1–6

Bianco A, Kostarelos K, Prato M (2005a) Applications of carbon nanotubes in drug delivery. Curr Opin Chem Biol 9(6):674–679. doi:10.1016/j.cbpa.2005.10.005

Bianco A, Kostarelos K, Partidos CD, Prato M (2005b) Biomedical applications of functionalised carbon nanotubes. Chem Commun (5):571–577. doi:10.1039/B410943K

Bobrie A, Théry C (2013) Exosomes and communication between tumours and the immune system: are all exosomes equal? Biochem Soc Trans 41(1):263–267. doi:10.1042/BST20120245

Boddapati SV, D'Souza GG, Erdogan S, Torchilin VP, Weissig V (2008) Organelle-targeted nanocarriers: specific delivery of liposomal ceramide to mitochondria enhances its cytotoxicity in vitro and in vivo. Nano Lett 8(8):2559–2563. doi:10.1021/nl801908y

Boomer JA, Qualls MM, Inerowicz HD, Haynes RH, Patri VS, Kim JM, Thompson DH (2009) Cytoplasmic delivery of liposomal contents mediated by an acid-labile cholesterol-vinyl ether-PEG conjugate. Bioconjug Chem 20(1):47–59. doi:10.1021/bc800239b

Boulikas T (2009) Clinical overview on lipoplatin: a successful liposomal formulation of cisplatin. Expert Opin Investig Drugs 18(8):1197–1218. doi:10.1517/13543780903114168

Cai W, Chen X (2008) Preparation of peptide-conjugated quantum dots for tumor vasculature-targeted imaging. Nat Protoc 3(1):89–96. doi:10.1038/nprot.2007.478

Camussi G, Deregibus MC, Bruno S, Cantaluppi V, Biancone L (2010) Exosomes/microvesicles as a mechanism of cell-to-cell communication. Kidney Int 78(9):838–848. doi:10.1038/ki.2010.278

Chapman S, Dobrovolskaia M, Farahani K, Goodwin A (2013) Nanoparticles for cancer imaging: the good, the bad, and the promise. Nano Today 8(5):454–460. doi:10.1016/j.nantod.2013.06.001

Cheng R, Meng F, Deng C, Klok HA, Zhong Z (2013) Dual and multi-stimuli responsive polymeric nanoparticles for programmed site-specific drug delivery. Biomaterials 34(14):3647–3657. doi:10.1016/j.biomaterials.2013.01.084

Choi HS, Frangioni JV (2010) Nanoparticles for biomedical imaging: fundamentals of clinical translation. Mol Imaging 9(6): 291–310

Cohen Y, Shoushan SY (2013) Magnetic nanoparticles-based diagnostics and theranostics. Curr Opin Biotechnol 24(4):672–681. doi:10.1016/j.copbio.2013.01.006

Croce CM (2009) Causes and consequences of microRNA dysregulation in cancer. Nat Rev Genet 10(10):704–714. doi:10.1038/nrg2634

Crommelin DJ, de Vlieger JS, Weinstein V, Mühlebach S, Shah VP, Schellekens H (2014) Different pharmaceutical products need similar terminology. AAPS J 16(1):11–14. doi:10.1208/s12248-013-9532-0

Davis ME (2009) The first targeted delivery of siRNA in humans via a self-assembling, cyclodextrin polymer-based nanoparticle: from concept to clinic. Mol Pharm 6(3):659–668. doi:10.1021/mp900015y

Davis ME, Zuckerman JE, Choi CHJ, Seligson D (2010) Evidence of RNAi in humans from systemically administered siRNA via targeted nanoparticles. Nature 464:1067–1070

Devalapally H, Duan Z, Seiden MV, Amiji MM (2007) Paclitaxel and ceramide co-administration in biodegradable polymeric nanoparticulate delivery system to overcome drug resistance in ovarian cancer. Int J Cancer 121(8):1830–1838. doi:10.1002/ijc.22886

Devalapally H, Duan Z, Seiden MV, Amiji MM (2008) Modulation of drug resistance in ovarian adenocarcinoma by enhancing intracellular ceramide using tamoxifen-loaded biodegradable polymeric nanoparticles. Clin Cancer Res 14(10):3193–3203. doi:10.1158/1078-0432.CCR-07-4973

Dickson MA, Carvajal RD, Merrill AH (2011) A phase I clinical trial of safingol in combination with cisplatin in advanced solid tumors. Clin Cancer Res 17(8):2484–2492. doi:10.1158/1078-0432.CCR-10-2323

Dobrovolskaia MA, McNeil SE (2007) Immunological properties of engineered nanomaterials. Nat Nanotechnol 2(8):469–478. doi:10.1038/nnano.2007.223

Dobrovolskaia MA, McNeil SE (2013) Understanding the correlation between in vitro and in vivo immunotoxicity tests for nanomedicines. J Control Release 172(2):456–466. doi:10.1016/j.jconrel.2013.05.025

Dobrovolskaia MA, Aggarwal P, Hall JB, McNeil SE (2008) Preclinical studies to understand nanoparticle interaction with the immune system and its potential effects on nanoparticle biodistribution. Mol Pharm 5(4):487–495. doi:10.1021/mp800032f

Dobrovolskaia MA, Patri AK, Zheng J (2009a) Interaction of colloidal gold nanoparticles with human blood: effects on particle size and analysis of plasma protein binding profiles. Nanomedicine 5(2):106–117. doi:10.1016/j.nano.2008.08.001

Dobrovolskaia MA, Germolec DR, Weaver JL (2009b) Evaluation of nanoparticle immunotoxicity. Nat Nanotechnol 4(7):411–414. doi:10.1038/NNANO.2009.175

Doherty GJ, McMahon HT (2009) Mechanisms of endocytosis. Annu Rev Biochem 78:857–902. doi:10.1146/annurev.biochem.78.081307.110540

Duncan R (2006) Polymer conjugates as anticancer nanomedicines. Nat Rev Cancer 6(9):688–701. doi:10.1038/nrc1958

Duncan R, Gaspar R (2011) Nanomedicine(S) under the microscope. Mol Pharm 8(6):2101–2141. doi:10.1021/mp200394t

Duncan R, Richardson SCW (2012) Endocytosis and intracellular trafficking as gateways for nanomedicine delivery: opportunities and challenges. Mol Pharm 9(9):2380–2402. doi:10.1021/mp300293n

Ehmann F, Sakai-Kato K, Duncan R, Hernán Pérez de la Ossa D, Pita R, Vidal JM, Kohli A et al (2013) Next-generation nanomedicines and nanosimilars: EU regulators' initiatives relating to the development and evaluation of nanomedicines. Nanomedicine (Lond) 8(5):849–856. doi:10.2217/nnm.13.68

EL Andaloussi S, Mäger I, Breakefield XO, Wood MJ (2013) Extracellular vesicles: biology and emerging therapeutic opportunities. Nat Rev Drug Discov 12(5):347–357. doi:10.1038/nrd3978

European Medicines Agency (2011) Reflection paper on non-clinical studies for generic nanoparticle iron medicinal product applications. EMA, London

European Medicines Agency (2013a) Joint MHLW/EMA reflection paper on the development of block copolymer micelle medicinal products. EMA, London

European Medicines Agency (2013b) Reflection paper on the data requirements for intravenous liposomal products developed with reference to an innovator liposomal product. EMA, London

Farokhzad OC, Langer R (2009) Impact of nanotechnology on drug delivery. ACS Nano 3(1): 16–20. doi:10.1021/nn900002m

Fasol U, Frost A, Büchert M, Arends J, Fiedler U, Scharr D, Scheuenpflug J, Mross K (2012) Vascular and pharmacokinetic effects of endoTAG-1 in patients with advanced cancer and liver metastasis. Ann Oncol 23(4):1030–1036. doi:10.1093/annonc/mdr300

Filipazzi P, Bürdek M, Villa A, Rivoltini L (2012) Recent advances on the role of tumor exosomes in immunosuppression and disease progression. Semin Cancer Biol 22(4):342–349

Gandon Y, Heautot JF, Brunet F, Guyader D, Deugnier Y, Carsin M (1991) Superparamagnetic iron oxide: clinical time-response study. Eur J Radiol 12(3):195–200

Gao X, Chung LW, Nie S (2007) Quantum dots for in vivo molecular and cellular imaging. Methods Mol Biol 374:135–145. doi:10.1385/1-59745-369-2:135

Gaspar R (2010) Therapeutic products: regulating drugs and medical devices. In: Hodge GA, Bowman DM, Maynard AD (eds) International handbook on regulating nanotechnologies. Edward Elgar, Cheltenham/Northampton, pp 291–320

Gaspar R, Duncan R (2009) Polymeric carriers: preclinical safety and the regulatory implications for design and development of polymer therapeutics. Adv Drug Deliv Rev 61(13):1220–1231. doi:10.1016/j.addr.2009.06.003

Gautier J, Allard-Vannier E, Munnier E, Soucé M, Chourpa I (2013) Recent advances in theranostic nanocarriers of doxorubicin based on iron oxide and gold nanoparticles. J Control Release 169(1–2):48–61. doi:10.1016/j.jconrel.2013.03.018

Ge R, Tan E, Sharghi-Namini S, Asada HH (2012) Exosomes in cancer microenvironment and beyond: have we overlooked these extracellular messengers? Cancer Microenviron 5(3): 323–332. doi:10.1007/s12307-012-0110-2

Gradishar WJ (2006) Albumin-bound paclitaxel: a next-generation taxane. Expert Opin Pharmacother 7(8):1041–1053. doi:10.1517/14656566.7.8.1041

Grandal MV, Madshus IH (2008) Epidermal growth factor receptor and cancer: control of oncogenic signalling by endocytosis. J Cell Mol Med 12(5a):1527–1534. doi:10.1111/j.1582-4934.2008.00298.x

Grimm D (2009) Small silencing RNAs: state-of-the-art. Adv Drug Deliv Rev 61(9):672–703. doi:10.1016/j.addr.2009.05.002

Gultepe E, Reynoso FJ, Jhaveri A, Kulkarni P, Nagesha D, Ferris C, Harisinghani M, Campbell RB, Sridhar S (2010) Monitoring of magnetic targeting to tumor vasculature through MRI and biodistribution. Nanomedicine (Lond) 5(8):1173–1182. doi:10.2217/nnm.10.84

Gupta Y, Jain A, Jain P, Jain SK (2008) Design and development of folate appended liposomes for enhanced delivery of 5-FU to tumor cells. J Drug Target 15(3):231–240. doi:10.1080/10611860701289719

Hawkins MJ, Soon-Shiong P, Desai N (2008) Protein nanoparticles as drug carriers in clinical medicine. Adv Drug Deliv Rev 60(8):876–885. doi:10.1016/j.addr.2007.08.044

Hervé K, Douziech-Eyrolles L, Munnier E, Cohen-Jonathan S, Soucé M, Marchais H, Limelette P et al (2008) The development of stable aqueous suspensions of PEGylated SPIONs for biomedical applications. Nanotechnology 19(46):465608. doi:10.1088/0957-4484/19/46/465608

Hillaireau H, Couvreur P (2009) Nanocarriers' entry into the cell: relevance to drug delivery. Cell Mol Life Sci 66(17):2873–2896. doi:10.1007/s00018-009-0053-z

Holloway C, Mueller-Berghaus J, Lima BS, Lee SL, Wyatt JS, Nicholas JM, Crommelin DJ (2012) Scientific considerations for complex drugs in light of established and emerging regulatory guidance. Ann N Y Acad Sci 1276(1):26–36. doi:10.1111/j.1749-6632.2012.06811.x

International Conference on Harmonisation of Technical Requirements for Registration of Pharmaceuticals for Human Use Q10 (2008) Pharmaceutical Quality System. International Conference on Harmonisation of Technical Requirements for Registration of Pharmaceuticals for Human Use, ICH Tripartite Guideline, ICH, Geneva

International Conference on Harmonisation of Technical Requirements for Registration of Pharmaceuticals for Human Use Q8 (2009) Pharmaceutical Development. International Conference on Harmonisation of Technical Requirements for Registration of Pharmaceuticals for Human Use, ICH Tripartite Guideline, ICH, Geneva

International Conference on Harmonisation of Technical Requirements for Registration of Pharmaceuticals for Human Use Q9 (ICH) (2005) Quality Risk Management. International

Conference on Harmonisation of Technical Requirements for Registration of Pharmaceuticals for Human Use, ICH Tripartite Guideline, ICH, Geneva
Iyer AK, Singh A, Ganta S, Amiji MM (2013) Role of integrated cancer nanomedicine in overcoming drug resistance. Adv Drug Deliv Rev 65(13–14):1782–1802
Jiang Y, DiVittore NA, Kaiser JM, Shanmugavelandy SS, Fritz JL, Heakal Y, Tagaram HR et al (2011) Combinatorial therapies improve the therapeutic efficacy of nanoliposomal ceramide for pancreatic cancer. Cancer Biol Ther 12(7):574–585
Jung D, Barbel R, Chowdhury N (2011) Is the European medical products authorisation regulation equipped to cope with the challenges of nanomedicines? Law Policy 33(2):276–303. doi:10.1111/j.1467-9930.2011.00339.x
Kan P (2007) A brief review on development of liposome in Taiwan. J Med Biol Eng 27(1):53–56
Kievit FM, Zhang M (2011) Cancer nanotheranostics: improving imaging and therapy by targeted delivery across biological barriers. Adv Mater 23(36):H217–47. doi:10.1002/adma.201102313
Kim DH, Vitol EA, Liu J, Balasubramanian S, Gosztola DJ, Cohen EE, Novosad V, Rozhkova EA (2013) Stimuli-responsive magnetic nanomicelles as multifunctional heat and cargo delivery vehicles. Langmuir 29(24):7425–7432. doi:10.1021/la3044158
Kocbek P, Kralj S, Kreft ME, Kristl J (2013) Targeting intracellular compartments by magnetic polymeric nanoparticles. Eur J Pharm Sci 50(1):130–138. doi:10.1016/j.ejps.2013.04.004
Kong M, Park H, Cheng X, Chen X (2013) Spatial-temporal event adaptive characteristics of nanocarrier drug delivery in cancer therapy. J Control Release 172(1):281–291. doi:10.1016/j.jconrel.2013.08.022
Koshkaryev A, Piroyan A, Torchilin VP (2012) Increased apoptosis in cancer cells in vitro and in vivo by ceramides in transferrin-modified liposomes. Cancer Biol Ther 13(1):50–60. doi:10.4161/cbt.13.1.18871
Krop IE, Beeram M, Modi S, Jones SF, Holden SN, Yu W, Girish S et al (2010) Phase I study of Trastuzumab-DM1, an HER2 antibody-drug conjugate, given every 3 weeks to patients with HER2-positive metastatic breast cancer. J Clin Oncol 28(16):2698–2704. doi:10.1200/JCO.2009.26.2071
Kwong GA, Radu CG, Hwang K, Shu CJ, Ma C, Koya RC, Comin-Anduix B et al (2009) Modular nucleic acid assembled P/MHC microarrays for multiplexed sorting of antigen-specific T cells. J Am Chem Soc 131(28):9695–9703. doi:10.1021/ja9006707
Leamon CP, Reddy JA (2004) Folate-targeted chemotherapy. Adv Drug Deliv Rev 56(8):1127–1141
Leopold PL, Crystal RG (2007) Intracellular trafficking of adenovirus: many means to many ends. Adv Drug Deliv Rev 59(8):810–821
Li SD, Huang L (2008) Pharmacokinetics and biodistribution of nanoparticles. Mol Pharm 5(4):496–504
Li SD, Huang L (2010) Stealth nanoparticles: high density but sheddable PEG is a key for tumor targeting. J Control Release 145(3):178–181. doi:10.1016/j.jconrel.2010.03.016
Li W, Nicol F, Szoka FC Jr (2004) GALA: a designed synthetic pH-responsive amphipathic peptide with applications in drug and gene delivery. Adv Drug Deliv Rev 56(7):967–985
Liu Y, Tan J, Thomas A, Ou-Yang D, Muzykantov VR (2012) The shape of things to come: importance of design in nanotechnology for drug delivery. Ther Deliv 3(2):181–194. doi:10.4155/tde.11.156
Low PS, Kularatne SA (2009) Folate-targeted therapeutic and imaging agents for cancer. Curr Opin Chem Biol 13(3):256–262. doi:10.1016/j.cbpa.2009.03.022
Lv Y, Ding G, Zhai J, Guo Y, Nie G, Li X (2013) A superparamagnetic Fe_3O_4-loaded polymeric nanocarrier for targeted delivery of evodiamine with enhanced antitumor efficacy. Colloids Surf B Biointerfaces 110(October 1):411–418. doi:10.1016/j.colsurfb.2013.04.038
Maeda H, Ueda M, Morinaga T, Matsumoto T (1985) Conjugation of poly(styrene-co-maleic acid) derivatives to the antitumor protein neocarzinostatin: pronounced improvements in pharmacological properties. J Med Chem 28(4):455–461
Main C, Bojke L, Griffin S, Norman G, Barbieri M, Mather L, Stark D, Palmer S, Riemsma R (2006) Topotecan, PEGylated liposomal doxorubicin hydrochloride and paclitaxel for second-line or subsequent treatment of advanced ovarian cancer: a systematic review and economic evaluation. Health Technol Assess 10(9):1–132. iii–iv

Mastrobattista E, Hennink WE, Schiffelers RM (2007) Delivery of nucleic acids. Pharm Res 24(8):1561–1563. doi:10.1007/s11095-007-9349-6

Matsumura Y, Maeda H (1986) A new concept for macromolecular therapeutics in cancer chemotherapy: mechanism of tumoritropic accumulation of proteins and the antitumor agent smancs. Cancer Res 46(12 Pt 1):6387–6392

McLeland CB, Dobrovolskaia MA, McNeil SE (2009) Nanoparticle interaction with plasma proteins as it relates to particle biodistribution, biocompatibility and therapeutic efficacy. Adv Drug Deliv 61(6):428–437. doi:10.1016/j.addr.2009.03.009

McNeil SE (Ed.), (2011) Characterization of Nanoparticles Intended for Drug Delivery. Humana Press, Series: Methods in Molecular Biology 697(12):269

Mercer J, Helenius A (2012) Gulping rather than sipping: macropinocytosis as a way of virus entry. Curr Opin Microbiol 15(4):490–499. doi:10.1016/j.mib.2012.05.016

Minotti G, Menna P, Salvatorelli E, Cairo G, Gianni L (2004) Anthracyclines: molecular advances and pharmacologic developments in antitumor activity and cardiotoxicity. Pharmacol Rev 56(2):185–229. doi:10.1124/pr.56.2.6

Moen MD, Lyseng-Williamson KA, Scott LJ (2009) Liposomal amphotericin B: a review of its use as empirical therapy in febrile neutropenia and in the treatment of invasive fungal infections. Drugs 69(3):361–392. doi:10.2165/00003495-200969030-00010

Moghimi SM, Andresen TL (2009) Complement-mediated tumour growth: implications for cancer nanotechnology and nanomedicines. Mol Immunol 46(8–9):1571–1572. doi:10.1016/j.molimm.2009.02.014

Moghimi SM, Hamad I (2008) Liposome-mediated triggering of complement cascade. J Liposome Res 18(3):195–209. doi:10.1080/08982100802309552

Moghimi SM, Hunter AC, Andresen TL (2012) Factors controlling nanoparticle pharmacokinetics: an integrated analysis and perspective. Annu Rev Pharmacol Toxicol 52(1):481–503. doi:10.1146/annurev-pharmtox-010611-134623

Morad SAF, Cabot MC (2013) Ceramide-orchestrated signalling in cancer cells. Nat Rev Cancer 13(1):51–65. doi:10.1038/nrc3398

Muro S (2012) Challenges in design and characterization of ligand-targeted drug delivery systems. J Control Release 164(2):125–137. doi:10.1016/j.jconrel.2012.05.052

Nel AE, Mädler L, Velegol D, Xia T, Hoek EM, Somasundaran P, Klaessig F, Castranova V, Thompson M (2009) Understanding biophysicochemical interactions at the nano-bio interface. Nat Mater 8(7):543–557. doi:10.1038/nmat2442

Oh YK, Park TG (2009) siRNA delivery systems for cancer treatment. Adv Drug Deliv Rev 61(10):850–862

Paillard A, Hindré F, Vignes-Colombeix C, Benoit JP (2010) The importance of endo-lysosomal escape with lipid nanocapsules for drug subcellular bioavailability. Biomaterials 31(29): 7542–7554. doi:10.1016/j.biomaterials.2010.06.024

Paliwal SR, Paliwal R, Agrawal GP, Vyas SP (2011) Liposomal nanomedicine for breast cancer therapy. Nanomedicine (Lond) 6(6):1085–1100. doi:10.2217/nnm.11.72

Parton RG, del Pozo MA (2013) Caveolae as plasma membrane sensors, protectors and organizers. Nat Rev Mol Cell Biol 14(2):98–112. doi:10.1038/nrm3512

Pastorino F, Brignole C, Di Paolo D (2006) Targeting liposomal chemotherapy via both tumor cell-specific and tumor vasculature-specific ligands potentiates therapeutic efficacy. Cancer Res 66(20):10073–10082. doi:10.1158/0008-5472.CAN-06-2117

Pecot CV, Calin GA, Coleman RL, Lopez-Berestein G, Sood AK (2011) RNA interference in the clinic: challenges and future directions. Nat Rev Cancer 11(1):59–67. doi:10.1038/nrc2966

Peer D, Karp JM, Hong S, Farokhzad OC, Margalit R, Langer R (2007) Nanocarriers as an emerging platform for cancer therapy. Nat Nanotechnol 2(12):751–760. doi:10.1038/nnano.2007.387

Plosker GL (2008) Pegylated liposomal doxorubicin: a review of its use in the treatment of relapsed or refractory multiple myeloma. Drugs 68(17):2535–2551

Ponce AM, Viglianti BL, Yu D, Yarmolenko PS, Michelich CR, Woo J, Bally MB, Dewhirst MW (2007) Magnetic resonance imaging of temperature-sensitive liposome release: drug dose painting and antitumor effects. J Natl Cancer Inst 99(1):53–63. doi:10.1093/jnci/djk005

Poon RT, Borys N (2011) Lyso-thermosensitive liposomal doxorubicin: an adjuvant to increase the cure rate of radiofrequency ablation in liver cancer. Future Oncol 7(8):937–945. doi:10.2217/fon.11.73

Puri A, Loomis K, Smith B, Lee J-H, Yavlovich A, Heldman E, Blumenthal R (2009) Lipid-based nanoparticles as pharmaceutical drug carriers: from concepts to clinic. Crit Rev Ther Drug Carrier Syst 26(6):523–580. doi:10.1615/CritRevTherDrugCarrierSyst.v26.i6.10

Rajendran L, Knölker HJ, Simons K (2010) Subcellular targeting strategies for drug design and delivery. Nat Rev Drug Discov 9(1):29–42. doi:10.1038/nrd2897

Record M, Carayon K, Poirot M (2014) Exosomes as new vesicular lipid transporters involved in cell–cell communication and various pathophysiologies. Biochim Biophys Acta 184(1):108–120. doi:10.1016/j.bbalip.2013.10.004

Reddy GR, Bhojani MS, McConville P, Moody J, Moffat BA, Hall DE, Kim G et al (2006) Vascular targeted nanoparticles for imaging and treatment of brain tumors. Clin Cancer Res 12(22): 6677–6686. doi:10.1158/1078-0432.CCR-06-0946

Reischl D, Zimmer A (2009) Drug delivery of siRNA therapeutics: potentials and limits of nanosystems. Nanomedicine 5(1):8–20. doi:10.1016/j.nano.2008.06.001

Reiss P, Protière M, Li L (2009) Core/shell semiconductor nanocrystals. Small 5(2):154–168. doi:10.1002/smll.200800841

Richardson SCW, Pattrick NG, Lavignac N (2010) Intracellular fate of bioresponsive poly(amidoamine) S in vitro and in vivo. J Control Release 142(1):78–88. doi:10.1016/j.jconrel.2009.09.025

Ritchie M, Tchistiakova L, Scott N (2013) Implications of receptor-mediated endocytosis and intracellular trafficking dynamics in the development of antibody drug conjugates. MAbs 5(1):13–21. doi:10.4161/mabs.22854

Rodriguez MA, Pytlik R, Kozak T, Chhanabhai M, Gascoyne R, Lu B, Deitcher SR, Winter JN; Marqibo Investigators (2009) Vincristine sulfate liposomes injection (Marqibo) in heavily pretreated patients with refractory aggressive non-Hodgkin lymphoma. Cancer 115(15): 3475–3482. doi:10.1002/cncr.24359

Roepstorff K, Grøvdal L, Grandal M, Lerdrup M, van Deurs B (2008) Endocytic downregulation of ErbB receptors: mechanisms and relevance in cancer. Histochem Cell Biol 129(5):563–578. doi:10.1007/s00418-008-0401-3

Ruenraroengsak P, Cook JM, Florence AT (2010) Nanosystem drug targeting: facing up to complex realities. J Control Release 141(3):265–276. doi:10.1016/j.jconrel.2009.10.032

Ruiz ME, Gantner ME, Talevi A (2014) Applications of nanosystems to anticancer drug therapy (Part II. Den-Drimers, micelles, lipid-based nanosystems). Recent Pat Anticancer Drug Discov 9:1–30

Santhosh PB, Ulrih NP (2013) Multifunctional superparamagnetic iron oxide nanoparticles: promising tools in cancer theranostics. Cancer Lett 336(1):8–17. doi:10.1016/j.canlet.2013.04.032

Schellekens H, Stegemann S, Weinstein V, de Vlieger JS, Flühmann B, Mühlebach S, Gaspar R, Shah VP, Crommelin DJ (2014) How to regulate nonbiological complex drugs (NBCD) and their follow-on versions: points to consider. AAPS J 16(1):15–21. doi:10.1208/s12248-013-9533-z

Schütz CA, Juillerat-Jeanneret L, Mueller H, Lynch I, Riediker M; NanoImpactNet Consortium (2013) Therapeutic nanoparticles in clinics and under clinical evaluation. Nanomedicine 8(3):449–467. doi:10.2217/nnm.13.8

Seetharamu N, Kim E, Hochster H, Martin F, Muggia F (2010) Phase II study of liposomal cisplatin (SPI-77) in platinum-sensitive recurrences of ovarian cancer. Anticancer Res 30(2):541–545

Segal E, Satchi-Fainaro R (2009) Design and development of polymer conjugates as antiangiogenic agents. Adv Drug Deliv Rev 61(13):1159–1176. doi:10.1016/j.addr.2009.06.005

Seymour LW, Ferry DR, Anderson D, Hesslewood S (2002) Hepatic drug targeting: phase I evaluation of polymer-bound doxorubicin. J Clin Oncol 20(6):1668–1676. doi:10.1200/JCO.20.6.1668

Simon A (2012) Phase I safety, pharmacokinetic and pharmacodynamic results for ALN-PCS – preliminary study results. Pharmaceuticals A. http://www.tekmirapharm.com/Programs/Products.asp - plk1

Singh SK, Hajeri PB (2009) siRNAs: their potential as therapeutic agents—Part II. Methods of delivery. Drug Discov Today 14(17–18):859–865. doi:10.1016/j.drudis.2009.06.002
Singh A, Sahoo SK (2013) Magnetic nanoparticles: a novel platform for cancer theranostics. Drug Discov Today 19(4):474–481. doi:10.1016/j.drudis.2013.10.005
Smith MQ, Staley CA, Kooby DA, Styblo T (2009) Multiplexed fluorescence imaging of tumor biomarkers in gene expression and protein levels for personalized and predictive medicine. Curr Mol Med 9(8):1017–1023
Staruch R, Chopra R, Hynynen K (2011) Localised drug release using MRI-controlled focused ultrasound hyperthermia. Int J Hyperthermia 27(2):156–171. doi:10.3109/02656736.2010.518198
Szebeni J, Moghimi SM (2009) Liposome triggering of innate immune responses: a perspective on benefits and adverse reactions. J Liposome Res 19(2):85–90. doi:10.1080/08982100902792855
Taylor DD, Gercel-Taylor C (2011) Exosomes/microvesicles: mediators of cancer-associated immunosuppressive microenvironments. Semin Immunopathol 33(5):441–454. doi:10.1007/s00281-010-0234-8
Tekade RK, Kumar PV, Jain NK (2008) Dendrimers in oncology: an expanding horizon. Chem Rev 109(1):49–87. doi:10.1021/cr068212n
Thakor AS, Gambhir SS (2013) Nanooncology: the future of cancer diagnosis and therapy. Cancer J Clin 63(6):395–418. doi:10.3322/caac.21199
Tokatlian T, Segura T (2010) siRNA applications in nanomedicine. Wiley Interdiscip Rev Nanomed Nanobiotechnol 2(3):305–315. doi:10.1002/wnan.81
Torchilin VP (2007) Micellar nanocarriers: pharmaceutical perspectives. Pharm Res 24(1):1–16. doi:10.1007/s11095-006-9132-0
Torchilin V (2009) Multifunctional and stimuli-sensitive pharmaceutical nanocarriers. Eur J Pharm Biopharm 71(3):431–444. doi:10.1016/j.ejpb.2008.09.026
Torchilin V (2011) Tumor delivery of macromolecular drugs based on the EPR effect. Adv Drug Deliv Rev 63(3):131–135. doi:10.1016/j.addr.2010.03.011
Tran MA, Smith CD, Kester M (2008) Combining nanoliposomal ceramide with sorafenib synergistically inhibits melanoma and breast cancer cell survival to decrease tumor development. Clin Cancer Res 14(11):3571–3581. doi:10.1158/1078-0432.CCR-07-4881
Vail DM, Kurzman ID, Glawe PC, O'Brien MG, Chun R, Garrett LD, Obradovich JE et al (2002) STEALTH liposome-encapsulated cisplatin (SPI-77) versus carboplatin as adjuvant therapy for spontaneously arising osteosarcoma (OSA) in the dog: a randomized multicenter clinical trial. Cancer Chemother Pharmacol 50(2):131–136. doi:10.1007/s00280-002-0469-8
van Lummel M, van Blitterswijk WJ, Vink SR, Veldman RJ (2011) Enriching lipid nanovesicles with short-chain glucosylceramide improves doxorubicin delivery and efficacy in solid tumors. FASEB J 25(1):280–289. doi:10.1096/fj.10-163709
van Vlerken LE, Duan Z, Seiden MV (2007) Modulation of intracellular ceramide using polymeric nanoparticles to overcome multidrug resistance in cancer. Cancer Res 67(10):4843–4850. doi:10.1158/0008-5472.CAN-06-1648
Vicent MJ, Ringsdorf H, Duncan R (2009) Polymer therapeutics: clinical applications and challenges for development. Adv Drug Deliv Rev 61(13):1117–1120. doi:10.1016/j.addr.2009.08.001
Videira M, Arranja A, Rafael D, Gaspar R (2013) Preclinical development of siRNA therapeutics: towards the match between fundamental science and engineered systems. Nanomedicine. doi:10.1016/j.nano.2013.11.018
Wagner V, Dullaart A, Bock AK (2006) The emerging nanomedicine landscape. Nat Biotechnol 24:1211–1217
Wahajuddin AS (2012) Superparamagnetic iron oxide nanoparticles: magnetic nanoplatforms as drug carriers. Int J Nanomedicine 7:3445–3471. doi:10.2147/IJN.S30320
Wang R, Billone PS, Mullett WM (2013) Nanomedicine in action: an overview of cancer nanomedicine on the market and in clinical trials. J Nanomater 2013(3):1–12. doi:10.1155/2013/629681
White SC, Lorigan P, Margison GP, Margison JM (2006) Phase II study of SPI-77 (sterically stabilised liposomal cisplatin) in advanced non-small-cell lung cancer. Br J Cancer 95(7): 822–888

Wood BJ, Poon RT, Locklin JK, Dreher MR, Ng KK, Eugeni M, Seidel G et al (2012) Phase I study of heat-deployed liposomal doxorubicin during radiofrequency ablation for hepatic malignancies. J Vasc Interv Radiol 23(2):248–255.e7. doi:10.1016/j.jvir.2011.10.018

Xie J, Jon S (2012) Magnetic nanoparticle-based theranostics. Theranostics 2(1):122–124. doi:10.7150/thno.4051

Yang KK, Kong M, Wei YN, Liu Y, Cheng XJ, Li J, Park HJ, Chen XG (2013) Folate-modified–chitosan-coated liposomes for tumor-targeted drug delivery. J Mater Sci 48(4):1717–1728. doi:10.1007/s10853-012-6930-0

Yessine MA, Leroux JC (2004) Membrane-destabilizing polyanions: interaction with lipid bilayers and endosomal escape of biomacromolecules. Adv Drug Deliv Rev 56(7):999–1021

Yessine MA, Lafleur M, Meier C, Petereit HU (2003) Characterization of the membrane-destabilizing properties of different pH-sensitive methacrylic acid copolymers. Biochim Biophys Acta 1613(1–2):28–38

Yu MK, Park J, Jon S (2012) Targeting strategies for multifunctional nanoparticles in cancer imaging and therapy. Theranostics 2(1):3–44. doi:10.7150/thno.3463

Zhang JA, Anyarambhatla G, Ma L, Ugwu S, Xuan T, Sardone T, Ahmad I (2005) Development and characterization of a novel Cremophor EL free liposome-based paclitaxel (LEP-ETU) formulation. Eur J Pharm Biopharm 59(1):177–187. doi:10.1016/j.ejpb.2004.06.009

Chapter 16
BO-110, a dsRNA-Based Anticancer Agent

Case Study: Gene Therapy Using Polymer Carriers

Mercedes Pozuelo-Rubio, Damià Tormo Carulla, Andrés P. Gagete, and María Soengas

16.1 Overview

Efficient treatment of aggressive cancers such as melanoma has been traditionally hampered by a plethora of (epi)genetic alterations that deactivate classical apoptotic death programs and potentiate multiple intrinsic mechanisms of cell survival. Targeted activation of tumor self-degradation by autophagy is emerging as an alternative strategy to deplete tumor cells of essential organelles and promote their demise. However, autophagy is frequently activated as a fail-safe mechanism, for example, as a catabolic process to sustain the high-energy demand of tumor cells. Moreover, autolysosome formation may also be induced to counteract reactive oxygen or other stress-inducing signals that may accompany drug treatment. An additional complication in the design of pro-autophagy treatments is that normal cells also undergo basal autophagy. Therefore, a key challenge in the field is to design compounds that are selectively incorporated by malignant cells and activate self-degradation without unwanted secondary effects to normal cell compartments. Here we describe the identification and functional validation of BO-110, a new double-stranded (ds)RNA-based nanocomplex that we found as a potent anti-cancer agent in vivo. BO-110 is a polyplex of viral dsRNA mimetics (improved version of the classical polyinosinic-polycytidylic acid, pIC), complexed with cationic carriers for enhanced cytosolic delivery. First tested in melanoma cells, BO-110 was found to

M. Pozuelo-Rubio • D.T. Carulla • A.P. Gagete
Bioncotech Therapeutics, Catedrático Agustín Escardino, 9, 46980 Paterna, Valencia, Spain

M. Soengas (✉)
Melanoma Laboratory, Molecular Oncology Programme, Spanish National Cancer Research Centre (CNIO), Melchor Fernández Almagro, 3, Madrid 28049, Spain
e-mail: msoengas@cnio.es

M.J. Alonso, M. Garcia-Fuentes (eds.), *Nano-Oncologicals: New Targeting and Delivery Approaches*, Advances in Delivery Science and Technology,
DOI 10.1007/978-3-319-08084-0_16

induce a tumor-cell selective autophagosome-lysosome fusion, in part via a selective induction of the innate immunity sensor Melanoma Differentiation-Associated gene-5 (MDA5), a cytosolic dsRNA helicase. Importantly, BO-110-driven activation of innate responses led to the upregulation of the pro-apoptotic protein NOXA in a manner not recapitulated by classical chemo or immunotherapeutic agents. BO-110 was also found to display a broad therapeutic effect against pancreatic cancer, bladder cancer, and glioblastoma, among others. These results revealed a novel link between cytosolic RNA helicases, autophagy, and apoptosis that can be exploited for therapeutic intervention in otherwise highly resistant human cancers.

16.2 The First Indication of BO-110: Malignant Melanoma

Melanoma is an aggressive cancer derived from the malignant transformation of pigment-producing cells, melanocytes, which predominantly reside in the skin (Chin et al. 2006). This tumor type is characterized by an intrinsically high metastatic potential (Gupta et al. 2005; Scott et al. 2011) and a persistent resistance to standard chemotherapy (Soengas and Lowe 2003).

In the last years, comprehensive genomic and functional analyses of human melanoma tumors (Berger et al. 2012; Hodis et al. 2012; Krauthammer et al. 2012) have resulted in the discovery of new promising anti-tumoral drugs (Chapman et al. 2011; Flaherty et al. 2012; Hodi et al. 2010). Nevertheless, a complex and heterogeneous histopathology (Quintana et al. 2010; Whiteman et al. 2011) and genetic background (Dutton-Regester and Hayward 2012; Lawrence et al. 2013) have complicated the development of long-term effective therapies. Despite great progress in the understanding of mechanisms underlying melanoma initiation and progression (see below), this tumor remains as a prototype of deadly cancer at advanced stages. Thus, the prognosis of patients with metastatic melanoma is bleak, with an average survival of 6–12 months after diagnosis. With an incidence in continuous increase for the last 30 years and being responsible for over 80 % of skin cancer related deaths, melanoma is becoming a key public health problem (Gray-Schopfer et al. 2007; Siegel et al. 2013).

Currently approved therapies against melanoma. The clinical management of patients with cutaneous melanoma is conditioned by the staging at the time of diagnosis. Specifically, standard care of localized melanoma is surgical excision. Patients with stage III disease (i.e., presence of tumor cells in draining lymph nodes) usually undergo complete sentinel lymph node dissection (Sondak and Zager 2014) and adjuvant therapy traditionally with interferon alpha (IFNα)-2b, which has substantial side effects (Davar et al. 2013). For metastatic stages, the alkylating agent dacarbazine was used since 1970 and until very recently, although the corresponding rates were traditionally of less than 10 % and generally transient (Eggermont and Kirkwood 2004). Interleukin 2 (IL-2) was also approved by the FDA in 1998, although responses were also limited (<0–8 % of patients) and associated with significant toxicity (Tarhini and Agarwala 2005). More recently, new strategies have revolutionized the treatment of melanoma. These correspond to (1) ipilimumab, a monoclonal antibody against the cytotoxic T-lymphocyte-associated antigen 4 (CTLA-4) aimed to potentiate intrinsic T-cell responses (Page et al. 2014); (2) vemurafenib and dabrafenib, selective inhibitors of activated BRAF mutations from valine to a glutamic acid at position 600 (V600E) characteristic of melanoma; (3) and trametinib, a blocker of MEK (a key effector in MAPK pathways) (Bollag et al. 2010; Chapman et al. 2011; Flaherty et al. 2010; Kim et al. 2013; Wright and McCormack 2013). Other ongoing efforts include vaccines against immunogenic melanoma antigens (Kazaks et al. 2008; Slingluff et al. 2008; Zarour and Kirkwood 2003); immunotherapy targeting PD1 or PDL1 (Hamid et al. 2013); targeted therapy against *KIT, MEK, PI3K, AKT, NRAS, mTOR,* and *CDK4* (Gray-Schopfer et al. 2007); and several combinatorial approaches (Flaherty et al. 2012), among many other strategies (see Tables 16.1 and 16.2).

A compilation of new drugs that are now in development is indicated below (Table 16.2).

Despite the new anticancer agents described above, the melanoma field still faces key challenges. First, durable anti-tumoral responses are transient or limited to specific subsets of melanoma patients. For example, BRAF inhibitors are invariably opposed by an array of resistance mechanisms (Prahallad et al. 2012; Shi et al. 2012; Sloot et al. 2014; Vultur and Herlyn 2009; Wilson et al. 2012). Similarly, in the case of the immunotherapy, relatively slow responses, serious side effects, and no consistent biomarkers also compromise clinical benefits (Fecher et al. 2013; Holzel et al. 2013; Ma and Armstrong 2014; Page et al. 2014; Pennock et al. 2012). Therefore, improved strategies to overcome the intrinsic and acquired resistance of melanoma cells are still needed.

16.3 Apoptosis and Autophagy in Cancer Therapy

Traditionally, most anticancer treatments have aimed to activate intrinsic programs of cell death in tumor cells and/or their accompanying stroma (Bunz 2001; Galluzzi et al. 2007; Maiuri et al. 2007). In this context, large efforts have been dedicated to the activation of programmed cell death by apoptotic cysteine-aspartate proteases (caspases) (Strasser et al. 2000). Dying apoptotic cells are recognized by macrophages and other phagocytic cells and thus represent a "clean" strategy to preclude the release of inflammatory responses typical of uncontrolled or necrotic death inducers (Johnstone et al. 2002). Different research teams, including ours, have been able to induce caspase activation in several mouse melanoma models. Targeted drugs against MEK (Verhaegen et al. 2006), anti-apoptotic factors such as Mcl-1 or Bcl-xL (Wolter et al. 2007), the proteasome (Fernandez et al. 2005, 2006), or MDM2 (Verhaegen et al. 2012), as well as small interfering RNAs against chromatin remodeling factors (DEK) (Khodadoust et al. 2009), have been shown to delay melanoma growth in mice. These compounds have identified novel candidate targets for second-generation drugs, but durable responses have yet to be achieved in vivo (Soengas and Lowe 2003). Given the partial efficacy of classical apoptotic

Table 16.1 Current therapies against melanoma

Drug	Mechanism of action	Response	Survival	Adverse events
Dacarbazine (Generic)	Alkylating agent	ORR 10–20 %	Median OS of 5.6–7.8 months	Vomiting, headaches, diarrheas, birth defects
Proleukin (Interleukin-2, Novartis)	Immunotherapy	Reported responses in 13–20 % of patients	Median OS 10–11.4 months	Mortality 1/50 treated. Capillary leak syndrome. CV, hepatic, neurological, endocrine, renal complications
Intron A (IFN-2a, Merck & Co.)	Adjuvant to surgical treatment, immunomodulator	ORR 15 % as single agent	PFS improvement 9 months. No OS improvement	Flu-like symptoms, GI, depression, neutropenia, fatigue anorexia, etc.
Yervoy (Ipilimumab, BMS)	Anti-CLTA-4 human monoclonal antibody, immunomodulator	ORR 10.9 % (MDX010-20 study)	Median OS of 10.1–11.2 months	Black-box warning of adverse immune reactions
Zelboraf (Vemurafenib, Plexxikon, Roche)	Raf B protein kinase inhibitor	ORR 48.4 % (BRIM3 study in untreated pats)	Median PFS 5.32 months. Median OS 13.2 months. In BRIM3	Rash, joint pain, photosensitivity, and fatigue. Cutaneous neoplasms

Source: Bioncotech Therapeutics
Legend: *ORR* objective response rate; *OS* overall survival, *PFS* progression free survival, *GI* gastrointestinal, *CV* cardiovascular, *AE* adverse event

Table 16.2 Drugs in development

Drug	Mechanism of action	Stage of development	Response	Survival	Adverse events
Dabrafenib GSK	Raf B protein kinase inhibitor	FDA approved, launch pending	CR 3 %; PR 50 %, for the dabrafenib arm. Compared to CR 0 % and PR 19 %, for the dacarbazine arm	PFS 5.1 months for dabrafenib and 2.7 months for dacarbazine ($p<0.0001$). Phase III BREAK3 study	Grade 3 events noted in trials include fatigue, pyrexia, and headache. 9 % of patients developed squamous cell carcinoma
Trametinib GSK	MEK-1 and MEK-2 protein kinase inhibitor	FDA approved, launch pending	40 % RR seen in BRAF mutated patients in Phase I in 162 patients	Median PFS in BRAF V600E mutant patients. 4.8 months in the trametinib DMSO arm vs. 1.4 months in the chemo arm ($p<0.0001$)	Rash and diarrhea
Nivolumab BMS	Anti-PD-1 antibody	Phase III	40 % ORR at 3 mg	20 months OS at 3 mg	21 % grade 3–4 AEs
Astuprotimut-r GSK	Melanoma associated antigen 3 (MAGE-3) modulator	Phase III (data in 2016)	ORR 11 % (Phase II, highest RR in AS15 adjuvant arm)	Median OS 33 months with AS15 adjuvant	Grade III toxicities included fatigue (8 %), anorexia (3 %), lymphopenia (6 %), and pyrexia (3 %)
T-VEC (OncoVex $^{GM\text{-}CSF}$, Amgen)	Herpes simplex virus type-1 (HSV-1) with the granulocyte-macrophage colony stimulating factor (GM-CSF) gene	Phase III (final OS data)	ORR 28 %. 92 % of responses lasted at least 6 months (Phase II). Met primary endpoint of DRR. OS trend	OS was 58 and 52 % at 1 and 2 years, respectively	On site of injection and grade 1 influenza-like symptoms
Abraxane Celgene	Nanoparticle albumin-bound paclitaxel. Microtubule stabilizer	Phase III (final OS data by YE)	Not reported	PFS 4.8 months, compared with 2.5 months for dacarbazine, $p=0.044$. Interim OS 12.8 and 10.7 months for ABI-007 and dacarbazine, $p=0.094$ (Phase III)	Bone marrow suppression is a dose limiting toxicity. Neutropenia, alopecia, sensory neuropathy, ECG abnormalities, and myalgia/arthralgia
Lambrolizumab Merck & Co.	Anti-PD-1 antibody	Phase II	ORR 52 % with 10 mg/kg dose (10 % CR)	N/A	10 % grade 3–4 AEs
MPDL-3280A Roche	Anti-PD-L1 antibody	Phase I	ORR 26 %	24-week PFS 35 %	33 % grade 3–4 AEs

Source: Bioncotech Therapeutics
Legend: *ORR* objective response rate, *OS* overall survival, *PFS* progression free survival, *AE* adverse event, *CR* complete response, *RR* response rate

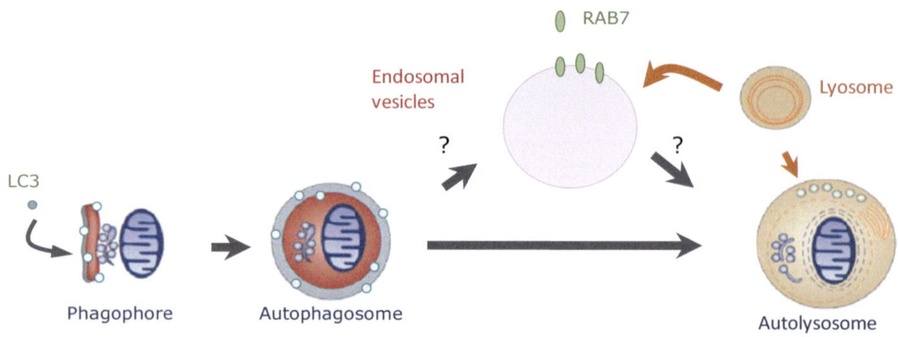

Fig. 16.1 Simplified overview of macroautophagy in cross talk with less characterized endo-lysosomal pathways. LC3 and the small GTPase RAB7 are shown as examples of autophagosome and late endosome markers

inducers, we decided to screen for compounds with different modes of action. In particular, we turned our attention to self-degradative mechanisms mediated by autophagy. Macroautophagy (herein refer to as autophagy for simplicity) involves the sequestration of bulk cytosolic components in autophagosomes, for subsequent lysosomal degradation (Yang and Klionsky 2010) (see schematic in Fig. 16.1). This degradative process is intrinsically active in a large range of cell types as a mechanism of survival to remove large protein aggregates, damaged organelles, and contribute to ATP generation in situations of starvation (Cuervo 2004; Mizushima et al. 2008). However, hyperactivation of autophagy can result on the depletion of essential organelles, and consequently, promote cell death (Galluzzi et al. 2008; Rubinsztein et al. 2007). How to shift the protective function of autophagy into tumor cell killing is an active area of research in oncology.

16.4 Discovery of BO-110

To identify novel autophagy inducers, we set up a screening based on the well-known autophagosome marker LC3. In resting conditions, LC3 is homogeneously distributed in the cytosol, but gets lipidated and inserted in the membrane of autophagosomes at early stages of autophagy (Xie and Klionsky 2007) (Fig. 16.1). Fluorescent derivatives of LC3 (i.e., fusion to green fluorescence protein, GFP) thus allow for a relatively straightforward scoring of non-autophagic vs. autophagosome containing cells (diffuse vs. focal staining) by fluorescence microscopy (Klionsky et al. 2012).

From that study we obtained 3 groups of compounds: (1) displaying no autophagic activity, (2) inducing autophagy as a survival mechanism, and finally, (3) promoting excessive or persistent autophagy leading to cellular collapse. Among all the compounds with tumor cell-selective activity in vitro, we focused on those allowing for systemic delivery without associated secondary toxicities. From those bioavailable compounds, the most potent were nanoparticles constituted of long synthetic dsRNA (pIC: polyinosinic-polycytidylic acid) packed with polycations (polyethylenimine) to improve cellular uptake (Tormo et al. 2009b). For simplicity, these nanocomplexes of pIC were named BO-110 (in reference to Bioncotech Therapeutics, the company now pursuing the clinical development of this compound).

16.5 Mode of Action of BO-110

16.5.1 Mobilization of the Endocytic Machinery

Biochemical and functional analyses of BO-110-treated cells revealed a complex mode of action (Alonso-Curbelo and Soengas 2010; Tormo et al. 2009a, b). Electron microscopy showed the uptake of the BO-110 nanocomplexes by vesicular structures at the plasma membrane (Fig. 16.2a, black arrows). Classical autophagosomes were also found by this technique (Fig. 16.2a, middle panels) and confirmed independently by monitoring foci formation by the autophagosome marker LC3 (Fig. 16.2b) or by depletion of key autophagy genes (not shown). These features

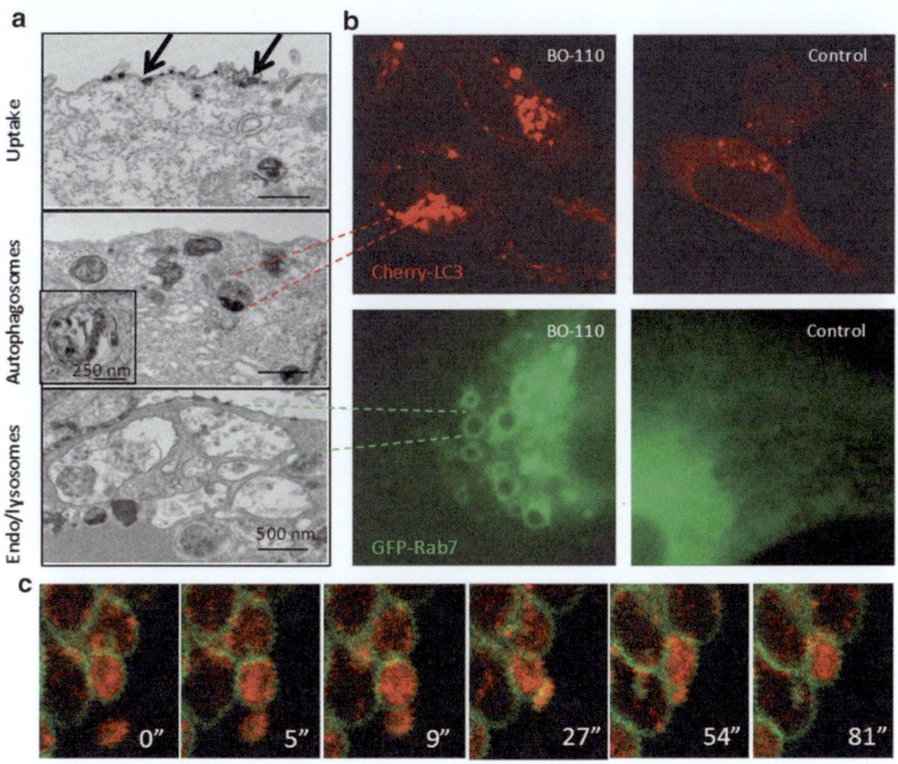

Fig. 16.2 Ultrastructural changes driven by BO-110. (**a**) Electron micrographs of BO-110 nanocomplexes (*black arrows*), and the autophagosomes and large endosomes generated thereafter. (**b**) Fluorescence micrographs of autophagosomes and endosomes by means of the visualization of LC3 and RAB7 fused to Cherry and GFP, respectively. (**c**) Fusion of lysosomes (labeled with LysoTracker Red) to late endosomes (decorated by GFP-RAB7) at the indicated times (*in seconds*) after treatment

were unexpected as dsRNA mimics had not been previously associated with autophagy in tumor cells. However, even more surprising was a massive and sustained mobilization of the endosomal machinery by BO-110. This could be visualized as large multivesicular structures by electron microscopy (Fig. 16.2b, c) and fluorescence-based imaging of the distribution of the small GTPase RAB7, a marker of late endosomes (Rink et al. 2005). Time-lapse fluorescence microscopy revealed a sequential order of fusion events whereby RAB7-decorated vesicles recruited the LC3 protein (i.e., to form hybrid endosome/autophagosome structures or amphisomes) to subsequently fuse with lysosomes (Fig. 16.3 and results not shown). These fusion events occurred for hours until cells finally autodigest. Importantly, these RAB7 > LC3 > lysosome fusion events mentioned were found to proceed an efficient cellular collapse independent of the mutational or functional status of BRAF, NRAS, PTEN, p14^{ARF}, or p16^{INK4a}, genes previously associated with melanoma progression. Moreover, the anti-tumoral activity of BO-110 was tumor cell

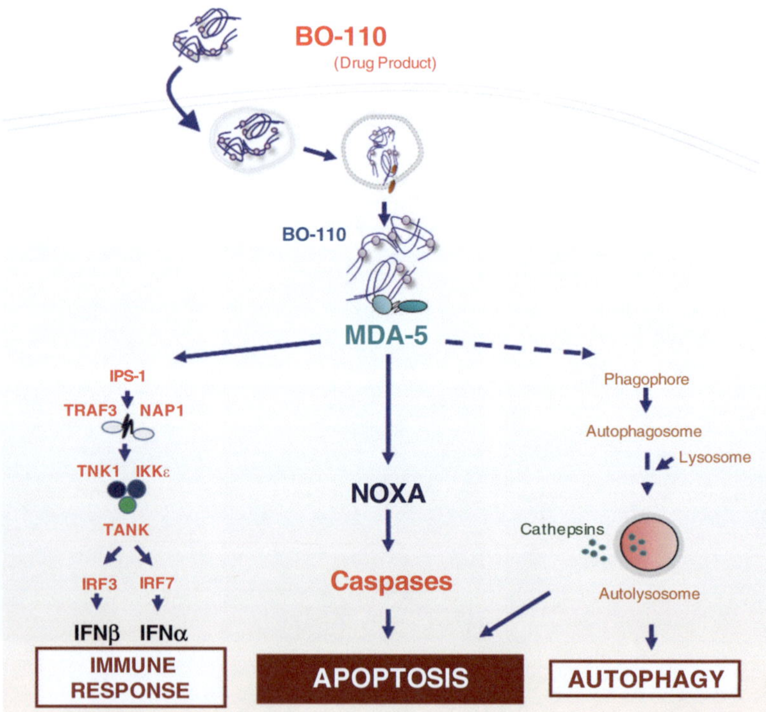

Fig. 16.3 Mechanism of action of BO-110. Model summarizing a new strategy to promote tumor cell death. BO-110 in the cytosol activates the helicase MDA-5 (*specific sensor of pIC, active ingredient of BO-110, in tumor cells). Accordingly, it induced a subsequent activation of apoptotic and autophagy-dependent cancer cell death. Within 2–3 h of incubation with BO-110, tumor cells undergo massive ultrastructural changes in the endosomal compartment. Although MDA-5 could facilitate autophagosome formation, MDA-5 actively induces the pro-apoptotic factor NOXA. Thus, the activation of apoptotic caspases accompanied by a sustained lysosomal-dependent degradative process can ultimately converge in an efficient tumor cell death in presence of BO-110

selective, as normal skin melanocytes, keratinocytes, or fibroblasts remained viable. Together, these results identified BO-110 as an efficient and selective anti-melanoma agent.

16.5.2 Biosensors of dsRNA: The MDA5 Helicase

Next, we questioned about the identity of the cellular sensors that underlie the efficacy and selectivity of BO-110. Specifically, we focused on the best known dsRNA-driven signaling cascades: (1) the Toll-like receptor 3 (TLR3), which is located in endosomes; (2) the Melanoma Differentiation-Associated gene-5 (MDA5), a cytosolic dsRNA helicase; and (3) the dsRNA-activated protein kinase (PKR) (Takeuchi and Akira 2008). To assess all these proteins simultaneously and to perform an unbiased analysis of alternative dsRNA sensors and effectors, we defined transcriptional profiles of BO-110-treated cells. Specifically, we compared changes in the transcriptome induced by BO-110 (which kills) but not by naked dsRNA (which does not affect viability). Melanocytes were also tested in parallel as controls for cells in which BO-110 lacked anticancer activity.

Interestingly, neither naked dsRNA (classical pIC) nor BO-110 induced significant changes in the expression profile of normal melanocytes. In melanoma cells, both naked dsRNA and BO-110 activated a transcriptional program consistent with an antiviral state. However, while changes in mRNA expression induced by naked dsRNA were transient, BO-110 induced a sustained response. These studies led to the cytosolic helicase MDA-5 (Melanoma Differentiation Associated gene-5) as the dsRNA sensor activated by BO-110. The functional role of MDA5 was independently assessed by generating xenografts with MDA-5 deficient cells and observing a reduced efficacy of BO-110. Of note, MDA-5 dependent induction of NOXA was confirmed in melanoma cells by an independent study (Besch et al. 2009). Nevertheless, although MDA5 is critical for the cytotoxic activity of BO-110, it is not absolutely essential, indicating that other dsRNA helicases may also be present in tumor cells for recognition of BO-110.

16.5.3 Dual Activation of Autophagy and Apoptosis by BO-110

cDNA arrays and protein expression studies identified a parallel wave of events induced by BO-110 involving the activation of a caspase-dependent apoptotic program (Fig. 16.3). Searching for caspase activators, we found that BO-110 promoted a marked upregulation of the proapoptotic protein NOXA. shRNA studies placed NOXA downstream of MDA5 and demonstrated BO-110's unique feature to activate both the endo-lysosomal pathways and the apoptotic machinery (Fig. 16.3). Our data suggest that the level of stress imposed by BO-110 on endocytic and apoptotic programs is of such magnitude that intrinsic protective mechanisms of autophagy are overridden, allowing for an effective tumor inhibition.

16.6 Efficacy and Tumor Cell Selectivity of BO-110 In Vivo

To validate the significance of the newly described mechanism of action of BO-110, three tumor models were investigated (Tormo et al. 2009b):

1. *Immunodeficient mice* in which human melanoma cell lines (showing resistance to other therapies) were implanted to pulmonary metastasis
2. *Immunocompetent mice* implanted with cell lines (B16) isolated from spontaneous melanoma originated in syngeneic animals
3. Immunocompetent genetically modified mice, in which melanoma formation is induced by oncogenic NRAS in IN4a/ARF$^{-/-}$ melanocytes (a *genetic background reproducing human melanoma*)

In all three models, BO-110 lead to a significantly stronger and sustained anti-melanoma activity than naked dsRNA (pIC), as determined by the inhibition in the number, onset, and growth of melanoma tumors, as well as by PET/CT-based analysis of metabolic activity. Importantly, BO-110 was found to extend progression-free survival at dosing schedules where no secondary toxicities are observed. Given the traditional resistance of melanomas to classical therapies, these results are significant as they offer new avenues for drug development.

16.7 Efficacy of BO-110 in Non-melanoma Tumors

To determine whether the therapeutic activity of BO-110 could benefit other tumor types, viability assays were performed on a battery of cell lines isolated from a variety of cancers. These included lines from cancer of the colon, bladder, prostate, lung, and ovary. Triple negative melanoma cells, as well as stem cells from glioma, were also tested for their response to BO-110. These studies confirmed a broad spectrum of action of BO-110, extending beyond melanoma (not shown).

16.8 Development of BO-110 as Antitumoral Drug

16.8.1 BO-110 Chemistry, Manufacturing, and Control

pIC has been reported to stimulate innate and adaptative immune responses against tumor cells and as well as to trigger apoptosis in some tumor types (Cheng and Xu 2010). However, naked pIC was found unsuccessful in clinical trials because of degradation by nucleases. Recent approaches to overcome the failure of naked pIC in the clinics are Ampligen (polyI:polyC12U) (Armstrong et al. 1992; Jasani et al. 2009; Navabi et al. 2009) and Oncovir's product Hiltonol (poly-ICLC) (Hartman et al. 2013; Levine et al. 1979; Ming Lim et al. 2013;

Rapoport et al. 2014; Sabado et al. 2013; Salazar et al. 2014). These drugs have mostly been described as adjuvants with immunostimulatory properties by modifying PolyI:C to increase its stability and efficacy in vivo.

BO-110 is as a pharmaceutical product formed by pIC as the active pharmacological ingredient (API), developed by Bioncotech under a proprietary manufacture process. In particular, proprietary chromatographic and electrophoretical analyses were optimized to guarantee size and high quality of the pIC derivative, according to regulatory agency guidelines. Regarding the carrier, BO-110 was formulated with a linear polyethylenimine (PEI). This particular coupling of pIC-PEI (1) overcame the nuclease degradation characteristic of naked dsRNA without affecting its inherent anti-tumoral properties and (2) favored the activation of dsRNA sensors in the cytosol, which as indicated above are otherwise maintained in a silent state. Other formulations in Ampligen (Armstrong et al. 1992; Jasani et al. 2009), Hiltonol (Nakamura et al. 1982; Salazar et al. 2014) or Poly-L-lysine polyplexes (Basarkar and Singh 2007), appear to require higher doses of pIC for efficacy in vivo, supporting a more efficient role of PEI as a packaging agent.

PEI is an organic macromolecule with high density of amine groups ideal to condense nucleic acids into particles of nanometer range (Dunlap et al. 1997). It has been reported that PEI promoted the intracellular delivery of nanoparticles and cell targeting ligands functioning as a viral delivery tool (Wagner 2004). PEI has been synthesized with a linear or a branched topology, available in a wide range of molecular weights (Kichler et al. 2001). The branched PEI was first described by Boussif and collaborators (Boussif et al. 1995) as an effective transfection reagent. Nevertheless, other authors (Wightman et al. 2001) found that linear low molecular weight (LMW)-PEI was more effective than branched high molecular weight (HMW)-PEI and with lower cytotoxicity (Kunath et al. 2003; Ledley 1996). Transfection efficiency of PEI has been found to be dependent on a multitude of factors such as molecular weight, degree of branching, nitrogen to phosphate (N/P) ratio, and complex size, among others (Thomas et al. 2005). In the case of BO-110, linear PEI became the best option to deliver synthetic dsRNA to the cytosol. PEI:dsRNA ratios were optimized to achieve a potent and sustained death response in tumor cells, by promoting autophagy initiation and apoptosis activation selectively in the malignant compartment.

Core stability data package for drug substance was established according to regulatory guidelines. Characterization tests included description of amorphous or crystal structures, potency, water content, impurity profile, microbiological content, and solubility of the lyophilized active ingredient. Comparisons between pIC batches showed reproducibility of the manufacturing process and the properties consistency between batches. Specifically, particle size distribution, osmolarity, and pH are highly controlled to produce a consistently optimal BO-110 manufacture for intravenous administration by bolus injections. Ultimately, BO-110 is provided by the manufacturer in vials containing 12 ml of product aimed to be stored at 4 °C.

16.8.2 Preclinical Evaluation of BO-110

BO-110 is intended to be tested initially in patients with solid cancers of poor prognosis, which are refractory to approved or available therapies. The BO-110 targets included melanoma, pancreatic cancer, bladder cancer, and triple negative cancers based on preclinical data. Guidelines will follow standard Nonclinical Evaluation for Anticancer Pharmaceuticals.

Minimum effective dose (MED) and maximum tolerated dose (MTD) defined in mouse xenografts were used to guide in toxicity studies in rats and dogs (granted by the expression and induction of MDA-5 and NOXA in tumor cells generated in these species). In particular, biochemical analyses demonstrated lack of apparent hematological toxicity in rodent species at doses of up to 25xMED. Exploratory pharmacokinetic analyses (PK) using phosphorus-32 (^{32}P) labelled-BO-110 indicated accumulation in the kidney and liver but with no apparent side effects (not shown). In addition, no adverse effects of BO-110 have been found in cardiac function through in vitro analyses for hERG ("human Ether-à-go-go-Related Gene") that codes for the alpha subunit of a potassium ion channel, to evaluate the electrical activity of the heart. In addition, no functional changes were detected for diverse cytochromes. BO-110 toxicity effects were in line with those found previously for pIC (Homan et al. 1972), where signs of toxicity were dose dependent. Doses close to the MTD promoted hypoactivity and diarrhea; in addition, clinical changes included affectation in red blood cells and elevation of liver transaminases and interferon. All these tox signs appeared to be reversible in all the preclinical species after cessation of dosing. Additional toxicity evaluation is being completed and Phase I clinical trials are planned for late-2014.

16.8.3 Clinical Development of BO-110

Melanoma accumulates many genetic and epigenetic defects that contribute to the limited efficacy of current anticancer treatments. This has been largely evidenced by a variety of resistance mechanisms which ultimately limit life expectancy. Nevertheless, melanoma tumors retain the ability to sense and respond to mimetics of viral dsRNA, inducing autophagy and apoptosis in melanoma and a variety of other tumor cell types. Bioncotech Therapeutics is pursuing the clinical development of BO-110.

16.9 Other dsRNA-Based Anticancer Agents

As mentioned above, other strategies have been developed to protect naked dsRNA (pIC) to take advantage of its pro-stimulatory effects in the immune system. Compounds in clinical trials are Hemispherx's Ampligen (Strayer et al. 2012), Oncovir's poly-ICLC (Butowski et al. 2009a, b; Morse et al. 2011; Rapoport et al. 2014; Rosenfeld et al. 2010; Sabbatini et al. 2012), and BioLineRx's BL-7030 (Schaffert et al. 2011).

Ampligen (polyI:polyC12U) is an experimental antiviral and immunomodulatory double stranded RNA drug. It has been primarily tested as a treatment for chronic fatigue syndrome (ME/CFS) and acquired immunodeficiency syndrome (AIDS). The drug has undergone regulatory clinical trials up to registration. Clinical trials in 760 patients show that the drug is generally safe and effective (Strayer et al. 2012). According to Hemispherx, the Mode of action (MoA) of Ampligen is related to the interaction with Toll-like receptors such as TLR3. In this context, Ampligen is an immune-stimulator similar to naked pIC as it is not complexed with any stabilizer.

Oncovir's product Hiltonol (poly-ICLC) is made of pIC stabilized with poly-lysine. The resulting compound is a stable dsRNA that is a potent interferon inducer (Krown 1986). Hiltonol has been proposed for wider indications such as the treatment of viral infections and the treatment of glioblastoma and other brain tumors (Butowski et al. 2009a, b; Morse et al. 2011; Rosenfeld et al. 2010; Sabbatini et al. 2012).

BioLineRx's BL-7030 is a polyIC complexed with PEI and Endothelial Growth Factor (pIC-pEI-EGF). The drug was intended for the treatment of EGFR-overexpressing cancers. BL-7030 eradicated human breast tumors and glioblastoma in preclinical models.

16.10 Conclusions

In summary, dsRNA-based therapies, here exemplified by BO-110, represent an alternative strategy to control aggressive cancers of poor prognosis by activating distinct mechanisms of cell death by apoptosis and autophagy, together with a potent induction of innate immunity programs. Clinical trials underway will demonstrate the efficacy of dsRNA nanocomplexes in long-term treatments, hopefully providing a significant improvement of the overall survival of affected patients.

Acknowledgments The authors thank our colleagues at the Melanoma Group of the CNIO, particularly to Direna Alonso-Curbelo, Agnieszka Chęcinska, Eva Perez-Guijarro, Estela Cañon, Erica Riveiro-Falkenbach, Tonantzin G. Calvo, and Lionel Larribere. We are also grateful to Ana Villanueva, Diego Megías, and Francisca Mulero (CNIO) for their technical help and support. M.S.S. is funded by grants from the Spanish Ministry of Economy and Innovation (projects SAF2011-28317 and Consolider RNAREG), and a Team Science Award by the Melanoma Research Alliance.

Conflict of Interest María S. Soengas and Damià Tormo Carulla are co-founders and members of the advisory board of BiOncotech Therapeutics.

References

Alonso-Curbelo D, Soengas MS (2010) Self-killing of melanoma cells by cytosolic delivery of dsRNA: wiring innate immunity for a coordinated mobilization of endosomes, autophagosomes and the apoptotic machinery in tumor cells. Autophagy 6:148–150

Armstrong JA, McMahon D, Huang XL, Pazin GJ, Gupta P, Rinaldo CR Jr, Schoenfeld DA, Gaccione P, Tripoli CA, Bensasi S et al (1992) A phase I study of ampligen in human immunodeficiency virus-infected subjects. J Infect Dis 166:717–722

Basarkar A, Singh J (2007) Nanoparticulate systems for polynucleotide delivery. Int J Nanomedicine 2:353–360

Berger M, Hodis E, Heffernan TP, Lissanu Deribe Y, Lawrence MS, Protopopov A, Ivanova E, Watson IR, Nickerson E, Ghosh P et al (2012) Melanoma genome sequencing reveals frequent PREX2 mutations. Nature. doi:10.1038/nature11071

Besch R, Poeck H, Hohenauer T, Senft D, Hacker G, Berking C, Hornung V, Endres S, Ruzicka T, Rothenfusser S, Hartmann G (2009) Proapoptotic signaling induced by RIG-I and MDA-5 results in type I interferon-independent apoptosis in human melanoma cells. J Clin Invest 119:2399–2411

Bollag G, Hirth P, Tsai J, Zhang J, Ibrahim PN, Cho H, Spevak W, Zhang C, Zhang Y, Habets G et al (2010) Clinical efficacy of a RAF inhibitor needs broad target blockade in BRAF-mutant melanoma. Nature 467:596–599

Boussif O, Lezoualc'h F, Zanta MA, Mergny MD, Scherman D, Demeneix B, Behr JP (1995) A versatile vector for gene and oligonucleotide transfer into cells in culture and in vivo: polyethylenimine. Proc Natl Acad Sci U S A 92:7297–7301

Bunz F (2001) Cell death and cancer therapy. Curr Opin Pharmacol 1:337–341

Butowski N, Chang SM, Junck L, DeAngelis LM, Abrey L, Fink K, Cloughesy T, Lamborn KR, Salazar AM, Prados MD (2009a) A phase II clinical trial of poly-ICLC with radiation for adult patients with newly diagnosed supratentorial glioblastoma: a North American Brain Tumor Consortium (NABTC01-05). J Neurooncol 91:175–182

Butowski N, Lamborn KR, Lee BL, Prados MD, Cloughesy T, DeAngelis LM, Abrey L, Fink K, Lieberman F, Mehta M et al (2009b) A North American brain tumor consortium phase II study of poly-ICLC for adult patients with recurrent anaplastic gliomas. J Neurooncol 91:183–189

Chapman PB, Hauschild A, Robert C, Haanen JB, Ascierto P, Larkin J, Dummer R, Garbe C, Testori A, Maio M et al (2011) Improved survival with vemurafenib in melanoma with BRAF V600E mutation. N Engl J Med 364:2507–2516

Cheng YS, Xu F (2010) Anticancer function of polyinosinic-polycytidylic acid. Cancer Biol Ther 10:1219–1223

Chin L, Garraway LA, Fisher DE (2006) Malignant melanoma: genetics and therapeutics in the genomic era. Genes Dev 20:2149–2182

Cuervo AM (2004) Autophagy: in sickness and in health. Trends Cell Biol 14:70–77

Davar D, Tarhini AA, Kirkwood JM (2013) Adjuvant immunotherapy of melanoma and development of new approaches using the neoadjuvant approach. Clin Dermatol 31:237–250

Dunlap DD, Maggi A, Soria MR, Monaco L (1997) Nanoscopic structure of DNA condensed for gene delivery. Nucleic Acids Res 25:3095–3101

Dutton-Regester K, Hayward NK (2012) Reviewing the somatic genetics of melanoma: from current to future analytical approaches. Pigment Cell Melanoma Res 25:144–154

Eggermont AM, Kirkwood JM (2004) Re-evaluating the role of dacarbazine in metastatic melanoma: what have we learned in 30 years? Eur J Cancer 40:1825–1836

Fecher LA, Agarwala SS, Hodi FS, Weber JS (2013) Ipilimumab and its toxicities: a multidisciplinary approach. Oncologist 18:733–743

Fernandez Y, Verhaegen M, Miller TP, Rush JL, Steiner P, Opipari AW Jr, Lowe SW, Soengas MS (2005) Differential regulation of noxa in normal melanocytes and melanoma cells by proteasome inhibition: therapeutic implications. Cancer Res 65:6294–6304

Fernandez Y, Miller TP, Denoyelle C, Esteban JA, Tang WH, Bengston AL, Soengas MS (2006) Chemical blockage of the proteasome inhibitory function of bortezomib: impact on tumor cell death. J Biol Chem 281:1107–1118

Flaherty KT, Puzanov I, Kim KB, Ribas A, McArthur GA, Sosman JA, O'Dwyer PJ, Lee RJ, Grippo JF, Nolop K, Chapman PB (2010) Inhibition of mutated, activated BRAF in metastatic melanoma. N Engl J Med 363:809–819

Flaherty KT, Hodi FS, Fisher DE (2012) From genes to drugs: targeted strategies for melanoma. Nat Rev Cancer 12:349–361

Galluzzi L, Maiuri MC, Vitale I, Zischka H, Castedo M, Zitvogel L, Kroemer G (2007) Cell death modalities: classification and pathophysiological implications. Cell Death Differ 14:1237–1243

Galluzzi L, Vicencio JM, Kepp O, Tasdemir E, Maiuri MC, Kroemer G (2008) To die or not to die: that is the autophagic question. Curr Mol Med 8:78–91

Gray-Schopfer V, Wellbrock C, Marais R (2007) Melanoma biology and new targeted therapy. Nature 445:851–857

Gupta PB, Kuperwasser C, Brunet JP, Ramaswamy S, Kuo WL, Gray JW, Naber SP, Weinberg RA (2005) The melanocyte differentiation program predisposes to metastasis after neoplastic transformation. Nat Genet 37:1047–1054

Hamid O, Robert C, Daud A, Hodi FS, Hwu WJ, Kefford R, Wolchok JD, Hersey P, Joseph RW, Weber JS et al (2013) Safety and tumor responses with lambrolizumab (anti-PD-1) in melanoma. N Engl J Med 369:134–144

Hartman LL, Crawford JR, Makale MT, Milburn M, Joshi S, Salazar AM, Hasenauer B, Vandenberg SR, Macdonald TJ, Durden DL (2013) Pediatric phase II trials of poly-ICLC in the management of newly diagnosed and recurrent brain tumors. J Pediatr Hematol Oncol 36:451–457

Hodi FS, O'Day SJ, McDermott DF, Weber RW, Sosman JA, Haanen JB, Gonzalez R, Robert C, Schadendorf D, Hassel JC et al (2010) Improved survival with ipilimumab in patients with metastatic melanoma. N Engl J Med 363:711–723

Hodis E, Watson IR, Kryukov GV, Arold S, Imielinski M, Thereurillat J-P, Nickerson E, Auclair D, Li L, Place C et al (2012) A landscape of driver mutations in melanoma. Cell 150: 251–263

Holzel M, Bovier A, Tuting T (2013) Plasticity of tumour and immune cells: a source of heterogeneity and a cause for therapy resistance? Nat Rev Cancer 13:365–376

Homan ER, Zendzian RP, Schott LD, Levy HB, Adamson RH (1972) Studies on poly I:C toxicity in experimental animals. Toxicol Appl Pharmacol 23:579–588

Jasani B, Navabi H, Adams M (2009) Ampligen: a potential toll-like 3 receptor adjuvant for immunotherapy of cancer. Vaccine 27:3401–3404

Johnstone RW, Ruefli AA, Lowe SW (2002) Apoptosis: a link between cancer genetics and chemotherapy. Cell 108:153–164

Kazaks A, Balmaks R, Voronkova T, Ose V, Pumpens P (2008) Melanoma vaccine candidates from chimeric hepatitis B core virus-like particles carrying a tumor-associated MAGE-3 epitope. Biotechnol J 3:1429–1436

Khodadoust MS, Verhaegen M, Kappes F, Riveiro-Falkenbach E, Cigudosa JC, Kim DS, Chinnaiyan AM, Markovitz DM, Soengas MS (2009) Melanoma proliferation and chemoresistance controlled by the DEK oncogene. Cancer Res 69:6405–6413

Kichler A, Leborgne C, Coeytaux E, Danos O (2001) Polyethylenimine-mediated gene delivery: a mechanistic study. J Gene Med 3:135–144

Kim KB, Kefford R, Pavlick AC, Infante JR, Ribas A, Sosman JA, Fecher LA, Millward M, McArthur GA, Hwu P et al (2013) Phase II study of the MEK1/MEK2 inhibitor Trametinib in patients with metastatic BRAF-mutant cutaneous melanoma previously treated with or without a BRAF inhibitor. J Clin Oncol 31:482–489

Klionsky DJ, Abdalla FC, Abeliovich H, Abraham RT, Acevedo-Arozena A, Adeli K, Agholme L, Agnello M, Agostinis P, Aguirre-Ghiso JA et al (2012) Guidelines for the use and interpretation of assays for monitoring autophagy. Autophagy 8:445–544

Krauthammer M, Kong Y, Ha BH, Evans P, Bacchiocchi A, McCusker JP, Cheng E, Davis MJ, Goh G, Choi M et al (2012) Exome sequencing identifies recurrent somatic RAC1 mutations in melanoma. Nat Genet 44:1006–1014

Krown SE (1986) Interferons and interferon inducers in cancer treatment. Semin Oncol 13:207–217

Kunath K, von Harpe A, Fischer D, Petersen H, Bickel U, Voigt K, Kissel T (2003) Low-molecular-weight polyethylenimine as a non-viral vector for DNA delivery: comparison of physicochemical properties, transfection efficiency and in vivo distribution with high-molecular-weight polyethylenimine. J Control Release 89:113–125

Lawrence MS, Stojanov P, Polak P, Kryukov GV, Cibulskis K, Sivachenko A, Carter SL, Stewart C, Mermel CH, Roberts SA et al (2013) Mutational heterogeneity in cancer and the search for new cancer-associated genes. Nature 499:214–218

Ledley FD (1996) Pharmaceutical approach to somatic gene therapy. Pharm Res 13:1595–1614

Levine AS, Sivulich M, Wiernik PH, Levy HB (1979) Initial clinical trials in cancer patients of polyriboinosinic-polyribocytidylic acid stabilized with poly-L-lysine, in carboxymethylcellulose [poly(ICLC)], a highly effective interferon inducer. Cancer Res 39:1645–1650

Ma C, Armstrong AW (2014) Severe adverse events from the treatment of advanced melanoma: a systematic review of severe side effects associated with ipilimumab, vemurafenib, interferon alfa-2b, dacarbazine and interleukin-2. J Dermatolog Treat 25:401–408

Maiuri MC, Zalckvar E, Kimchi A, Kroemer G (2007) Self-eating and self-killing: crosstalk between autophagy and apoptosis. Nat Rev Mol Cell Biol 8:741–752

Ming Lim C, Stephenson R, Salazar AM, Ferris RL (2013) TLR3 agonists improve the immunostimulatory potential of cetuximab against EGFR head and neck cancer cells. Oncoimmunology 2:e24677

Mizushima N, Levine B, Cuervo AM, Klionsky DJ (2008) Autophagy fights disease through cellular self-digestion. Nature 451:1069–1075

Morse MA, Chapman R, Powderly J, Blackwell K, Keler T, Green J, Riggs R, He LZ, Ramakrishna V, Vitale L et al (2011) Phase I study utilizing a novel antigen-presenting cell-targeted vaccine with Toll-like receptor stimulation to induce immunity to self-antigens in cancer patients. Clin Cancer Res 17:4844–4853

Nakamura O, Shitara N, Matsutani M, Takakura K, Machida H (1982) Phase I–II trials of poly(ICLC) in malignant brain tumor patients. J Interferon Res 2:1–4

Navabi H, Jasani B, Reece A, Clayton A, Tabi Z, Donninger C, Mason M, Adams M (2009) A clinical grade poly I:C-analogue (Ampligen) promotes optimal DC maturation and Th1-type T cell responses of healthy donors and cancer patients in vitro. Vaccine 27:107–115

Page DB, Postow MA, Callahan MK, Allison JP, Wolchok JD (2014) Immune modulation in cancer with antibodies. Annu Rev Med 65:185–202

Pennock GK, Waterfield W, Wolchok JD (2012) Patient responses to ipilimumab, a novel immunopotentiator for metastatic melanoma: how different are these from conventional treatment responses? Am J Clin Oncol 35:606–611

Prahallad A, Sun C, Huang S, Di Nicolantonio F, Salazar R, Zecchin D, Beijersbergen RL, Bardelli A, Bernards R (2012) Unresponsiveness of colon cancer to BRAF(V600E) inhibition through feedback activation of EGFR. Nature 483:100–103

Quintana E, Shackleton M, Foster HR, Fullen DR, Sabel MS, Johnson TM, Morrison SJ (2010) Phenotypic heterogeneity among tumorigenic melanoma cells from patients that is reversible and not hierarchically organized. Cancer Cell 18:510–523

Rapoport AP, Aqui NA, Stadtmauer EA, Vogl DT, Xu YY, Kalos M, Cai L, Fang HB, Weiss BM, Badros A et al (2014) Combination immunotherapy after ASCT for multiple myeloma using MAGE-A3/poly-ICLC immunizations followed by adoptive transfer of vaccine-primed and costimulated autologous T cells. Clin Cancer Res 20:1355–1365

Rink J, Ghigo E, Kalaidzidis Y, Zerial M (2005) Rab conversion as a mechanism of progression from early to late endosomes. Cell 122:735–749

Rosenfeld MR, Chamberlain MC, Grossman SA, Peereboom DM, Lesser GJ, Batchelor TT, Desideri S, Salazar AM, Ye X (2010) A multi-institution phase II study of poly-ICLC and radiotherapy with concurrent and adjuvant temozolomide in adults with newly diagnosed glioblastoma. Neuro Oncol 12:1071–1077

Rubinsztein DC, Gestwicki JE, Murphy LO, Klionsky DJ (2007) Potential therapeutic applications of autophagy. Nat Rev Drug Discov 6:304–312

Sabado RL, Miller E, Spadaccia M, Vengco I, Hasan F, Bhardwaj N (2013) Preparation of tumor antigen-loaded mature dendritic cells for immunotherapy. J Vis Exp. doi: 10.3791/50085

Sabbatini P, Tsuji T, Ferran L, Ritter E, Sedrak C, Tuballes K, Jungbluth AA, Ritter G, Aghajanian C, Bell-McGuinn K et al (2012) Phase I trial of overlapping long peptides from a tumor self-antigen and poly-ICLC shows rapid induction of integrated immune response in ovarian cancer patients. Clin Cancer Res 18:6497–6508

Salazar AM, Erlich RB, Mark A, Bhardwaj N, Herberman RB (2014) Therapeutic in-situ autovaccination against solid cancers with intratumoral poly-ICLC: case report, hypothesis, and clinical trial. Cancer Immunol Res 2:720–724

Schaffert D, Kiss M, Rodl W, Shir A, Levitzki A, Ogris M, Wagner E (2011) Poly(I:C)-mediated tumor growth suppression in EGF-receptor overexpressing tumors using EGF-polyethylene glycol-linear polyethylenimine as carrier. Pharm Res 28:731–741

Scott KL, Nogueira C, Heffernan TP, van Doorn R, Dhakal S, Hanna JA, Min C, Jaskelioff M, Xiao Y, Wu CJ et al (2011) Proinvasion metastasis drivers in early-stage melanoma are oncogenes. Cancer Cell 20:92–103

Shi H, Moriceau G, Kong X, Lee MK, Lee H, Koya RC, Ng C, Chodon T, Scolyer RA, Dahlman KB et al (2012) Melanoma whole-exome sequencing identifies (V600E)B-RAF amplification-mediated acquired B-RAF inhibitor resistance. Nat Commun 3:724

Siegel R, Naishadham D, Jemal A (2013) Cancer statistics, 2013. CA Cancer J Clin 63:11–30

Slingluff CL Jr, Petroni GR, Olson W, Czarkowski A, Grosh WW, Smolkin M, Chianese-Bullock KA, Neese PY, Deacon DH, Nail C et al (2008) Helper T-cell responses and clinical activity of a melanoma vaccine with multiple peptides from MAGE and melanocytic differentiation antigens. J Clin Oncol 26:4973–4980

Sloot S, Fedorenko IV, Smalley KS, Gibney GT (2014) Long-term effects of BRAF inhibitors in melanoma treatment: friend or foe? Expert Opin Pharmacother 15:589–592

Soengas MS, Lowe SW (2003) Apoptosis and melanoma chemoresistance. Oncogene 22:3138–3151

Sondak VK, Zager JS (2014) Melanoma: MSLT-1-putting sentinel lymph node biopsy into context. Nat Rev Clin Oncol 11:246–248

Strasser A, O'Connor L, Dixit VM (2000) Apoptosis Signaling. Annu Rev Biochem 69:217–245

Strayer DR, Carter WA, Stouch BC, Stevens SR, Bateman L, Cimoch PJ, Lapp CW, Peterson DL, Mitchell WM (2012) A double-blind, placebo-controlled, randomized, clinical trial of the TLR-3 agonist rintatolimod in severe cases of chronic fatigue syndrome. PLoS One 7:e31334

Takeuchi O, Akira S (2008) MDA5/RIG-I and virus recognition. Curr Opin Immunol 20:17–22

Tarhini AA, Agarwala SS (2005) Interleukin-2 for the treatment of melanoma. Curr Opin Investig Drugs 6:1234–1239

Thomas M, Ge Q, Lu JJ, Chen J, Klibanov AM (2005) Cross-linked small polyethylenimines: while still nontoxic, deliver DNA efficiently to mammalian cells in vitro and in vivo. Pharm Res 22:373–380

Tormo D, Alonso-Curbelo D, Soengas MS (2009a) Cytosolic delivery of dsRNA triggers MDA-5 mediated autonomous cell death in malignant melanomas. Clin Transl Oncol 11:39–42

Tormo D, Checinska A, Alonso-Curbelo D, Perez-Guijarro E, Canon E, Riveiro-Falkenbach E, Calvo TG, Larribere L, Megias D, Mulero F et al (2009b) Targeted activation of innate immunity for therapeutic induction of autophagy and apoptosis in melanoma cells. Cancer Cell 16:103–114

Verhaegen M, Bauer JA, Martin de la Vega C, Wang G, Wolter KG, Brenner JC, Nikolovska-Coleska Z, Bengtson A, Nair R, Elder JT et al (2006) A novel BH3 mimetic reveals a mitogen-activated protein kinase-dependent mechanism of melanoma cell death controlled by p53 and reactive oxygen species. Cancer Res 66:11348–11359

Verhaegen M, Checinska A, Riblett MB, Wang S, Soengas MS (2012) E2F1-dependent oncogenic addiction of melanoma cells to MDM2. Oncogene 31:828–841

Vultur A, Herlyn M (2009) Cracking the system: melanoma complexity demands new therapeutic approaches. Pigment Cell Melanoma Res 22:4–5

Wagner E (2004) Strategies to improve DNA polyplexes for in vivo gene transfer: will "artificial viruses" be the answer? Pharm Res 21:8–14

Whiteman DC, Pavan WJ, Bastian BC (2011) The melanomas: a synthesis of epidemiological, clinical, histopathological, genetic, and biological aspects, supporting distinct subtypes, causal pathways, and cells of origin. Pigment Cell Melanoma Res 24:879–897

Wightman L, Kircheis R, Rossler V, Carotta S, Ruzicka R, Kursa M, Wagner E (2001) Different behavior of branched and linear polyethylenimine for gene delivery in vitro and in vivo. J Gene Med 3:362–372

Wilson TR, Fridlyand J, Yan Y, Penuel E, Burton L, Chan E, Peng J, Lin E, Wang Y, Sosman J et al (2012) Widespread potential for growth-factor-driven resistance to anticancer kinase inhibitors. Nature 487:505–509

Wolter KG, Verhaegen M, Fernandez Y, Nikolovska-Coleska Z, Riblett M, de la Vega CM, Wang S, Soengas MS (2007) Therapeutic window for melanoma treatment provided by selective effects of the proteasome on Bcl-2 proteins. Cell Death Differ 14:1605–1616

Wright CJ, McCormack PL (2013) Trametinib: first global approval. Drugs 73:1245–1254

Xie Z, Klionsky DJ (2007) Autophagosome formation: core machinery and adaptations. Nat Cell Biol 9:1102–1109

Yang Z, Klionsky DJ (2010) Eaten alive: a history of macroautophagy. Nat Cell Biol 12:814–822

Zarour HM, Kirkwood JM (2003) Melanoma vaccines: early progress and future promises. Semin Cutan Med Surg 22:68–75

Index

A

Active targeting, 3–30, 78, 80, 84, 87, 98, 100, 102, 116, 141, 222, 270–274, 280, 343, 356, 432, 441
Allergy, 406, 407
Apoptosis, 22, 53, 127–129, 138, 147, 178, 179, 193, 199, 201, 202, 214, 230, 245, 253–255, 260, 269, 284, 309, 319, 341, 352, 353, 379, 395, 396, 399, 412, 454–458, 461–465
Autophagy, 128, 129, 453–458, 460, 461, 463–465

B

Biocompatibility, 5, 70, 96–97, 149, 150, 254, 276, 298, 312, 317–319, 367, 369, 373, 416, 427, 428, 430
Bio-corona, 396, 398–402, 404, 412, 413, 415, 416
Biodegradation, 403, 408, 409, 416
Biological identity, 393, 399–402, 404, 414
BO-110, 453–465

C

Cancer
- nano-immunotherapy, 205, 224–234
- therapy, 19, 28, 67–88, 95–117, 128, 147, 195, 216, 220, 221, 224, 225, 258–261, 277, 278, 282, 298, 299, 307, 308, 311, 314, 330, 351–383, 431, 432, 436, 443, 455–458
- vaccines, 163–180, 193, 195, 196, 204, 225, 234

Chemotherapy, 28, 30, 37, 67, 80, 85–87, 113, 116, 137–139, 147, 153, 171, 177, 194, 196, 199–202, 216, 228, 260, 270, 304, 310, 315, 339, 346, 351, 352, 380, 382, 404, 435, 454
CHMP Reflection Paper, 444
Combination therapy, 137, 147, 436
Communication, 48–52, 57, 58, 433

D

Delivery, 3, 54, 69, 96, 125, 164, 197, 217, 245, 269, 296, 333, 356, 394, 431, 453
Dendritic cells (DCs), 49, 53, 55, 75, 108, 127, 131, 164–167, 169–180, 188, 189, 192–194, 198, 199, 203, 204, 216, 217, 220–221, 225, 227, 229–234, 250, 278, 403, 405, 407, 408
Diagnosis, 58, 69, 71, 74, 85, 87, 110, 112, 152, 153, 270, 296, 298–304, 320, 346, 351–383, 394, 405, 425, 454
Drug delivery, 3, 54, 69, 96, 132, 205, 215, 270, 296, 333, 356, 391, 425

E

Enhanced permeability retention (EPR), 3–30, 69, 74, 77, 83, 96, 111, 112, 141, 146, 148, 202, 271–273, 275, 279, 313, 345, 356, 369, 381, 430, 432, 444
EPR factors, 4, 6–12
Exosomes, 47–58, 201, 382, 433

M.J. Alonso, M. Garcia-Fuentes (eds.), *Nano-Oncologicals: New Targeting and Delivery Approaches*, Advances in Delivery Science and Technology,
DOI 10.1007/978-3-319-08084-0

G

Gene therapy, 80, 87, 88, 217, 229, 233, 276–278, 296, 307–313, 319, 336, 339–341, 345
Gold nanoparticles, 298, 313, 314, 318, 334, 369, 372, 378, 395, 396, 399, 401, 404, 413

H

High-throughput screening (HTS), 393, 410–412

I

Imaging, 4, 48, 69, 108, 125, 272, 295, 329, 351, 415, 430, 460
Immune modulation, 187–206
Immunosuppression, 130, 163, 164, 177, 178, 188, 189, 191, 193–195, 200, 205, 216, 402, 405–407, 428
Immunosuppressive cells, 178, 187–206, 219, 229–234
Immunotherapy, 163, 166, 172, 174, 176, 180, 195–196, 202–205, 215–234, 455, 456
Immunotoxicity, 403, 427
Inflammation, 54, 127, 131, 188, 189, 192, 199, 215, 216, 344, 370, 393, 398, 403, 406, 407, 414

L

Liposome(s), 3, 48, 50, 69, 95, 108, 125, 142, 196, 222, 254, 270, 272, 295, 329, 336, 351, 369, 394, 415, 425 4, 430, 460

M

Marketed nanosystems, 426, 431, 434, 439, 440
MDSCs. *See* Myeloid-derived suppressor cells (MDSCs)
Melanoma, 20, 53, 54, 80, 105, 114, 136, 144, 147, 150, 165, 168–171, 174, 177, 179, 195, 196, 205, 226, 227, 231, 233, 234, 255, 283, 453–457, 460–462, 464, 465
Melanoma Differentiation-Associated gene-5 (MDA5), 454, 461
Metastatic cancer, 84–87
Micelle(s), 3, 68, 95, 142, 204, 222, 270, 336, 373, 394, 434
Microbubbles, 115, 329–346, 364, 367
Microenvironment, 6–8, 13, 19, 29, 48, 54, 58, 69, 88, 108–111, 115, 124, 164, 177, 180, 188, 191–194, 196–198, 203, 216, 217, 219–222, 229, 230, 234, 269, 270, 351, 353, 356, 358, 365, 382, 391, 409, 432
MicroRNA (miRNAs), 47, 48, 55, 56, 58, 229, 232, 245–261, 274, 275, 281–285, 308, 311–313, 320, 380, 434, 439
Multifunctional nanoparticles, 70, 270
Myeloid-derived suppressor cells (MDSCs), 178, 188–193, 196, 197, 199–206, 216, 218, 219, 221, 229, 232–234

N

Nanocarrier(s), 3, 67, 98, 140, 200, 221, 295, 357, 398, 430
Nanomedicine, 18, 22, 28–30, 85, 88, 96, 98, 202, 205, 206, 222, 234, 286, 351, 392–395, 397–402, 404–406, 409, 411, 414–416, 426, 427, 429–434, 437–440, 443, 444
Nanoparticles (NPs), 4, 55, 67, 142, 179, 197, 222, 254, 270, 296, 334, 356, 392, 425, 457
Nanopharmaceuticals, 425–431, 434, 437, 438, 441, 442
Nanotechnology, 67, 72, 85, 96, 133, 140, 153, 179–180, 194, 205, 215–234, 285, 286, 295, 297, 301, 304, 307, 311, 317, 320, 351–358, 371, 393, 394, 398, 403, 427, 429, 431, 434, 439, 440, 444
Nanotoxicity, 317–319
Nanotoxicology, 391–416

P

Passive targeting, 3–30, 74, 75, 84, 141, 222, 274, 343, 356, 432, 441
Photodynamic therapy, 123–153, 367
Photosensitizers, 72, 123, 132, 135, 379
Physico-chemical properties, 128, 138, 393, 394, 396, 400, 405, 412
Polymer, 6, 10, 11, 13, 28, 68, 70–73, 75, 77, 78, 82, 99, 111, 114, 147, 222, 253, 254, 256, 274, 276, 277, 280, 281, 284–286, 309, 334, 337, 343, 365, 367, 369–371, 373, 379, 397, 427, 430, 434, 437, 439, 453–465
Positive target, 102, 196, 276
Preclinical and clinical studies, 130
Primary cancer, 83–84

Q

Quantitative structure-activity relationship (QSAR), 412

R

Regulatory requirements, 425–431, 439

S

Sequence-defined oligomers, 256–258, 262
Small Interfering RNA (siRNA), 21, 55, 68, 100, 139, 198, 223, 245, 272, 308, 340, 380, 434, 455
Soluble *N-ethylmaleimide-sensitive* factor attachment protein receptors (SNAREs), 49–52, 58
Sonoporation, 337, 339, 341, 345, 346
Systems biology, 391, 413, 414

T

TAMs. *See* Tumor-associated macrophages (TAMs)
Targeted delivery, 15, 22, 229, 269–286, 356, 369, 370, 372
Targeted therapy, 14, 21, 220, 342–345, 455
Targeting, 3, 55, 67, 98, 134, 172, 189, 222, 246, 270, 296, 330, 356, 392, 425, 455
T cells, 53, 55, 131, 135, 163–168, 170–180, 189–196, 202–205, 216, 217, 219–221, 223–226, 233, 455
Theranostics, 30, 84, 151, 153, 295–321, 329–347, 365, 367, 369–382, 440, 441, 444
Therapy, 12, 26, 28, 71, 85, 86, 110, 112, 134, 136, 137, 150–153, 179, 180, 195, 217, 218, 228–230, 260, 261, 274, 285, 296, 297, 304, 307, 311, 313, 314, 319, 320, 330, 336, 338, 342, 352, 354, 355, 358–360, 363, 364, 367, 370, 376, 378, 380, 382, 435, 436
Triggered release, 100, 115, 143, 326, 370, 373
Triggering, 103, 109–113, 115, 116, 151, 219, 220, 233, 308, 312, 334, 392, 398, 443
Tumor-associated macrophages (TAMs), 189–191, 196–200, 202, 203, 205, 216, 218, 220, 227, 229–234
Tumor microenvironment, 7, 8, 19, 29, 58, 69, 88, 110, 111, 177, 180, 191–194, 196–198, 203, 216, 217, 219–222, 229, 230, 234, 351, 353, 356, 358, 382, 391

U

Ultrasound contrast, 330, 333, 334, 344, 346

MIX
Papier aus verantwortungsvollen Quellen
Paper from responsible sources
FSC® C105338

If you have any concerns about our products,
you can contact us on
ProductSafety@springernature.com

In case Publisher is established outside the EU,
the EU authorized representative is:
Springer Nature Customer Service Center GmbH
Europaplatz 3, 69115 Heidelberg, Germany

Printed by Libri Plureos GmbH
in Hamburg, Germany